FLORA ZAMBESIACA

Flora terrarum Zambesii aquis conjunctarum

OCHNA PUBERULA

FLORA ZAMBESIACA

MOZAMBIQUE
FEDERATION OF RHODESIA AND NYASALAND
BECHUANALAND PROTECTORATE

VOLUME TWO: PART ONE

Edited by

A. W. EXELL, A. FERNANDES and H. WILD

on behalf of the Editorial Board:

Published on behalf of the Governments of
Portugal
The Federation of Rhodesia and Nyasaland
and the United Kingdom by the
Crown Agents for Oversea Governments and Administrations,
4, Millbank, London, S.W.1
May 23, 1963

FLORA ZAMBESIACA
MANAGING COMMITTEE

★

Printed at the University Press Glasgow
by Robert MacLehose & Company Limited

LIST OF NEW NAMES PUBLISHED IN THIS WORK

Grewia flavescens var. *olukondae* (Schinz) Wild, comb. nov. *page* 46

Triumfetta pilosa var. *effusa* (E. Mey. ex Harv.) Wild, comb. nov. 77

Monsonia ovata subsp. *glauca* (Knuth) Bowden & T. Müll., stat. nov. 137

Thamnosma rhodesica (Bak. f.) Mendonça, stat. nov. 183

Lepidotrichilia volkensii (Gürke) Leroy, comb. nov. 305

ABBREVIATIONS ADDITIONAL TO THOSE ON PAGE 21 OF VOLUME ONE

B.C.L.: Check-lists of the Forest Trees and Shrubs of the British Empire No. 6, Bechuanaland Protectorate (1948).

F.F.N.R.: Forest Flora of Northern Rhodesia.

LIST OF FAMILIES INCLUDED IN VOL. II, PART 1

ANGIOSPERMAE

31. Tiliaceae
32. Linaceae
33. Ixonanthaceae
34. Erythroxylaceae
35. Malpighiaceae
36. Zygophyllaceae
37. Geraniaceae
38. Oxalidaceae
39. Balsaminaceae
40. Rutaceae
41. Simaroubaceae
42. Irvingiaceae
43. Balanitaceae
44. Ochnaceae
45. Burseraceae
46. Meliaceae
47. Dichapetalaceae
48. Olacaceae
49. Opiliaceae
50. Icacinaceae

CLAVE DAS FAMÍLIAS E DOS GRUPOS SUPERIORES

por J. E. Dandy

Tradução de E. J. Mendes*

Esta clave foi elaborada com o intuito de permitir a identificação preliminar até ao nível da família a quem utilizar a *Flora Zambesiaca*. É fundamentalmente semelhante à clave artificial para as famílias publicada por F. W. Andrews na obra *The Flowering Plants of the Anglo-Egyptian Sudan*, vol. 1 (1950), que, por sua vez, foi adaptada à flora do Sudão a partir da que J. Hutchinson apresentou para as famílias de todo o mundo em *The Families of Flowering Plants*, vols. 1 e 2 (1926, 1934). As famílias aqui consideradas são aquelas que se sabe estarem representadas na área da *Flora Zambesiaca* por espécies indígenas ou naturalizadas e, na medida do previsível, esta clave incluirá todas as fanerogâmicas que virão a ser tratadas na *Flora Zambesiaca*. É óbvio, no entanto, que se tornarão necessárias algumas modificações, não só devido à descoberta de taxa a aditar ao inventário da área, mas também a alterações quer na delimitação quer na nomenclatura das famílias (a eventualidade de algumas destas alterações é considerada no índice respectivo). Não parece provável, contudo, que tais modificações diminuam de forma sensível a utilidade da clave.

(A versão original inglesa desta clave, publicada no vol. 1, sofreu algumas adaptações. A família *Irvingiaceae*, recentemente assinalada na área, foi acrescentada e as *Averrhoaceae* foram agora incluídas nas *Oxalidaceae*.—J. E. Dandy.)

Óvulos ou óvulo não incluídos em ovários fechados; flores (estróbilos) unissexuadas, sem perianto - - - - - - - - - GYMNOSPERMAE:

Plantas simulando palmeiras, com folhas pinadas formando uma coroa no topo do caule CYCADALES (ver abaixo)

Árvores com folhas simples dispostas ao longo dos ramos CONIFERAE (ver abaixo)

Óvulos ou óvulo incluídos em ovários fechados; flores bissexuadas ou unissexuadas, com ou sem perianto - - - - - - - - - ANGIOSPERMAE:

Embrião com 2 (raramente mais) cotilédones; feixes vasculares do caule dispostos geralmente em um cilindro único; folhas quase sempre com nervação reticulada; flores em geral 5- ou 4-meras - - - - DICOTYLEDONES (ver p. 8)

Embrião com 1 só cotilédone; feixes vasculares do caule dispersos; folhas geralmente de nervação paralela; flores em geral 3-meras - MONOCOTYLEDONES (ver p. 26)

CYCADALES

Família única - - - - - - - - - - - - **Cycadaceae**

CONIFERAE

Folhas bem desenvolvidas, loriformes, nem escamiformes nem subuladas, quase sempre alternas; óvulos solitários em cada carpelo; sementes 1–2, grandes e drupáceas, desenvolvidas sobre um receptáculo (frequentemente dilatado e carnudo) originado da base do estróbilo - - - - - - - - - - - **Podocarpaceae**

Folhas pequenas, escamiformes, oposto-cruzadas (ou subuladas e ternadas ou alternas nos ramos jóvens); óvulos solitários ou 2 ou mais em cada carpelo; sementes incluídas num falso fruto bacáceo indeiscente ou numa gálbula lenhosa deiscente **Cupressaceae**

* À Ex.ma Sr.a D. Rosette Fernandes, ilustre Naturalista do Instituto Botânico Dr. Júlio Henriques, Coimbra, apresento o meu mais vivo agradecimento pelas inestimáveis sugestões e cuidada revisão desta tradução. O Ex.mo Sr. A. W. Exell é credor do meu mais profundo reconhecimento pela ligação que estabeleceu, com inexcedível interesse, entre o Autor e o Tradutor.

DICOTYLEDONES

Gineceu de 2 ou mais carpelos livres com os estiletes e os estigmas também livres (por vezes, em *Nymphaeaceae*, os carpelos estão ± imersos no tecido dilatado do receptáculo):
 Pétalas presentes, livres - - - - - - - - *Grupo 1* (ver p. 8)
 Pétalas presentes, ± unidas - - - - - - - - *Grupo 2* (ver p. 9)
 Pétalas nulas - - - - - - - - - *Grupo 3* (ver p. 9)
Gineceu de 1 carpelo, ou de 2 ou mais carpelos unidos com estiletes livres ou unidos, ou de carpelos livres na base mas, então, com os estiletes ou com os estigmas unidos:
 Óvulos 2 ou mais, inseridos na parede do ovário (placentação parietal):
  Ovário súpero:
   Pétalas presentes, livres - - - - - - - - *Grupo 4* (ver p. 9)
   Pétalas presentes, ± unidas - - - - - - - *Grupo 5* (ver p. 10)
   Pétalas nulas - - - - - - - - - *Grupo 6* (ver p. 11)
  Ovário ± ínfero:
   Pétalas presentes, livres - - - - - - - - *Grupo 7* (ver p. 11)
   Pétalas presentes, ± unidas - - - - - - - *Grupo 8* (ver p. 11)
   Pétalas nulas - - - - - - - - - *Grupo 9* (ver p. 12)
 Óvulos 1 ou mais e, quando mais de 1, inseridos ou no eixo central ou na base ou no ápice do ovário:
  Ovário súpero:
   Pétalas presentes, livres - - - - - - - - *Grupo 10* (ver p. 12)
   Pétalas presentes, ± unidas - - - - - - - *Grupo 11* (ver p. 16)
   Pétalas nulas - - - - - - - - - *Grupo 12* (ver p. 20)
  Ovário ± ínfero:
   Pétalas presentes, livres - - - - - - - - *Grupo 13* (ver p. 23)
   Pétalas presentes, ± unidas - - - - - - - *Grupo 14* (ver p. 25)
   Pétalas nulas - - - - - - - - - *Grupo 15* (ver p. 26)

Grupo 1

Folhas opostas, sem estípulas; carpelos em número igual ao das pétalas; ervas, frequentemente suculentas - - - - - - - - - - - **Crassulaceae**
Folhas alternas, por vezes todas basilares:
 Plantas herbáceas, aquáticas, com folhas flutuantes longamente pecioladas e limbos ± peltados ou não; flores solitárias, longamente pedunculadas, frequentemente grandes e vistosas:
  Carpelos ± imersos no tecido dilatado do receptáculo; pétalas numerosas; limbo da folha cordado e ± peltado ou não - - - - **Nymphaeaceae**
  Carpelos não como acima; pétalas 3; limbo da folha peltado mas não cordado **Cabombaceae**
 Plantas terrestres; limbo da folha não peltado:
  Estames periginicos, inseridos na fauce do tubo do cálice; folhas pinadas ou digitadas, com estípulas; árvores, arbustos ou ervas e, por vezes, silvas - - **Rosaceae**
  Estames hipogínicos não inseridos no cálice:
   Sépalas 6–24 em 2 ou mais séries, livres ou as internas unidas; pequenos arbustos ou trepadeiras lenhosas com pequenas flores dióicas - **Menispermaceae**
   Sépalas 2–5, em 1 só série, livres ou unidas:
    Ervas; folhas crenadas, lobadas ou profundamente recortadas; sépalas livres, herbáceas ou petalóides, sendo a mediana (posterior) por vezes esporoada **Ranunculaceae**
    Árvores ou arbustos, por vezes escandentes:
     Folhas pinadas ou 3-folioladas:
      Flores bissexuadas; estames em número duplo do das pétalas; carpelos com 2 óvulos - - - - - - - - - - **Connaraceae**
      Flores unissexuadas; estames em número igual ao das pétalas; carpelos com 1 só óvulo - - - - - - - - - **Simaroubaceae**
     Folhas simples:
      Sépalas 2–3, valvadas, livres ou unidas; anteras frequentemente mais compridas do que os filetes, apresentando em geral um largo prolongamento do conectivo acima das tecas - - - - - - **Annonaceae**
      Sépalas 5, imbricadas, livres ou quase; anteras muito menores do que os filetes e sem prolongamento do conectivo:
       Estames em número igual ou duplo do das pétalas; carpelos com 2 óvulos, estilete sub-basal - - - - - **Simaroubaceae**
       Estames numerosos, em número maior que o duplo do das pétalas; carpelos com alguns óvulos (mais de 2), estilete terminal **Dilleniaceae**

Grupo 2

Folhas opostas; pétalas unidas em tubo alongado; ervas ou pequenos arbustos suculentos **Crassulaceae**
Folhas alternas; pétalas curtamente unidas na base ou acima desta; árvores ou arbustos, por vezes escandentes:
  Folhas 3-folioladas; sépalas 5, livres ou quase; pétalas 5 - - **Connaraceae**
  Folhas simples; sépalas 2–3, livres ou unidas; pétalas 6 ou 4 - - **Annonaceae**

Grupo 3

Sépalas unidas na base; estames periginicos, inseridos na fauce do tubo do cálice; folhas pinadas, digitadas ou palmatilobadas, com estípulas; árvores, arbustos ou ervas e, por vezes, silvas - - - - - - - - - - - - - **Rosaceae**
Sépalas livres; estames hipogínicos:
  Folhas compostas, por vezes recompostas; sépalas petalóides; ervas ou plantas fracamente lenhosas escandentes ou prostradas - - - - - - **Ranunculaceae**
  Folhas simples:
    Flores dióicas; trepadeiras lenhosas com folhas alternas:
      Sépalas 6–18, em 2 ou mais séries; estames 3–6, livres ou em sinândrio; flores em cimeiras ou racimos; folhas inteiras ou palmatilobadas **Menispermaceae**
      Sépalas 5, em 1 só série; estames 10–15, livres; flores em racimos alongados; folhas inteiras - - - - - - - - - - **Phytolaccaceae**
    Flores bissexuadas; ervas com folhas opostas ou subopostas:
      Folhas inteiras; cálice herbáceo; flores pequenas, em cimeiras ou panículas **Aizoaceae**
      Folhas dentadas ou lobadas; cálice petalóide; flores grandes e vistosas, solitárias ou em inflorescências paucifloras - - - - - **Ranunculaceae**

Grupo 4

Estames 6, tetradinâmicos (4 internos longos, 2 externos curtos); sépalas 4; pétalas 4 ou menos; folhas sem estípulas; ervas - - - - - - - **Cruciferae**
Estames não tetradinâmicos:
  Gineceu de 1 só carpelo (1 única placenta):
    Flores zigomórficas; pétalas 5 ou menos, por vezes 1 só; folhas simples ou compostas; árvores, arbustos ou ervas, por vezes escandentes - **Leguminosae**
    Flores actinomórficas:
      Folhas 2-pinadas, com estípulas; árvores, arbustos ou ervas, por vezes escandentes; flores em espigas ou capítulos - - - - - - **Leguminosae**
      Folhas simples, 1-pinadas ou 1-folioladas:
        Folhas pinadas, folíolos providos de pontuações glandulares translúcidas pelo menos na margem; árvores ou arbustos - - - - - - **Rutaceae**
        Folhas simples ou 1-folioladas, sem glândulas translúcidas:
          Folhas dentadas, com estípulas; estames em número igual ao das pétalas, a elas opostos e unidos na base; ervas ou pequenos arbustos **Sterculiaceae**
          Folhas inteiras, sem estípulas; estames em número maior que o das pétalas; árvores ou arbustos, por vezes escandentes:
            Sépalas 5; flores em racimos paucifloros - - - **Connaraceae**
            Sépalas 3; pétalas 6; flores solitárias, axilares - - **Annonaceae**
  Gineceu de 2 ou mais carpelos unidos (2 ou mais placentas):
    Folhas opostas ou verticiladas, simples, nunca todas basilares; ervas:
      Estames em número maior que o duplo do das pétalas, frequentemente agrupados em 3 ou mais feixes; plantas terrestres; folhas opostas providas de glândulas translúcidas - - - - - - - - - - **Guttiferae**
      Estames em número igual ao das pétalas, livres; plantas aquáticas; folhas verticiladas, com costelos terminais - - - - - - - **Droseraceae**
    Folhas alternas ou todas basilares:
      Flores unissexuadas:
        Estames em número igual ao das pétalas:
          Pétalas menores que as sépalas; trepadeiras herbáceas ou lenhosas, com gavinhas; folhas por vezes digitadas ou palmatilobadas - **Passifloraceae**
          Pétalas maiores que as sépalas; árvores ou arbustos sem gavinhas; folhas simples, não lobadas - - - - - - - - - **Pittosporaceae**
        Estames em número duplo do das pétalas ou maior; árvores ou arbustos sem gavinhas:
          Folhas palmati-3–7-lobadas; pétalas livres nas flores femininas, unidas nas masculinas - - - - - - - - - - - - **Caricaceae**
          Folhas não palmati-3–7-lobadas; pétalas sempre livres - **Flacourtiaceae**

Flores bissexuadas ou, por vezes, poligâmicas:
Ovário sobre um ginóforo nítido:
Flores com uma coroa fimbriada ou pilosa, exterior aos estames; estiletes 3–6; folhas simples; árvores ou arbustos, por vezes escandentes **Passifloraceae**
Flores sem coroa; estilete 1 ou estigma séssil; folhas simples ou digitadas; árvores, arbustos ou ervas, por vezes escandentes - - **Capparidaceae**
Ovário séssil ou subséssil:
Flores ± zigomórficas:
Folhas 2–3-pinadas; árvores; flores em panículas axilares; estames 5, alternando com 5 estaminódios - - - - - - **Moringaceae**
Folhas simples ou digitadas; ervas ou, por vezes, arbustos ou pequenas árvores:
Conectivo apendiculado acima das tecas; flores quase sempre solitárias; fruto uma cápsula 3-valve; folhas simples - - - **Violaceae**
Conectivo não prolongado acima das tecas; flores em racimos ou espigas terminais:
Fruto uma cápsula 2-valve, deiscente a todo o comprimento; folhas simples ou digitadas - - - - - - - **Capparidaceae**
Fruto uma cápsula ± aberta no cimo; folhas simples, inteiras a pinatipartidas - - - - - - - **Resedaceae**
Flores actinomórficas:
Estames em número superior ao das pétalas, ou igual e a elas opostos:
Anteras de deiscência por poros ou por curtas fendas semelhantes a poros:
Anteras em forma de ferradura, deiscentes no topo da curvatura; sépalas 4–6, caducas, deixando na base da flor grandes glândulas persistentes; fruto uma cápsula setosa 2-valve - - **Bixaceae**
Anteras rectas, deiscentes no ápice; sépalas 3, fruto uma cápsula lisa ou aculeada, 5–8-valve - - - - - **Flacourtiaceae**
Anteras deiscentes por fendas longitudinais:
Flores com uma coroa fimbriada, exterior aos estames; pequenos arbustos escandentes, com gavinhas axilares - **Passifloraceae**
Flores sem coroa; plantas sem gavinhas:
Ervas espinhosas; folhas pinatilobadas; sépalas 2–3, caducas; fruto uma cápsula com deiscência apical por 4–6 curtas valvas **Papaveraceae**
Árvores ou arbustos; folhas não pinatilobadas:
Fruto indeiscente, rodeado pelas 5 sépalas acrescentes, aliformes e persistentes; folhas inteiras - - - **Dipterocarpaceae**
Fruto capsular ou bacáceo, não rodeado por sépalas aliformes persistentes; folhas inteiras ou dentadas - **Flacourtiaceae**
Estames em número igual ao das pétalas e alternos com elas:
Estiletes 2–5, livres ou curtamente unidos na base:
Folhas com numerosas glândulas estipitadas e pegajosas; ervas insectívoras - - - - - - - - - **Droseraceae**
Folhas sem glândulas estipitadas pegajosas; plantas não insectívoras:
Flores com uma coroa vistosa, fimbriada ou pilosa, exterior aos estames; ervas ou arbustos, frequentemente escandentes com gavinhas - - - - - - - **Passifloraceae**
Flores sem coroa, ou com uma coroa muito pequena e pouco aparente; ervas sufruticosas sem gavinhas - - - **Turneraceae**
Estilete 1:
Estames alternando com estaminódios petalóides e frequentemente circundados por uma série exterior de estaminódios filiformes; ervas com estípulas incisas - - - - - **Ochnaceae**
Estames não acompanhados de estaminódios; árvores ou arbustos:
Conectivo apendiculado acima das tecas; folhas com estípulas **Violaceae**
Conectivo não ultrapassando as tecas; folhas sem estípulas **Pittosporaceae**

Grupo 5

Gineceu de 1 só carpelo (1 única placenta); folhas simples ou compostas, com estípulas ou com espinhos estipulares; árvores, arbustos ou ervas, por vezes escandentes **Leguminosae**
Gineceu de 2 ou mais carpelos unidos (2 ou mais placentas); folhas sem estípulas:
Flores zigomórficas; estames 4–2, em número menor que o dos lóbulos da corola e inseridos no tubo desta:

Folhas pinadas; árvores; estames 4; fruto grande, subcilíndrico e indeiscente **Bignoniaceae**
Folhas simples, por vezes reduzidas a escamas; ervas:
Plantas parasitas de raízes; folhas reduzidas a escamas; estames 4 **Orobanchaceae**
Plantas não parasitas; folhas (ou a única folha) bem desenvolvidas; estames 2 **Gesneriaceae**
Flores actinomórficas; estames em número igual ou maior que o dos lóbulos da corola:
Flores dióicas, pétalas unidas nas flores masculinas mas livres nas femininas; folhas palmati-3–7-lobadas; árvores - - - - - - - **Caricaceae**
Flores bissexuadas; folhas não palmatilobadas:
Estames não inseridos na corola e em número superior ao dos lóbulos desta; lóbulos da corola em número duplo do das sépalas; árvores ou arbustos **Annonaceae**
Estames inseridos no tubo da corola e em número igual ao dos lóbulos desta; lóbulos da corola tantos quantas as sépalas:
Folhas alternas, de limbo cordado-circular; lóbulos da corola induplicado-valvados; ervas aquáticas - - - - - **Menyanthaceae**
Folhas opostas; lóbulos da corola contortos; plantas terrestres:
Arbustos ou trepadeiras lenhosas; fruto uma baga grande ou uma cápsula espinhosa - - - - - - - - - **Apocynaceae**
Ervas; fruto uma cápsula septicida, não espinhosa - - **Gentianaceae**

Grupo 6

Ervas aquáticas, submersas, simulando fetos pelas suas folhas profundamente pinatissectas; flores dióicas em espigas alongadas e pedunculadas; cálice nulo **Hydrostachyaceae**
Plantas terrestres, não simulando fetos:
Folhas pinadas ou 2-folioladas; árvores - - - - - **Leguminosae**
Folhas simples ou digitado-3-folioladas, por vezes reduzidas a escamas:
Flores dióicas; folhas simples; árvores ou arbustos:
Cálice nulo; flores em amentilhos; fruto uma cápsula 2-valve; sementes com um tufo basal de pêlos finos e longos - - - - **Salicaceae**
Cálice presente; flores solitárias, fasciculadas ou em racimos curtos; fruto uma baga; sementes glabras ou com indumento lanoso - **Flacourtiaceae**
Flores bissexuadas:
Folhas opostas; fruto uma cápsula circuncisa; ervas ou pequenos arbustos **Aizoaceae**
Folhas alternas:
Estames em número igual ao das sépales e a elas opostos; flores com uma coroa fimbriada exterior aos estames; ervas ou arbustos, frequentemente escandentes com gavinhas - - - - - **Passifloraceae**
Estames em número superior ao das sépalas, ou igual e com elas alternos; árvores ou arbustos, sem gavinhas:
Ovário sobre um ginóforo nítido - - - - - **Capparidaceae**
Ovário séssil ou subséssil - - - - - - **Flacourtiaceae**

Grupo 7

Flores unissexuadas; estames 2–5, anteras frequentemente curvas, flexuosas ou plicadas; ervas ou arbustos, geralmente prostrados ou escandentes com gavinhas; folhas frequentemente palmati-lobadas ou profundamente recortadas - **Cucurbitaceae**
Flores bissexuadas; estames com anteras pequenas e rectas; plantas sem gavinhas:
Folhas nulas ou reduzidas a escamas; plantas suculentas, por vezes epifíticas **Cactaceae**
Folhas presentes, bem desenvolvidas:
Árvores ou arbustos; folhas alternas; estames opostos às pétalas, isolados ou em feixes; pétalas 5–9, persistentes - - - - - - - **Flacourtiaceae**
Ervas suculentas; folhas opostas; estames numerosos não dispostos em feixes; pétalas numerosas - - - - - - - - - - **Aizoaceae**

Grupo 8

Flores unissexuadas; folhas alternas, frequentemente palmatilobadas ou profundamente recortadas; estames 2–5, por vezes unidos, anteras frequentemente curvas, flexuosas ou plicadas; ervas, prostradas ou escandentes, ou arbustos com gavinhas **Cucurbitaceae**

Flores bissexuadas; folhas opostas ou verticiladas, inteiras, com estípulas interpeciolares; estames com anteras rectas; árvores ou arbustos, por vezes escandentes mas sem gavinhas - - - - - - - - - - - **Rubiaceae**

Grupo 9

Ervas ou arbustos, frequentemente escandentes, não parasitas; folhas bem desenvolvidas; flores bissexuadas; cálice com o tubo alongado e bojudo na base e com a fauce 3-lobulada ou $\pm$ desenvolvida unilateralmente em lóbulo único - **Aristolochiaceae**

Plantas parasitas sobre ramos de árvores; folhas reduzidas a escamas; flores unissexuadas, pequenas, solitárias ou por vezes geminadas, sésseis sobre o ritidoma do hospedeiro, rodeadas por folhas escamiformes e com cálice de 4 ou mais sépalas livres **Rafflesiaceae**

Grupo 10

Ovário com 1 só lóculo, por vezes septado na base:
 Sépalas 1–2, livres, por vezes caducas:
  Flores dióicas; sépalas 1–2; pétalas 1–4; estames unidos em sinândrio; limbo das folhas peltado ou subpeltado; trepadeiras lenhosas - **Menispermaceae**
  Flores bissexuadas; sépalas 2; pétalas 4–5; estames livres ou unidos pelos filetes; limbo das folhas não peltado; ervas:
   Flores zigomórficas; ovário com 1 só óvulo; fruto uma núcula; folhas muito recortadas - - - - - - - - - - - **Fumariaceae**
   Flores actinomórficas; ovário com óvulos numerosos; fruto uma cápsula; folhas inteiras, carnudas - - - - - - - - - - **Portulacaceae**
 Sépalas (ou lóbulos do cálice) 3 ou mais:
  Folhas opostas ou verticiladas, nunca todas basilares:
   Folhas com estípulas; ervas ou pequenos arbustos:
    Ovário com 2 óvulos; estames 1–2; fruto indeiscente, rodeado por brácteas carnudas - - - - - - - - - - - **Illecebraceae**
    Ovário com 3 ou mais óvulos; número dos estames não excedendo o duplo do das pétalas; fruto uma cápsula 3–5-valve - - **Caryophyllaceae**
   Folhas sem estípulas:
    Pétalas e estames perigínicos, inseridos no tubo do cálice; estilete 1:
     Ovário com 1 só óvulo apical pêndulo; estames em número igual ao das pétalas; arbustos ericóides - - - - - - **Thymelaeaceae**
     Ovário com óvulos numerosos em placenta basal livre ou central livre; estames em número duplo do das pétalas ou menor; ervas, arbustos ou árvores - - - - - - - - - - **Lythraceae**
    Pétalas e estames hipogínicos ou apenas ligeiramente perigínicos, não inseridos no cálice:
     Ervas; ovário com óvulos numerosos em placenta basal livre ou central livre; fruto uma cápsula; estiletes 2–5, livres - **Caryophyllaceae**
     Árvores ou arbustos; ovário com 1–2 óvulos basais ou apicais:
      Estames em número maior que o das pétalas (indefinidos), folhas com nervuras laterais numerosas e muito aproximadas - **Guttiferae**
      Estames em número igual ao das pétalas:
       Filetes unidos na base em tubo ou em taça - - **Salvadoraceae**
       Filetes livres:
        Óvulo 1; estiletes 3, livres ou unidos na base; flores dióicas ou poligâmicas - - - - - - - **Anacardiaceae**
        Óvulos 2; estilete 1 só ou estigma séssil; flores bissexuadas:
         Óvulos pêndulos do ápice do ovário - - - **Icacinaceae**
         Óvulos erectos na base do ovário - - - - **Celastraceae**
  Folhas alternas, por vezes todas basilares:
   Folhas compostas, de 3 ou mais folíolos; árvores ou arbustos, por vezes escandentes:
    Folhas 2-pinadas; flores bissexuadas; fruto alado, indeiscente **Leguminosae**
    Folhas 1-pinadas ou 3-folioladas; flores unissexuadas ou poligâmicas:
     Ovário com 2 óvulos; folíolos providos de pontuações glandulares translúcidas pelo menos na margem - - - - - - **Rutaceae**
     Ovário com 1 só óvulo; folíolos sem glândulas translúcidas **Anacardiaceae**
   Folhas simples ou 1-folioladas:
    Ovário com 1 só óvulo:
     Flores zigomórficas, com as 2 sépalas interiores maiores que as restantes, a pétala inferior (mediana) aquilhada e as 2 pétalas superiores vestigiais ou nulas; fruto longamente alado, indeiscente; árvores ou arbustos, por vezes escandentes - - - - - - - - **Polygalaceae**
     Flores actinomórficas; fruto não alado mas por vezes rodeado de sépalas persistentes e aliformes:

Folhas com estípulas; estames em número igual ao das pétalas; ervas **Illecebraceae**

Folhas sem estípulas:

Sépalas unidas em tubo alongado; estames em número duplo do das pétalas, inseridos no tubo do cálice; arbustos, por vezes ericóides **Thymelaeaceae**

Sépalas livres ou quase:

Estames unidos em sinândrio; limbo das folhas peltado ou subpeltado; trepadeiras lenhosas com flores dióicas - **Menispermaceae**

Estames livres ou quase; limbo das folhas não peltado:

Flores poligâmicas ou dióicas; estames 1–10; árvores ou arbustos **Anacardiaceae**

Flores bissexuadas; estames em número igual ao das pétalas e a elas opostos; trepadeiras lenhosas - - - **Opiliaceae**

Ovário com 2 ou mais óvulos:

Anteras deiscentes por duas valvas que se encurvam para cima; arbustos com espinhos 3-partidos dispostos nos nós; fruto uma baga **Berberidaceae**

Anteras deiscentes por fendas longitudinais:

Folhas com estípulas:

Pétalas e estames hipogínicos; ervas ou pequenos arbustos com pêlos estrelados; estames em número igual ao das pétalas, a elas opostos e unidos na base - - - - - - - - **Sterculiaceae**

Pétalas e estames perigínicos, inseridos na fauce do tubo do cálice; árvores ou arbustos:

Estames em número igual ao das pétalas e a elas opostos; óvulos 4, pêndulos do ápice do ovário - - - - **Flacourtiaceae**

Estames em número maior que o das pétalas; óvulos 2:

Ovário implantado lateralmente na parede do tubo do cálice; estilete inserido junto da base do ovário - - **Chrysobalanaceae**

Ovário centrado, não implantado no tubo do cálice; estilete terminal **Rosaceae**

Folhas sem estípulas; árvores ou arbustos, por vezes escandentes:

Pétalas imbricadas:

Estames em número igual ao das pétalas e a elas opostos; óvulos em placenta basal livre - - - - - - - **Myrsinaceae**

Estames em número igual ao das pétalas e com elas alternos; óvulos pêndulos do ápice ou de perto do ápice do ovário - **Rutaceae**

Pétalas valvadas:

Óvulos 2, pêndulos do ápice do ovário; estames em número igual ao das pétalas e com elas alternos - - - **Icacinaceae**

Óvulos 2–5, pêndulos do ápice da placenta central; estames em número igual ao das pétalas e a elas opostos, ou não excedendo o número duplo do das pétalas, ou reduzidos a 3 e acompanhados de estaminódios - - - - - - - - **Olacaceae**

Ovário com 2 ou mais lóculos:

Estames em número igual ao das pétalas e a elas opostos:

Filetes ± unidos em tubo ou em taça, por vezes alternando com estaminódios; segmentos do cálice valvados; ervas ou arbustos, frequentemente com pêlos estrelados - - - - **Sterculiaceae**

Filetes livres, não alternando com estaminódios:

Ovário com óvulos numerosos em cada lóculo; fruto uma cápsula loculicida; flores em panículas terminais; árvores com folhas providas de pontuações glandulares **Heteropyxidaceae**

Ovário com 1–2 óvulos em cada lóculo; fruto drupáceo ou bacáceo:

Inflorescências opostas às folhas; plantas herbáceas ou lenhosas, muitas vezes escandentes com gavinhas; folhas simples (frequentemente lobadas ou ± profundamente recortadas) ou digitadas; óvulos 2 em cada lóculo; fruto uma baga - - - - - - - - - - **Vitaceae**

Inflorescências axilares; árvores ou arbustos, frequentemente espinhosos, sem gavinhas; folhas simples, não profundamente recortadas; óvulos 1 em cada lóculo; fruto drupáceo - - - - - - **Rhamnaceae**

Estames em número igual ao das pétalas e com elas alternos, ou em número maior ou menor:

Folhas compostas, de 2 ou mais folíolos:

Inflorescências com gavinhas; arbustos ou ervas, escandentes; folhas pinadas ou 2-ternadas - - - - - - - - - - **Sapindaceae**

Inflorescências sem gavinhas:

Ervas; estames em número duplo do das pétalas:

Folhas opostas (as do mesmo par frequentemente desiguais), pinadas ou 2-folioladas, com estípulas; estilete 1 só - - - **Zygophyllaceae**
Folhas alternas (por vezes todas basilares), pinadas ou 3-folioladas, sem estípulas; estiletes 5 - - - - - **Oxalidaceae**
Árvores ou arbustos, por vezes escandentes:
Anteras com 1 só teca; folhas digitado-3–9-folioladas; estames 10 ou mais, com os filetes unidos na base - - - - **Bombacaceae**
Anteras com 2 tecas; folhas pinadas ou 2-pinadas ou 2–3-folioladas:
Folhas opostas ou subopostas:
Folhas imparipinadas ou 3-folioladas, folíolos providos de pontuações glandulares translúcidas pelo menos na margem; fruto bacáceo ou separável em 2–4 mericarpos - - - - - **Rutaceae**
Folhas imparipinadas, folíolos sem glândulas translúcidas; fruto uma cápsula lateralmente comprimida, sementes aladas - **Ptaeroxylaceae**
Folhas alternas:
Folhas com estípulas laterais ou intrapeciolares, imparipinadas; flores ± zigomórficas, em racimos; estames 4–5 - - **Melianthaceae**
Folhas sem estípulas:
Filetes unidos em tubo; folhas pinadas ou 2-pinadas - **Meliaceae**
Filetes livres ou apenas curtamente unidos na base:
Folíolos providos de pontuações glandulares translúcidas; folhas pinadas ou 3-folioladas - - - - - - **Rutaceae**
Folíolos sem pontuações glandulares translúcidas:
Ovário com 2 ou mais óvulos em cada lóculo:
Óvulos numerosos em cada lóculo; estiletes 5; fruto bacáceo; folhas pinadas - - - - - **Oxalidaceae**
Óvulos 2 em cada lóculo; estilete 1; fruto drupáceo, com exocarpo por vezes deiscente; folhas pinadas ou 3-folioladas; plantas resiníferas - - - - - **Burseraceae**
Ovário com 1 óvulo em cada lóculo:
Estiletes 3–5, livres e afastados na base; flores poligâmicas ou dióicas - - - - - - - **Anacardiaceae**
Estilete ou estiletes centrais ou terminais, não afastados na base:
Óvulos erectos ou ascendentes; folhas paripinadas, 2-pinadas ou 2–3-folioladas - - - - - **Sapindaceae**
Óvulos pêndulos:
Folhas 2-folioladas; ramos providos de espinhos rectos **Balanitaceae**
Folhas imparipinadas ou 3-folioladas; ramos inermes ou providos de acúleos curvos e curtos **Simaroubaceae**
Folhas simples ou 1-folioladas, por vezes profundamente recortadas:
Perianto acentuadamente zigomórfico; sépalas 3 ou, por vezes, 5, a mediana esporoada; pétalas 3; estames 5, anteras unidas em torno do ovário; fruto uma cápsula explosiva; ervas - - - - - **Balsaminaceae**
Perianto actinomórfico ou apenas ligeiramente zigomórfico:
Folhas opostas ou verticiladas, nunca todas basilares:
Estames em número superior ao duplo do das pétalas:
Folhas com estípulas:
Árvores; estípulas interpeciolares; pétalas franjadas no ápice; estames 15–45, filetes não unidos em feixes - - **Rhizophoraceae**
Ervas; estípulas não interpeciolares; pétalas não franjadas; estames 15, filetes unidos na base em 5 feixes - - - - **Geraniaceae**
Folhas sem estípulas:
Sépalas unidas na base; estames e pétalas periginicos, inseridos na fauce do tubo do cálice; filetes livres; estilete 1, comprido; folhas sem glândulas; árvores - - - - - **Sonneratiaceae**
Sépalas livres; estames e pétalas hipogínicos; filetes dispostos irregularmente ou ± unidos em feixes; estiletes 1–5 ou nulos; folhas em geral providas de pontuações ou linéolas glandulares translúcidas ou opacas; árvores, arbustos ou ervas - - - - - - **Guttiferae**
Estames não excedendo o número duplo do das pétalas:
Sépalas unidas na base, constituindo um tubo:
Pétalas periginicas, inseridas na fauce do tubo do cálice; ovário com óvulos numerosos em cada lóculo:
Anteras deiscentes por poros apicais; conectivo frequentemente apendiculado abaixo da antera; folhas com 3 ou mais nervuras longitudinais paralelas; ervas ou arbustos - **Melastomataceae**
Anteras deiscentes por fendas longitudinais; conectivo não apendi-

culado; folhas sem nervuras longitudinais paralelas; ervas, arbustos ou árvores - - - - - - - **Lythraceae**

Pétalas hipogínicas, não inseridas no tubo do cálice; ovário com 1–2 óvulos em cada lóculo; árvores ou arbustos:

Estames em número duplo do das pétalas; folhas com estípulas interpeciolares; pétalas franjadas no ápice; plantas inermes **Rhizophoraceae**

Estames em número igual ao das pétalas; folhas sem estípulas; pétalas não franjadas; plantas espinhosas - - - **Salvadoraceae**

Sépalas livres ou quase:

Ovário com 1 óvulo em cada lóculo fértil (por vezes 1 ou 2 dos lóculos vazios); estames em número duplo do das pétalas; fruto alado; arbustos, frequentemente escandentes - - - **Malpighiaceae**

Ovário com 2 ou mais óvulos em cada lóculo:

Folhas lobadas ou profundamente recortadas; ovário rostrado; ervas ou pequenos arbustos - - - - - - **Geraniaceae**

Folhas não lobadas nem recortadas; ovário não rostrado:

Estiletes 3–5; ervas ou pequenos arbustos:

Ovário com óvulos numerosos em cada lóculo; sépalas inteiras; folhas com estípulas - - - - - **Elatinaceae**

Ovário com 2 óvulos em cada lóculo; sépalas dentadas ou lobadas no ápice; folhas sem estípulas - - - - **Linaceae**

Estilete 1; árvores ou arbustos, por vezes escandentes:

Estames em número igual ao das pétalas e alternando com estaminódios petalóides; folhas providas de pontuações glandulares translúcidas - - - - - **Rutaceae**

Estames em número igual ou inferior ao das pétalas, não alternando com estaminódios; folhas sem pontuações glandulares translúcidas - - - - - - - **Celastraceae**

Folhas alternas, por vezes todas basilares:

Ovário com 1 óvulo em cada lóculo fértil (por vezes 1 ou 2 dos lóculos vazios):

Flores unissexuadas ou poligâmicas:

Folhas sem estípulas; árvores ou arbustos sem pêlos estrelados nem escamas; flores pequenas, em racimos ou panículas estreitas semelhantes a racimos - - - - - - **Sapindaceae**

Folhas com estípulas; plantas frequentemente com pêlos estrelados ou escamas:

Estilete 1, curto, com grande estigma 2–3-lobado; fruto com 4 asas longitudinais; estames numerosos; árvores - - **Tiliaceae**

Estiletes 2–5, livres ou unidos na base, frequentemente lobados ou ramosos; fruto não alado; estames 5 ou mais; árvores, arbustos ou ervas - - - - - - - - **Euphorbiaceae**

Flores bissexuadas:

Anteras com 1 só teca; estames numerosos com os filetes ± unidos em tubo; sépalas valvadas, com ou sem epicálice de bractéolas; ervas ou pequenos arbustos, frequentemente com pêlos estrelados **Malvaceae**

Anteras com 2 tecas:

Folhas sem estípulas:

Sépalas 4; estilete 1; folhas inteiras, dentadas ou ± pinatissectas; fruto uma silícula comprimida ou divisível em 2 cocas; ervas - - - - - - - - - **Cruciferae**

Sépalas 5; estiletes 2; folhas inteiras; fruto divisível em 2 cocas:

Arbustos, frequentemente escandentes; ovário 3-locular ou 2-locular por aborto; estiletes muito compridos; cocas com asa subterminal - - - - - - - **Malpighiaceae**

Ervas; ovário 2-locular; estiletes curtos; cocas não aladas ou aladas ao longo do dorso - - - - - **Aizoaceae**

Folhas com estípulas; fruto drupáceo; árvores ou arbustos:

Pétalas e estames periginicos, inseridos na fauce do tubo do cálice; ovário implantado lateralmente na parede do tubo do cálice; estilete inserido lateralmente junto da base do ovário **Chrysobalanaceae**

Pétalas e estames hipogínicos; ovário centrado, não implantado no cálice; estilete ou estiletes centrados e terminais:

Ovário profundamente lobado, com 1 só estilete, separando-se os carpelos na frutificação; estames em número duplo do das pétalas ou maior, livres - - - - **Ochnaceae**

Ovário inteiro, com 1 ou 3 estiletes, de carpelos não separáveis na frutificação; estames em número duplo do das pétalas:
Filetes unidos na base em tubo ou em taça; ovário com 2 lóculos férteis ou com 1 lóculo fértil e 2 vazios; estiletes 3 ou 1 só - - - - - **Erythroxylaceae**
Filetes livres; ovário com 2–6 lóculos férteis; estilete 1 só **Irvingiaceae**

Ovário com 2 ou mais óvulos em cada lóculo:
Estames em número igual ou inferior ao das pétalas:
Óvulos numerosos em cada lóculo; flores em umbelas pedunculadas e axilares; árvores ou arbustos - - - - - **Brexiaceae**
Óvulos 2 em cada lóculo:
Ovário 5-locular:
Pétalas imbricadas; folhas ± lobadas ou ± profundamente recortadas, com estípulas; ovário e fruto rostrados; ervas ou pequenos arbustos - - - - - - **Geraniaceae**
Pétalas contortas; folhas não lobadas nem profundamente recortadas; ovário e fruto não rostrados:
Ervas; pétalas caducas; estiletes 5 - - - - - **Linaceae**
Árvores ou arbustos; pétalas persistentes; estilete 1 só **Ixonanthaceae**
Ovário 2–4-locular; árvores ou arbustos, por vezes escandentes:
Pétalas 2-lobadas; fruto drupáceo; folhas inteiras **Dichapetalaceae**
Pétalas não 2-lobadas:
Flores unissexuadas; estiletes 2–3, livres ou unidos na base, frequentemente lobados ou ramosos - - **Euphorbiaceae**
Flores bissexuadas; estilete 1 só - - - **Celastraceae**
Estames em número superior ao das pétalas:
Folhas sem estípulas:
Filetes unidos em tubo; pétalas alongadas; árvores ou arbustos **Meliaceae**
Filetes livres:
Folhas providas de pontuações glandulares translúcidas, por vezes profundamente recortadas; ovário com 2 ou mais óvulos em cada lóculo; árvores, arbustos ou ervas - - - - **Rutaceae**
Folhas sem pontuações glandulares translúcidas; ovário com 2 óvulos em cada lóculo; árvores ou arbustos, resiníferos **Burseraceae**
Folhas com estípulas:
Fruto indeiscente, rodeado pelas 5 sépalas aliformes acrescentes; estames numerosos, livres; ovário 3-locular com 2 óvulos em cada lóculo; árvores ou arbustos - - - **Dipterocarpaceae**
Fruto não rodeado por sépalas aliformes persistentes:
Sépalas imbricadas; plantas sem pêlos estrelados; ovário 5-locular com 2 óvulos em cada lóculo:
Pétalas imbricadas; ovário rostrado, com estilete 5-ramoso; ervas ou pequenos arbustos - **Geraniaceae**
Pétalas contortas; ovário não rostrado, com 5 estiletes livres; árvores ou arbustos, por vezes escandentes - - **Linaceae**
Sépalas (ou lóbulos do cálice) valvadas; plantas frequentemente com pêlos estrelados:
Filetes livres ou quase; árvores, arbustos ou ervas - **Tiliaceae**
Filetes ± unidos em tubo ou em grupos de 2 ou 3:
Anteras com 2 tecas; árvores ou arbustos - **Sterculiaceae**
Anteras com 1 só teca; ervas ou arbustos ou, por vezes, árvores **Malvaceae**

Grupo 11

Ovário com 1 só lóculo, por vezes septado na base:
Flores dióicas, as masculinas com pétalas unidas, as femininas com 1–4 pétalas livres; estames unidos em sinândrio; limbo da folha peltado ou subpeltado; trepadeiras lenhosas - - - - - - - - - - - - - **Menispermaceae**
Flores bissexuadas; estames livres ou com filetes unidos:
Ovário com 1 só óvulo:
Flores zigomórficas, papilionáceas, com as 2 pétalas inferiores unidas em quilha; estames 10, diadelfos, sendo o superior livre e os 9 restantes unidos pelos filetes; ervas - - - - - - - - - - - - **Leguminosae**

Flores actinomórficas; estames não diadelfos:
Estames em número duplo do dos lóbulos do cálice, inseridos no tubo alongado do cálice; corola anular, inserida na fauce do tubo do cálice; óvulo pêndulo do ápice do ovário; arbustos, por vezes escandentes **Thymelaeaceae**
Estames em número igual ao dos lóbulos do cálice, não inseridos no cálice; corola 4–5-lobulada, não anular; óvulo erguendo-se da base:
Folhas alternas; estames opostos aos lóbulos da corola; cálice revestido por glândulas estipitadas; ervas ou arbustos, por vezes escandentes **Plumbaginaceae**
Folhas opostas; estames alternos com os lóbulos da corola; cálice sem glândulas estipitadas; arbustos ou árvores - - - **Salvadoraceae**
Ovário com 2 ou mais óvulos:
Estames em número menor que o dos lóbulos da corola; flores ± zigomórficas; ervas, frequentemente aquáticas:
Estames 2; corola acentuadamente zigomórfica, com tubo esporoado; folhas inteiras ou laciniadas e muitas vezes com ascídias carnívoras **Lentibulariaceae**
Estames 4; corola levemente zigomórfica, com tubo não esporoado; folhas não laciniadas, sem ascídias - - - - - - **Scrophulariaceae**
Estames em número igual ao dos lóbulos da corola; flores actinomórficas ou quase:
Estames opostos aos lóbulos da corola:
Árvores ou arbustos; fruto indeiscente, geralmente monospérmico; folhas alternas - - - - - - - - - **Myrsinaceae**
Ervas; fruto uma cápsula circuncisa ou 5-valve, polispérmico, ou, por vezes, indeiscente e monospérmico; folhas opostas ou alternas **Primulaceae**
Estames alternos com os lóbulos da corola; árvores ou arbustos:
Folhas opostas; lóbulos da corola e estames 4; óvulos 4, pêndulos do ápice duma placenta basal livre 4-alada; fruto uma cápsula 2-valve comprimida **Verbenaceae**
Folhas alternas; lóbulos da corola e estames 5; óvulos 2, pêndulos do ápice do ovário; fruto uma drupa - - - - - - **Icacinaceae**
Ovário com 2 ou mais lóculos:
Lóbulos da corola numerosos (10 ou mais):
Estiletes 5; fruto uma cápsula 5-valve; ervas com folhas carnudas - **Aizoaceae**
Estilete 1; fruto uma baga; árvores ou arbustos, por vezes escandentes:
Lóbulos da corola em 2–3 séries, imbricados; estames em número duplo do dos lóbulos internos da corola, ou em número igual e a eles opostos e, então, alternando com estaminódios; ovário com 1 óvulo em cada lóculo; folhas alternas **Sapotaceae**
Lóbulos da corola em 1 só série; estames em número igual ou inferior ao do dos lóbulos da corola e com eles alternos; ovário com 2 ou mais óvulos em cada lóculo:
Estames 10 ou mais, em número igual ao dos lóbulos da corola; lóbulos da corola contortos; óvulos numerosos em cada lóculo; folhas opostas **Potaliaceae**
Estames 2; lóbulos da corola imbricados; óvulos 2–4 em cada lóculo; folhas opostas ou, por vezes, alternas - - - - - - - **Oleaceae**
Lóbulos da corola menos de 10:
Estames em número superior ao dos lóbulos da corola:
Folhas com estípulas, frequentemente digitadas ou palmatilobadas; flores unissexuadas; árvores, arbustos ou ervas - - - - **Euphorbiaceae**
Folhas sem estípulas:
Flores zigomórficas, com a pétala inferior (mediana) aquilhada; filetes unidos em baínha fendida superiormente; ovário 2-locular com 1 óvulo em cada lóculo; ervas ou arbustos - - - - - - **Polygalaceae**
Flores actinomórficas; filetes não unidos em baínha; árvores ou arbustos:
Ovário com 1–2 óvulos em cada lóculo; fruto indeiscente, bacáceo; flores bissexuadas ou unissexuadas - - - - - **Ebenaceae**
Ovário com alguns ou numerosos óvulos em cada lóculo; fruto uma cápsula loculicida; flores bissexuadas:
Estames grupados em feixes de 3, alternos com os lóbulos da corola; anteras deiscentes por poros apicais; estilete 5-ramoso - - **Theaceae**
Estames não grupados em feixes; anteras deiscentes por poros apicais ou por fendas longitudinais; estilete indiviso; frequentemente plantas lembrando as urzes - - - - - - - - - **Ericaceae**
Estames em número igual ou inferior ao dos lóbulos da corola:
Estames 2–4, em número menor que o dos lóbulos da corola:
Ovário com mais de 4 óvulos em cada lóculo; flores zigomórficas;

Folhas pinadas, opostas ou ternadas; estames 4; fruto uma cápsula loculicida com sementes aladas; árvores ou arbustos, por vezes escandentes **Bignoniaceae**

Folhas simples, por vezes profundamente recortadas ou reduzidas a escamas:

Ovário com 4 compartimentos completos ou incompletos (os 2 lóculos originais subdivididos em 2 por falsos septos); estames 4; ervas ou arbustos - - - - - - - - **Pedaliaceae**

Ovário 2-locular, com os lóculos não subdivididos por falsos septos:

Óvulos dispostos em mais de 2 fiadas em cada placenta; estames 4 ou 2; fruto uma cápsula ou, por vezes, uma baga; funículos das sementes não uncinados nem endurecidos; folhas alternas, opostas ou verticiladas, por vezes reduzidas a escamas; ervas ou arbustos ou, por vezes, árvores - - - - - - - **Scrophulariaceae**

Óvulos dispostos em 1–2 fiadas em cada placenta; folhas opostas ou, por vezes, alternas:

Fruto indeiscente ou tardiamente deiscente, provido de pontas com ganchos; funículos das sementes não uncinados nem endurecidos; flores solitárias, axilares; estames 4; folhas ± sinuadas ou lobadas; ervas - - - - - - - - - **Pedaliaceae**

Fruto uma cápsula loculicida, sem espinhos; funículos das sementes frequentemente uncinados e endurecidos; flores solitárias ou grupadas em inflorescências; estames 4 ou 2; folhas não lobadas; ervas ou arbustos - - - - - - **Acanthaceae**

Ovário com 1–4 óvulos em cada lóculo:

Pedúnculos adnados aos pecíolos das folhas axilantes; estames 2; árvores ou arbustos com folhas alternas - - - - - **Dichapetalaceae**

Pedúnculos não adnados aos pecíolos:

Perianto actinomórfico; estames 2; folhas simples ou compostas, opostas ou, por vezes, ternadas ou alternas; árvores ou arbustos, por vezes escandentes - - - - - - - - **Oleaceae**

Perianto zigomórfico:

Ovário ± profundamente 4-lobado, com o estilete ginobásico; fruto divisível em 4 pseudo-aquénios (ou menos por aborto); folhas simples, opostas ou verticiladas ou, por vezes, alternas; ervas ou arbustos, frequentemente aromáticos - - - - **Labiatae**

Ovário não profundamente 4-lobado, com o estilete não ginobásico:

Fruto uma cápsula loculicida ou, por vezes, indeiscente, funículos das sementes frequentemente uncinados e endurecidos; ovário 2-locular, com 2–4 óvulos em cada lóculo; ervas ou arbustos com folhas simples e opostas - - - - - **Acanthaceae**

Fruto indeiscente ou divisível em 2 ou mais pirenos ou cocas, sementes presas a funículos não uncinados nem endurecidos:

Flores solitárias, axilares; fruto indeiscente, armado com 2 ou 4 pontas ou longitudinalmente 4-alado; estames 4; folhas opostas ou subopostas, frequentemente dentadas ou pinatilobadas; ervas - - - - - - - - - **Pedaliaceae**

Flores reunidas em inflorescências; fruto sem pontas nem asas:

Anteras com 1 só teca; folhas quase sempre alternas, simples, estreitas; ovário 2-locular com 1 óvulo apical pêndulo em cada lóculo; ervas ou arbustos - - - - - **Selaginaceae**

Anteras com 2 tecas; folhas opostas ou verticiladas, simples ou digitadas; ovário 2–8-locular com 1 óvulo basal ou axial em cada lóculo; ervas, arbustos ou árvores - **Verbenaceae**

Estames 4 ou mais, em número igual ao dos lóbulos da corola:

Folhas nulas ou reduzidas a escamas:

Plantas filamentosas, volúveis e parasitas; ovário com 2 óvulos em cada lóculo; fruto uma cápsula; corola com ou sem escamas infra-estaminais **Convolvulaceae**

Plantas suculentas ou arbustos, não parasitas; ovário com numerosos óvulos em cada lóculo; fruto formado por 2 folículos separados (ou 1 por aborto); flores com uma coroa - - - - - - - - - **Asclepiadaceae**

Folhas bem desenvolvidas:

Estames opostos aos lóbulos da corola; árvores ou arbustos com folhas alternas:

Folhas 2-pinadas; lóbulos da corola valvados; estames não acompanhados de estaminódios, filetes unidos em tubo - - - - - **Leeaceae**

Folhas simples, inteiras; lóbulos da corola imbricados; estames por vezes alternando com estaminódios - - - - - - **Sapotaceae**

Estames alternos com os lóbulos da corola:

Folhas opostas ou verticiladas, nunca todas basilares:
Estames hipogínicos, não inseridos na corola; anteras deiscentes por fendas apicais semelhantes a poros; arbustos com folhas pequenas e verticiladas, lembrando as urzes - - - - - **Ericaceae**
Estames inseridos no tubo da corola:
Lóbulos da corola imbricados:
Estilete 2-ramoso, com os ramos por sua vez 2-ramulosos; ovário 2-locular com 2 óvulos em cada lóculo; fruto uma cápsula loculicida 2-lobada e comprimida; arbustos - **Loganiaceae**
Estilete indiviso ou com 2–3 ramos apicais curtos:
Ovário com 1–2 óvulos em cada lóculo; fruto drupáceo ou divisível em pirenos ou pseudo-aquénios:
Lóbulos da corola e estames 5; ervas frequentemente híspidas com pêlos de base bolbosa - - - **Boraginaceae**
Lóbulos da corola e estames 4; ervas, arbustos ou árvores **Verbenaceae**
Ovário com óvulos numerosos em cada lóculo; fruto uma cápsula ou, por vezes, bacáceo; lóbulos da corola e estames 4:
Flores solitárias ou geminadas na axila das folhas; ervas **Scrophulariaceae**
Flores em cimeiras, racimos ou panículas:
Ervas; flores zigomórficas; fruto uma cápsula loculicida **Acanthaceae**
Árvores ou arbustos; flores actinomórficas; fruto uma cápsula septicida ou bacáceo - **Buddlejaceae**
Lóbulos da corola contortos ou valvados:
Flores com uma coroa; pólen aglutinado em grânulos ou em massas cerosas; ervas, arbustos ou árvores, frequentemente escandentes **Asclepiadaceae**
Flores sem coroa; pólen não aglutinado:
Lóbulos da corola valvados; folhas frequentemente com 3 ou mais nervuras longitudinais; árvores ou arbustos, por vezes escandentes, frequentemente com espinhos axilares ou com gavinhas - - - - **Strychnaceae**
Lóbulos da corola contortos:
Ervas; fruto uma cápsula septicida - - **Gentianaceae**
Árvores ou arbustos, por vezes escandentes; fruto uma baga, uma drupa ou com os carpelos separáveis na frutificação:
Lóbulos da corola e estames 6–9; filetes unidos em tubo; fruto uma baga - - **Potaliaceae**
Lóbulos da corola e estames 4–5; filetes livres **Apocynaceae**
Folhas (pelo menos as inferiores) alternas, por vezes todas basilares:
Folhas todas basilares; flores pequenas, em espigas ou capítulos pedunculados; fruto uma cápsula circuncisa; ervas - - **Plantaginaceae**
Folhas nunca todas basilares:
Ovário com mais de 2 óvulos em cada lóculo:
Lóbulos da corola contortos; fruto formado por 2 carpelos separados; arbustos ou árvores - - - - **Apocynaceae**
Lóbulos da corola não contortos; fruto uma cápsula ou uma baga:
Estiletes 2–3; lóbulos da corola imbricados; fruto uma cápsula; ervas - - - - **Hydrophyllaceae**
Estilete 1:
Lóbulos da corola plicados ou valvados; ervas, arbustos ou árvores - - - - - **Solanaceae**
Lóbulos da corola imbricados:
Fruto uma cápsula com sementes aladas; folhas simples ou 3-folioladas; arbustos espinhosos - - **Bignoniaceae**
Fruto uma baga ou uma cápsula com sementes não aladas; folhas simples; ervas ou arbustos espinhosos **Solanaceae**
Ovário com 1–2 óvulos em cada lóculo:
Filetes unidos em baínha fendida superiormente; flores zigomórficas, com o lóbulo inferior (mediano) da corola aquilhado; fruto drupáceo; arbustos ou árvores - - **Polygalaceae**
Filetes não unidos em baínha:
Tubo da corola fendido anteriormente, 4-lobulado; anteras com 1 só teca; ovário 2-locular com 1 óvulo apical pêndulo em cada lóculo; ervas ou pequenos arbustos com flores em espigas - - - - - - - - **Selaginaceae**

Tubo da corola não fendido anteriormente; anteras com 2 tecas:
Fruto drupáceo; árvores ou arbustos:
Estilete 2-ramoso (com os ramos indivisos ou por sua vez 2-ramulosos) ou terminando num anel horizontal provido de um estigma 2-lobado; folhas sem estípulas; flores bissexuadas, em cimeiras ou panículas - **Boraginaceae**
Estilete nulo, estigma séssil; folhas com estípulas; flores dióicas, em cimeiras ou fascículos axilares **Aquifoliaceae**
Fruto não drupáceo; ervas ou pequenos arbustos, por vezes escandentes:
Fruto divisível em 4 pseudo-aquénios (ou menos por aborto); lóbulos da corola imbricados ou, por vezes, contortos **Boraginaceae**
Fruto capsular ou, por vezes, indeiscente; lóbulos da corola plicados ou valvados; plantas frequentemente volúveis ou prostradas - - - - - - **Convolvulaceae**

Grupo 12

Ovário com 2 ou mais óvulos em cada lóculo:
Plantas aquáticas, simulando hepáticas ou musgos, crescendo sobre rochas; flores pequenas e pouco aparentes; estames 1–2; ovário com óvulos numerosos; fruto uma cápsula - - - - - - - - - - - - **Podostemaceae**
Plantas terrestres, não simulando hepáticas nem musgos:
Folhas reduzidas a escamas formando baínhas dentadas envolvendo os nós; árvores com flores unissexuadas, as masculinas em espigas e as femininas em capítulos; estame 1; ovário 1-locular com 2 óvulos colaterais; estilete com 2 ramos alongados **Casuarinaceae**
Folhas bem desenvolvidas, não reduzidas a escamas:
Folhas opostas, verticiladas ou todas basilares:
Flores unissexuadas; árvores ou arbustos:
Cálice nulo; folhas dentadas no ápice, flabeladas; flores em espigas simulando amentilhos; óvulos numerosos em cada lóculo - **Myrothamnaceae**
Cálice presente; folhas inteiras, não flabeladas; flores solitárias ou fasciculadas na axila das folhas; óvulos 2 em cada lóculo:
Folhas com estípulas; flores dióicas; estames muito numerosos, dispostos em espiral ao longo do receptáculo alongado - - **Euphorbiaceae**
Folhas sem estípulas; flores monóicas; estames 4–6 - - **Buxaceae**
Flores bissexuadas:
Cálice provido de um esporão adnato ao pedicelo; ovário rostrado; folhas dentadas ou lobadas; ervas - - - - - - - **Geraniaceae**
Cálice sem esporão; ovário não rostrado; folhas inteiras:
Sépalas livres ou quase; estames hipogínicos; ervas:
Ovário 3–5-locular com placentas axiais - - - **Aizoaceae**
Ovário 1-locular com placenta basal livre ou central livre **Caryophyllaceae**
Sépalas unidas na base; estames perigínicos, inseridos no tubo do cálice:
Estiletes 2–5 ou estilete 1 só e, então, ovário 1-locular e com poucos óvulos; fruto uma cápsula circuncisa; ervas - - - - - **Aizoaceae**
Estilete 1; ovário com 1 ou mais lóculos, óvulos numerosos em cada lóculo:
Árvores do litoral; estames numerosos em número maior que o duplo do dos lóbulos do cálice; ovário com 10 ou mais lóculos; fruto uma baga - - - - - - - **Sonneratiaceae**
Ervas, por vezes aquáticas; estames em número que não excede o duplo do dos lóbulos do cálice, por vezes 1 único; ovário completamente ou incompletamente 2–5-locular; fruto uma cápsula **Lythraceae**
Folhas alternas, nunca todas basilares:
Folhas pinadas; árvores com flores unissexuadas ou poligâmicas; ovário 2-locular com 2 óvulos em cada lóculo; fruto indeiscente - **Sapindaceae**
Folhas simples ou digitadas:
Ovário com 1 só lóculo; flores em racimos ou espigas:
Flores bissexuadas; cálice escarioso; fruto uma cápsula circuncisa; ervas **Amaranthaceae**
Flores dióicas; cálice não escarioso; fruto uma drupa; árvores ou arbustos, por vezes escandentes:
Folhas com estípulas; cálice imbricado; estames opostos às sépalas; estiletes 3, geralmente 2-lobados - - - - **Euphorbiaceae**

Folhas sem estípulas; cálice valvado; estames alternos com as sépalas; estilete nulo, estigma séssil e multirradiado - - **Icacinaceae**
Ovário com 2 ou mais lóculos:
Gineceu de 3 ou mais carpelos ligeiramente unidos mas separáveis na frutificação em outros tantos folículos (ou menos por aborto); árvores ou arbustos com flores unissexuadas ou poligâmicas; folhas simples ou digitadas; sépalas valvadas, unidas na base; estames unidos em coluna **Sterculiaceae**
Gineceu de carpelos completamente unidos, formando na frutificação uma cápsula ou um fruto indeiscente ou divisível em cocas aladas:
Ovário sobre um ginóforo nítido; flores bissexuadas; arbustos com flores pedunculadas, solitárias e axilares - - **Capparidaceae**
Ovário séssil ou subséssil; flores unissexuadas ou poligâmicas:
Folhas com estípulas; fruto uma cápsula ou um fruto indeiscente ou divisível em 2 cocas aladas; árvores, arbustos ou ervas **Euphorbiaceae**
Folhas sem estípulas; fruto uma cápsula septicida com 2 ou mais asas longitudinais membranosas; arbustos ou árvores **Sapindaceae**
Ovário com 1 óvulo em cada lóculo:
Ovário com 2 ou mais lóculos:
Folhas pinadas; árvores ou arbustos com flores poligâmicas ou dióicas; fruto drupáceo ou bacáceo - - - - - - - - - - **Sapindaceae**
Folhas simples, por vezes lobadas ou muito recortadas, ou reduzidas a escamas ou a espinhos estipulares:
Flores unissexuadas ou poligâmicas:
Ovário 5-locular com os carpelos ligeiramente unidos e separando-se na frutificação; cálice presente, sépalas valvadas e unidas na base; estames 5–10, unidos em coluna; folhas alternas; árvores ou arbustos - **Sterculiaceae**
Ovário 2–4-locular com os carpelos completamente unidos e não se separando na frutificação; cálice presente ou nulo; flores masculinas por vezes reduzidas a 1 estame único e dispostas em um invólucro comum em torno de 1 flor feminina pedicelada; folhas alternas ou opostas, por vezes reduzidas a escamas ou a espinhos estipulares; árvores, arbustos ou ervas, por vezes escandentes ou suculentos - - - - - - **Euphorbiaceae**
Flores bissexuadas:
Sépalas unidas em tubo; estames em número duplo do dos lóbulos do cálice, periginicos e inseridos no tubo do cálice; ovário 2-locular; fruto uma drupa; arbustos ou árvores - - - **Thymelaeaceae**
Sépalas livres; estames hipogínicos ou quase:
Folhas $\pm$ dentadas ou pinatilobadas a profundamente recortadas; estilete 1; fruto uma silícula comprimida ou divisível em 2 cocas; ervas **Cruciferae**
Folhas inteiras; estiletes 2 ou mais, livres ou unidos na base:
Ovário 7–10-locular; fruto uma baga; flores em racimos simulando espigas; ervas ou pequenos arbustos - - - **Phytolaccaceae**
Ovário 2–5-locular; fruto uma cápsula loculicida ou divisível em 2 cocas, estas por vezes aladas ao longo do dorso; flores em cimeiras laxas ou densas; ervas - - - - - - - - - - **Aizoaceae**
Ovário com 1 só lóculo:
Folhas nulas ou reduzidas a escamas; flores em espigas:
Plantas filamentosas, volúveis e parasitas; flores não incluídas em escavações do eixo da espiga; estames 6–9, acompanhadas de estaminódios; anteras deiscentes por valvas - - - - - - - - - **Lauraceae**
Plantas herbáceas suculentas, dos salgadiços, com ramos articulados, não parasitas; flores incluídas em escavações do eixo da espiga; estames 1–2; anteras deiscentes por fendas longitudinais - - - - - - **Chenopodiaceae**
Folhas bem desenvolvidas:
Folhas com estípulas que, por vezes, formam uma bainha (ócrea) a rodear o caule:
Folhas 3–4-pinadas; cálice petalóide; fruto de aquénios estipitados na extremidade de um pedículo delgado; ervas - - - - - **Ranunculaceae**
Folhas simples ou digitadas:
Óvulo pêndulo do ápice ou de perto do ápice do ovário; flores unissexuadas ou poligâmicas:
Flores em espigas ou capítulos densos, ou dispostas em grande número sobre um receptáculo aberto ou plano ou no interior de um receptáculo oco quase fechado (figo), as femininas por vezes $\pm$ imersas no tecido do receptáculo; cálice por vezes nulo; árvores, arbustos ou ervas **Moraceae**
Flores solitárias, fasciculadas, em cimeiras, racimos ou panículas; nas flores femininas cálice por vezes vestigial ou nulo:

Ervas anuais; folhas opostas ou alternas, todas ou apenas as inferiores ± profundamente recortadas; flores masculinas em panículas alongadas, flores femininas em espigas, envolvidas por brácteas; fruto seco, indeiscente - - - - - - **Cannabaceae**

Árvores ou arbustos; folhas alternas:

Estames em número igual ao dos segmentos do cálice, 4–5; estilete 2-ramoso com estigmas simples ou divididos; fruto uma drupa; árvores ou arbustos - - - - - - - **Ulmaceae**

Estames em número maior que o dos segmentos do cálice; estilete indiviso, por vezes muito curto:

Flores solitárias ou geminadas na axila das folhas; folhas 1–3-folioladas, com folíolos pequenos e estreitos; arbustos ericóides **Rosaceae**

Flores em racimos ou panículas axilares; folhas simples; árvores **Euphorbiaceae**

Óvulo ascendente na base ou perto da base do ovário:

Folhas opostas, com marcas de cistólitos; flores unissexuadas; ervas ou arbustos - - - - - - - - - - **Urticaceae**

Folhas alternas:

Cálice nulo; flores muito pequenas, em espigas densas; arbustos, por vezes escandentes - - - - - - - **Piperaceae**

Cálice presente:

Estiletes 2–3, livres ou unidos na base; estames 4–8; fruto uma núcula; estípulas formando frequentemente uma ócrea que rodeia o caule; ervas ou arbustos, por vezes escandentes - **Polygonaceae**

Estilete 1 ou nulo; estames 5 ou menos:

Flores bissexuadas; sépalas unidas na base, com as partes livres alternas com os lóbulos de um epicálice; estilete inserido lateralmente junto da base do ovário; ervas com folhas palmatilobadas **Rosaceae**

Flores unissexuadas; sépalas livres ou unidas, sem epicálice; estilete, ou estigma séssil, terminal:

Folhas digitadas ou palmatilobadas, sem cistólitos; flores dióicas, as masculinas muito numerosas em espigas ramificadas, as femininas em capítulos; árvores - - - - **Moraceae**

Folhas simples, por vezes pinatilobadas, frequentemente com marcas de cistólitos; flores monóicas ou dióicas, em cimeiras frequentemente densas ou capituliformes; ervas, arbustos ou árvores, por vezes escandentes ou epifíticos - **Urticaceae**

Folhas sem estípulas:

Ervas aquáticas, submersas; folhas verticiladas, profunda e bifurcadamente recortadas, com os segmentos lineares ou filiformes; flores unissexuadas, solitárias e sésseis na axila das folhas - - - **Ceratophyllaceae**

Plantas não aquáticas; folhas alternas ou opostas, não recortadas ou, por vezes, pinatissectas:

Cálice nulo; flores em espigas:

Flores unissexuadas; estames 3–12; estiletes 2, livres ou curtamente unidos na base; folhas alternas; árvores ou arbustos - **Myricaceae**

Flores bissexuadas; estames 2; estilete nulo, estigma séssil; folhas alternas, opostas ou verticiladas; ervas, por vezes prostradas, escandentes ou epifíticas - - - - - - - - - **Piperaceae**

Cálice presente (por vezes nulo nas flores femininas):

Estames em número duplo do dos segmentos do cálice, ou maior:

Óvulo erecto na base do ovário; fruto encerrado no tubo do cálice persistente e longitudinalmente 4-alado; arbustos espinhosos com folhas pequenas e lineares, frequentemente fasciculadas; flores polígamo-dióicas - - - **Nyctaginaceae**

Óvulo pêndulo do ápice do ovário; fruto não encerrado em um tubo alado do cálice; plantas não espinhosas:

Flores bissexuadas; sépalas unidas em tubo alongado; estames inseridos no tubo do cálice; estilete alongado, delgado; arbustos (frequentemente ericóides) ou, por vezes, árvores, com folhas alternas ou opostas - - - - - - - - **Thymelaeaceae**

Flores dióicas; sépalas curtamente unidas na base; estames não inseridos no tubo do cálice; estilete nulo; estigma séssil; arbustos ou árvores com folhas opostas ou subopostas - **Monimiaceae**

Estames em número menor que o duplo do dos segmentos do cálice, por vezes acompanhados de estaminódios:

Anteras deiscentes por valvas; cálice 6-lobulado; estames 6–9 acompanhados de estaminódios; árvores - - - **Lauraceae**

Anteras deiscentes por fendas longitudinais; cálice com 3–5 segmentos ou lóbulos:

Folhas opostas ou subopostas:

Sépalas ± petalóides, unidas em tubo constricto acima do ovário; fruto incluído na parte inferior do tubo do cálice, persistente e frequentemente glandulosa no exterior; estames não acompanhados de estaminódios; ervas ou arbustos escandentes **Nyctaginaceae**

Sépalas ± escariosas, livres ou curtamente unidas na base; estames em número igual ao dos segmentos do cálice e a eles opostos, frequentemente alternando com estaminódios; ervas ou arbustos **Amaranthaceae**

Folhas alternas:

Ervas volúveis; sépalas unidas na base em um tubo com 2 bractéolas adnadas; flores em espigas pedunculadas e axilares **Basellaceae**

Plantas não volúveis; sépalas livres ou unidas, sem bractéolas:

Cálice com 2–4 segmentos valvados; árvores ou arbustos:

Flores dióicas, em capítulos pequenos solitários ou aglomerados (por vezes paniculados); estames 3–5 unidos em coluna; fruto com pericarpo espesso, carnudo e deiscente; semente com arilo - - - - - - - - **Myristicaceae**

Flores bissexuadas, em grandes capítulos involucrados, espigas alongadas ou racimos; estames 4, livres, inseridos nos segmentos do cálice; fruto uma noz; sementes sem arilo **Proteaceae**

Cálice com 3–5 segmentos imbricados ou, por vezes, quase completamente tubuloso e com segmentos indistintos; ervas ou arbustos:

Estames em número menor que o dos segmentos do cálice e a maioria alternos com eles; folhas pequenas, lineares; flores em espigas simples e alongadas - **Phytolaccaceae**

Estames em número igual ou inferior ao dos segmentos do cálice e a eles opostos:

Sépalas ± escariosas, livres ou curtamente unidas na base; folhas inteiras - - - - - **Amaranthaceae**

Sépalas herbáceas, livres ou ± unidas em tubo; folhas inteiras, ± dentadas ou pinatissectas **Chenopodiaceae**

Grupo 13

Arbustos parasitas sobre outros arbustos ou árvores; óvulo dificilmente distinguível do tecido do ovário que o envolve; cálice truncado ou obsoleto; estames em número igual ao das pétalas e nelas inseridos; folhas simples, inteiras, opostas ou alternas, por vezes reduzidas a escamas - - - - - - - - - - - **Loranthaceae**

Plantas não parasitas; óvulo ou óvulos nìtidamente distinguíveis:

Ovário com 1 só lóculo:

Árvores ou arbustos, por vezes escandentes; estilete 1, não ramoso:

Óvulo 1 só, pêndulo do ápice do ovário; flores em cimeiras axilares; estames em número igual ao das pétalas; folhas alternas, frequentemente assimétricas na base **Alangiaceae**

Óvulos 2 ou mais:

Óvulos pêndulos do ápice do ovário; fruto indeiscente, com 2–6 asas ou arestas longitudinais; flores em racimos, espigas ou capítulos; estames em número duplo do das pétalas ou, por vezes, igual; folhas opostas ou, por vezes, verticiladas ou alternas - - - - - - - - - - - **Combretaceae**

Óvulos em placenta basal livre ou central livre; fruto bacáceo; flores em cimeiras, fascículos ou umbelas axilares; estames em número duplo do das pétalas; folhas opostas - - - - - - - - **Melastomataceae**

Ervas; estiletes ou ramos do estilete 2–6:

Óvulos numerosos em placenta basal livre; sépalas 2, frequentemente caducas; fruto uma cápsula circuncisa - - - - - - - - **Portulacaceae**

Óvulo ou óvulos pêndulos do ápice do ovário ou em placentas apicais pêndulas; sépalas ou lóbulos do cálice 2–5; fruto não circunciso:

Óvulos numerosos em placentas pêndulas; fruto uma cápsula de deiscência apical; folhas opostas; flores bissexuadas, geminadas, axilares; pétalas e estames 5 - - - - - - - - - - - **Vahliaceae**

Óvulos 1–4, pêndulos do ápice do ovário; fruto indeiscente; folhas alternas ou opostas, por vezes todas basilares; flores bissexuadas ou unissexuadas, em espigas paniculadas ou em glomérulos axilares; pétalas e estames 2–4 **Haloragaceae**

Ovário com 2 ou mais lóculos:

Óvulo 1 em cada lóculo:

Estames em número duplo do das pétalas; estiletes 3–10:

Folhas verticiladas, pinatissectas com segmentos filiformes; ervas aquáticas com flores em espigas interrompidas - - - - - **Haloragaceae**

Folhas alternas; plantas não aquáticas:

Ervas com flores solitárias axilares; ovário 10-locular; fruto uma cápsula espinhosa; folhas ± pinatilobadas - - - - - - **Rosaceae**

Árvores ou arbustos com flores em espigas; ovário 3–5-locular; fruto uma drupa; folhas inteiras - - - - - - - **Rhizophoraceae**

Estames em número igual ao das pétalas:

Estames opostos às pétalas; arbustos, muitas vezes escandentes com gavinhas; fruto frequentemente divisível em cocas, por vezes alado - **Rhamnaceae**

Estames alternos com as pétalas; plantas sem gavinhas:

Flores solitárias na axila das folhas; ervas aquáticas, flutuantes; folhas alternas, aproximadas em roseta, com pecíolo ± inflado; fruto grande, indeiscente, com endocarpo duro, armado com 2 ou 4 pontas **Trapaceae**

Flores grupadas em inflorescências; plantas não aquáticas:

Fruto divisível em 2 cocas; ervas, por vezes arborescentes; flores em umbelas simples ou compostas, por vezes em capítulos; folhas frequentemente muito e profundamente recortadas ou compostas **Umbelliferae**

Fruto drupáceo ou capsular, não divisível em cocas; árvores ou arbustos:

Folhas pinadas, digitadas ou palmatissectas; flores em umbelas, racimos ou espigas; fruto drupáceo - - - **Araliaceae**

Folhas simples, indivisas:

Flores unissexuadas, em capítulos; pétalas e estames 5; fruto uma cápsula 2-valve; folhas inteiras - - - **Hamamelidaceae**

Flores bissexuadas ou unissexuadas, em panículas ou cimeiras simulando umbelas; pétalas e estames 4; fruto uma drupa; folhas opostas, inteiras ou dentadas - - - - - **Cornaceae**

Óvulos 2 ou mais em cada lóculo:

Folhas alternas:

Folhas com estípulas, frequentemente assimétricas; flores monóicas; estames numerosos; óvulos muito numerosos em cada lóculo; ervas, por vezes epifíticas - - - - - - - - - - - **Begoniaceae**

Folhas sem estípulas:

Estiletes 2; flores unissexuadas ou poligâmicas, em panículas axilares; estames em número igual ao das pétalas; ovário 2-locular; arbustos **Escalloniaceae**

Estilete 1, com estigma inteiro ou lobado; flores bissexuadas ou, por vezes, poligâmicas:

Estames em número igual ou duplo do das pétalas; fruto uma cápsula de deiscência valvar ou irregular ou, por vezes, indeiscente; ervas ou arbustos, por vezes aquáticos - - - - - **Onagraceae**

Estames numerosos, em número maior que o duplo do das pétalas; fruto uma baga ou uma drupa:

Flores solitárias ou geminadas na axila das folhas; ovário 2-locular; folhas com pontuações glandulares; pequenos arbustos **Myrtaceae**

Flores em racimos terminais; ovário 4-locular; folhas sem pontuações glandulares; árvores - - - - - - - **Lecythidaceae**

Folhas opostas:

Estames em número igual ao das pétalas; pétalas alternando com escamas incurvadas; árvores ou arbustos - - - - - - **Oliniaceae**

Estames em número duplo do das pétalas ou maior; pétalas não alternando com escamas:

Estames numerosos, em número maior que o duplo do das pétalas; árvores ou arbustos com folhas providas de pontuações glandulares **Myrtaceae**

Estames em número duplo do das pétalas:

Folhas com estípulas interpeciolares; óvulos 2 em cada lóculo; árvores ou arbustos de mangal, vivíparos - - - - **Rhizophoraceae**

Folhas sem estípulas; óvulos numerosos em cada lóculo; plantas não vivíparas:

Anteras deiscentes por um poro apical; conectivo frequentemente apendiculado abaixo da antera; folhas com 3 ou mais nervuras longitudinais

paralelas; sementes sem tufo de pêlos; ervas, arbustos ou pequenas árvores - - - - - - - - **Melastomataceae**
Anteras deiscentes por fendas longitudinais; conectivo inapendiculado; folhas sem nervuras longitudinais paralelas; sementes com um tufo de pêlos apical; ervas - - - - - - - **Onagraceae**

Grupo 14

Arbustos parasitas sobre outros arbustos ou árvores; óvulo dificilmente distinguível do tecido do ovário que o envolve; cálice truncado ou curtamente lobulado; estames em número igual ao dos lóbulos da corola e neles inseridos; folhas simples, inteiras, opostas, alternas ou por vezes ternadas - - - - - - - **Loranthaceae**
Plantas não parasitas; óvulo ou óvulos nìtidamente distinguíveis:
Ovário com 1 só lóculo:
Óvulo 1 só; flores actinomórficas ou zigomórficas, em capítulos involucrados; fruto indeiscente, frequentemente coroado pelo cálice persistente e transformado em papilho de pêlos, aristas ou escamas:
Anteras conadas em tubo, rodeando o estilete; óvulo erecto na base do ovário; ervas, arbustos ou árvores; corola das flores marginais frequentemente diferindo da das internas - - - - - - **Compositae**
Anteras livres; óvulo pêndulo do ápice do ovário; ervas com folhas opostas **Dipsacaceae**
Óvulos numerosos em placenta basal livre; flores actinomórficas, não dispostas em capítulos involucrados:
Cálice formado por 2 sépalas frequentemente caducas; fruto uma cápsula circuncisa; estames em número igual ao dos lóbulos da corola e com eles alternos, ou mais numerosos; ervas - - - - - - **Portulacaceae**
Cálice 4–5-lobulado; fruto não circunciso; estames em número igual ao dos lóbulos da corola e a eles opostos:
Árvores ou arbustos; flores em panículas ou racimos axilares; fruto uma baga **Myrsinaceae**
Ervas; flores em racimos terminais; fruto uma cápsula 5-valve **Primulaceae**
Ovário com 2 ou mais lóculos (por vezes 3-locular com 1 lóculo fértil e 2 vazios):
Pétalas unidas em uma peça caduca (caliptra); estames muito numerosos; árvores ou arbustos com folhas providas de pontuações glandulares - **Myrtaceae**
Pétalas ± unidas em tubo, mas não em caliptra; estames em número que não excede o duplo do dos lóbulos da corola:
Ovário 3-locular com 2 lóculos vazios, o lóculo fértil com 1 só óvulo apical pêndulo; estames 3; fruto indeiscente, frequentemente coroado pelo cálice persistente e transformado em papilho plumoso; ervas com folhas opostas **Valerianaceae**
Ovário com 2 ou mais lóculos, sem lóculos vazios:
Ervas ou arbustos, prostrados ou escandentes com gavinhas; flores unissexuadas; estames 3–5, com as anteras por vezes curvas, flexuosas ou plicadas; folhas frequentemente palmatilobadas, pedatilobadas ou profundamente recortadas **Cucurbitaceae**
Plantas sem gavinhas; flores bissexuadas ou, por vezes, unissexuadas:
Folhas opostas ou verticiladas, com estípulas (por vezes foliáceas) interpeciolares ou intrapeciolares e com margens inteiras; árvores, arbustos ou ervas, por vezes escandentes - - - - - - - **Rubiaceae**
Folhas alternas ou opostas, sem estípulas, por vezes reduzidas a escamas:
Estames em número duplo do dos lóbulos da corola; anteras deiscentes por poros apicais; flores em racimos axilares; árvores ou arbustos com folhas alternas - - - - - - - - - **Ericaceae**
Estames em número igual ao dos lóbulos da corola; anteras abrindo por fendas longitudinais:
Óvulo 1 em cada lóculo; fruto uma drupa; flores zigomórficas, tubo da corola fendido dorsalmente; anteras livres; arbustos do litoral **Goodeniaceae**
Óvulos numerosos em cada lóculo; fruto uma cápsula:
Lóbulos da corola imbricados; cápsula circuncisa; flores actinomórficas, em espigas terminais; anteras livres; ervas anuais **Sphenocleaceae**
Lóbulos da corola valvados; cápsula deiscente por valvas; flores actinomórficas ou zigomórficas, tubo da corola por vezes fendido dorsalmente; anteras livres ou conadas em tubo que rodeia o estilete; ervas ou arbustos, por vezes escandentes ou atingindo grande estatura **Campanulaceae**

Grupo 15

Ovário com 2 ou mais lóculos:
Flores unissexuadas; folhas com estípulas:
Flores em capítulos; estames em número igual ao dos segmentos do cálice; ovário com 1 óvulo em cada lóculo; árvores ou arbustos com folhas inteiras **Hamamelidaceae**
Flores em cimeiras ou panículas; estames numerosos; ovário com óvulos muito numerosos em cada lóculo; sépalas petalóides; ervas por vezes epifíticas, com folhas alternas e frequentemente assimétricas - - - - **Begoniaceae**
Flores bissexuadas; folhas sem estípulas:
Folhas opostas; flores solitárias, subsésseis na axila das folhas; ervas **Onagraceae**
Folhas alternas:
Sépalas 5, livres ou curtamente unidas na base; estames acompanhados de numerosos estaminódios, os da série exterior unidos simulando uma corola; anteras não adnadas ao estilete; árvores ou arbustos - - - - **Lecythidaceae**
Sépalas unidas em tubo alongado e bojudo na base; estames não acompanhados de estaminódios; anteras adnadas ao estilete; ervas ou arbustos frequentemente escandentes - - - - - - - - **Aristolochiaceae**
Ovário com 1 só lóculo:
Óvulos 2 ou mais:
Plantas afilas, parasitas de raízes; flores solitárias, subsésseis sobre o rizoma; óvulos muito numerosos em placentas apicais pêndulas - - - **Hydnoraceae**
Plantas com folhas por vezes muito pequenas; flores solitárias na axila das folhas ou reunidas em inflorescências; óvulos 2–4:
Estames em número duplo do dos lóbulos do cálice; flores em racimos, espigas ou capítulos; fruto indeiscente, frequentemente com 2–5 asas longitudinais; árvores ou arbustos - - - - - - - **Combretaceae**
Estames em número igual ao dos lóbulos do cálice:
Óvulos 2–4, pêndulos duma placenta basal livre; estilete 1 com o estigma inteiro ou dividido; flores solitárias ou em cimeiras ou racimos, bissexuadas ou unissexuadas; ervas ou arbustos, frequentemente parasitas de raízes e por vezes com folhas reduzidas - - - - - - - **Santalaceae**
Óvulos 4, pêndulos do ápice do ovário; estiletes 4; flores em glomérulos na axila das folhas, unissexuadas ou poligâmicas; ervas não parasitas **Haloragaceae**
Óvulo 1 só:
Ervas suculentas, parasitas de raízes; folhas reduzidas a escamas; flores unissexuadas dispostas em capítulos involucrados, ou as flores masculinas paniculadas e as femininas grupadas em capítulos paniculados - - - **Balanophoraceae**
Plantas não parasitas; folhas bem desenvolvidas:
Flores unissexuadas, dispostas em capítulos densos ou sobre um receptáculo aberto ou plano, as femininas por vezes solitárias no receptáculo; árvores, arbustos ou ervas - - - - - - - - - - **Moraceae**
Flores em cimeiras ou espigas paniculadas:
Árvores com folhas inteiras ou palmati-3–5-lobadas; flores em cimeiras, unissexuadas ou poligâmicas; fruto uma noz coroada por 2 asas alongadas originadas pelos lóbulos acrescentes e persistentes do cálice **Hernandiaceae**
Ervas com folhas basilares longamente pecioladas, reniformes e dentadas; flores em espigas paniculadas, bissexuadas ou unissexuadas; fruto pequeno, não alado, drupáceo - - - - - - - **Haloragaceae**

MONOCOTYLEDONES

Gineceu de 2 ou mais carpelos livres, com estiletes e estigmas também livres *Grupo 1* (ver p. 27)
Gineceu de 1 carpelo ou de 2 ou mais carpelos unidos, com estiletes e estigmas livres ou unidos:
Ovário súpero:
Perianto presente, composto de 4 ou mais segmentos livres ou unidos, não reduzidos a sedas:
Perianto composto de cálice e corola separados, o cálice frequentemente herbáceo, a corola em geral petalóide ou de algum modo diferindo do cálice, as sépalas e as pétalas quer livres quer conadas mas nunca unidas em um tubo comum *Grupo 2* (ver p. 27)
Perianto composto de segmentos semelhantes ou quase, dispostos em 2 séries, em geral petalóides mas por vezes herbáceos ou escariosos e glumáceos, quer livres quer todos unidos em um tubo comum - - - *Grupo 3* (ver p. 27)

Perianto nulo ou reduzido a sedas ou a 1–3 escamas:
Flores unissexuadas, não dispostas em pequenas espigas com brácteas escamiformes *Grupo 4* (ver p. 28)
Flores bissexuadas ou unissexuadas, dispostas em pequenas espigas (espiguetas) com brácteas escamiformes (glumas e lemas ou glumelas inferiores) e por vezes unifloras - - - - - - - - *Grupo 5* (ver p. 29)
Ovário ± ínfero; perianto presente:
Perianto composto de cálice e corola separados, o cálice herbáceo ou de algum modo diferindo da corola petalóide, as sépalas e as pétalas quer livres quer conadas mas nunca unidas em um tubo comum - - - - - - *Grupo 6* (ver p. 29)
Perianto composto de segmentos semelhantes ou quase, em regra 6 em 2 séries, por vezes 3 em 1 só série, em geral petalóides, quer livres quer todos unidos em um tubo comum - - - - - - - - - - *Grupo 7* (ver p. 29)

Grupo 1

Folhas pinadas; palmeiras - - - - - - - - - **Palmae**
Folhas simples; plantas herbáceas, aquáticas ou palustres:
Perianto nulo ou cupular; estames 1 ou 2 unidos; plantas submersas, marinhas ou de água doce, com folhas estreitas; flores unissexuadas, axilares, solitárias ou em cimeiras - - - - - - - - **Zannichelliaceae**
Perianto composto de 1–6 segmentos livres; estames 3 ou mais; plantas de água doce ou de pântanos:
Flores com brácteas, dispostas em verticilos, umbelas simples ou compostas ou, por vezes, espigas; perianto de 3 sépalas e 3 pétalas ou, por vezes, pétalas nulas; plantas aquáticas ou terrestres - - - - - - **Alismataceae**
Flores sem brácteas, dispostas em espigas; perianto de 1–4 segmentos semelhantes; plantas aquáticas:
Folhas basilares; espigas simples ou 2-ramosas, longamente pedunculadas, de início encerradas numa espata; segmentos do perianto 1–3; estames em geral 6, com filetes alongados; óvulos 2 ou mais em cada carpelo **Aponogetonaceae**
Folhas dispostas ao longo de caules compridos; espigas simples, pedunculadas, axilares e sem espata; segmentos do perianto 4; estames, em geral, 4, anteras sésseis; 1 óvulo em cada carpelo - - - - **Potamogetonaceae**

Grupo 2

Folhas pinadas ou flabeladas; palmeiras - - - - - - - **Palmae**
Folhas simples, não flabeladas; plantas herbáceas:
Ovário 1-locular com placentas parietais com óvulos numerosos:
Folhas filiformes, dispostas ao longo de caules compridos; flores solitárias, longamente pedunculadas e axilares; pétalas livres; anteras deiscentes por um poro apical; estilete indiviso; plantas aquáticas - - - - - - - **Mayacaceae**
Folhas quase todas basilares; flores em espigas ou capítulos longamente pedunculados; pétalas unidas na base em tubo; anteras deiscentes por fendas longitudinais; estilete 3-ramoso - - - - - - - - - - **Xyridaceae**
Ovário 2–3-locular com placentas axiais ou apicais com 1 ou mais óvulos:
Flores monóicas, pequenas e actinomórficas, em capítulos involucrados longamente pedunculados; corola não vistosa, de cores pouco vivas, frequentemente muito pequena nas flores masculinas; estilete ramoso; folhas estreitas, ± basilares ou congestas - - - - - - - - - **Eriocaulaceae**
Flores bissexuadas ou poligâmicas, actinomórficas ou zigomórficas, em cimeiras, panículas ou fascículos laxos ou densos e frequentemente subtendidos por brácteas plicadas ou cimbiformes; corola vistosa, em geral de cores vivas, frequentemente azul ou amarela; estilete indiviso - - - - - - **Commelinaceae**

Grupo 3

Segmentos do perianto 6, escariosos e glumáceos ou o mais interno hialino; folhas estreitas ou reduzidas à baínha:
Flores dióicas; ovário 2-locular com 1 óvulo apical pêndulo em cada lóculo; estiletes 3 **Restionaceae**
Flores bissexuadas; ovário 1- ou 3-locular com 3 ou mais óvulos em cada lóculo; estilete 1, 3-ramoso - - - - - - - - - - - **Juncaceae**
Segmentos do perianto petalóides ou herbáceos:
Plantas herbáceas aquáticas; flores em espigas ou racimos:
Segmentos do perianto 6, petalóides, unidos na base em tubo; ovário 1–3-locular com óvulos numerosos em cada lóculo; fruto uma cápsula; folhas com limbo alargado-ovado ou lineares e, então, submersas - - - - - - **Pontederiaceae**

Segmentos do perianto 4, herbáceos, livres; ovário 1-locular com 1 só óvulo; fruto uma pequena drupa; folhas estreitamente lineares ou filiformes, submersas **Potamogetonaceae**

Plantas terrestres:

Inflorescência uma espádice protegida por uma espata; flores monóicas, muito pequenas, as femininas dispostas na base da espadice, as masculinas na parte superior; ovário 1–2-locular; folhas (ou a folha única) basilares, 1–4-pinadas ou com limbo sagitado ou hastado; plantas herbáceas - - **Araceae**

Inflorescência não como acima; flores bissexuadas ou dióicas; ovário 3-locular:

Folhas reduzidas a escamas ou a espinhos, sendo as funções foliares desempenhadas frequentemente por ramos foliáceos (cladódios) aciculares ou achatados; plantas herbáceas ou pequenos arbustos, frequentemente escandentes **Liliaceae**

Folhas (ou a folha única) bem desenvolvidas, por vezes aparecendo depois das flores:

Flores dióicas, pequenas, dispostas em umbelas axilares; folhas com nervação reticulada e com gavinhas estipulares; fruto uma baga; arbustos escandentes, frequentemente com caules aculeados - - - - **Smilacaceae**

Flores bissexuadas; folhas sem gavinhas estipulares:

Flores em umbelas protegidas por uma ou mais brácteas espatáceas, pedúnculos nus; plantas herbáceas com folhas basilares; fruto uma cápsula loculicida - - - - - - - - **Amaryllidaceae**

Flores solitárias ou em racimos, espigas ou panículas:

Segmentos do perianto unidos na base em um tubo:

Anteras deiscentes por um poro apical; filetes curtos, inseridos na fauce do tubo do perianto; plantas herbáceas - - **Tecophilaeaceae**

Anteras deiscentes por fendas longitudinais; plantas herbáceas, arbustos ou árvores, por vezes suculentos ou escandentes - - **Liliaceae**

Segmentos do perianto livres:

Estames com anteras sésseis; fruto divisível em 3 cocas; perianto herbáceo; flores sem brácteas, em racimos longamente pedunculados; ervas com folhas basilares lineares ou filiformes **Juncaginaceae**

Estames com filetes ± compridos; fruto uma cápsula, uma baga ou uma drupa; perianto ± petalóide:

Ovário com 1 óvulo em cada lóculo; fruto uma drupa; perianto subpetalóide; trepadeiras herbáceas cujas folhas terminam em gavinhas; flores pequenas, em panículas - - **Flagellariaceae**

Ovário com 2 ou mais óvulos em cada lóculo; fruto uma cápsula ou uma baga; perianto petalóide; plantas herbáceas, por vezes escandentes, cujas folhas por vezes terminam em gavinhas **Liliaceae**

Grupo 4

Plantas aquáticas, sem folhas, muito pequenas, cujo corpo se reduz a uma "fronde" talóide sem ou com 1 ou mais raízes pequenas e pendentes - - - **Lemnaceae**

Plantas com folhas (ou a folha única) bem desenvolvidas:

Ervas aquáticas submersas com caule ± alongado e folhas lineares estreitas:

Folhas opostas ou verticiladas; flores solitárias ou poucas e reunidas nas axilas das folhas; plantas de água doce - - - - - - - - **Najadaceae**

Folhas alternas; inflorescência uma espiga de flores alternadamente masculinas e femininas, protegida por uma espata; plantas marinhas - - **Zosteraceae**

Plantas terrestres ou aquáticas mas não submersas:

Plantas herbáceas aquáticas, flutuantes, acaules; folhas sésseis dispostas em roseta **Araceae**

Plantas não flutuantes:

Árvores ou arbustos, frequentemente com raízes aéreas; folhas lineares ou ensiformes, com dentes espinhosos na margem e frequentemente também sobre a nervura central na página inferior - - - - **Pandanaceae**

Plantas herbáceas ou trepadeiras lenhosas; folhas sem dentes espinhosos:

Folhas estreitas, lineares; inflorescência uma espiga densa e cilíndrica, feminina na base, masculina na parte superior; ovário 1-locular com 1 só óvulo apical pêndulo; ervas palustres, frequentemente com a base submersa **Typhaceae**

Folhas (ou a folha única) largas, frequentemente profundamente recortadas; inflorescência uma espádice subtendida ou envolvida por uma espata, feminina na base, masculina na parte superior e, por vezes, com um apêndice terminal estéril; ovário 1–4-locular com 1 ou mais óvulos em placentas basais, axiais ou parietais; plantas herbáceas ou trepadeiras, terrestres - - **Araceae**

Grupo 5

Flores da espigueta situadas, cada uma, entre uma bráctea externa (lema ou glumela inferior) e uma bractéola interna (pálea ou glumela superior), espiguetas em regra com 2 brácteas estéreis (glumas) na base; perianto nulo ou representado por 2–3 escamas (lodículas ou glumélulas); caules em geral com entrenós ocos, cilíndricos ou comprimidos; baínha das folhas em geral aberta; semente em geral adnata ao pericarpo; plantas herbáceas ou, por vezes, lenhosas, arbustivas ou arborescentes (bambús) **Gramineae**

Flores da espigueta situadas, cada uma, na axila de uma única bráctea (gluma), sem bractéola ou as femininas por vezes encerradas em uma bractéola aberta apenas no cimo (utrículo); perianto nulo ou representado por sedas ou escamas; caules em geral sólidos, frequentemente trigonais; baínha das folhas em geral fechada; semente não adnata ao pericarpo; plantas herbáceas - - - - - - - **Cyperaceae**

Grupo 6

Plantas herbáceas aquáticas; flores actinomórficas, bissexuadas ou dióicas, solitárias ou 2 ou mais, protegidas por uma espata tubulosa; espatas sésseis ou longamente pedunculadas - - - - - - - - - - - **Hydrocharitaceae**

Plantas herbáceas terrestres ou epifíticas; flores ± zigomórficas ou assimétricas:

Estames férteis 5; flores unissexuadas ou bissexuadas; plantas herbáceas de estatura elevada e folhas grandes - - - - - - - - - **Musaceae**

Estame fértil 1; flores bissexuadas:

Estame não acompanhado de estaminódios petalóides; pólen aglutinado em massas (polinídias); ovário 1-locular com óvulos numerosos em placentas parietais; flores zigomórficas cuja pétala mediana (labelo) difere ± das laterais; plantas terrestres ou epifíticas, por vezes escandentes ou saprofíticas **Orchidaceae**

Estame acompanhado de 1 ou mais estaminódios petalóides; pólen não aglutinado em massas; plantas terrestres:

Sépalas unidas em tubo; flores zigomórficas; anteras com 2 tecas; ovário 3-locular com óvulos numerosos em cada lóculo - - - **Zingiberaceae**

Sépalas livres; flores assimétricas; antera com 1 só teca:

Ovário 3-locular com óvulos numerosos em cada lóculo - **Cannaceae**

Ovário 1–3-locular com 1 óvulo em cada lóculo - - - **Marantaceae**

Grupo 7

Plantas herbáceas aquáticas, por vezes marinhas; flores actinomórficas, dióicas, solitárias (ou as masculinas por vezes grupadas) protegidas por uma espata tubulosa ou 2-lobada; espatas sésseis ou longamente pedunculadas - - - - **Hydrocharitaceae**

Plantas não aquáticas:

Flores unissexuadas, pequenas; plantas volúveis; folhas simples ou digitadas, com nervação reticulada; fruto uma cápsula 3-alada ou 3-gonal - - **Dioscoreaceae**

Flores bissexuadas:

Folhas profundamente recortadas com segmentos pinatipartidos, basilares; flores em umbelas involucradas com as brácteas exteriores largas e as interiores compridas e filamentosas; plantas herbáceas, tuberosas - - - - - - **Taccaceae**

Folhas simples e não recortadas ou, por vezes, nulas:

Estame 1; pólen aglutinado em massas (polinídias); ovário 1-locular com óvulos numerosos em placentas parietais; segmento mediano interno (labelo) do perianto ± diferente dos restantes; plantas herbáceas, terrestres ou epifíticas, por vezes escandentes ou saprofíticas - - - - - **Orchidaceae**

Estames 3 ou 6; pólen não aglutinado em massas:

Estames 3:

Estames opostos aos segmentos internos do perianto; plantas saprofíticas com algumas folhas estreitas basilares ou com as folhas reduzidas a escamas; segmentos do perianto unidos em tubo; flores actinomórficas, em racimos ou cimeiras - - - - - - - - **Burmanniaceae**

Estames opostos aos segmentos externos do perianto; plantas herbáceas não saprofíticas, com folhas estreitas e com bolbo sólido ou rizoma; segmentos do perianto livres ou unidos em tubo; flores actinomórficas ou zigomórficas, em espigas ou panículas ou, por vezes, solitárias ou algumas grupadas e protegidas por uma espata - - - - - - - - **Iridaceae**

Estames 6:

Plantas saprofíticas, com folhas reduzidas a escamas; ovário 1-locular com placentas parietais; segmentos do perianto unidos em tubo urceolado; flores zigomórficas - - - - - - - **Burmanniaceae**

Plantas não saprofíticas, com folhas bem desenvolvidas; ovário 3-locular:

Flores em umbelas (por vezes 1-floras), subtendidas por 1 ou mais brácteas espatáceas, pedúnculos nus; plantas herbáceas, bolbosas, com folhas basilares - - - - - - - - **Amaryllidaceae**
Flores solitárias ou em racimos ou corimbos, não subtendidas por brácteas espatáceas; plantas não bolbosas:
Plantas lenhosas de estatura elevada ou não; caule ramoso ou não, densamente revestido pelas bases persistentes das folhas velhas; flores solitárias ou fasciculadas - - - - - - **Velloziaceae**
Plantas herbáceas, com tubérculo ou bolbo sólido:
Anteras deiscentes por um poro apical; ovário com 2 óvulos em cada lóculo; flores em racimos ou panículas; plantas glabras **Tecophilaeaceae**
Anteras deiscentes por fendas longitudinais; ovário com vários óvulos em cada lóculo; flores solitárias ou em racimos ou corimbos; plantas ± pilosas - - - - - - - - **Hypoxidaceae**

ÍNDICE DAS FAMÍLIAS DA CLAVE

31. TILIACEAE

By H. Wild

Small trees, lianes, shrubs, or annual or perennial herbs, often stellately hairy; leaves usually alternate, simple or rarely digitate, entire, toothed, or lobed; stipules usually small and deciduous. Inflorescence usually cymose, with the cymes often leaf-opposed, sometimes in corymbs or panicles, usually axillary, sometimes terminal. Flowers actinomorphic, usually bisexual. Sepals 5 or sometimes 2–4, free or occasionally connate, usually valvate. Petals free, equalling the number of sepals and alternating with them, rarely absent, often with a glandular claw or appendage at the base. Stamens 4–∞ (usually ∞), often on a raised torus or androgynophore, free or connate at the base, all fertile or the outer ones sterile; anthers dehiscing longitudinally or by terminal pores. Ovary superior, 2–10-locular, with 1 to many ovules per loculus. Style entire or lobed at the apex or the stigmas almost free. Fruit a dry or somewhat fleshy drupe or a schizocarp, 2–10-locular or 1-locular by abortion, sometimes transversely septate between the seeds. Seeds with endosperm.

Sepals free, 4–5; fruit not winged or very narrowly so; more than 1 ovule per loculus, or, if apparently 1 per loculus by the intrusion of false septa, then ovary falsely 10-locular:
- Fruit indehiscent, without bristles, drupaceous or a (1) 2–4-lobed berry, or longitudinally ribbed and woody:
  - Anthers dehiscing by terminal pores; fruit woody, longitudinally ribbed — **2. Glyphaea**
  - Anthers dehiscing by slits; fruit drupaceous or a (1) 2–4-lobed berry, usually only slightly fleshy - - - - - - - - - - - **3. Grewia**
- Fruit usually a dehiscent (if indehiscent then densely bristly), glabrous, hairy or bristly capsule or schizocarp:
  - Flowers white, pinkish or mauve; some stamens sterile with no anthers or, if sterile stamens very few or apparently absent, then filaments nodose:
    - Fertile stamens 12 or fewer; flowers large, 5 cm. in diam. or more, pinkish-mauve, rarely white; ovary and fruit 4–8-locular; seeds discoid **6. Clappertonia**
    - Fertile stamens many; flowers white, c. 2 cm. in diam.; ovary and fruit 4–5-locular; seeds ellipsoid-biconical - - - - - **5. Sparrmannia**
  - Flowers yellow; all the stamens fertile, filaments never nodose:
    - Fruits globose to ovoid, bristly, spiny, or with conical spine-tipped tubercles; ovules 2 per loculus - - - - - - - - - - - **4. Triumfetta**
    - Fruits usually long and pod-like, rarely ellipsoid or ovoid; ovules more than 2 per loculus, or, if only 2, then capsule smooth; leaves often tailed at the base — **7. Corchorus**

Sepals connate into a 2–3-lobed calyx; fruit a capsule with very broad wings; ovary 2-locular, with 1 or sometimes 2 pendulous ovules per loculus - **1. Carpodiptera**

1. CARPODIPTERA Griseb.

Carpodiptera Griseb. in Mem. Acad. Amer. Sci. Art. N. S. 8, **1**: 163 (1860).

Small trees; leaves oblong or ovate, with margins entire or repand, petiolate, stipulate. Inflorescence of axillary pedunculate cymes. Flowers sometimes unisexual. Calyx campanulate, 2–3-lobed. Petals 5, narrowed to the base, glandless. Androgynophore obsolete; stamens many, united at the base. Ovary sessile, 2-lobed, 2-locular, with 1 (2) pendulous ovules per loculus; style obsolete or very short, stigma large and spreading, 2-lobed. Fruit a 2-valved capsule, usually with 1 seed; each valve prolonged by 2 unequal, horizontally spreading, foliaceous wings in a vertical plane. Seed large, villous or hairy, at least at the apex; testa coriaceous; endosperm fleshy.

Carpodiptera africana Mast. in Oliv., F.T.A. **1**: 241 (1868).—Sim, For. Fl. Port. E Afr.: 21 (1909).—Engl., Pflanzenw. Afr. **3**, 2: 345, t. 167 (1921).—Brenan, T.T.C.L.: 613 (1949). TAB. **1.** Type from Tanganyika (Rovuma R.).

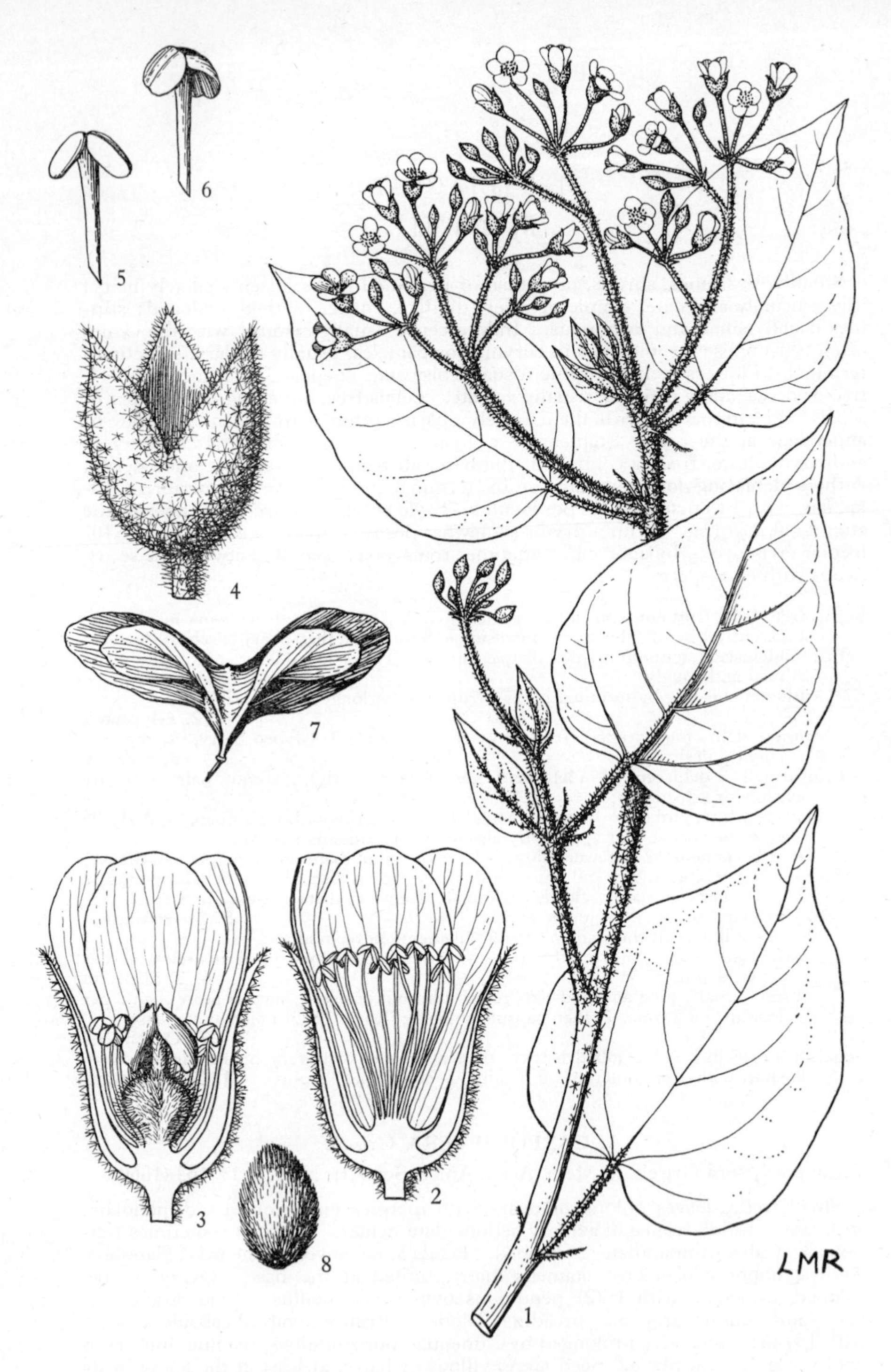

Tab. 1. CARPODIPTERA AFRICANA. 1, flowering branch with male flowers (×1); 2, vertical section of male flower (×7); 3, vertical section of female flower (×7); 4, calyx (×7); 5, fertile stamen (×25); 6, sterile stamen (×25); 7, fruit (×⅔); 8, seed (×2). 1–6 from *Kirk* s.n., 7 and 8 from *Hornby* 2718.

Small tree up to c. 8 m. tall; young branchlets sparsely stellately hairy, soon glabrous. Leaf-lamina up to 20 × 13 cm., ovate to oblong, acute or obtuse at the apex, with margin entire or repand, rounded or cordate at the base, stellate-pubescent when young but very soon becoming glabrous, 3–5-nerved at the base; petiole up to 6 cm. long, stellate-pubescent; stipules up to 8 mm. long, setaceous, stellate-pubescent, very caducous. Inflorescence of axillary cymes; peduncles up to 3·5 cm. long and as long as or longer than the petioles, stellate-pubescent; primary branches c. 5, c. 2·5 cm. long, each terminating in a cymule of c. 5–7 flowers; pedicels 0·3–1·0 cm. long, stellate-pubescent; bracts c. 2 mm. long, subulate, stellate-pilose, caducous. Calyx campanulate, divided c. $\frac{3}{4}$ of the way down into 2–3 lobes; lobes 3–4 mm. long, broadly triangular, acute or acuminate, often unequal, densely stellate-pubescent outside. Petals white, 5–6 mm. long, narrowly obovate to obovate, narrowed to the base. Flowers unisexual, dioecious or more rarely monoecious. Male flowers with many stamens c. 4 mm. long; filaments joined at the base; ovary absent. Female flowers with many short sterile stamens 2–3 mm. long and with a sessile 2-lobed stellately pilose ovary c. 1·3 mm. in diam., with 1 or sometimes 2 ovules per loculus; style c. 1 mm. long, pubescent; stigma 2–3-lobed, 1·5 mm. wide, conical. Capsule up to 1·5 cm. long, ellipsoid, 2-valved, each valve with 2 unequal obtuse foliaceous wings 2–5 × 1–2 cm. spreading horizontally in a vertical plane, stellate-pubescent. Seed usually solitary, 8 × 5 mm., ellipsoid, brown, finely reticulate, villous.

Mozambique. N: between Liúpo and Mogincual, fr., *Pedro & Pedrógão* 4661 (LMJ; SRGH). SS: Macovane, fr. 1.vi.1947, *Hornby* 2718 (K; SRGH).

Also in Kenya, Tanganyika and Zanzibar. Coastal species in *Brachystegia* woodland, mixed woodland or thicket. Masters (loc. cit.), described both genus and species as having a 5-lobed calyx. This is an error as the calyx of this species is 2–3-lobed as given in Grisebach's description of the genus (based on a Cuban species). The genus has an unusual distribution as it is absent from West Africa and is known only from Central America, the West Indies and East Africa. There is however, a related genus, *Berrya* Roxb. with species distributed in tropical Asia and the Pacific Is. *Berrya* differs from *Carpodiptera* only in its long slender styles, punctiform stigmas and in having 2–c. 6 ovules per loculus. These differences are rather small and La Maza (in Ann. Soc. Esp. Hist. Nat. **19**: 215 (1890)) transferred *Carpodiptera cubensis* Griseb. to *Berrya*. Nevertheless, his example is not followed here for the stigma character provides a constant difference in all the known species of the two genera, so that it seems inadvisable to unite them, although this would admittedly render the phytogeography of the group more easily understood. Our species is very like the West Indian *C. cubensis* Griseb. but differs markedly in its stigma, which is sessile and crenate-multilobulate in the latter species. *Carpodiptera minor* Sim (loc. cit.) is *Hymenocardia acida* Tul. in the *Euphorbiaceae*.

2. GLYPHAEA Hook. f. ex Planch.

Glyphaea Hook. f. ex Planch. in Hook., Ic. Pl. **8**: t. 760 (1848).

Shrubs or trees, with at least the young parts stellate-pubescent. Leaf-lamina oblong or obovate-oblong, finely toothed, with 3 strong nerves from the base; petiole slightly widened just below the leaf-lamina; stipules caducous, very small, subulate. Inflorescence of leaf-opposed or extra-axillary cymes; cymes 1–several-flowered; bracts very small, caducous. Sepals 5, linear-oblong, stellately hairy outside. Petals 5, bright yellow, glandless, linear-oblong. Androgynophore obsolete. Stamens ∞, all fertile; anther-thecae linear, adnate throughout their length, dehiscing by terminal pores and with the connective prolonged into a small crest. Ovary sessile, 6–10-locular; loculi many-ovuled. Style simple; stigma simple, not wider than the style. Fruit oblong-cylindric or fusiform, sulcate, indehiscent, with hard fibrous walls; endocarp with pits containing the seeds. Seeds discoid with a fleshy endosperm; cotyledons cordate and roundish.

Glyphaea tomentosa Mast. in Oliv., F.T.A. **1**: 267 (1868).—Sim, For. Fl. Port. E. Afr.: 21 (1909). TAB. **2.** Syntypes: Mozambique, Chupanga, *Meller* (K); Morrumbala, *Waller* (K).

Oncoba sulcata Sim, tom. cit.: 12 (1909). Type: Mozambique, Maganja da Costa, *Sim* 5900 (PRE, holotype).

Glyphaea grewioides sensu Bak. f. in Journ. Linn. Soc., Bot. **40**: 33 (1911).

Small tree up to 4 m. tall; young shoots shortly stellately tomentellous. Leaf-lamina 4–9 × 2–4 cm., oblong to obovate-oblong, acuminate or acute at the apex, with

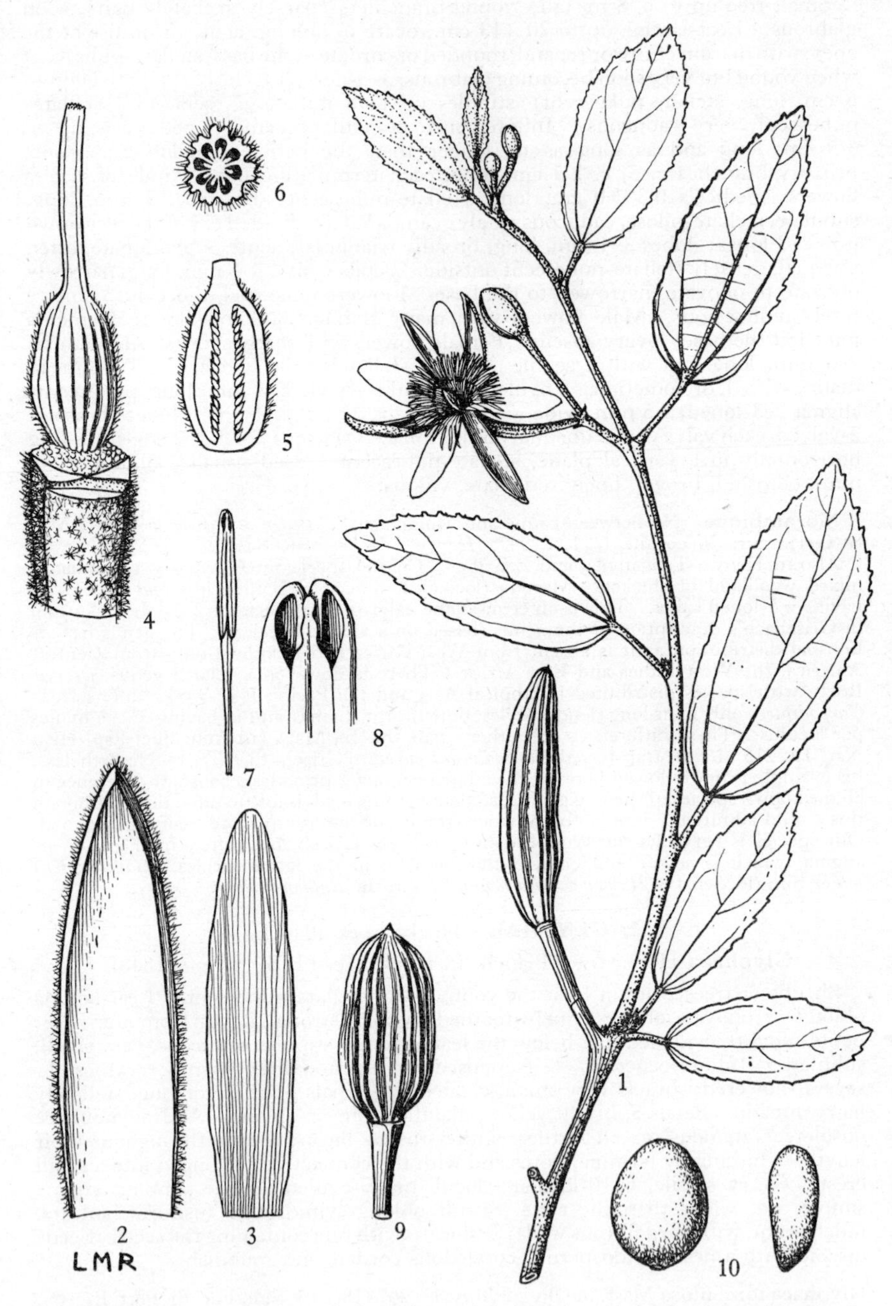

Tab. 2. GLYPHAEA TOMENTOSA. 1, flowering and fruiting branch (× ⅔); 2, sepal (× 3); 3, petal (× 3); 4, gynoecium (× 4); 5, longitudinal section of ovary (× 4); 6, cross section of ovary (× 4); 7, stamen (× 4); 8, tip of anther with pores (× 15); 9, fruit (× ½); 10, seeds (× 4). All from *Meller* s.n.

margin irregularly crenate-dentate, slightly cordate or rounded at the base, strongly 3-nerved and sometimes asymmetrical at the base, stellate-pubescent on both surfaces and with tufts of hairs in the axils of the nerves; petiole tomentellous, widening just below the leaf-blade; stipules c. 2 mm. long, very caducous, tomentellous, subulate. Inflorescence of leaf-opposed 1–3-flowered cymes towards the ends of the branches; peduncles up to 1 cm. long, tomentellous; pedicels similar, up to 1·7 cm. long, thickening considerably in fruit up to a diameter of c. 4 mm.; bracts c. 2 mm. long, subulate, tomentellous. Sepals up to 2·2 × 0·4 cm., narrowly oblong, rather blunt at the apex, tomentellous outside, glabrous and yellow inside, with margin slightly thickened. Petals golden-yellow, slightly shorter than the sepals, narrowly oblong, glabrous, blunt or 2-dentate at the apex. Stamens very numerous, about half the length of the sepals; anthers linear with two narrow parallel thecae, dehiscing by terminal pores, with the connective prolonged at the apex by a very short crest and with a pair of very short blunt horns prolonging the bases of the thecae. Ovary subsessile, villous, subcylindric, sulcate, with many ovules in each loculus; style glabrous, rather flattened, blunt at the apex; stigmatic surface not wider than the style, c. 5 mm. long. Fruit 4–7·5 × 2–3 cm., hard and woody, oblong-cylindric, blunt or very abruptly apiculate at the apex, abruptly truncate at the base, with 8–10 paired deeply incised ridges running longitudinally down the outside, smoothly stellate-tomentellous. Seeds numerous, c. 4 × 3 mm., discoid, brown, with a rather roughened testa.

Mozambique. N: Nampula, fl. 1.i.1942, *Mendonça* 1198 (LISC; SRGH). Z: Morrumbala, fl., *Waller* (K). MS: Cheringoma, Inhaminga, fr. 25.v.1948, *Mendonça* 4385 (BM; LISC).

Apparently confined to Mozambique and occurring in deciduous woodland.

Closely related to *G. brevis* (Spreng.) Monachino, the only other member of the genus, but differing in the following respects: *G. brevis* is a forest species, its fruiting pedicels remain slender and c. 2 mm. in diam. or less in fruit, the number of loculi in the ovary is usually fewer, the fruit is only sparsely stellate-pubescent and is usually gradually narrowed to both ends. The flowers of *G. tomentosa* are usually larger and the texture of the leaves thicker but in flowering material the separation of the two species is not so easy. No doubt this is a case where the area of distribution of one widespread species, under the influence of adverse conditions, probably in this case of decreasing rainfall, has contracted and left a relic population in Mozambique. Once cut off, this isolated population developed differences which now enable us to separate it at the specific level. It is in fact a borderline case and the differences from *G. brevis* would only have to be a little less definite before it would be best to consider *G. tomentosa* a subspecies.

3. GREWIA L.

Grewia L., Sp. Pl. **2**: 964 (1753); Gen. Pl. ed. 5: 412 (1754); Burret in Engl., Bot. Jahrb. **44**: 198–238 (1910); op. cit. **45**: 156–203 (1910).

Shrubs or smallish trees. Leaves alternate, simple, serrate or very rarely entire, 3–7-nerved at the base, petiolate; stipules lateral. Flowers in terminal or axillary panicles or in leaf-opposed or axillary umbel-like cymes. Sepals 5, linear-oblong or linear-spathulate, stellately hairy outside, often coloured within like the petals. Petals 5, shorter than the sepals, yellow, purplish-blue or white, usually with a nectariferous claw or gland at the base. Stamens indefinite, free, usually raised on a torus or androgynophore which is short and glabrous or produced above into a pubescent extension. Ovary 2–4-locular, entire or 2–4-lobed, with 2 to many axile ovules in each loculus; style longer than the ovary, with subulate or flattened lobes or almost entire. Fruit a 1–4-lobed drupe or berry with 1–4 pyrenes; mesocarp somewhat fleshy or fibrous; endocarp hard and woody.

Note 1. In common with other recent treatments of this genus in Africa, such as those by Brenan (T.T.C.L. (1949)) and Exell and Mendonça (C.F.A.), I have decided not to follow Burret's later division of the African *Grewia* spp. into the following smaller genera; *Grewia* L. in a more restricted sense, *Microcos* L. and *Vincentia* Boj. With our material at least, there is no real justification for this division. Further to this, I have not thought it necessary to quote the long list of synonyms thus created by Burret in subdividing the genus in his later work (Burret in Notizbl. Bot. Gart. Berl. **9**: 592–880 (1926) and **12**: 715 (1935)).

Note 2. None of our representatives of this genus is of any real commercial value since, although their very hard wood makes good assegai handles or walking-sticks, it never grows to sufficient size to make its exploitation worth while. The fruits of most species are edible, but have too little flesh to make them really attractive. However, the large number of species and their wide distribution in all kinds of habitats make them useful to the ecologist as indicator species.

Stigma not lobed; inflorescence terminal or axillary, of several to many-flowered panicles; fruit never lobed - - - - - - - - - Sect. *Microcos* (below)
Stigmas lobed; inflorescences axillary or terminal, sometimes leaf-opposed, often in rather umbel-like cymes:
 Inflorescences leaf-opposed, of rather dense heads with very short peduncles, several to many-flowered; leaves usually subcircular - - Sect. *Glomeratae* (p. 40)
 Inflorescences rather lax, 1 to several-flowered, if leaf-opposed then leaves not subcircular:
  Ovary and fruit very rarely lobed; ovules 10–20 per loculus; stigma-lobes subulate; flowers yellow - - - - - - - - Sect. *Pluriovulatae* (below)
  Ovary and fruit 2–4-lobed, or 1-lobed by abortion (when the style is lateral); stigma-lobes flattened; ovules 4–6 (8) in each loculus:
   Inflorescence axillary; ovary shallowly 2(4)-lobed or 1-lobed by abortion; flowers yellow - - - - - - - - - - - Sect. *Axillares* (p. 39)
   Inflorescences terminal or ± leaf-opposed; ovary and fruit deeply 4-lobed; flowers white, blue or mauve - - - - - - - Sect. *Grewia* (p. 39)

Sect. **Microcos** (L.) Wight & Arn.

Leaves almost glabrous on both sides, usually less than 5 cm. long; stipules always entire - - - - - - - - - - - - - 1. *microthyrsa*
Leaves noticeably pubescent, tomentose or tomentellous below, usually more than 5 cm. long; stipules usually divided:
 Leaves whitish-tomentose below, with the main nerves brownish and subglabrous 5. *gilviflora*
 Leaves not whitish-tomentose below or, if whitish-tomentose, uniformly so:
  Inflorescence with ferruginous tufts of hairs; scrambler or liane - 3. *barombiensis*
  Inflorescence not ferruginously hairy; shrubs or small trees:
   Inflorescence and young branchlets with short velvety tomentum mixed with longer patent hairs; leaves pilose below - - - - 2. *transzambesica*
   Inflorescence and young branchlets with grey velvety tomentum only; leaves shortly greyish-pubescent below - - - - - - 4. *conocarpa*

Sect. **Pluriovulatae** Burret

Androgynophore produced well beyond the basal glabrous portion (noticeable in fruit as well as in flower):
 Leaves glabrous or very sparsely stellate-pubescent below; stipules not auriculate 13. *caffra*
 Leaves greyish-white-tomentose below; stipules auriculate at the base 12. *falcistipula*
Androgynophore not produced beyond the basal glabrous portion:
 Bracts undivided; stipules always entire:
  Fruits warted or subspinose; coastal species - - - - - 10. *forbesii*
  Fruits not warted or subspinose:
   Leaves lanceolate to oblong; peduncles less than 1 cm. long in the flowering stage:
    Pubescence of all parts rather fine and appressed, or lacking; fruits always quite entire:
     Leaves oblong or elliptic-oblong, rounded or acute at the apex, finely reticulate above; low bush of c. 1 m. tall; older branches ± cylindric 8. *retinervis*
     Leaves acute to acuminate at the apex, nerves not reticulate; spreading shrubs or lianes; older branches becoming 4-angled:
      Androgynophore-apex glabrous and shallowly cupular, clasping the ovary base; liane of forest and forest-edges or constituent of coastal thickets 6. *holstii*
      Androgynophore pubescent at the apex, not clasping the base of the ovary; shrub of dry woodlands - - - - - - - - 7. *gracillima*
    Pubescence of buds and stems rather coarse and harsh; fruit often shallowly 2–4-lobed - - - - - - - - - - - - 9. *flavescens*
   Leaves broadly oblong or almost circular; flowering peduncles normally more than 1 cm. long - - - - - - - - - - - - 9. *flavescens* var. *olukondae*
 Bracts bifid or trifid; stipules mostly divided; low shrub usually branching from ground level - - - - - - - - - - - 11. *decemovulata*

Sect. **Axillares** Burret

Leaves discolorous, whitish-, greyish- or yellowish-tomentose below:
Nerves on under surface of leaf subglabrous or with tufts of long brownish hairs standing out against the white tomentum of the areas between them:
Sepals more than 2 cm. long; leaves with a yellowish-white very dense woolly tomentum below - - - - - - - - - - 22. *hexamita*
Sepals less than 2 cm. long; leaves, if whitish-tomentose below, finely and closely so:
Young branches floccose-tomentose; leaves oblong to oblong-obovate, lacking longer tufts of hairs on the nerves beneath, up to 18 cm. long; petals orbicular 21. *inaequilatera*
Young branches not floccose-tomentose; leaves oblong to elliptic, apex rounded or acute, somewhat shiny and rugose above, nerves below with tufts of longer brownish hairs; petals oblong to obovate - - - - - 20. *micrantha*
Nerves on under surface of leaf not differentiated in colour, whitish-tomentose:
Sepals usually 1 cm. long or less, rarely up to 1·3 cm.; lobes of fruit c. 0·75 cm. in diam.; androgynophore not produced:
Flowering branches elongated, virgate, rather flattened - - 14. *mollis*
Flowering branches not markedly elongate, cylindric in section:
Leaves asymmetric at the base; peduncles usually 3-flowered:
Bracts entire:
Leaves asymmetrically cordate, oblong-ovate to ovate or broadly obovate-oblong:
Leaf-margin often irregularly and coarsely serrate, leathery:
Petals broadly obovate to circular; leaves normally c. 8 × 5·5 cm. 23. *schinzii*
Petals oblong to narrowly obovate; leaves normally c. 5 × 3·5 cm. 16. *monticola*
Leaf-margin regularly serrate, thin-textured; lamina not rugose above 17. *subspathulata*
Leaves elliptic, elliptic-oblong or lanceolate - - - - 15. *bicolor*
Bracts filamentous, bifid or trifid - - - - - - 18. *hornbyi*
Leaves symmetric at the base; peduncles commonly 1-flowered, if more than 1-flowered, nerves rather raised and reticulate below - 19. *flava*
Sepals normally more than 2 cm. long; fruit lobes c. 1·5 cm. in diam.; androgynophore with a produced pubescent portion above - - 22. *hexamita*
Leaves pubescent below, not discolorous - - - - - - - 24. *microcarpa*

Sect. **Grewia** (Sect. *Oppositiflorae* Burret)

Flowers appearing with the mature leaves; stipules various:
Inflorescences all 1-flowered - - - - - - - - - 29. *tenax*
Inflorescences 2–several-flowered:
Petals and inside of sepals purple, pinkish or mauve (very rare albino specimens occur); petals oblong-lanceolate - - - - - - 28. *occidentalis*
Leaves coriaceous, obovate, 3–7·5 × 2–4 cm., tertiary nerves reticulate below; coastal dunes plant - - - - - - - - - - 28. *occidentalis* var. *litoralis*
Leaves not coriaceous, rhomboid-elliptic, acute at the apex, cuneate at the base 28. *occidentalis* var. *occidentalis*
Petals and inside of sepals white, if pinkish or purple then petals circular and fruit ferruginously hairy or villous (*G. glandulosa*):
Leaves obovate, rounded at the base and apex; inflorescences mostly terminal, mostly more than 3-flowered; petals narrowed from base and apex and with the lamina several times longer than the nectariferous claw 25. *sulcata*
Leaves ovate, ovate-oblong, rhombic or obovate-oblong; petals circular, elliptic or, if narrowed from base to apex, then lamina not more than about twice the length of the nectariferous claw:
Petals narrowed from base to apex with straight sides and not exceeding the nectariferous claw in width:
Bracteoles at the apex of the peduncles entire; young branchlets ferruginously hairy; marginal teeth of leaves glandular - - - - 26. *stolzii*
Bracteoles trifid; young branchlets yellowish-brown-pubescent; leaves pubescent above - - - - - - - - - 27. *avellana*
Petals circular or ovate-lanceolate with lamina wider than the nectariferous claw:
Leaves truncate or retuse at the apex - - - - - 31. *truncata*
Leaves acute or acuminate at the apex:
Fruits rufous-tomentose; flowers pink or purple - - 32. *glandulosa*

Fruits not tomentose; flowers usually white:
Sepals smoothly tomentellous outside:
Sepals 3·5 mm. wide or more - - - - - 33. *pachycalyx*
Sepals less than 3 mm. wide - - - - - 34. *lepidopetala*
Sepals densely pubescent with short stellate hairs and longer fasciculate-stellate hairs - - - - - - - - - - 35. *limae*
Flowers appearing with the young leaves; stipules persisting on second-year branches and becoming dark brown and cucullate - - - - - - - 30. *praecox*

Sect. **Glomeratae** Burret

Stipules and bracts ovate or broadly lanceolate; stigma-lobes laciniate; low shrub 36. *villosa*
Stipules and bracts lanceolate; stigma-lobes flattened, broad, margins denticulate; branches virgate and springing from a woody rootstock - - - - 37. *herbacea*

1. **Grewia microthyrsa** K. Schum. ex Burret in Engl., Bot. Jahrb. **45**: 163 (1910). TAB. **3** fig. C. Syntypes: Mozambique, Lourenço Marques, *Schlechter* 11632 (B†, holotype; BM; K); *Schlechter* 11616 (B†; BM; K).

Shrub 2–3 m. tall, with pale grey bark. Leaf-lamina 2–5·5 × 1·2–2·5 cm., oblong to oblong-elliptic, apex obtuse, margin almost entire or subserrate in the upper half, base rounded, both surfaces glabrous or with minute scattered stellate hairs; lateral nerves fairly prominent below, in 4–5 pairs, looping within the margin; petiole c. 5 mm. long, stellate-pubescent; stipules subulate, very caducous. Inflorescences terminal on the branches, of smallish panicles; lateral branches of panicle c. 5 mm. long, whitish- or brownish-tomentellous, each bearing 2–3 flowers; pedicels similar and c. 2·5 mm. long; bracts c. 5 mm. long, deeply trifid, densely puberulous, very caducous. Flower-buds obovoid, greyish-brown-tomentellous. Sepals 6–9 mm. long, revolute and hooded in the upper half, slipper-shaped, puberulous within, tomentellous without. Petals yellow, c. 3 mm. long, usually retuse or bifid at the apex and densely puberulous within above the nectary which does not form a scale-like ledge above. Androgynophore 1–1·5 mm. long, glabrous, not produced above. Ovary not lobed, glabrous, usually 3-locular and the loculi 3–4-ovulate; style 5 mm. long, glabrous; stigma hardly wider than the style. Fruit c. 13 × 7 mm., pendulous on a recurved pedicel, narrowly ovoid, never lobed.

Mozambique. SS: Panda, fr. 25.ii.1955, *E. M. & W.* 599 (BM; LISC; SRGH). LM; Lourenço Marques, fl. 6.xii.1897, *Schlechter* 11632 (B†; BM; K).
Restricted to the southern part of Mozambique and the north-eastern Transvaal. Dry bushland or dry mixed woodland at low altitudes.

2. **Grewia transzambesica** Wild in Bol. Soc. Brot., Sér. 2, **31**: 81, t. 1 fig. A (1957). TAB. **4** fig. A. Type: Mozambique, Beira, *Simão* 1260 (LM; SRGH, holotype).

Shrub or small tree up to 7 m. tall; young branches pale brownish-stellate-pubescent with some caducous longer hairs, later glabrescent and grey-brown. Leaf-lamina 4–10 × 2·5–5·5 cm., lanceolate to ovate-lanceolate; apex acuminate or acute, margin repand-dentate to shallowly crenate (especially towards the apex), base rounded or slightly cordate, ± symmetrical, upper surface with sparsely scattered minute stellate hairs and with a pubescent midrib, lower surface patently pilose with rather prominent nerves; petiole up to 6 mm. long, pubescent. Inflorescences paniculate, terminal or axillary, all parts except the pedicels and flowers themselves with a fine brownish or cream tomentum interspersed with much longer patent hairs; peduncles c. 4 mm. long; pedicels c. 3 mm. long; basal bracts deeply bifid or trifid. Sepals 6–7·5 mm. long, oblong-spathulate, involute in the upper half, boat-shaped, with slightly undulate margins, creamy-velvety outside, pubescent within. Petals white with a triangular lamina 1·75 mm. long, pilose inside and outside, and with a basal nectariferous claw tomentose at the back and around the margin within. Androgynophore 1 mm. long, glabrous, not produced. Ovary ovoid, tomentose, not lobed but somewhat triangular in section, 3-locular with c. 4 ovules per loculus; style 3·5 mm. long, stellate-pilose except towards the apex; stigma no wider than the style.

Mozambique. Z: between Inhamacurra and Maganja da Costa, fl. 25.iii.1943, *Torre* 4987 (K; LISC; SRGH). MS: Cheringoma, at mile 15 on the Transzambesi Railway, fl. 20.ii.1946, *Simão* 1260 (LM; SRGH).

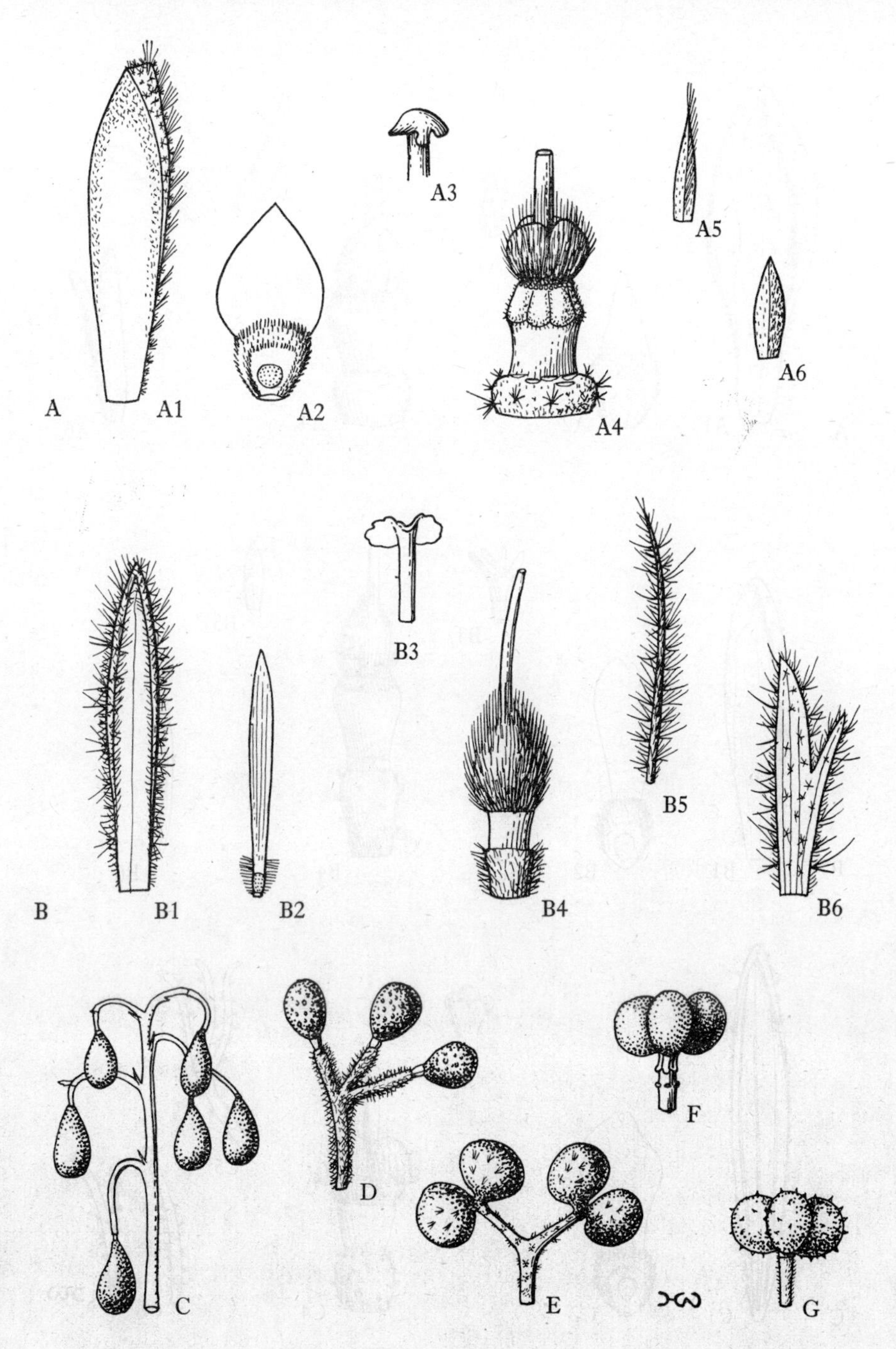

Tab. 3. A.—GREWIA LIMAE (SECT. GREWIA). A1, sepal (×3); A2, petal (×3); A3, stigma (×3); A4, androgynophore and ovary (×5); A5, bract (×5); A6, stipule (×5), all from *Pires de Lima* 60. B.—GREWIA HERBACEA (SECT. GLOMERATAE). B1, sepal (×4); B2, petal (×4); B3, stigma (×10); B4, androgynophore and ovary (×5); B5, bract (×5); B6, stipule (×5). B1, 3, 5 and 6 from *Purves* 31, B2 and 4 from *Eyles* 2. C.—GREWIA MICROTHYRSA (SECT. MICROCOS), fruits ($\times \frac{2}{3}$) *E. M. & W.* 599. D.—GREWIA DECEMOVULATA (SECT. PLURIOVULATAE), fruits (×1) *Pardy* in GHS 13295. E.—GREWIA INAEQUILATERA (SECT. AXILLARES), fruits ($\times \frac{2}{3}$) *Wild* 3352. F.—GREWIA PACHYCALYX (SECT. GREWIA), fruit ($\times \frac{2}{3}$) *Davies* 1451. G.—GREWIA HERBACEA (SECT. GLOMERATAE). fruit (×1) *Rattray* 575.

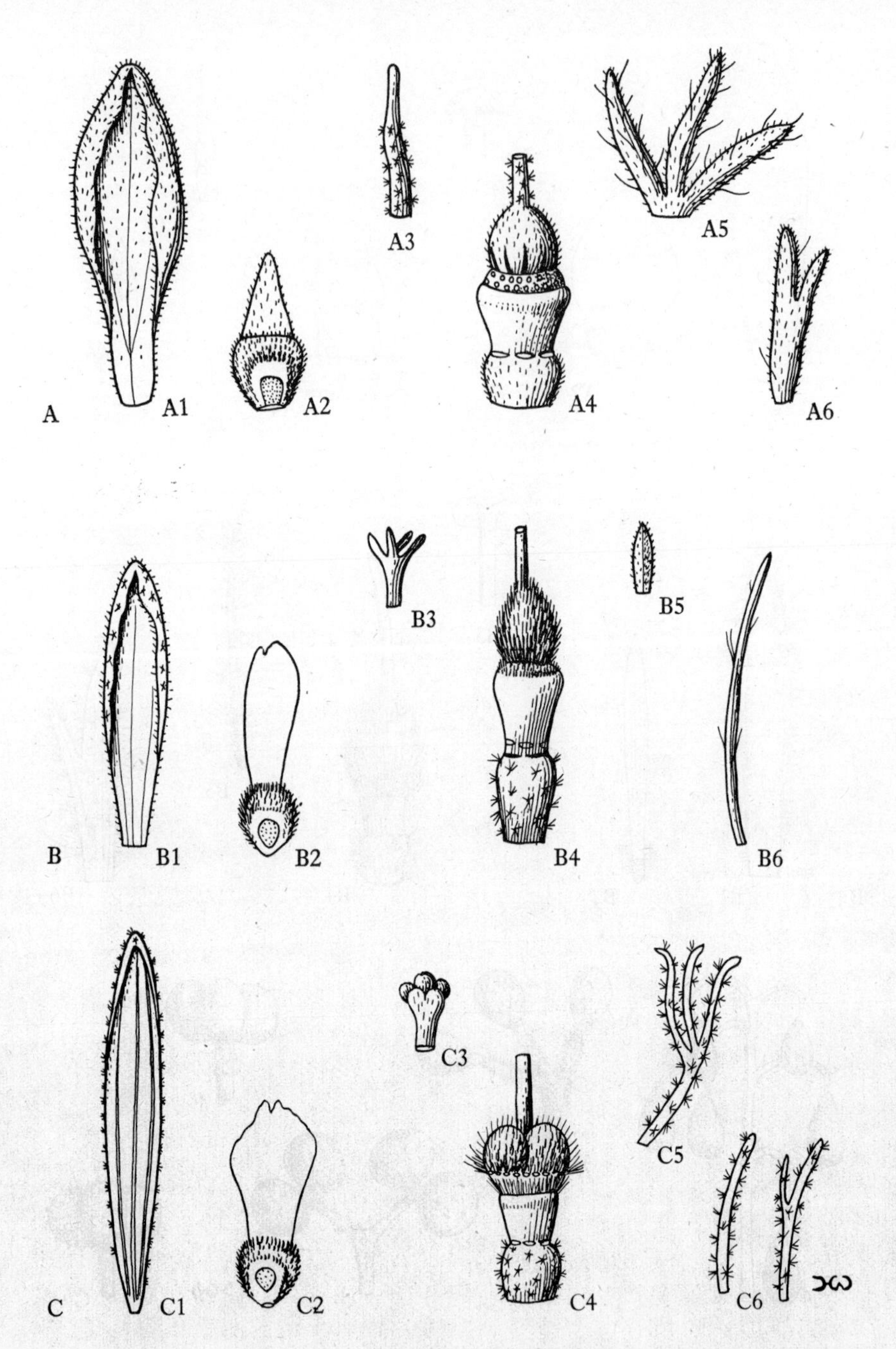

Tab. 4. A.—GREWIA TRANSZAMBESICA (SECT. MICROCOS). A1, sepal (× 7); A2, petal (× 7); A3, stigma (× 7); A4, androgynophore and ovary (× 7); A5, bract (× 7); A6, stipule (× 7), all from *Simão* 1304. B.—GREWIA GRACILLIMA (SECT. PLURIOVULATAE). B1, sepal (× 7); B2, petal (× 7); B3, stigma (× 7); B4, androgynophore and ovary (× 10); B5, bract (× 7); B6, stipule (× 7), all from *Mylne* 42/51. C.—GREWIA HORNBYI (SECT. AXILLARES). C1, sepal (× 5); C2, petal (× 5); C3, stigma (× 5); C4, androgynophore and ovary (× 5); C5, bract (× 5); C6, stipules (× 5), all from *Hornby* 2497.

Apparently confined to Mozambique but it may be the same as *G. heterotricha* Burret (1914) from Dar es Salaam. The type of this latter species is destroyed but, in any case, the name is invalidated by *G. heterotricha* Mast. (1875) an Indian species. Shrub or small tree of open *Brachystegia* woodland on the coastal plains to the north and south of the Zambezi.

3. **Grewia barombiensis** K. Schum. in Engl., Bot. Jahrb. **15**: 124 (1892).—Burret in Engl., Bot. Jahrb. **45**: 165 (1910).—Exell & Mendonça, C.F.A. **1**, 2: 214 (1951).—Keay, F.W.T.A. ed 2, **1**, 2: 303 (1958). Type from Cameroons.

Scrambling shrub or liane, with the young branches ferruginously stellate-hairy. Leaf-lamina 6–15 × 2·5–7 cm., oblong-lanceolate to ovate-lanceolate, apiculate to acute at the apex, margin entire or subentire, rounded to subcordate at the base, glabrous or pubescent on the midrib above, stellate-pubescent below, slightly discolorous, nerves reticulate and somewhat prominent; petiole up to 8 mm. long, with a ferruginous tomentum; stipules up to 4 mm. long, subulate, digitate or simple, with 1–4 lobes, ferruginously pubescent. Inflorescences terminal or axillary, paniculate, all parts with a tufted ferruginous tomentum; peduncles up to 13 mm. long; pedicels up to 6 mm. long. Sepals c. 7 mm. long, oblong-spathulate, involute in the upper part, cream-coloured and velvety outside with scattered tufted ferruginous hairs, pubescent within. Petals with a triangular blade up to 1 mm. long, pubescent at the base; nectariferous claw up to 1·5 mm. long, densely villous at the back and around the margin within. Androgynophore 1 mm. long, not produced, glabrous. Ovary densely tomentose, not lobed; style 3–4 mm. long, slightly pubescent at the base; stigma not wider than the style. Fruit bright red, rather fleshy, up to 3 × 1·5 cm., ± ellipsoid but tapering to both ends, sparsely stellately hairy.

N. Rhodesia. N: Kawambwa, fr. 22.viii.1957, *Fanshawe* 3516 (K). W: Chingola, fl. 1.iii.1956, *Fanshawe* 2807 (K; SRGH).
Also in Angola, Cameroons, Nigeria, Gaboon and the Congo. Forest liane or climber at forest edges. Reaching the limit of its range of distribution in N. Rhodesia and found in small forest patches or " mateshi ".

Our material is rather more pubescent on the lower surface of the leaves than most West African material but matches Angolan material such as *Gossweiler* 13999 (BM; K) from the Lunda province.

4. **Grewia conocarpa** K. Schum. in Engl., Pflanzenw. Ost-Afr. **C**: 264 (1895).—Burret in Engl., Bot. Jahrb. **45**: 164 (1910).—Brenan, T.T.C.L.: 614 (1949). Type from Tanganyika (Usaramo).
Grewia salamensis Sprague in Kew Bull. **1909**: 67 (1909).—Burret in Engl., Bot. Jahrb. **45**: 164 (1910). Type from Zanzibar.

Shrub or small tree up to 5 m. tall, often branching near the base; young branches shortly grey-tomentose, becoming glabrous and greyish-brown. Leaf-lamina 3·5–13 × 1·5–7 cm., obovate or obovate-oblong, minutely denticulate or subentire, subacute at the apex, rounded or subcordate at the base, with minute scattered stellate hairs above and a short grey indumentum of stellate hairs below; lateral nerves 4–5, fairly prominent; petiole up to 1 cm. long, tomentellous; stipules up to 8 mm. long, subulate, bifid, finely pubescent. Inflorescences of terminal or axillary many-flowered panicles; inflorescence branches 2–5 mm. long, shortly brownish- or greyish-tomentose; pedicels similar, 1–3 mm. long; bracts c. 4 mm. long, bifid or trifid, densely and shortly greyish-velvety. Sepals 0·7–0·9 mm. long, margins inrolled, hooded at the apex, greyish-velvety outside, pubescent within. Petals pinkish with a narrowly triangular lamina 1·5 mm. long, pubescent on both sides; nectariferous claw 1 mm. long, tomentose on the margins within and on the back. Androgynophore 1 mm. tall, glabrous, not produced. Stamens pubescent at the base. Ovary tomentose, ovoid, not lobed; style 4 mm. long, pubescent at the base; stigma not widened at the apex. Fruit c. 10 × 7 mm., rather dry, tomentellous, ovoid.

Mozambique. N: R. Messalo, fl. & fr. ii.1912, *Allen* 121 (K). Also in Tanganyika. Coastal districts, either in thickets or in forest patches.

5. **Grewia gilviflora** Exell in Journ. of Bot. **77**: 167 (1939).—Brenan, T.T.C.L.: 614 (1949).—White, F.F.N.R.: 236 (1962). Type from Tanganyika (Mkwemi).

Shrub up to 5 m. tall; young branches ferruginous-tomentose but soon glabrous, becoming dark purple-brown. Leaf-lamina 4·5–11 × 2–4·5 cm., lanceolate-elliptic to obovate-elliptic or oblong-elliptic, apex acuminate to subobtuse, margin serrulate or irregularly denticulate, base rounded to subcordate, ± equal-sided, stellate-puberulous above, densely but shortly white-velvety below but glabrescent later; primary nerves brownish, glabrescent; petiole up to 1 cm. long, stellate-pubescent. Inflorescences paniculate at the ends of the branchlets, flowers 2–3 together on each lateral branch of the panicle; panicle branches 2–3 mm. long, densely ferruginously pubescent; pedicels similar, up to 5 mm. long; bracts 7–9 × 6–7 mm., obovate with a bifid or trifid apex, densely tomentellous. Sepals pale brown-velvety, 9–11 × 2 mm., oblong, margins inrolled. Petals whitish, 3·5 × 1 mm., oblong, with a basal nectary which is pilose on its margins. Androgynophore 1 mm. long, glabrous, not elongated above the node. Ovary subglobose, puberulous, not lobed; style c. 6 mm. long, glabrous; stigma entire, wider than the style. Fruit yellowish, up to 15 × 8 mm., obovoid to obovoid-ellipsoid, not lobed, at first stellate-puberulous, later glabrescent.

N. Rhodesia. N: Lake Tanganyika coast between Kalambo mouth and Mpulungu, fr. 18.v.1936, *Burtt* 6035 (BM; K).
Also in Tanganyika. Associated with *Brachystegia* spp. or forming thickets on the shingle beaches of Lake Tanganyika.
Flowering from January to April. Fruit edible. Used for making hut poles.

6. **Grewia holstii** Burret in Engl., Bot. Jahrb. **45**: 167 (1910).—Brenan, T.T.C.L.: 618 (1949). Type from Zanzibar.

Shrub or small tree up to 8 m. tall or liane climbing over tall trees; young branches pubescent, soon becoming glabrous; older stems 3–4-angled, blackish and sometimes developing stout spine-like bosses. Leaf-lamina 3–6 × 1·5–3 cm., elliptic or lanceolate-elliptic, acute to acuminate at the apex, margin serrate, rounded or subcordate at the base, subglabrous, with minute scattered appressed-stellate hairs on both sides; petiole 0·2–0·3 cm. long, pubescent; stipules up to 5 mm. long, subulate to linear, slightly pubescent. Inflorescences all axillary; peduncles c. 5 mm. long, slightly pubescent; pedicels usually 3 together, c. 5 mm. long, slightly pubescent. Flower-buds oblong, hardly swollen at the base. Sepals up to 1·2 cm. long, linear-oblong, sparsely pubescent and greenish outside, yellow and glabrous within. Petals yellow, slightly shorter than the sepals, glabrous except for the nectariferous base, which is up to 1 mm. long circumvillous within and setose-pubescent on the back. Androgynophore 1 mm. long, glabrous and shallowly cup-like at the apex, clasping the ovary. Ovary ovoid, pubescent, not at all lobed; style c. 7 mm. long, glabrous, with 4 subulate stigma-lobes. Fruit 5–8 mm. in diam., globose, not lobed, yellowish, very minutely stellate-pubescent.

Mozambique. N: Cabo Delgado, between Porto Amélia and S. Paulo Mission, fl. 27.x.1942, *Mendonça* 1084 (LISC; SRGH).
Also in both coastal and inland districts of Tanganyika. Shrub in open woodland or grassland, sometimes forming thickets but also growing as a liane under forest conditions.
Related to *G. carpinifolia* Juss. but with consistently smaller leaves and flowers.

7. **Grewia gracillima** Wild in Bol. Soc. Brot., Sér. 2, **31**: 82, t. 1 fig. B (1957). TAB. **4** fig. B. Type: S. Rhodesia, Ndanga, Chipinda Pools, *Mylne* 42/51 (SRGH, holotype).

Shrub c. 2 m. tall, becoming scandent if crowded with other trees or bushes; young branches very delicate, sparsely pubescent, soon becoming glabrous; older branches 4-angled. Leaf-lamina 1·5–6 × 0·8–2·5 cm., lanceolate, apex acute to subacute, margin serrate or serrate-crenate, base rounded or subcordate, glabrous above, very sparsely appressed-stellate-pubescent below, especially on the nerves; lateral nerves 4–5 pairs; petiole 1–3 mm. long, sparsely setulose-pubescent; stipules c. 6 mm. long, subulate, sparsely setulose. Inflorescences all axillary; peduncles 3–10 mm. long, 1–3-flowered, slender, sparsely setulose; pedicels similar, 2·5–6 mm. long; basal bracts 1 mm. long, oblong, minutely pubescent. Flower-buds oblong, slightly wider towards the apex. Sepals c. 6·5 mm. long, narrowly oblong-spathulate, involute towards the apex, pubescent outside, 3-nerved, apex somewhat hooded. Petals yellow, about half the length of the sepals, often incised at the apex, with a basal nectary which is circumvillous on the margins within, with a slight ledge above and glabrous at the back. Androgynophore 0·75 mm. long, glabrous, barely elongated at the pubescent apex. Ovary never lobed, appressed-setose; style c. 3 mm. long, glabrous; stigmas very small, subulate. Fruit c. 8 mm. in diam., globose, yellowish, very sparsely stellate-setulose.

S. Rhodesia. N: Urungwe, Sanyati-Chiroti, fl. 6.i.1958, *Goodier* 534 (K; SRGH). C: Beatrice, fl. 4.i.1946, *Wild* 603 (K; SRGH). E: Umtali, Zimunya's Reserve, fr. 8.iv.1956, *Chase* 6063 (SRGH). S: Ndanga, Chipinda Pools, fl., *Mylne* 42/51 (SRGH). **Mozambique.** MS: between Mavita and Dombe, fr. 18.vi.1942, *Torre* 4344 (LISC; SRGH).

Also on the Soutpansberg in the Transvaal. Bush or scrambler of lower-altitude woodland up to 700 m. but most commonly found, not in more open woodland, but in somewhat protected situations such as stream banks and on rocky hills where vegetation is relatively dense.

Like *G. decemovulata* Merxm. this species is rather difficult to place in Burret's classification. Superficially it appears to be nearest to *G. carpinifolia* Juss. but it lacks the shallow hollowing out of the apex of the androgynophore, which clasps the base of the ovary in that species. Instead there is the slightest suggestion of an elongation of the androgynophore and the apex is pubescent. Its androgynophore is, therefore, of a type intermediate between subsection *Apodogynae* Burret and subsection *Podogynae* Burret and in fact this species is in all other respects remarkably near *G. caffra* Meisn. of the *Podogynae*.

8. **Grewia retinervis** Burret in Engl., Bot. Jahrb. **45**: 170 (1910) pro parte quoad specim. Baum. excl. specim. Dinter.—Burtt Davy, F.P.F.T. **1**: 255 (1926).—O. B. Mill. in Journ. S. Afr. Bot. **18**: 53 (1952). Type from Angola (Bié).
Grewia pilosa sensu Eyles in Trans. Roy. Soc. S. Afr. **5**: 411 (1916).
Grewia flavescens sensu Exell and Mendonça, C.F.A. **1**, 2: 221 (1951) pro parte.

Small shrub 1–2 m. tall, usually branching low down; young stems brownish-pubescent or glabrescent; older branches not becoming 4-angled and grooved. Leaf-lamina 2–5 × 1–2·75 cm., elliptic-oblong, acute or rounded at the apex, with margins serrate, rounded at the base, sparsely pubescent above or glabrous; nerves reticulate on both sides, slightly pubescent below; petiole 1–3 mm. long, pubescent; stipules c. 2·5 mm. long, subulate, pubescent. Inflorescences all axillary; peduncles 3–6 mm. long, pubescent; pedicels 2–3 together, 2–4 mm. long, pubescent; basal bracts c. 2 mm. long, entire, lanceolate-triangular, pubescent. Flower-buds oblong, somewhat sulcate. Sepals 6–8 × 1 mm., lorate, appressed-pubescent outside, glabrous within. Petals yellow, 5–7 × 1·5 mm., narrowly oblong or lorate, with a basal nectariferous claw 1 mm. long and circumvillous within. Androgynophore 1 mm. long, glabrous, not extended above the node, rather cupular at the apex and clasping the ovary-base. Ovary appressed-pubescent, never lobed; style 8–9 mm. long, glabrous; stigmas usually 4, subulate. Fruit shining, reddish-brown, c. 8 mm. in diam., globose, never lobed, minutely and sparsely stellate-pubescent.

Caprivi Strip. Katima Mulilo area, fl. 24.xii.1958, *Killick & Leistner* 3089 (PRE; SRGH). **Bechuanaland Prot.** N: Chobe-Zambezi confluence, fl. 11.iv.1955, *E.M. & W.* 1464 (BM; LISC; SRGH). SE: Artesia, fr. 12.iv.1931, *Pole Evans* 3168 (PRE; SRGH). SW: Ghanzi, fr. 11.vi.1955, *Story* 4875 (K; PRE). **N. Rhodesia.** B: Sesheke, fl. i.1922, *Borle* 329 (PRE; SRGH). S: Namwala, fr. 11.vi.1949, *Hornby* 3014 (SRGH). **S. Rhodesia.** W: Victoria Falls. fr. 4–6.v.1948, *Rodin* 4488 (K; PRE; SRGH). C: Vungu Native Purchase Area, Gwelo Distr., fl. i.1960, *Davies* 2669 (SRGH). S: Nuanetsi, fl. xii.1955, *Davies* 1856 (SRGH).

Also in the Transvaal, SW. Africa and Angola. Open woodland, common on Kalahari Sand formations.

This species has, by a number of botanists, been treated as a form of *G. flavescens* Juss. However, the fact that the fruits are always 1-lobed, the leaves almost glabrous and the older branches rounded rather than 4-sided relates this species rather to *G. carpinifolia* Juss. than to *G. flavescens*, if one follows the criteria used by Keay in F.W.T.A. ed. 2 (1958) to separate these last two species. Ecologically, as a small bush of dry Kalahari conditions, it is readily separated from *G. carpinifolia*, which is a forest species, but, like a number of other species such as *Combretum platypetalum* Welw. and *Dichapetalum cymosum* (Hook.) Engl., which are also plants of the open woodlands of our area, it may have evolved from a liane-like forest ancestor. All these plants are most nearly related to other species in the same genera which are forest scramblers or lianes. For these reasons *G. retinervis* is once more treated as a distinct species and, under our conditions at least, its rather dwarf habit, apart from the other characters mentioned, renders it readily distinguishable from the larger *G. flavescens*. The *Dinter* specimen from SW. Africa cited by Burret as *G. retinervis* is *G. flavescens*.

9. **Grewia flavescens** Juss. in Ann. Mus. Nation. Hist. Nat. Par. **4**: 91 (1804).—Burret in Engl., Bot. Jahrb. **45**: 168 (1910).—Eyles in Trans. Roy. Soc. S. Afr. **5**: 410 (1916).—Burtt Davy, F.P.F.T. **1**: 254 (1926).—Brenan, T.T.C.L.: 618 (1949).—Exell & Mendonça, C.F.A. **1**, 2: 221 (1951) pro parte.—O.B.Mill. in Journ. S. Afr. Bot. **18**: 53 (1953).—Keay, F.W.T.A. ed. 2, **1**, 2: 305 (1958). Type from India.
Grewia flava sensu Mast. in Oliv., F.T.A. **1**: 250 (1868) pro parte quoad specim. Kirk.
Grewia pilosa sensu O.B.Mill. in Journ. S. Afr. Bot. **18**: 53 (1953).

Shrub 2–5 m. tall; young branches stellate-pubescent, becoming glabrous; older stems 4-angled and somewhat sulcate, sometimes scandent. Leaf-lamina 4–12 × 2–8·5 cm., oblanceolate, obovate or oblong-lanceolate, occasionally almost circular, acute or acuminate at the apex, margins irregularly serrate, rounded or subcordate at the base, harshly pubescent, particularly below; petiole up to 7 mm. long but usually shorter, pilose; stipules up to 1 cm. long, subulate to oblong-lanceolate, somewhat keeled, pubescent on both sides. Inflorescences all axillary on pubescent peduncles up to 1·5 cm. long; pedicels up to 2 cm. long, 2–3 per peduncle, pubescent; bracts entire, c. 3 mm. long, lanceolate to ovate-lanceolate, pubescent, especially on the back. Flower-buds oblong with a slight swelling and then a constriction just above the base. Sepals yellow inside, yellowish-tomentose outside, 12–20 mm. long, lorate. Petals yellow, 6–8 × 1·5 mm., narrowly oblong, with a nectariferous claw villous on the margins within and at the base outside. Androgynophore 1·5–2 mm. long, glabrous, not elongated above the node. Ovary closely setose-pubescent, entire or slightly 2-lobed; style c. 1 cm. long, glabrous; stigma-lobes usually 4, subulate. Fruit 8–14 × 13–15 mm., depressed-globose, sometimes entire but usually shallowly 2- or occasionally 4-lobed, yellowish-brown with short, appressed-stellate hairs often mixed with longer stellate hairs, somewhat shining when ripe.

Widespread throughout tropical Africa, the Transvaal and SW. Africa; also in Arabia and India. Open woodland in all but the high-altitude high-rainfall areas. Above 1400 m. tending to occur more commonly on termite mounds.

Var. **flavescens**

Leaf-lamina oblanceolate, oblong-lanceolate or obovate, usually subtruncate at the base. Flowering peduncles usually less than 1 cm. long.

Bechuanaland Prot. N: Ngamiland, Kwebe Hills, fl. 21.i.1898, *Lugard* 121 (K). SE: Mochudi, fr. i–iv.1914, *Harbor* (BM). SW: 98 km. from Ghanzi on road to Maun, fr. 27.vii.1955, *Story* 5053 (PRE). **N. Rhodesia.** B: Sesheke, fl. 23.xii.1952, *Angus* 1007 (FHO; K). N: Abercorn, fl. 29.xii.1951, *Richards* 188 (K). W: Mwinilunga, fl. 13.xii.1937, *Milne-Redhead* 3638 (BM; K). C: Lusaka, fl. 1953, *Puffet* (K; PRE). E: between Sasare and Petauke, fl. 9.xii.1958, *Robson* 880 (K; SRGH). S: Mazabuka, fl. 14.i.1952, *White* 1908 (FHO; K). **S. Rhodesia.** N: Mtoko, fr. 14.iv.1951, *Lovemore* 2 (K; SRGH). W: Bubi, fr. 4.vi.1947, *Keay* in FHO 21308 (FHO; K; SRGH). C: Salisbury, fl. 2.i.1946, *Wild* 654 (K; SRGH). E: Inyanga North, fr. 19.ii.1954, *West* 3327 (SRGH). S: Ndanga, fl. 26.i.1949, *Wild* 2769 (K; SRGH). **Nyasaland.** N: Karonga Distr. Waye R., fl. 17.iii.1954, *Jackson* 1244 (FHO; K). S: Lower Shire Valley, fl. i.1862, *Kirk* (K). **Mozambique.** N: without precise locality, fl. & fr. ii–iv.1937, *Torre* 1560 (COI). Z: between Mocuba and Milange, fl. & fr. 18.iii.1943, *Torre* 4953 (LISC; SRGH). T: Tete, fr. *Kirk* (K). MS: Vila Machado, fr. 14.iv.1948, *Mendonça* 3919 (LISC). LM: Goba, fl. 24.x.1940, *Torre* 1853 (LISC; SRGH).

Distribution and ecology as for the species as a whole.

Var. **olukondae** (Schinz) Wild, comb. nov.
Grewia olukondae Schinz in Bull. Herb. Boiss., Sér. 2, **8**: 701 (1908).—Burret in Engl., Bot. Jahrb. **45**: 166 (1910).—Weim. in Bot. Notis. **1936**: 39 (1936).—O.B. Mill. in Journ. S. Afr. Bot. **18**: 53 (1953). Type: SW. Africa, Amboland, Olukonda, *Schinz* 1114 (Z, holotype).

Leaf-lamina broadly oblong to almost circular, rounded or often ± cordate at the base. Flowering peduncles often more than 1 cm. long.

Bechuanaland Prot. N: Chobe R., Serondela, fl. xii.1951, *Miller* 1270 (K). SE: Eastern Bamangwato country, Tamasetzi, fl. 5.iii.1875, *Holub* (K). **N. Rhodesia.** B: Sesheke, fl. 26.xii.1952, *Angus* 1036 (FHO; K). S: Sinachirundu, fl. & fr. 29.xii.1958, *Robson* 994 (K; SRGH). **S. Rhodesia.** N: Sebungwe, fl. & fr. ix.1955, *Davies* 1508

(K; SRGH). W: Wankie, fl. 15.ii.1956, *Wild* 4731 (K; SRGH). C: Rusape, fl. 6.i.1931, *Norlindh & Weimarck* 4148 (LD). S: Victoria, fl. 1909, *Monro* 794 (BM). **Mozambique.** T: between Mandié and Tete, fr. 26.vi.1941, *Torre* 2935 (LISC).

Also in SW. Africa and the Transvaal. Often in *Baikiaea plurijuga* woodland, i.e. on Kalahari Sand; also on the alluvial banks of rivers just above flood level.

The young leaves of this variety are indistinguishable in shape from those of var. *flavescens*, so it is not practicable to retain it as a distinct species.

10. **Grewia forbesii** Harv. ex Mast. in Oliv., F.T.A. **1**: 250 (1868).—Burret in Engl., Bot. Jahrb. **45**: 171 (1910).—Brenan, T.T.C.L.: 618 (1949). Type: Mozambique, *Forbes* (K, holotype).

Grewia pilosa var. *grandifolia* Kuntze, Rev. Gen. **3**, 2: 26 (1898).—Burret, loc. cit. Type: Mozambique, without precise locality, *Kuntze* (B†, holotype; K; NY).

Grewia latiunguiculata K. Schum. in Notizbl. Bot. Gart. Berl. **3**: 102 (1901). Type: Nyasaland, Zomba, *Whyte* (K, holotype).

Much-branched shrub or small tree; young branches coarsely and ferruginously hairy. Leaf-lamina 3·5–12 × 2·5–7 cm., ovate-oblong or obovate-oblong to broadly elliptic, scabrid above with the nervation noticeably impressed, stellate-pubescent below with the nerves strongly reticulate; petiole up to 4 mm. long, rufous-hairy; stipules up to 5 mm. long, subulate and often falcate, pubescent. Inflorescences all axillary; peduncles 0–3 mm. long, ferruginously hairy; pedicels similar, about 3 together, 1–3 mm. long; bracts 3–5 × 1·5–3 mm., lanceolate to ovate-lanceolate, entire, pubescent on the back. Flower-buds oblong but slightly swollen at the base and usually slightly constricted above the swelling. Sepals 14–20 × 2·5 mm., closely stellate-pubescent outside, glabrous and yellow inside. Petals yellow, slightly shorter than the sepals, narrowly elongate, with a basal nectariferous claw c. 1·5 mm. long setose-pubescent at the back circumvillous within and produced into a ledge above. Androgynophore 1·5 mm. long, glabrous, not produced above the node, slightly lobed on the upper margin producing a shallow cup clasping the ovary. Ovary ellipsoid, densely setose, entire or subentire; style up to 1·3 cm. long, setose-pubescent except at the apex; stigma-lobes 4, subulate. Fruit entire or shallowly 2–4-lobed, c. 1·6 cm. in diam., sparsely setose and covered with warty rough protuberances.

Nyasaland. S: Ncheu, fl. & fr. iv.1956, *Jackson* 1850 (FHO; K; SRGH). **Mozambique.** N: R. Rovuma, Mocímboa, fr., *Stocks* (K).

Also on the coasts of Tanganyika, Kenya and Zanzibar. Mainly coastal species related to *G. flavescens* Juss., but a robust shrub larger in all parts than this latter species and readily recognized by its warted fruits.

11. **Grewia decemovulata** Merxm. in Proc. & Trans. Rhod. Sci. Ass. **43**: 104 (1951). —White, F.F.N.R.: 239, fig. 42E (1962). TAB. **3** fig. D. Type: S. Rhodesia, Marandellas, *Dehn* 270 (M, holotype; SRGH).

Grewia flavescens var. *longipedunculata* Burret in Engl., Bot. Jahrb. **45**: 169 (1910). Type from Angola.

Grewia trothai sensu Hutch., Botanist in S. Afr.: 485 (1946).

Grewia flavescens sensu Exell and Mendonça, C.F.A. **1**, 2: 221 (1951) pro parte.

Procumbent or spreading shrub, branching from ground level or occasionally along the stems; young branches yellowish-pubescent. Leaf-lamina 3·5–9 × 2·5–7·5 cm., very broadly oblong, rounded, acute or acuminate at the apex, margin serrate, truncate or subcordate at the base, upper surface bronzed when dry, sparsely appressed-pubescent, more densely pubescent beneath; nerves rather prominent below; petiole 2–3 mm. long, densely pubescent; stipules c. 5 mm. long, subulate, entire or more usually bifid, pubescent. Inflorescences all axillary with 3–5 cymes often crowded in one axil, on thinly pubescent peduncles up to 1·5 cm. long; pedicels similar, up to 5 mm. long; bracts deeply bifid, trifid or occasionally simple, pubescent. Flower-buds oblong, not swollen at the base, somewhat longitudinally sulcate. Sepals c. 7 mm. long, linear-oblong, densely and closely pubescent on the back, glabrous and reddish-yellow within. Petals yellow, somewhat shorter than the sepals, linear-oblong, often bidentate or incised at the apex, with a basal nectariferous claw circumvillous on the margins within setose-pubescent on the back and with a slight ledge above on the inside. Androgynophore 1 mm. long, glabrous, barely elongated above the node but tomentose at the apex and not clasping the ovary-base. Ovary ovoid, not lobed, setose-pubescent; style up to 6 mm. long, glabrous; stigma-lobes 4, short, subulate. Fruit c. 8 mm.

in diam., yellowish when ripe, globose or obovoid, finely appressed-stellate-hairy, not lobed.

N. Rhodesia. B: near Kabompo pontoon, fr. 23.iii.1961, *Drummond & Rutherford-Smith* 7235 (K; PRE; SRGH). N: Shiwa Ngandu, fl. 3.i.1937, *Ricardo* 173 (BM). W: Ndola, fl. & fr. 4.xi.1955, *Fanshawe* 2586 (K; SRGH). C: Chisamba, fl. & fr. 8.iv.1933, *Michelmore* 663 (K). S: Mazabuka, fl. 15.i.1952, *White* 1925 (FHO; K). **S. Rhodesia.** W: Matopos, fl. 18.i.1930, *Brain* 30 (SRGH). C: Charter, fl. 28.xii.1926, *Eyles* 4571 (K; SRGH).

Also in Angola (Huila). Small bush of open *Brachystegia* woodland on the central plateaux of N. and S. Rhodesia.

Although this species has been confused with *G. flavescens* Juss., its dwarf habit is quite distinctive and a further minute but important difference is that the androgynophore is in fact very slightly elongated. It therefore falls rather artificially in Burret's Subsect. *Apodogynae* with *G. flavescens* etc. when it would be better placed with *G. falcistipula* K. Schum. in the *Podogynae*. The divided bracts of this species point to the same affinity.

Grewia trothai Burret to which Hutchinson (loc. cit.) refers *Hutchinson & Gillett* 3560 (K) is an unpublished MS name based on *von Trotha* 25a from Windhoek, SW. Africa. Burret later (in Engl., Bot. Jahrb. **45**: 166 (1911)) referred this latter specimen to *G. olukondae* Schinz.

12. **Grewia falcistipula** K. Schum. in Warb., Kunene-Samb.-Exped. Baum. 296: (1903).—Burret in Engl., Bot. Jahrb. **45**: 172 (1910).—Exell & Mendonça, C.F.A. **1**, 2: 216 (1951).—White, F.F.N.R.: 238, fig. 41D (1962). Type from Angola (R. Cubango).

Shrub with branches spreading from the base, up to 1 m. tall; young branches greyish-tomentose. Leaf-lamina 2–7·5 × 1·3–4 cm., elliptic, broadly elliptic or ovate, acute at the apex, margin serrate, asymmetrically rounded or subcordate at the base, finely reticulate and sparsely appressed-stellate-hairy above, densely greyish-white-tomentellous below; petiole c. 5 mm., greyish- or greyish-brown-tomentose; stipules up to 1 cm. long, entire, auriculate at the base, subulate above, pubescent, fairly persistent. Inflorescences all axillary; peduncles c. 1 cm. long, tomentellous; pedicels normally 3 together, c. 2 mm. long; bracts c. 3 mm. long, normally bifid or trifid ⅓- or ½-way, greyish-tomentellous. Flower-buds oblong. Sepals greenish-grey, up to 1·2 cm. long, linear-oblong, tomentellous on the back, glabrous within. Petals bright yellow, slightly shorter than the sepals, linear, often bifid at the apex, widening into a nectariferous base which is circumvillous within ledged above and tomentellous on the back. Androgynophore with a glabrous base, c. 1 mm. long, extended above into a densely pubescent column up to 2 mm. long. Ovary ovoid, not lobed, setose-pubescent; style 5–6 mm. long, pubescent at the base; stigma-lobes c. 4, short, subulate. Fruit up to 1 cm. in diam., never lobed, globose, shortly and sparsely stellate-pubescent.

Caprivi Strip. Katima Mulilo area, fl. 24.xii.1958, *Killick & Leistner* 3055 (PRE; SRGH). **N. Rhodesia.** B: Sesheke, fr. 1.ii.1952, *White* 2000 (FHO; K). S: Nanga Forest, N. of Machili, fl. 19.xii.1962, *Angus* 961 (FHO).

Also widely distributed in Angola. Often on Kalahari Sand formations and so frequently found in *Baikiaea plurijuga* woodland or accompanying such species as *Pterocarpus angolensis*, *Diplorhynchus condylocarpon*, *Burkea africana*, etc.

Fruit edible.

13. **Grewia caffra** Meisn. in Hook., Lond. Journ. Bot. **2**: 53 (1843).—Harv. in Harv. & Sond., F.C. **1**: 225 (1860).—Burret in Engl., Bot. Jahrb. **45**: 171 (1910).—Burtt Davy, F.P.F.T. **1**: 255 (1926). Type from Natal.

Grewia fruticetorum J. R. Drummond ex Bak. f. in Journ. Linn. Soc., Bot. **40**: 31 (1911). Type: Mozambique, Manica and Sofala, Lower Chibabava, *Swynnerton* 1222 (BM, holotype; K).

Scandent many-stemmed shrub; branchlets sparsely setose-pubescent, soon becoming glabrous; older stems quadrangular. Leaf-lamina 2–5 × 1–2·5 cm., oblong-ovate to lanceolate, acute to acuminate at the apex, broadly cuneate to rounded at the base, finely serrate or serrulate, glabrous or very sparsely appressed-stellate-pubescent on both sides; petiole 2–3 mm. long, sparsely setose-pubescent; stipules c. 6 mm. long, setaceous, very sparsely setulose-pubescent or glabrous. Inflorescences all axillary, on slender setulose-pubescent peduncles up to 7 mm. long; pedicels similar, 2–3 together, up to 1 cm. long. Flower-buds oblong.

Sepals 6–9 mm. long, linear, finely pubescent on the back, glabrous within. Petals yellow, slightly shorter than the sepals, with a basal nectariferous gland circumvillous within, produced into a ledge above and setose-pubescent on the back. Androgynophore glabrous below for 1 mm., produced above into a closely pubescent extension c. 2 mm. long. Ovary ovoid, not lobed, setose-pubescent; style c. 7 mm. long, glabrous; stigma with 4 subulate lobes. Fruit 7·5–10 mm. in diam., globose, never lobed, very sparsely stellate-pubescent or glabrous.

S. Rhodesia. S: Junction of Sabi and Lundi Rivers, fr. 6.vi.1950, *Wild* 3366 (K; SRGH). **Mozambique.** Z: Chamo, Shire-Zambezi Junction, fr. 15.i.1863, *Kirk* (K). MS: Chibabava, Lower R. Buzi, fl. 28.xi.1906, *Swynnerton* 1222 (BM; K). SS: Inhachenga, fr. 26.ii.1955, *E.M. & W.* 622 (BM; LISC; SRGH). LM: Magude, fl & fr. 16.i.1948, *Torre* 7144 (LISC; SRGH).

Also in the Transvaal and Natal. Common along the banks of largish rivers at low altitudes, either as a thicket-forming species or as a scrambler or liane in riverine fringes. Fruit edible.

14. **Grewia mollis** Juss. in Ann. Mus. Nation. Hist. Nat. Par. **4**: 91 (1804).—Mast. in Oliv., F.T.A. **1**: 248 (1868).—Burret in Engl., Bot. Jahrb. **45**: 174 (1910).—Brenan, T.T.C.L.: 621 (1949).—Keay, F.W.T.A. ed. 2, **1**, 2: 304 (1958).—White, F.F.N.R.: 238, fig. 42C (1962). Type from Nigeria.

Grewia venusta Fresen. in Mus. Senckenb. **2**: 159, t. 10 (1837). Type from Ethiopia.

Shrub or small tree up to 4 m. tall; bark thick and flaking, leaving a yellow-brown surface exposed; flowering branches elongated, up to 2 m. long, flattened somewhat towards their tips, brownish- or greyish-tomentose. Leaf-lamina 5–15 × 2·5 cm., oblong, oblong-lanceolate or lanceolate, apex acute or subacute, margin serrate, coarsely and irregularly so in large leaves, rounded to broadly cuneate at the base, green and minutely stellate-pubescent above, greyish- or greyish-brown-tomentose below on both nerves and spaces between them; petiole up to 1 cm. long but usually less, brownish-tomentose, more densely so above than below, somewhat channelled above; stipules up to 1 cm. long, linear-subulate, tomentellous, caducous. Inflorescences all axillary; peduncles sometimes 2–3 together in one axil, c. 1 cm. long, greyish-tomentose; pedicels similar, usually 3 but sometimes up to 6 together, c. 5 mm. long; bracts up to 5 mm. long, tomentellous, entire or rarely 2- or 3-dentate at the apex, very caducous. Flower-buds oblong-ovoid. Sepals 6–9 mm. long, linear-oblong, greyish-tomentose outside, yellow and glabrous inside. Petals bright yellow, about $\frac{2}{3}$ the length of the sepals, oblong-spathulate or obovate, often 2-dentate at the apex, with a basal nectariferous claw circumvillous within, with a short ledge above, and glabrous on the back. Androgynophore c. 1 mm. long, glabrous, villous at the apex, not elongated above. Ovary villous, shallowly 2-lobed or 1-lobed by abortion of 1 loculus and then style eccentric; style 6–7 mm. long, glabrous, with 4 flattened stigma-lobes. Fruit c. 7 mm. in diam., very deeply 2-lobed or more often 1-lobed by abortion of 1 loculus, globose, greyish-yellow, puberulous.

N. Rhodesia. N: Mkupa-Chiengi, fl. 10–12.x.1949, *Bullock* 1209 (K; SRGH).

Widespread to the north of our area from Tanganyika to Abyssinia in the east and through the Congo to Senegambia in the west. A species of open woodland and riverine vegetation, or sometimes a constituent of semi-evergreen thicket.

The wood is used for axe-handles and the fruit is edible.

15. **Grewia bicolor** Juss. in Ann. Mus. Nation. Hist. Nat. Par. **4**: 90, t. 50 fig. 2 (1804).—Burret in Engl., Bot. Jahrb. **45**: 176 (1910).—Brenan, T.T.C.L.: 621 (1949).—Exell & Mendonça, C.F.A. **1**, 2: 217 (1951).—O.B. Mill. in Journ. S. Afr. Bot. **18**: 52 (1953).—Keay, F.W.T.A. ed. 2, **1**, 2: 304 (1958).—White, F.F.N.R.: 238, fig. 42D (1962). Type from Senegal.

Grewia miniata Mast. ex Hiern, Cat. Afr. Pl. Welw. **1**: 95 (1896).—R.E.Fr. Wiss. Ergebn. Schwed. Rhod.-Kongo-Exped. **1**: 141 (1914).—Burtt Davy, F.P.F.T. **1**: 254 (1926). Type from Angola.

Grewia grisea N.E.Br. in Kew Bull. **1909**: 94 (1909). Type: Bechuanaland Prot., Kwebe, *Lugard* 54 (K, holotype).

Grewia kwebensis N.E.Br., tom. cit.: 95 (1909). Type: Bechuanaland Prot., Kwebe, *Lugard* 92 (K, holotype).

Grewia mossambicensis Burret in Engl., Bot. Jahrb. **45**: 178 (1910). Type: Mozambique, Ressano Garcia, *Schlechter* 11930 (B, holotype †; BM; K).

Grewia madandensis J. R. Drummond ex Bak. f. in Journ. Linn. Soc., Bot. **40**: 30 (1911). Type: Mozambique, Manica and Sofala, Madanda Forest, *Swynnerton* 1221 pro parte (BM, holotype; K).

Shrub or occasionally moderate-sized tree up to 9 m. tall; bark dark grey in large specimens, deeply fissured longitudinally and exfoliating in long strips, in smaller specimens bark grey and smooth; young branches grey- or brown-tomentellous. Leaf-lamina 1·5–7 × 1–3·2 cm., elliptic, elliptic-oblong or lanceolate, acute or rounded at the apex, margins finely serrate, sometimes almost entire, rounded or subcordate or broadly cuneate at the base, glabrous and green to very shortly greenish-tomentellous above, shortly and densely white-tomentose below; petiole c. 2 mm. (rarely up to 4 mm.) long; stipules c. 6 mm. long, linear or subulate, grey- or brownish-tomentellous. Inflorescences all axillary; peduncles brownish- or grey-tomentellous, 5–10 mm. long; pedicels similar, 2–3 together, 3–10 mm. long; bracts similar to the stipules, c. 4 mm. long, caducous. Flower-buds oblong-ovoid. Sepals up to 1·2 cm. long, linear-oblong or spathulate-oblong, slightly hooded at the apex, grey- or brownish-green-tomentellous outside, yellow and glabrous inside. Petals bright yellow, $\frac{2}{3}$–$\frac{3}{4}$ the length of the sepals, linear-oblong, sometimes 2-dentate at the apex, with a basal nectariferous claw circumvillous within and with a ledge above and glabrous on the back, or with the nectary entirely lacking. Androgynophore with a glabrous basal portion up to 1 mm. long when nectaries present, or completely absent, in the latter case the ovary almost sessile on a short pubescent cushion carrying the anthers. Ovary shortly stellate-pubescent, shallowly 2-lobed or 1-lobed with an eccentric style by abortion of 1 loculus; style c. 4 mm. long, glabrous; stigma-lobes c. 4, broad and flattened. Fruit deeply divided into two globose lobes, each c. 6 mm. in diam. or 1-lobed, sparsely stellate-pubescent, finally purple-black.

Bechuanaland Prot. N: Ngamiland, Kwebe Hills, fl. xii.1896, *Lugard* 92 (K). SE: Gaberones, fl. 1.xii.1954, *Codd* 8903 (PRE; SRGH). **N. Rhodesia.** B: Sesheke, fl. 20.xii.1952, *Angus* 973 (FHO; K). N: Abercorn, fl. 16.xi.1952, *Angus* 767 (FHO; K). W: Ndola, fr. 7.v.1954, *Fanshawe* 1170 (K). C: Chisamba, fr. 8.iv.1933, *Michelmore* 665 (K). E: Petauke, fl. 19.iv.1952, *White* 2412 (FHO; K). S: Chirundu, fl. 17.xii.1947, *Whellan* 333 (K; SRGH). **S. Rhodesia.** N: Mtoko, fl. 29.xii.1950, *Whellan* 489 (K; SRGH). C: Hartley, fr. vii.1927, *Jack* in Herb. Eyles 4960 (K; SRGH). E: Inyanga, Lawley's Concession, fr. 19.ii.1954, *West* 3351 (SRGH). S: Nuanetsi, Malipati, fl. 2.xi.1955, *Wild* 4694 (K; SRGH). **Nyasaland.** N: Karonga Distr., Rumpi to Karonga, c. 24 km. from Karonga, 480 m., fl. i.1959, *Richards* 10566 (K). S: Zomba, fl. 1934, *Townsend* 73 (FHO; K). **Mozambique.** N: c. 10 km. from Macomia, fl. 6.xi.1959, *Gomes e Sousa* 4501 (PRE; SRGH). T: near Chicoa, Zambezi valley, fr. 9.vi.1947, *Hornby* 2735 (K; SRGH). MS: Mossurize, Madanda Forest, fl. 5.xii.1906, *Swynnerton* 1221, partly (BM; K). SS: Vilanculos, fr. 21.v.1941, *Torre* 2704 (BM; LISC; SRGH). LM: Ressano Garcia, fl. 25.xii.1897, *Schlechter* 11930 (B†; BM; K).

Widespread from S. Africa to Ethiopia, Angola and West Africa; also in Arabia and India. On the whole more common in the drier types of deciduous woodland but also penetrating areas with an annual rainfall above 60 cm. in special habitats such as termite mounds. Often associated with *Colophospermum mopane* but also found in many other types of mixed woodland.

The wood of larger specimens is used for making axe-handles and walking-sticks. Fruit edible.

An extremely variable species, perhaps at least partly due to the fact that it hybridizes freely with *G. monticola* Sond., particularly in Bechuanaland, S. Rhodesia, the Transvaal and southern Mozambique. North of our area it seems to hybridize with *G. mollis* Juss. and it appears likely that *G. bicolor* var. *tephrodermis* (K. Schum.) Burret of Kenya, Tanganyika and Uganda owes its origin to hybridization between these two parents. The possibility of its hybridization with *G. micrantha* Boj. ex Mast. is discussed under that species.

16. **Grewia monticola** Sond. in Linnaea, **23**: 20 (1850).—Harv. in Harv. & Sond., F.C. **1**: 226 (1860).—Burret in Engl., Bot. Jahrb. **45**: 179 (1910).—Eyles in Trans. Roy. Soc. S. Afr. **5**: 410 (1916).—Burtt Davy, F.P.F.T. **1**: 254 (1926).—Steedman, Trees etc. S. Rhod.: 47 (1933).—Weim. in Bot. Notis. **1936**: 42 (1936).—Brenan, T.T.C.L.: 621 (1949).—O.B.Mill. in Journ. S. Afr. Bot. **18**: 53 (1953).—White, F.F.N.R.: 238, fig. 42B–C (1962). Syntypes from the Transvaal.

Grewia cordata N.E.Br. in Kew Bull. **1909**: 96 (1909).—Hutch., Botanist in S. Afr.: 480 (1946).—O.B.Mill. in Journ. S. Afr. Bot. **18**: 52 (1953). Type: Bechuanaland Prot., Kwebe Hills, *Lugard* 102 (K).

Grewia obliqua Weim. in Bot. Notis. **1936**: 39, fig. 10 (1936).—Suesseng. & Merxm. in Proc. & Trans. Rhod. Sci. Ass. **43**: 105 (1951) non *Grewia obliqua* Juss. (1804). Type: S. Rhodesia, Inyanga, *Norlindh & Weimarck* 4300 (BM; LD, holotype; SRGH).

Shrub or small tree up to 5 m. tall; young branches densely ferruginously tomentose. Leaf-lamina 2·5–9 × 1–5 cm., obliquely elliptic-oblong to ovate, acute at the apex, margin irregularly serrate, asymmetrically cordate or rounded at the base, sparsely pubescent or glabrous and somewhat rugose above, densely white-tomentose below; petiole 2–5 mm. long, ferruginously tomentose; stipules c. 8 mm. long, linear-lanceolate, apiculate. Inflorescences all axillary; peduncles up to 1 cm. long, often 2–3 together in one axil, ferruginous-tomentose; pedicels c. 5 mm. long, about 3 per peduncle, ferruginously tomentose; bracts c. 5 mm. long, linear to linear-lanceolate, ferruginously tomentose. Flower-buds ovoid or ovoid-oblong. Sepals ferruginously tomentose outside, glabrous and yellow inside, c. 10 mm. long, narrowly oblong. Petals bright yellow, $\frac{1}{2}$–$\frac{2}{3}$ the length of the sepals, oblong, narrowly obovate or obovate, often 2-dentate at the apex, either with a basal nectariferous claw which is circumvillous within ledged above and glabrous on the back, or with the gland entirely lacking. Androgynophore 1 mm. long, glabrous but villous at the apex producing a cushion that bears the stamens, or with the basal glabrous portion lacking in specimens with no nectaries. Ovary villous, shallowly 2-lobed; style c. 4 mm. long, glabrous, with broad flattened stigma-lobes. Fruit either deeply 2-lobed or 1-lobed by abortion of 1 loculus, each lobe c. 8 mm. in diam., setulose-pubescent, yellowish when ripe.

Bechuanaland Prot. N: Ngamiland, Kwebe Hills, fl. 12.i.1898, *Lugard* 102 (K). SE: Mahalapye, fl. 31.i.1912, *Rogers* 6096 (BM; K). **N. Rhodesia.** C: Marandellas, fl. 3.xii.1941, *Dehn* 395 (M; SRGH). C: Mt. Makulu Research Station, near Chilanga, st. 15.xi.1957, *Angus* 1734 (FHO; K). E: Ft. Jameson, fr. iii.1955, *E.M. & W.* 1142 (BM; LISC; SRGH). S: Livingstone, fl. i.1910, *Rogers* 7265 (K; SRGH). **S. Rhodesia.** N: Lomagundi, fl. x.1920, *Eyles* 2681 (K; SRGH). W: Bulawayo, fr. 27.xi.1920, *Borle* 9 (K; PRE; SRGH). E: Inyanga, Cheshire, fl. 14.i.1931, *Norlindh & Weimarck* 4300 (LD; SRGH). S: South of Lundi R., Beit Bridge Road, fr. 15.ii.1955, *E.M. & W.* 367 (BM; LISC; SRGH). **Mozambique.** N: Marrupa, Mecopo Mts., fl. 15.x.1942, *Mendonça* 869 (LISC; SRGH). MS: Vila Gouveia, Mungári, fl. 27.x.1943, *Torre* 6094 (K; LISC; SRGH). LM: Goba, fl. 23.xii.1944, *Mendonça* 3475 (LISC; SRGH).

Widespread in S. Africa and SW. Africa and reaching Tanganyika in the north. Often occurring in our area with *G. bicolor* Juss. but also found in *Brachystegia* woodland at annual rainfall levels of c. 100 cm. where *G. bicolor* does not penetrate, except perhaps on termitaria. A species of open woodland and found on many soil types.

The wood is commonly used for making walking-sticks and assegai handles.

Like *G. bicolor* this is a very variable species, possibly because a whole range of hybrids seems to exist between these two species and it is most difficult to deal with them in a conventional taxonomic manner. Mrs. Lugard, who collected the type specimens of *G. cordata*, *G. kwebensis* and *G. grisea*, remarked on the label accompanying the type of *G. cordata* that every hybrid seemed to exist in the Kwebe Hills between this species and *G. bicolor*. This was in 1898 and her comment evidently received little attention at the time. Now that much more material has been collected it is more difficult than ever to explain this complex of forms in any other way, and all the above named species, as Mrs. Lugard thought, seem to form part of a hybrid series.

As in *G. bicolor*, *G. micrantha* and *G. microcarpa*, forms lacking the glabrous androgynophore are common. In this species both forms seem about equally common. A galling of the fruits in *G. monticola* causes them to develop a characteristic brown woolly appearance, which, if not recognized as a gall, could easily lead one into assuming that a distinct species was involved.

17. **Grewia subspathulata** N.E.Br. in Kew Bull. **1909**: 96 (1909).—O.B.Mill. in Journ. S. Afr. Bot. **18**: 53 (1953). Type: Ngamiland, Kwebe Hills, *Lugard* 92a (K, holotype).

Grewia monticola sensu White, F.F.N.R.: 238 (1962).

This species has a rather greyish short pubescence, like that of *G. bicolor* Juss., but a leaf-shape quite unlike the elliptic-oblong or lanceolate leaves of that species and agreeing instead with *G. monticola* Sond. Its leaves are asymmetrically cordate and oblong-ovate to ovate. The leaf texture is not so thin as in *G. bicolor*, nor so leathery as in *G. monticola*. In all respects it seems morphologically to be about half-way between *G. monticola* and *G. bicolor*.

Bechuanaland Prot. N: Ngamiland, Kwebe Hills, fl. xii.1896, *Lugard* 92a (K). SE: Kanye, fl. xi.1948, *Miller* 715 (K; PRE). **N. Rhodesia.** E: Petauke, fl. 3.xii.1958, *Robson* 813 (K; SRGH). S: Livingstone, fr. 10.iii.1952, *White* 2229 (FHO; K). **S. Rhodesia.** N: Bindura, fl. 30.xii.1928, *Young* 917 (SRGH). W: Bulawayo, fr. 8.iv.1917, *Zealley* 30 (SRGH). E: Umtali, Odzani R., 1915, *Teague* 361 (K). S: Gwanda, fr. 5.i.1955, *Oates* 1751 (K; SRGH). **Nyasaland.** C: Lilongwe, Chitedze, fr. 22.iii.1955, *E.M. & W.* 1123 (BM; LISC; SRGH).

Also in the Transvaal and SW. Africa.

It is suggested that this species may be of hybrid origin having *G. bicolor* Juss. and *G. monticola* Sond. as parents. It is known throughout the common range of these two species (except for Mozambique) and does not occur beyond it. It is apparently fertile but morphologically intermediate between these two species. The only other way to deal with this problem would be to consider the whole complex, including both *G. bicolor* and *G. monticola*, as one species under *G. bicolor*, the oldest name. These two species, however, are so different that even the most ardent lumper would hesitate to unite them and, as the evidence for hybridization is so strong, the present treatment is felt to be more logical. This is not proof of the hypothesis, however, which must await the outcome of breeding experiments which in these species will take some years to carry out. Meanwhile we may retain the name of *G. subspathulata* as a provisional measure without indicating definitely, in the absence of proof, that we are dealing with a hybrid population.

18. **Grewia hornbyi** Wild in Bol. Soc. Brot., Sér. 2, **31**: 84, t. 1 fig. C (1957). TAB. **4** fig. C. Type: Mozambique, R. Save, *H. E. Hornby* 2497 (K; SRGH, holotype).

Shrub 3–4 m. tall; young branches with a short white tomentum and scattered tufts of longer ferruginous hairs. Leaf-lamina 1·2–4 × 0·7–2·2 cm., oblong-elliptic to oblong-ovate, apex obtuse, margin serrate and somewhat revolute, base obliquely subcordate or rounded, somewhat shining, sparsely stellate-pubescent or glabrescent above, densely whitish-tomentose below including the nerves, which have however additional scattered tufts of longer brownish hairs and are rather prominent; petiole 2–3 mm. long, with an indumentum as in the young branches; stipules 3·5–4 mm. long, subulate, pubescent. Inflorescences all axillary; peduncles c. 1 cm. long, tomentum as for young branches; pedicels normally 3 together, 3–4 mm. long; bracts 3–4 mm. long, deeply trifid or bifid, very narrow, almost filamentous, tomentose. Flower-buds oblong, sulcate. Sepals 9–10 mm. long, linear, 3-nerved, tomentellous outside, yellow and glabrous within. Petals bright yellow, 4–4·5 × 1·75–2 mm., oblong-obovate, shortly 2-dentate at the apex, with a nectariferous claw circumvillous within ledged above and glabrous on the back. Androgynophore up to 1·5 mm. long, glabrous except for a villous apex, not elongate above the node. Ovary shallowly 2-lobed, villous; style c. 7 mm. long, glabrous; stigma-lobes 4–5, broad. Fruit usually deeply 2-lobed, lobes c. 4 mm. in diam., globose, yellowish when ripe, sparsely setulose-pubescent.

Mozambique. SS: R. Save, 100 km. E. of Massangena, fl. 12.x.1946, *H. E. Hornby* 2497 (K; SRGH). LM: Maputo, fr. 19.ii.1947, *R. M. Hornby* 2528 (K; PRE; SRGH).

Endemic, as far as is known, in the southern part of Mozambique. Occurring in deciduous woodland with such species as *Sclerocarya caffra*, *Guibourtia coleosperma*, *Afzelia quanzensis*, etc.

This species is related to *G. monticola* but is remarkable for its very small leaves and fruits and is the only species of this affinity with almost filamentous divided bracts.

19. **Grewia flava** DC., Cat. Pl. Hort. Bot. Monsp.: 113 (1813).—Harv. in Harv. & Sond., F.C. **1**: 225 (1860).—Burret in Engl., Bot. Jahrb. **45**: 179 (1910).—Eyles in Trans. Roy. Soc. S. Afr. **5**: 410 (1916).—Burtt Davy, F.P.F.T. **1**: 254 (1926).—O.B.Mill. in Journ. S. Afr. Bot. **18**: 52 (1953). Type a specimen from a cultivated plant of S. African origin.

Grewia cana Sond. in Linnaea, **23**: 20 (1850).—Eyles in Trans. Roy. Soc. S. Afr. **5**: 410 (1916).—Steedman, Trees etc. S. Rhod.: 46 (1933). Syntypes from the Transvaal.

Grewia hermannioides Harv. in Harv. & Sond., F.C. **1**: 226 (1860). Type from the Transvaal.

Compact shrub c. 2 m. tall; young branchlets greyish or greyish-brown, tomentellous; older branches dark purplish-black. Leaf-lamina 1·4–7 × 0·75–2·5 cm., elliptic or oblanceolate, rounded at the apex, margin finely serrulate to dentate, cuneate and equal-sided at the base, very finely and closely tomentellous above, somewhat paler and more densely tomentellous below, venation fairly prominent and reticulate; petiole c. 2 mm. long, tomentellous; stipules c. 5 mm. long, subulate, tomentellous. Inflorescences all axillary; peduncles 7·5–10 mm. long, tomentellous; pedicels normally 1 per peduncle, up to 1 cm. long, tomentellous; basal bracts 3–4 mm. long, very caducous, subulate, tomentellous. Flower-buds obovoid, slightly sulcate. Sepals c. 8 mm. long, linear-lanceolate to linear-oblong, greenish-grey-tomentellous without, yellow and glabrous within, 3-nerved. Petals yellow, c. $\frac{2}{3}$ the length of the sepals, linear-oblong to oblanceolate, with a basal nectariferous claw circumvillous within ledged above and sparsely pilose outside. Androgynophore c. 1 mm. tall, glabrous except at the apex, not extended above the node. Ovary villous, shallowly 2-lobed or 1-lobed by abortion when the style is eccentrically placed on the ovary; style c. 4 mm. long, glabrous, with flattened, broad, stigma-lobes. Fruit reddish when ripe, c. 8 mm. in diam., globose or 2-lobed, sparsely setulose, glabrescent.

Bechuanaland Prot. N: Francistown, fr. 25.xii.1911, *Rogers* 607 (PRE; SRGH). SW: Chukudu Pan, fr. 20.vi.1955, *Story* 4941 (K; PRE). SE: Eastern Bamangwato country, Sosongu, fl., *Holub* (K). **S. Rhodesia.** W: Matopos, fl. xii.1905, *Gibbs* 226 (BM; K). C: Chilimanzi, fl. & fr. 7.ii.1951, *McGregor* 29/51 (K; SRGH). S: Gwanda, Shashi Plain, fl. xii.1954, *Davies* 879 (K; SRGH).

Widespread also in S. Africa and SW. Africa. In the drier types of deciduous woodland and bushland found towards the west of our area. Recorded from alluvial granite and schist soils.

In the Transvaal the fruits are used for making an intoxicating drink and the tough bark for basket making. The fruit is edible.

Most of the available material of this species has 1-flowered peduncles but specimens with 2- and even 3-flowered peduncles occur, e.g. *Gibbs* 226 (BM) from the Matopos. In this they match the type of *G. hermannioides* (*Burke* in Herb. Kew, from the Magaliesberg, Transvaal) which Burret considers synonymous with *G. flava*. His opinion is followed here but with such material its separation from *G. bicolor* presents some difficulty. In these cases the coarser texture of the leaves, the more prominent lateral nerves, the reticulation of the tertiary nerves and above all the symmetrically cuneate leaf-bases can be used to separate *G. flava* from *G. bicolor*. These are all relative characters but appear to give a reliable method of separation. The *Kirk* specimen from Tete in Mozambique (*Kirk* in Herb. Kew) placed under *G. flava* DC. by Masters (in Oliv., F.T.A. **1**: 250 (1868)) is not this species, but a poor fruiting specimen of *G. flavescens*. Baines's specimen (*Baines* (K)) from S. Central Africa (Bechuanaland or S. Rhodesia) quoted in the same place is true *G. flava*.

The plant illustrated under this name by Martineau (Rhod. Wild Fl.: 47, t. 16 (1954). is not this species. The drawing is not a good one but probably represents *G. flavescens*.

20. **Grewia micrantha** Boj. in Proc.-Verb. Soc. Hist. Nat. Maurice, **1842–5**: 28 (1846).—Mast. in Oliv., F.T.A. **1**: 244 (1868).—Burret in Engl., Bot. Jahrb. **45**: 181 (1910).—Brenan, T.T.C.L.: 620 (1949).—White, F.F.N.R.: 434 (1962). Type from Madagascar.

Grewia woodiana K. Schum. in Engl., Bot. Jahrb. **33**: 307 (1904). Syntypes: Nyasaland, Shire Highlands, *Buchanan* (K); Zomba, *Whyte* (K).

Grewia aurantiaca Weim. in Bot. Notis. **1936**: 41, fig. 11 (1936). Type: S. Rhodesia, Umtali, *F.N. & W.* 2861 (LD, holotype; SRGH).

Shrub or small tree up to 8 m. tall; young branchlets ferruginously tomentose, becoming grey or brownish with paler lenticels. Leaf-lamina 2·5–9 × 1·2–4·8 cm., ovate-oblong or elliptic, apex rounded or acute, margins serrate, rounded or asymmetrically cordate at the base, sparsely stellate-pubescent above or glabrous, finely reticulate-rugose, older leaves somewhat shiny, closely appressed-whitish-tomentose between the nerves below, the nerves themselves brownish and subglabrous, except for tufts of scattered longer and darker hairs; petiole up to 7 mm. long, ferruginously pubescent; stipules c. 5 mm. long, subulate, coarsely brown-hirsute. Inflorescences all axillary, on coarsely brown-hairy peduncles up to 1·5 cm. long, 1 or sometimes 2–3 peduncles per axil; pedicels normally 3 per peduncle, c. 0·5 cm. long, coarsely brown-hairy; bracts 3–4 mm. long, linear-oblong to oblong, entire or bifid or trifid at the apex, tomentose. Flower-buds oblong-ovoid. Sepals up to 0·8 cm. long, linear-oblong, coarsely stellately hairy outside, yellow and glabrous within. Petals yellow, about half the length of the sepals, oblong to obovate, often 2-dentate at the apex, basal claw either with circumvillous nectary within, or often absent and replaced by a small tuft of hairs at the cuneate petal base. Androgynophore, in specimens with nectaries, glabrous and 1 mm. long with a villous apex, in specimens with nectaries absent glabrous portion of androgynophore absent or almost so and ovary almost sessile on a villous cushion. Ovary shallowly 2-lobed, villous; style c. 5 mm. long, glabrous; stigma with about 4 broad lobes. Fruit yellowish, deeply 2-lobed or 1-lobed by abortion, each lobe c. 7 mm. in diam., pubescent.

N. Rhodesia. E: Nsadyu–Fort Jameson road, fl. 25.xi.1958, *Robson* 705 (K; SRGH). **S. Rhodesia.** N: Shamva, fr. 26.xi.1952, *Pemberthy* in GHS 40295 (K; SRGH). E: Umtali, fl. 11.xi.1930, *F.N. & W.* 2861 (LD; SRGH). S: Nuanetsi, fl. xii.1955. *Davies* 1722 (K; SRGH). **Nyasaland.** C: Chongoni Forest, Dedza, fl. 18.xii.1957, *Adlard* 264 (K; SRGH). S: Shire Highlands, *Buchanan* 218 (K). **Mozambique.** N: between Mucojo and Ingoane, fl. 11.ix.1948, *Pedro & Pedrógão* 5148 (LMJ; SRGH). Z: Mocuba, fl. 12.xii.1942, *Torre* 4796 (LISC; SRGH). T: Zobue Mts., fl. 21.x.1941, *Torre* 3695 (LISC; SRGH). MS: Mossurize, Madanda Forest, fl. 5.xii.1906, *Swynnerton* 1222, pro parte (K). SS: R. Save, Meringuas, fr. 20.vi.1950, *Chase* 2527 (K; SRGH). LM: between Moamba and Boane, fl. 2.xii.1940, *Torre* 2179 (LISC).

Also in Tanganyika, Kenya and Madagascar. Small tree of *Brachystegia* woodland or mixed bushland with such genera as *Afzelia*, *Sclerocarya*, *Strychnos*, etc. and occasionally found on termite mounds.

A group of Mozambique specimens represented by *Gomes e Sousa* 2307 (K; PRE; SRGH) and *Barbosa* 2507 (LISC) from Niassa Prov., *Hornby* 2568 (K; PRE; SRGH) from Lourenço Marques, *Myre & Balsinhas* 591 (LM; SRGH) from the same area, *Faulkner* 79 (BM; K; PRE) from Zambesia, and *Swynnerton* 1221 pro parte (K) from Manica and Sofala require a little further comment. They have noticeably smaller leaves than usual and the veins beneath are not differentiated as well as in typical *G. micrantha.* It would not be worth while to create a separate taxon for this group as they intergrade with normal specimens; but they suggest the possibility of hybridization between *G. micrantha* and *G. bicolor*. Further to this, *Swynnerton* 1221 is the type of *G. madandensis* which in turn is a synonym of *G. bicolor*, as can be seen if the holotype in the British Museum is examined. There are two duplicate sheets of this gathering at Kew, however, and one of them is the smaller-leaved form of *G. micrantha*. It looks as if this mixed gathering was taken from adjacent bushes and this would tend to reinforce the argument in favour of hybridization.

Finally, *G. micrantha* is like *G. monticola* and a number of other species in that the presence or absence of the lower glabrous portion of the androgynophore and the accompanying nectariferous claw of the petal which fits against it is of no taxonomic significance and both forms occur throughout its range. Specimens with both torus and nectaries lacking, however, are by far the most common in this species.

The *Bojer* specimen quoted by Masters (loc. cit.) cannot be traced at Kew and he may have written " Mombas Is." in error for " Madagascar " which is written on the type sheet. There are no other Madagascar specimens of this species at Kew so there could be an alternative explanation i.e. that Masters was correct and the type in fact came from Mombasa. Further evidence from Madagascar is needed to confirm or reject this possibility.

21. **Grewia inaequilatera** Garcke in Peters, Reise Mossamb. Bot. **1**: 134 (1861).—Mast. in Oliv., F.T.A. **1**: 245 (1868).—Burret in Engl., Bot. Jahrb. **45**: 183 (1910).—Bak. f. in Journ. Linn. Soc., Bot. **40**: 32 (1911).—Burtt Davy, F.P.F.T. **1**: 255 (1926).—Gomes e Sousa, Dendrol. Moçamb. **4**: 97, 99 cum. tab. (1958). —White, F.F.N.R.: 434 (1962). TAB. **3** fig. E. Type: Mozambique, Sena, *Peters* 28 (B, holotype †; K).

Shrub or small tree up to 7 m. tall, with spreading or scandent branches; young branchlets stellate-pilose, soon becoming brown and glabrous; older branches pale grey. Leaf-lamina 5–18 × 2·5–7·5 cm., oblong to oblong-ovate, acute or abruptly acuminate at the apex, margin serrate but less so towards the base, rounded and markedly asymmetric at the base, green and glabrous above, with a fine white or greyish tomentum between the nerves below; nerves reticulate and sparsely hispid; petiole up to 1 cm. long, with a tufted stellate-pubescence and widening somewhat near the lamina; stipules 3·5 mm. long with a 2·5 mm. long apiculus, greyish-tomentose, semi-circular, very caducous, and present only in bud. Inflorescences always axillary; peduncles up to 1 cm. long with scattered tufts of brown stellate hairs; pedicels similar, 2–3 together, c. 6 mm. long; bracts c. 5 mm. long, obovate to oblong, stellate-pubescent. Flower-buds ellipsoid. Sepals c. 1·6 cm. long, linear-oblong, with tufted stellate hairs outside, bright yellow and glabrous inside. Petals yellow, 7·5 mm. long, subcircular to very broadly obovate, margins slightly undulate, with a basal nectary less than 1 mm. long, very small for the genus, pubescent all over its surface within ledged above glabrous on the back. Androgynophore equal in length to the nectary, c. 0·5 mm. long in flower, glabrous, but with tufts of hair at its base. Stamens very numerous, c. 4-seriate. Ovary 2-lobed, pubescent; style c. 7 mm. long, glabrous; stigma with c. 4 broad lobes. Fruit yellow when ripe, divided almost to the peduncle into 2 globose lobes c. 13 mm. in diam., very sparsely stellate-pubescent.

N. Rhodesia. E: between Changwe and Luangwa R., fl. & fr. 16.xii.1958, *Robson* 965 (K; SRGH). **S. Rhodesia.** N: Mtoko, fl. i.1953, *Phelps* 13 (SRGH). W: Victoria Falls, fl. 12.ii.1912, *Rogers* 5590 (K). C: Headlands, fl. xii. 1959, *Davies* 2663 (SRGH). E: Melsetter, Odzi R., fl. 29.xii.1948, *Chase* 1412 (BM; K; SRGH). S: Sabi-Lundi Junction, fr. 4.vi.1950, *Wild* 3352 (K; SRGH). **Nyasaland.** S: Lower Shire, fl. 30.x.1858, *Kirk* (K). **Mozambique.** N: Mutuáli, fl. 25.ii.1954, *Sousa* 4207 (K; PRE; SRGH). Z: Morrumbala Mts., fr. 13.v.1943, *Torre* 5305 (K; LISC; SRGH). T: opposite Sena, fl. i.1859, *Kirk* (K). MS: Chibabava, Lower Buzi R., fl. 1.xii.1906, *Swynnerton* 1223 (BM; K). SS: Guijá, between Macarretane and Majejamela, fr. 6.v.1957, *Carvalho* 155 (BM; LM).

Also in the Transvaal. In the low-altitude river valleys draining into Mozambique; either a thicket-forming species on fertile alluvial soils or a scrambler in riverine fringes or in deciduous bushland.

Fruit edible.

One sheet, *Van Son* (PRE; SRGH) from the Chobe R., Bechuanaland, offers some diffculty here. It may belong to *G. inaequilatera* but it might also be assigned to *G. cyclopetala* Wawra, an Angolan and West African species. Unfortunately the specimen is in fruit and until flowering material from this area is available the problem cannot be resolved with confidence. The two species are in any case closely related. This is the specimen referred to *G. inaequilatera* by Miller (in Journ. S. Afr. Bot. **18**: 53 (1953)).

22. **Grewia hexamita** Burret in Engl., Bot. Jahrb. **45**: 184 (1910).—Burtt Davy, F.P.F.T. **1**: 254 (1926). Type from the Transvaal.

Grewia messinica Burtt Davy & Greenway in Burtt Davy, tom. cit.: 41 (1926). Type from the Transvaal.

Grewia schweickerdtii Burret in Bothalia, **3**: 244 (1937). Type from the Transvaal.

Shrub 2–5 m. tall; young branches ferruginously woolly; older branches with a reddish-brown bark and pale lenticels. Leaf-lamina 3–10 × 2·5–6 cm., elliptic to oblong-elliptic, rounded or acute at the apex, margin serrate and somewhat revolute, asymmetrically cordate at the base, glabrous and somewhat rugose above, rather shining; nerves impressed and very densely pale yellowish-tomentose below but sometimes less hairy and brownish; petiole 2–3 mm. long, or rarely up to 6 mm. Inflorescences all axillary, on tomentose peduncles 5–10 mm. long; pedicels c. 5 mm. long, tomentose, 2–3 together; bracts up to 5 mm. long, ovate to lanceolate, glabrous inside, pubescent outside. Flower-buds oblong-ovoid. Sepals 1·8–2·6 cm. long, linear-oblong, golden-brown tomentose on the back, glabrous and yellow within. Petals with a subcircular lamina up to 0·7 cm. in diam. and with a basal nectariferous claw up to 2 mm. long with a short ledge above circumvillous on the inner margin pilose at the back. Androgynophore with a very stout glabrous lower portion up to 2 mm. long, produced into a pilose upper portion up to 2 mm. long. Ovary 2-lobed or 1-lobed by abortion, silky-villous; style up to 10 mm. long, glabrous; stigma-lobes c. 4, rounded and dilated. Fruit reddish, deeply divided into 2 globose lobes up to 2 cm. in diam. or 1-lobed by abortion, pilose, shining.

S. Rhodesia. S: Nuanetsi, Mateke Hills, fr. 4.v.1958, *Drummond* 5556 (K; SRGH). **Mozambique.** SS: between Caniçado and Mapai, fr. 9.v.1944, *Torre* 6601 (LISC; SRGH). LM: Magude, fl. 29.xi.1944, *Mendonça* 3128 (K; LISC; SRGH).

Also in the Transvaal and Tanganyika. Common in deciduous bushland in the valleys of the Komati and the Limpopo on both the Transvaal and Mozambique sides of the border.

A very fine species with large handsome flowers.

23. **Grewia schinzii** K. Schum. in Engl., Bot. Jahrb. **15**: 124 (1892).—Burret in Engl., Bot. Jahrb. **45**: 183 (1910).—Exell & Mendonça, C.F.A. **1**, 2: 217 (1951).—White, F.F.N.R.: 238, fig. 42A (1962). Type from SW. Africa.

Grewia velutinissima Dunkley in Kew Bull. **1935**: 256 (1935).—O.B.Mill. in Journ. S. Afr. Bot. **18**: 53 (1953). Type: N. Rhodesia, Bombwe, *Martin* 350/32 (FHO; K, holotype; SRGH).

Shrub or small tree, usually branching from the base; main stems up to c. 12 cm. in diam.; young branches ferruginously tomentose, glabrous with age and becoming brown with pale lenticels; crown rounded. Leaf-lamina 4–14 × 2·5–9 cm., obliquely oblong to broadly obovate-oblong, apex acute or obtuse or broadly rounded, margin coarsely and irregularly serrate, base asymmetrically rounded or subcordate, green and shortly stellate-tomentose above, densely greyish-brown-tomentose below; lateral nerves in c. 4 pairs; petiole up to 1 cm. long, ferruginously hairy; stipules c. 6 mm. long, obliquely lanceolate-subulate. Inflorescences all axillary, with densely ferruginous-tomentose peduncles up to 15 mm. long; pedicels 2–3 together, up to 5 mm. long, ferruginously tomentose; bracts c. 5 mm. long, entire, oblong-lanceolate, ferruginously tomentose on the back, almost glabrous within. Flower-buds oblong, slightly sulcate. Sepals 10–13 mm. long, narrowly oblong, with a tufted stellate tomentum on the back, glabrous inside. Petals yellow, 6–8 mm. long, obovate to circular, with a basal nectariferous claw circumvillous within ledged above and glabrous on the back. Androgynophore 1·5 mm. long, glabrous except at the villous apex, not elongated above the node. Ovary densely pilose, 2-lobed; style up to 10 mm. long, glabrous, with c. 4 broad stigmatic lobes. Fruit yellowish, normally very deeply 2-lobed, each lobe c. 7·5 mm. in diam., sparsely stellately pilose.

Bechuanaland Prot. N: Ngamiland, Gomare, fr. vi.1940, *Miller* 433 (PRE; SRGH). **N. Rhodesia.** B: Sesheke, fl. & fr. 20.xii.1952, *Angus* 970 (FHO; K). S: Bombwe, fl. 3.xi.1932, *Martin* 350/32 (FHO; K; SRGH).

Also in Angola and SW. Africa. Open bush or woodland; also occurring on termite mounds and showing a tendency to grow near rivers.

24. **Grewia microcarpa** K. Schum. in Notizbl. Bot. Gart. Berl. **2**: 190 (1898).—Burret in Engl., Bot. Jahrb. **45**: 185 (1910).—Brenan, T.T.C.L.: 619 (1949). Type from Tanganyika.

Grewia parvifolia sensu Mast. in Oliv., F.T.A. **1**: 251 (1868) pro parte quoad specim. Kirk.

Grewia swynnertonii J. R. Drummond ex Bak. f. in Journ. Linn. Soc., Bot. **40**: 30 (1911). Type: Mozambique, Manica and Sofala, Lower Umswirizwi R., *Swynnerton* 201 (BM, holotype).

Grewia utilis Exell in Journ. of Bot. **72**: 316 (1934). Type from Tanganyika.

Shrub or tree up to 9 m. tall; young branchlets very slender, pubescent, soon becoming glabrous and plum-coloured. Leaf-lamina 1·5–6 × 0·7–3 cm., elliptic, ovate-elliptic or lanceolate, acute at the apex, margin finely serrate, rounded at the base, minutely pubescent above or glabrous, very minutely reticulate and more densely pubescent below; petiole 1–2 mm. long, pubescent; stipules c. 3 mm. long, subulate, pubescent. Inflorescences all axillary; peduncles up to 1 cm. long, slender, pubescent; pedicels similar, 2–3 together, up to 1 cm. long; bracts 3–4 mm. long, subulate, pubescent. Flower-buds oblong or oblong-ovoid. Sepals 6 mm. long, oblong-elliptic, tomentellous on the back, yellow and glabrous inside. Petals yellow, about ⅓ the length of the sepals, oblong-lanceolate, basal nectariferous claw quite lacking or represented by a cluster of hairs, or present and then circumvillous inside with a minute ledge above and glabrous at the back. Androgynophore without a basal glabrous portion if nectaries lacking and then ovary almost sessile on a pubescent cushion bearing the stamens, if nectaries present then basal glabrous portion 0·5–0·75 mm. long. Ovary 2-lobed and villous; style c. 3 mm. long, glabrous; stigma-lobes c. 4, flattened. Fruit yellowish, deeply 2-lobed or 1-lobed by abortion, lobes c. 4 mm. in diam., globose, sparsely setulose-pubescent.

S. Rhodesia. E: Chipinga, Makoho, 8 km. E. of Hippo Mine, fr. 13.ii.1957, *Goodier* 112 (K; SRGH). **Nyasaland.** S: Zomba, fl. xii.1900, *Purves* 61 (K). **Mozambique.** N: between Memba and Geba, fl. 14.x.1948, *Barbosa* 2397 (LISC; SRGH). MS: Lower Umswirizwi R., fl. 23.xi.1906, *Swynnerton* 201 (BM).

Also in Tanganyika and the Congo. Low-altitude thicket-forming species, also found on the margins of riverine fringes.

As in *G. bicolor*, *G. monticola* and *G. micrantha*, the petals may be nectariferous or not and the basal glabrous portion of the androgynophore can also be either present or absent.

25. **Grewia sulcata** Mast. in Oliv., F.T.A. **1**: 252 (1868).—Burret in Engl., Bot. Jahrb. **45**: 188 (1910).—Burtt Davy, F.P.F.T. **1**: 255 (1926).—Brenan, T.T.C.L.: 616 (1949); in Mem. N.Y.Bot. Gard. **8**: 228 (1953). Syntypes: Mozambique, Chupanga, *Kirk* 358 (K); R. Luabo, *Kirk* (K); S. Rhodesia or Bechuanaland, S. of Zambezi R., *Baines* (K).

Grewia obovata K. Schum. [in Abh. Preuss. Akad. Wiss. **1894**: 18 (1894) *nom. nud.*] in Engl., Pflanzenw. Ost-Afr. **C**: 263 (1895).—Bak. f. in Journ. Linn. Soc., Bot. **40**: 32 (1911).—Eyles in Trans. Roy. Soc. S. Afr. **5**: 410 (1916). Type from Tanganyika.

Grewia sulcata var. *obovata* (K. Schum.) Burret in Engl., Bot. Jahrb. **45**: 188 (1910). Type as for *G. obovata*.

Erect or straggling shrub 2–3 m. tall; young branches and inflorescences ferruginously tomentose. Leaf-lamina 2–7 × 1·5–4·5 cm., obovate or occasionally oblong, rounded or bluntly acuminate at the apex, margin crenate-dentate, rounded or slightly cordate at the slightly asymmetrical base, slightly scabrous above, hispid or tomentose below, secondary venation rather prominent below; petiole up to 0·5 cm. long, tomentellous; stipules c. 2 mm. long, linear, caducous. Inflorescences usually at the ends of the branchlets, usually more than 3-flowered; peduncles 1–1·5 cm. long, ferruginously hairy; pedicels similar, c. 1 cm. long; bracts up to 2·5 mm. long, hairy on the outside, entire, linear-lanceolate. Flower-buds oblong, slightly inflated at the base, slightly constricted above the base. Sepals 12–18 mm. long, linear-oblong, rufous-tomentose outside, glabrous and white within. Petals white, 7·5–13 mm. long, narrowly triangular, tapering to the apex, with a basal nectariferous claw circumvillous within and ledged above. Androgynophore up to 1·5 mm. long, glabrous below, sulcate, elongated above the node into a tomentellous upper portion 2–3·5 mm. tall. Ovary globose, 4-lobed; stigma with 4–5 broad lobes. Fruit 1·5–2 cm. in diam., 4-lobed, sparingly hispid, slightly fleshy.

S. Rhodesia. E: Chipinga, Sabi-Lundi confluence, fl. & fr. 29.vi.1955, *Mowbray* 45 (K; SRGH). S: Nuanetsi R., fl. xi.1955, *Davies* 1667 (K; SRGH). **Nyasaland.** S: Chikwawa District, Lower Mwanza R., fl. 6.x.1946, *Brass* 17999 (K; NY; SRGH). **Mozambique.** N: Cabo Delgado, Montepuez, fr. 2.ix.1948, *Barbosa* 1975 (LISC; SRGH). Z: Mopeia, fl. & fr. 14.x.1941, *Torre* 3652 (LISC; SRGH). T: between Mungari and Tambara, fl. & fr. 2.ix.1943, *Torre* 5822 (LISC; SRGH). MS: R. Buzi, fl. 20.viii.1947, *Pimenta* 33 (SRGH). SS: Vilanculos, fl. & fr. 1944, *Mendonça* 1899 (LISC; SRGH). LM: Lourenço Marques, 10.xii.1940, *Torre* 2309 (LISC; SRGH).

From the Transvaal to Tanganyika. In low-altitude river valleys and in the coastal plain, common on sandy river-banks. Fruit edible.

26. **Grewia stolzii** Ulbr. in Engl., Bot. Jahrb. **51**: 347 (1914).—Weim. in Bot. Notis. **1936**: 43 (1936).—White, F.F.N.R.: 236 (1962). Type from Tanganyika (Lake Nyasa).

Grewia hopkinsii Suesseng. & Merxm. in Proc. & Trans. Rhod. Sci. Ass. **43**: 105 (1951). Type: S. Rhodesia, Marandellas, *Dehn* 578 (M, holotype; SRGH).

Shrub or small tree or semi-scandent with purplish branches; young branchlets coarsely ferruginously pubescent. Leaf-lamina up to 12 × 5·5 cm., elliptic-oblong or ovate-lanceolate, acute to acuminate at the apex, margin finely serrate with glands on the serratures, broadly cuneate or truncate and sometimes oblique at the base, glabrous above, sparsely stellate-pubescent below; nerves fairly prominent on both sides; petiole up to 7·5 mm. long, hispid-pubescent; stipules 0·5 mm. long, linear, pubescent outside, caducous. Inflorescences terminal or leaf-opposed, 2–5-flowered; peduncles up to 15 mm. long, pubescent; pedicels similar, c. 1 cm. long; bracts up to 4 mm. long, linear or lanceolate-triangular, pubescent on the back, entire. Flower-buds oblong. Sepals 1·2–2 cm. long, linear-oblong, greenish or ferruginously pubescent outside with a dense short stellate tomentum interspersed with longer fasciculate hairs, glabrous and white inside. Petals white, 5–9 mm. long, narrowly triangular; lamina not wider than the nectariferous claw and not more than twice its length; claw circumvillous within and ledged above. Androgynophore up to 1·5 mm. long, glabrous, not produced above. Ovary 4-lobed, villous; stigma-lobes broad. Fruit reddish when ripe, up to 2.3 cm. in diam., 4-lobed, sparsely hispid or glabrous.

N. Rhodesia. N: Abercorn, fl. 7.ii.1955, *Richards* 4345 (K). W: Chingola, fl. 18.i.1956, *Fanshawe* 2739 (K; SRGH). **S. Rhodesia.** C: Marandellas, Cave Farm, fl. 30.i.1942, *Dehn* 578 (M; SRGH). E: Inyanga, fl. 18.xii.1930, *F.N. & W.* 3835 (BM; LD; SRGH). S: Zimbabwe, fr. *Pole Evans* 2715 (PRE; SRGH). **Nyasaland.** N: Champika, fl. i.1956, *Chapman* 278 (FHO). C: between Kongwe and Mwera Hill, fl. 21.ii.1959, *Robson* 1700 (K; SRGH). S: Shire Highlands, *Buchanan* (K). **Mozambique.** MS: Mossurize, Mt. Maruma, fr. 12.ix.1906, *Swynnerton* 1349 (BM; K).

Also in Tanganyika, Kenya and Uganda. In small patches of forest or on rocky ground with some fire protection in *Brachystegia* country.
Fruit edible.

27. **Grewia avellana** Hiern, Cat. Afr. Pl. Welw. **1**: 94 (1896).—Burret in Engl., Bot. Jahrb. **45**: 190 (1910).—Exell & Mendonça, C.F.A. **1,** 2: 218 (1951).—O.B.Mill. in Journ. S. Afr. Bot. **18**: 52 (1953).—White, F.F.N.R.: 236, fig. 41B (1962). Type from Angola.
Grewia guazumifolia sensu Mast. in Oliv., F.T.A. **1**: 245 (1868).—Burret in Engl., Bot. Jahrb. **45**: 201 (1910).—Eyles in Trans. Roy. Soc. S. Afr. **5**: 410 (1916).
Grewia calycina N.E. Br. in Kew Bull. **1909**: 97 (1909). Type: Bechuanaland Prot., Botletle Valley, *Lugard* 237 (K, holotype).

Rounded bush with yellowish-brown densely pubescent branchlets, glabrescent and dark brown with age. Leaf-lamina 2–4·5 mm. × 1·5–3 cm., elliptic to ovate-elliptic, obtuse or acute at the apex, margin crenate-dentate, rounded at the base, softly pubescent on both surfaces; petiole up to 2·5 mm. long, densely pubescent. Inflorescences terminal and leaf-opposed, 2–3-flowered; peduncles up to 10 mm. long, densely pubescent; pedicels similar, 2–3 mm. long; bracts up to 10 mm. long, pubescent, divided into 2–3 filiform segments. Sepals up to 17 mm. long, linear-lanceolate, acute at the apex, greenish and tomentellous outside, white and glabrous inside. Petals white, c. 2–3 mm. long, with a shortly triangular lamina and a nectariferous claw larger and wider than the lamina circumvillous within and ledged above. Androgynophore c. 1·5 mm. long, glabrous at the base, pubescent at the apex at the insertion of the stamens but not elongated. Ovary 4-lobed, villous; style c. 10 mm. long, glabrous; stigma-lobes broad. Fruit blackish when ripe, up to 2·5 cm. in diam., deeply 4-lobed, rather fleshy, very sparsely setulose.

Bechuanaland Prot. N: Ngamiland, Botletle Valley, fl. xii.1898, *Lugard* 237 (K). SW: near Chukudu Pan, fr. 22.vi.1955, *Story* 4959 (K; PRE; SRGH). SE: Lobatsi, fl. 1912, *Rogers* 6110 (BM). **N. Rhodesia.** B: Sesheke, fl. 19.xii.1952, *Angus* 952 (FHO; K). N: Fort Rosebery, fr. 26.viii.1952, *Angus* 318 (FHO; K). S: Bombwe, fl. 7.x.1932, *Martin* 310/32 (FHO; K; SRGH). **S. Rhodesia.** W: Nyamandhlovu, Gwaai Forest, fl. i.1906, *Allen* 236 (K; SRGH).
Also in Angola and SW. Africa. Constituent of rather low-rainfall (up to 60 cm. p.a.) woodland and often found with *Baikiaea plurijuga* on the Kalahari Sands of the Rhodesias and Bechuanaland.
Fruit edible.

28. **Grewia occidentalis** L., Sp. Pl. **2**: 964 (1753).—Harv. in Harv. & Sond., F.C. **1**: 225 (1860).—Burret in Engl., Bot. Jahrb. **45**: 191 (1910).—Bak. f. in Journ. Linn. Soc., Bot. **40**: 32 (1911).—Eyles in Trans. Roy. Soc. S. Afr. **5**: 411 (1916).—Burtt Davy, F.P.F.T. **1**: 255 (1926).—Weim. in Bot. Notis. **1936**: 43 (1936). Type from S. Africa (Cape Province).
Grewia chirindae Bak. f. in Journ. Linn. Soc., Bot. **40**: 31 (1911).—Eyles in Trans. Roy. Soc. S. Afr. **5**: 410 (1916).—Steedman, Trees etc. S. Rhod.: 47, t. 46 (1933). Type: S. Rhodesia, Chirinda Forest, *Swynnerton* 131 (BM, holotype).
Grewia microphylla Weim. in Bot. Notis. **1936**: 46 (1936). Type: S. Rhodesia, Makoni District, *F.N. & W.* 3401 (BM; LD, holotype; SRGH).

Shrub or small tree (rarely climbing) up to 3 m. tall with rather slender glabrescent branchlets. Leaf-lamina 2–7·5 × 1·5–4 cm., lanceolate, rhombic-lanceolate or obovate-lanceolate, acute or rounded at the apex, margin crenate or crenate-dentate, subcuneate or rounded or slightly cordate at the base, glabrous or slightly pubescent on both surfaces, rather thin-textured; petiole up to 1·3 cm. long, glabrescent; stipules up to 4 mm. long, linear, pubescent, caducous. Inflorescences leaf-opposed, 1–3-flowered; peduncles slender, up to 15 mm. long, glabrescent; pedicels similar, up to 12 mm. long. Sepals up to 18 mm. long, linear-oblong, pubescent and greenish outside, purplish or pink inside. Petals purple, mauve, pinkish or very rarely white, up to 14 mm. long, oblong-lanceolate, with a basal nectariferous claw 1 mm. long circumvillous within and ledged above. Androgynophore glabrous below for 1 mm. prolonged above into a densely pubescent portion up to 3 mm. long. Ovary 4-lobed, densely hairy; style up to 10 mm. long, glabrous; stigma-lobes broad. Fruit reddish-purple, up to 2·5 cm. in diam., 4-lobed, shining, glabrescent, somewhat fleshy.

Var. **occidentalis**

Small tree. Leaf-lamina usually acute at the apex, margin crenate-dentate.

S. Rhodesia. C: Salisbury, Greenwood Park, st. v.1950, *Pardy* 24/50 (SRGH), cultivated. E: Chipinga, Chirinda Forest, xii.1908, *Swynnerton* 6628 (BM; K). **Mozambique.** SS: Zavala, fl. & fr. x.1937, *Gomes e Sousa* 2049 (K). LM: Lourenço Marques, Costa do Sol, fl. 15.xi.1944, *Mendonça* 2862 (BM; LISC; SRGH).

Widespread in the Cape, Natal, Transvaal and Swaziland. Climber in closed forest or a shrub or small tree at forest edges and in forest relics; occurring only in the wetter and eastern parts of our area south of the Zambezi.

Var. **litoralis** Wild in Bol. Soc. Brot., Sér. 2, **31**: 85 (1957). Type: Mozambique, Sul do Save, Praia de Zavora, *E.M. & W.* 690 (BM, holotype of var.; LISC; SRGH).

Spreading low bush up to 1 m. tall or with its branches semi-prostrate. Leaf-lamina 2–7·5 × 1·4 cm., oblong to obovate, margin slightly revolute, rounded at the apex, rounded to subcuneate at the base, entire or shallowly repand, glabrous on both sides, tertiary venation visibly reticulate below. Otherwise as in var. *occidentalis*.

Mozambique. SS: Praia de Zavora, fl. & fr. 27.ii.1955, *E.M. & W.* 690 (BM; LISC; SRGH). LM: Maputo, fl. 20.ii.1952, *Barbosa & Balsinhas* 4769 (LM).

This variety is apparently endemic to the coast of Mozambique and is known only from Sul do Save and Lourenço Marques Provinces. It is entirely confined to the high dunes next the shore-line and has a most characteristic habit. It does not differ in floral characters from *G. occidentalis* var. *occidentalis*.

29. **Grewia tenax** (Forsk.) Fiori in Agric. Colon. 5, Suppl.: 23 (1912).—Burret in Engl., Bot. Jahrb. **45**: 200 (1910).—Brenan, T.T.C.L.: 615 (1949).—Keay, F.W.T.A. ed. 2, **1**, 2: 305 (1958). Type from Arabia.
Chadara tenax Forsk., Fl. Aegypt.-Arab.: CXIV, 105 (1775). Type as above.
Grewia populifolia Vahl, Symb. Bot. **1**: 33 (1790).—Mast. in Oliv., F.T.A. **1**: 246 (1868).—Burret, tom. cit: 192 (1910). Type as for *Grewia tenax*.
Grewia betulifolia Juss. in Ann. Mus. Nation. Hist. Nat. Par. **4**: 92, t. 50 fig. 1 (1804). Type from Senegal.

Small shrub up to 2 m. tall; branches glabrescent. Leaf-lamina up to 3 × 2 cm., orbicular to obovate, rounded at the apex, margin rather coarsely dentate, rounded or abruptly cuneate at the base, slightly scabrous-pubescent especially below, or glabrescent, greyish-green, coriaceous; petiole up to 1 cm. long, but usually much less, pubescent; stipules up to 4 mm. long, filiform, pubescent. Flowers always borne singly, leaf-opposed; peduncles c. 10 mm. long, almost glabrous, slender; pedicels similar, up to 10 mm. long, the whole giving the appearance of an articulated pedicel; bracts minute, glabrescent, c. 1·5 mm. long. Sepals 10–18 mm. long, greenish and shortly pubescent outside, white and glabrous inside, linear-oblong. Petals white, with a linear and often 2-dentate lamina almost as long as the sepals and narrower than the basal nectariferous claw which is circumvillous within ledged above and up to 1·5 mm. long. Androgynophore with a basal glabrous portion up to 1·5 mm. long and a densely pubescent upper portion up to 1·5 mm. long. Ovary 4-lobed, glabrous or glabrescent; style c. 10 mm. long, glabrous; stigma-lobes broad. Fruit c. 10 mm. in diam., 4-lobed, shining, glabrous.

S. Rhodesia. S: Gwanda Distr., Shashi Plain, fl. xii.1954, *Davies* 876 (K; SRGH).

Widespread in Africa from the Transvaal and SW. Africa to Ethiopia and Arabia in the North-East and through West Africa to Senegal. Only found in the driest types of woodland or semi-desert scrub.

The presence of this species in the extreme south-west of S. Rhodesia suggests that it should be found in Bechuanaland also.

30. **Grewia praecox** K. Schum. in Engl., Bot. Jahrb. **15**: 117 (1892).—Burret in Engl., Bot. Jahrb. **45**: 195 (1910).—Brenan, T.T.C.L.: 615 (1949).—White, F.F.N.R.: 238, fig. 41C (1962). Type from Tanganyika.
Grewia albiflora K. Schum. in Engl., Bot. Jahrb. **28**: 428 (1900).—Burret, tom. cit.: 196 (1910).—Brenan, T.T.C.L.: 615 (1949). Syntypes from Tanganyika.
Grewia congesta Weim. in Bot. Notis. **1936**: 43, t. 12 (1936). Type: S. Rhodesia, Inyanga, *Norlindh & Weimarck* 4286 (BM; LD, holotype; SRGH).

Bush or small tree up to 4 m. tall, with rather pale pubescent branches. Leaves incompletely expanded at flowering time; leaf-lamina c. 1–2 × 1–1·25 cm. expanding to 5–6 × 2–5 cm. in fruit, oblong, ovate-oblong or subobovate-rhombic, acute at the apex, margin denticulate and becoming irregularly serrate with age, rounded or broadly cuneate at the base, subtomentose on both surfaces but not discolorous when young, sparsely pubescent in fruit; petiole 1–4 mm. long, pubescent; stipules subulate, exceeding the petiole at flowering time; bud-scales persistent, characteristically brown-black and cucullate. Inflorescences 1-, rarely 2-flowered, extra-axillary; peduncles very short or obsolete; pedicels up to 18 mm. long, densely pubescent; bracts similar to the young stipules. Sepals up to 17 mm. long, linear-oblong, greenish and densely pubescent outside, white and glabrous inside. Petals white, c. 0·7 cm. long, broadly lanceolate to almost circular, with a narrower basal claw 1 mm. tall, circumvillous within and ledged above. Androgynophore glabrous below for 1 mm., shortly tomentellous above for c. 1 mm. Ovary 4-lobed, villous; style c. 8 mm. long, pubescent except in the upper third; stigma-lobes broad. Fruit c. 15 mm. in diam., very deeply 4-lobed, with scattered setulose hairs often borne in pairs on a tuberculate base.

N. Rhodesia. C: Lusaka, fr. 20.iii.1952, *White* 2311 (FHO; K). E: between Chadiza turn-off and Nsadzu R., fl. 27.xi.1958, *Robson* 738 (K; SRGH). S: Livingstone District, Ngwesi R., fl. 8.xii.1932, *Martin* 426/32 (FHO; SRGH). **S. Rhodesia.** N: Urungwe District, Naodsa R., fl. 26.xi.1953, *Wild* 4267 (K; SRGH). W: Victoria Falls, fl. xi.1905, *Sykes* in GHS 221 (K; SRGH). C: Gatooma, fl. 22.xii.1927, *Eyles* 5082 (K; SRGH). E: Inyanga, Cheshire, fr. 14.i.1931, *Norlindh & Weimarck* 4286 (BM; LD; SRGH). **Nyasaland.** N: Mzimba R., fr. 26.ii.1959, *Robson* 1720 (K; SRGH).

Also in Tanganyika. In drier areas with annual rainfall of c. 60 cm., often in *Colophospermum mopane* woodland and with *Terminalia, Commiphora* spp. etc. The apparently anomalous record from Inyanga, usually looked upon as a high-rainfall area, is explained by the fact that the farm Cheshire is in the rain-shadow area to the west of Mt. Inyangani and is very hot and dry for Inyanga, so much so that occasional Baobab (*Adansonia digitata* L.) specimens are to be found not far away.

31. **Grewia truncata** Mast. in Oliv., F.T.A. **1**: 244 (1868).—Burret in Engl., Bot. Jahrb. **45**: 197 (1910).—Brenan, T.T.C.L.: 617 (1949). Type: Mozambique, between Lupata and Tete, *Kirk* (K, holotype).

Shrub or small tree up to 6 m. tall, with pubescent branchlets. Leaf-lamina 4–18 × 3–10 cm., oblong or obovate-oblong, retuse or truncate at the apex with the midrib often produced as a short mucro, margin crenate-dentate, rounded and often asymmetric at the base, glabrescent above, softly pubescent below; petiole up to 8 mm. long, pubescent; stipules c. 6 mm. long, lanceolate-acuminate, dorsally keeled, pubescent. Inflorescences opposite the leaves, 3–9-flowered; peduncles pubescent, very short; pedicels c. 15 mm. long, pubescent; bracts similar to the stipules, c. 0·7 cm. long. Sepals 1·5–2·5 cm. long, linear-oblong with a thickened and swollen apex, green and tomentellous outside, white and glabrous inside. Petals white, 10–16 mm. long with an ovate acute lamina much wider than the nectariferous claw which is circumvillous within ledged above and up to 2 mm. long. Androgynophore with a glabrous lower portion up to 2 mm. long and a pubescent elongation above 2·5–3 mm. long. Ovary 4-lobed, setulose-pubescent; style c. 1 cm. long, glabrous; stigma-lobes broad. Fruit c. 1·5 cm. in diam., deeply 4-lobed, with the lobes somewhat conical and sparsely and minutely tuberculate-setulose.

Nyasaland. S: Blantyre, fl. 10.i.1956, *Jackson* 1794 (K; SRGH). **Mozambique.** T: between Lupata and Tete, fl., *Kirk* (K).

Also in Tanganyika. Riverine species, sometimes thicket-forming.

32. **Grewia glandulosa** Vahl, Symb. Bot. **1**: 34 (1790).—Mast. in Oliv., F.T.A. **1**: 246 (1868).—Burret in Engl., Bot. Jahrb. **45**: 196 (1910).—Brenan, T.T.C.L.: 617 (1949). Type from Mauritius.

Shrub or small tree up to 6 m. tall; young branches pubescent. Leaf-lamina 4·5–12·5 × 3–7 cm., ovate to ovate-oblong, acutely acuminate at the apex, margins finely crenate-serrate with the serratures somewhat glandular, rounded or slightly cordate at the base, very sparsely pubescent or glabrous on both surfaces, tertiary venation reticulate; petiole up to 10 mm. long, pubescent; stipules up to 0·75 cm. long, linear, finely pubescent. Inflorescences 2–3-flowered, opposite the leaves; peduncles 2–4 mm. long, pubescent; pedicels up to 5 mm. long, pubescent; bracts similar to the stipules, up to 7 mm. long. Sepals 1·5–2·5 cm. long, linear-oblong, tomentellous outside, glabrous inside, not thickened at the apex. Petals up to 10 mm. in diam., pinkish or purple, circular, with a much narrower nectariferous claw c. 1 mm. long circumvillous within and ledged above. Androgynophore with a lower glabrous portion c. 1 mm. long and a pubescent upper portion 1 mm. long or often shorter. Ovary 4-lobed, villous; style c. 10 mm. long, glabrous; stigma-lobes broad. Fruit up to 2·8 cm. in diam., 4-lobed, densely ferruginously hairy or villous.

Mozambique. N: Cabo Delgado, fr. 21.x.1942, *Mendonça* 1040 (LISC; SRGH). Without precise locality, *Forbes* (K).

Common along the coasts of Tanganyika, Kenya and Zanzibar; occurring also in Madagascar and Mauritius. Entirely confined to coastal habitats and often thicket-forming.

33. **Grewia pachycalyx** K. Schum. in Engl., Bot. Jahrb. **15**: 123 (1892).—Burret in Engl., Bot. Jahrb. **45**: 196 (1910).—Brenan, T.T.C.L.: 617 (1949).—White, F.F.N.R.: 238, fig. 41E (1962). TAB. **3** fig. F. Type from Tanganyika.
Grewia truncata sensu Burret in Engl., Bot. Jahrb. **45**: 197 (1910) pro parte quoad specim. Kirk. ex Tete, ii.1859.
Grewia occidentalis sensu O.B.Mill. in Journ. S. Afr. Bot. **18**: 53 (1953).

Shrub, sometimes scandent, up to 5 m. tall; branchlets rather pale or reddish-pubescent, soon becoming glabrous. Leaf-lamina 6–19 × 2·5–9 cm., oblong-lanceolate, subrhombic or narrowly obovate, acute or shortly acuminate at the apex, margin finely crenate-dentate, rounded or subcuneate at the base, finely pubescent above when young but soon glabrous, finely but more densely pubescent below and also eventually glabrous; petiole up to 0·8 cm. long, pubescent; stipules up to 1·4 cm. long, lanceolate-acuminate, with a dorsal keel, pubescent. Inflorescences opposite the leaves, 3–9-flowered; peduncles up to 2 mm. long, pubescent; pedicels up to 15 mm. long, pubescent; bracts c. 2 mm. long, shortly triangular. Sepals 1·5–2·2 × 0·35–0·5 cm., linear-oblong, hooded but not thickened at the apex, greenish-pubescent outside, white and glabrous inside. Petals white, up to 15 × 10 mm., ovate-lanceolate; lamina much wider than the nectariferous claw which is circumvillous and ledged above on the inside and up to 1·5 mm. long. Androgynophore with a lower glabrous portion up to 1·5 mm. long and an elongated pubescent upper portion up to 4·5 mm. long. Ovary 4-lobed, pubescent; style up to 10 mm. long, finely stellate-pubescent or glabrous except at the base; stigma-lobes broad. Fruit orange-red when ripe, c. 2 cm. in diam., 4-lobed, shining, glabrescent.

Caprivi Strip. Lisikili, 24 km. E. of Katima Mulilo, fr. 17.vii.1952, *Codd* 7099 (BM; K; PRE). **Bechuanaland.** N: Ngamiland, fl. xii.1930, *Curson* 93 (PRE). **N. Rhodesia.** B: Sesheke, fr. 20.xii.1952, *Angus* 972 (FHO; K). N: Fort Rosebery, near Lake Bangweulu shore, st. 27.viii.1952, *Angus* 320 (FHO; K). C: Lusaka, fl. ix.1954, *Gilges* 403 (K; SRGH). S: Victoria Falls, fl. & fr. 11.iii.1952, *White* 2240 (FHO; K). **S. Rhodesia.** N: Urungwe District, Msukwe R., fl. 19.xi.1953, *Wild* 4211 (K; SRGH). W: Wankie, fl. 18.ii.1956, *Wild* 4741 (K; SRGH). E: Sabi Valley, Rupisi Hot Springs, fr. 28.i.1948, *Wild* 2374 (K; SRGH). S: Birchenough Bridge, fl. i.1938, *Obermeyer* 2417 (PRE; SRGH). **Nyasaland.** N: Rumpi Distr., 1110 m., fl. 19.ii.1961, *Richards* 14421 (K). **Mozambique.** T: Tete, fl. & fr. ii.1859, *Kirk* (K).

Also in Tanganyika. Common thicket-forming species in the Zambezi valley; apparently rather rare in the Sabi valley.

34. **Grewia lepidopetala** Garcke in Peters, Reise Mossamb. Bot. **1**: 135 (1861).—Mast. in Oliv., F.T.A. **1**: 245 (1868).—Burret in Engl., Bot. Jahrb. **45**: 196 (1910).—Brenan, T.T.C.L.: 617 (1949). Type: Mozambique, Sena, *Peters* 19 (B, holotype †; K).

Scrambling or erect shrub or small tree up to 6 m. tall; branches thinly pubescent. Leaf-lamina 2·5–11 × 1·6–5 cm., obovate-oblong to elliptic-oblong, shortly and acutely acuminate at the apex, margin finely serrate-dentate, abruptly cuneate or rounded at the base, glabrous above, very finely but densely stellately whitish-pubescent below or finally glabrescent; petiole up to 5 mm. long, thinly pubescent; stipules 3–4 mm. long, keeled, linear-lanceolate. Inflorescences leaf-opposed but usually concentrated towards the ends of the branches, 3–9-flowered; peduncles up to 15 mm. long, finely pubescent; pedicels similar, up to 12 mm. long; bracts 2–3 mm. long, pubescent. Sepals up to 15 × 2 mm., linear-oblong, not thickened at the apex, greenish-tomentellous outside, white and glabrous within. Petals white or pinkish, 4–5 mm. in diam., subcircular, with a much narrower nectariferous claw c. 1 mm. long circumvillous inside and ledged above. Androgynophore with a glabrous lower portion c. 1 mm. long and a pubescent upper portion c. 1·5 mm. long. Ovary 4-lobed, densely setulose-pubescent; style c. 6 mm. long, glabrous; stigma-lobes broad. Fruit c. 15 mm. in diam., deeply 4-lobed, sparsely hispid.

S. Rhodesia. N: Mtoko, Mkota Reserve, fl. 29.xii.1950, *Whellan* 486 (K; SRGH). E: Umtali, Wengesi R., fl. 18.xii.1954, *Chase* 5356 (K; SRGH). S: Nuanetsi, Lundi Hotel, fl. xii.1955, *Davies* 1760 (K; SRGH). **Nyasaland.** S: Shire R., fl. i.1894, *Scott Elliott* 8801 (BM; K). **Mozambique.** T: Tete, fl. & fr. xi.1858, *Kirk* (K). MS: Madanda Forest, fl. 5.xii.1906, *Swynnerton* 1220 (BM; K).

Also in Tanganyika. Confined to the eastern parts of our area and apparently to the low-altitude dry valleys of the Zambezi and Limpopo; occasionally found on termitaria.

35. **Grewia limae** Wild in Bol. Soc. Brot., Sér. 2, **31**: 86, t. 1 fig. D (1957). TAB. **3** fig. A. Type: Mozambique, Cabo Delgado, Palma, *Pires de Lima* 60 (PO).

Shrub or small tree; branchlets at first brownish-stellate-pubescent, later becoming glabrous; lenticels greyish and slightly tubercular. Leaf-lamina 2·5–9 × 1·5–5 cm., narrowly obovate to narrowly oblong-obovate, apex abruptly acuminate to cuspidate, margin denticulate to denticulate-crenate, the base rounded and more or less symmetrical, minutely stellate-pilose above or glabrous, densely brownish-stellate-pubescent below, at least when young; petiole up to 5 mm. long; stipules c. 3 mm. long, linear-lanceolate, caducous, setulose-pubescent. Inflorescences terminal or opposite the leaves; peduncle 2–3 mm. long, 3–5-flowered, stellate-pubescent; pedicels up to 8 mm. long, with short stellate hairs and some longer fasciculate-stellate hairs; bracts 2–3 mm. long, narrowly lanceolate, setulose-pubescent on the back. Flower-buds ovoid. Sepals c. 14 × 3·5 mm., linear-oblong, apex acute, somewhat thickened, tomentose outside with dense, shortly-stellate hairs and scattered longer fasciculate-stellate hairs, tomentellous within towards the apex and margins. Petals white; lamina c. 5 × 4 mm., ovate, subacute at the apex, with a circumvillous nectary at the base 1·5 × 1 mm. ledged above and glabrous at the back. Androgynophore with a glabrous basal portion c. 1·5 mm. long, prolonged above into a pubescent upper portion c. 1 mm. long. Ovary 4-lobed, densely villous, with 4 ovules per loculus. Style 7–8 mm. long, glabrous, stigmatic lobes broad and flattened. Fruit not known.

Mozambique. N: Cabo Delgado Distr., Palma, fl. 17.i.1917, *Pires de Lima* 60 (PO).

Only known from northern Mozambique but will no doubt be found in southern Tanganyika. Small tree of coastal woodlands (as far as is known at present).

36. **Grewia villosa** Willd. in Ges. Nat. Fr. Berl. Neue Schr. **4**: 205 (1803).—Mast. in Oliv., F.T.A. **1**: 249 (1868).—Burret in Engl., Bot. Jahrb. **45**: 198 (1910).—Burtt Davy, F.P.F.T. **1**: 254 (1926).—Brenan, T.T.C.L.: 614 (1949).—Exell & Mendonça, C.F.A. **1,** 2: 223 (1951).—O.B.Mill. in Journ. S. Afr. Bot. **18**: 54 (1953). Type from India.

Much-branched shrub 1–3 m. tall; young branchlets with yellowish silky hairs. Leaf-lamina up to 12 cm. in diam., subcircular to broadly elliptic, apex rounded, margin serrate, slightly cordate at the base and sometimes asymmetric, sparsely pubescent, green and finely reticulate above, grey-pubescent or villous below; veins prominent and reticulate; petiole up to 4 cm. long, setulose-pubescent; stipules 5–12 mm. long, ovate to broadly oblanceolate, pubescent, submembranous. Inflorescences leaf-opposed or frequently arising some distance from the node; peduncle up to 5 mm. long, pubescent; pedicels similar, up to 3 mm. long; bracts c. 7 mm. long, ovate, similar to the stipules. Flower-buds globose to cylindric-globose. Sepals 5–9 mm. long, linear-lanceolate, silky-pubescent outside, slightly pubescent and yellow inside. Petals yellow, about half the length of the sepals, oblong or obovate-oblong, with a basal nectariferous claw circumvillous within but not extended above into a narrow ledge, nectary and lamina of about equal length, pubescent behind the nectary and along the mid-line of the lamina. Androgynophore up to 0·75 mm. long, glabrous, with a more or less membranous undulate rim. Ovary not lobed, densely villous; style c. 3 mm. long, pubescent, especially towards the base; stigma divided into many laciniate segments. Fruit reddish, c. 15 mm. in diam., globose, shallowly 4-lobed, with small scattered tubercles bearing rather long caducous setulose hairs.

Bechuanaland Prot. N: Ngamiland, Kwebe Hills, fl. xii.1896, *Lugard* 46 (K). **S. Rhodesia.** N: Sebungwe, Chikwatata Spring, fl. 7.xi.1958, *Phipps* 1394 (K; SRGH).

W: Matobo, Prospect Ranch, fr. 22.iii.1950, *Orpen* 4/50 (SRGH). S: Birchenough Bridge, Sabi valley, fl. i.1938, *Obermeyer* 2402 (PRE; SRGH). **Mozambique.** LM: between Sábiè and Machatuíne, fl. 2.v.1953, *Myre & Balsinhas* 1674 (K; LM).

Widely distributed through the drier parts of Africa and also in the Cape Verde Is., Arabia and India. Often found in *Acacia* bush or woodland or in mixed *Commiphora-Terminalia* bushland.

37. **Grewia herbacea** Welw. ex Hiern, Cat. Afr. Pl. Welw. **1**: 96 (1896).—Burret in Engl., Bot. Jahrb, **45**: 198 (1910).—Brenan, T.T.C.L.: 614 (1949).—Exell & Mendonça, C.F.A. **1,** 2: 224 (1951).—White, F.F.N.R.: 263, fig. 41A (1962). TAB. **3** fig. B, G. Type from Angola.

Grewia leucodiscus K. Schum. in Notizbl. Bot. Gart. Berl. **3**: 101 (1901). Syntypes: Nyasaland, Zomba, *Whyte* (K); without precise locality, *Buchanan* 133 (B†), 171 (B†), 653 (B†; K).

Very like *G. villosa* Willd. but with long virgate branches arising from a woody rootstock. Stipules linear or filiform, somewhat keeled and up to 13 mm. long. Bracts up to 10 mm. long, narrowly lanceolate to linear-lanceolate or sometimes bifid, pubescent. Stigma-lobes broad and flat with toothed or lobed margins. Flowers smaller than in *G. villosa* and more per inflorescence. Otherwise as in *G. villosa.*

N. Rhodesia. S: Mazabuka, fl. 23.vii.1952, *Angus* 16 (FHO; K). **S. Rhodesia.** N: Darwin, fl. 10.vi.1923, *Swynnerton* 4027 (K). W: Matopos, fl. 29.xi.1951, *Plowes* 1343 (K; SRGH). C: Charter, fr. 22.ii.1933, *Rattray* 575 (BM; SRGH). **Nyasaland.** C: Kasunga-Bua road, fl. 13.i.1951, *Robson* 1123 (K; SRGH). S: Shire Highlands, fr. 1891, *Buchanan* 171 (B†; BM; K). **Mozambique.** N: Cuamba, fr. 13.v.1948, *Pedro & Pedrógão* 3365 (LMJ; SRGH).

Also in Nigeria, the Congo, Angola and Tanganyika. Dwarf shrub of deciduous woodland and often in *Brachystegia boehmii* woodland; apparently occurring in somewhat higher rainfall conditions than *G. villosa.*

4. TRIUMFETTA L.

Triumfetta L., Sp. Pl. **1**: 444 (1753); Gen. Pl. ed. 5: 203 (1754); Sprague & Hutch. in Journ. Linn. Soc., Bot. **39**: 231–276 (1909).

Annual or perennial herbs, shrublets or shrubs, sometimes with annual stem from a woody rootstock. Leaves alternate, petiolate, simple or digitate, often lobed, serrate or crenate, often several-nerved from the base; stipules lateral. Flowers actinomorphic, borne in cymes in terminal inflorescences or at the nodes. Sepals 5, usually linear and with a short horn just behind the apex, usually stellately hairy without. Petals 5, yellow or orange, linear to obovate, narrowed to the base, and often hairy at the base or just above it. Stamens 4–40, raised on a short glabrous androgynophore or torus with a glandular patch just above each petal base; apex of the androgynophore produced into a ciliate or pubescent or villous disk or annulus between which and the ovary the stamens are inserted. Ovary often tubercled or echinulate, each tubercle surmounted by one or more minute bristles, 2–5-locular with 2 pendulous collateral ovules in each loculus or falsely 10-locular by the intrusion of longitudinal false septa; style terete, about as long as the stamens; stigma entire or very shortly 2–5-lobed. Fruit a capsule, dividing into 3–5-valves with 1–2 seeds per loculus or indehiscent, usually globose, sometimes ovoid, echinate or setose. Seeds obovoid or subreniform; testa rather leathery and brown; embryo straight; cotyledons flat, suborbicular; endosperm fleshy, scanty.

Note: Many species of this genus are weeds of cultivation and waste places or early colonisers of fallows or abandoned cultivation; others are frequent in forest clearings and forest margins; many of them are used for the production of native fibres.

Leaves palmately divided or lobed to at least half-way or digitately 5–7-foliolate:
- Plant creeping, with slender stems and 3–5-lobed leaves; inflorescences solitary at the nodes - - - - - - - - - - - - - - - 2. *kirkii*
- Plant erect, or, if prostrate or trailing, then inflorescences terminal or axillary and of dense cymes:
  - Leaves 3–5-lobed to the middle or slightly beyond - - - - - 4. *trifida*
  - Leaves 5–9-lobed almost to the base or with 5–9 leaflets:
    - Stipules oblong-lanceolate, persistent; leaves rugose, softly pubescent above 5. *grandistipulata*

Stipules linear, ± caducous; leaves not rugose but scabrous above - 3. *digitata*
Leaves not lobed or at the most shallowly 3-lobed to much less than half-way:
Sepals lepidote outside; fruit with broad-based conical tubercles - 1. *amuletum*
Sepals not lepidote; fruits with slender aculei:
Herbs with annual stems from a woody rootstock:
Leaves up to 9 × 3·5 cm.; young stems and leaves at least with a very dense spreading golden-yellow tomentum; stipules lanceolate-oblong, c. 0·8 × 0·5 cm. 7. *heliocarpa*
Leaves narrower, 3–11 × 0·4–2 cm.; young parts appressed-tomentellous or glabrescent; stipules subulate-lanceolate, 0·7 × 0·1–0·3 cm.:
Young leaves stellately tomentellous, at least beneath, and retaining some vestiges of tomentum with age - - 6. *welwitschii* var. *welwitschii*
Young leaves not tomentellous:
Pubescence, if any, of minute appressed-stellate hairs 6. *welwitschii* var. *descampsii*
Pubescence sparsely setose with some scattered simple hairs 6. *welwitschii* var. *hirsuta*
Herbs, shrubs or suffrutices but not having annual stems from a woody rootstock:
Aculei of fruit or the majority of them with a terminal crown of several minute bristles or setae:
Leaves tomentose or tomentellous, at least below; rigidly erect small shrubs:
Upper surface of leaves stellate-pubescent or glabrescent, not tomentellous 9. *setulosa*
Upper surface of leaves tomentellous:
Terminal setae of aculei spreading, 6–8; aculei stellately plumose 8. *dekindtiana*
Terminal setae pointing forwards, 1–3; aculei broad-based, sparsely puberulous - - - - - - - - - - 12. *reticulata*
Leaves not tomentose or tomentellous:
Sepals glabrous towards the base; aculei of fruit very sparsely pubescent:
Stems with patent simple hairs; leaves ovate to lanceolate 11. *paradoxa*
Stems with appressed-stellate hairs; leaves narrowly lanceolate 13. *intermedia*
Sepals stellate-pubescent along their whole length; aculei of fruit markedly pilose 10. *angolensis*
Spines (or aculei) of fruit terminated by a single bristle or spine:
Bristle or spine straight:
Stems prostrate; leaves ovate to subcircular, obtuse at the apex 15. *glechomoides*
Stems erect; leaves narrowly lanceolate to ovate, apex acute:
Peduncles very slender, up to 2·5 cm. long; leaves narrowly elliptic-lanceolate to narrowly lanceolate; capsule and aculei glabrous 16. *tenuipedunculata*
Peduncles not slender, c. 2 mm. long; leaves ovate or lanceolate-ovate; capsule and aculei pilose or only occasionally almost glabrous 14. *tomentosa*
Bristle or spine hooked or falcate:
Fruit c. 5 mm. in diam. including the aculei:
Stamens 5–13; fruit ovoid, aculei ciliate on one side - 18. *pentandra*
Stamens c. 15; fruit globose, aculei glabrous, body of fruit tomentose 17. *rhomboidea*
Fruit 10 mm. in diam. or more:
Aculei glabrous or very nearly so:
Indumentum of simple hairs; leaves membranous; sepals 2·5–5 mm. long - - - - - - - - - 19. *annua* forma *annua*
Indumentum stellate-tomentose, sepals c. 10 mm. long:
Buds appressed-tomentellous; leaves discolorous, whitish-tomentellous beneath - - - - - - - 21. *cordifolia* var. *tomentosa*
Buds tomentose with spreading hairs; leaves not noticeably discolorous, with a coarse yellowish tomentum 22. *pilosa* var. *effusa*
Aculei conspicuously pilose:
Leaves membranous with simple hairs:
Aculei densely pilose, not conspicuously broadened at the base, about 120 per capsule - - - - - - - - - 20. *trichocarpa*
Aculei sparsely pilose, broadened towards the base, about 60 per capsule - - - - - - - - - 19. *annua* forma *piligera*
Leaves not membranous, densely stellately hairy or tomentose:
Aculei with falcate spines at the tip - - - - 14. *tomentosa*
Aculei strongly uncinate:
Leaves tomentellous on both sides; subapical horn of sepals inconspicuous - - - - 21. *cordifolia* var. *tomentosa*

Leaves tomentose to glabrescent but never thinly tomentellous; subapical horn of sepals conspicuous:
Indumentum densely tomentose on both sides of leaf; cauline leaves ovate, 6–12 × 3–7·5 cm. - 22. *pilosa* var. *tomentosa*
Indumentum of dense coarse yellowish hairs or sparsely pilose:
Buds and leaves sparsely pilose; leaf-lamina 8–14 × 1·5–3·5 cm. 22. *pilosa* var. *glabrescens*
Buds and leaves densely and coarsely yellowish-pubescent; leaf-lamina 9–18 × 2·5–7 cm. - 22. *pilosa* var. *nyasana*

1. **Triumfetta amuletum** Sprague in Bull. Herb. Boiss., Sér. 2, **5**: 702 (1905).—Sprague & Hutch. in Journ. Linn. Soc., Bot. **39**: 245, t. 17 fig. 1 (1909).—Eyles in Trans. Roy. Soc. S. Afr. **5**: 411, (1916) "*annuletum*".—Weim. in Bot. Notis. **1936**: 47 (1936). TAB. **5** fig. A. Syntypes: Nyasaland, Lake Chirwa, *Kirk* 164 (K); Mozambique, Missale, *Nicholson* (K).

Prostrate perennial with stellate-pubescent trailing stems up to 1 m. long or ascending to 1·2 m. Leaf-lamina 7–10 × 3·5–11 cm., very broadly ovate or subcircular, densely stellate-pubescent above, subtomentose below, apex subacute, rounded or broadly cuneate at the base, margin irregularly crenate, 5-nerved from the base; petiole up to 5 cm. long, somewhat channelled above, stellate-pubescent, with 3–5 pairs of discoid glands c. 1 mm. in diam. just below the lamina; stipules 5–6 mm. long, lanceolate-acuminate to broadly triangular, setose-pubescent. Inflorescences of leaf-opposed cymes 1–2 together on short stout pubescent peduncles up to 5 mm. long. Flowers crowded several together, on short pedicels or subsessile; bracts up to 4 mm. long, broadly lanceolate to lanceolate, setose pubescent. Sepals 10–12 mm. long, linear, with crowded scales outside, and a subapical blackish 2-dentate horn 1–1·5 mm. long. Petals yellow, c. 1 × 0·5 cm., obovate, narrowing rather abruptly into a basal claw which is ciliate at the margins. Androgynophore with suborbicular glands; annulus with a shortly ciliate margin. Ovary tuberculate with the tubercles terminated by 2 or 3 short setae, apparently 10-locular by the intrusion of 5 additional false septa. Stamens c. 25, c. 8 mm. long; style up to 5 mm. long, slender, glabrous. Fruit blackish when ripe, c. 2 cm. in diam., globose, covered with hard conical glabrous processes or spinose bosses c. 5 mm. long.

N. Rhodesia. N: Mpika, fl. 30.i.1955, *Fanshawe* 1891 (K). C: Chilanga, fl. xi.1909, *Rogers* 8665 (K). S: Choma, Mapanza Mission, fl. 10.v.1953, *Robinson* 207 (K). **S. Rhodesia.** N: Banket, fr. 15.iv.1947, *Wild* 1887 (K; SRGH). W: Bulawayo, fl. iii.1915, *Rogers* 13630 (BM; PRE). C: Salisbury, fl. & fr. 8.iii.1925, *Eyles* 4486 (K; SRGH). E: Inyanga, Cheshire, fl. 15.i.1930, *F.N. & W.* 4355 (BM; LD). S: Victoria, fl. 1909, *Monro* 902 (BM). **Nyasaland.** C: Dedza, Dzenza Forest Reserve, fl. 21.iii.1955, *E.M. & W.* 1103 (BM; LISC; SRGH). S: Lake Chirwa (Chilwa), fr. iv.1859, *Kirk* 164 (K). **Mozambique.** T: Vila Mousinho, fr. 16.vii.1949, *Barbosa & Carvalho* 3641 (K; LM; SRGH).

Known only from our area. Usually growing in open grasslands at swamp (vlei) edges, etc. or in open woodland on rather heavy soils subject to some waterlogging. Persists in fallows and as a weed.
Reported to be grazed by cattle.

2. **Triumfetta kirkii** Mast. in Oliv., F.T.A. **1**: 259 (1868).—K. Schum. in Engl., Pflanzenw. Ost-Afr. **C**: 265 (1895).—Sprague & Hutch. in Journ. Linn. Soc., Bot. **39**: 246, t. 17 fig. 2 (1909). Type from Tanganyika (Rovuma R.).

Creeping annual with slender thinly stellate-pubescent stems. Leaf-lamina 0·75–1·5 × 0·7–3 cm., cordate, divided to $\frac{2}{3}$–$\frac{3}{4}$ of the way into 3–5 lobes; lobes oblong, obtuse, crenate, thinly stellate-pubescent on both sides; petiole up to 1 cm. long, thinly pubescent; stipules c. 1·5 mm. long, linear, setose-pubescent. Inflorescences of 1–3-flowered cymes, solitary at the nodes; peduncle up to 15 mm. long, filiform; bracts up to 1·5 mm. long, linear, similar to the stipules; pedicels c. 5 mm. long, filiform. Sepals up to 4·5 mm. long, linear, hooded in the upper half, pubescent outside. Petals yellow, 3·5 mm. long, narrowly oblanceolate. Androgynophore c. 0·3 mm. long with subquadrate glands; annulus ciliate on the margin. Stamens 12; filaments up to 3 mm. long. Ovary bristly, 3-locular. Fruit 15 mm. in diam. including the aculei; aculei very slender and very numerous, plumose above, glabrous below, terminated by a single seta.

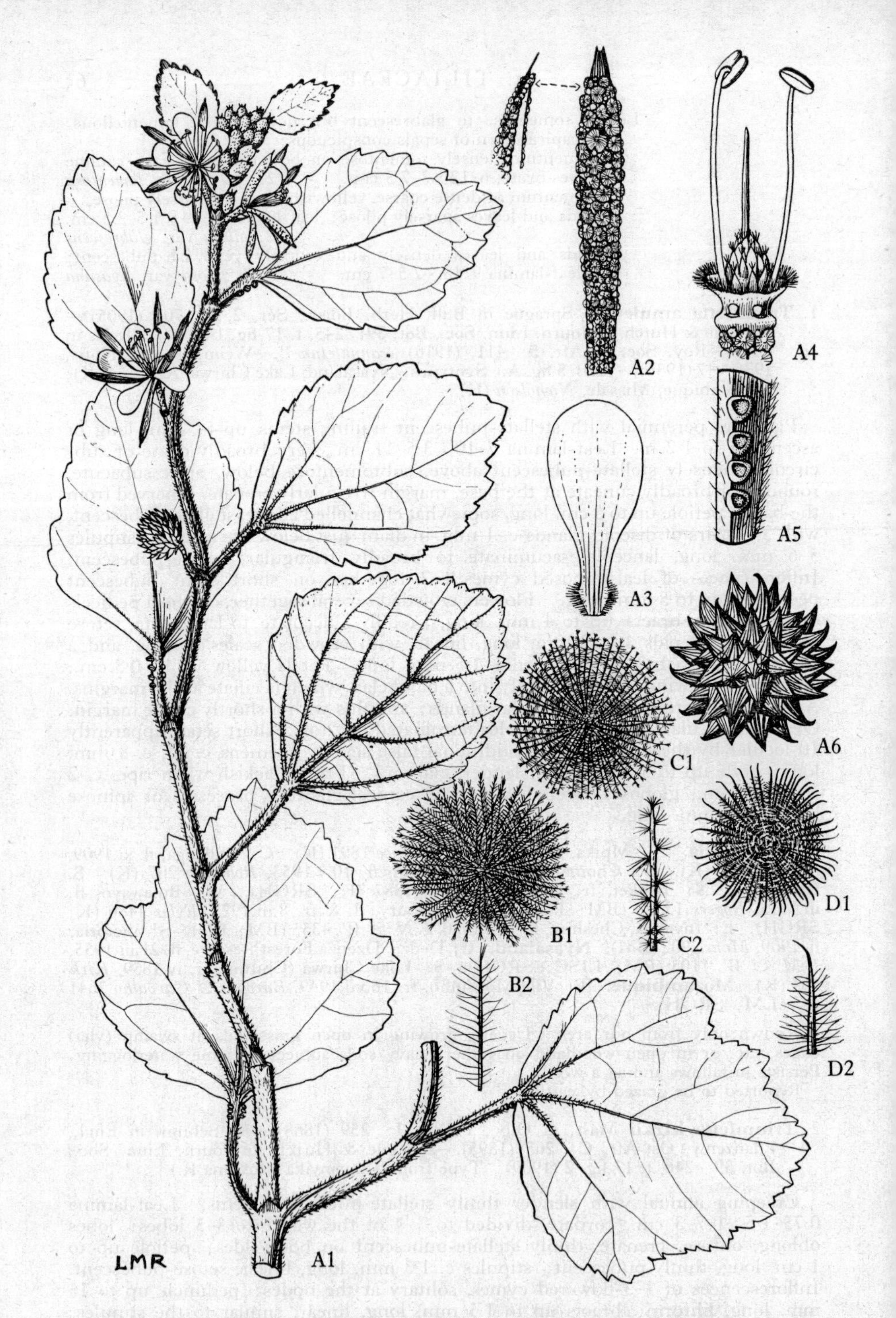

Tab. 5. A.—TRIUMFETTA AMULETUM. A1, flowering branch (× ⅔); A2, sepal with side view of apex (× 5); A3, petal (× 3); A4, gynoecium with all but 2 stamens removed (× 5); A5, portion of petiole showing glands (× 3); A6, fruit (× 1), all from *Robinson* 532. B.—TRIUMFETTA WELWITSCHII VAR. WELWITSCHII. B1, fruit (× 2); B2, bristle (× 3), both from *Milne-Redhead* 2589. C.—TRIUMFETTA DEKINDTIANA. C1, fruit (× 2); C2, bristle (× 4), both from *Cruse* 367. D.—TRIUMFETTA TOMENTOSA. D1, fruit (× 1); D2, bristle (× 2), both from *Richards* 1642.

Mozambique. N: Nacala, fl. & fr. 17.v.1937, *Torre* 1424 (COI; LISC). MS: Inhaminga, fl. & fr. 22.v.1942, *Torre* 4177 (LISC; SRGH).
Also in Tanganyika Territory.
Apart from the type known only from Torre's two gatherings. As it could be easily overlooked the species may not be so rare as this would indicate.

3. **Triumfetta digitata** (Oliv.) Sprague & Hutch. in Journ. Linn. Soc., Bot. **39**: 247 (1909).—R.E.Fr., Wiss. Ergebn. Schwed. Rhod.-Kongo-Exped. **1**: 141 (1914). Type: N. Rhodesia, Abercorn District, Kawimbe (Fwambo), *Carson* 1 (K, holotype).
Ceratosepalum digitatum Oliv. in Hook., Ic. Pl. **24**: t. 2307 (1894). Type as above.

Low shrub about 0·6 m. tall or sometimes with the branches trailing and almost prostrate, branching mostly from the base; branches densely ferruginously stellate-hairy when young. Leaf-lamina 1·5–5 × 0·7–1·4 cm., deeply 5–7-partite or -digitate; segments or leaflets oblanceolate, with appressed-stellate hairs of two kinds, some small and some much larger above, whitish- or pale-brownish-tomentellous below, acute or subacute at the apex, margin serrate, cuneate at the base; petiole up to 2·5 cm. long, ferruginously hairy. Inflorescences ferruginous, of rather dense cymes opposite the bracts ± aggregated into small terminal panicles; bracts similar to the leaves but smaller and often displaced upwards so that the inflorescences appear axillary; peduncles up to 2·5 cm. long, ferruginous; pedicels up to 5 mm. long but usually less, ferruginous and bearing linear ferruginously hairy bracteoles up to 8 mm. long. Sepals 12–14 mm. long, linear, narrowing to the apex, ferruginously stellate-hairy outside, with a horn 4–5 mm. long inserted 0·75–1 mm. below the apex. Petals yellow, 9–10 mm. long, obovate and narrowed into a linear basal claw which is densely villous just above the base. Androgynophore very short but with a villous annulus clasping the base of the ovary. Stamens 35–40, 0·8 cm. long. Ovary depressed-globose, villous, 5–6-locular. Fruit up to 3·5 cm. in diam. including the aculei, globose; aculei extremely numerous, densely plumose, with a single fine terminal seta.

N. Rhodesia. N: Abercorn Distr., Kawimbe (Fwambo), fl. ii.1893, *Carson* 1 (K). W: Solwezi, fr. 16.iv.1960, *Robinson* 3487 (K; SRGH).
Also in Tanganyika as far north as Lake Victoria and in the Congo near Elisabethville. Locally very common and a constituent of *Brachystegia* woodland.

4. **Triumfetta trifida** Sprague & Hutch. in Journ. Linn. Soc., Bot. **39**: 248 (1909).—R.E.Fr., Wiss. Ergebn. Schwed. Rhod.-Kongo-Exped. **1**: 141 (1914).—White, F.F.N.R.: 240 (1962). Type from the Congo (Mwero).
Triumfetta palmatiloba Dunkley in Kew Bull. **1933**: 170 (1933). Type: N. Rhodesia, Abercorn, *Miller* 100 (FHO; K, holotype).

Shrub or small tree up to c. 5 m. tall; branches rather thick and with a ferruginous tufted scabrous stellate tomentum. Leaf-lamina 3–10 × 4–14 cm., palmately lobed to rather more than half-way, the base broadly cordate to truncate; lobes 3–5, 2–8 × 1–4 cm., oblong, acute at the apex, irregularly serrate, sparsely appressed-stellate-pilose and somewhat scabrid above, brownish-tomentose below, with raised nerves; petiole up to 8 cm. long, stellately ferruginous-tomentose. Inflorescences terminal panicles or corymbs, of shortly pedunculate c. 10-flowered cymes; peduncles of cymes 1–5 mm. long, ferruginously scabrous-tomentose; pedicels similar, 1–3 mm. long, bracts 4–8 mm. long, lanceolate to subulate, setulose-pubescent and with a short appressed-stellate pubescence also. Sepals c. 10 mm. long, broadened at the base giving a swelling in the bud, linear above, appressed-stellately tomentellous outside; subapical horn c. 1 mm. long. Petals orange-yellow, c. 10 mm. long, spathulate-oblong, rounded or truncate and serrate at the apex, narrowed to a falcate claw below, tomentose at the base. Androgynophore 1–1·5 mm. long, with large closely spaced glands; annulus villous. Stamens c. 30, 10 mm. long. Ovary depressed-globose, villous, 3-locular. Fruit up to 6·5 cm. in diam. including the aculei; aculei very numerous, subulate-filiform, softly stellate-pubescent and terminating in a fine long seta which is easily rubbed off and then apparently absent.

N. Rhodesia. N: Abercorn, fl. & fr. 20.x.1947, *Brenan & Greenway* 8155 (FHO; K).
Also in the Congo. In *Brachystegia* woodlands and often on red soils.
The bark is used for rope and the roots as a native stomach-medicine.

5. **Triumfetta grandistipulata** Wild in Bol. Soc. Brot., Sér. 2, **32**: 49 (1958). Type: N. Rhodesia, Kawambwa, *Fanshawe* 2959 (K, holotype; SRGH).

Shrub or small tree up to 4 m. tall; young branches with a dense tufted golden-stellate tomentum. Leaf-lamina up to 11 × 14 cm., subcircular or oblate in outline, deeply 3–9-lobed or some branches with narrowly elliptic undivided leaves; lobes 8 × 2 cm., narrowly elliptic, apex acute, margin serrate, green and rugose above, densely and shortly white-tomentose below, nerves and veins prominent below, reticulate; petiole up to 7 cm. long; stipules up to 2 × 0·6 cm., persistent, oblong-lanceolate, acute at the apex, densely and shortly tomentose. Inflorescences of dense, terminal or leaf-opposed, leafless cymes; peduncles up to 1 cm. long, densely stellate-tomentose; pedicels similar, up to 4 mm. long; bracts caducous, oblong-elliptic, apex acute, entire or 1–2-dentate, shortly and densely grey-tomentose on both sides. Flower-buds cylindric, widened below, densely golden-tomentose outside, crowned with apical horns up to 4 mm. long. Sepals up to 2 × 0·3 cm., linear, widened below, apex acuminate, glabrous within. Petals yellow, clawed; blade 7 × 9 mm., circular or oblate, rounded at the apex; claw c. 5 mm. long, tomentose at the base. Androgynophore 1 mm. long, with 5 reniform glands; annulus undulate and ciliate at the margin. Stamens c. 40, filaments 15 mm. long. Ovary globose, villous, 5–7-locular; style c. 13 mm. long. Fruit up to 3 cm. in diam. including the aculei, globose, woody, indehiscent; aculei c. 3·5 mm. long, very numerous, collected into conical fascicles, minutely stellate-tomentose.

N. Rhodesia. N: Kawambwa, fl. & fr. 30.i.1957, *Fanshawe* 2959 (K; SRGH) Known so far only from the type collection. On sandy flats in rather scrubby woodland.

6. **Triumfetta welwitschii** Mast. in Oliv., F.T.A. **1**: 255 (1868).—Sprague & Hutch. in Journ. Linn. Soc., Bot. **39**: 253 (1909).—Eyles in Trans. Roy. Soc. S. Afr **5**: 412 (1916).—Exell & Mendonça, C.F.A. **1,** 2: 229 (1951).—Suesseng. & Merxm. in Proc. & Trans. Rhod. Sci. Ass. **43**: 106 (1951).—Brenan in Mem. N.Y. Bot. Gard. **8,** 3: 228 (1953).—Martineau, Rhod. Wild Fl.: 47, t. 16 (1954). Type from Angola.

Triumfetta rehmannii Szyszyl., Polypet. Thalam. Rehm.: 151 (1887). Type from the Transvaal.

Triumfetta mastersii Bak. f. in Trans. Linn. Soc., Ser. 2, Bot. **4**: 6 (1894).—Sprague & Hutch. in Journ. Linn. Soc., Bot. **39**: 252 (1909).—Bak. f. in Journ. Linn. Soc., Bot. **40**: 33 (1911).—R.E.Fr., Wiss. Ergebn. Schwed. Rhod.-Kongo-Exped. **1**: 249 (1914).—Eyles in Trans. Roy. Soc. S. Afr. **5**: 411 (1916).—Weim. in Bot. Notis. **1936**: 46 (1936). Type: Nyasaland, Mt. Mlanje, *Whyte* 163 (K, holotype).

Triumfetta laxiflora Engl. in Engl., Bot. Jahrb. **39**: 579 (1907). Type: S. Rhodesia, Salisbury, *Engler* 3025 (B†).

Triumfetta welwitschii var. *rehmannii* (Szyszyl.) Sprague & Hutch. in Journ. Linn. Soc., Bot. **39**: 253 (1909).—Eyles in Trans. Roy. Soc. S. Afr. **5**: 412 (1916).—Burtt Davy, F.P.F.T. **1**: 256 (1926).—Weim. in Bot. Notis. **1936**: 48 (1936). Type as for *T. rehmannii*.

Perennial herb sending up annual stems from a woody rootstock; flowers developing before the leaves on the first shoots and some purely vegetative leafy shoots appearing later in the growing season; stems 20–45 cm. tall, yellowish- or greyish-tomentellous, glabrescent below. Leaf-lamina 3–11 × 0·4–2 cm., linear-lanceolate, oblanceolate or obovate-oblong, tomentellous or softly pubescent, more densely so below, or almost entirely glabrous except when very young, apex obtuse, rounded, acute or apiculate, narrowly cuneate at the base, margin entire or serrulate or conspicuously serrate; petiole c. 0·5 cm. long, tomentellous; stipules c. 0·75 cm. long, subulate-triangular, pubescent. Inflorescences terminal, up to 15 cm. long, of small dense cymes with the internodes elongating in fruit, closely covered with a yellowish or golden velvety tomentum; bracts c. 2·5 mm. long, linear, silky-pubescent outside. Sepals 6–10 mm. long, linear-oblong, spathulate, golden-tomentellous outside, with a subapical horn 0·25–0·5 mm. long. Petals yellow, 5–9·5 × 3–5 mm., spathulate-oblong to obovate, rounded or slightly emarginate at the apex, villous just above the base. Androgynophore c. 0·5 mm. long; annulus with a ciliate rim. Stamens 20–50. Ovary globose, villous. Fruit c. 2–2·5 cm. in diam. including the aculei, globose; aculei very numerous, c. 7 mm. long, very slender, plumose, with 1–3 fine setae at the apex.

Var. **welwitschii.** TAB. 5 fig. B.

Leaves closely stellately tomentellous, especially beneath, and with their under surfaces retaining some vestiges of tomentum even when older.

N. Rhodesia. N: Abercorn Distr., Kawimbe (Fwambo), fr., *Carson* 131 (K). W: Mwinilunga, fr. 5.x.1937, *Milne-Redhead* 2589 (BM; K). S: Kalomo, fl. xii.1920, *Rogers* 26036 (K). **S. Rhodesia.** N: Sinoia, fl. & fr. x.1926, *Rand* 295 (BM). W: Bulawayo, fl. ix.1902, *Eyles* 1078 (BM; SRGH). C: Salisbury, fl. 14.ix.1911, *Rogers* 4013 (BM; K; SRGH). E: Melsetter, fr. 28.xii.1948, *Chase* 1386 (SRGH). S: Victoria, fl., *Monro* 502 (BM; K). **Nyasaland.** S: Mt. Mlanje, fl. 1891, *Whyte* 163 (BM; K). **Mozambique.** N: Mandimba, fl. & fr. 5.xi.1941, *Hornby* 3469 (K; PRE). Z: Milange, fr. 13.xi.1942, *Mendonça* 1428 (LISC; SRGH). MS: Mavita, R. Messambuzi, fl. 13.xiii.1949, *Pedro & Pedrógão* 7986 (LMJ; SRGH).

Known from SW. Tanganyika to the Transvaal and Angola. Very common in *Brachystegia* woodland and grassland; most commonly flowering just before the rainy season and particularly obvious after the grass is burnt. The leaves expand and more leafy shoots appear in the course of the rainy season.

Var. **descampsii** (De Wild. & Dur.) Brenan in Mem. N.Y.Bot. Gard. **8,** 3: 229 (1953). Type from the Congo.

Triumfetta descampsii De Wild. & Dur. in Bull. Soc. Roy. Bot. Belg. **39**: 95 (1901). Type as above.

Triumfetta mastersii var. *descampsii* (De Wild. & Dur.) Sprague & Hutch. in Journ. Linn. Soc., Bot. **39**: 252 (1909). Type as above.

Leaves rapidly becoming quite glabrous below and, even when young, not densely tomentellous as in var. *welwitschii*.

N. Rhodesia. N: Luvingo, Malolo, fl. & fr. x.1911, *Fries* 2436 (UPS). E: Mbozi, fl. 28.ix.1936, *Burtt* 6127 (K). **S. Rhodesia.** N: Concession, fr. 24.x.1951, *Greatrex* in GHS 34629 (K; SRGH). C: Marandellas, fl. & fr. 26.xi.1941, *Dehn* 54 (SRGH). **Nyasaland.** C: Dedza aerodrome, fl. & fr. 6.x.1949, *Wiehe* N/263 (K). **Mozambique.** N: Massangulo, fl. xii.1932, *Gomes e Sousa* 1094 (COI). T: Angónia, fl. 29.ix.1942, *Torre* 546 (LISC; SRGH).

Also from the Congo.

Var. **hirsuta** (Sprague & Hutch.) Wild in Bol. Soc. Brot. Ser., 2, **31**: 87 (1957). Syntype from the Transvaal.

Triumfetta hirsuta Sprague & Hutch. in Journ. Linn. Soc., Bot. **39**: 251 (1909).—Weim. in Bot. Notis. **1936**: 47 (1936). Type as above.

Triumfetta mastersii var. *descampsii* sensu Burtt Davy, F.P.F.T. **1**: 256 (1926) quoad specim. Transv. & Swazil.

Like var. *descampsii* but having on the leaves and stems a proportion of scattered simple setose hairs.

N. Rhodesia. S: Victoria Falls, Kandahar I., fl. 11.x.1911, *Rogers* 5472 (BM; K). **S. Rhodesia.** C: Marandellas, Coquetdale, fl. xi.1931, *Myres* 40 (K). E: Inyanga, Mare R., fl. 21.x.1946, *Wild* 1448 (K; SRGH).

Also in the Transvaal and Swaziland. Ecologically vars. *hirsuta* and *descampsii* behave in the same way as var. *welwitschii*.

7. **Triumfetta heliocarpa** K. Schum. in Engl., Bot. Jahrb. **15**: 131 (1892).—Brenan in Mem. N.Y.Bot. Gard. **8**, 3: 228 (1953). Type from the Congo (Lulu R).

Triumfetta mastersii var. *heliocarpa* (K. Schum.) Sprague & Hutch. in Journ. Linn. Soc., Bot. **39**: 253 (1909).—Hutch., Botanist in S. Afr.: 483 (1946). Type as above.

Triumfetta humilis N.E.Br. in Kew Bull. **1921**: 290 (1921). Type: N. Rhodesia, Mazabuka, *Rogers* 26138 (K, holotype).

Perennial with a similar habit and general appearance to *T. welwitschii* but differing in having broader lanceolate-oblong stipules c. 8 × 5 mm.; stems and lower surfaces of leaves with a dense spreading golden-yellow tomentum and the upper surfaces dark green becoming bronzed when dry. Leaves on the whole larger than in *T. welwitschii*, i.e. up to 11 × 6 cm., more prominently veined and with densely crenate-serrate margins.

N. Rhodesia. N: Mporokoso Distr., Mweru-Wantipa, Kabwe Plain, 1000 m., fl. & fr. 14.xii.1960, *Richards* 13696 (K). W: Bwana Mkubwa, fl. 31.x.1907, *Kassner* 2272 (BM). C: Lusaka, Mt. Makulu, fl. 4.iv.1955, *E.M. & W.* 1400 (BM; LISC; SRGH). S: Mazabuka, fl. vii.1909, *Rogers* 8329 (BM; K).

Also in the Congo. Like *T. welwitschii*, a species of *Brachystegia* woodlands and open grassy areas. Flowers as a rule just before the rains but only expands its leaves fully during the rainy season.

A few specimens, *Fanshawe* 1524 (K) from Ndola, *Macaulay* 321 (K) from Mumbwa, and *Rogers* 8529 (K; SRGH) from Chilanga are perhaps hybrids with *T. welwitschii* since they are intermediate between the two species. On the other hand it must be admitted that these specimens may be evidence for reducing *T. heliocarpa* to the rank of a variety of *T. welwitschii* as was done by Sprague and Hutchinson. This must, for the time being, remain a matter of opinion but *T. heliocarpa* does seem so distinct in the field that Brenan's opinion is followed here and the species considered a good one.

8. **Triumfetta dekindtiana** Engl., Bot. Jahrb. **39**: 580 (1907).—Sprague & Hutch. in Journ. Linn. Soc., Bot. **39**: 254 (1909).—Bak. f. in Journ. Linn. Soc., Bot. **40**: 33 (1911).—R.E.Fr., Wiss. Ergebn. Schwed. Rhod.-Kongo-Exped. **1**: 142 (1914).—Eyles in Trans. Roy. Soc. S. Afr. **5**: 411 (1916).—Weim. in Bot. Notis. **1936**: 48 (1936).—Hutch., Botanist in S. Afr.: 485 (1946).—Brenan, T.T.C.L.: 622 (1949).—Exell & Mendonça, C.F.A. **1**, 2: 230 (1951).—White, F.F.N.R.: 240 (1962). TAB. **5** fig. C. Type from Angola.

Small shrub up to 1·3 m. tall with a long tap-root; young branches greyish-brown, tomentellous, becoming scurfy pubescent and later glabrescent and reddish-brown. Leaf-lamina 1–6·5 × 0·5–2·5 cm., ovate, ovate-oblong or oblong-lanceolate, acute, obtuse or rounded at the apex, margins ± irregularly serrate, rounded at the base, greyish- or greyish-brown-tomentellous on both sides; petiole 5–15 mm. long, tomentellous; stipules 2–5 mm. long, linear, tomentellous. Inflorescences of small cymes crowded at the nodes; cymes 1–3 or rarely more-flowered, with a tomentellous peduncle 5–8 mm. long; pedicels similar, 1–3 mm. long; bracts at the base of the pedicels 2–3 mm. long, forming an involucre, tomentellous, subulate to lanceolate. Sepals 6–9 mm. long, c. 1 mm. wide at the base, 0·5 mm. wide above, linear, reflexed just above the base, tomentellous outside, with a horn 0·5 mm. long almost at the apex. Petals yellow, 4–5 × 1–1·5 mm., oblong-oblanceolate, apex rounded or retuse, base ciliate. Androgynophore c. 0·5 mm. long, glands obovate; annulus ciliate, reflexed. Stamens 9–10; filaments c. 6 mm. long. Ovary depressed-globose, 4-locular, tomentellous. Fruit up to 16 mm. in diam. including the aculei; aculei very numerous, slender, stellately plumose, with c. 6–8 terminal setae.

N. Rhodesia. B: Balovale, fl. & fr. 18.iv.1954, *Gilges* 336 (K; PRE; SRGH). N: Mpika, fl. 4.ii.1955, *Fanshawe* 1970 (K). W: Bwana Mkubwa, fr. viii.1911, *Fries* 331 (UPS). C: Mkushi, Fiwila, fl. 1932, *Hewitt* 41 (BM). S: 27 km. SE. of Choma, fr. 11.vii.1930, *Hutchinson & Gillett* 3535 (BM; K; SRGH). **S. Rhodesia.** W: Shangani, fl, iii.1955. *Goldsmith* 107/55 (K; SRGH). C: Salisbury, fl. ii.1920, *Eyles* 2066 (K; PRE; SRGH). E: Inyanga, fl. 6.ii.1931, *Norlindh & Weimarck* 4891 (BM; LD). S: Belingwe, fl. 25.ii.1931, *Norlindh & Weimarck* 5123 (BM; LD). **Mozambique.** N: Mandimba, fr. 14.v.1948, *Pedro & Pedrógão* 3410 (LMJ; SRGH). Z: between Ile & Gurué, fl. & fr. 5.iv.1943, *Torre* 5070 (LISC; SRGH).

Also in Angola and Tanganyika. Fairly common species in *Brachystegia* woodlands.

9. **Triumfetta setulosa** Mast. in Oliv., F.T.A. **1**: 259 (1868).—Sprague & Hutch. in Journ. Linn. Soc., Bot. **39**: 256 (1909).—Exell & Mendonça, C.F.A., **1**, 2: 231 (1951). Type from Angola.

Small bushy annual herb up to 1 m. tall; branchlets yellowish-tomentose when young, stellate-pubescent later. Leaf-lamina 2·5–5 × 1–3 cm., lanceolate to ovate-lanceolate, acute at the apex, margin serrate, rounded at the base, sometimes irregularly so, stellate-pubescent above or glabrescent with stellate hairs of different sizes or with some longer simple hairs, greyish-tomentose below; petiole up to 6 mm. long, densely pubescent; stipules 2–3 mm. long, subulate, pubescent. Inflorescences leafy, with 2–5 cymes per node; peduncles of cymes 1–3 mm. long, slender, pubescent; pedicels similar, 0–2 mm. long; bracts 2–3 mm. long, subulate, ciliolate. Sepals reddish-brown, 6–7 mm. long, stellate-pubescent outside but almost glabrous towards the base; subapical horn setulose with a few bristles. Petals yellow, c. 5 mm. long, linear-oblanceolate, ciliate just above the base. Androgynophore c. 0·25 mm. long, with glands; annulus very finely ciliolate. Stamens 10–12, c. 5 mm. long. Ovary subglobose, densely pubescent, 4-locular. Fruit 7–8 mm. in diam. including the aculei; aculei stellately pubescent with 2–6 spreading setae at the apex.

N. Rhodesia. W: Kitwe, fl. 19.ii.1959, *Mutimushi* 43 (K). C: 10 km. E. of Lusaka, fl. & fr. 22.ii.1956, *King* 325 (K).

Also in Angola, the Congo and Uganda. Species of deciduous woodland.

10. **Triumfetta angolensis** Sprague & Hutch. in Journ. Linn. Soc., Bot. **39**: 256 (1909).—Eyles in Trans. Roy. Soc. S. Afr. **5**: 411 (1916).—Weim. in Bot. Notis. **1936**: 49 (1936).—Exell & Mendonça, C.F.A. **1**, 2: 231 (1951). Type from Angola.
Triumfetta sonderi sensu Eyles in Trans. Roy. Soc. S. Afr. **5**: 412 (1916).

Erect or rather lax perennial herb c. 0·6 m. tall, branching from the base; stems stellate-pilose. Leaf-lamina up to 6 × 3 cm., elliptic, ovate-elliptic or lanceolate-oblong, apex subacute or rounded, margin serrate or crenate-serrate, base subcordate or rounded or broadly cuneate, both surfaces sparsely stellate-pilose with some simple hairs, a little more densely so below, nerves rather raised beneath; petiole up to 1·3 cm., stellate-pilose; stipules c. 3 mm. long, subulate, setose-pilose. Inflorescences leafy, of c. 2–5 small cymes per node; peduncles of cymes c. 3 mm. long, pubescent; pedicels similar, c. three together, 2–3 mm. long, pilose; bracts up to 2·5 mm. long, linear-lanceolate, forming an involucre, setose-pilose. Sepals reddish-brown, 5 mm. long, linear, stellate-pubescent outside, subapical horn 0·5 mm. long, bearing simple setose hairs. Petals yellow, c. 5 mm. long, oblanceolate, rounded at the apex, ciliate just above the base. Androgynophore 0·5 mm. long, with transversely elliptic-oblong glands; annulus with a ciliate margin. Stamens 9–10; filaments c. 5 mm. long. Ovary subglobose, 4-locular. Fruit c. 8 mm. in diam. including the aculei; aculei broadening towards the base, shortly pilose, with 2–5 spreading setae at the apex.

N. Rhodesia. B: near Kabompo pontoon, fr. 23.iii.1961, *Drummond & Rutherford-Smith* 7291 (K; PRE; SRGH). N: Abercorn, fl. & fr. 9.iii.1952, *Richards* 1056 (K). W: Ndola, fl. 5.ii.1954, *Fanshawe* 779 (K). E: Chadiza, fl. i.xii.1958, *Robson* 802 (K; SRGH). S: Livingstone, fl. & fr. 12.ii.1956, *Gilges* 589 (K; SRGH). **S. Rhodesia.** N: Lomagundi Distr., Trelawney, fl. & fr. 22.ii.1943, *Jack* 117 (SRGH). W: Victoria Falls, fl. 9.ii.1912, *Rogers* 5707 (K). C: Salisbury, fl. ii.1918, *Eyles* 927 (BM; K; SRGH). E: Inyanga, fl. 21.i.1931, *Norlindh & Weimarck* 4541 (BM; LD). **Nyasaland.** C: Dzalanyama Forest Reserve, fl. 9.ii.1959, *Robson* 1534 (K; SRGH). **Mozambique.** MS: without precise locality, fl. 6.ii.1948, *Barbosa* 972 (LISC; SRGH).
Also in Angola and the Transvaal. Most commonly in partial shade and often by rivers.

11. **Triumfetta paradoxa** (Welw. ex Hiern) Sprague & Hutch. in Journ. Linn. Soc., Bot. **39**: 257 (1909).—Exell & Mendonça, C.F.A. **1**, 2: 232 (1951). Type from Angola.
Triumfetta setulosa var.? *paradoxa* Welw. ex Hiern, Cat. Afr. Pl. Welw. **1**: 100 (1896). Type as above.

Very similar to *T. angolensis* but differs in that it is an annual and strictly erect herb up to c. 60 cm. tall with patent simple hairs on the stems, with its sepals quite glabrous towards the base and with black capsules whose aculei are only very sparsely pubescent or glabrescent and 3 mm. long rather than 1·5–2 mm. long.

N. Rhodesia. N: Abercorn, Lunzua Falls, fr., *Nash* 139 (BM).
Also in Angola and the Congo. Open woodland.

12. **Triumfetta reticulata** Wild in Bol. Soc. Brot., Sér. 2, **31**: 88 (1957). Type: N. Rhodesia, Ndola, *Fanshawe* 21 (K, holotype; SRGH).

Small shrub c. 0·6 m. tall; branches densely ferruginously pubescent with appressed-stellate hairs which are somewhat tubercular at the base. Leaf-lamina 3–12 × 0·8–5 cm., lanceolate to narrowly ovate or narrowly trapeziform, apex acute, margin irregularly serrate, base cuneate or narrowly truncate, appressed-stellate-tomentose above with impressed nerves, whitish-tomentose below with the nerves raised; petiole up to 2·5 cm. long, shortly tomentose. Inflorescences leafy with up to 5 cymes per node; peduncles of cymes up to 4 mm. long, densely pubescent; pedicels 2–3 together, c. 2 mm. long; bracts 2 mm. long, narrowly lanceolate, ciliate. Sepals 6–7 mm. long, linear, 3-nerved, subacute at the apex with a very short subapical horn c. 0·25 mm. long, setulose-pubescent outside, particularly towards the apex. Petals yellow, 4–4·5 × 0·75 mm., linear-oblanceolate, obtuse at the apex, ciliate just above the base. Androgynophore less than 0·5 mm. tall, with subquadrate glands; annulus with a shortly ciliate margin. Stamens 8–9, c. 5 mm. long. Ovary depressed-globose, 4-locular; style 4·5 mm. long, glabrous. Fruit (not quite mature) c. 7·5 mm. in diam. including the aculei; aculei numerous, minutely and sparsely puberulous, widened below, apex with 1–3 forward-pointing setae.

N. Rhodesia. W: Ndola, fl. & fr. 20.v.1953, *Fanshawe* 21 (K; SRGH). Known so far only from the type collection. Waste places and woodlands.

13. **Triumfetta intermedia** De Wild. in Ann. Mus. Congo, Sér. 5, **1**: 56 (1903).—Sprague & Hutch. in Journ. Linn. Soc., Bot. **39**: 258 (1909). Type from the Congo.

Perennial herb or small shrub with slender stellate-pubescent stems becoming glabrous later. Leaf-lamina 4–8 × 0·8–2 cm., oblong-lanceolate, acute to subacute at the apex, margin serrate-denticulate, the teeth rather widely spaced (3–6 mm. apart), broadly cuneate at the base, upper surface with a thinly scattered simple setulose pubescence, lower surface with a thinly scattered simple and stellate-setulose pubescence, veins rather prominent; petiole 4–15 mm. long, stellate-pubescent; stipules 4–5 mm. long, subulate, setulose-pubescent. Inflorescences leafy, of small cymes up to 7 at a node; peduncles of cymes 3–4 mm. long, pubescent; pedicels similar, 1–3 together, 1–2 mm. long; bracts 2 mm. long, subulate, setulose. Sepals 4–5 mm. long, linear, glabrous towards the base, sparsely stellate-pilose towards the apex; subapical horn 0·5 mm. long, terminated by a single seta. Petals yellow, 4–4·5 mm. long, oblanceolate, apex rounded, sparsely pilose just above the base. Androgynophore 0·3 mm. tall, with transversely oblong glands; annulus very shortly ciliate. Stamens 10; filaments 3–4 mm. long. Ovary subglobose, 4-locular; style 3 mm. long. Fruit 4–5 mm. in diam. (hardly mature), globose, with many aculei which are very sparsely setulose-pubescent and with 1–4 forward-pointing terminal setae.

Nyasaland. S: without precise locality, fl. & fr., *Buchanan* 216 & 614 (BM; K).

Also known from Kisantu in the Congo and from Southern Tanganyika.

Very little is known of this species and it badly needs re-collecting with mature fruits and more copious collectors' data.

14. **Triumfetta tomentosa** Boj. [Hort. Maurit.: 43 (1837) *nom. nud.*] in Bouton, Douz. Rapp. Ann. Maur.: 19 (1842).—Mast. in Oliv., F.T.A. **1**: 258 (1868).—Sprague & Hutch. in Journ. Linn. Soc., Bot. **39**: 260, t. 17 fig. 5 (1909).—Brenan, T.T.C.L.: 622 (1949).—Exell & Mendonça, C.F.A. **1**, 2: 232 (1951).—Keay, F.W.T.A. ed. 2: **1**, 2: 309 (1958).—White, F.F.N.R.: 240 (1962). TAB. **5** fig. D. Syntypes from Mauritius, cult. ex Kenya, Mombasa I.

Small shrub occasionally up to 3 m. tall; branches with a brown woolly indumentum. Leaf-lamina 5–12 × 2·5–7 cm., ovate or lanceolate-ovate, sometimes slightly 3-lobed, acute at the apex, margin irregularly serrate, cordate or subcordate at the base, greyish-brown-tomentose, more densely so beneath; petiole up to 6 cm. long, woolly; stipules c. 7 mm. long, linear-lanceolate, densely pubescent. Inflorescences leafy, with the leaves narrower than the cauline ones and progressively smaller upwards, of small cymes crowded at the nodes; peduncles of cymes and pedicels very short, c. 2 mm. long or less, densely pilose or tomentose; bracts 3–4 mm. long, linear, tomentose. Sepals 4·5–8·5 mm. long, linear, pubescent outside towards the base and tomentose towards the apex; subapical horn 0·2–0·5 mm. long, densely stellate-pubescent. Petals yellow, 4–7 mm. long, linear-oblanceolate, rounded at the apex, villous at the base. Androgynophore 0·5 mm. long, with suborbicular glands; annulus densely ciliate. Stamens 8–10; filaments 6–7 mm. long; ovary 4-locular, globose, densely setose. Fruit 1–1·5 cm. in diam. including the aculei; aculei numerous, pilose with simple hairs or almost glabrous and with the terminal setae single, straight or slightly curved.

N. Rhodesia. N: Abercorn, fr. 6.v.1952, *Richards* 1642 (K). W: Ndola, fr. 5.v.1954, *Fanshawe* 1167 (K; SRGH). C: Chilanga, fl. & fr. 7.ix.1929, *Sandwith* 146 (K; SRGH). S: Mapanza Mission, fl. 18.ii.1954, *Robinson* 538 (K). **S. Rhodesia.** N: Urungwe, Karoi Experimental Farm, fl. & fr. 5.iii.1947, *Wild* 1853 (SRGH). C: Marandellas, Ruzawi, fr. 20.iv.1924, *Eyles* 7525 (SRGH). E: Melsetter, fr. 24.ii.1907, *Johnson* 187 (K). **Nyasaland.** C: Dedza, fr. 18.v.1960, *Adlard* 336 (SRGH). S: Blantyre Distr., Chota, fr. 5.xii.1898, *Cameron* 15 (K). **Mozambique.** N: Massangulo, fr. 15.v.1948, *Pedro & Pedrógão* 3531 (LMJ; SRGH). Z: Mocuba, Namagoa, fl. & fr. 1.vii.1949, *Faulkner* 456 (K; PRE; SRGH). T: Missale, fl. ii.1897, *Nicholson* (K).

Throughout tropical Africa and also in southern tropical America. Often in fallows or old native cultivations.

This species is very difficult to distinguish from some varieties of *T. pilosa* Roth unless it is in fruit. *Norlindh & Weimarck* 4510 (BM; LD) collected at Inyanga, S. Rhodesia, is an example of this. Placed by Weimarck (in Bot. Notis. **1936**: 49 (1936)) under *T. tomentosa*, it is a young flowering specimen and may belong here; but I would place it under *T. pilosa* var. *nyasana*.

15. **Triumfetta glechomoides** Welw. ex Mast. in Oliv., F.T.A. **1**: 258 (1868).—Sprague & Hutch. in Journ. Linn. Soc., Bot. **39**: 261 (1909).—Exell & Mendonça, C.F.A. **1**, 2: 233 (1951). Type from Angola.

Prostrate perennial with a woody rootstock; stems up to 1 m. long, stellate-pubescent. Leaf-lamina 3·5–7 × 2·5–5·5 cm., ovate to very broadly ovate or sub-circular, obtuse or rounded at the apex, margin crenate-serrate, cordate and 7–9-nerved at the base, sparingly hairy with simple appressed-setulose hairs above, scabrous or scaberulous below with mostly stellate hairs; petiole up to 1·5 cm. long, stellate-pubescent; stipules up to 0·8 cm. long, linear-lanceolate, setulose. Inflorescences at the nodes of the creeping stems or on erect flowering branches with reduced leaves up to 1·5 cm. long; cymes 2–3 per node; peduncles of cymes up to 5 mm. long, sparsely pubescent; pedicels similar, 2–3 together, c. 5 mm. long; bracts up to 2·5 mm. long, forming a small involucre, lanceolate, setulose at least on the margins. Sepals c. 9 mm. long, ciliate at the base, linear, glabrous in the middle, setulose with simple or bifurcate hairs in the upper half; subapical horn up to 0·5 mm. long, usually with a single apical bristle. Petals yellow, almost as long as the sepals, obovate-oblong, ciliate at the base. Androgynophore 0·5 mm. tall, glands orbicular; annulus with a ciliate margin. Stamens 10, as long as the petals. Ovary globose, setulose, 4-locular. Fruit 10 mm. in diam. including the aculei, pink, glabrous; aculei short, 1·5 mm. long with a single terminal seta or sometimes 2–3 setae.

N. Rhodesia. N: Abercorn, fl. & fr. 3.v.1955, *Richards* 5490 (K). W: Luanshya, fr. 24.iv.1955, *Fanshawe* 2251 (K; SRGH). S: Mumbwa, fl. & fr., *Macaulay* 620 (K). **Nyasaland.** N: Mzimba, Mbawa Exp. Sta., fl. 6.iv.1955, *Jackson* 1603 (K; SRGH). S: Kirk Range, Zonze Hill, fl. & fr. 17.iii.1955, *E.M. & W.* 985 (BM; LISC; SRGH). **Mozambique.** N: Mandimba, fr. 14.v.1948, *Pedro & Pedrógão* 3418 (LMJ; SRGH).
Also in Angola. Usually on sandy soils.

16. **Triumfetta tenuipedunculata** Wild in Bol. Soc. Brot., Sér. 2, **31**: 89 (1957).—White, F.F.N.R.: 434 (1962). Type: N. Rhodesia, Abercorn Distr., *Richards* 5255 (K, holotype).

Perennial herb or small shrub c. 0·6–1·5 m. tall; with slender erect pubescent branches, pubescence of scattered dark brown appressed-stellate hairs. Leaf-lamina 3–10 × 0·5–3 cm., narrowly elliptic-lanceolate to narrowly lanceolate, apex acute, margin serrate or serrate-crenate, base cuneate or narrowly rounded, both surfaces with a sparse pubescence of simple setulose hairs and minute appressed-stellate hairs principally on the nerves, very young leaves densely stellate-tomentellous on both sides; petiole up to 15 mm. long, stellately pilose. Inflorescences leafy, of small 1–3-flowered cymes 1–3 per node of cymes; peduncles of cymes very slender, up to 2·5 cm. long, minutely stellate-pubescent along one side; pedicels similar, c. 3 mm. long; bracts c. 1·5 mm. long, linear, sparsely ciliate. Sepals 6–6·5 × 0·5 mm., linear, glabrous; subapical horn 0·2 mm. long. Petals yellow, 3·5 cm. long, linear-oblanceolate, obtuse at the apex, ciliate at the base. Androgynophore 0·25 mm. long, glands subquadrate; annulus reflexed, shortly recurved, ciliate. Stamens 8–9; filaments c. 4 mm. long. Ovary globose, 4-locular; style 4 mm. long. Fruit globose, 4 mm. in diam. (immature), glabrous; aculei numerous, widened at the base, each terminated by a single seta.

N. Rhodesia. N: Abercorn District, Chishima Falls, Luombe R., fl & fr. 31.iii.1955, *Richards* 5255 (K). W: Chifubwa Gorge, Solwezi Distr., fl. 20.iii.1961, *Drummond & Rutherford-Smith* 7114 (K; LISC; PRE; SRGH).
Known so far only from N. Rhodesia. Damp shady woodland.
An unusual species with a fruit like that of *T. glechomoides* but a completely different habit.

17. **Triumfetta rhomboidea** Jacq., Enum. Pl. Carib.: 22 (1760).—Mast. in Oliv., F.T.A. **1**: 257 (1868).—Harv. in Harv. & Sond., F.C. **1**: 227 (1860).—Ficalho, Pl. Ut. Afr. Port.: 111 (1884).—Sprague & Hutch. in Journ. Linn. Soc., Bot. **39**: 266 (1909).—Bak. f. in Journ. Linn. Soc., Bot. **40**: 33 (1911).—R.E.Fr., Wiss. Ergebn. Schwed. Rhod.-Kongo-Exped. **1**: 142 (1914).—Eyles in Trans. Roy. Soc.

S. Afr. **5**: 412 (1916).—Burtt Davy, F.P.F.T. **1**: 257 (1926).—Weim. in Bot. Notis. **1936**: 49 (1936).—Brenan, T.T.C.L.: 623 (1949).—Mendonça & Torre, Contr. Conh. Fl. Moçamb. **1**: 21 (1950).—Suesseng. & Merxm. in Proc. & Trans. Rhod. Sci. Ass. **43**: 106 (1951).—Exell & Mendonça, C.F.A. **1,** 2: 234 (1951).—Brenan in Mem. N.Y.Bot. Gard. **8,** 3: 229 (1953).—Keay, F.W.T.A. ed. 2: **1,** 2: 309 (1958).—White, F.F.N.R.: 240 (1962). Type from the West Indies.

Bartramia indica L., Sp. Pl. **1**: 389 (1753) non *Triumfetta indica* Lam. (1791). Type from Ceylon.

Triumfetta bartramia L., Syst. Nat. ed. 10, **2**: 1044 (1759) *nom. illegit.* Type as for *Bartramia indica*.

Triumfetta diversifolia E. Mey. in Drège, Zwei Pflanz.-Docum.: 227 (1823) *nom. nud.*—Eyles in Trans. Roy. Soc. S. Afr. **5**: 411 (1916).

Very polymorphic species, usually annual but there seem to be perennial forms. Stems almost glabrous to velvety or tomentose, up to c. 2 m. tall. Leaf-lamina 2·5–15 × 2–10 cm., ovate to ovate-lanceolate, acute at the apex, often 3-lobed, cordate or truncate at the base, 3–7-nerved from the base, irregularly serrate; petiole up to 5 cm. long; stipules c. 4 mm. long, linear-lanceolate, setose-pubescent. Inflorescences leafy, with the inflorescence leaves smaller and narrower than the cauline ones; cymes crowded at the nodes; peduncles of cymes and pedicels short, 1·5–3 mm. long; bracts c. 3 mm. long, linear, pubescent. Sepals 4–5 mm. long, linear, hooded towards the apex, stellate-pubescent outside or almost glabrous; subapical horn c. 0·5 mm. long, setulose-pubescent. Petals yellow, slightly shorter than the sepals, linear-oblanceolate, villous at the base. Androgynophore 0·25 mm. tall with circular glands: annulus villous on its upper margin. Stamens c. 15. Ovary 2–3-locular, closely setulose or echinulate. Fruit 4–5 mm. in diam. including the aculei, globose or ovoid-globose, its body densely tomentose; aculei uncinate at the apex, glabrous.

N. Rhodesia. N: Abercorn, fl. & fr. 4.iv.1952, *Richards* 1280 (K). W: Kitwe, fl. & fr. 18.iii.1955, *Fanshawe* 2155 (K; SRGH). C: Lusaka, fr. vi.1925, *Distr. Vet. Officer* 5 (PRE). S: 22 km. west of Pemba, fl. & fr. 21.iv.1954, *Robinson* 698 (K). **S. Rhodesia.** W: Matobo, fl. iii.1953, *Miller* 1606 (SRGH). C: Salisbury, fl. & fr. iii.1920, *Eyles* 2131 (K; PRE; SRGH). E: Chipinga Distr., Chirinda, fr. 18.v.1906, *Swynnerton* 472 (BM; K; SRGH). S: Ndanga, fr. 26.i.1949, *Wild* 2750 (K; SRGH). **Nyasaland.** N: Nyika Plateau, fl. & fr. ii.–iii.1903, *McClounie* 126 (K). C: Ncheu, fl. 12.iii.1955, *Adlard* 145 (FHO; K). S: Cholo Mt., fl. & fr. 19.ix.1946, *Brass* 17645 (K; SRGH). **Mozambique.** N: Mandimba, fl. 14.v.1948, *Pedro & Pedrógão* 3413 (LMJ; SRGH). Z: Mocuba, Namagoa, fl. & fr., *Faulkner* 353 (COI; K; PRE; SRGH). T: Zóbuè, fr. 9.v.1948, *Mendonça* 4145 (LISC). MS: Chimoio, Amatongas Missn., fl. & fr. 25.iv.1948, *Mendonça* 4048 (BM; LISC; SRGH). SS: Massinga, fl. & fr. vii.1936, *Gomes e Sousa* 1778 (COI; K). LM: Maputo, between Bela Vista and Santaca, fl. 12.iv.1949, *Myre & Balsinhas* 576 (LM; SRGH).

Known throughout the tropics and subtropics (including S. Africa). Commonly occurs as a weed of cultivation.

Extremely variable except for the fruits.

18. **Triumfetta pentandra** A. Rich. in Guill. & Perr. & Rich., Fl. Senegamb. Tent. **1**: 93, t. 19 (1831).—Mast. in Oliv., F.T.A. **1**: 255 (1868).—Sprague & Hutch. in Journ. Linn. Soc., Bot. **39**: 267, t. 17 fig. 9 (1909).—Mendonça & Torre, Contr. Conh. Fl. Moçamb. **1**: 22 (1950).—Keay, F.W.T.A. ed. 2, **1,** 2: 309 (1958). Syntypes from Senegal.

Triumfetta trichocarpa sensu Eyles in Trans. Roy. Soc. S. Afr. **5**: 412 (1916).

Annual herb very like *T. rhomboidea* in general appearance but much more sparingly hairy, never being more than stellate-pubescent, the leaves as a rule thinner-textured, the stamens usually fewer and often about 5, but the most important differences are in the fruit which is ovoid not globose, glabrous not tomentose, except for the aculei which are ascending and densely ciliate on their upper sides below their uncinate apices.

Bechuanaland Prot. N: Kwebe Hills, fl. & fr. 3.iii.1898, *Lugard* 206 (K). SE: Mahalapye, fl. & fr. 7.ii.1958, *de Beer* 597 (K; SRGH). **N. Rhodesia.** N: Abercorn Distr., Lake Tanganyika, Kasaba Game Reserve, st. 17.ii.1959, *McCallum-Webster* 631 (K). S: Livingstone, fl. & fr. 26.vii.1909, *Rogers* 7021 (K; SRGH). **S. Rhodesia.** N: Urungwe, 1·5 km. east of Kariba, fl. & fr. 3.iii.1956, *Goodier* 46 (K; SRGH). W: Bulalima-Mangwe, fl. & fr. 6.v.1942, *Feiertag* (SRGH). S: Gwanda, fl. & fr. ii.1955, *Plowes* 1761 (K; SRGH). **Nyasaland.** N: Nyika Plateau, fl. ii.1903, *McClounie* 161 (K). S: Port Herald, between

Muona R. and Shire R., fl. 20.iii.1960, *Phipps* 2584 (K; SRGH). **Mozambique.** N: Nampula, fl. & fr. 2.iv.1936, *Torre* 696 (COI). Z: " Zambeziland ", fl. & fr. 12.xii.1865, *Kirk* (K). T: Boroma, fl. & fr. ii.1891, *Menyharth* 523 (K). MS: Cheringoma, Inhamitanga, fl. 5.iv.1945, *Simão* 331 (LISC; LM). SS: Inhambane, fl. & fr. iv.1926, *Gomes e Sousa* 1723 (COI). LM: Rikatla, fl. & fr. iv.1919, *Junod* (PRE; SRGH).

Widely spread through tropical Africa, the Transvaal, SW. Africa, India and Formosa. Weed or plant of waste places.

19. **Triumfetta annua** L., Mant. Pl. **1**: 73 (1767).—Mast. in Oliv., F.T.A. **1**: 256 (1868).—Sprague & Hutch. in Journ. Linn. Soc., Bot. **39**: 268, t. 17 fig. 10 (1909).—Bak. f., in Journ. Linn. Soc., Bot. **40**: 32 (1911).—Eyles in Trans. Roy. Soc. S. Afr. **5**: 411 (1916).—Burtt Davy, F.P.F.T. **1**: 257 (1926).—Weim. in Bot. Notis. **1936**: 50 (1936).—Exell & Mendonça, C.F.A. **1, 2**: 236 (1951). Type a figure in Mill., Fig. Pl. **2**: t. 298 (1760) drawn from a plant cultivated in London from seed collected in India.

Annual 10–60 cm. tall; young stems with a sparse pubescence of simple hairs and a short crisped pubescence in a single line or two opposite lines down the stem. Leaf-lamina 3–12 × 2–7 cm., ovate, becoming narrower upwards, acutely acuminate at the apex, rounded at the base, coarsely serrate or crenate, membranous, sparingly setulose-pubescent above and below; petiole up to 7·5 cm. long, with a line of crisped pubescence on the upper side; stipules c. 5 mm. long, subulate or lanceolate setulose. Inflorescences of small cymes 1–5 together clustered at the nodes; peduncles of cymes usually very short, up to 2 mm. long, puberulous along one side; pedicels similar, 0·5 to 2 mm. long; bracts c. 2 mm. long, lanceolate to subulate, setulose-pubescent at least on the margins. Sepals 2·5–5 mm. long, linear, sparingly setulose-pubescent particularly towards the apex or almost glabrous; subapical horn up to 0·5 mm. long, with one or several setulose hairs. Petals yellow, slightly shorter than the sepals, oblanceolate or linear-oblanceolate, minutely ciliate at the base. Androgynophore very short, c. 0·2 mm. long, glands subquadrate; annulus with its margin reflexed and sparingly ciliate. Stamens 4–12. Ovary depressed-globose, echinulate, 4-locular. Fruit c. 15 mm. in diam. including the aculei, depressed-globose, glabrous or with some weak hairs on the body of the fruit and the base of the aculei, fruit body deeply reticulate; aculei c. 60 per fruit, 3–5 mm. long, broadening at the base, uncinate and slightly flexuous particularly when young.

Forma **annua**

Fruits and aculei quite glabrous.

N. Rhodesia. B: Shangombo, fr. 7.viii.1952, *Codd* 7426 (K; PRE). N: Abercorn, fl. 4.iii.1952, *Richards* 923 (K). W: Mufulira, fl. & fr. 8.ii.1948, *Cruse* 186 (K). S: Mumbwa, fr., *Macaulay* 714 (K). **S. Rhodesia.** N: Urungwe, Karoi Exp. Farm, fl. 5.iii.1947, *Wild* 1720 (K; SRGH). W: Shangani-Bubi Districts, Gwampa For. Res., fr. v.1956, *Goldsmith* 95/56 (SRGH). C: Salisbury, fl. 26.ii.1927, *Eyles* 4715 (K; SRGH). E: Chirinda, fl. & fr. 26.v.1906, *Swynnerton* 473 (BM; K; SRGH). S: Belingwe, fl. 25.ii.1931, *Norlindh & Weimarck* 5131 (BM; LD). **Nyasaland.** C: Dedza, Mua-Livulezi For. Res., fl. 19.iii.1955, *E.M. & W.* 1049 (BM; LISC; SRGH). **Mozambique.** MS: Inhamatinga, fl. & fr. 2.vii.1947, *Simão* 1334 (LM; SRGH). LM: Namaacha, fl. & fr. 21.v.1957, *Carvalho* 222 (BM; LM).

Widely distributed in tropical Africa, Transvaal, SW. Africa, Madagascar, and also in India, China and Malaya. In rather shady situations at forest edges and in bushy shady places. Often appearing as a ruderal or weed of cultivation.

Forma **piligera** Sprague & Hutch. in Journ. Linn. Soc., Bot. **39**: 268 (1909). Syntypes from Natal and Madagascar.

Triumfetta trichocarpa sensu Mendonça & Torre, Contr. Conhec. Fl. Moçamb. **1**: 22 (1950).

Forms with the body and base of the aculei inconspicuously pilose with long weak hairs.

N. Rhodesia. N: Abercorn, Kalambo Falls, fl. & fr. 29.iii.1955, *E.M. & W.* 1287 (BM; LISC; SRGH). W: Mufulira, fl. & fr. 8.ii.1948, *Cruse* 186 (K). **S. Rhodesia.** N: Trelawney, fl. & fr. 27.iii.1943, *Jack* 149 (SRGH). W: Matopos, fr. iv.1955, *Miller* 2736 (K; SRGH). **Mozambique.** MS: Vila Machado, Serra do Chiluvo, fr. 16.iv.1948, *Mendonça* 3977 (BM; LISC).

Also in Natal, Madagascar and E. Africa. Does not appear to differ in habit and ecology from forma *annua*.

20. **Triumfetta trichocarpa** Hochst. ex A. Rich., Tent. Fl. Abyss. **1**: 84 (1847).—Mast. in Oliv., F.T.A. **1**: 259 (1868).—Sprague & Hutch. in Journ. Linn. Soc., Bot. **39**: 269 (1909). Type from Ethiopia.

Annual up to 1·5 m. tall, very like *T. annua* but differing in its fruits, which have aculei about 120 in number not so conspicuously broadened at the base and densely setose-ciliate except in the upper third. The fruit body is also puberulous and not so deeply honeycombed.

N. Rhodesia. N: Isoka, fl. & fr. 23.xii.1960, *Lawton* 697 (K). W: Ndola, fl. & fr. 22.iii.1954, *Fanshawe* 1018 (K). C: 11 km. E. of Lusaka, fl. & fr. 23.iii.1956, *King* 365 (K). **Nyasaland.** C: Lilongwe, fl. & fr. 31.iii.1955, *Jackson* 1543 (K; SRGH). S: Shire Highlands, fl. & fr., *Buchanan* 379 (K). **Mozambique.** MS: Chimoio, Vanduzi, fr. 29.iii.1948, *Garcia* 801 (LISC; SRGH).
Also northwards through East Africa to Ethiopia.

21. **Triumfetta cordifolia** A. Rich. in Guill., Perr. & Rich., Fl. Senegamb. Tent. **1**: 91, t. 18 (1831).—Sprague in Kew Bull. **1908**: 231 (1908).—Sprague & Hutch. in Journ. Linn. Soc., Bot. **39**: 270 (1909).—Keay, F.W.T.A. ed. 2, **1**, 2: 310 (1958).—White, F.F.N.R.: 240 (1962). Type from Cape Verde.

Shrub 1–2 m. tall; stems shortly stellate-pubescent or glabrescent, often rather angled or 4-sided. Leaf-lamina 5–14 × 2–13 cm., ovate, acuminate at the apex, cordate at the base, 3–5-lobed, irregularly serrate, stellate-puberulous on the upper surface, sparingly stellate-pubescent on the lower or densely stellate-tomentellous on both sides, 5–7-nerved at the base, or in some varieties oblong-lanceolate and unlobed; petiole up to c. 9 cm. long; stipules c. 1 cm. long, subulate, pubescent to tomentellous. Inflorescences large, scarcely leafy; cymes very numerous and dense at the nodes; peduncles of cymes c. 5 mm. long; pedicels 3–5 together; bracts 2–3 mm. long, linear. Sepals up to 11 mm. long, linear, greyish-tomentellous to almost glabrous outside; subapical horn very short as a rule but occasionally up to 0·75 mm. long. Petals yellow, a little shorter than the sepals, linear, subacute or obtuse at the apex, ciliate at the base. Androgynophore c. 0·4 mm. tall, glands rotund; annulus with a shortly villous upper margin. Stamens 10–12. Ovary 4–5-locular, depressed-globose, minutely echinulate. Fruit c. 10 mm. or a little more in diam. (including the aculei); aculei uncinate at the tip, pilose or occasionally glabrous.

Var. **tomentosa** Sprague in Kew Bull. **1908**: 232 (1908).—Sprague & Hutch. in Journ. Linn. Soc., Bot. **39**: 371 (1909). Syntypes: a cultivated specimen grown from seed from Angola, and specimens from Angola and the Cameroons.

Leaves subcircular, 3-lobed or shortly 5-lobed, densely stellate-tomentellous on both sides.

N. Rhodesia. N: Abercorn, D'hulmuti, fl. 6.v.1955, *Richards* 5572 (K).
Throughout tropical West Africa, in Angola and southern Tanganyika. In our area restricted to the fairly high-rainfall areas at the northern extremity of N. Rhodesia around Abercorn but there apparently fairly common.

22. **Triumfetta pilosa** Roth, Nov. Pl. Sp.: 223 (1821).—Mast. in Oliv., F.T.A. **1**: 257 (1868).—Harv. in Harv. & Sond., F.C. **1**: 227 (1860).—Sprague & Hutch. in Journ. Linn. Soc., Bot. **39**: 273, t. 17 fig. 12 (1909). Type from India.

Small shrub c. 1·3 m. tall; stems often quadrangular, sparingly stellate-pubescent or densely tomentose, hair-bases often tubercled. Leaf-lamina 5–12 × 3–7 cm., ovate to oblong-lanceolate, acutely acuminate or acute at the apex, margin coarsely serrate, rounded to cordate at the base, puberulous or pubescent or densely tomentose especially below; petiole up to c. 5 cm. long; stipules up to 0·8 cm. long, subulate. Inflorescences leafy, of small cymes crowded at the nodes; peduncles of cymes and pedicels very short, up to 5 mm. long, but often less; bracts 2–4 mm. long, linear or lanceolate-linear, setulose-pubescent. Sepals 7–10 mm. long, linear, pubescent or tomentose; subapical horn very short, c. 1 mm. long. Petals yellow, slightly shorter than the sepals, linear to oblanceolate-linear, ciliate at the base. Androgynophore up to 0·75 mm. tall, glands orbicular; annulus densely hirsute above. Stamens 8–10. Ovary globose, echinulate, 3–4-locular. Fruit 1·5–2·7 cm. in diam. including the aculei; aculei 0·5–1 cm. long, glabrous or pilose, uncinate at the apex.

Var. **tomentosa** [Szyszyl., Polypet. Thalam. Rehm.: 58 (1887) *nom. nud.*] ex Sprague & Hutch. in Journ. Linn. Soc., Bot. **39**: 273 (1909).—Burtt Davy, F.P.F.T. **1**: 257 (1926).—Mendonça & Torre, Contr. Conhec. Fl. Moçamb. **1**: 23 (1950). Syntypes from Natal.

Triumfetta pilosa sensu Hutch., Botanist in S. Afr.: 464 (1946).

Stems densely brown-tomentose. Leaf-lamina 6–12 × 3–7·5 cm., usually ovate, densely and shortly stellate-tomentose above, more densely so below. Buds tomentose.

S. Rhodesia. W: Matobo Distr., Besna Kobila, fl. & fr. i.1954, *Miller* 2092 (SRGH). C: Gwelo, fl. ii.1957, *Evans* 1 (K; SRGH). E: Chipinga, Upper Buzi R., fr. 19.iv.1907, *Swynnerton* 2105 (BM; K; SRGH). S: kopjes near Lundi R., fr. 30.vi.1930, *Hutchinson & Gillett* 3293 (K). **Nyasaland.** S: Ncheu Distr., Lower Kirk Range, fl. & fr. 17.iii.1955, *E.M. & W.* 946 (BM; LISC; SRGH). **Mozambique.** Z: between Milange and Mocuba, fl. 25.v.1949, *Barbosa & Carvalho* 2800 (LM; SRGH). MS: between Vila Pery and Munhinga, fr. 3.vi.1949, *Pedro & Pedrógão* 6139 (LMJ; SRGH).

Also in the Transvaal and Natal.

Stems used for cordage.

Var. **nyasana** Sprague & Hutch. in Journ. Linn. Soc., Bot. **39**: 273 (1909).—Bak. f. in Journ. Linn. Soc., Bot. **40**: 32 (1911).—Brenan in Mem. N.Y.Bot. Gard. **8**, 3: 229 (1953). Syntypes: Kenya, Nairobi; Nyasaland, Masuku Plateau, *Whyte* (K); Namasi, *Cameron* 21 (K); Shire Highlands, *Buchanan* 380, 381 (K).

Triumfetta morrumbalana De Wild., Pl. Nov. Herb. Hort. Then. **5**: 167, t. 37 (1905).—Bak. f. in Journ. Linn. Soc., Bot. **40**: 33 (1911). Type a cultivated specimen, *Luja* 428 (BR, holotype) grown from seed from Mozambique (Morrumbala).

Triumfetta pilosa sensu Eyles in Trans. Roy. Soc. S. Afr. **5**: 412 (1916).

Differs from var. *tomentosa* in having a sparser and more strigose pubescence on the upper leaf-surface, the lower surface with a dense pubescence but not tomentose, and in the leaf-lamina being 9–18 × 2·5–7 cm., lanceolate rather than ovate even in the cauline leaves.

N. Rhodesia. N: Abercorn, fl. iii.1934, *Gamwell* 197 (BM). **S. Rhodesia.** E: Chipinga Distr., Chirinda, Gungunyana Farm, fl. ii.1907, *Johnson* 120 (K). **Nyasaland.** N: Masuku Plateau, near Karonga, *Whyte* (K). C: Kota Kota Distr., Nchisi Mt., fr. 29.vii.1946, *Brass* 17019 (K; SRGH). S: Namasi, fl. & fr. 5.xii.1898, *Cameron* 21 (K). **Mozambique.** N: Vila Cabral, fl. & fr. 24.v.1934, *Torre* 112 (COI; LISC). T: between Vila Vasco da Gama and Fíngoè, fr. 27.vi.1949, *Barbosa & Carvalho* 3346 (LM; SRGH). MS: Vumba Mt., fl. & fr. 18.iii.1948, *Garcia* 658 (BM; LISC).

Also in Kenya and Tanganyika. According to Brass a plant of forest margins.

Var. **glabrescens** Sprague & Hutch. in Journ. Linn. Soc., Bot. **39**: 274 (1909).—Brenan, T.T.C.L.: 623 (1949).—Syntypes: Tanganyika; Nyasaland, without precise locality, *Buchanan* 726 (K); between Mpata and Tanganyika Plateau, *Whyte* (K); Nyika Plateau, *Whyte*; Mt. Chiradzulu, *Whyte* (K).

Leaves and buds only sparsely pilose, and leaves narrower than in var. *nyasana*, 8–14 × 1·5–3·5 cm.

N. Rhodesia. N: Abercorn, Shamba Milimani, fl. & fr. 27.v.1952, *Richards* 1841 (K). **S. Rhodesia.** E: Vumba Mt., fr. 27.vi.1948, *Fisher* 1623 (PRE). **Nyasaland.** N: Nyika Mts., between Mpata and commencement of Tanganyika Plateau, fr. vii.1896, *Whyte* (K). S: Mt. Chiradzulu, fr. *Whyte* (K). **Mozambique.** Z: Quelimane, Gúruè, fl. 28.v.1937, *Torre* 1540 (COI). T: Angónia, Posto Zootécnico, fl. & fr. 12.v.1948, *Mendonça* 4174 (BM; LISC). MS: Vila de Manica, fr. 9.vi.1948, *Mendonça* 4476 (LISC).

Also in Tanganyika. Ecology probably similar to that of var. *nyasana*.

Var. **effusa** (E. Mey. ex Harv.) Wild, comb. nov. Syntypes from Natal.

Triumfetta effusa E. Mey. ex Harv. in Harv. & Sond., F.C. **1**: 228 (1860).—Sprague & Hutch. in Journ. Linn. Soc., Bot. **39**: 275 (1909).—Bak. f. in Journ. Linn. Soc., Bot. **40**: 32 (1911).—R.E.Fr., Schwed. Rhod.-Kongo-Exped. **1**: 142 (1914).—Eyles in Trans. Roy. Soc. S. Afr. **5**: 411 (1916).—Burtt Davy, F.P.F.T. **1**: 257 (1926).—Brenan in Mem. N.Y.Bot. Gard. **8**, 3: 229 (1953).—White, F.F.N.R.: 240 (1962). Syntypes as above.

Triumfetta melanocarpa Suesseng. in Proc. & Trans. Rhod. Sci. Ass. **43**: 105 (1951). Type: S. Rhodesia, Marandellas, *Dehn* 80 (M, holotype; SRGH).

Similar to var. *nyasana* but the aculei of the fruit are glabrous.

N. Rhodesia. N: Mokawe by Lake Bangweulu, fl. & fr., *Fries* 842 (UPS). S: Livingstone, fr. iv.1909, *Rogers* 7083 (K). **S. Rhodesia.** C: Marandellas, fl. & fr. 26.ii.1942, *Dehn* 80 (M, holotype; SRGH). E: Chirinda Forest, fr. 14.iv.1907, *Swynnerton* 1156 (BM; K; SRGH). **Nyasaland.** S: Mlanje Mt., fr. 18.vii.1946, *Brass* 16865 (K; SRGH). **Mozambique.** LM: Lourenço Marques, fr. 2.v.1920, *Borle* 477 (PRE).

Also from Natal and the Transvaal. Forest margins or river banks mostly in the higher-rainfall areas.

The cortex is used in Rhodesia to make a native fibre.

Teague 231 (K) from the Odzani valley, S. Rhodesia, is intermediate between this variety and var. *nyasana* as it has a few hairs at the base of the aculei of the fruits.

The plant illustrated as *T. effusa* in Martineau, Rhod. Wild Fl.: 47, t. 16 fig. 4 (1954) is, from the size of its fruits, more likely to be *T. rhomboidea* but the illustration is not good enough to be certain.

5. SPARRMANNIA L. f.

Sparrmannia L. f., Suppl. Plant.: 41 (1781), *nom. conserv.*

Large shrubs with petiolate alternate leaves; stipules deciduous, subulate or setaceous; all parts stellate-pubescent or glabrescent, sometimes with intermixed simple hairs. Leaf-lamina 3–7-angled or -lobed, palmately nerved, crenate-dentate or crenate or serrate. Inflorescences of extra-axillary or leaf-opposed umbels on longish peduncles; bracts similar to the stipules. Sepals 4, lanceolate, deciduous. Petals 4, oblanceolate, glandless. Stamens numerous, outer ones sometimes sterile and moniliform. Androgynophore obsolete. Ovary 4–5-locular; ovules numerous; style slender, glabrous, with a 4–5-toothed stigma. Capsule globose, oblong-ovoid or ellipsoid, 4–5-valved, covered with rigid bristles.

Note: the correctness of the above spelling as against "*Sparmannia*" is explained by Brenan in Mem. N.Y.Bot. Gard. **8,** 3: 229 (1953).

Sparrmannia ricinocarpa (Eckl. & Zeyh.) Kuntze, Rev. Gen. Pl. **3,** 2: 26 (1898).—Brenan in Mem. N.Y.Bot. Gard. **8,** 3: 229 (1953).—White, F.F.N.R.: 239 (1962). TAB. **6.** Type from S. Africa (Eastern Cape).

Urena ricinocarpa Eckl. & Zeyh., Enum. Pl. Afr. Austr. Extratrop.: 37 (1835).—C. Presl, Bot. Bemerk.: 19 (1844). Type as above.

Sparrmannia palmata E. Mey. ex Harv. in Harv. & Sond., F.C. **1**: 224 (1860) *nom. illegit.*—Bak. f. in Journ. Linn. Soc., Bot. **40**: 33 (1911).—Eyles in Trans. Roy. Soc. S. Afr. **5**: 410 (1916). Type as for *S. ricinocarpa*.

Sparrmannia abyssinica var. *micrantha* Burret in Mildbr., Wiss. Ergebn. Deutsch. Zentr. Afr.-Exped. 1907–8, **2**: 494 (1910). Type from Tanganyika.

Sparrmannia ricinocarpa subsp. *micrantha* (Burret) Weim. in Svensk Bot. Tidskr. **27**: 404 & 408 (1933); in Bot. Notis. **1936**: 37 (1936).—Exell & Mendonça, C.F.A. **1,** 2: 238 (1951). Type as above.

Slender shrub or scrambler, up to 3 m. tall; stems slender, with spreading simple hairs and shorter stellate hairs. Leaf-lamina 3–13 × 1·5–10 cm., 3–7-lobed, cordate or cordate-sagittate and 5–7-nerved at the base, lobed about half-way to three-quarters, middle lobe the longest; lobes, or at least the middle one, long-acuminate, inciso-sinuate, crenate-dentate and often with secondary lobing; with stellate or simple pubescence on both surfaces or with a mixture of both; petiole up to 8 cm. long, pubescent; stipules up to 8 mm. long, setaceous, pubescent. Inflorescences of extra-axillary or leaf-opposed 6–20-flowered umbels; peduncles elongated and up to 8 cm. long, pubescent; pedicels pubescent, articulated in the upper half and more densely pubescent above the articulation; bracts up to 7 mm. long, subulate to lanceolate, pubescent. Sepals c. 10 mm. long, lorate, subacute at the apex, pubescent and green on the back, white or purplish inside. Petals white, the same length as the sepals or slightly longer, oblanceolate to oblanceolate-oblong. Stamens many (c. 50), filamentous, c. 8 mm. long, the majority with numerous nodose swellings along the filament, a few outer ones sometimes sterile. Ovary ovoid, echinulate-setulose; style 7–8 mm. long, slender, glabrous; stigma shortly 5-toothed. Capsule 2–2·5 × 1·5–2 cm. including the bristles, ellipsoid, 4–5-valved, covered with rigid bristles, sparsely stellate-pubescent; loculi opening from the apex. Seeds dark brown, 2·5 × 1·5 mm. ellipsoidally biconical.

N. Rhodesia. E: Nyika Plateau, near source of Chire R., fl. & fr. 3.v.1952, *White* 2562 (FHO). **S. Rhodesia.** C: Marandellas, Wedza, fl. & fr. iii.1955, *Davies* 986 (SRGH). E:

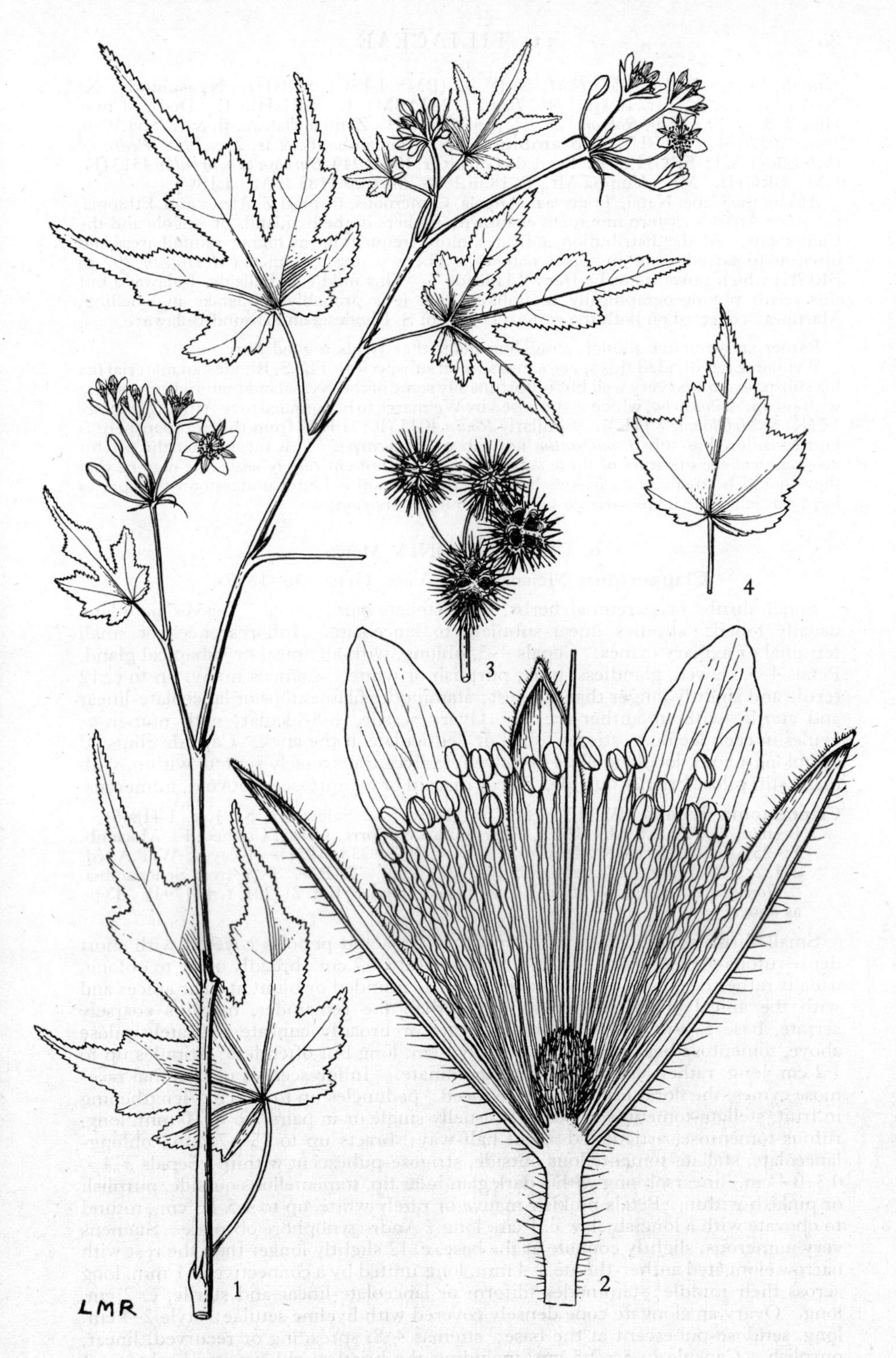

Tab. 6. SPARRMANNIA RICINOCARPA. 1, flowering branch (× ⅔) *Lawrence* A13; 2, longitudinal section of flower (× 7) *Lawrence* A13; 3, fruits (× ⅔) *Kirk* s.n.; 4, shallowly lobed leaf (× ⅔) *Eyles* 1161.

Umtali, Engwa, fl. 1.ii.1955, *E.M. & W.* 42 (BM; LISC; SRGH). **Nyasaland.** N: Nyika Plateau, fl. & fr. 16.viii.1946, *Brass* 17262 (BM; K; SRGH). C: Dedza, Ciwas Hill, fl. & fr. 19.i.1959, *Robson* 1267 (K; SRGH). S: Zomba Plateau, fl. & fr. 5.vi.1946, *Brass* 16270 (K; SRGH). **Mozambique.** N: Maniamba, fl. & fr. 29.v.1948, *Pedro & Pedrógão* (LMJ; SRGH). Z: Serra do Gúruè, fr. 18.x.1949, *Barbosa & Carvalho* 4513 (K; LM; SRGH). MS: Vumba Mt., fl. 18.iii.1948, *Barbosa* 1183 (BM; LISC).

Also in the Cape, Natal, Transvaal, Angola, Cameroons, tropical E. Africa and Ethiopia, i.e. along Africa's eastern mountain chain with outliers in the highlands of Angola and the Cameroons. As the distribution indicates, more frequent in the higher-rainfall areas and often at forest edges or in forest patches. There is one specimen (*Martineau* 297 in SRGH) which purports to be from " Hillside ". This might be Hillside, Bulawayo but this seems phytogeographically so unlikely that it is probably a mistake in labelling. Martineau collected on both the eastern border of S. Rhodesia and around Bulawayo.

Rather an attractive though straggling plant that yields a good fibre.

Weimarck has divided this species into several subspecies. The S. Rhodesian material fits his subsp. *micrantha* very well but unfortunately some of the Nyasaland material agrees best with subsp. *ricinocarpa,* which is supposed by Weimarck to be confined to S. Africa (*Lawrence* 13 (K) & *McClounie* 80 (K)). Similarly *Rodin* 4045 (K; PRE) from the Soutpansberg is best considered as subsp. *micrantha* not subsp. *ricinocarpa.* This means that the phytogeographical classification of these subspecific taxa is not entirely reliable. It may be that they should be treated as varieties but until the position is better understood the species has been left as a single variable taxon without subdivisions.

6. CLAPPERTONIA Meisn.

Clappertonia Meisn., Plant. Vasc. Gen.: 36 (1837).

Small shrubs or perennial herbs, with stellate pubescence. Leaves petiolate, usually lobed; stipules linear-subulate to lanceolate. Inflorescences of small terminal or axillary cymes. Sepals 4–5, oblong, with an apical or subapical gland. Petals 4–5, clawed, glandless, pink, purplish or white. Stamens many, up to c. 12 fertile and slightly longer than the rest; staminodes filamentous or lanceolate-linear and sterile, without anther-thecae. Ovary sessile, 4–8-locular, with numerous ovules in each loculus; stigma lobed or denticulate at the apex. Capsule ellipsoid or oblong-ovoid, loculicidally 4–8-valved; valves transversely septate within, with many stiff pilose bristles outside. Seeds discoid or compressed-obovoid, numerous.

Clappertonia ficifolia (Willd.) Decne. in Deless., Ic. Select. Pl. **5**: 1, t. 1 (1846).—Brenan, T.T.C.L.: 613 (1949).—Mendonça & Torre, Contr. Conhec. Fl. Moçamb. **1**: 25 (1950).—Exell & Mendonça, C.F.A. **1,** 2: 238 (1951).—Keay, F.W.T.A. ed. 2, **1,** 2: 310 (1958).—White, F.F.N.R.: 232 (1962). TAB. **7.** Type from Senegambia.

Honckenya ficifolia Willd. in Usteri, Delect. Opusc. Bot. **2**: 201, t. 4 (1793). Type as above.

Small shrub up to c. 2 m. tall, with the stems and petioles covered with short dense rufous-stellate hairs. Leaf-lamina up to 13 × 7 cm., broadly ovate to oblong, usually rather deeply 3–7-lobed, with the lobes rounded or blunt at their apices and with the apical one broader and larger than the remainder, margins coarsely serrate, base 5–9-nerved, cordate, truncate or broadly cuneate, stellately pilose above, tomentose beneath; petiole up to 4 cm. long but often less; stipules up to 1·2 cm. long, rather leafy, lanceolate-acuminate. Inflorescences of terminal racemose cymes, the floral leaves much reduced; peduncles up to 2·5 cm., lengthening in fruit, stellate-tomentose; pedicels usually single or in pairs, up to 10 mm. long, rufous-tomentose, articulated about half-way; bracts up to 13 × 7 mm., oblong-lanceolate, stellate-tomentellous outside, strigose-pubescent within. Sepals 3–4 × 0·3–0·4 cm., linear-oblong with a dark glandular tip, tomentellous outside, purplish or pinkish within. Petals pinkish-mauve or rarely white, up to 4·5 × 3 cm., rotund to obovate with a longish claw c. 1 cm. long. Androgynophore obsolete. Stamens very numerous, slightly connate at the base, c. 12 slightly longer than the rest with narrow elongated anther-thecae 3–4 mm. long united by a connective c. 1 mm. long across their middle; staminodes filiform or lanceolate-linear and sterile, c. 2 cm. long. Ovary an elongate cone densely covered with hyaline setulae; style 2–4 cm. long, setulose-pubescent at the base; stigmas 4–8, spreading or recurved, linear, purplish. Capsule c. 5 × 2·5 cm. including the bristles, oblong-ovoid, obtuse at both ends, pubescent, covered outside with numerous stiff ciliate bristles, each terminated by a deciduous hyaline straight or bent seta. Seeds c. 2·8 × 1·6 mm., discoid, with a pale brown often rather loose testa.

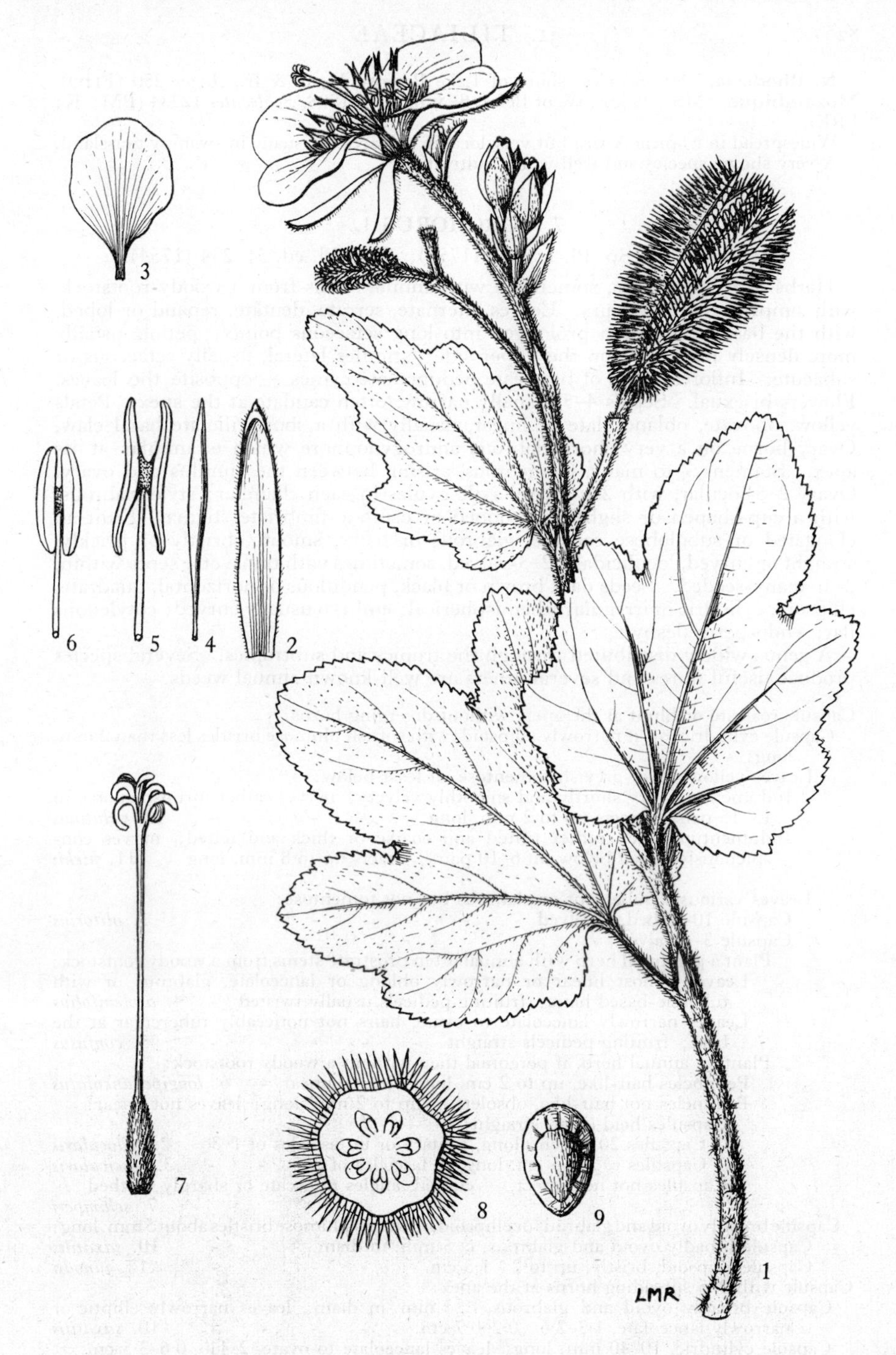

Tab. 7. CLAPPERTONIA FICIFOLIA. 1, flowering and fruiting branch ($\times\frac{2}{3}$); 2, sepal ($\times 2$); 3, petal ($\times\frac{2}{3}$); 4, staminode ($\times 5$); 5, anthers of longer stamens ($\times 5$); 6, anthers of shorter stamens ($\times 5$); 7, gynoecium ($\times 2$); 8, transverse section of ovary ($\times 10$); 9, seed ($\times 5$). 1–8 from *Andrews* A 1579, 9 from *Purseglove* P 1726.

N. Rhodesia. N: Samfya, shore of Lake Bangweulu, fl. & fr., *Angus* 250 (FHO). **Mozambique.** MS: 40 km. W. of Beira, fl. & fr. 10.iv.1898, *Schlechter* 12234 (BM; K; PRE).

Widespread in tropical Africa but very local in our area. Usually in swampy grassland. A very showy species and well worth cultivation.

7. CORCHORUS L.

Corchorus L., Sp. Pl. **1**: 529 (1753); Gen. Pl. ed. 5: 234 (1754).

Herbs or small shrubs, sometimes with annual stems from a woody rootstock, with simple or stellate hairs. Leaves alternate, serrate, dentate, repand or lobed, with the basal teeth often prolonged into long setaceous points; petiole usually more densely pubescent on the upper side; stipules lateral, usually setaceous or subacute. Inflorescences of bracteate pedunculate cymes ± opposite the leaves. Flowers bisexual. Sepals 4–5, usually narrow, often caudate at the apex. Petals yellow, obovate, oblanceolate or linear, usually with a short ciliolate basal claw. Ovary borne on a very short glabrous androgynophore which is annular at its apex. Stamens 7 to many, filamentous, arising between the annulus and ovary. Ovary 2–5-locular, with 2 to many axile ovules in each chamber; style glabrous, with a cup-shaped or slightly 2–6-lobed or capitate-fimbriate stigma. Fruit an elongated or subglobose capsule, glabrous or hairy, smooth, bristly or prickly, straight or curved, loculicidally 2–5-valved, sometimes with transverse septa within, 2- to many-seeded. Seeds dark brown or black, pendulous or horizontal, quadrate, ellipsoid, cylindric or irregularly hemispherical; embryo usually curved; cotyledons flat; endosperm fleshy.

A genus widely distributed through the tropics and subtropics. Several species produce useful fibres and several others are well-known annual weeds.

Capsule rostrate or blunt at the apex, without diverging horns:
- Capsule cylindric, or, if narrowly ellipsoid, with softish plumose bristles less than 2 mm. long:
  - Leaves yellowish- or greyish-tomentose, at least below:
    - Indumentum very shortly and smoothly velvety; nerves rather inconspicuous, in 13–15 pairs; bracts up to 2 mm. long - - - - - 12. *velutinus*
    - Indumentum often rather tufted and coarse or thick and felted; nerves conspicuously raised below, in 6–10 pairs; bracts up to 8 mm. long 11. *kirkii*
  - Leaves variously pubescent or glabrous but not tomentose:
    - Capsule 10-ribbed, 5-valved - - - - - - - 1. *olitorius*
    - Capsule 3–4-valved:
      - Plant a perennial herb with annual often prostrate stems from a woody rootstock:
        - Leaves almost linear or narrowly oblong or lanceolate, glabrous or with tubercle-based hairs; fruiting pedicels usually twisted - 4. *asplenifolius*
        - Leaves narrowly lanceolate to ovate, hairs not noticeably tubercular at the base; fruiting pedicels straight - - - - - - 5. *confusus*
      - Plant an annual herb, if perennial then without a woody rootstock:
        - Peduncles hair-like, up to 2 cm. long; leaves linear - 6. *longipedunculatus*
        - Peduncles not hair-like, obsolete or up to 2 mm. long; leaves not linear:
          - Capsules held erect, straight:
            - Capsules 20–70 cm. long, solitary or in fascicles of 1–3 2. *trilocularis*
            - Capsules c. 1–1·5 cm. long, in fascicles of 2–5 - - 3. *fascicularis*
          - Capsules not held erect, ± curved, angles muricate or sharply toothed 7. *schimperi*
- Capsule broadly ovoid and glabrous or ellipsoid, with stiff plumose bristles about 5 mm. long:
  - Capsule broadly ovoid and glabrous, c. 5 mm. in diam. - - 10. *saxatilis*
  - Capsule ellipsoid, bristly, up to 2 × 1·5 cm. - - - - - 13. *junodii*

Capsule with 3–5 spreading horns at the apex:
- Capsule broadly ovoid and glabrous, c. 5 mm. in diam., leaves narrowly elliptic or narrowly lanceolate, 1·3–2·6 × 0·2–0·7 cm. - - - - 10. *saxatilis*
- Capsule cylindric, 10–40 mm. long; leaves lanceolate to ovate, 2–11 × 0·6–3·5 cm.:
  - Capsule winged, 10–15 × 3–4 mm. - - - - - - 8. *aestuans*
  - Capsule not winged, 25–40 mm. long - - - - - 9. *tridens*

1. **Corchorus olitorius** L., Sp. Pl. **1**: 529 (1753).—Mast. in Oliv., F.T.A. **1**: 262 (1868). —Mendonça & Torre, Contr., Conhec. Fl. Moçamb. **1**: 24 (1950).—Exell & Mendonça, C.F.A. **1,** 2: 240 (1951). TAB. **8** fig. B. Lectotype a cultivated specimen in Herb. Hort. Cliff. (BM).

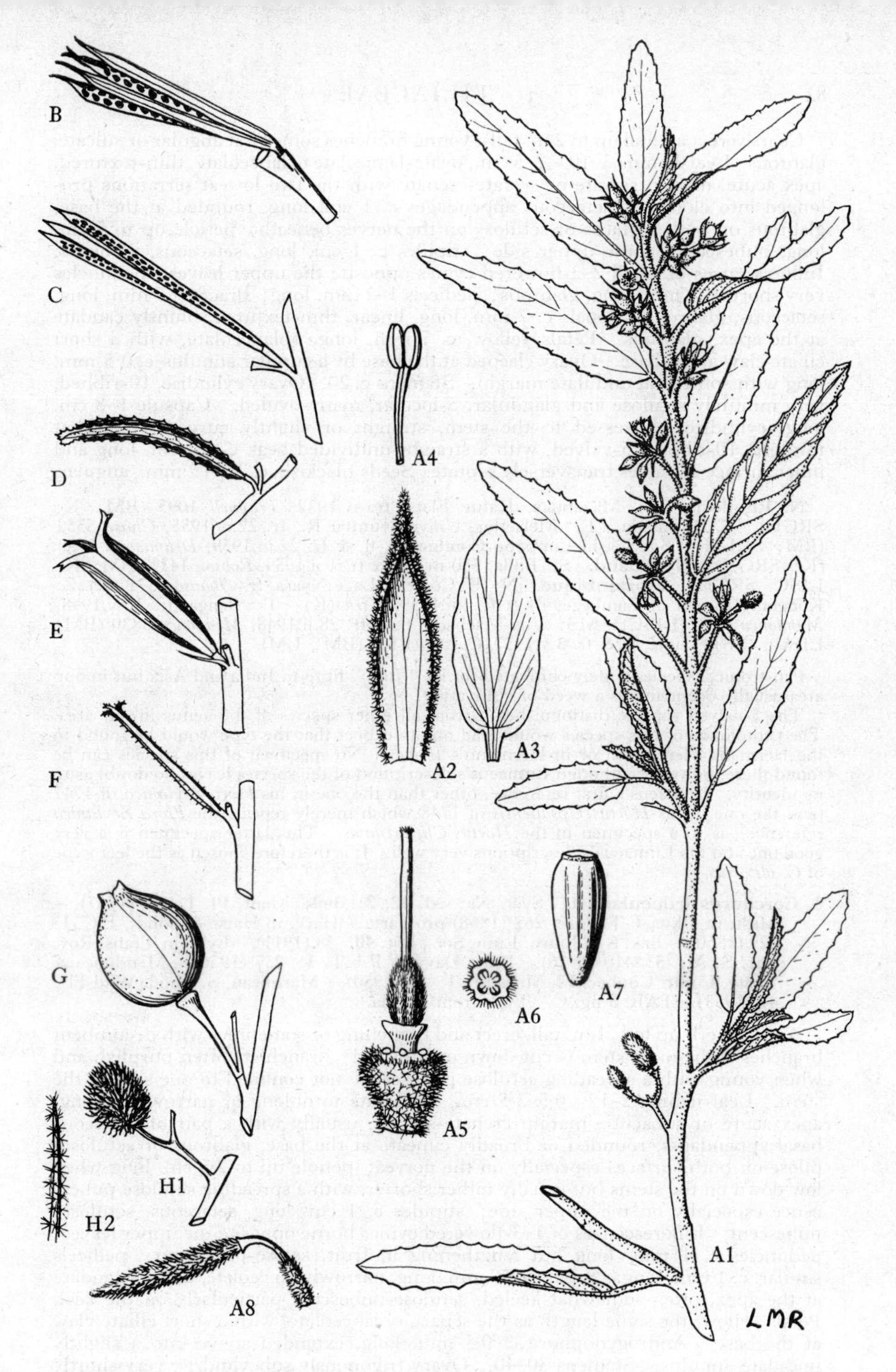

Tab. 8. A.—CORCHORUS KIRKII. A1, flowering branch (× ⅔); A2, sepal (× 5); A3, petal (× 5); A4, stamen (× 16); A5, gynoecium and torus (× 7·5); A6, transverse section of ovary (× 7·5); A7, seed (× 7·5); A8, fruit (× ⅔), all from *Kirk* s.n. B.—CORCHORUS OLITORIUS, fruit (× ⅔) *Kirk* s.n. C.—CORCHORUS TRILOCULARIS, fruit (× ⅔) *Cecil* 29. D.—CORCHORUS SCHIMPERI, fruit (× 1) *Sandwith* 155. E.—CORCHORUS AESTUANS, fruit (× 1) *Wakefield* s.n. F.—CORCHORUS TRIDENS, fruit (× ⅔) *Robinson* 262. G.—CORCHORUS SAXATILIS, fruit (× 4) *Robinson* 486. H.—CORCHORUS JUNODII. H1, fruit (× ⅔); H2, fruit bristle (× 7), both from *Gomes e Sousa* 1759.

Coarse erect annual up to 2 m. tall; young branches somewhat angular or sulcate, glabrous. Leaf-lamina 3–10 × 2–5 cm., ovate-lanceolate to lanceolate, thin-textured, apex acute, margin serrate or serrate-crenate with the two lowest serrations prolonged into elongated setaceous appendages c. 1 cm. long, rounded at the base, glabrous on both surfaces or setulose on the nerves beneath; petiole up to 5 cm. long, pubescent on the upper side; stipules c. 1 cm. long, setaceous, glabrous. Inflorescences of small 2–3-flowered cymes opposite the upper leaves; peduncles very short, c. 1 mm. long, glabrous; pedicels 1–2 mm. long; bracts c. 5 mm. long, setaceous, glabrous. Sepals c. 7 mm. long, linear, thin-textured, bluntly caudate at the apex, glabrous. Petals yellow, c. 7 mm. long, oblanceolate, with a short ciliate claw at the base. Ovary clasped at the base by a cupular annulus c. 0·5 mm. long with somewhat undulate margin. Stamens c. 20. Ovary cylindric, 10-ribbed, very minutely setulose and glandular, 5-locular, many-ovuled. Capsule 1–8 cm. long, cylindric, appressed to the stem, straight or slightly curved, somewhat torulose, 10-ribbed, 5-valved, with a straight undivided beak c. 1·2 cm. long and inner surface of valves transversely septate. Seeds blackish, c. 2·3 × 2 mm., angular.

N. Rhodesia. S: Mazabuka, Kafue Flats, fr. iv.1932, *Trapnell* 1095 (BM; K; SRGH). **S. Rhodesia.** E: Melsetter, Umvumvumvu R., fr. 22.iv.1955, *Chase* 5553 (BM; SRGH). S: Shashi-Limpopo Confluence, fl. & fr. 22.iii.1959, *Drummond* 5953 (K; SRGH). **Nyasaland.** S: Bilila, 550 m., fl. & fr. 1.ii.1959, *Robson* 1411 (BM; K; LISC; SRGH). **Mozambique**. N: E. Coast of Lake Nyasa, fr., *Johnson* 371 (K). Z: Kongone, mouth of Zambeze, fl. & fr. xii.1859, *Kirk* (K). T: Benga, fr. 16.v.1948, *Mendonça* 4270 (LISC). MS: Vila Machado, fl. & fr. 28.ii.1948, *Mendonça* 3830 (BM; LISC). SS: Guijá, fl. & fr. 3.v.1957, *Carvalho* 145 (BM; LM).

Pan-tropical species widely cultivated for its " Jute " fibre in India and Asia but in our area usually occurring as a weed of cultivation.

The 5-valved capsule distinguishes it from all other species of the genus in our area. The protologue of this species would lead one to expect that the type would be found in the Hermann Herbarium or in Hermann's Icones. No specimen of this species can be found there, however, although Linnaeus's description of the species leaves no doubt as to its identity. Linnaeus's first reference, other than the one in his *Flora Zeylanica* of 1747 (and the one in his *Hortus Upsaliensis* of 1748 which merely repeats the *Flora Zeylanica* reference), is to a specimen in the *Hortus Cliffortianus*. This latter specimen is a very good one and fits Linnaeus's descriptions very well. It is therefore chosen as the lectotype of *C. olitorius*.

2. **Corchorus trilocularis** L., Syst. Nat. ed. 12, **2**: 369; Mant. Pl. **1**: 77 (1767).—Mast. in Oliv., F.T.A. **1**: 262 (1868) pro parte.—Harv. in Harv. & Sond., F.C. **1**: 229 (1860).—Bak. f. in Journ. Linn. Soc., Bot. **40**: 33 (1911).—Eyles in Trans. Roy. Soc. S. Afr. **5**: 410 (1916).—Burtt Davy, F.P.F.T. **1**: 257 (1926).—Mendonça & Torre, Contr. Conhec. Fl. Moçamb. **1**: 24 (1950).—Martineau, S. Rhod. Wild Fl.: 47 (1953). TAB. **8** fig. C. Type from Arabia.

Annual herb up to c. 1 m. tall, erect and branching or sometimes with decumbent branches if the main stem is cut down or browsed; branchlets often purplish and when young with a spreading setulose pubescence not confined to one side of the stem. Leaf-lamina 2–12 × 0·5–3·5 cm., lanceolate to oblong or narrowly oblong, apex acute or subacute, margin crenate-serrate, usually with a pair of setaceous basal appendages, rounded or broadly cuneate at the base, glabrous or setulose-pilose on both surfaces especially on the nerves; petiole up to 2·5 cm. long when low down on the stems but usually rather shorter, with a spreading setulose pubescence especially on the upper side; stipules c. 1 cm. long, setaceous, setulose-pubescent. Inflorescences of 1–3-flowered cymes borne opposite the upper leaves; peduncles c. 1 mm. long but lengthening in fruit, setose-pubescent; pedicels similar, c. 1 mm. long. Sepals 6–10 mm. long, narrowly lanceolate, usually caudate at the apex, often somewhat keeled, setulose-pubescent particularly on the keel. Petals yellow, the same length as the sepals, oblanceolate, with a short ciliate claw at the base. Androgynophore c. 0·5 mm. long, extended above into a slightly undulate annulus. Stamens 30–40. Ovary trigonously subcylindric, very shortly pubescent, 3(4)-locular; style 1–1·5 mm. long, glabrous. Capsule many-seeded, 2·5–7 cm. long, held erect, straight or slightly curved, 3–4-angled, 3–4-valved; valves scabrous and sometimes somewhat torulose outside, with a series of hollows inside fitting the seeds. Seeds dark brown, 1–1·3 × 0·7–1 mm., oblong-ovoid.

N. Rhodesia. B: Sesheke, fl. iv., *Macaulay* 485 (K). C: Chilanga, fl. & fr. 31.viii.1929,

Sandwith 156 (K). S: Mazabuka, fl. & fr. v.1915, *Rogers* 8732 in part (BM). **S. Rhodesia.** E: Chipinga, Chipete Forest, fl. & fr. 23.v.1906, *Swynnerton* 469 (BM; K). S: Gwanda, fl. 17.xii.1956, *Davies* 2368 (K; SRGH). **Nyasaland.** C: Lilongwe, fl. & fr. 1.iv.1955, *Jackson* 1551 (K; PRE; SRGH). S: Upper Shire R., fl. & fr. 1893–4, *Scott Elliot* 8425 (BM). **Mozambique.** N: Nampula, fl. & fr. 27.ii.1937, *Torre* 1275 (COI; LISC). Z: Mocuba, Namagoa, fl. & fr. 1945, *Faulkner* 179 (K; PRE). T: Tete, fl. i.1932, *Guerra* 82 (COI). MS: Chimoio, Garuso, fl. & fr. 8.iv.1948, *Mendonça* 3880 (BM; LISC; SRGH). SS: Guijá, fl. & fr. 3.v.1957, *Carvalho* 148 (BM; LM). LM: Umbeluzi, fl. & fr. 20.ii.1951, *Myre* 1073 (LM; SRGH).

Widespread as an annual weed in Africa and Asia.

3. **Corchorus fascicularis** Lam., Encycl. Méth., Bot. 2: 104 (1786).—Mast. in Oliv., F.T.A. **1**: 263 (1868).—Hiern, Cat. Afr. Pl. Welw. **1**: 101 (1896).—Exell & Mendonça, C.F.A. **1**, 2: 239 (1951). Type from India.

Annual (? or perennial) herb with prostrate or ascending stems up to c. 0·6 m. long; branches glabrous or with a line of scattered setulose hairs on one side only. Leaf-lamina up to 6 × 0·9 cm., lorate to narrowly lanceolate-oblong, apex obtuse or acute, base rounded or broadly cuneate, margin crenate-serrate, basal setae absent, glabrous or glabrescent; petiole c. 7 mm. long, pubescent; stipules c.5 mm. long, setaceous. Inflorescences of 2–5-flowered fascicles opposite or subopposite the upper leaf axils; pedicels 0–1 mm. long, pubescent; bracts c. 5 mm. long, setaceous. Sepals 2–3 mm. long, linear, apex acuminate but not caudate, not keeled. Petals yellow, slightly longer than the sepals, oblanceolate, tapering but not clawed at the base. Androgynophore scarcely developed; annulus undulate. Stamens 5–10. Ovary trigonous-cylindric, pubescent, 3-locular; style c. 0·5 mm. long, glabrous. Capsule c. 1–1·5 cm. long, on an erect fruiting pedicel, many-seeded, ±cylindric, with a short beak ±1 mm. long. Ripe seeds not seen.

N. Rhodesia. B: Machili, fl. & fr. 3.iii.1961, *Fanshawe* 6374 (SRGH). **S. Rhodesia.** N: Chirundu, fr. 31.iii.1961, *Drummond & Rutherford-Smith* 7513 (K; PRE; SRGH). **Mozambique.** MS: Lake Gambue, c. 22 km. S. of Tica, coastal plain, fl. & fr. 20.vi.1961, *Leach & Wild* 11115 (K; SRGH).

Widespread in tropical Africa. Also in India and Australia. Damp places.

Very variable in size, leaf-shape, pubescence and, particularly if it is grazed or trampled, becoming semi-prostrate when it can be confused with *C. angolensis*. It is, however, always an annual, whilst *C. angolensis* is always a perennial.

4. **Corchorus asplenifolius** Burch., Trav. Int. S. Afr. **1**: 400 (1822).—Harv. in Harv. & Sond., F.C. **1**: 229 (1860).—Eyles in Trans. Roy. Soc. S. Afr. **5**: 409 (1916).—Burtt Davy, F.P.F.T. **1**: 257, fig. 39 (1926).—Martineau, S. Rhod. Wild Fl.: 46, t. 16 fig. 3 (1953). Type from Cape Province (Vaal River).

Corchorus serrifolius Burch., tom. cit.: 537 (1822).—Harv. in Harv. & Sond., F.C. **1**: 229 (1860).—Bak. f. in Journ. Linn. Soc., Bot. **40**: 33 (1911).—Eyles in Trans. Roy. Soc. S. Afr. **5**: 410 (1916).—Burtt Davy, F.P.F.T. **1**: 257 (1926).—Weim. in Bot. Notis. **1936**: 36 (1936).—Mendonça & Torre, Contr. Conhec. Fl. Moçamb. **1**: 25 (1950). Type from Cape Province (Asbestos Mts.).

Corchorus mucilagineus Gibbs in Journ. Linn. Soc., Bot. **37**: 433 (1906).—Eyles in Trans. Roy. Soc. S. Afr. **5**: 409 (1916). Type: S. Rhodesia, Matopos, *Gibbs* 8 (BM, holotype).

Corchorus trilocularis sensu Eyles, tom. cit.: 410 (1916) pro parte quoad specim. Eyles.

Perennial herb, with prostrate or suberect annual stems from a woody rootstock; stems glabrous or with a line of short curly hairs on one side only, or with spreading hairs all round the stem as well as the line of short curly hairs. Leaf-lamina 1·5–8 × 0·2–1·5 cm., lanceolate, narrowly oblong or almost linear, apex acute or subacute, rounded or broadly cuneate at the base, dentate-crenate or serrate, toothing very variable in size, sometimes irregular or biserrate and from very coarsely to rather finely toothed, basal setae absent, glabrous on both surfaces or sparsely to densely hispid with tubercle-based hairs; petiole up to 1 cm. long, pubescent at least on the upper side; stipules up to 1 cm. long, setaceous, setulose-pubescent. Inflorescences of solitary 1–3-flowered cymes opposite or subopposite the upper leaf-axils; peduncle obsolete or up to 4 mm. long, pubescent or glabrous; pedicels similar, up to 5 mm. long; bracts c. 2·5 mm. long, setaceous, pubescent. Sepals 6–10 mm. long, linear to linear-oblanceolate, glabrous on both sides or setulose-pubescent at the back, acuminate but not caudate at the apex, not keeled.

Petals yellow, the same length as the sepals, oblanceolate to obovate, with a basal ciliate claw c. 0·75 mm. long. Androgynophore c. 0·3 mm. long, extended above into a slightly undulate annulus. Stamens very numerous. Ovary trigonously subcylindric, very shortly setulose-pubescent, 3-locular; style 2–7 mm. long, glabrous. Capsule 2–3 cm. long, many-seeded, often on a rather twisted pedicel and therefore variable in its presentation, subcylindric not 3-angled, sparsely setulose-scabrid, attenuated to a blunt undivided apex. Seeds dark brown, 1·3–2 × 0·75–1·0 mm., shortly cylindric.

Bechuanaland Prot. N: Ngamiland, Kwebe Hills, fl. & fr. 18.i.1898, *Lugard* 115 (K). SE: Mahalapye, fl. & fr. 3.i.1912, *Rogers* 6095 (BM; K; PRE; SRGH). **N. Rhodesia.** S: Mazabuka, fl. 15.xii.1931, *Vet. Dept. CRS* 544 (PRE). **S. Rhodesia.** N: Urungwe, Kariba, fl. i.1956, *Goodier* 21 (K; SRGH). W: Shangani, fl. & fr. iii.1918, *Eyles* 946 (BM; K; SRGH). C: Enkeldoorn, fl. 21.iii.1937, *Eyles* 8955 (SRGH). E: Inyanga, Cheshire, fl. & fr. 14.i.1931, *Norlindh & Weimarck* 4293 (BM; LD). S: Beitbridge, fl. & fr. 15.ii.1955, *E.M. & W.* 427 (BM; LISC; SRGH).

Also in the Cape Prov., Transvaal and SW. Africa. This species is one of those which, without protection, are regularly burnt down to ground level in the dry season. It occurs in open woodland and at the margins of seasonal swamps (vleis or dambos).

Like so many species of this habit and ecology it is extremely polymorphic, but all the variations intergrade so freely that it hardly seems worth while to subdivide it even at the varietal level. As regards pubescence, for instance, every stage exists between the form with numerous bulbous-based hairs on the leaves, represented by the type of *C. mucilagineus*, and the quite glabrous leaves of the type of *C. serrifolius*. Unlike most species of this type, which flower mainly in the pre-rainy season, *C. asplenifolius* flowers mainly between December and March in the mid-rainy season.

5. **Corchorus confusus** Wild in Bothalia, **7**, 2: 422 (1960). Type from the Transvaal (Kruger National Park).

Corchorus trilocularis sensu Burtt Davy, F.P.F.T. **1**: 257 (1926) pro parte excl. specim. *Thorncroft* 2058 et *Nelson* 381.

Perennial herb with prostrate or spreading stems up to 0·6 m. long; rootstock woody; branchlets, at least when young, with a spreading pubescence all round the stem. Leaf-lamina up to 7 × 2·6 cm., narrowly lanceolate to ovate, apex acute, margin crenate or crenate-serrate, base rounded or slightly cordate, sometimes with a pair of setaceous basal lobes, pilose on both surfaces, especially on the nerves, hairs not tubercle-based, strongly 3–5-nerved from the base; petiole up to 8 mm. long, setulose-pilose all round; stipules up to 6 mm. long, setaceous, pubescent. Inflorescences of small (1)2–3-flowered cymes opposite the upper leaves; peduncles 0·4–2·5 cm., with a patent pubescence all round; pedicels similar, up to 8 mm. long; bracts setaceous, similar to the stipules. Sepals up to 10 × 1·5 mm., linear-lanceolate to narrowly lanceolate, apex caudate-acuminate, setulose-pilose outside. Petals yellow, slightly shorter than the sepals, oblanceolate to obovate, with a short basal ciliate claw. Androgynophore c. 0·5 mm. long, extended above into a slightly undulate glabrous annulus. Stamens c. 50. Ovary trigonously cylindric, densely pubescent; style c. 2·5 mm. long, slender, glabrous. Capsule 2·5–5 cm. long, on a straight fruiting pedicel, undivided at the apex, glabrous or sparsely scabrous on the angles; valves hollowed out to contain the seeds within. Seeds numerous, dark brown-grey, c. 2 × 1·2 mm.

S. Rhodesia. W: Shangani, fl. & fr. iii.1943, *Feiertag* in GHS 4551 (K; SRGH). C: Gwelo, *Gardner* 40 (K). E: Hippo Mine, fl. 12.iii.57, *Phipps* 586a (K; SRGH). S: Gwanda, fl. 17.xii.1956, *Davies* 2368 (K; SRGH). **Mozambique.** LM: Incanine, fl. & fr. 13.i.1898, *Schlechter* 12024 (K). SS: Inharrime, Ponta Závora, fr. 4.iv.1959, *Barbosa & Lemos* (K; LMJ; SRGH).

Also in the Transvaal, Cape Prov., Natal and Swaziland. In low-altitude woodlands in our area.

This species has often been confused in herbaria with *C. trilocularis* but it cannot be included in that species as it is a perennial. It is, in fact, more nearly related to *C. asplenifolius*.

6. **Corchorus longipedunculatus** Mast. in Oliv., F.T.A. **1**: 262 (1868).—Burtt Davy, F.P.F.T. **1**: 257 (1926). Type: Mozambique, opposite Sena, i.1860, *Kirk* (K, holotype).

Annual herb up to 0·6 m. tall, branching low down, with many ascending slender stems; branchlets angular or compressed at first, glabrous. Leaf-lamina 1–10 × 0·2–0·6 cm., linear, acuminate at the apex, margin denticulate, cuneate or narrowly truncate at the base, glabrous or minutely setulose-pubescent on the midrib below; petiole up to 5 mm. long, very shortly pubescent on the upper side; stipules 5–7 mm. long, setaceous, glabrous. Inflorescences of 1–3-flowered cymes opposite the upper leaves; peduncles up to 2 cm. long, hair-like, glabrous; pedicels similar, up to 1 cm. long; bracts 2–3 mm. long, setaceous, glabrous. Sepals c. 4 mm. long, very narrowly elliptic to linear, caudate or setaceous at the apex, glabrous, margins somewhat inrolled. Petals yellow, 3–4 mm. long, linear to very narrowly oblanceolate, not clawed or ciliate at the base. Androgynophore and annulus almost or quite obsolete. Stamens c. 12. Ovary trigonously subcylindric, 3-locular, very minutely setulose-pubescent; style 1–1·5 mm. long, glabrous. Capsule 2–5 cm. long, many-seeded, trigonous, with an undivided beak, narrowed to the base, glabrous, 3-valved; valves hollowed out to contain the seeds within. Seeds brown, c. 0·8 × 1·5 mm., subcylindric.

S. Rhodesia. E: Sabi valley, Hot Springs, fl. ii.1931, *Myres* 719 (K; SRGH). S: Ndanga, Mkwasini R., fl. 28.i.1957, *Phipps* 199 (SRGH). **Mozambique.** T: opposite Vila de Sena, R. Zambeze, fl. & fr. 5.i.1860, *Kirk* (K). LM: Ressano Garcia, fl. & fr. 26.xii.1897, *Schlechter* 11943 (BM; K).
Also in the Transvaal.

A most characteristic but inconspicuous plant, probably commoner and more widely distributed than the known gatherings indicate.

6. **Corchorus schimperi** Cufod. in Bull. Jard. Bot. Brux. **28**: 516 (1958). TAB. **8** fig. D. Type from Ethiopia.
Corchorus muricatus sensu Hochst. ex A. Rich., Tent. Fl. Abyss. **1**: 81 (1847).—Mast. in Oliv., F.T.A. **1**: 263 (1868).—Eyles in Trans. Roy. Soc. S. Afr. **5**: 410 (1916).—Weim. in Bot. Notis. **1936**: 36 (1936).

Annual herb branching from low down, at first erect but later branches often prostrate; branches with a line of short curly hairs on one side. Leaf-lamina 1–5 × 0·7–1·5 cm., lanceolate or oblong, apex rounded, margin crenate or crenate-serrate, base broadly cuneate or rounded, sometimes rather asymmetric, sparsely pilose at least on the nerves beneath and at the margins; petiole up to 7 mm. long, pubescent on the upper side; stipules 2–3 mm. long, setaceous, sparsely ciliate-pubescent. Inflorescences of 1–3-flowered extra-axillary fascicles, the peduncles usually obsolete; pedicels up to 2 mm. long, pubescent, twisted in fruit; bracts 1–2 mm. long, setaceous. Sepals c. 3–5 mm. long, linear to very narrowly elliptic, shortly acuminate at the apex, pubescent on the back. Petals yellow, the same length as the sepals, narrowly obovate to obovate, with a very short minutely ciliolate basal claw less than 0·5 mm. long or obsolete. Androgynophore and annulus very minute, c. 0·25 mm. long. Stamens numerous. Ovary trigonously cylindric, 3-locular, very minutely setulose-pubescent; style 2·5 mm. long, glabrous, very slender. Capsule up to 2·5 cm. long, straight or often curved, trigonous, with the angles muricate or sharply toothed, 3-valved, with many seeds; beak very short, undivided and blunt. Seeds c. 1·4 × 0·75 mm., dark brown, subcylindric.

N. Rhodesia. C: Chilanga, fl. & fr. 24.viii.1929, *Sandwith* 155 (K). **S. Rhodesia.** N: Sinoia, fr. 17.v.1943, *McKinnon* in GHS 10076 (SRGH). W: Victoria Falls, fl. & fr. xii.1905, *Allen* 87 (K; SRGH). C: Salisbury Experimental Station, fl. & fr. 3.v.1960, *Drummond* 6867 (K; SRGH). E: Inyanga, Cheshire, fl. & fr. 15.i.1931, *Norlindh & Weimarck* 4337 (BM; LD).
Known also from Ethiopia, Transvaal, Orange Free State, northern Cape Province, Natal and SW. Africa.

Usually a weed of cultivation in our area (but *Allen* 87 comes from the edge of a seasonal swamp or vlei). An inconspicuous plant, and no doubt the large gap in its distribution in East Africa is only apparent.

7. **Corchorus aestuans** L., Syst. Nat. ed. 10, **2**: 1079 (1759).—Mendonça & Torre, Contr. Conhec. Fl. Moçamb. **1**: 24 (1950).—Exell & Mendonça, C.F.A. **1**, 2: 241 (1951). TAB. **8** fig. E. Type from Jamaica.
Corchorus acutangulus Lam., Encycl. Méth. Bot. **2**: 104 (1786).—Mast. in Oliv., F.T.A. **1**: 264 (1868). Type from India.

Branching annual up to 50 cm. tall; first branches more or less erect, the later ones tending to be prostrate; branches pilose, usually more densely so on one side, compressed at first, later cylindric. Leaf-lamina 2–8 × 1–3·5 cm., ovate, narrowly ovate or broadly elliptic, acute or subacute at the apex, margin serrate-crenate, rounded at the base, usually with two basal setaceous appendages, sparsely pilose on both surfaces, especially on the nerves; petiole up to 3 cm. long, pilose, more densely so on the upper side; stipules up to 1 cm. long, setaceous, sparsely setulose. Inflorescences of 1–3-flowered cymes, opposite the upper leaves; peduncles very short or obsolete, up to 1 mm. long, pubescent; pedicels similar, c. 1 mm. long; bracts up to 3 mm. long, setaceous, sparsely setulose-pubescent. Sepals c. 4 mm. long, linear, somewhat hooded and caudate-acuminate at the apex, glabrous. Petals yellow, the same length as the sepals, narrowly obovate, slightly undulate at the rounded apex, with a very short ciliate basal claw 0·5 mm. long. Androgynophore and annulus 0·3 mm. long. Stamens c. 10. Ovary 3–5-angled, cylindric, puberulous; style c. 1 mm. long, glabrous. Capsule up to 4 cm. long, usually straight, erect, 3–5-sided with narrow membranous wings on the angles, glabrous, terminating in 3–5 spreading horns 1·5–3 mm. long; horns entire or sometimes bifid; valves only shallowly hollowed out to contain the seeds on the inner surface. Seeds c. 0·8 × 0·8 mm., numerous, shortly cylindric, brown.

N. Rhodesia. N: Lake Tanganyika, Mpulungu, fl. & fr. 17.iv.1955, *Richards* 5446 (K). S: Gwembe valley, fr. iv.1932, *Trapnell* 1476 (K). **S. Rhodesia.** N: Urungwe, fl. 21.ii.1954, *Lovemore* 383 (K; SRGH). **Nyasaland.** S: Port Herald, fl. 28.iii.1933, *Lawrence* 37 (K). **Mozambique.** N: Palma, fl. 26.iii.1917, *Pires de Lima* 151 (PO). Z: Mocuba, fr. 19.v.1949, *Barbosa & Carvalho* 2724 (K; LM; SRGH). T: between Lupata and Tete, fl. ii.1859, *Kirk* (K). MS: Vila Machado, fl. & fr. 16.v.1948, *Mendonça* 3953 (BM; LISC).

Widely distributed throughout the tropics. In our area apparently confined to altitudes below 750 m.

Angola plants according to Welwitsch (quoted by Masters in Oliv., F.T.A. **1**: 264 (1868)) are often apetalous; but apetaly has not been noted so far in plants from our area.

9. **Corchorus tridens** L., Mant. Pl. **2**: 566 (1771).—Mast. in Oliv., F.T.A. **1**: 264 (1868).—Ficalho, Pl. Ut. Afr. Port.: 111 (1884).—Eyles in Trans. Roy. Soc. S. Afr. **5**: 410 (1916).—Burtt Davy, F.P.F.T. **1**: 257 (1926).—Mendonça & Torre, Contr. Conhec. Fl. Moçamb. **1**: 25 (1950).—Exell & Mendonça, C.F.A. **1**, 2: 241 (1951). TAB. **8** fig. F. Type from India.

Annual herb up to c. 0·6 m. tall, usually erect but with the older branches rather spreading; branchlets at first rather compressed or angular, glabrous or sparsely pilose. Leaf-lamina 2·5–11 × 0·6–4·5 cm., oblong to lanceolate, acute at the apex, margin serrate or serrate-crenate, rounded at the base, glabrous or sparsely setulose-pubescent on the nerves; petiole up to 2 cm. long, pilose on the upper side; stipules up to 10 mm. long, setaceous, glabrous. Inflorescences of 1–3-flowered leaf-opposed cymes; peduncles up to 1 mm. long, glabrous; pedicels similar, up to 1 mm. long; bracts 1–2 mm. long, setaceous, glabrous. Sepals c. 5 mm. long, linear, slightly broader in the upper half, bluntly acuminate, glabrous. Petals yellow, the same length as the sepals, very narrowly oblanceolate, with a basal minutely ciliolate claw c. 0·3 mm. long. Androgynophore almost obsolete, but annulus visible and clasping the base of the ovary. Stamens 8–10. Ovary trigonously cylindric, papillose, 3-locular; style c. 2 mm. long, glabrous. Capsule up to c. 4 cm. long, held more or less erect, straight or slightly curved, usually somewhat ribbed, sparsely and minutely setulose-scabrid, terminated by 3 spreading horns c. 1 mm. long; valves only shallowly pitted inside to contain the seeds. Seeds c. 1·4 × 0·8 mm., numerous, dark brown, cylindric or somewhat quadrangular-cylindric.

Caprivi Strip. Mpilila I., 18.i.1959, *Killick & Leistner* 3334 (PRE; SRGH). **Bechuanaland Prot.** N: Ngamiland, Kwebe, fl. & fr. 5.iii.1898, *Lugard* 209 (K). **N. Rhodesia.** N: Abercorn Distr., Mpulungu, fl. & fr. 6.iv.1952, *Richards* 1452 (K). W: Ndola, fl. & fr. 30.iii.1954, *Fanshawe* 1039 (K). C: Broken Hill, Chirukutu, fr. 7.viii.1911, *Fries* 263 (UPS). E: Luangwa R., fl. 17.iii.1959, *Robson* 1753 (K; SRGH). S: Kalomo, fl. & fr. v.1909, *Rogers* 8208 (K; SRGH). **S. Rhodesia.** N: Urungwe, Zambezi valley, fl. & fr. 24.ii.1953, *Wild* 4100 (K; SRGH). W: Victoria Falls, fl. & fr. 12.ii.1912, *Rogers* 5537 (BM; K; PRE; SRGH). C: Marandellas, fl. & fr. 26.ii.1942, *Dehn* 623

(M; SRGH). E: Lower Sabi R., Mtema, fl. 28.i.1948, *Wild* 2397 (K; SRGH). S: Victoria, fl. & fr. 1908, *Monro* 974 (BM; SRGH). **Nyasaland.** ?N or C: Pirazinge, fl. & fr. 17.iii.1947, *Lowe* 471 (BM). S: near Blantyre, fr. 1887, *Last* (K). **Mozambique.** N: Nampula, fl. 2.iv.1936, *Torre* 814 (COI; LISC). T: Tete, fl. & fr. 5.v.1948, *Mendonça* 4080 (BM; LISC). MS: Mavita, Macequece road, fl. & fr. 7.iv.1948, *Barbosa* 1402 (BM; LISC). SS: Massingir, fl. & fr. 7.v.1957, *Carvalho* 166 (BM; LM). LM: Delagoa Bay, fl. & fr. 5.i.1898, *Schlechter* 11996 (COI; PRE; K).

Widespread in the tropics and subtropics of the old world. Common weed of cultivation in our area.

The leaves are commonly eaten as a spinach by natives and the plant has been used for fibre in Nyasaland.

10. **Corchorus saxatilis** Wild in Bol. Soc. Brot., Sér. 2, **31**: 90, t. 2 (1957). TAB. **8** fig. G. Type: N. Rhodesia, Mapanza, *Robinson* 486 (K, holotype; SRGH).

Annual herb up to 30 cm. tall; stems at first erect but later branches bending downwards and becoming semi-prostrate; branchlets very slender, very sparsely pubescent or glabrous, purplish-brown and with a raised line decurrent from the petioles. Leaf-lamina 1·3–3·8 × 0·2–1·5 cm., narrowly elliptic, narrowly oblong, or narrowly lanceolate, apex acute, margin crenate-dentate, base narrowly rounded to cuneate, sparsely setulose on the nerves; petiole 2–10 mm. long, pilose on the upper side; stipules c. 2 mm. long, setaceous, glabrous. Inflorescences of 1–3-flowered cymes subopposite the leaves; peduncles up to 1 mm. long, glabrous; pedicels up to 1·5 mm. long, glabrous; bracts c. 1 mm. long, setaceous. Sepals 3 mm. long, linear, acute, slightly inrolled at the edges, glabrous. Petals yellow, the same length as the sepals, linear-oblanceolate, minutely ciliolate at the base. Androgynophore almost obsolete, c. 0·1 mm. long; annulus present. Stamens 7–8. Ovary globose, slightly sulcate, 3-locular; loculi 2-ovulate; style c. 1·5 mm. long, glabrous. Capsule c. 5 mm. in diam., broadly ovoid, apex very shortly and bluntly 3-horned. Seeds c. 1·5 mm. in diam., roughly hemispherical, blackish.

N. Rhodesia. W: Solwezi, fl. & fr. 20.iii.1961, *Drummond & Rutherford-Smith* 7091 (K; LISC; PRE; SRGH). E: 32 km. E. of Kachalola, fl. & fr. 24.iii.1955, *E.M. & W.* 1166 (BM; LISC; PRE; SRGH). S: Mapanza, fl. & fr. 24.i.1954, *Robinson* 486 (K; SRGH).

So far as is known at present, confined to N. Rhodesia. Appears to be confined to very shallow soil over and in the cracks of rocky outcrops.

The affinities of this species are with *C. capsularis* L., a much larger plant with more ovules per loculus and *C. hochstetteri* Milne-Redh. which has echinate fruits.

11. **Corchorus kirkii** N.E.Br. in Kew Bull. **1908**: 288 (1908).—Eyles in Trans. Roy. Soc. S. Afr. **5**: 409 (1916).—Steedman, Trees etc. S. Rhod.: 46 (1933).—Weim. in Svensk Bot. Tidskr. **30**: 478, t. 16 (1936); in Bot. Notis. **1936**: 37 (1936). TAB. **8** fig. A. Syntypes: Mozambique, Tete, *Kirk* (K); between Tete and Lupata, *Kirk* (K); between Tete and the sea coast, *Kirk* (K).

Corchorus hirsutus sensu Mast. in Oliv., F.T.A. **1**: 264 (1868).—Eyles in Trans. Roy. Soc. S. Afr. **5**: 409 (1916).—Steedman, loc. cit.

Corchorus pongolensis Burtt Davy & Greenway in Burtt Davy, F.P.F.T. **1**: 37 (1926).—Weim. in Svensk Bot. Tidskr. **30**: 477, t. 1 fig. a (1936); in Bot. Notis. **1936**: 37 (1936).—O.B.Mill. in Journ. S. Afr. Bot. **18**: 52 (1952).—Type from the Transvaal.

Small shrub up to 2·5 m. tall; branches densely greyish- or yellowish-tomentose. Leaf-lamina 2·4–10 × 0·8–2·7 cm., oblong to oblong-lanceolate, rounded or acute at the apex, margin coarsely serrate or repand or finely serrate, rounded or broadly cuneate at the base, thick textured, densely grey-green- or yellowish-tomentose on both surfaces, nerves impressed above, somewhat prominent below; petiole up to 1·3 cm. long, greyish- or yellowish-tomentose; stipules up to 5 mm. long, subulate, tomentose on the back. Inflorescences of leaf-opposed 3–8-flowered cymes; peduncles up to 1 cm. long, but often much shorter, tomentose; pedicels similar, up to 3 mm. long; bracts up to 8 mm. long, subulate, tomentose or pilose on the back. Sepals up to 10 mm. long, lanceolate-acuminate to linear, pilose or tomentose on the back. Petals yellow, the same length as the sepals, narrowly obovate to linear-oblong, with a short basal claw ciliolate on its margins. Androgynophore 0·4 mm. long, with an annulus above. Stamens 20–30. Ovary cylindric, 3(4)-locular, many-seeded, densely pilose; style glabrous. Capsule up to 3 cm. long, not beaked, narrowly ovoid or linear-cylindric, densely clothed with soft stellately hairy bristles. Seeds c. 2·5 × 1·3 mm., compressed-ellipsoid.

Bechuanaland Prot. N: Shashi R., fl. & fr. 27.iii.1876, *Holub* 1429 (K). SE: Kanye, fl. & fr. iii.1944, *Miller* 307 (PRE). **S. Rhodesia.** N: Sebungwe, fl. & fr. 5.iv.1951, *Lovemore* 148 (SRGH). W: Matopos, fl. & fr. 14.ii.1912, *Rogers* 5650 (BM; K; SRGH). C: Marandellas, fl. 22.iv.1942, *Dehn* 649a (M; K; SRGH). E: Umtali, fl. 18.xi.1951, *Chase* 4192 (BM; K; SRGH). S: Belingwe, fl. & fr. 26.ii.1931, *Norlindh & Weimarck* 5150 (BM; LD; K; SRGH). **Mozambique.** T: Tete, fl. ii.1859, *Kirk* (K).

Also in the Transvaal. In open woodland, often occurring on the slopes of rocky hills.

The extreme form represented by the type of *C. pongolensis* would at first sight appear to be distinct. Its characteristic greyish tufted indumentum, its narrowly ovoid fruits, bracts usually longer than the flowers, and leaves wider in proportion to their length, all seem to separate it from *C. kirkii*, which has a smooth velvety yellowish indumentum, cylindric fruits and bracts shorter than the flowers. When the full range of material is examined, however, each one of these characters of typical *C. pongolensis* can be found in what is otherwise typical *C. kirkii*, uncorrelated with the remainder. What is more, there is no phytogeographical segregation of the two entities. Plants with the elongate fruits of *C. kirkii* are found near the type locality of *C. pongolensis*, in the Transvaal, and plants with the indumentum of the latter but otherwise agreeing with *C. kirkii* are found on the northern borders of S. Rhodesia, not far from the type locality of *C. kirkii*. It is possible that two distinct species once existed but, owing to the overlapping of their two ranges of distribution, widespread hybridization took place and a hybrid swarm was produced with a continuous range of variation between the two extremes.

It is reported that the bark of this plant is used in the Matopos for basket making.

12. **Corchorus velutinus** Wild in Bol. Soc. Brot., Sér. 2, **31**: 92 (1957). Type from the Transvaal.

Small shrub up to 1 m. tall; all parts except the corolla with a short dense velvety greyish-yellow indumentum; old stems becoming glabrous, with a brownish bark. Leaf-lamina 2–5 × 0·6–2·5 cm., narrowly oblong or narrowly elliptic, submucronate-obtuse at the apex, dentate or repand-dentate at the margin, widely cuneate at the base, with 13–15 pairs of rather inconspicuous nerves slightly impressed above and slightly raised below; petiole c. 1 cm. long; stipules 1–2 mm. long, subulate, caducous. Inflorescences of small 3–11-flowered cymes opposite the upper leaves or borne some distance from the nodes; peduncles up to 1·5 cm. long; pedicels similar, up to 1·5 cm. long; bracts c. 2 mm. long, subulate. Sepals up to 10 × 2·5 mm., narrowly elliptic or very narrowly elliptic, somewhat keeled dorsally, caudate at the apex, velvety outside, glabrous within. Petals yellow, up to 10 × 3 mm., narrowly oblanceolate-oblong, with a small claw c. 1·2 mm. long at the base with its margins very shortly stellately pubescent. Androgynophore 0·5 mm. tall, with an annulus above. Stamens numerous, filiform. Ovary 3-locular, cylindric, shortly stellately villous; style c. 4 mm. long, glabrous. Capsule c. 3 cm. long, cylindric, not beaked, sometimes slightly torulose, greyish- or yellowish-velvety-tomentose with tufted longer stellate hairs scattered among the shorter indumentum. Seeds numerous, very dark brown, c. 2·5 × 1·2 mm., irregularly ellipsoid.

S. Rhodesia. S: Nuanetsi, fl. & fr. 1.xi.1955, *Wild* 4685 (K; SRGH). **Mozambique.** SS: 100 km. S. of R. Save, fl. 21.xii.1946, *Hornby* 2489 (SRGH).

Also in the Northern Transvaal in localities quite near to those in our area. Dry deciduous woodland on sandy soils in the valleys of the Sabi and Limpopo rivers.

12. **Corchorus junodii** (Schinz) N.E.Br. in Kew Bull. **1908**: 287 (1908).—Weim. in Svensk Bot. Tidskr. **30**: 480, t. 1 fig. d (1936). TAB. **8** fig. H. Type: Mozambique, Delagoa Bay, *Junod* (K; PRE; Z, holotype).

Triumfetta junodii Schinz in Mém. Herb. Boiss. **10**: 49 (1900). Type as above.

Small shrub up to 1·6 m. tall; young branches greyish-pubescent, older branches glabrescent and brown. Leaf-lamina 1·5–6 × 0·6–2·0 cm., lanceolate, oblong or elliptic, rounded or acute at the apex, margin coarsely dentate, rounded or broadly cuneate at the base, shortly and densely greenish-tomentose or glabrescent above, densely whitish- or yellowish-tomentose below; petiole up to 1 cm. long, densely grey-pubescent; stipules 2–4 mm. long, subulate, densely grey-pubescent. Inflorescences of leaf-opposed 3–5-flowered cymes; peduncles up to 1 cm. long, densely grey-pubescent; pedicels similar, up to 5 mm. long; bracts c. 3 mm. long, subulate, grey-pubescent. Sepals up to 1 cm. long, linear, with a caudate apex up to 4·5 mm. long but often less, densely pubescent outside, glabrous within. Petals yellow, 7–8 mm. long, oblanceolate, with a claw at the base up to 1 mm. long and with ciliolate margins. Androgynophore 0·5 mm. long; annulus with its margin somewhat undulate. Stamens numerous. Ovary ovoid, densely pubescent, 5-locular; ovules very numerous; style 3–4 mm. long, glabrous. Capsule up to 2 × 1·5 cm. including the bristles, ovoid with a dense covering of rather stiff stellately plumose bristles up to 5 mm. long. Seeds brown, 2 × 1·3 mm., irregularly ellipsoid.

Mozambique. MS: Chiloane Is., fl. & fr. 12.viii.1887, *Scott* (K). SS: Inhachengo, fr. 26.ii.1955, *E.M. & W.* 620 (BM; LISC; SRGH). LM: Lourenço Marques, fl. & fr. 29.xi.1897, *Schlechter* 11561 (BM; COI; K; PRE).

Also in Natal and the Transvaal. Mainly in coastal areas.

32. LINACEAE

By N. K. B. Robson

Trees, shrubs, lianes or herbs, often with tendrils on climbing shoots, glabrous or with an indumentum of simple hairs. Leaves alternate or opposite, simple, penninerved or 1-nerved; stipules divided or entire or gland-like, deciduous (rarely absent). Flowers in terminal or axillary cymes (or rarely solitary), actinomorphic, bisexual, usually heterostylic. Sepals (4) 5, imbricate, free or partially united. Petals (4) 5, contorted in bud, free or very rarely united at the base, often unguiculate, fugacious. Stamens twice (rarely three times) as many as the petals, or with the antipetalous whorl staminodial or absent; filaments ± united in the lower part, sometimes glandular at the base; anthers introrse, dorsifixed, dehiscing longitudinally. Ovary superior, with 2–5 2-ovulate loculi sometimes subdivided nearly to the placentae or alternating with an equal number of empty loculi; ovules collateral, pendulous; styles free or united at the base, slender, with simple capitate or clavate stigmas. Fruit a capsule dehiscing septicidally into (4) 5 2-seeded valves, or septicidally and loculicidally into (8) 10 1-seeded valves, or a drupe usually with fewer seeds than the 2–5 originally fertile loculi. Seeds ± compressed, shining, exarillate, with or without endosperm; embryo straight or slightly curved, with flat cotyledons.

A family of about 25 genera and about 300 species distributed over the tropical and temperate regions of both hemispheres. In the present treatment it is equivalent to subfamily *Linoideae* of Winkler (in Engl. & Prantl, Nat. Pflanzenfam., ed. 2, **19a**: 107 (1931)) excluding the tribe *Nectaropetaleae*, which has been transferred to the *Erythroxylaceae*.

Stamens 10, all fertile; fruit drupaceous; woody plants - - - **1. Hugonia**
Stamens 4–5, sometimes alternating with staminodes; fruit capsular; annual or perennial herbs, rarely shrublets:
Sepals undivided; flowers 5-merous - - - - - - **2. Linum**
Sepals 3(4)-fid; flowers 4-merous - - - - - - **3. Radiola**

1. HUGONIA L.

Hugonia L., Sp. Pl. 2: 675 (1753); Gen. Pl. ed. 5: 305 (1754).

Trees, shrubs (rarely shrublets), or lianes climbing by recurved tendrils (usually in subopposite pairs). Leaves petiolate, alternate, entire or with a crenulate to serrate or dentate margin, penninerved, sometimes with domatia in the nerve axils and with ± densely reticulate tertiary venation; stipules pinnatifid or palmatifid, often caducous. Flowers solitary or in pairs or few-flowered cymes in the leaf-axils, more rarely in condensed terminal bracteate cymes, (always?) trimorphically heterostylic. Sepals 5, quincuncial, usually unequal, entire, free, persistent. Petals 5, usually yellow, darker at the base, shortly and narrowly unguiculate with the claw inserted on the dorsal side, free. Stamens 5 + 5, all fertile, with filaments united at the base to form a cup. Ovary with 2–5 fertile loculi alternating with an equal number of sterile loculi; styles 2–5, free or rarely partially united; stigmas capitate, grooved. Fruit a drupe with 3–5 or fewer seeds (not more than one per originally fertile loculus), with the hard endocarp vertically grooved. Seeds smooth, flattened, endospermic; embryo straight or slightly curved.

A genus of about 35 species in palaeotropical and subtropical regions. The recurved cirrhi on the climbing shoots (when present) provide a useful field character.

Flowers in 3–4-flowered pedunculate cymes; leaves elliptic - - 1. *elliptica*
Flowers solitary or paired, not pedunculate; leaves oblong to oblanceolate or obovate, rarely elliptic:
Sepals 9–10 mm. long; petals 20–30 mm. long; pedicels 4–8 mm. long; indumentum chocolate-brown - - - - - - - - - - 2. *grandiflora*
Sepals 4–8 (9) mm. long; petals 10–25 mm. long; pedicels of varying lengths; indumentum fulvous to orange-brown:
Plant a shrub, tree or liane over 1 m. high; pedicels at least 8 mm. long:
Drupe usually obovoid when mature; pedicels and sepals usually pubescent 3. *busseana*
Drupe globose; pedicels and sepals glabrous - - - - 4. *orientalis*
Plant a shrublet up to c. 60 cm. high; pedicels c. 1 mm. long - 5. *gossweileri*

1. **Hugonia elliptica** N. Robson in Bol. Soc. Brot., Sér. 2, **36**: 5 (1962). Type: Mozambique, Bajone, between Murroa and Namuera, *Barbosa & Carvalho* 4276 (LISC, holotype; LMJ).

Shrub 3 m. high, climbing; branches striate and brown-sericeo-pubescent when young, eventually terete, with cream-brown bark and numerous raised whitish lenticels. Leaves petiolate; lamina 8–13·5 × 2·5–4·3 cm., elliptic to oblong-elliptic, acute to subacute at the apex, broadly to narrowly cuneate at the base, entire or remotely and very shallowly serrate, chartaceous, entirely glabrous when mature or appressed-pubescent at base of midrib below, with the midrib, 9–13 pairs of lateral nerves and densely reticulate tertiary venation prominent on both sides, especially below; petiole 6–12 mm. long, shortly subappressed-pubescent; stipules 3–4·5 mm. long, 4–5-digitate or subpinnatifid with linear lobes, long-appressed-pubescent outside, glabrous within, ribbed, soon deciduous. Flowers 3–4 in condensed pedunculate cymes, in axils of foliage leaves; peduncles c. 3 mm. long; pedicels 5–6 mm. long, articulated, appressed-brown-pubescent. Sepals 6–8 mm. long, ovate, the outer ones subacute, the inner ones rounded, shallowly vertically ribbed, appressed-brown-pubescent outside, glabrous within. Petals yellow, c. 10 mm. long, obovate, rounded at the apex, with margin pubescent, otherwise glabrous. Stamens glabrous; longer filaments linear, shorter ones triangular; anthers oblong. Ovary c. 2 mm. long, ovoid, glabrous; styles 2–3, united for a short distance at the base, glabrous; stigmas broadly and obliquely capitate. Drupe (immature) globose, with thin brown pericarp and 2–3 fertile 1-seeded loculi.

Mozambique. Z: Bajone, between Murroa and Namuera, fl. & fr. 2.x.1949, *Barbosa & Carvalho* 4276 (LISC; LMJ).

Known from only the above locality. Habitat unrecorded, c. 150 m.

H. elliptica appears to be allied to *H. reticulata* Engl., a species of which the type material (from the Congo—Kasai) is destroyed, but differs in the number of styles and the length of the pedicels. It also has affinities with *H. castanea* Baill. from Madagascar.

2. **Hugonia grandiflora** N. Robson in Bol. Soc. Brot., Sér. 2, **36**: 6 (1962). Type: Mozambique, Macondes, Mueda, *Mendonça* 945 (LISC, holotype).

Shrub or small liane; branches striate and with short ± spreading chocolate-brown pubescence when young, eventually (3rd year) terete, glabrescent, with cream-brown bark and numerous raised whitish lenticels. Leaves petiolate; lamina (5) 7–12 × (2·5) 3·5–5 cm., obovate-rhombic to oblong, rounded to apiculate or shortly acuminate at the apex, broadly to narrowly cuneate or decurrent at the base, remotely and shallowly serrate, subcoriaceous, ± densely appressed- or subappressed-pubescent along the midrib above and below, on lateral nerves below and at the base of the lamina, otherwise glabrous (more rarely wholly glabrous), with the midrib and 9–12 pairs of lateral nerves impressed above and prominent below, and the densely reticulate tertiary venation prominent on both sides; petiole (5) 6–10 (11) mm. long, subappressed-chocolate-pubescent; stipules 5–10 mm. long, 3-fid or 4–7-pinnatifid with subulate lobes, long-appressed-pubescent outside, glabrous and ribbed within, persistent. Flowers solitary or paired, in axils of foliage leaves; peduncles 4–8 mm. long, stout, subappressed-chocolate-pubescent. Sepals 9–10 mm. long, ovate to lanceolate-elliptic, the outer ones acute or subacute, the inner ones rounded, shallowly vertically ribbed, appressed-chocolate-pubescent outside, glabrous within. Petals yellow, 20–30 mm. long, obovate, emarginate at the apex, with margin pubescent, otherwise glabrous. Stamens with linear glabrous filaments; anthers ovate-oblong, with apical tuft of hairs. Ovary c. 4 mm. long, ovoid, glabrous; styles 3–5, glabrous, free; stigmas broadly and obliquely capitate. Drupe c. 1·5 cm. in diameter, subglobose, with thin brown pericarp and 3–5 fertile 1-seeded loculi. Seeds c. 8 mm. long, whitish.

Mozambique. N: Macondes, between Mueda and Chomba, fl. & fr. 25.ix.1948, *Barbosa* 2247 (LISC).

Apparently confined to SE. Tanganyika (Rondo Plateau) and NE. Mozambique (Mueda region). Evergreen forest, c. 500 m.

H. grandiflora is allied to *H. castaneifolia* Engl. from the coastal region of Kenya and northern Tanganyika, but differs by its larger flowers on stout pedicels and its obovate-rhombic to oblong leaves with 9–12 paired nerves.

3. **Hugonia busseana** Engl., Bot. Jahrb. **40**: 45 (1907); Pflanzenw. Afr. **3,** 1: 722 (1915) ("bussei").—Brenan, T.T.C.L.: 266 (1949).—White, F.F.N.R.: 167 (1962). TAB. **9** fig. A. Type from Tanganyika (S. Highlands).

Hugonia buchananii De Wild., Pl. Bequaert. **4**: 269 (1926). Type: Nyasaland, *Buchanan* 367 (BM, holotype; E).

Hugonia arborescens Mildbr. in Notizbl. Bot. Gart. Berl. **12**: 513 (1935).—Brenan, loc. cit. Type from Tanganyika (Lindi).

Hugonia arborescens var. *schliebenii* Mildbr., tom. cit.: 514 (1935).—Brenan, loc. cit. Type from Tanganyika (Lindi).

Hugonia faulknerae Meikle in Kew Bull. **1950**: 338, fig. 1 (1951). Type: Mozambique, Lugela-Mocuba, Namagoa, *Faulkner* K 80 (BM; COI; K, holotype; PRE; SRGH).

Shrub or small tree 2–7 (10) m. high, often climbing. Branches sometimes arching or pendulous, striate and with dense fulvous to orange-brown spreading villous indumentum when young, eventually terete and tomentellous to glabrescent, with whitish-cream spongy striate or grooved bark on the older shoots. Leaves variable in shape and indumentum, petiolate; lamina 4·5–13 (17·5) × 1·6–4·3 (4·8) cm., oblong or oblanceolate to obovate (rarely elliptic), acute or shortly acuminate to rounded at the apex, cuneate to truncate or subcordate at the base, subentire to strongly dentate with triangular or ciliate teeth, chartaceous, rusty- or fulvous-pilose along the midrib above and below and on the nerves below, with the other parts varying from densely pubescent to glabrous, often with domatial tufts in the nerve angles below, with midrib and 12–13 nerves impressed above and raised below, and densely reticulate venation less prominent on both sides; petiole 4–8 (11) mm. long, spreading-rusty- to fulvous-pubescent; stipules 4–7 mm. long, pinnatifid with long or short lobes, densely sericeo-pubescent on both sides, deciduous. Flowers solitary or rarely paired, in axils of foliage leaves; peduncles 8–26 mm. long, ± slender, ± densely spreading- or appressed- rusty- or fulvous-pubescent (rarely glabrous). Sepals 6–8 mm. long, ovate or ovate-lanceolate to elliptic, obtuse to rounded, shallowly vertically ribbed, appressed-rusty- or fulvous-pubescent outside, pubescent near the margin or wholly glabrous within (rarely glabrous on both sides except round the margin). Petals yellow, 16–25 mm. long, triangular-obovate, rounded to retuse at the apex, with pubescent margin, otherwise glabrous. Stamens with linear glabrous filaments; anthers ovate-oblong, usually with an apical tuft of hairs. Ovary c. 2·5 mm. long, ovoid-pyriform, glabrous; styles (2) 3 (4), free, glabrous; stigmas broadly capitate. Drupe up to 3 cm. long, obovoid, apiculate, with (2) 3 (4) fertile 1-seeded loculi and thin orange-yellow pericarp. Seeds c. 10 mm. long.

N. Rhodesia. C: Lusaka Distr., fr. 29.v.1958, *Fanshawe* 4463 (K; SRGH). E: Fort Jameson Distr., Chadiza, fl. 25.xi.1958, *Robson* 693 (BM; K; LISC; SRGH). **Nyasaland.** N: Rumpi Distr., fr. 13.v.1952, *White* 2858 (BM; FHO; K). C: Dowa Distr., Lake Nyasa Hotel, fr. 30.vii.1951, *Chase* 3878 (BM; COI; K; SRGH). S: Fort Johnston, Monkey Bay Road, fl. 25.xi.1954, *Jackson* 1406 (BM; FHO; K). **Mozambique.** N: Nampula, fl. 1.xi.1936, *Torre* 904 (COI; LISC). Z: Quelimane Distr., Lugela, Namagoa, fl. x. & fr. xi.1946 & 1948, *Faulkner* K 80 (BM; COI; K; PRE; SRGH). T: between Matundo and Massanga, fr. 5.vii.1949, *Barbosa & Carvalho* 3444 (LISC; SRGH).

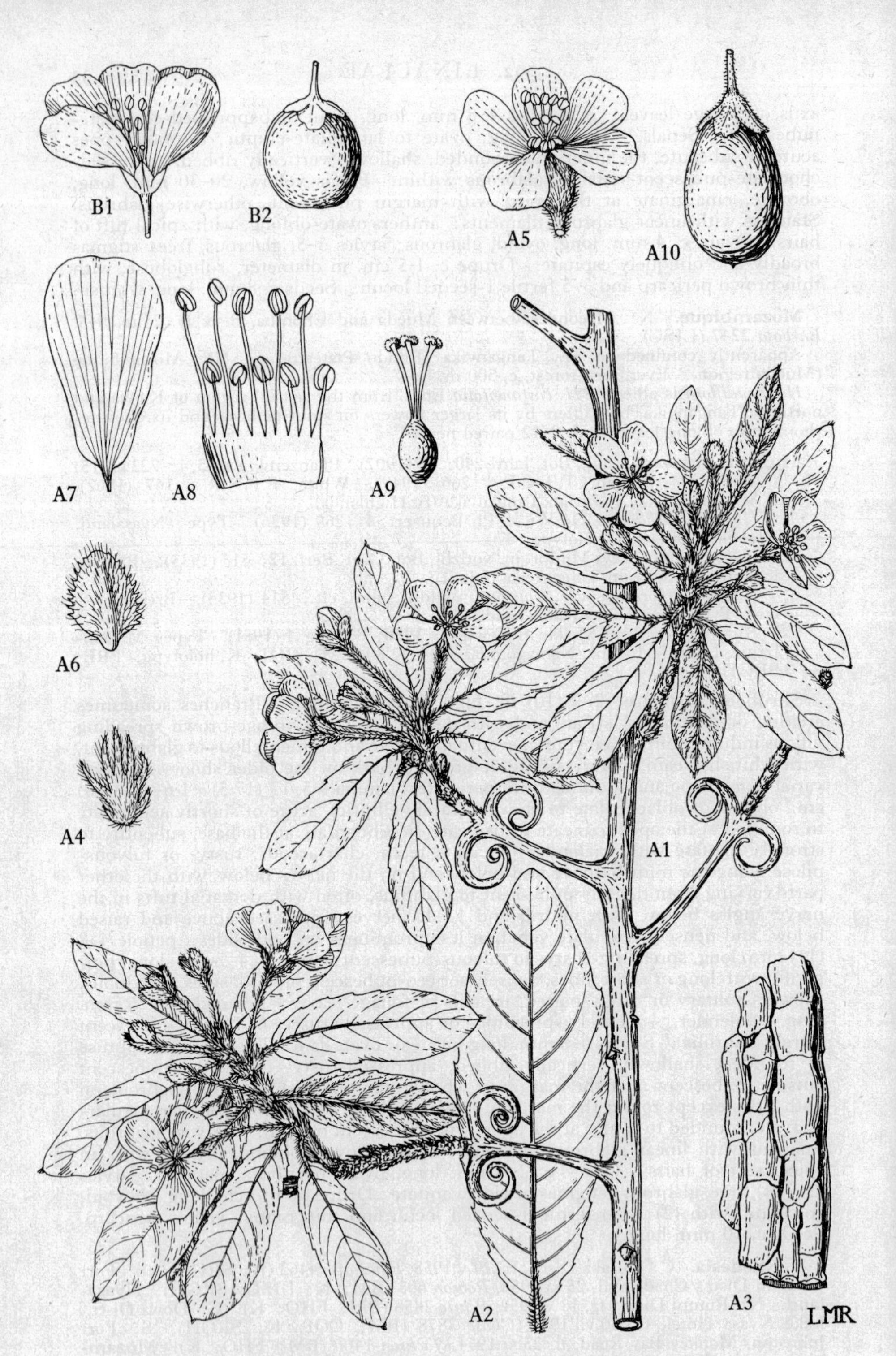

Tab. 9. A.—HUGONIA BUSSEANA. A1, flowering branch (× ⅔) *Fanshawe* 4131; A2, mature leaf (× ⅔) *Brenan* 7853; A3, portion of bark (× ⅔) *Mendonça* 851; A4, bract (× 2); A5, flower (× 1); A6, sepal (× 2); A7, petal (× 2); A8, stamen tube opened (× 2); A9, gynoecium (× 2), all from *Fanshawe* 4131; A10, fruit (× 1) *Milne-Redhead & Taylor* 8505. B.—HUGONIA ORIENTALIS. B1, flower (× 1) *Andrada* 1399; B2, fruit (× 1) *Wild* 3371.

In southern Tanganyika, eastern N. Rhodesia, Nyasaland and northern Mozambique. Open deciduous woodland on dry sandy soils or among rocks, 130–1100 m.

A very variable species in which forms tend to be distributed geographically. Thus the western specimens tend to have oblong or ovate-oblong leaves which are truncate or subcordate at the base, shortly petiolate and without domatial tufts; whereas in those from the eastern parts of its range the leaves are usually oblong or oblanceolate, cuneate at the base, with longer petioles and with domatial tufts. These distinctions, however, are insufficiently constant to allow taxonomic separation of the two forms.

4. **Hugonia orientalis** Engl., Bot. Jahrb. **32**: 107 (1902); Pflanzenw. Afr. **3**, 1: 722 (1915). TAB. **9** fig. B. Type: Mozambique, " Sofala-Gasa-Land ", Matola, *Schlechter* 11724 (B, holotype †).

Hugonia trigyna Summerh. in Kew Bull. **1926**: 238 (1926).—De Wild., Pl. Bequaert. **4**: 285 (1927). Type: Mozambique, without locality, *Allen* 76 (K).

Hugonia swynnertonii De Wild., loc. cit. Type: Mozambique, Madanda forests, *Swynnerton* 1335 (BM, holotype; K).

Shrub or rarely small tree 1–4 (6) m. high, often climbing. Branches long, trailing or climbing, striate and with fulvous to golden or rusty pubescence (dense and ± spreading on short lateral shoots, ± sparse and appressed on elongate climbing shoots) when young, eventually terete and glabrescent, with yellowish-white corky bark and raised lenticels on the older shoots. Leaves petiolate; lamina 2·5–11·6 (14) × 1–3·5 (4·5) cm., oblong or oblanceolate to obovate or elliptic, subacute (or rarely apiculate) to rounded at the apex, cuneate to truncate at the base, entire to strongly triangular-dentate, chartaceous, ± sparsely appressed-fulvous-pubescent or -pilose along the midrib and nerves below and usually on the midrib above, the remainder glabrous (or very rarely sparsely appressed-pubescent below), occasionally with small domatial tufts in the nerve-angles below, with midrib and 6–12 (14) nerves prominent on both sides (not impressed above and raised below), and densely reticulate venation prominent on both sides or not; petiole 3–11 mm. long, appressed- or subappressed-fulvous- or golden-pubescent; stipules 4–8 mm. long, 3–4 (5)-fid with ± long subulate lobes, appressed-pubescent or sericeous on both sides, deciduous. Flowers solitary, in axils of foliage leaves; peduncles 8–21 (30) mm. long, slender, glabrous. Sepals 4–7 mm. long, ovate to elliptic, obtuse to rounded, smooth or vertically ribbed, glabrous apart from a usually pubescent margin. Petals yellow, 12–13 mm. long, obovate to oblanceolate, rounded at the apex, glabrous apart from the pubescent margin. Stamens with linear glabrous filaments; anthers ovate-oblong, usually with an apical tuft of hairs. Ovary c. 2 mm. long, cylindric-ellipsoid, glabrous; styles (2) 3, free or united at the base, glabrous; stigmas broadly capitate. Drupe up to c. 1·5 cm. long, globose, flattened on top, with thin yellow pericarp and (2) 3 fertile 1-seeded loculi. Seeds c. 4–5 mm. long.

S. Rhodesia. E: Chipinga Distr., Mt. Makossa, fl. 3.xii.1956, *Goodier* 100 (K; SRGH). S: Nuanetsi Distr., Tswiza Camp, fl. xi.1955, *Davies* 1627 (K; SRGH). **Mozambique.** N: between Mocímboa da Praia and Palma, fr. 15.ix.1948, *Barbosa* 2132 (LISC). MS: Gorongosa Game Reserve, fl. 25.ix.1953, *Chase* 5077 (BM; K; LISC; PRE; SRGH). SS: Gaza, Chongoene to Chibuto, fr. 15.viii.1957, *Barbosa* 7851 (K; LISC). LM: Magude, near Mapulanguene on road to Massingir, fr. 4.v.1944, *Torre* 6559 (LISC).

In southern Mozambique (with an outlying population in north-eastern Mozambique), south-eastern S. Rhodesia and north-eastern Transvaal. Open deciduous woodland on dry sandy soil or at forest margins, 30–580 m.

A variable species with climbing shoots which may differ considerably from the non-climbing ones. It is closely related to *H. busseana*, and the population in Manica e Sofala Province tends to approach the latter species in some respects. On the other hand the four collections from Niassa Province along the coast between Mecúfi and Palma are similar to the coastal collections from the south of Mozambique. Their entire or sub-entire rhombic leaves are characteristic, but can be matched by those of the type specimen of *H. trigyna* among others.

Both *H. orientalis* and *H. swynnertonii* were originally described as having 5 styles, but only 3 are present in the type specimens of the latter. Although the holotype of *H. orientalis* no longer exists, Engler's description agrees well in other respects with specimens from the south of Mozambique. It seems justifiable, therefore, to assume that Engler, too, was mistaken in thinking that his specimen had 5 styles—the only number known in the genus up to that time.

5. **Hugonia gossweileri** Bak. f. & Exell ex De Wild., Pl. Bequaert. **4**: 262, 272 (Apr., 1927); apud Exell in Journ. of Bot. **65,** Suppl. Polypet.: 49 (Aug., 1927).—Exell & Mendonça, C.F.A. **1,** 2: 244 (1951).—White, F.F.N.R.: 167 (1962). Type from Angola (Bié).

Shrub or shrublet up to c. 60 cm. high with woody underground rootstock and erect caespitose shoots. Branches angular with raised lines decurrent from the stipules and fulvous-tomentellous when young, eventually terete and glabrous, with cream-brown or yellowish bark but without obvious lenticels. Leaves petiolate; lamina (3·5) 4–6·5 × 1·3–3·5 cm., elliptic to oblong, acute or shortly acuminate at the apex, cuneate to rounded or truncate at the base, coriaceous, glabrescent and slightly glossy above, ± densely fulvous-tomentellous below, with c. 9–13 nerves and reticulate venation prominent on both sides; petiole 2–4 mm. long, fulvous-tomentellous; stipules 4–6 mm. long, 3-fid or narrowly 4–7-pinnatifid with subulate lobes, appressed-pubescent or tomentellous outside, glabrous within, ribbed, eventually deciduous. Flowers solitary, in axils of foliage leaves or bracts; peduncles c. 1 mm. long, stout, fulvous-tomentellous. Sepals 6–9 mm. long, ovate to lanceolate, the inner ones clawed, acute, not ribbed, fulvous-tomentellous outside, glabrous within. Petals yellow, 10–16 mm. long, obovate, rounded or subtruncate at the apex, wholly glabrous. Stamens glabrous, with linear filaments; anthers ovate-orbicular. Ovary c. 2·5 mm. long, ellipsoid, glabrous; styles 3, 3–4 times as long as the ovary, free, glabrous; stigmas narrowly and obliquely capitate. Drupe 1 × 0·5 cm., ellipsoid, with thin brown pericarp and 1 fertile and 2 abortive 1-seeded loculi. Seeds c. 6 mm. long, whitish.

N. Rhodesia. N: Kawambwa Distr., Chipili Mission to Luwingu, 5·25 km., fl. 10.x.1947, *Brenan* 8087 (FHO; K).

Known from only Angola (Bié) and two localities in N. Rhodesia (Chipili and Chinsali). In open *Brachystegia* woodland and stony or sandy soil, c. 1300 m.

H. gossweileri appears to be very local, although abundant where it does occur. It should therefore be sought in dry open woodlands in the intervening areas between Bié and Fort Rosebery.

2. LINUM L.

Linum L., Sp. Pl. **1**: 277 (1753); Gen. Pl. ed. 5: 135 (1754).

Shrublets or, more frequently, perennial or annual herbs, sometimes woody at the base. Leaves sessile, alternate (or sometimes opposite or whorled at the base of the stem), entire or with a denticulate or ciliate margin, 1- to several-nerved; stipules glandular or absent. Flowers terminal, few to many in monochasial or subdichasial cymes or more rarely solitary, homostylic or heterostylic. Sepals 5, quincuncial, entire or with glandular-ciliate margin, free, persistent. Petals 5, yellow, blue, red, pink or white, free or very rarely united at the base, shortly and narrowly unguiculate with a terminal claw. Stamens 5, alternating with the petals and with the filaments ± united in the lower part to form a tube bearing 5 nectary glands outside; staminodes alternating with the stamens and forming filiform processes from the staminal tube, or absent. Ovary 5-locular,* each loculus partially divided by a false septum; styles 5,* free or rarely united in the lower half; stigmas capitate or obliquely clavate. Fruit a 5-locular* 10-valved capsule with 2-seeded loculi, the seeds in each loculus partially separated by a well-developed false septum. Seeds smooth, flat, with slime-epidermis and little or no endosperm; embryo straight.

A genus of about 200 species, almost cosmopolitan but especially abundant in the Mediterranean region and in south-western North America.

In addition to the following species, *L. usitatissimum* L. (flax) is grown in our area and sometimes becomes established as an escape. It can be immediately distinguished from our native species by its larger blue (rarely white) flowers.

Leaves opposite and relatively broader towards the base of the stem; sepals with uniform midrib usually extending to the acute or acuminate apex - - - 1. *thunbergii*

Leaves alternate and uniform in shape throughout the length of the stem; sepals with midrib thicker in the lower half, more slender above, or usually not extending to the acuminate or caudate apex - - - - - - - - - 2. *holstii*

* *L. digynum* A. Gray, from N. America, has a 2-merous gynoecium.

1. **Linum thunbergii** Eckl. & Zeyh., Enum. Pl. Afr. Austr. Extratrop.: 35 (1835) excl. syn.—Planch. in Hook., Lond. Journ. Bot. **7**: 495 (1848).—Sond. in Harv. & Sond., F.C. **1**: 310 (1860).—Eyles in Trans. Roy. Soc. S. Afr. **5**: 386 (1916).—Burtt Davy, F.P.F.T. **1**: 186, t. 21 (1926).—Exell & Mendonça, C.F.A. **1**, 2: 242 (1951). TAB. **10** fig. A. Syntypes from S. Africa (Cape Prov.).
Linum gallicum var. *abyssinicum* sensu Suesseng. in Proc. & Trans. Rhod. Sci. Ass. **43**: 107 (1951).

Herb 15–60 cm. high or sometimes higher, erect or suberect, perennial (or sometimes annual?), glabrous or very rarely with the lower leaves and parts of the stem sparsely pubescent. Stems single or in groups, unbranched for most of their length or more rarely branched in the lower half, ridged above, becoming terete towards the base. Leaves opposite (rarely whorled) at the stem base, becoming alternate towards the inflorescence; lamina of average leaves 9–20 (25) × (1) 2–5 (6) mm., oblong-lanceolate to oblong-elliptic to linear (rarely oblanceolate), acute or mucronulate at the apex, with margin entire sometimes involute, those towards the stem base broader and often more obtuse or rounded at the apex, 1 (3)-nerved; stipular glands globose or ellipsoid, dark brown, sometimes absent. Inflorescence a lax monochasial or subdichasial paniculate or corymbose cyme, usually many-flowered; bracts small, linear; pedicels 1–2 mm. long, elongating in fruit. Flowers homostylic. Sepals 2–3 mm. long, elliptic to ovate-elliptic, acute to sharply acuminate, with the margin entire or, more usually, ciliate to glandular-dentate or glandular-ciliate, and the midrib extending to the apex. Petals chrome-yellow, sometimes tipped purplish-red in bud, 5–10 mm. long, obovate to obovate-oblanceolate, rounded or retuse at the apex. Stamens c. 3–4 mm. long, with filaments united in the lower quarter; anthers c. 0·5 mm. long, oblong; staminodial processes absent. Ovary c. 0·75 mm. long, ovoid-ellipsoid; styles c. 4 times as long as the ovary and just exceeding the stamens, free, linear; stigmas slightly capitate. Capsule about as long as the sepals, globose or depressed-globose, yellow-brown, sometimes purplish towards the apex. Seeds c. 1·2 mm. long, ellipsoid, plano-convex, reddish-brown to yellowish-brown, smooth, shining.

S. Rhodesia. C: Macheke, fl. xii.1919, *Eyles* 2003 (K; PRE; SRGH). E: Inyanga, Mare R., fl. 17.i.1951, *Chase* 3671 (BM; SRGH). **Nyasaland.** N: Nyika Plateau, 2300 m., fl. 14.iii.1961, *Robinson* 4506 (K). S: Kirk Range, Goche, fl. 31.i.1959, *Robson* 1379 (BM; K; LISC; SRGH).

In Tanganyika (W. Usambaras & S. Highlands), Nyasaland, S. Rhodesia, Angola, and S. Africa from Transvaal and Natal to the Cape of Good Hope. Grasslands and marshes, 1500–2150 m. (in our area).

L. thunbergii is related to *L. africanum* L. from Cape Province, a shrubby species with partially united styles and sepals which exceed the capsule. *L. thesioides* Bartl., from the same region, is herbaceous, but its leaves are always alternate.

2. **Linum holstii** Engl. [Bot. Jahrb. **17**: 168 (1893) *nom. nud.*] ex Wilczek in Bull. Jard. Bot. Brux. **25**: 311 (1955); F.C.B. **7**: 38, t. 4 (1958).—Cufod. in Bull. Jard. Bot. Brux. **26**, Suppl.: 354 (1956). TAB. **10** fig. B. Type from Tanganyika (Usambara).
Linum gallicum var. *holstii* Engl. in Phys. Abh. Königl. Akad. Wiss. Berl. **1894,** 1: 58 (1894) *nom. nud.*
Linum gallicum var. *abyssinicum* sensu Engl., Bot. Jahrb. **30**: 336 (1901); in Mildbr., Deutsch. Z.-Afr.-Exped. 1907–1908, **2**: 241 (1912).—Staner in Rev. Zool. Bot. Afr. **23**: 219 (1933); op. cit. **24**: 217 (1933).
Linum abyssinicum sensu Robyns, Fl. Parc Nat. Alb. **1**: 404, t. 40 (1948).

Herb (10) 20–75 (90) cm. high, erect or somewhat spreading, perennial or annual, glabrous (or rarely glabrescent?). Stems single and branched or grouped and unbranched for most of their length, ridged above, becoming terete towards the base. Leaves alternate throughout; lamina (5) 10–25 × 1–3 mm., uniformly linear-lanceolate, sharply acuminate to apiculate at the apex, with margin denticulate or shortly ciliate (rarely entire) sometimes involute, 1 (rarely 3)-nerved; stipular glands globular or ellipsoid, dark brown, persistent. Inflorescence a lax or ± dense monochasial paniculate to corymbose cyme, usually many-flowered; bracts linear to narrowly triangular, with the margin usually ± glandular; pedicels 1–1·5 (2) mm. long, not elongating in fruit. Flowers homostylic. Sepals 2–3 mm. long, ovate-elliptic, sharply acuminate to caudate at the apex, with the margin gland-fringed and the midrib not extending beyond the midpoint (or rarely reaching the

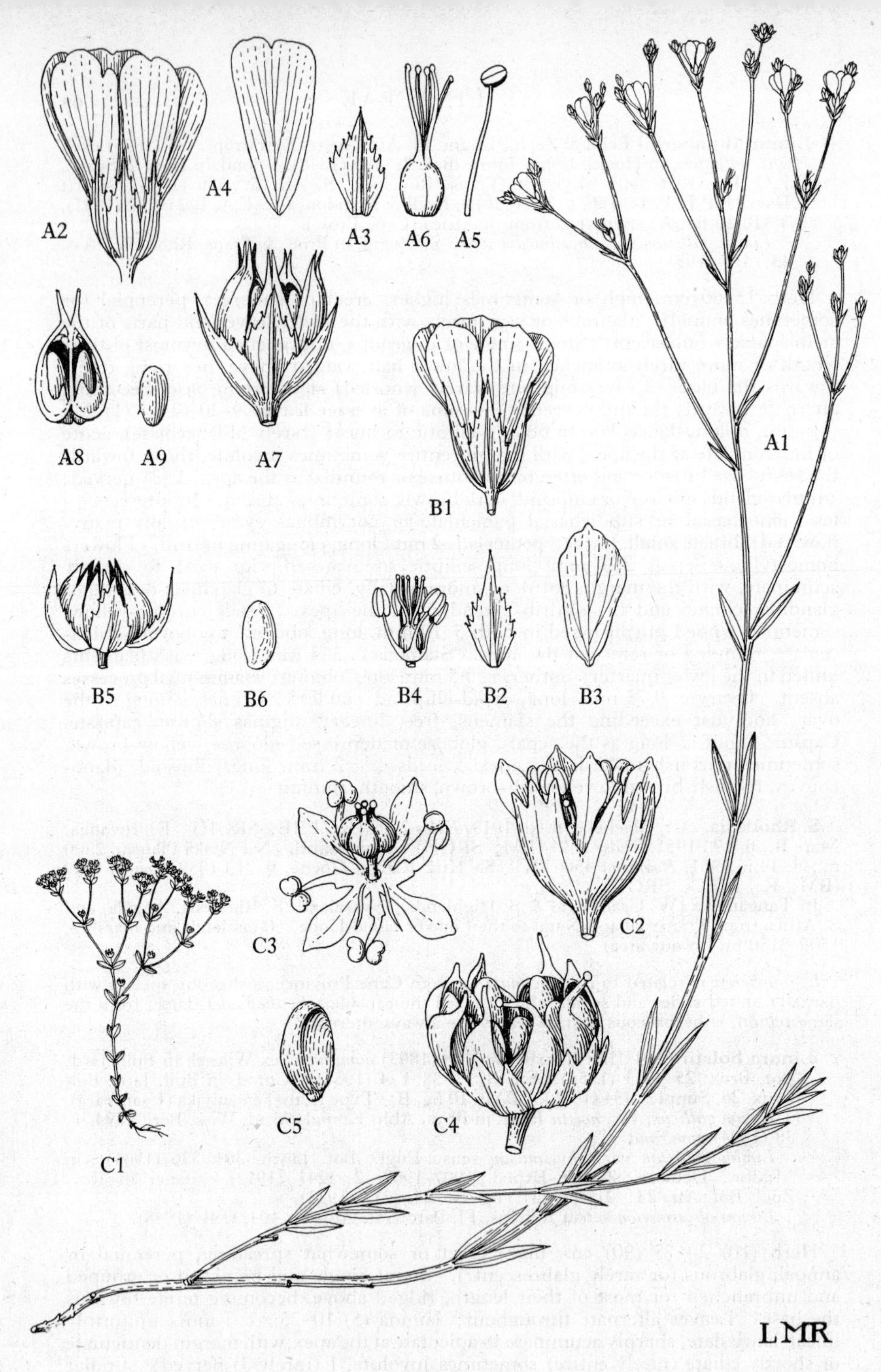

Tab. 10. A.—LINUM THUNBERGII. A1, flowering shoot (× ⅔) *Chase* 3671; A2, flower (× 4); A3, sepal (× 4); A4, petal (× 4); A5, stamen (× 4); A6, gynoecium (× 4); A7, capsule (× 6); A8, capsule valve with seeds (× 6); A9, seed (× 6), all from *Robson* 1379. B.—LINUM HOLSTII. B1, flower (× 4); B2, sepal (× 4); B3, petal (× 4); B4, androecium and gynoecium (× 4); B5, capsule (× 6); B6, seed (× 6), all from *Whyte* s.n. C.—RADIOLA LINOIDES. C1, plant (× ⅔) *Mann* 1334; C2, flower (× 18); C3, petals, staminodes and stamens united at the base (after separation from base of gynoecium) (× 18), both from *Richards* s.n.; C4, capsule (× 18); C5, seed (× 18), both from *Mann* 1334.

apex, when the lower third is more prominent than the rest). Petals dull yellow, sometimes with reddish veins, 5–8 mm. long, obovate, rounded at the apex. Stamens 2·5–3·5 mm. long, with filaments united in the lower quarter; anthers c. 0·75 mm. long, oblong-ovate; staminodial processes very short. Ovary c. 1 mm. long, globose; styles c. 2–3 times as long as the ovary, exceeding the stamens, free, linear; stigmas capitate-globose. Capsule about as long as the sepals, globose, yellow-brown, sometimes purplish towards the apex. Seeds c. 1 mm. long, oblong-ellipsoid, plano-convex, reddish-brown, smooth, shining.

Nyasaland. N: Masuku Plateau, fl. & fr. vii.1896, *Whyte* (K). **Mozambique.** N: Vila Cabral, fr. 10.v.1934, *Torre* 65 (COI; LISC).
In S. Ethiopia, E. Congo, tropical East Africa, Nyasaland and Mozambique. Grasslands and stony areas, sometimes beside streams, c. 1800–2100 m. (in our area).

L. holstii can be distinguished from *L. thunbergii* by the wholly alternate and ± uniform leaves, and by the usually more elongate sepals in which the midrib is most prominent in the lower half.

3. RADIOLA Hill

Radiola Hill, Brit. Herbal: 227 (1756).—Roth, Tent. Fl. Germ. **1**: 71 (1788).—Druce, Rep. B.E.C. **3**: 438 (1914).

Herb, annual. Leaves sessile, opposite, entire, 1-nerved, exstipulate. Flowers in terminal regularly dichotomous dichasia, homostylic. Sepals 4, (2) 3 (4)-fid, united at the base, persistent. Petals 4, white, united with the stamen filaments to form a very short tube, very shortly unguiculate. Stamens 4, with filaments united at the base only; staminodes usually absent; nectaries obscure. Ovary 4-locular, each loculus with 2 pendulous ovules and partially divided by a false septum; styles 4, free; stigmas capitate. Fruit a 4-locular 8-valved capsule with 2-seeded loculi, the seeds in each loculus partially separated by an enlarged false septum. Seeds small, with scanty endosperm; embryo straight.
A monotypic genus closely related to *Linum*.

Radiola linoides Roth, Tent. Fl. Germ. **1**: 71 (1788).—Gmel., Syst. Nat. **2, 1**: 289 (1791).—Planch. in Hook., Lond. Journ. Bot. **7**: 165 (1848).—Mansfeld in Fedde, Repert. **46**: 301 (1939).—Brenan in Mem. N.Y. Bot. Gard. **8, 3**: 230 (1953).—Keay, F.W.T.A. ed. 2, **1**, 2: 361 (1958). TAB. **10** fig. C. Type from Europe.
Linum radiola L., Sp. Pl. **1**: 281 (1753). Type as above.
Linum multiflorum Lam., Fl. Fr. **3**: 70 (1778) *nom. illegit.* Type as above.
Radiola millegrana Sm., Fl. Brit. **1**: 202 (1800); English Bot.: t. 893 (1801).—Oliv., F.T.A. **1**: 268 (1868). Type as above.
Radiola multiflora (Lam.) Aschers., Fl. Brand.: 106 (1860). Type as above.
Millegrana radiola (L.) Druce, Fl. Berks.: 114 (1897).—Fernald, Gray's Man. ed. 8: 943 (1950). Type as above.

Herb (0·2) 1–7 (10) cm. high, ± spreading, annual, glabrous. Stem as long as the inflorescence or usually shorter, terete, often purplish-red. Leaf-lamina 1–3 × 0·75–2 mm., elliptic or the lower ones obovate, acute to rounded at the apex, cuneate to rounded at the base, 1-nerved. Inflorescence a regular repeatedly dichasial many-flowered cyme, densely corymbose, flat or forming a rounded cushion; bracts foliar. Sepals c. 0·5–0·75 mm. long, (2) 3 (4)-fid, with acute teeth, united at the base. Petals white, as long as sepals, with obovate-circular lamina rounded at the apex and claw equalling the lamina or slightly longer. Stamens as long as the sepals, with anthers c. 0·1 mm. long, globose-oblong; filaments expanded at the base; staminodes subulate (when present). Ovary shallowly 8-lobed; styles free, short; stigma capitate. Capsule almost as long as the sepals, straw-coloured, 8-seeded. Seeds c. 0·3 mm. long, ± ovoid, pale brown, smooth, shining.

Nyasaland. N: Nyika Plateau, fr. 17.viii.1946, *Brass* 17292 (K).
In Europe from Ireland to S. Russia and from Scandinavia south to Spain; also in Madeira and Teneriffe, and with a scattered distribution in Africa (Morocco, Tunisia, Cameroons Mt., Ethiopia, S. Tanganyika and N. Nyasaland). Tussocky grassland, 1800–2300 m. (in our area).

33. IXONANTHACEAE

By N. K. B. Robson

Trees, shrubs, or rarely suffrutices, glabrous or with an indumentum of simple hairs. Leaves alternate, simple, penninerved, with small stipules or exstipulate. Flowers in axillary cymes or racemes, sometimes paniculate, actinomorphic, bisexual, homostylic. Sepals 5, imbricate, free or united at the base. Petals 5, contorted in bud, free, not unguiculate, persistent, often becoming indurated. Stamens 20, 10 or 5, all fertile; filaments free, inserted on the outer side of an eglandular annulus or cup; anthers dorsifixed, dehiscing longitudinally. Ovary superior, 3–5-locular, with each loculus 2-ovulate and often ± subdivided by a false septum; ovules collateral, pendulous; styles completely united, with 5 stigmas, free and radiating or united. Fruit a capsule dehiscing septicidally (sometimes also partially loculicidally) into 5 valves. Seeds slightly compressed, shining, almost surrounded by a fleshy aril or winged, with fleshy endosperm; embryo often oblique or lateral.

A family comprising 2 genera, *Ochthocosmus* (Africa and Guiana-Brazil) and *Ixonanthes* (Malaysia). The persistent petals, united styles and arillate or winged seeds distinguish it from *Linaceae*.

OCHTHOCOSMUS Benth.

Ochthocosmus Benth. in Hook., Lond. Journ. of Bot. **2**: 366 (1843).
Phyllocosmus Klotzsch in Abh. Akad. Wiss. Berl.: 232, t. 1 (1857).

Trees, shrubs, or rarely suffrutices. Leaves petiolate or subsessile, alternate, entire or with a serrate or dentate margin, frequently with caducous black glands on the teeth, usually ± coriaceous, penninerved, with densely reticulate tertiary venation; stipules small and caducous or absent. Flowers in axillary racemes, sometimes paniculate. Sepals 5, quincuncial, equal, entire or denticulate, united at the base, persistent. Petals 5, white or greenish-white to yellow, free, persistent in fruit, elongating and often becoming indurated, vertically ribbed, and appressed to the capsule. Stamens 5 or 5 + 5, with filaments inserted on the outer side of a shallow annulus; anthers dorsifixed. Ovary 5-locular, loculi 2-ovulate, with an incomplete false septum; styles terminal, united; stigmas 5, free and radiating or united to form a globular mass. Capsule globose to cylindric or ovoid, coriaceous or ± woody, septicidal, each valve frequently splitting at the apex along the false septum, 1–3(5)-seeded. Seeds smooth, endospermic, almost surrounded by a fleshy aril.

A genus of about 10 species in tropical Africa and eastern tropical South America.

Ochthocosmus lemaireanus De Wild. & Dur. in Bull. Soc. Roy. Bot. Belg. **40**: 16 (1901).—Hall. f. in Beih. Bot. Centralbl. **39**, 2: 19 (1921).—De Wild., Pl. Bequaert. **4**: 243 (1927).—Wilczek, F.C.B. **7**: 33 (1958).—White, F.F.N.R.: 167 (1962). TAB. **11.** Type from Katanga.

Phyllocosmus senensis Klotzsch ex Engl., Bot. Jahrb. **32**: 110 (1902) pro parte quoad specim. ex Senna. Type: Mozambique, Sena, 1846, *Peters* (B†).

Phyllocosmus senensis forma *latifolia* Engl., loc. cit. Type from Tanganyika (S. Highlands).

Phyllocosmus candidus Engl. & Gilg in Warb., Kunene-Samb.-Exped. Baum: 237 (1903).—R.E.Fries, Schwed. Rhod.-Kongo-Exped. **1**: 109 (1914)—White, loc. cit. Type from Angola (Bié).

Phyllocosmus lemaireanus (De Wild. & Dur.) T. & H. Dur., Syll. Fl. Cong.: 76 (1909). Type as for *Ochthocosmus lemaireanus*.

Ochthocosmus candidus (Engl. & Gilg) Hall. f., loc. cit.—Burtt Davy & Hoyle, N.C.L.: 48 (1936).—Exell & Mendonça, C.F.A. **1,** 2: 247 (1951).—White, F.F.N.R.: 167 (1962). Type as for *Phyllocosmus candidus*.

Ochthocosmus senensis (Klotzsch ex Engl.) Hall. f., loc. cit.—Brenan, T.T.C.L.: 266 (1949). Type as for *Phyllocosmus senensis*.

Ochthocosmus gillettae Hutch. in Kew Bull. **1931**: 249 (1931); Botanist in S. Afr.:

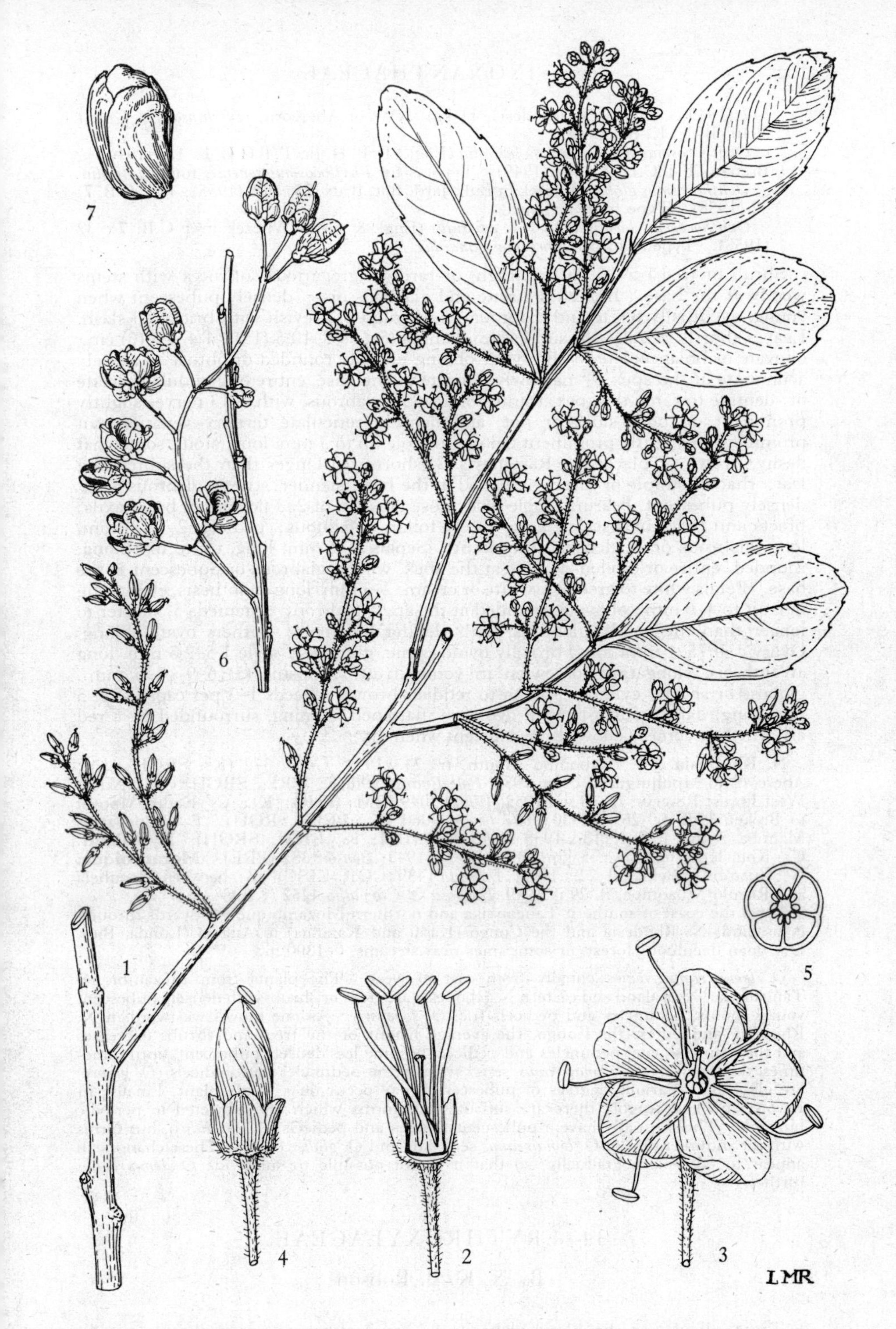

Tab. 11. OCHTHOCOSMUS LEMAIREANUS. 1, flowering branch (× ⅔); 2, young flower with petals and 2 sepals removed (× 4); 3, flower at anthesis (× 4); 4, flower closed after anthesis (× 4); 5, transverse section of ovary (× 8), all from *Richards* 6359B; 6, fruiting branch (× ⅔) *Gilges* 174; 7, seed showing aril (× 4) *Fanshawe* 1653.

512 (1949). Type: N. Rhodesia, 11 km. NW. of Abercorn, *Hutchinson & Gillett* 3910 (BM; K, holotype).

Ochthocosmus senensis forma *latifolia* (Engl.) F. B. Hora, T.T.C.L. **1**: 134 (1940).—Brenan, T.T.C.L. **2**: 267 (1949). Type as for *Phyllocosmus senensis* forma *latifolia*.

Ochthocosmus glaber Wilczek in Bull. Jard. Bot. Brux. **25**: 308 (1955); in F.C.B. **7**: 32 (1958). Type from Katanga.

Ochthocosmus lemaireanus var. *candidus* (Engl. & Gilg) Wilczek in F.C.B. **7**: 32 (1958). Type as for *Phyllocosmus candidus*.

Shrub or tree 1·5–10 (20) m. high, or rarely a gregarious suffrutex with stems woody at the base. Branches striate and glabrous or ± densely pubescent when young, eventually terete and glabrous; bark rugose, greyish with brick-red slash. Leaves shortly petiolate or subsessile; lamina (5·2) 6·5–10·5 (15) × 1·4–4·5 (6) cm., obovate or oblanceolate to elliptic or oblong-elliptic, rounded or obtuse to acutely acuminate at the apex, ± narrowly cuneate at the base, entire or glandular-serrate or -dentate towards the apex, thinly coriaceous, glabrous, with 6–13 nerves slightly prominent on both sides or not, and densely reticulate tertiary venation not prominent or slightly prominent below; petiole up to 3 mm. long, stout, somewhat fleshy, brown or blackish. Racemes 1–3, shorter to longer than the subtending leaf; rhachis simple or rarely branched at the base, slender, striate, glabrous or ± densely pubescent, bearing single or cymose clusters of 2–3 flowers in bract-axils; bracts and bracteoles scale-like or rarely foliar, deciduous; pedicels 2–5 (8) mm. long, glabrous or ± densely pubescent. Sepals c. 2 mm. long, ovate to oblong, rounded, entire or eroded-dentate at the apex, wholly glabrous or pubescent at the base. Petals white to greenish-white or cream, 2–3 mm. long at anthesis, elongating in fruit to 4–6 mm., obovate, rounded at the apex, glabrous. Stamens 5, shorter to longer than the petals, glabrous, with slender filaments; anthers ovate-oblong. Ovary c. 0·75–1 mm. long, broadly ovoid-conic, glabrous; style 1·5–2·5 mm. long at anthesis, elongating somewhat in young fruit. Capsule (5) 6–9 × 4–6 mm., globose or shortly cylindric, green to reddish-brown. Seeds 1–3 per capsule, c. 5 mm. long, dark reddish-brown, ovoid, ± flattened, shining, surrounded by a red aril which becomes yellowish-translucent when dry.

N. Rhodesia. B: Kabompo Mouth, fr. 24.v.1954, *Gilges* 372 (K; SRGH). N: Abercorn to Mpulungu, fl. 19.vii.1930, *Pole Evans* 2996 (K; PRE; SRGH). W: Ndola West Forest Reserve, fr. 13.viii.1952, *White* 3049 (BM; FHO; K). C: Kapiri Mposhi to Broken Hill, fr. 26.vii.1930, *Pole Evans* 3060 (K; PRE; SRGH). E: Lundazi to Mzimba, 3 km., fl. & fr. 18.x.1958, *Robson* 144 (BM; K; LISC; SRGH). **Nyasaland.** C: Kota Kota to Dowa, 5 km., fl. & fr. 3.x.1943, *Benson* 382 (PRE). **Mozambique.** N: Cuamba to Belém, fl. 22.x.1948, *Andrada* 1439 (COI; LISC). Z: between Mocubela and Régulo Muaquiua, fl. 29.ix.1949, *Barbosa & Carvalho* 4252 (LISC).

From the coast of southern Tanganyika and northern Mozambique westwards through Nyasaland, N. Rhodesia and the Congo (Kasai and Katanga) to Angola (Lunda, Bié). Dry open deciduous forest, or sometimes near streams, 0–1300 m.

O. lemaireanus varies clinally from east to west. The plants from Mozambique, Tanganyika, Nyasaland and eastern N. Rhodesia are trees or shrubs with densely pubescent young shoots, peduncles and pedicels (i.e. *O. senensis*). As one moves westward in N. Rhodesia and enters the Congo, the average height of the trees and shrubs decreases and the young shoots, peduncles and pedicels become less densely pubescent, until sometimes the pedicels (*O. lemaireanus* sens. str.) or the peduncles and pedicels (*O. glaber*) are glabrous. Various degrees of pubescence may occur on a single plant. Finally, in Barotseland and Angola there are suffrutescent forms which are subjected to periodic burning. These usually have a pubescent rhachis and pedicels (*O. candidus*), but forms with the indumentum of *O. lemaireanus* sens. str. and *O. glaber* occur. These changes all appear to take place gradually, so that it is not possible to subdivide *O. lemaireanus* further.

34. ERYTHROXYLACEAE

By N. K. B. Robson

Trees, shrubs or shrublets, glabrous. Leaves alternate (or rarely opposite*), simple, entire, penninerved, with stipules ± united and intrapetiolar (or rarely interpetiolar*). Flowers axillary, solitary or in fascicles (rarely pedunculate), actinomorphic, bisexual or rarely subdioecious, heterostylic. Sepals (4) 5, valvate,

± united. Petals (4) 5, contorted in bud, free, unguiculate, caducous, usually with a ventral ligule-like nectariferous appendage. Stamens 5 + 5, all fertile, with filaments united at the base to form a deep urceole or shallow rarely glandiferous cup; anthers basifixed, dehiscing longitudinally. Ovary superior, 2–3 (4)-locular, with each loculus 1-ovulate (or rarely 2-ovulate*); ovules pendulous; styles 2–3 (4), free or ± united, with stigmas clavate to obliquely capitate or depressed-capitate rarely acute. Fruit a 1-seeded fleshy drupe (or rarely a 3 (4)-locular 3 (4)-seeded capsule dehiscing longitudinally*). Seeds not compressed, exarillate, without or with scanty endosperm; embryo straight, with flat or semiconvex cotyledons.

A family of 4 genera in tropical and warm-temperate regions of both hemispheres. The genus *Aneulophus*, from the lower Congo and Gaboon, is atypical in several respects (see characters marked * above).

Styles 3 (very rarely 2), free or partially united; stigmas capitate or obliquely capitate; petal-nectaries large, convolute, exceeding the sepals; flowers opening on leafy shoots **1. Erythroxylum**
Styles 2, completely united; stigmas flattened, spreading or reflexed; petal-nectaries small, hidden by the sepals; flowers (in our species) opening on leafless shoots **2. Nectaropetalum**

1. ERYTHROXYLUM Browne

Erythroxylum Browne, Hist. Jam.: 278 (1756).

Trees, shrubs or shrublets, with young shoots usually ± compressed. Leaves petiolate, alternate, simple, entire, involute when young, with venation varying in prominence and degree of reticulation; stipules intrapetiolar, ± united, persistent or partially to completely deciduous or caducous. Flowers axillary, in fascicles or solitary, heterostylic. Sepals 5, valvate, triangular, connate and broadened at the base, ± coriaceous. Petals white to cream-yellow, free, unguiculate; nectary appendage (ligule) large, usually exceeding the calyx, entire or 2–3-lobed. Stamens 5 + 5, with filaments equal or unequal (antisepalous shorter than antipetalous), united in the lower part to form a deep cup with margin entire or ± denticulate. Ovary (2) 3-locular with 1-ovulate loculi; styles free or partially united, divergent; stigmas obliquely capitate, rarely acute. Fruit a 1-seeded drupe, with sterile loculi ± apparent. Seed oblong-elliptic, usually with fleshy endosperm.

A genus of c. 200 species distributed throughout the tropics and in warm temperate regions, but most abundant in America and Madagascar.

Stipules 2–3-setose, with fimbriate margin and persistent base; young shoots angular, scarcely compressed; leaves with midrib usually more prominent above than below 1. *gerrardii*
Stipules entire, persistent or deciduous; young shoots ± compressed; leaves with midrib equally prominent on both sides or more prominent below:
Leaves with prominent venation (usually on both sides), always emarginate; drupe ampullaceous, sometimes becoming broadly ellipsoid; styles free in both forms of flower; stipules persistent - - - - - - - - - 2. *emarginatum*
Leaves with reticulate venation visible but not prominent (or slightly prominent above), emarginate or not; drupe cylindric or cylindric-ellipsoid, usually becoming 3-gonal; styles free or partly united; stipules persistent or not:
Styles free in both forms of flower; leaves with apex mucronulate and rounded to shallowly emarginate, venation below densely reticulate and clearly visible (at least near the midrib); stipules deciduous - - - - - 3. *platclyadum*
Styles partially united in both forms of flower; leaves with apex rounded to shallowly emarginate but not apiculate, venation below lax and rather indistinct; stipules persistent or not:
Petiole 1·5–3 mm. long; lamina oblanceolate (rarely obovate), not decurrent; stipules deciduous - - - - - - - - - - - 4. *delagoense*
Petiole 5–12 mm. long; lamina obovate (more rarely oblong or elliptic), usually decurrent along the petiole; stipules persistent - - - - 5. *zambesiacum*

1. **Erythroxylum gerrardii** Bak. in Journ. Linn. Soc., Bot. **20**: 109 (1883).—O.E. Schulz, Pflanzenr. IV, 134: 128 (1907).—Perrier in Mém. Inst. Sci. Madag., Sér. B: 247 (1950); Fl. Madag., Erythrox.: 11 (1952) excl. var. *sylvicola* Perrier. TAB. **12** fig. C. Type from Madagascar.

Shrub (1) 1·5–3 (5) m. high, much branched, evergreen, glabrous; branches slender, red or blackish, scarcely flattened and angular with raised lines decurrent

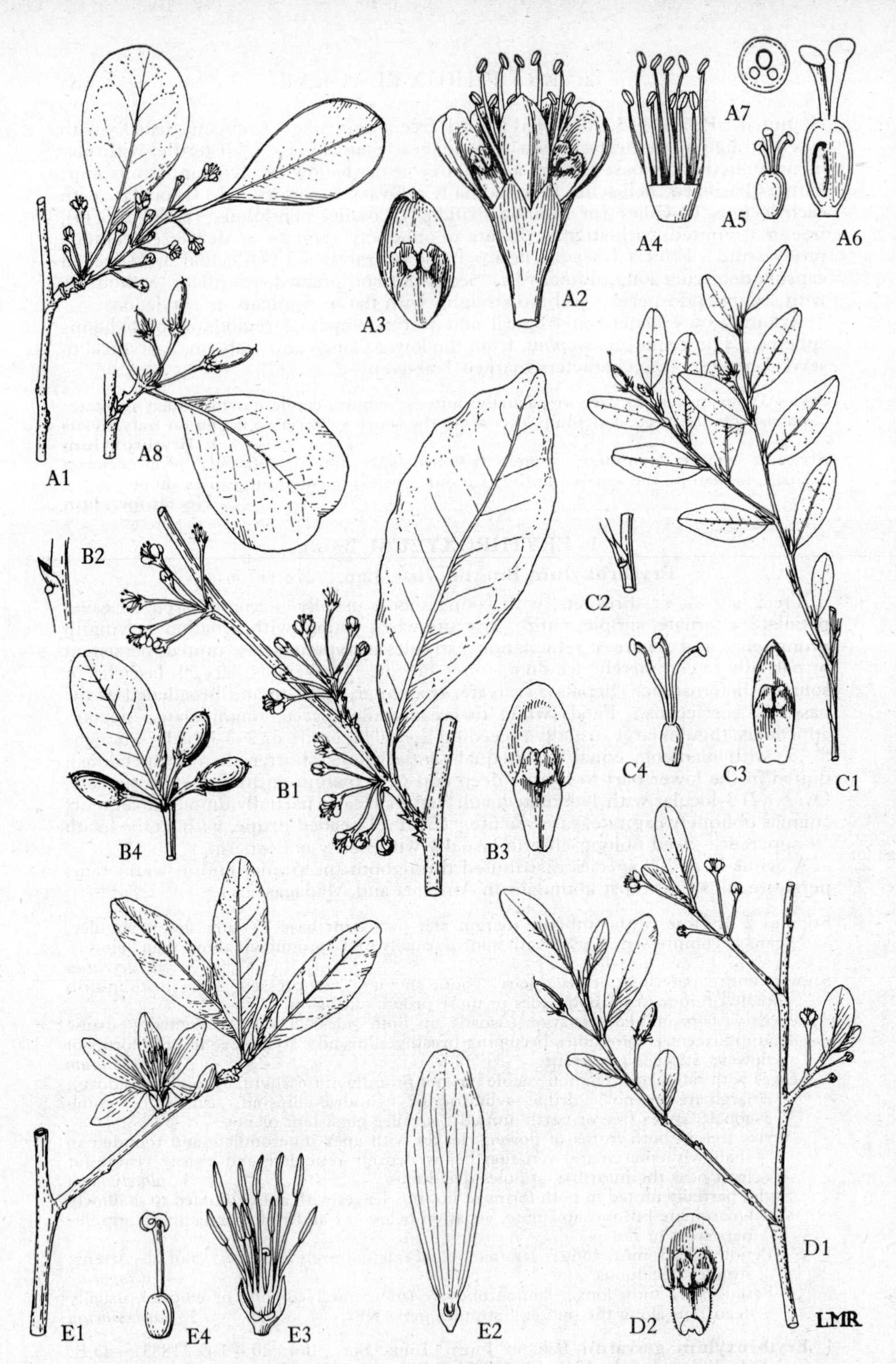

Tab. 12. A.—ERYTHROXYLUM ZAMBESIACUM. A1, flowering branch ($\times \frac{2}{3}$); A2, flower ($\times 6$); A3, petal ($\times 6$); A4, stamen-tube ($\times 6$); A5, gynoecium ($\times 6$); A6, longitudinal section of gynoecium ($\times 12$); A7, transverse section of ovary ($\times 12$), all from *Phelps* 101; A8, fruiting branch ($\times \frac{2}{3}$) *Drummond* 5396. B.—ERYTHROXYLUM EMARGINATUM. B1, flowering branch ($\times \frac{2}{3}$); B2, stipule ($\times 2$); B3, petal ($\times 6$), all from *Faulkner* 221; B4, fruiting branch ($\times \frac{2}{3}$) *Kirk* s.n. C.—ERYTHROXYLUM GERRARDII. C1, fruiting branch ($\times \frac{2}{3}$); C2, stipule ($\times 2$); C3, petal ($\times 6$); C4, gynoecium ($\times 6$), all from *Gomes e Sousa* 4353. D.—ERYTHROXYLUM DELAGOENSE. D1, flowering and fruiting branch ($\times \frac{2}{3}$); D2, petal ($\times 6$), both from *Barbosa* 7931. E.—NECTAROPETALUM CARVALHOI. E1, flowering branch with young leaves ($\times \frac{2}{3}$); E2, petal ($\times 2$); E3, flower with petals and 2 sepals removed ($\times 2$); E4, gynoecium ($\times 4$), all from *Carvalho* s.n.

from the petiole bases when young, becoming terete and brown; bark smooth, finely vertically fissured, pale grey-brown. Leaves petiolate; lamina 0·8–3·5 (4·3) ×0·3–1·4 (2·1) cm., elliptic or rhombic to oblanceolate or rarely obovate, not or scarcely acuminate, rounded or shallowly (rarely deeply) emarginate at the apex, cuneate at the base, chartaceous, dark- or brownish-green above, paler below, sublucent on both surfaces or dull (rarely glaucous) below, with midrib usually ± prominent above, but nerves and venation inconspicuous or slightly prominent above, more rarely also visible below; petiole 0·75–1·5 mm. long, rather slender, grooved above; stipules partially united, 1·5–5 mm. long, triangular-amplexicaul with fimbriate margin, each free upper part bearing a long slender fimbriate seta, sometimes also with a short median entire seta, or the stipules almost completely united, with three short terminal setae, the setae and fimbriate margin frequently breaking off leaving the persistent entire base. Flowers solitary, axillary; peduncle 2–5 mm. long, slender, angular. Sepals 0·75–1·5 mm. long, triangular, acute, united for c. ½ their length. Petals white, 2–2·5 mm. long, oblong, rounded at the apex, shortly unguiculate; ligule c. 1 mm. long, with eroded margin, inserted ¼ of the length above the base. Short-styled flowers: stamens equal, 2·5 mm. long; stamen-cup equal to the sepals, with 10-crenulate margin; ovary c. 1 mm. long, as long as the stamen-cup, cylindric; styles 3, 1–2·5 mm. long, united at the base. Long-styled flowers: stamens unequal, the antisepalous 1·75 mm. and antipetalous 2·5 mm. long; stamen-cup as above; styles 3, 2·75–3 mm. long, united in the lower half. Drupe 6–7·5 mm. long, red, obliquely ovoid.

Mozambique. N: Ribáuè, Serra de Ribáuè, fl. 13.x.1935, *Torre* 655 (COI; LISC). MS: Chinizíua, near Gama village, fl. & fr. 12.iv.1957, *Gomes e Sousa* 4353 (K; PRE; SRGH).

Also in western and southern Madagascar. In damp shady places, 0–1200 m.

A variable species, each population of which tends to differ from the others. It does not seem possible, however, to maintain Perrier's subspecies and varieties except var. *sylvicola* which even appears to belong to another section of the genus. The leaves of the type specimen are more glaucous below than those of any subsequent gathering, but the other characters of this specimen can be matched in later collections. The setose fimbriate stipules distinguish *E. gerrardii* from all other species of *Erythroxylum* on the African mainland.

2. **Erythroxylum emarginatum** Thonn. in Schum. & Thonn., Beskr. Guin. Pl.: 224 (1827) (" Erytroxylon emarginatus ").—Oliv., F.T.A. **1**: 274 (1868).—O.E.Schulz, Pflanzenr., IV, 134: 135 (1907).—Steedman, Trees, etc. S. Rhod.: 27, t. 23 (1933).—Burtt Davy & Hoyle, N.C.L.: 42 (1936).—Brenan, T.T.C.L.: 194 (1949); in Mem. N.Y. Bot. Gard. **8**, 3: 230 (1953).—Exell & Mendonça, C.F.A. **1**, 2: 248 (1951).—Keay, F.W.T.A. ed. 2, **1**, 2: 356 (1958).—White, F.F.N.R.: 168 (1962). TAB. **12** fig. B. Type from Ghana.

Erythroxylum caffrum Sond. in Linnaea, **23**: 22 (1850); in Harv. & Sond., F.C. **1**: 233 (1860). Type from Natal.

Erythroxylum emarginatum var. *caffrum* (Sond.) O.E.Schulz, tom. cit.: 136 (1907).—Bak. f. in Journ. Linn. Soc., Bot. **40**: 34 (1911).—Phillips in S. Afr. Journ. Sci. **32**: 310 (1935).—Burtt Davy & Hoyle, loc. cit.—Suesseng. in Proc. & Trans. Rhod. Sci. Ass. **43**: 108 (1951). Type as for *E. caffrum*.

Erythroxylum emarginatum var. *angustifolium* O.E.Schulz, tom. cit.: 136 (1907).—Burtt Davy & Hoyle, loc. cit. Type: Nyasaland, *Buchanan* 971 (K).

Erythroxylum monogynum sensu Sim, For. Fl. Port. E. Afr.: 21 (1909) pro parte.

Erythroxylum monogynum var. *caffrum* (Sond.) Eyles in Trans. Roy. Soc. S. Afr. **5**: 387 (1916). Type as for *E. caffrum*.

Shrub or small tree (1) 1·5–9 m. high, occasionally taller (up to 18 m.), evergreen, glabrous; branches rather stout, somewhat flattened and with raised lines decurrent from the petiole-bases when young, soon terete, pale grey or grey-brown, almost smooth; bark vertically fissured, dark grey or grey-brown, sometimes pink-tinged. Leaves petiolate; lamina (1·5) 2–12·5 (15) ×(0·6) 1–4·5 (5) cm., varying from oblanceolate or rhombic (rarely obovate) to elliptic (rarely subcircular), rounded or shortly acuminate towards the apex but always ± shallowly emarginate at the tip, narrowly to broadly cuneate (rarely rounded or truncate) at the base, coriaceous, bright to dark green or bluish-green and shining above, paler green or pale reddish-brown below, with midrib and ± loosely reticulate venation prominent on both sides; petiole 1·5–6 mm. long, rather slender, grooved above; stipules completely united, 1–2 mm. long, broadly triangular, acute, with entire membranous margin, persistent. Flowers solitary or in fascicles of 2–8 in the leaf-axils, usually borne in clusters on stem regions with short internodes; pedicels 5–14 mm. long, rather slender, angular. Sepals c. 1 mm. long, triangular, acute, united for about ¼ of their length. Petals white, c. 4 mm. long, oblong, rounded at the apex, shortly unguiculate; ligule 1–1·5 mm. long, shortly fimbriate, inserted ¼ of the length above the base. Short-styled flowers: stamens equal, 4 mm. long; stamen-cup equal to the sepals, usually with 10-denticulate margin; ovary c. 1·5 mm. long, slightly longer than stamen-cup, obovoid; styles 3, 1·5 mm. long, free. Long-styled flowers: stamens unequal, the antisepalous 1·75 mm. and antipetalous 3·5 mm. long; stamen-cup as above; ovary as above; styles 3, 3–3·5 mm. long, free. Drupe 8–12 mm. long, bright red, ampullaceous at first but occasionally becoming broadly ellipsoid.

N. Rhodesia. N: Abercorn to Tunduma, 40 km., fl. 22.i.1947, *Brenan & Greenway* 8182 (BM; FHO; K). W: Mwinilunga Distr., Zambezi R. 6·4 km. N. of Kalene Hill Mission, fl. 20.ix.1952, *Angus* 495 (BM; FHO; K; PRE). S: Mazabuka Distr., Choma to Namwala, km. 48, Ngongo R., fr. 22.vi.1952, *White* 2957 (FHO; K). **S. Rhodesia.** N: Urungwe, Nyagugutu R., fr. 21.xi.1957, *Goodier* 398 (K; SRGH). W: Matopos, Diana's Pool, fl. ix.1947, *Hodgson* 8/47 (SRGH). C: Chilimanzi, 6·4 km. Umvuma-Gwelo, fr. 6.ii.1951, *Greenhow* 615/51 (SRGH). E: Chimanimani Mts., fl. 26.ix.1906, *Swynnerton* 1364 (BM; K). S: Fort Victoria, Mtilitswa R., fl. xi.1956, *Davies* 2181 (SRGH). **Nyasaland.** C: Dowa Distr., Nchisi Mt., fr. 19.ii.1959, *Robson* 1670 (BM; K; LISC; SRGH). S: Cholo Distr., Nswadzi R., fl. 29.ix.1946, *Brass* 17866 (BM; K; SRGH). **Mozambique.** N: Niassa, Unango Mission, fl. 1.xi.1934, *Torre* 598 (COI; LISC). Z: Lugela, Namagoa, fl. & fr. xi.1946, *Faulkner* K.117 (COI; K; SRGH). T: Moatize, Mt. Zóbuè, fl. 18.x.1943, *Torre* 6057 (LISC). MS: Manica, Dombe-Sambanhe km. 11, Matindire, fl. 29.x.1953, *Pedro* 4512 (K; LISC; PRE). SS: Zavala, Quissico, fl. 6.xii.1944, *Mendonça* 3283 (LISC). LM: Namaacha, near Fonte de Goba, fl. 25.xi.1944, *Mendonça* 3054 (LISC).

Also from Guinea to Kenya and Tanganyika, and in Angola, Natal, Transvaal and Cape Province. In the understorey of evergreen forest, forest margins, thickets and fringing forest, and also sometimes on rocky outcrops, sandy soils or sand dunes, 0–c. 1900 m.

The leaves of *E. emarginatum* are very variable in shape and size. Narrow-leaved and small-leaved forms have been distinguished as var. *angustifolium* and var. *caffrum* respectively; but although these extreme forms appear distinct, they are connected with the typical ones by continuous series of intermediates and cannot therefore be given separate taxonomic rank.

E. fischeri Engl., an East African species, is found in the south of Tanganyika and may well occur also in our area. It can be easily distinguished from *E. emarginatum* by the longer anthers (1–1·25 mm. as against 0·5 mm.) and the cylindric or triquetrous fruits.

3. **Erythroxylum platycladum** Boj. in Ann. Sci. Nat., Sér. 2, Bot. **18**: 186 (1842).—O.E.Schulz, Pflanzenr. IV, 134: 151 (1907).—Brenan, T.T.C.L.: 195 (1949).—Perrier in Mem. Inst. Sci. Madag., Sér. B, **2**: 258 (1950); Fl. Madag., Erythrox.: 33, t. 5 figs. 1–7 (1952). Type from Madagascar.
Erythroxylum crassipes Baill. in Bull. Soc. Linn. Par. **1**: 605 (1886). Type from Madagascar.

Shrub or small tree 2–6 m. high, glabrous; branches numerous, spreading, rather stout, ± flattened and with raised lenticels when young, eventually terete; bark striate, reddish-brown to dark grey or whitish. Leaves petiolate; lamina 2·5–7 (8) × 1·8–4·2 (4·8) cm., obovate or rarely obcordate, mucronulate and rounded to truncate or shallowly emarginate at the apex, cuneate or angustate at the base, subcoriaceous, bright to brownish-green but not shining above, paler or ferrugineous below, ± plane on both sides except for the slightly prominent midrib below, with very densely reticulate venation clearly visible below especially towards the midrib; petiole 2–5 mm. long, slender, grooved above; stipules completely united, 2–5 mm. long, triangular-lanceolate, acute to somewhat obtuse, with entire margin, bicarinate, deciduous. Flowers solitary or in fascicles of 2–6 in the axils of foliage or scale leaves; pedicels (2) 5–10 (14) mm. long, slender, subterete. Sepals 1–2 mm. long, lanceolate to oblong-lanceolate, acute, united for c. ¼ of their length. Petals white, 2·5–3 mm. long, oblong, rounded and entire at the apex, dorsally carinate, shortly and narrowly unguiculate; ligule 0·5–0·8 mm. long, emarginate, inserted c. ⅓ of the length above the base. Short-styled flowers: stamens equal, c. 3·5 mm. long; stamen-cup much shorter than the sepals, with plane margin;

ovary c. 1 mm. long, ovoid-cylindric; styles 3, c. 1–1·5 mm. long, free. Long-styled flowers: stamens subequal, 2–2·5 mm. long; stamen-cup as above; ovary as above; styles 3, c. 2·5 mm. long, free (? or sometimes shortly connate at the base). Drupe 5–6 mm. long, red, cylindric-ellipsoid or obtusely 3-gonal when dry.

Mozambique. N: Tecomaze I., fr. 29.iii.1961, *Gomes e Souza* 4676 (K).

In coastal areas from northern Kenya to northern Mozambique and in the Comoro Is. and northern Madagascar. Coral rock and sand dunes near sea level, but occurs up to 150 km. inland and at altitudes up to 600 m. in Madagascar where it also grows in woodland.

Schulz (l.c.) and Perrier de la Bâthie (l.c.) state that the long-styled flowers have styles very shortly united at the base (unlike those of other species in Sect. *Venelia*, in which they are free). In all specimens which I have examined, however, they are completely free.

4. **Erythroxylum delagoense** Schinz in Bull. Herb. Boiss., Sér. 2, **1**: 876 (1901).—O.E.Schulz, Pflanzenr. IV, 134: 148 (1907). TAB. **12** fig. D. Type: Mozambique, Delagoa Bay, *Junod* 207 (Z, holotype).

Erythroxylum monogynum sensu Harv., F.C. **2**: 591 (1862).—Sim, For. Fl. Port. E. Afr.: 21 (1909) pro parte.

Erythroxylum pulchellum Engl., Bot. Jahrb. **34**: 149 (1904) pro parte quoad specim. Mozamb. Lectotype: Mozambique, Lourenço Marques, *Schlechter* 11600 (B†).

Erythroxylum brownianum Burtt Davy in Kew Bull. **1924**: 230 (1924); F.P.F.T. **2**: 286, t. 43 A (1932).—Phillips in S. Afr. Journ. Sci. **32**: 311 (1935). Type from Natal.

Shrub or small tree (0·6) 1–6 (8) m. high, glabrous; branches numerous, erect, slender, smooth and only slightly flattened when young, soon terete; bark grey, verrucose. Leaves petiolate; lamina 1·2–3·2 (4·5) × 0·75–1·6 cm., oblanceolate or rarely obovate, rounded or more rarely obtuse or subretuse at the apex, narrowly cuneate at the base, chartaceous, pale green, paler below, plane on both sides except for the slightly prominent midrib below, with very loosely reticulate venation scarcely visible below; petiole 1·5–3 mm. long, slender, grooved above; stipules completely united, 1·5–4 mm. long, lanceolate, acute or retuse, with entire margin, bicarinate, deciduous. Flowers solitary or paired (rarely in fascicles of 3–4) in the axils of fallen leaves; pedicels (3) 5–9 (11) mm. long, slender, angular. Sepals 1–1·5 mm. long, ovate-triangular, acute, united for $\frac{1}{4}$–$\frac{1}{2}$ of their length. Petals white or yellowish- to greenish-white, c. 3 mm. long, oblong, rounded and crenulate to fimbriate at the apex, dorsally carinate, shortly and narrowly unguiculate; ligule 1–1·25 mm. long, entire, inserted $\frac{1}{3}$–$\frac{1}{4}$ of the length above the base. Short-styled flowers: stamens equal, c. 1·5 mm. long; stamen-cup shorter than (or rarely equal to) the sepals, with slightly undulate margin; ovary c. 1 mm. long, ovoid-conic; styles (2) 3, c. 0·5 mm. long, united at the base. Long-styled flowers: stamens slightly unequal, the antisepalous 1·25 mm. and antipetalous 1·75 mm. long; stamen-cup as above; ovary as above; styles (2) 3, c. 2 mm. long, united for $\frac{2}{3}$–$\frac{3}{4}$ their length (or rarely almost up to the stigmas). Drupe 8–10 mm. long, red, cylindric-ellipsoid or 3-gonal when dry.

Mozambique. SS: Muchopes, 30 km. from Manjacaze, fr. 26.i.1941, *Torre* 2531 (LISC). LM: Marracuene, near Bobole, fl. & fr. 2.x.1957, *Barbosa* 7931 (COI; K; LISC; SRGH).

Also occurs in Swaziland, Natal and eastern Transvaal. In deciduous forest or scrub, usually on dry stony or sandy soils, 0–1350 m.

One syntype of *E. pulchellum* (*Schlechter* s.n.) came from Zanzibar but, like the lectotype, it no longer exists. It was almost certainly a specimen of *E. platycladum* Boj., small-leaved forms of which resemble *E. delagoense* to some extent.

5. **Erythroxylum zambesiacum** N. Robson in Bol. Soc. Brot., Sér. 2, **36**: 7 (1962). TAB. **12** fig. A. Type: S. Rhodesia, Sebungwe, Kariangwe Hill, *Lovemore* 213 (K, holotype; SRGH).

Erythroxylum monogynum sensu Eyles in Trans. Roy. Soc. S. Afr. **5**: 387 (1916).
Erythroxylum sp.—O.B.Mill. in Journ. S. Afr. Bot. **18**: 37 (1952).
Erythroxylum sp. 1.—White, F.F.N.R.: 168 (1962).

Shrub or small tree 3–7·5 (9) m. high, glabrous; branches slender, somewhat flattened and with raised lines decurrent from the petiole-bases when young, eventually terete; bark smooth, grey. Leaves petiolate; lamina 1·8–6·7 × 1·3–4·2 cm., obovate or more rarely elliptic or oblong, rounded or truncate to emarginate at the apex, narrowly cuneate or decurrent at the base, subchartaceous, pale green, paler below, with nerves usually slightly prominent above and loosely to rather densely reticulate venation visible below; petiole 5–10 (12) mm. long, slender, grooved above; stipules completely united, 1·5–3 mm. long, broadly triangular, obtuse, bicarinate, with entire margin, persistent. Flowers solitary or in fascicles of 2–5 in the leaf-axils; pedicels 4–8 mm. long, slender, angular. Sepals c. 1 mm. long, ovate-triangular, acute to obtuse or apiculate, united for $\frac{1}{2}$–$\frac{2}{3}$ of their length. Petals white or yellowish-white, c. 4 mm. long, oblong, rounded and ± shallowly crenulate at the apex, shortly unguiculate; ligule 1–1·5 mm. long, shortly fimbriate, inserted $\frac{1}{6}$–$\frac{1}{4}$ of the length above the base. Stamens equal; stamen-cup longer than the sepals but shorter than the ovary, with a slightly undulate margin. Short-styled flowers: stamens c. 5 mm. long; ovary c. 1·5 mm. long, cylindric; styles c. 1 mm. long, united for c. $\frac{3}{5}$ of their length. Long-styled flowers: stamens c. 2 mm. long; ovary as above; styles 2·5–3 mm. long, united for c. $\frac{5}{6}$ of their length. Drupe 9 mm. long, cylindric or 3-gonal when dry.

Bechuanaland Prot. N: Chobe Distr., Kasane, fl. xii.1948, *Miller* B/825 (K; PRE). **N. Rhodesia.** S: Mazabuka Distr., c. 3 km. from Chirundu Bridge on Lusaka road, fr. 1.ii.1958, *Drummond* 5411 (K; SRGH). **S. Rhodesia.** N: Urungwe, half-way down Zambezi escarpment on main road, fl. & fr. 21.ii.1957, *McGregor* 55/57 (K; SRGH). W: Victoria Falls, fl. 12.ii.1912, *Rogers* 5539 (K; SRGH).

Apparently confined to the Zambezi valley area, from Kariba to above the Victoria Falls. In *Colophospermum* woodland and on rocky hillsides, 600–1100 m.

E. zambesiacum is related to the W. African *E. mannii* Oliv., but differs *inter alia* by the more slender and less strongly flattened branches and the shape of the leaves.

2. NECTAROPETALUM Engl.

Nectaropetalum Engl., Bot. Jahrb. **32**: 109 (1902).—Stapf & Boodle in Kew Bull. **1909**: 188 (1909).

Peglera Bolus in Kew Bull. **1907**: 362 (1907).

Shrubs or small trees. Young shoots terete or ± compressed. Leaves petiolate, alternate, simple, entire, ± coriaceous, penninerved, with densely reticulate tertiary venation; stipules intrapetiolar, usually completely united, deciduous or caducous. Flowers axillary, solitary or in few-flowered fascicles, heterostylic. Sepals (4) 5, valvate, connate at the base, ± coriaceous. Petals (4) 5, white, free, unguiculate, deciduous; nectary appendage small (more rarely absent). Stamens 5 + 5, with filaments of equal or unequal length united below to form a shallow cup sometimes bearing 5 glands outside. Ovary 2-locular, loculi 1-ovulate; styles united; stigmas 2, free, spreading or reflexed. Fruit probably drupaceous.

A genus of 6 species, 5 in coastal regions from northern Kenya to the Cape and 1 in the Congo.

Phillips (S. Afr. Journ. Sci. **32**: 308 (1935)) decided that *Nectaropetalum* was not distinct from *Erythroxylum*; but, although a few species of *Erythroxylum* have styles partially united, and 1 or 2 are sometimes found with 2-carpellary ovaries, the occurrence of 2-carpellary ovaries together with completely united styles, relatively large petals with short nectary scales (or rarely without these), and a shallow filament cup, is sufficient to distinguish *Nectaropetalum*. *N. lebrunii* G. Gilbert, from the Congo, is unusual in having small flowers with large nectary appendages and a deep filament cup, and may have been independently derived from species of *Erythroxylum*. *Pinacopodium* (Cabinda and Gaboon) differs from *Nectaropetalum* in having a pedunculate inflorescence and a very short stylar column.

Nectaropetalum carvalhoi Engl., Bot. Jahrb. **32**: 109 (1902); Pflanzenw. Afr. **3**, 1: 723, t. 335, fig. G-O (1915). TAB. **12** fig. E. Type: Mozambique, between Mossuril and Cabeceira, (" Mussori le Cabeceira "), *Carvalho* (B, holotype †; COI).

Erythroxylum carvalhoi (Engl.) Phillips in S. Afr. Journ. Sci. **32**: 308 (1935). (" carvahoi "). Type as above.

Shrub 1–2 m. high, glabrous; branches flattened when young, becoming terete, smooth or faintly striate, with reddish- to greyish-brown bark. Leaves petiolate; lamina (young) c. 4 × 1·3 cm., elliptic to oblanceolate, rounded at the apex, cuneate at the base; petiole 1·5–2 mm. long; stipules completely united, 4·5–5 mm. long, linear-lanceolate, reddish-brown, soon deciduous. Flowers solitary in the axils of fallen leaves, subsessile. Sepals 4 mm. long, lanceolate, acute. Petals 20–28 × 6–7 mm., narrowly oblong-oblanceolate, rounded at the apex; nectary depression shallow, with entire margin. Stamens with linear filaments united below to form a shallow cup; anthers 2–3 mm. long, linear. Ovary c. 1 mm. long, subglobose to cylindric; stylar column c. 2 mm. long (in brachystylic flower); stigmas recurved spirally. Fruit not yet collected.

Mozambique. N: between Mossuril and Cabeceira, fl. 1884–1885, *Carvalho* (COI).

In northern Mozambique and probably also southern Tanganyika, 0–230 m. *Schlieben* 5784 from Mlingura (Tanganyika, Lindi Distr.) probably also belongs to this species, although the floral parts are smaller.

35. MALPIGHIACEAE

By E. Launert

Mostly woody climbers, sometimes shrubs or small trees, with unicellular appressed (sometimes V-shaped) medifixed ± stiff hairs. Leaves opposite, ternate or alternate, simple and entire, often with glands near the base of the lamina or on the petiole; stipules present or absent. Inflorescence terminal or axillary, usually many-flowered and racemose (or more rarely flowers solitary); bracts and bracteoles present. Flowers actinomorphic or zygomorphic, bisexual (in African genera). Sepals 5, free or connate at the base, persistent, often with dorsal glands. Petals 5, frequently unguiculate, free, imbricate, entire or with fringed or dentate margin. Stamens 10, obdiplostemonous, with filaments often connate at the base; anthers dehiscing longitudinally, introrse, basifixed or dorsifixed, 2-thecous. Ovary superior, syncarpous, 3 (rarely 2, 4, or 5) -locular and -lobed, with 1 pendulous axile ovule in each loculus; styles as many as the carpels, with usually entire stigmas. Fruit a schizocarp, usually winged (samara), rarely a fleshy drupe. Seeds with a large usually straight embryo, without endosperm.

A large family of about 60 genera and 800 species, with a mainly pantropical distribution, but extending into the subtropics, most abundant in the New World.

Leaves alternate - - - - - - - - - **1. Acridocarpus**
Leaves opposite or ternate:
Anthers linear, 3–4 mm. long; samara with the lateral wing divided into 5–7 narrow stellately arranged lobes - - - - - - - **2. Tristellateia**
Anthers ovate, oblong or ovate-oblong, 1–1·8 (2·5) mm. long; wing of the samara entire:
Ovary glabrous - - - - - - - - - **3. Triaspis**
Ovary hairy:
Styles shorter than, as long as, or only slightly exceeding the stamens; petals usually auriculate near the base - - - - - - - **4. Caucanthus**
Styles distinctly exceeding the stamens; petals not auriculate:
Flowers actinomorphic or subactinomorphic; samara with an oblique dorsal wing only - - - - - - - - - **5. Sphedamnocarpus**
Flowers distinctly zygomorphic; samara with a shield-like lateral wing and a dorsal crest (sometimes absent) - - - - - - **3. Triaspis**

1. ACRIDOCARPUS Guill. & Perr.

Acridocarpus Guill. & Perr. in Guill., Perr. & Rich., Fl. Senegamb. Tent.: 123, t.29 (1831) *nom. conserv. prop.*

Erect suberect trailing or climbing shrubs, rarely small trees. Leaves alternate, petiolate, entire, usually with glands on the under surface at the base, and sometimes with two rows of smaller glands parallel to the margins, exstipulate. Inflorescences few- to many-flowered corymbs, racemes or panicles, axillary or terminating leafy branches; bracts present and persistent, small; bracteoles present at the base of the pedicels, sometimes with a circular gland at the base. Flowers actinomorphic or nearly so. Calyx ± coriaceous, with one or more subcircular sessile or sunken glands; lobes 5, equal or subequal, obtuse. Petals 5, white or yellow, usually unguiculate, longer than the sepals, entire or fimbriate or lacerate. Stamens 10; anthers basifixed, glabrous; filaments somewhat thickened and ± broadened and connate at the base, glabrous. Ovary 3-locular, but usually with 1 loculus abortive, usually densely sericeous or tomentose-sericeous; styles 2, terete, curved inwards. Samara with a straight or oblique dorsal wing.

A genus of about 30 species, occurring mainly in tropical Africa; 1 species in Madagascar, 1 in New Caledonia.

Bracteoles with a circular gland at the base - - - - - - - 1. *katangensis*
Bracteoles eglandular:
Bracts subulate, 3–5 mm. long; younger leaves glabrous or only sparsely hairy, older ones quite glabrous, usually cuneate, very rarely rounded at the base; wing of the samara oblique-ovate, rarely ± obovate, up to 2·8 cm. long, but usually shorter 2. *natalitius*
Bracts triangular-lanceolate, 1–2 mm. long; younger leaves ferrugineous-tomentose on both surfaces, older ones with the tomentum usually persisting beside the midrib beneath, rounded at the base; wing of the samara oblique-obovate, up to 4 cm. long 3. *chloropterus*

1. **Acridocarpus katangensis** De Wild. in Ann. Mus. Cong., Bot. Sér. 4, **1**: 27, t. 1 fig. 11–21 (1902); in Ann. Soc. Sci. Brux. **38,** 2: 16 (1914); Contr. Fl. Katang.: 105 (1921).—Sprague in Journ. of Bot. **44**: 205 (1906).—T. & H. Dur., Syll. Fl. Cong.: 77 (1909).—Niedenzu in Arb. Bot. Inst. Akad. Braunsb. **7**: 10 (1921); in Engl., Pflanzenr. IV, 141: 269 (1928).—Wilczek, F.C.B. **7**: 230, fig. 5 (1958).—White, F.F.N.R.: 433 (1962). Type from the Congo.
Acridocarpus rufescens Hutch., Botanist in S. Afr.: 520 (1946). Type: N. Rhodesia, Lake Tanganyika, near Mpulungu, fl. 21.vii.1930, *Hutchinson & Gillett* 3952 a (BM; K, holotype).
Acridocarpus natalitius sensu White, tom. cit.: 183 (1962).

A small tree or much-branched upright shrub, up to 6 m. tall; branches with the younger parts ± densely rufous-pubescent, very soon glabrescent, lenticellate. Leaf-lamina 3 (5)–9 (12) × 1·5–3·5 cm., obovate, oblanceolate, oblong-obovate to oblong-elliptic, obtuse, retuse to emarginate and usually finely apiculate at the apex, attenuate-cuneate at the base, coriaceous, brownish- or rufous-tomentose on both surfaces (more densely beneath) when very young, later usually glabrous (except sometimes the midrib beneath), with a pair of small glands near the insertion of the petiole underneath, and with 8–13 pairs of ± prominent lateral nerves; petiole 0·3–0·5 cm. long, glabrous or somewhat pubescent. Inflorescences 2–5 (7) cm. long, of racemes terminating leafy branches, few- to many-flowered, ± loose, with a stout ± densely rufous-pubescent to tomentose rhachis; bracts up to 2 mm. long, ovate-triangular, persistent, rufous-tomentose; bracteoles c. 1 mm. long, with a circular gland at the base. Flowers ± 2 cm. in diam. Calyx with 2–3 glands; lobes 3–3·5 mm. long, oblong to ovate-oblong, usually glabrous. Petals 10–15 × 6–10 mm., yellowish, unguiculate. Stamens 3–5 mm. long, with ovate-oblong glabrous anthers; filaments c. 0·5 mm. long. Ovary densely tomentose; styles 2, 7–10 mm. long, slender. Samara with wing 2·5–3·5 × 1·3–1·8 cm., oblique-oblong-obovate, finely pubescent or glabrous, often tinged with purple.

N. Rhodesia. N: Lake Tanganyika, near Mpulungu, fl. 21.vii.1930, *Pole Evans* 3028 (K; PRE; SRGH). W: Mwinilunga Distr., NW. of Kalene Hill Mission Station, fl. 21.ix.1952, *Holmes* 902 (FHO).

Also in Katanga. In open woodland and in clearings in denser woodland.

2. **Acridocarpus natalitius** A. Juss. in Arch. Mus. Par. **3**: 486 (1843).—Sond. in Harv. & Sond., F.C. **1**: 231 (1860).—Sprague in Journ. of Bot. **44**: 202 (1906).—Niedenzu in Arb. Bot. Inst. Akad. Braunsb. **6**: 54 (1915); op. cit. **7**: 8 (1921); in Engl., Pflanzenr. IV, 141: 266 (1928). Type from Natal.
Banisteria kraussiana Hochst. in Flora, **27**: 296 (1844). Type as above.
Acridocarpus natalitius var. *acuminatus* Niedenzu in Arb. Bot. Inst. Akad. Braunsb. **7**: 8 (1921); in Engl., tom. cit.: 267 (1928) pro parte excl. specim. ex Tanganyika.
Acridocarpus natalitius var. *obtusa* Niedenzu, tom. cit.: 9 (1921); in Engl., tom. cit.: 266 (1928). Type as above.
Acridocarpus reticulatus Burtt Davy, F.P.F.T. **1**: 36 (1926). Type from the Transvaal.

Small tree or shrub up to 5 m. in height, sometimes scrambling or climbing;

stems with younger parts bright yellow- or reddish-sericeous, older ones glabrous, lenticellate. Leaf-lamina 7–12 (15) × 0·5–4·5 cm., oblong, oblanceolate, oblong-ovate, oblong-oblanceolate, linear or linear-lanceolate, apex obtuse, subobtuse, or acuminate, base cuneate, rarely rounded, mostly quite glabrous on both surfaces (younger ones sometimes ± pubescent), coriaceous, with 7–11 pairs of lateral nerves beneath, somewhat reticulate, with usually 2 (4) glands at the very base beneath; petiole stout, 2 (5)–10 (13) mm. long, canaliculate, usually glabrous. Inflorescences 5–15 (25) cm. long, of many-flowered racemes terminating leafy branches, very rarely axillary, pyramidal in outline; bracts up to 5 mm. long, subulate, persistent, eglandular; bracteoles 1–2·5 mm. long, subulate, eglandular. Flowers 2–3·3 cm. in diam. Calyx-lobes 3·5–5 mm. long, 2–2·5 mm. broad, broadly ovate to ovate, glabrous or rarely sericeous outside, the 2 anterior ones usually with 2 glands affixed near the edges above the base. Petals 10–15 mm. long, subcircular, yellowish, ± lacerate, shortly unguiculate, claw c. 2 mm. long. Stamens with anthers c. 5 mm. long, lanceolate or lanceolate-oblong; filaments 2–2·5 mm. long, glabrous. Ovary densely hairy; styles ± 1·5 mm. long, curved inwards. Samara with wing up to 3 (4) × 1·7–2·2 cm., broadly obliquely ovate, with the upper margin straight, rarely slightly curved, distally obtuse, surrounding the nut to the base.

Leaf-lamina oblong, oblanceolate, oblong-obovate or oblong-oblanceolate, always broader than 1 cm. - - - - - - - - - - - - var. *natalitius*
Leaf-lamina linear or linear-oblanceolate, 0·5–1 (1·3) cm. broad - - var. *linearifolius*

Var. **natalitius**

Mozambique. SS: Gaza, between Chibuto and Gomes da Costa, fl. 14.xi.1957, *Barbosa & Lemos* 8113 (LISC). LM: Goba, fl. & fr. 23.ii.1955, *E.M. & W.* 561 (BM; LISC; SRGH).

Also in Natal and the Transvaal. Forest edges and clearings, on rocky shallow red clay soils rich in organic matter.

Var. **linearifolius** Launert in Bol. Soc. Brot., Sér. 2, **35**: 32 (1961). TAB. **13.** Type: Mozambique, Magude, Mapulanguene, *Torre* 6564 (LISC).

Acridocarpus pondoensis Engl. ex Niedenzu in Arb. Bot. Inst. Akad. Braunsb. **7**: 7 (1921); in Engl., Pflanzenr. IV, 141: 265 (1928) ex descr. Type from S. Africa (Pondoland).

Mozambique. SS: Mau-é-ele, fl. & fr. xii.1936, *Gomes e Sousa* 1947 (K; LISC). LM: Magude, Mapulanguene, fl. & fr. 4.v.1944, *Torre* 6564 (LISC).

Known only from Mozambique. Ecology as for the type variety.

3. **Acridocarpus chloropterus** Oliv., F.T.A. **1**: 279 (1868).—Sprague in Journ. of Bot. **44**: 205 (1906).—Niedenzu in Arb. Bot. Inst. Akad. Braunsb. **7**: 10 (1921); in Engl., Pflanzenr. IV, 141: 269 (1928).—Brenan in T.T.C.L.: 295 (1949). Type: Mozambique, Zambézia, R. Chire, *Meller* (K).

Scandent shrub or tall woody climber with branches up to 15 m. in length, younger ones ± densely rusty pubescent, older ones quite glabrous, lenticellate. Leaf-lamina 7–15 (19) × 2·5–5·5 (7) cm., oblong or oblong-elliptic, apex acute to subobtuse or acuminate, base rounded or rarely subcordate or broadly cuneate, coriaceous, usually with one pair of glands near the insertion of the petiole beneath and with 7–13 pairs of prominent lateral nerves, ferrugineous-tomentose (more densely underneath) when young, later usually glabrescent except usually beside the midrib underneath; petiole 2–6 mm. long, stout, canaliculate, pubescent or glabrescent. Inflorescences (5) 8–15 (25) cm. long, usually racemes terminating leafy branches, sometimes axillary, many-flowered, pyramidal in outline; bracts 1–2 mm. long, triangular-lanceolate, densely ferrugineous-tomentose, eglandular; bracteoles very small, eglandular. Flowers c. 2·25 cm. in diam. Calyx with 2–3 circular glands; sepals 3–3·5 × ±2·5 mm., ovate, sericeous or glabrescent outside. Petals up to 11 × 8 mm., elliptic to ovate, unguiculate, yellow. Stamens with anthers c. 4 mm. long, oblong-ovate, glabrous; filaments ± 1·5 mm. long, thick, ± ligulate. Ovary densely ferrugineous-tomentose; styles 2, 8–10 mm. long, stout, terete, curved inwards. Samara with wing 3–4 (5) × 1·5–2·3 cm., obliquely obovate, sometimes somewhat constricted in the lower half, often distally oblique-truncate or obtuse, not clasping the nut, often purplish, usually glabrous.

Tab. 13. ACRIDOCARPUS NATALITIUS VAR. LINEARIFOLIUS. 1, branch with flowers (×⅔); 2, flower-bud (×⅔); 3, flower (×2); 4, anther (×4); 5, fruit (×⅔); 6, samara (×1), all from *Hornby* 2561.

Nyasaland. S: Chikwawa, fl. & fr. 4.vii.1955, *Jackson* 1716 (FHO). **Mozambique.** N: Mogovolas between Corrane and Nametil, fl. 18.viii.1935, *Torre* 947 (LISC). Z: between Mucubela and Muidebo, 7·1 km. from Mucubela, fl. 29.ix.1949, *Barbosa & Carvalho* 4249 (K; LISC).

Also in Tanganyika. Climbing on trees on forest edges, in openings in forests and in thickets, often close to streams.

2. TRISTELLATEIA Thou.

Tristellateia Thou., Gen. Nov. Madag. **14,** no. 47 (1806).

Woody climbers, usually glabrous. Leaves opposite, entire, usually with two glands on the margin of the lamina near the base (in *T. africana* glands near the apex of the petiole); stipules very small. Inflorescence racemose, terminal; pedicels articulated, 2-bracteolate. Flowers actinomorphic, bisexual. Sepals persistent, with dorsal glands in some species. Petals oblong, unguiculate, carinate, entire. Stamens with glabrous basifixed anthers; filaments of the outer whorl longer and broader at the base. Ovary globose; only 1 (very seldom 2) styles fully developed, terete. Fruit subligneous; lateral wing divided into 4–10 narrow stellately arranged lobes; median wing usually absent; samara sometimes with a dorsal crest. Seeds subglobose, with a short acumen formed by the radicle.

A genus of 22 species, mainly in Madagascar, but with 1 species known from the African continent and 1 from SE. Asia.

Tristellateia africana S. Moore in Journ. of Bot. **6**: 289 (1877).—Niedenzu in Engl., Pflanzenr. IV, 141: 64 (1928).—Arènes in Mém. Mus. Nation. Hist. Nat. Par., N. Sér. **21**: 311 (1947).—Brenan, T.T.C.L.: 297 (1949).—Cufod. in Bull. Jard. Bot. Brux. **26,** Suppl.: 404 (1956). TAB. **14.** Type from Tanganyika.

Woody climber up to 5 m. or more in length; younger stems usually greyish-pubescent, older ones glabrous, densely lenticellate. Leaves petiolate; lamina 4·5–9·5 × 3·5–5 cm., ovate or broadly elliptic, apex acute or obtuse, base rounded or cordate, coriaceous to chartaceous, glabrous (younger ones sometimes ± sericeous); petiole 1·5–3 cm. long, with a pair of glands near the apex. Inflorescence 5–12 cm. long, racemose, many-flowered; pedicels 5–18 mm. long, decussate, distinctly articulate; bracts 2–3 mm. long, linear-lanceolate; bracteoles very short, subulate. Flowers 2·3–2·5 cm. in diam. Sepals 4–5 mm. long, oblong, sericeous outside. Petals 10–12 mm. long, bright yellow, shortly unguiculate, oblanceolate to oblong, apex rounded, base cordate or subsagittate. Anthers 3–4 mm. long, linear, orange; filaments glabrous, somewhat incurved. Samara 1·5–2 cm. in diam; lateral wing divided into 6 oblanceolate-linear, usually denticulate lobes; dorsal crest with a spine 8–10 mm. long.

Mozambique. N: mouth of R. Messalo, fr. 7.viii.1912, *Allen* 12 (K); Goa I., fl. 31.x.1953, *Gomes e Sousa* 4152 (LISC). SS: Praia de Závora, fl. 27.ii.1955, *E.M. & W.* 685 (BM; LISC; SRGH).

Also in Kenya and Tanganyika. Climbing on bushes or small trees, locally common on damp sandy or alkaline soils near the sea-shore, on coral rocks or in thickets in coastal districts.

3. TRIASPIS Burch.

Triaspis Burch., Trav. Int. S. Afr., **2**: 280, t. 290 (1824).

Small trees, woody climbers, or scandent or semiscandent shrubs. Leaves opposite or subopposite or rarely ternate, usually with 2–4 glands on the under surface near the base, petiolate or sessile, with or without interpetiolar stipules. Inflorescences terminal or axillary, usually forming many-flowered corymbs, umbels or panicles; bracts and bracteoles usually present, usually deciduous. Flowers actinomorphic or zygomorphic, bisexual. Sepals 5, almost always without glands. Petals 5, unguiculate, usually with fringed or denticulate margins; pedicels as long as, or longer than, the peduncle and articulated with it. Stamens 10; anthers basifixed, usually glabrous; filaments subulate, glabrous or finely-pubescent. Ovary hairy or glabrous, 3-locular (or sometimes 2-locular outside our region); styles 3 (2), somewhat curved, with incurving stigmas. Samara with a circular or ovate membranous or coriaceous lateral wing; dorsal wing shorter and narrower or absent.

An African genus of 15 species.

Tab. 14. TRISTELLATEIA AFRICANA. 1, branch with flowers ($\times \frac{2}{3}$); 2, branch with flowers and fruits ($\times \frac{2}{3}$); 3, flower ($\times 3$); 4, fruit ($\times 1$), all from *E. M. & W.* 685.

Ovary glabrous; flowers actinomorphic - - - - - - - 1. *lateriflora*
Ovary densely hairy; flowers distinctly zygomorphic:
  Anthers and filaments finely pubescent; stipules rather large, 3–15 mm. long
  2. *macropteron*
  Anthers and filaments glabrous; stipules small or absent:
    Petiole less than 5 mm. long, or leaves nearly sessile - - - 6. *nelsonii*
    Petiole more than 5 mm. long:
      Indumentum on younger branches and leaves ferrugineous; anthers 1–1·25 mm. long; filaments 2·5–3 mm. long - - - - - - 3. *mozambica*
      Indumentum on younger branches and leaves greyish or absent; anthers 1·5–2 mm. long; filaments 4–5·6 mm. long:
        Leaves of the main stem and the vegetative branches 3·4–4 cm. long, ovate-oblong to oblong, apex obtuse, different in shape and size from those of the inflorescence; petiole 0·5–1·2 cm. long - - - - - 5. *suffulta*
        Leaves of the main stem and the vegetative branches not different from those of the inflorescence, (3) 4·5–10 cm. long, ovate and acute at the apex; petiole (1·2) 2–2·5 cm. long - - - - - - - 4. *dumeticola*

1. **Triaspis lateriflora** Oliv. in F.T.A. **1**: 281 (1868).—Hiern, Cat. Afr. Pl. Welw. **1**: 104 (1896).—Exell in Journ. of Bot. **65,** Suppl. Polypet.: 51 (1927).—Niedenzu in Engl., Pflanzenr. IV, 141: 42 (1928).—Gossweiler & Mendonça, Cart. Fitogeogr. Angola: 90 (1939). Type from Angola.
Triaspis angolensis Niedenzu in Arb. Bot. Inst. Akad. Braunsb. **6**: 22 (1915); in Verz. Vorl. Akad. Braunsb. S.-Sem. **1924**: 5 (1924). Type from Angola.
Flabellaria sp. 1.—White, F.F.N.R.: 183 (1962).

Woody climber; young stems densely covered with a ferrugineous ± hispid indumentum, very soon glabrescent. Leaves petiolate; lamina 5·5–10 × 3–6·5 cm., broadly elliptic, or elliptic-oblong, apex shortly apiculate, base rounded, coriaceous, sparsely pubescent when young, later usually glabrous, deciduous; petiole 1·5–2 cm. long, somewhat pubescent or glabrous, canaliculate. Inflorescences terminal and axillary, of loose, few- to many-flowered umbellate panicles, densely covered with reddish spreading hairs. Flowers 1–1·25 cm. in diam., actinomorphic. Sepals 1·5–2 mm. long, oblong, rounded at the apex, densely hairy outside. Petals 5–6 mm. long, white or pale yellow, oblong-elliptic to obovate, entire or somewhat irregularly dentate at the margins. Stamens with anthers c. 1 mm. long, ovoid; filaments 2 mm. long, glabrous. Ovary glabrous; styles 3, c. 3 mm. long. Samara up to 3 cm. in diam.; lateral wing subcircular to ovate, entire or ± retuse (rarely emarginate); dorsal crest 5–9 × 2–3 mm., oblique-lanceolate.

N. Rhodesia. N: Samfya, fl. & fr. 2.x.1953, *Fanshawe* 340 (EA; FHO; K).
Climbing on small trees and bushes at forest margins.

2. **Triaspis macropteron** Welw. ex Oliv., F.T.A. **1**: 281 (1868). Type from Angola (Cuanza Norte).

A tall woody climber (sometimes creeping) up to 4 m. or more in length; younger stems densely ferrugineous-pubescent when young, becoming glabrous when older. Leaves petiolate; lamina 6–13 × 2–6·5 cm., ovate, broadly lanceolate, lanceolate-oblong or oblong, apex acute or apiculate, or sometimes subobtuse, base cordate, rounded or cuneate, densely ferrugineous-pubescent on both surfaces when young, later usually glabrous, dark green above, grey-green beneath, secondary and tertiary nerves ± strongly developed; petiole 0·5–2 cm. long, canaliculate, usually glabrous; stipules 2–15 × 1–8 mm., broadly lanceolate, obovate or oblong, ± elliptic, deciduous or ± persistent, usually glabrous. Inflorescences terminal or axillary, forming many-flowered corymbose panicles; bracts 0·5–2·5 mm. long, deciduous; bracteoles ± 1 mm. long, ± persistent. Flowers 1·5–2·5 cm. in diam., zygomorphic. Sepals 2–3 × 1–1·5 mm., oblong-elliptic, sericeous outside, glabrescent. Petals 6–9 × 4–5 mm., white to cream or yellow-orange (red?), distinctly unguiculate, deeply fimbriate at the margins. Stamens with anthers 1–1·5 mm. long, ovate-oblong; filaments 4–7 mm. long, minutely pubescent. Ovary densely sericeous; styles 4–8 mm. long, glabrous. Samara with lateral wing 4–5 cm. in diam., circular or broadly ovate, glabrous, somewhat undulate, dorsal wing 1·5–3 × 0·4–0·6 cm., crest-like.

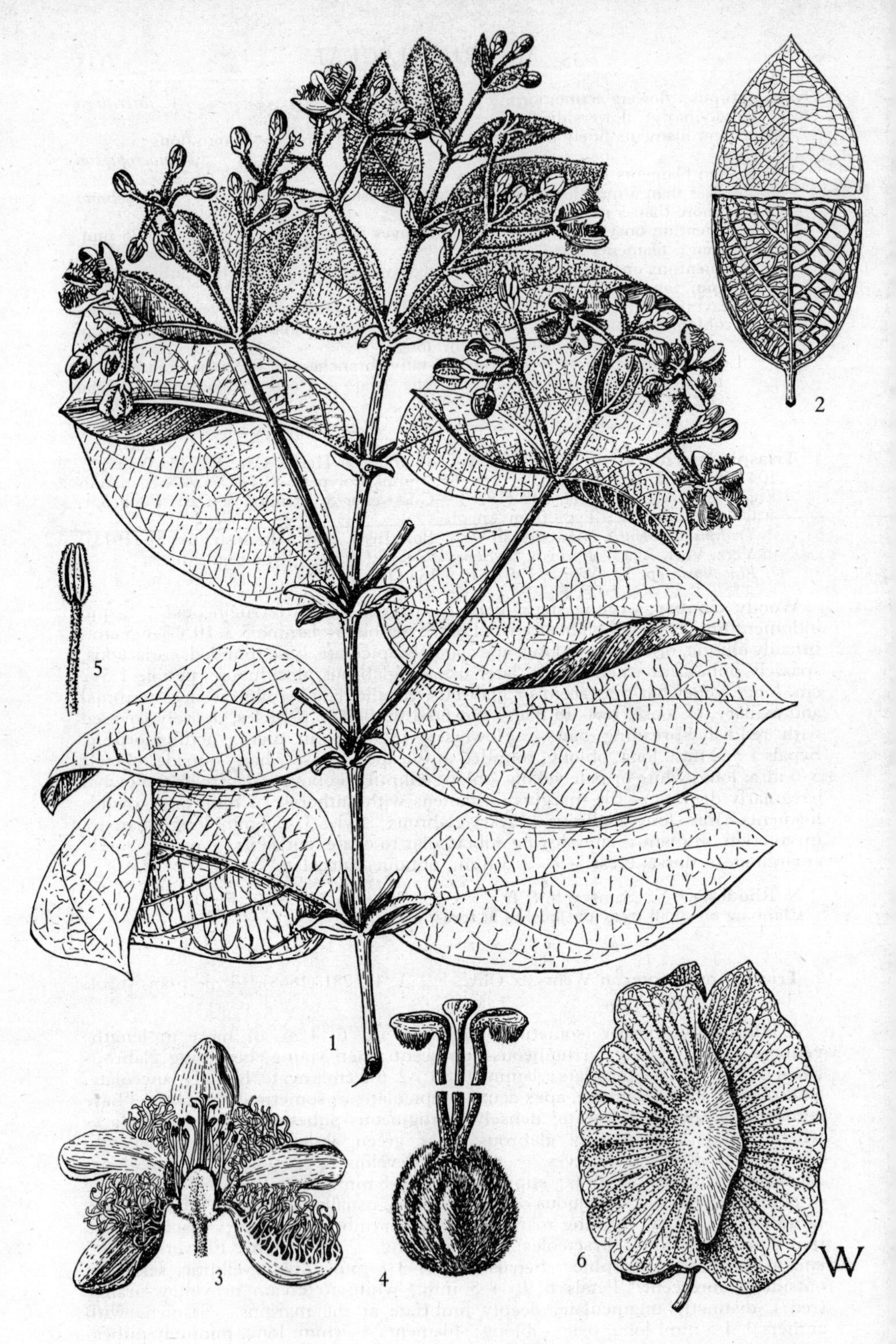

Tab. 15. TRIASPIS MACROPTERON SUBSP. MASSAIENSIS. 1, branch with flowers (×⅔); 2, leaf (×⅔); 3, flower (×2); 4, ovary (×8); 5, stamen (×8); 6, fruit (×⅔), all from *Chase* 5200.

Subsp. **massaiensis** (Niedenzu) Launert in Bol. Soc. Brot., Sér. 2, **35**: 31 (1961). TAB. **15.** Type from Tanganyika.
Triaspis speciosa Niedenzu in Engl., Pflanzenw. Ost-Afr. **C**: 232 (1895). Syntypes from Tanganyika.
Triaspis stipulata Engl., Bot. Jahrb. **43**: 382 (1909) non Oliv. (1868). Type from Tanganyika.
Triaspis macropteron var. *speciosa* (Niedenzu) Niedenzu in Verz. Vorl. Akad. Braunsb. S.-Sem. **1924**: 7 (1924).—White, F.F.N.R.: 183 (1962). Type as for *T. speciosa*.
Triaspis macropteron var. *speciosa* forma *massaiensis* (Engl.) Niedenzu in Verz. Vorl. Akad. Braunsb. S.-Sem. **1924**: 7 (1924); in Engl., Pflanzenr. IV, 141: 52 (1928).—Brenan, T.T.C.L.: 296 (1949). Type as for the subspecies above.
Triaspis macropteron var. *speciosa* forma *brevistipulata* Niedenzu in Engl., Pflanzenr. IV. 141: 52 (1928).—Brenan, loc. cit. Syntypes from Tanganyika.
Triaspis massaiensis Engl. ex Niedenzu in Engl., Pflanzenr. IV, 141: 52 (1928) *in syn.*

Leaf-lamina broadly lanceolate, lanceolate-oblong or oblong, base cuneate, rarely rounded or very rarely slightly cordate; tertiary nerves strongly developed; stipules 3–15 × 3–8 mm., broadly lanceolate to obovate, ± persistent.

N. Rhodesia. N: Crocodile I., Mpulungu, fl. 6.iv.1952, *Richards* 1284 (K). S: Mapanza, Choma, 1065 m., fl. 27.ii.1958, *Robinson* 2772 (K). **S. Rhodesia.** N: Urungwe Distr., near tributary of Nyakasanga R., 610 m., fl. 31.i.1958, *Drummond* 5400 (SRGH). E: Umtali, 1095 m., fl. 27.ii.1954, *Chase* 5200 (BM; K; SRGH). **Nyasaland.** S: without precise locality (probably Shire Highlands), *Buchanan* 1268 (K). **Mozambique.** T: Baroda, Msusa, banks of R. Mkanya, fr. 25.vii.1950, *Chase* 2215 (BM; K; SRGH).
Also in Tanganyika. In deciduous scrub, in riverine and ravine thickets and on thicketed rocky hills.

Triaspis macropteron subsp. *macropteron* does not occur in the Flora Zambesiaca area, but in N.E. Angola and western Congo. It differs from subsp. *massaiensis* mainly in having leaves with a usually cordate base and fewer tertiary nerves strongly developed underneath, and very small caducous stipules.

3. **Triaspis mozambica** A. Juss. in Ann. Sci. Nat., Sér. 2, Bot., **13**: 268 (1840); in Archiv. Mus. Par. **3**: 505 (1843).—Oliv., F.T.A. **1**: 281 (1868).—Niedenzu in Arb. Bot. Inst. Akad. Braunsb. **6**: 23 (1915); in Verz. Vorl. Akad. Braunsb. S.-Sem. **1924**: 6 (1924); in Engl., Pflanzenr. IV, 141: 46 (1928).—Brenan, T.T.C.L.: 297 (1949). Type: Mozambique, Delagoa Bay, *Forbes* (K).

Small climber, up to 3 m. long, with younger stems and inflorescences ± densely ferrugineo-sericeous, older ones glabrescent. Leaves petiolate; lamina 2·5–8 × 2·4 cm., lanceolate, ovate, ovate-lanceolate or elliptic, apex obtuse or acute, apiculate, base rounded or subcordate, membranous, ferrugineo-sericeous on both surfaces when young, soon glabrescent; petiole 0·7–1·2 cm. long, canaliculate, usually sericeous; stipules very small, deciduous. Inflorescence corymbose, ± lax, terminal or axillary, generally many-flowered; bracts 3·5 mm. long, linear-lanceolate; bracteoles 1–2 mm. long, linear. Flowers c. 1–1·5 cm. in diam., zygomorphic. Sepals c. 2 mm. long, ovate, sericeous outside. Petals 4·5–6 mm. long, green or yellowish-green, obovate and ± cucullate, spreading or recurved, shortly unguiculate, with the margins shortly fimbriate or crenulate. Stamens with the anthers 1–1·25 mm. long, ovate or oblong-ovate; filaments 2·5–3 mm. long, glabrous. Ovary densely sericeous; styles glabrous. Samara with lateral wing 2–2·7 cm. in diam., subcircular, retuse or ± emarginate at the apex; dorsal wing 0·8 × 0·25 cm., crest-like, subcordate at the base.

Mozambique. LM: Delagoa Bay, *Forbes* (K).
Also in Kenya and Tanganyika. Climbing on shrubs in coastal evergreen bush.

4. **Triaspis dumeticola** Launert in Bol. Soc. Brot., Sér. 2, **35**: 29, t. 1 (1961). Type: S. Rhodesia, Matobo Distr., Malema Dam, *Miller* 7013 (K, holotype; SRGH, isotype).

A small climber with younger stems and inflorescences greyish-sericeous or -pubescent, soon glabrescent. Leaves petiolate; lamina (3) 4·5–10 cm. long and (1·8 2·5 × 6 cm. broad, ovate, apex acute, base rounded to cordate, thickly herbaceous, usually glabrous on both surfaces, rarely sparsely hairy, dark green on the upper, bright green on the lower surface; petiole (1·2) 2–2·5 cm. long, subcanaliculate, slender, usually glabrous; stipules absent. Inflorescence loosely paniculate, axillary, many-flowered, open; peduncle 1·5–5 cm. long; bracts 1·5 mm. long, triangular-subulate, acute, pubescent or glabrous; bracteoles c. 1 mm. long, subulate. Flowers zygomorphic, c. 1·8 cm. in diam. Sepals c. 2·8 mm. long, ovate, ovate-oblong or broadly elliptic, usually glabrous, rarely somewhat hairy. Petals unguiculate, white with pink tinge, spreading or recurved; the lower one c. 8·8 mm. long, boat-shaped, fimbriate all along the margin, the others c. 8 mm. long, ovate or broadly elliptic, fimbriate at the lower half, rarely at the whole margin. Anthers ± 2 mm. long, oblong; filaments 5–5·6 mm. long, glabrous. Ovary ± densely sericeous; styles 3, c. 4·8 mm. long, glabrous, slightly curved outwards. Mature samara not known; dorsal wing crest-like or absent.

S. Rhodesia. W: Matobo Distr., Malema Dam, c. 1300 m., fl. xii.1959, *Miller* 7013 (K; SRGH).

Known only from S. Rhodesia. On edges and in openings of woodland.

5. **Triaspis suffulta** Launert in Bol. Soc. Brot., Sér. 2, **35**: 30 (1961). Type: Mozambique, Vilanculos, *Barbosa & Balsinhas* 5006 (BM, holotype; LMJ).

A delicate woody climber up to 5 m. or more in length, many-branched. Stems terete, glabrous or sparsely strigose. Leaves petiolate; lamina 3·5–4 × 2·2–2·5 cm., ovate-oblong to oblong, apex obtuse sometimes slightly retuse, base rounded to cordate, glabrous on both surfaces or ± sericeous when young, bright green on the upper, greyish green on the lower surface, eglandular or with two small circular glands at the base beneath near the insertion of the petiole; petiole 0·5–1·2 cm. long, slightly canaliculate, sericeous or glabrous. Inflorescence corymbose, many-flowered; bracts up to 2 mm. long, ovate-lanceolate; bracteoles very short, lanceolate. Flowers zygomorphic, 1·2–1·5 cm. in diam. Sepals 2·4–2·9 × c. 1·6 mm. broad, ovate-oblong to oblong, sericeous outside. Petals c. 7·5 mm. long, obovate-cochleate, the lower one boat-shaped, unguiculate, with the margins partially fimbriate. Anthers c. 1·6 mm. long, oblong; filaments c. 4 mm. long, glabrous. Ovary densely sericeous-villous; styles 3, 4–5 mm. long, glabrous, slightly curved outwards. Samara with lateral wing 2·5–3·3 × 1·7–2·3 cm., ovate to ovate-triangular, apex obtuse or subobtuse, rarely subacute, finely puberulous when young, later glabrous; dorsal wing completely reduced.

Mozambique. SS: Vilanculos, fl. & fr. 27.iii.1952, *Barbosa & Balsinhas* 5006 (BM; LMJ).

Known only from Mozambique. Climbing on trees or shrubs on dunes near the coast.

6. **Triaspis nelsonii** Oliv. in Hook., Ic. Pl.: t. 1418 (1883).—Niedenzu in Arb. Bot. Inst. Akad. Braunsb. **6**: 251 (1915); in Verz. Vorl. Akad. Braunsb. S.-Sem. **1924**: 6 (1924); in Engl., Pflanzenr. IV, 141: 47 (1928).—Burtt Davy in F.P.F.T. **2**: 286 (1932). Type from the Transvaal.
Triaspis rehmannii Szyszyl. in Polypet. Discifl. Rehm.: 3 (1888).—Niedenzu in Engl., Pflanzenr. IV, 141: 52 (1928). Type from the Transvaal.

Erect to suberect, scandent or straggling shrub, up to 1 m. high. Younger stems and inflorescences densely sericeous to villous, older ones canescent or glabrescent. Leaves sessile or shortly petiolate; lamina 1·5–3·5 (5) × 1–2 (3·4) cm., ovate or oblong-ovate, apex subacute to acute and usually finely apiculate, base obliquely rotundate or subcordate, subcoriaceous, pale green or glaucous, ± densely appressed-pilose on both surfaces when young, later villous-canescent or glabrescent, but usually only the midrib beneath remaining sericeous or pubescent, usually with 2–4 small glands underneath near the base; petiole 1–2 (4) mm. long, stout, densely hairy. Flowers 1·5 mm. long, ovate. Sepals 3–4 mm. long, ovate-elliptic, subglabrous. Petals 8–10 mm. long, obovate, ± spathulate, mauve, pinkish or purplish-blue, fimbriate, shortly unguiculate. Stamens with anthers 1·5–2 mm. long, oblong-elliptic; filaments glabrous, somewhat curved. Ovary densely sericeous; styles about 6 mm. long, glabrous. Samara with the lateral wing 2·5–4 cm. in diam., circular, slightly retuse at the apex; dorsal wing ± developed or much reduced, up to 2·5 × 0·8 cm., obliquely lanceolate.

Leaf-lamina ovate, subacute, soon glabrescent (except on the midrib beneath); hairs with trabeculae 0·6–1 mm. long - - - - - - - - Subsp. *nelsonii*
Leaf-lamina oblong-ovate or triangular-ovate, acute, canescent or eventually older ones glabrescent; hairs with trabeculae 1·2–2·1 mm. long - - - Subsp. *canescens*

Subsp. **nelsonii**

S. Rhodesia. W: Matobo Distr., fr. 22.iii.1950, *Orpen* 05/50 (SRGH). Also in the

Transvaal and SW. Africa. In exposed dry places, in open grassland, deciduous scrub or on rocky hillsides.

Subsp. **canescens** (Engl.) Launert in Bol. Soc. Brot., Sér. 2, **35**: 32 (1961). Type: Mozambique, Ressano Garcia, fl. 18.xii.1897, *Schlechter* 11827 (B†; K, lectotype).
Triaspis canescens Engl., Bot. Jahrb. **36**: 249 (1905).—Niedenzu in Arb. Bot. Inst. Akad. Braunsb. **6**: 26 (1915); in Verz. Vorl. Akad. Braunsb. S.-Sem. **1924**: 7 (1924); in Engl., Pflanzenr. IV, 141: 47 (1928). Type as above.

Mozambique. SS: Ressano Garcia, fl. & fr. 20.ii.1948, *Torre* 7379 (LISC). Known only from Mozambique. Ecology as for the type subspecies.

4. CAUCANTHUS Forsk.

Caucanthus Forsk., Fl. Aegypt.-Arab.: CXI, 91 (1775).

Woody climbers or upright or semiscandent shrubs; stems with younger parts usually ± densely appressed-pubescent or sericeous. Leaves spirally arranged or opposite, with 2 glands on the margin near the base or eglandular; stipules very small, deciduous. Inflorescences racemose or corymbose-paniculate, axillary or terminal. Flowers actinomorphic, bisexual. Sepals without glands. Petals unguiculate (sometimes very shortly), sometimes auriculate or hastate at the base, glabrous, the margins wholly or partially fimbriate. Stamens glabrous with dorsifixed anthers. Ovary densely sericeous; styles truncate, shorter than, as long as, or slightly exceeding the stamens. Fruit with lateral wing completely surrounding the nut, circular or broadly elliptic; dorsal wing small, obliquely lanceolate or absent.

A genus of 3 species, confined to eastern Africa from Somaliland and Ethiopia to Mozambique and S. Rhodesia.

Caucanthus auriculatus (Radlk.) Niedenzu in Arb. Bot. Inst. Akad. Braunsb. **4**: 18 (1915); in Verz. Vorl. Akad. Braunsb. S.-Sem. **1924**: 2 (1924); in Engl., Pflanzenr. IV, 141: 35 (1928).—Brenan, T.T.C.L.: 296 (1949).—Cufod. in Bull. Jard. Bot. Brux. **26**, Suppl.: 403 (1956). TAB. **16.** Type from Kenya.
Triaspis auriculata Radlk. in Abhandl. Naturw. Ver. Bremen, **8**: 379 (1884).—Engl., Pflanzenw. Ost-Afr. **A**: 57 (1895) in obs. Type as above.
Caucanthus argenteus Niedenzu in Bull. Herb. Boiss., Sér. 2, **4**: 1010 (1904). Type: Mozambique, Boroma, *Menyhart* 964 (B†).

Climber up to 5 m. in length; younger stems densely covered with short soft white sericeous hairs, older ones very finely pubescent or glabrescent. Leaves petiolate; lamina 6–12 × 4–9·5 cm., ovate-cordate, apex acute to shortly acuminate, membranous, pubescent or glabrescent above, grey-tomentose beneath and with 2 large glands near the base (usually concealed by the indumentum); petiole 1–3 cm. long, densely sericeous with usually 2 small glands above the middle. Flowers c. 1·5 cm. in diam., in dense axillary and terminal corymbs, evil-smelling; peduncles and rhachis sericeous; pedicels 1–1·5 cm. long, sericeous; bracts ovate; bracteoles lanceolate or linear-subulate. Sepals 2–2·5 mm. long, broadly ovate with a narrowed base, sericeous outside. Petals 6–7 mm. long, pale yellow, shortly unguiculate, ovate, carinate, subhastate at the base, usually reflexed. Stamens with anthers subversatile, 2·3–2·5 mm. long, oblong; filaments somewhat fleshy. Ovary densely sericeous; styles 2·5–3 mm. long, fairly stout, sericeous. Samara with lateral wing 1–2 cm. in diam., oblong-elliptic or oblong-obovate, with entire margins; dorsal wing absent.

S. Rhodesia. N: Darwin Distr., between Chiswiti and Musengezi, c. 600 m., fl. 28.i.1960, *Phipps* 2474 (BM; K; SRGH). **Nyasaland.** S: Port Herald, c. 150 m., fl. 25.iii.1960, *Phipps* 2732 (BM; K; SRGH). **Mozambique.** T: Between Mutarara and Doa, fr. 16.vi.1949, *Barbosa & Carvalho* 3108 (EA; LISC).
Also in Ethiopia, Kenya and Tanganyika. Climbing in thickets and on small trees outside forests.

5. SPHEDAMNOCARPUS Planch. ex Hook. f.

Sphedamnocarpus Planch. ex Hook. f. in Benth. & Hook., Gen. Pl., **1**: 256 (1862).

Climbers, upright or trailing shrubs or undershrubs. Leaves opposite, ternate or rarely subopposite, with 2–4 sessile or stalked glands on the margin of the lamina

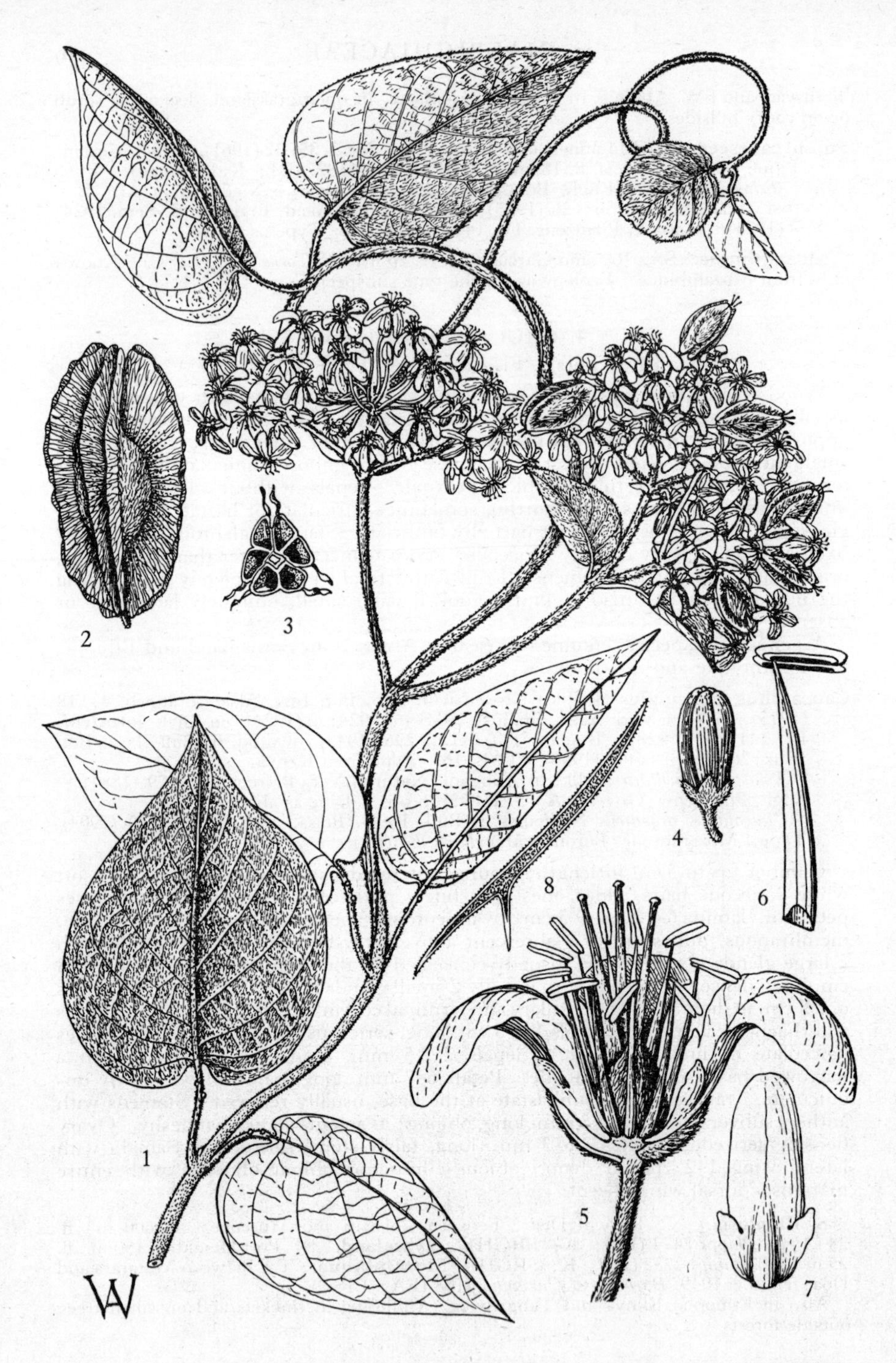

Tab. 16. CAUCANTHUS AURICULATUS. 1, branch with flowers and fruits (×⅔) *Barbosa & Carvalho* 3108; 2, fruit (×⅔); 3, cross-section of fruit (×⅔) both from *Dale* 3854; 4, flower bud (×2); 5, flower with sepals and petals partly removed (×4); 6, stamen (×6); 7, petal (×3); 8, hair taken from a leaf (×30), all from *Anderson* 345.

near the base or near the apex of the petiole; stipules present or absent. Inflorescences umbellate, few-flowered, terminating leafy branches or branchlets, or axillary panicles, very rarely flowers solitary; bracts and bracteoles present. Flowers subactinomorphic, bisexual. Sepals 5, equal or subequal, without glands. Petals 5, shortly unguiculate, glabrous, usually yellow, entire or crisped-denticulate. Stamens 10; anthers basifixed, glabrous; filaments subulate, somewhat dilated and united at the very base. Ovary 3–4-locular, usually densely sericeous; styles 3 (4), free, slender, with somewhat thickened terminal stigmas. Samara with an oblique dorsal wing only.

A genus of 18 species of which 10 are in Madagascar and Mauritius and the remainder in southern tropical Africa.

Leaves sericeous-tomentose on both surfaces (usually less so on the upper one) or only underneath and glabrescent above:
- Plant an erect or suberect, sometimes trailing shrub; leaves some ternate, some opposite (very rarely subopposite), sericeous-tomentose on both surfaces, less so on the upper surface and occasionally glabrescent - - - - - - - 1. *angolensis*
- Plant a climber; leaves strictly opposite, sericeous-tomentose underneath, greyish-tomentose or arachnoid or glabrous or nearly so on the upper surface 2. *pruriens*

Leaves glabrous or sparsely hairy on both surfaces, or glabrous above and hairy (but never tomentose) beneath:
- Wing of the fruit broadly ovate, extending to the base of the nut; lax shrub or shrublet with twining branchlet tips - - - - - - - - 4. *transvalicus*
- Wing of the fruit oblong or oblong-obovate, not extending to the base of the nut, woody climber or creeper - - - - - - - 3. *galphimiifolius*

1. **Sphedamnocarpus angolensis** (A. Juss.) Planch. ex Oliv., F.T.A. **1**: 279 (1868).—Hiern, Cat. Afr. Pl. Welw. **1**: 104 (1896).—Niedenzu in Arb. Bot. Inst. Akad. Braunsb. **6**: 48 (1916); in Verz. Vorl. Akad. Braunsb. S.-Sem. **1924**: 17 (1924); in Engl., Pflanzenr. IV, 141: 255 (1928).—Burtt Davy, F.P.F.T. **2**: 284 (1932).—Gossweiler & Mendonça, Carta Fitogeogr. Angola: 160, 161 (1939).—Arènes in Notul. Syst. **11**: 121 (1943).—Exell & Mendonça, C.F.A. **1**, 2: 252 (1951).—White, F.F.N.R.: 183 (1962). TAB. **17**. Type from Angola (Cuanza Norte).

Acridocarpus (?) *angolensis* A. Juss. in Ann. Sci. Nat., Sér. 2, Bot. **13**: 272 (1840); in Archiv. Mus. Hist. Nat. Par. **3**: 490 (1843). Type as above.

Sphedamnocarpus pulcherrimus Engl. & Gilg in Warb., Kunene-Samb.-Exped. Baum: 272 (1903). Type from Angola (Bié).

Sphedamnocarpus angolensis var. *pulcherrimus* (Engl. & Gilg) Niedenzu in Verz. Vorl. Akad. Braunsb. S.-Sem. **1924**: 17 (1924); in Engl., Pflanzenr. IV, 141: 255 (1928). Type as above.

Erect or suberect shrub or shrublet, or woody herb, sometimes trailing and tending to twine at the branchlet tips; younger stems greyish- or yellowish-sericeous-tomentose, older ones ± densely pubescent or sometimes glabrescent. Leaves usually in whorls of three or opposite, rarely subopposite, petiolate; lamina 3·5–6 × 1·5–2·5 cm., lanceolate or oblong-lanceolate, apex obtuse to subobtuse, apiculate, base always rounded, with 2 glands on the margin beneath near the insertion of the petiole (or often near the apex of the petiole), usually greyish- or yellowish-sericeous-tomentose on both surfaces (denser underneath), sometimes glabrescent above; petiole 2–5 mm. long, stout, usually tomentose; stipules 3–10 mm. long, ovate, oblong or rarely ovate-lanceolate, apex obtuse, retuse or ± apiculate, sericeous-tomentose, often absent. Flowers 1·5–2 cm. in diam., in few-flowered axillary or terminal umbels. Sepals about 5 mm. long, oblong, sericeous outside, sometimes glabrescent. Petals 8–10 mm. long, obovate to obovate-oblong, bright yellow. Stamens with anthers 1–1·5 mm. long, glabrous; filaments c. 4 mm. long, slender. Ovary densely sericeous; styles about 6 mm. long. Samara with wing 1·75–2 × 0·9–1·5 cm., obliquely ovate or obovate-oblong, very often reddish, sericeous to nearly glabrous.

N. Rhodesia. B: Sesheke, 20 miles SW. of Machili on the Mwandi road, 21.xii.1952, *Angus* 1001 (FHO, K). N: Abercorn, fl. xii.1908, *Merkel* (BM). S: Batoka Plateau, fl. ii.1907, *Allen* 444 (K; SRGH). **S. Rhodesia.** W: Nyamandlhovu, fl. 19.i.1957, *West* 3180 (K; SRGH). C: Rusape, fl. 23.i.1939, *Hopkins* in GHS 7043 (K; SRGH). **Nyasaland.** C: Lilongwe, fl. & fr. 3.iii.1958, *Jackson* 2153 (SRGH).

Also in Angola and northern Transvaal. On dry sandy ground, in open grassland or deciduous woodland.

2. **Sphedamnocarpus pruriens** (A. Juss.) Szyszyl., Polypet. Discifl. Rehm.: 2 (1888).—Niedenzu in Arb. Bot. Inst. Akad. Braunsb. **6**: 49 (1915); in Verz. Vorl. Akad. Braunsb. S.-Sem. **1924**: 18 (1924); in Engl., Pflanzenr. IV, 141: 257 (1928).—Burtt Davy, F.P.F.T. **2**: 284 (1932).—Arènes, Notul. Syst. **11**: 119 (1943). Type from Natal.

Banisteria pruriens E. Mey. ex Drège, Cat. Pl. Exsicc. Afr. Austr.: 19 (1838) *nom. nud.*

Acridocarpus (?) *pruriens* A. Juss., Malpigh. Synops.: 272 (1840); in Arch. Mus. Par. **3**: 492 (1843).—Sond. in Harv. & Sond., F.C. **1**: 232 (1860). Type as above.

S. pruriens var. *platypterus* Arènes in Notul. Syst. **11**: 120 (1943). Syntypes: Mozambique, Delagoa Bay, *Junod* 497 (P?), and Lourenço Marques, *Borle* 350 (SRGH).

A herbaceous or woody climber or scrambler. Younger stems densely ± appressed greyish-silvery- or yellowish-sericeous, older ones ± pubescent or nearly glabrous. Leaves petiolate; lamina 3·5–7·5 (11) × 1·7–5 (10) cm., ovate, ovate-subcircular or ovate-oblong, rarely oblong-elliptic, ovate-lanceolate or lanceolate, apex obtuse to acute, mostly distinctly apiculate, base subcordate to rounded or cuneate, greyish- or yellowish-sericeous-tomentose underneath, greyish-pubescent or arachnoid or glabrescent on the upper surface; petiole (0·5) 0·8–2·5 (4·5) cm. long, canaliculate, sericeous, with 2 glands in the upper half; stipules 2–4 mm. long, obovate, usually absent. Umbels 2–6-flowered; bracts ± 2·5 mm. long, lanceolate; bracteoles 1·5–2 mm. long, oblong. Flowers 2–2·5 cm. in diam. Sepals ± 2·5 mm. long, ovate or obovate-oblong, sericeous outside; petals 8–12 mm. long, obovate to obovate-oblong, yellow, unguiculate, usually entire. Stamens with anthers c. 1 mm. long, oblong, glabrous; filaments c. 4 mm. long, glabrous. Ovary densely sericeous; styles c. 5 mm. long. Samara with wing 1·5–2 × 1·1–1·5 cm., sericeous to nearly glabrous.

Leaves glabrous or glabrescent on the upper surface, ovate, subcircular, or ovate-oblong, rarely elliptic, apex obtuse to acute, base cordate or rounded:
  Leaf-lamina subcircular, ovate-elliptic, ovate-oblong or elliptic, apex obtuse sometimes slightly retuse, apiculate, coriaceous - - - - - var. *pruriens*
  Leaf-lamina ovate, rarely ovate-elliptic, apex acute or subacute, apiculate, subcoriaceous to stiffly herbaceous - - - - - - - - var. *latifolius*
Leaves tomentellous or arachnoid (the older ones rarely becoming glabrous) on the upper surface, lanceolate, rarely ovate-lanceolate, apex acute, base rounded to broadly cuneate var. *lanceolatus*

Var. **pruriens.** TAB. **17.**

Bechuanaland Prot. SE: Gaberones, fl. 10.iii.1930, *van Son* in Herb. Transv. Mus. 28943 (BM); Molepolole, fr. 11.vi.1955, *Story* 4874 (BM; PRE). **Nyasaland.** S: Isonge Hill, *Banda* 91 (K). **S. Rhodesia.** N: Shamva, 915 m., fl. & fr. v.1920, *Eyles* 2247 (SRGH). W: Bubi, fl. ii.1954, *Goldsmith* 30/54 (K; SRGH). E: Umtali, fl. & fr. 2.iv.1950, *Chase* 2126 (K; BM; SRGH). S: Lundi Drift, fr. vii.1929, *Smuts* in Nat. Herb. Pret. 28708 (PRE). **Mozambique.** Z: Morrumbala, fr. 13.vi.1949, *Barbosa & Carvalho* 3059 (LISC; SRGH). MS: Chimoio, Bandula, Chibata, fl. 26.iii.1948, *Barbosa* 1225 (BM; LISC). LM: Namaacha, Goba, fl. 17.iii.1945, *Sousa* 83 (LISC; PRE).

Also in the Transvaal, northern Cape Province and SW. Africa. A climber upon shrubs or small trees or twining round grasses, sometimes trailing; in open woodland, open grassland and on the forest floor.

Var. **lanceolatus** Launert in Bol. Soc. Brot., Sér. 2., **35**: 42, t. 4 fig. 2 (1961). Type: S. Rhodesia, Matopos, fl. 14.xii.1912, *Rogers* 5651 (BM, holotype; K; PRE).

S. wilmsii Engl., Bot. Jahrb. **36**: 249 (1905).—Burtt Davy, F.P.F.T. **2**: 284 (1932).—Arènes, Notul. Syst. **11**: 120 (1943).—Type from the Transvaal.

S. pruriens forma *wilmsii* (Engl.) Niedenzu in Verz. Vorl. Akad. Braunsb. S.-Sem. **1924**: 18 (1924); in Engl., Pflanzenr. IV, 141: 257 (1928). Type as above.

S. Rhodesia. N: Darwendale, 1370 m., fl. iv.1917, *Eyles* 702 (PRE; SRGH). W: Matobo, fl. ii.1957, *Gourlay* 121 (K; SRGH). C: Marandellas, fl. 30.i.1942, *Dehn* 85 (M; SRGH).

Also in the Transvaal. Ecology as for the type variety.

Var. **latifolius** Engl., Bot. Jahrb. **36**: 249 (1905). Type from the Transvaal.

Sphedamnocarpus latifolius (Engl.) Niedenzu in Arb. Bot. Inst. Akad. Braunsb. **6**: 48 (1915); in Verz. Vorl. Akad. Braunsb. S.-Sem. **1924**: 17 (1924); in Engl., Pflanzenr. IV, 141: 256 (1928).—Burtt Davy, F.P.F.T. **2**: 284 (1932).—Arènes in Notul. Syst. **11**: 120 (1943).

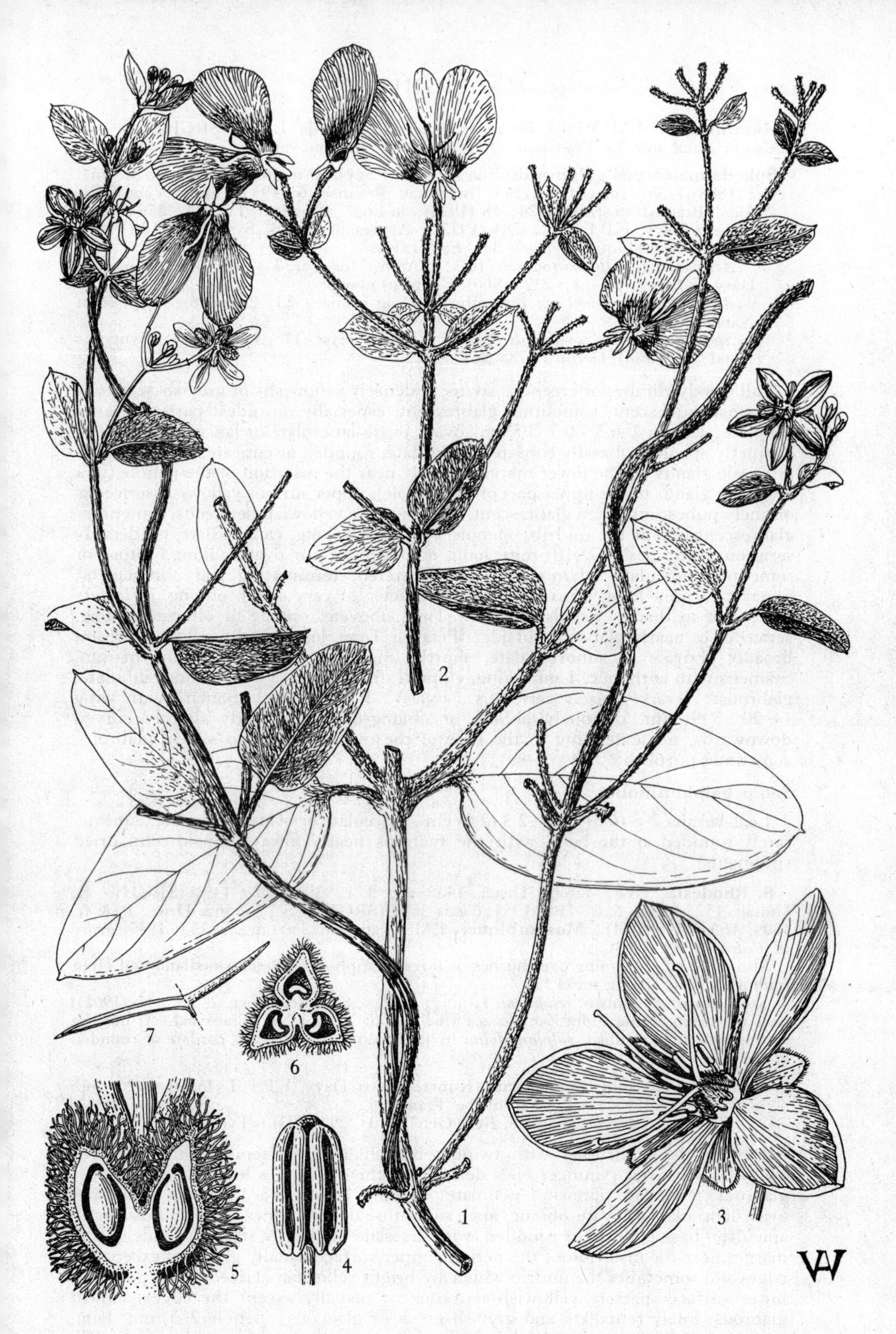

Tab. 17. SPHEDAMNOCARPUS PRURIENS VAR. PRURIENS. 1, branch with flowers ($\times \frac{2}{3}$); 2, branch with fruits ($\times \frac{2}{3}$); 3, flower ($\times 3$); 4, anther ($\times 14$); 5, longitudinal section of ovary ($\times 4$); 6, cross-section of ovary ($\times 4$); 7, hair taken from a leaf ($\times 30$), all from *Rand* 431.

Mozambique. LM: Goba, *Barbosa & Lemos* 8260 (BM; LISC; SRGH).
Also in Natal and the Transvaal. Ecology as for the type variety.

3. **Sphedamnocarpus galphimiifolius** (A. Juss.) Szyszyl. in Polypet. Discifl. Rehm.: 2 (1888).—Niedenzu in Arb. Bot. Inst. Akad. Braunsb. **6**: 49 (1915); in Verz. Vorl. Akad. Braunsb. S.-Sem. **1924**: 18 (1924); in Engl., Pflanzenr. IV, 141: 256 (1928). —Burtt Davy, F.P.F.T. **2**: 284 (1932).—Arènes in Notul. Syst. **11**: 118 (1943). Type: Mozambique, Delagoa Bay, *Forbes* (K).
Acridocarpus galphimiifolius A. Juss. in Arch. Mus. Par. **3**: 491 (1843).—Sond. in Harv. & Sond., F.C. **1**: 232 (1860). Type as above.
Acridocarpus pruriens var. *laevigatus* Sond. in Linnaea, **23**: 22 (1850). Type from Natal.
Sphedamnocarpus woodianus Arènes in Notul. Syst. **11**: 118 (1943). Syntype: Natal (Zululand), *Gerrard* 1788 (K).

Tall woody climber or creeper; stems ± densely yellowish- or greyish-sericeous or greyish-pubescent, sometimes glabrescent, especially on oldest parts. Leaves petiolate; lamina 2–6·5 × 0·7–3·5 cm., ovate, ovate-lanceolate or lanceolate, acute or distinctly apiculate, basally cordate, subcordate, rounded or cuneate, with 2 stalked or sessile glands on the lower margin, mostly near the insertion of the petiole (in a few cases glands in the upper part of the petiole), upper surface yellowish-sericeous or finely pubescent, often glabrescent, lower surface yellowish-sericeous, sometimes glabrescent except the midrib; petiole 0·5–1·7 cm. long, canaliculate, ± densely sericeous; stipules 3–7 (10) mm. long, oblong, ovate or ovate-oblong, obtuse or sometimes apiculate. Umbels 2–4 (6)-flowered, terminating leafy branchlets; bracts c. 3 mm. long, linear-lanceolate; bracteoles very small, oblong. Flowers 2–2·3 cm. in diam. Sepals c. 4 mm. long, obovate, oblong or obovate-oblong, sericeous or nearly glabrous outside. Petals c. 1 cm. long, bright yellow or cream, broadly obovate or suborbiculate, shortly unguiculate, crisped at the margin. Stamens with anthers c. 1 mm. long, elliptic; filaments 3–3·5 mm. long, subulate, glabrous. Ovary densely sericeous; styles 3–5 mm. long. Samara with wing 14–20 × 5–9 mm., obliquely oblong or oblong-obovate, mostly slightly curved downwards, not extending to the base of the nut, sericeous to almost glabrous, sometimes purplish.

Subsp. **galphimiifolius**

Leaf-lamina 2–5 (6·5) × 0·7–2·3 (2·8) cm., lanceolate or ovate-lanceolate, cuneate, rarely rounded at the base, with the margins nearly always involute (in dried specimens).

S. Rhodesia. W: Matobo Distr., 1435 m., fl. i.1939, *Miller* 1491 (SRGH). E: Umtali, 1525 m., fl. & fr. 11.iii.1951, *Chase* 3617 (SRGH). S: Victoria Distr., fl. & fr. 1909, *Monro* 875 (BM). **Mozambique.** LM: Namaacha, 600 m., fr. 25.iv.1947, *Pedro & Pedrógão* 760 (LISC).
Also in Natal. Twining over bushes or forest margins or in open woodland and often trailing in grass among rocks.
S. galphimiifolius subsp. *rehmannii* Launert (in Bol. Soc. Brot., Sér. 2, **35**: 45 (1961)) does not occur in the Flora Zambesiaca area, but in N. and NE. Transvaal. It may be distinguished from subsp. *galphimiifolius* by its broader ovate leaves, cordate or rounded at the base.

4. **Sphedamnocarpus transvalicus** (Kuntze) Burtt Davy, F.P.F.T. **1**: 50 (1926); op. cit. **2**: 284 (1932). Type from the Transvaal.
Triaspis transvalica Kuntze, Rev. Gen. Pl. **3**: 29 (1893). Type as above.

Lax shrub or shrublet with twining branchlet tips; stems whitish-grey- or yellowish-sericeous (younger ones densely so, the older ones less dense or almost glabrous). Leaves opposite, petiolate; lamina 1·2–3 × 0·5–1·5 cm., lanceolate, ovate-lanceolate or ovate-oblong, apex subobtuse or sometimes acute and distinctly apiculate, base cuneate or rounded, with 2 sessile or shortly stalked glands on the margin near the insertion of the petiole, upper surface usually glabrous except the edges and sometimes the midrib which are bright yellowish-sericeous and purplish, lower surface sparsely yellowish-sericeous or (usually except the edges) almost glabrous, finely reticulate and greyish-green or glaucous; petiole 2–7 mm. long, ± densely sericeous; stipules 1·5–5 mm. long, obovate or obovate-oblong, shortly apiculate, deciduous. Umbels 2–6-flowered; bracts broadly ovate, 2–3 mm. long; bracteoles linear, c. 1 mm. long. Flowers 1·5–2 cm. in diam. Sepals 2–3 mm.

long, ovate-lanceolate, obtuse, sericeous outside. Petals 7–10 mm. long, canary yellow, subcircular or ovate, shortly unguiculate, entire or ± crisped. Stamens with anthers c. 1 mm. long, broad-elliptic, glabrous; filaments subulate, c. 2·5 mm. long. Ovary densely sericeous; styles about 4–5 mm. long. Samara with wing 1·6–2 × 1·2–1·5 cm., obliquely and broadly ovate, extending to the base of the nut, sericeous or almost glabrous.

Bechuanaland Prot. SE: Kanye, 1250 m., fr. iii.1946, *Miller* B/455 (PRE), fl. i.1950, *Miller* B/978 (K; PRE).

Also in northern Transvaal. Recorded from rock crevices in open grassland.

36. ZYGOPHYLLACEAE

By E. Launert

Annual or perennial herbs (sometimes with a woody base) or shrublets or shrubs, with simple hairs. Leaves usually opposite, 2- (rarely 1- or 3-) foliolate or paripinnate; leaflets entire. Stipules present, persistent. Flowers solitary or rarely in cymes, actinomorphic, bisexual or rarely unisexual, hypogynous. Sepals (4) 5, free or connate at the very base, persistent or deciduous, usually imbricate. Petals (4) 5, imbricate, rarely contorted or valvate, free, sessile or unguiculate; disk usually present, sometimes provided with extra- and intrastaminal glands. Stamens obdiplostemonous, usually twice as many as the petals, with free filaments and 2-thecous introrse anthers. Ovary usually sessile, syncarpous, 4–5-locular with 2 or more pendulous ovules in each loculus; style usually simple, terete or angular or furrowed. Fruit a capsule or at length breaking up into indehiscent cocci, sometimes winged or provided with tubercules or spines. Seeds 1 or more in each loculus; embryo straight or curved; endosperm present (often very scanty) or absent.

A family of 27 genera and about 200 species, distributed throughout the drier tropics and subtropics of both hemispheres, mainly on the alkaline soil of deserts, where they form a characteristic constituent of the flora.

Leaves 2-foliolate, those of each pair equal, succulent; ovary glabrous 1. *Zygophyllum*
Leaves paripinnate, one of each pair usually longer than the other, herbaceous; ovary densely covered with stiff setose hairs - - - - - - - 2. *Tribulus*

ZYGOPHYLLUM L.

Zygophyllum L., Sp. Pl. **1**: 385 (1753); Gen. Pl. ed. 5: 182 (1754).

Small shrubs or shrublets, rarely perennial or annual herbs, often armed. Leaves opposite, (1) 2-foliolate, succulent, often caducous; leaflets very variable in size and shape, subcircular, spathulate, elliptic, linear, lanceolate or lanceolate-elliptic. Stipules present, often spiny. Flowers solitary or sometimes 2-nate, terminal. Sepals 4–5, imbricate, persistent or deciduous, sometimes connate at the base, somewhat fleshy. Petals 4–5, sometimes unguiculate, usually longer than the sepals, imbricate, whitish or yellow and often with a red spot at the base, spreading. Disk cup-shaped, fleshy, 8–10-angled. Stamens 8–10, inserted at the base of the disk; filaments subulate, provided with a usually fimbriate or pectinate, bifid or bipartite appendage at the base; anthers oblong or oblong-ovate. Ovary sessile on the disk, 4–5-locular, with 2 or more pendulous ovules in each loculus; style simple, terete, with a small stigma. Fruit a capsule or samara, 4–5-locular, 4–5-angled or winged, usually with several seeds. Seeds crustaceous, with scanty endosperm.

A genus of c. 120 species, mostly in the drier parts of Africa and Australia; about 50 species in S. Africa, mainly in the Karroo, but also in SW. Africa, the Transvaal and Cape Province.

Zygophyllum cuneifolium Eckl. & Zeyh., Enum. Pl. Afr. Austr. Extratrop.: 97 (1835).—Sond. in Harv. & Sond., F.C. **1**: 359 (1860).—Engl. in Engl. & Prantl, Nat.

Pflanzenfam. **3,** 4: 81 (1896); op. cit. ed. 2, **19**a: 164 (1931). Tab. **18.** Type from Cape Prov.

Z. rigescens E. Mey. in Drège, Zwei Pflanz.-Docum.: 96 (1843). Type from Cape Prov.

A small rigid shrub or suffrutex with a greyish or brownish rimose or rugose bark on older parts, ultimate branches vividly green and subsucculent, subangular and distinctly striate; internodes 5–15 mm. long. Leaves sessile or very shortly petiolate, 2-foliolate, stipulate; leaflets sessile, 5–12 × 1·2–2·3 mm., linear-cuneate or rarely narrowly oblanceolate, apex acute or subobtuse, glabrous, succulent; stipules very small, broadly triangular, membranous, acute to subobtuse. Flowers solitary, c. 18 mm. in diam.; pedicels c. 6 mm. long, glabrous. Sepals 3·75–4·25 × 2·9–3·25 mm., broadly ovate to (rarely) subcircular, apex obtuse or rarely slightly retuse, margins ± scarious, base broadly cuneate, glabrous, green. Petals bright yellow, 8–9 × 4–5 mm., obovate to obovate-oblong, apex obtuse, base cuneate, patent to slightly reflexed. Stamens with the anthers c. 2 mm. long, ovate to ovate-oblong; filaments 4·5–5 mm. long, slender, terete, glabrous; appendage of the filaments c. 2·4 × 1·5 mm., obovate-oblong or obcuneate, apex obtuse to truncate, margins shortly pectinate, glabrous. Ovary obovoid, glabrous; style c. 1·6 mm. long, terete, narrowly conical, glabrous; disk shallowly cup-shaped, succulent, glabrous. Fruit a 5-winged samara, 7–10 mm. long, 6–8 mm. in diam., glabrous, the wings dark green.

Bechuanaland Prot. SW: Takatshwane Pan, fl. 20.ii.1960, *Wild* 5083 (BM; SRGH). SE: Letlaking, fl. v.1958, *Patterson* 5 (SRGH).

Also in S. Africa. Growing in pans and on limestone outcrops.

There is one gathering from southeast Bechuanaland Prot. (Mone Valley near Letlaking, *Wild* 4960) which might represent another species. The plant differs from *Z. cuneifolium* by its larger leaflets which are distinctly petiolulate (petiolule 4–8 mm. long). A decision about its status can only be made when better and at least fruiting material becomes available.

TRIBULUS L.

Tribulus L., Sp. Pl. **1**: 386 (1753); Gen. Pl. ed. 5: 183 (1754).

Annual or perennial xerophilous herbs, sometimes with a woody base, very rarely shrubby, with usually prostrate or ascending branches. Leaves opposite, one of each pair usually longer than the other, paripinnate; leaflets opposite, sessile or very shortly petiolate, entire and somewhat oblique. Stipules herbaceous. Flowers solitary, axillary. Sepals 5, deciduous or sometimes persistent. Petals 5, usually yellow, rarely white, shorter to longer than the sepals, spreading. Stamens 10 (very rarely 5, but not in species of our area); filaments subulate, with extra- and intrastaminal glands at the base; anthers cordate or oblong-cordate. Ovary 5-lobed, consisting of 5 concrescent carpels with 3–5 ovules in each carpel, densely covered with stiff erect hairs; stigma 5-angled, pyramidal or hemispheric, formed by fusion of the 5 stigmatic lobes. Fruit at length breaking up into 5 cocci; cocci dorsally tuberculate and provided with spines or wings, rarely unarmed, 1–5-seeded. Seeds without endosperm.

A genus of about 25 species, found in all tropical and warm-temperate countries of the world.

Peduncles usually shorter than or as long as the subtending leaf; intrastaminal glands free; stigma hemispheric, often asymmetric, nearly sessile on the ovary 1. *terrestris*

Peduncles usually longer than the subtending leaf; intrastaminal glands ± connate and forming a shallow cup at the base of the ovary; stigma slender and usually pyramidal or hemispheric and then with a rather elongated style:

- Stigma pyramidal, nearly sessile on the ovary or style only slightly elongated; leaflets oblong to ovate-oblong, usually rather acute - - - - - 2. *zeyheri*
- Stigma hemispheric (often much reduced) but also sometimes ± pyramidal, always with a ± elongated style; leaflets oblong to obovate-oblong, rarely elliptic, usually obtuse and mucronulate but sometimes ± acute - - - - - 3. *cistoides*

1. **Tribulus terrestris** L., Sp. Pl. **1**: 387 (1753).—Eckl. & Zeyh., Enum. Pl.: 95 (1835) —Harv., Gen. S. Afr. Pl.: 46 (1838).—Drège in Zwei Pflanz.—Docum.: 58, 73, 131

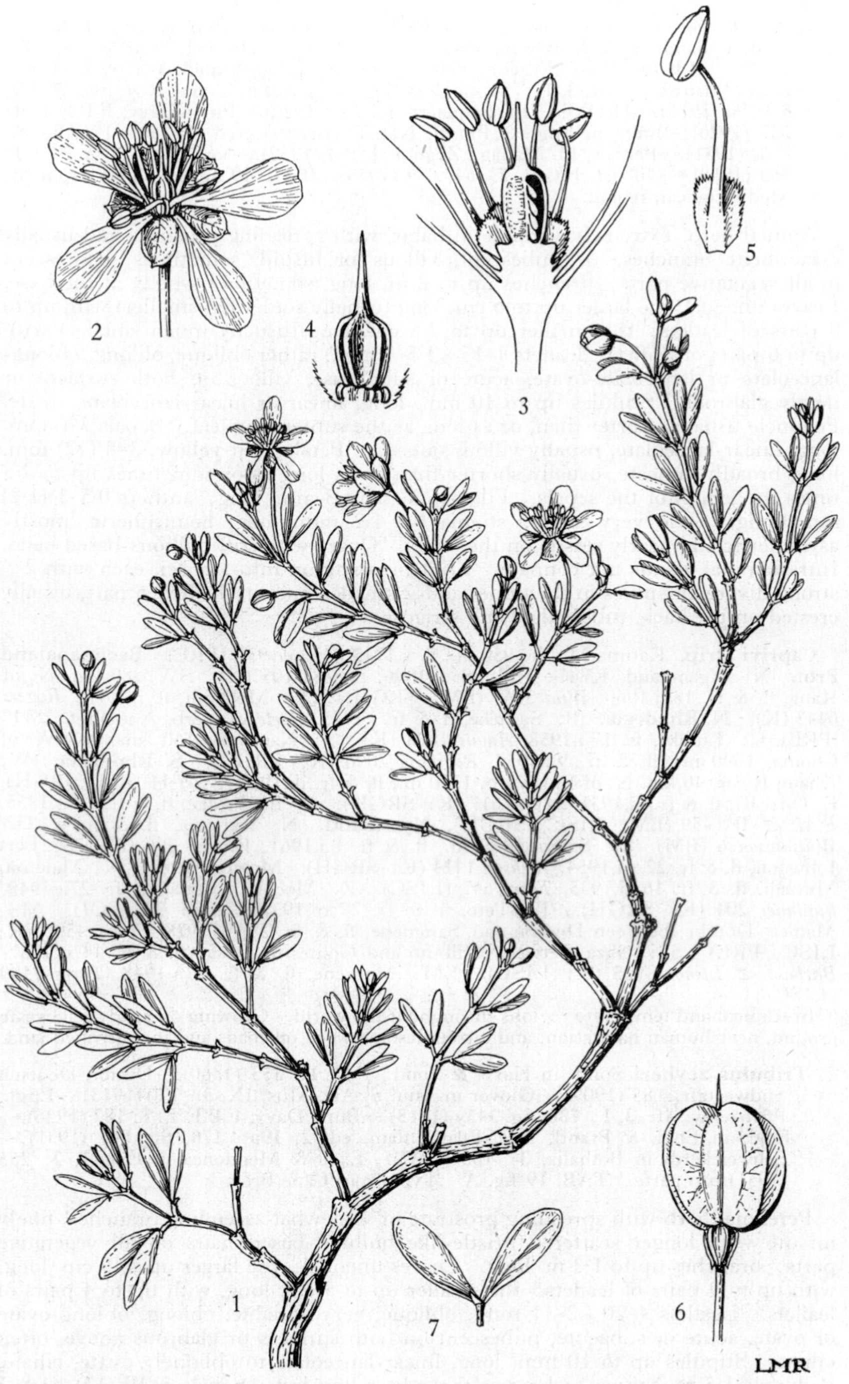

Tab. 18. ZYGOPHYLLUM CUNEIFOLIUM. 1, flowering branch ($\times\frac{2}{3}$) *Ferrar* 5680; 2, flower in surface view ($\times 2$) *Rodin* 3573; 3, vertical section of flower ($\times 4$) *Rodin* 3573; 4, ovary and disk ($\times 4$) *Rodin* 3573; 5, stamen ($\times 6$) *Rodin* 3573; 6, fruit ($\times 2$) *Ferrar* 5680; 7, leaf ($\times 1$) *Ferrar* 5680.

(1843).—Sond. in Harv. & Sond., F.C. **1**: 352 (1860).—Engl. & Gilg in Warb., Kunene-Samb.-Exped. Baum: 269 (1903).—Dinter, Deutsch Südwestafr.: 85 (1909).—R.E.Fr., Wiss. Ergebn. Schwed. Rhod.-Kongo-Exped. **1**: 109 (1914).—Engl., Pflanzenw. Afr. **3,** 1: 736, fig. 343 (1915).—Eyles in Trans. Roy. Soc. S. Afr. **5**: 390 (1916).—De Wild., Pl. Bequaert. **1**: 233 (1922).—Burtt Davy, F.P.F.T. **1:** 187 (1926).—Engl. in Engl. & Prantl, Nat. Pflanzenfam. ed. 2, **19a**: 176, fig. 84 E–L (1931).—Perrier, Fl. Madag., Zygophyll.: 4 (1952).—Keay, F.W.T.A. ed 2, **1:** 363 (1958).—Gilbert, F.C.B. **7**: 64, t. 9 (1958). TAB. **19** fig. B. Type from the Mediterranean region.

Annual herb, extremely variable in habit, with spreading prostrate and usually decumbent branches, ± pubescent, villous or hispid, sometimes glabrescent in all vegetative parts; branches up to 2 m. long, sometimes woody at the base. Leaves unequal, the larger up to 6 cm. long (usually somewhat smaller) with up to 8 pairs of leaflets; the smaller up to 3·5 cm. long (usually much smaller) with up to 6 pairs of leaflets; leaflets 4–15 × 1·5–7 mm., rather oblique, oblong, oblong-lanceolate or lanceolate-ovate, acute or subobtuse, villous on both surfaces or nearly glabrous. Stipules up to 10 mm. long, linear or linear-lanceolate, acute. Peduncle usually shorter than, or as long as the subtending leaf. Sepals 3–6 mm. long, linear-lanceolate, usually villous outside. Petals light yellow, 3–8 (12) mm. long, broadly cuneate, usually shorter than or as long as or sometimes up to 2·3 times the length of the sepals. Filaments about 3 mm. long; anthers 0·5–1 (1·2) mm. long. Style very short; stigma 0·8–1·2 mm. long, hemispheric, mostly asymmetric and nearly sessile on the ovary. Ovary with stiff bulbous-based hairs. Intrastaminal glands not connate. Fruit breaking up into 5 cocci, each with 2 ± strong divergent spines in the upper and 2 smaller ones in the lower part, usually crested on the back, tuberculate and strigose.

Caprivi Strip. Katima Mulilo, 950 m., fl. x.1945, *Curson* 951 (PRE). **Bechuanaland Prot.** N: Ngamiland, Kwebe Hills, fl. ii.1896, *Lugard* 105 (K). SW: 80 km. N. of Kang, fl. & fr. 18.ii.1960, *Wild* 5064 (BM; SRGH). SE: Mochudi, fl. iii.1914, *Rogers* 6443 (K). **N. Rhodesia.** B: Sesheke, fl. & fr. i.1925, *Borle* in Herb. Nat. Pret. 28717 (PRE). C: Lusaka, fr. 17.i.1958, *Angus* 1823 (K). S: Namwala, 160 km. NNW. of Choma, 1000 m., fl. & fr. 9.i.1957, *Robinson* 2096 (K; SRGH). **S. Rhodesia.** W: Khami River, 40 km. N. of Bulawayo, 1370 m., fl. & fr. iii.1904, *Eyles* 11 (BM; SRGH). E: Odzi R., fl. & fr. xi.1931, *Myres* 617 (K; SRGH). S: Beitbridge, fl. & fr. 16.ii.1955, *E.M. & W.* 459 (BM; LISC; SRGH). **Nyasaland.** N: Karonga, fl. & fr. v.1952, *Williamson* 6 (BM). C: Salima, 450 m., fl. & fr. 8.i.1961, *Wright* 296 (K). S: Fort Johnston, fl. & fr. 22.xii.1954, *Jackson* 1424 (K; SRGH). **Mozambique.** N: Malema, Mutuali, fl. & fr. 16.vi.1935, *Torre* 647 (LISC). Z: Mocuba, Namagoa, fr. 27.i.1948, *Faulkner* 204 (K; SRGH). T: Tete, fl. & fr. 27.xii.1931, *Guerra* 28 (COI). MS: Manica, Dombe, between Dombe and Sanguene, fl. & fr. 28.viii.1928, *Pedro* 4500 (K; LISC; PRE). SS: Gaza, between Chibuto and Gomes da Costa, fl. & fr. 14.xi.1957, *Barbosa & Lemos* 8125 (K; LISC). LM: Macaene, fl. & fr. 23.i.1948, *Torre* 7199 (LISC).

In tropical and temperate regions throughout the world. Growing on sandy soil, waste ground, near human habitation, and a troublesome weed on roads and in cultivated land.

2. **Tribulus zeyheri** Sond. in Harv. & Sond., F.C. **1**: 353 (1860).—Dinter, Deutsch Südwestafr.: 85 (1909).—Glover in Ann. S. Afr. Mus. IX, **3**: 170 (1913).—Engl., Pflanzenw. Afr. **3,** 1: 736, fig. 343y (1915).—Burtt Davy, F.P.F.T. **1**: 187 (1926).—Engl. in Engl. & Prantl, Nat. Pflanzenfam. ed. 2, **19a**: 176, fig. 84y (1931).—Schweickerdt in Bothalia, **3**: 168 (1937).—Exell & Mendonça, C.F.A. **1,** 2: 255 (1951) pro parte. TAB. **19** fig. A. Type from Cape Prov.

Perennial herb with spreading prostrate or somewhat ascending branches, finely hirsute with longer scattered bristle-like bulbous-based hairs on all vegetative parts; branches up to 1·2 m. long. Leaves unequal; the larger up to 9 cm. long, with up to 9 pairs of leaflets; the smaller up to 5 cm. long, with up to 4 pairs of leaflets. Leaflets 4–20 × 2–11 mm., oblique, very variable, oblong, oblong-ovate or ovate, acute or subacute, pubescent on both surfaces or glabrous above, often ciliate. Stipules up to 10 mm. long, linear-lanceolate to obliquely ovate, ciliate. Peduncle 1·5 to 2 times as long as the subtending leaf. Sepals 5–10 (12) × 1·8–2 mm., linear-lanceolate, acute, densely pubescent. Petals bright yellow, 10–20 (25) mm. long, up to 2·5 times the length of the sepals, broadly cuneate. Filaments 3–4 mm. long, anthers (1·2) 1·5–2·5 (3) mm. long. Style short; stigma 2–3 mm. long, slender, pyramidal. Ovary with stiff bristle-like tubercule-based hairs. Intra-

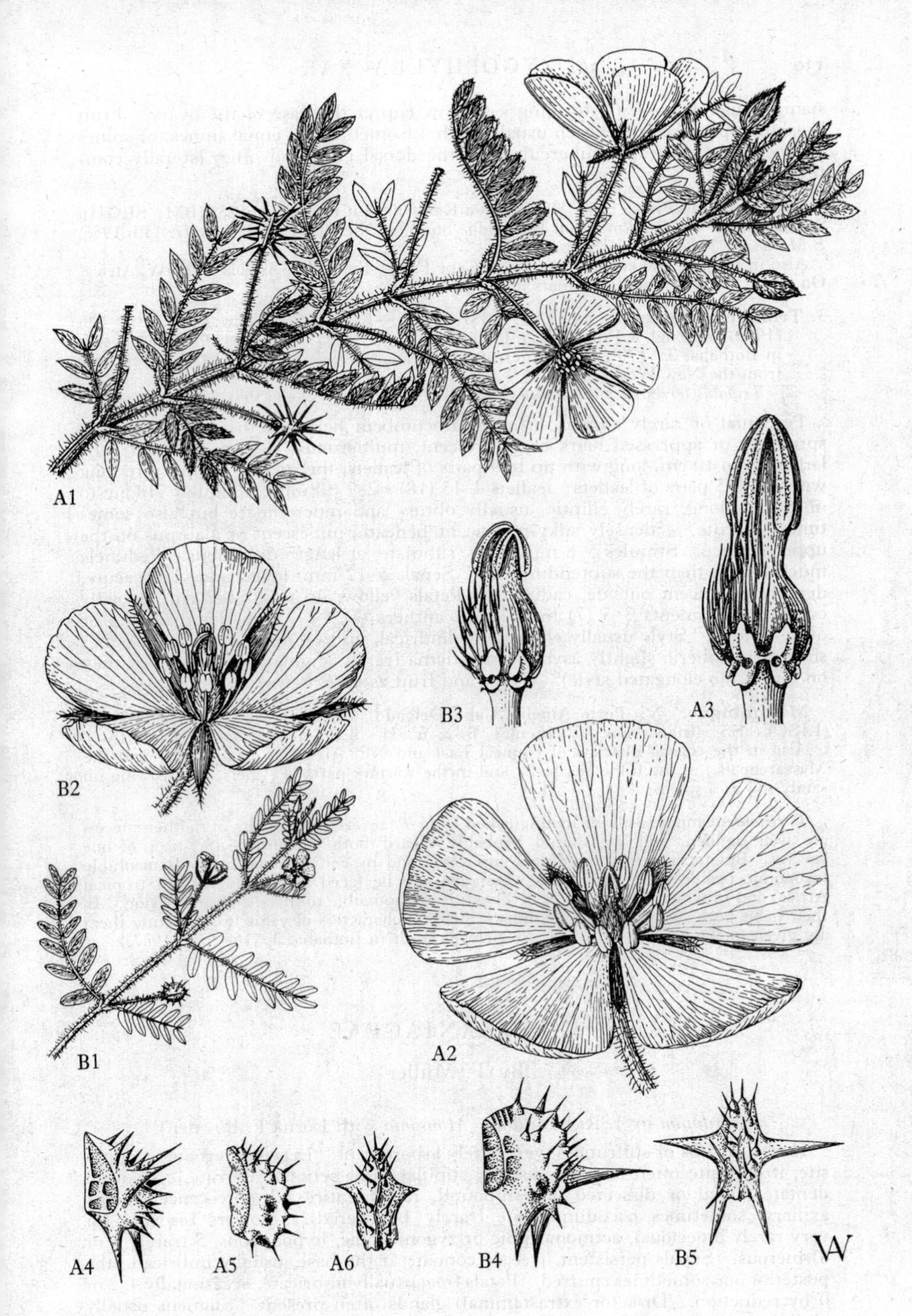

Tab. 19. A.—TRIBULUS ZEYHERI. A1, branch with flowers (×⅔); A2, flower (×3); A3, ovary (×6); all from *E.M. & W.* 394; A4, 5, 6, fruit (×2) after Schweickerdt. B.—TRIBULUS TERRESTRIS. B1, branch with flowers (×⅔); B2, flower (×4); B3, ovary (×6), all from *Schlechter* 11896; B4, 5, fruit (×2) after Schweickerdt.

staminal glands connate, forming a shallow cup at the base of the ovary. Fruit breaking up into 5 cocci, each usually with 4 (sometimes 6) equal spines, or spines reduced to thick warts, tuberculate on the dorsal crest and often laterally compressed.

Bechuanaland Prot. SW: 80 km. N. of Kang, fl. 18.ii.1960, *Wild* 5057 (BM; SRGH). **S. Rhodesia.** S: 3·2 km. from Beitbridge on West Nicholson road, fl. & fr. 15.ii.1955, *E.M. & W.* 394 (BM; LISC; SRGH).

Also in the arid and subarid regions of Cape Prov., Transvaal, Angola and SW. Africa. On sandy soil, forming dense mats in overgrazed areas.

3. **Tribulus cistoides** L., Sp. Pl. **1**: 387 (1753).—Engler, Pflanzenw. Afr. **3,** 1: 736 (1915); in Engl. & Prantl., Nat. Pflanzenfam. ed. 2, **19a**: 176 (1931).—Schweickerdt in Bothalia, **3**: 170 (1937).—Perrier, Fl. Madag., Zygophyll.: 4 (1952). Syntypes from the New World.

Tribulus terrestris var. *cistoides* (L.) Oliv., F.T.A. **1**: 284 (1868). Type as above.

Perennial or rarely annual diffusely procumbent herb; stems pubescent with spreading or appressed hairs or glabrescent, multistriate. Leaves unequal; the larger up to 10 cm. long with up to 9 pairs of leaflets, the smaller up to 6 cm. long with up to 5 pairs of leaflets; leaflets 4–15 (18) × 2–7 (9) mm., obliquely oblong to obovate-oblong, rarely elliptic, usually obtuse and mucronulate but also sometimes ± acute, ± densely silky-pubescent beneath, pubescent or glabrous on the upper surface. Stipules 5–8 mm. long, subulate, at length deciduous. Peduncle much longer than the subtending leaf. Sepals 5–12 mm. long, lanceolate, acute, densely pubescent outside, caducous. Petals yellow, up to 2 × 1–2 cm., broadly cuneate. Filaments 3–5 (7) mm. long; anthers (1) 1·2–3 mm. long, oblong or ovate-oblong. Style usually elongate, cylindrical, up to 4 mm. long with a rather short hemispheric slightly asymmetric stigma (rarely stigma pyramidal, but then on top of the elongated style). Ovary and fruit as in *T. zeyheri*.

Mozambique. N: Porto Amélia, Cabo Delgado, Mecufi, fl. 26.ix.1935, *Torre* 829 (LISC). SS: Inhambane; Vilanculos, fl. & fr. 31.viii.1944, *Mendonça* 1902 (LISC).

Also in the coastal districts of tropical East and NE. Africa, in Madagascar and the Mascarene Is., in the Cape Verde Is. and in the warmer parts of America. Growing on sandy soil as a weed.

T. cistoides cannot easily be distinguished from *T. zeyheri*; the limits of neither species are well defined, and we should probably regard both as infraspecific taxa of one species. But in view of the highly complicated and by conventional methods insoluble taxonomic problems in this genus which have to be faced (especially in NE. tropical Africa) and in which *T. cistoides* is involved, it is impossible to make a clear decision. In their habit both species are very similar; the only characters of value in separating them are given in the key above. (See also Schweickerdt in Bothalia, **3**: 169–171 (1937)).

37. GERANIACEAE

By T. Müller

(*Geranium* by J. R. Laundon; *Monsonia* with Lorna F. Bowden)

Herbs, shrubs or suffrutices, very rarely arborescent. Leaves alternate or opposite, if opposite often unequal, usually stipulate and petiolate, serrate, crenate or dentate, lobed or dissected or compound, rarely entire. Inflorescence usually axillary, sometimes pseudumbellate (rarely 1-flowered). Flowers bisexual (or very rarely dioecious), actinomorphic or zygomorphic, hypogynous, 5 (rarely 4 or 8)-merous. Sepals persistent, free or connate at the base, usually imbricate, the posterior one sometimes spurred. Petals free, usually imbricate, occasionally 4, 2 or 0 by reduction. Disk (or extrastaminal) glands often present. Stamens usually obdiplostemonous, twice as many as the sepals, more rarely 3 times as many (some sometimes sterile); filaments usually ± connate at the base, sometimes in 5 bundles of 3 each; anthers 2-thecous, dehiscing longitudinally, introrse. Ovary superior, syncarpous, 3–5 (rarely 2 or 8)-locular, usually lobed, usually rostrate (always in our area); style present or absent; stigmas ligulate, clavate or filiform, rarely capitate; loculi 1–2 ovulate; ovules pendulous, anatropous, superposed, placentation axile. Fruit a schizocarp or sometimes a 3–5 (rarely 8)-lobed capsule; lobes or mericarps usually 1-seeded and dehiscing septicidally; mericarps (cocci) rostrate, breaking away from a persistent central column. Seeds smooth or minutely reticulate; embryo curved, rarely straight; endosperm scanty or absent.

Flowers actinomorphic; spur absent; filaments 10 or 15, all bearing anthers (in our species); disk or extra-staminal glands present:
- Stamens 10; fruit up to 3 cm. long (in our species); rostrum of mericarps rolling upwards at dehiscence - - - - - - - - - - 1. *Geranium*
- Stamens 15; fruit 5–11 cm. long (in our species); rostrum of mericarps twisting helically at dehiscence - - - - - - - - - - 2. *Monsonia*

Flowers zygomorphic; posterior sepal with a spur adnate to the pedicel; filaments 10 (only 2–7 bearing anthers); no disk or extra-staminal glands present 3. *Pelargonium*

1. GERANIUM L.

By J. R. Laundon

Geranium L., Sp. Pl. **2**: 676 (1753); Gen. Pl. ed. 5: 306 (1754).

Annual or perennial herbs. Leaves alternate or opposite, stipulate, palmately lobed or dissected. Flowers actinomorphic, solitary or in pairs, axillary. Sepals and petals 5. Disk-glands present. Stamens 10 (rarely 5), usually all antheriferous; filaments slightly connate at the base or sometimes free. Ovary 5-lobed, 5-locular; loculi 2-ovulate, with 1 ovule aborting. Fruit a capsule, dehiscing by the splitting of the rostrum, the segments of which roll upwards (remaining attached at the apex) releasing the cocci from the base of the calyx. Seeds with little or no endosperm.

Stems with sharp reflexed prickles; leaves divided to about three-quarters; stipules divided - - - - - - - - - - - - 1. *aculeolatum*

Stems without prickles:
- Stipules entire, ovate:
  - Lobes of leaves rather shallowly divided (to less than half-way); inflorescence (1-)2-flowered; petals with conspicuous red veins - - - 2. *arabicum*
  - Lobes of leaves pinnatilobed to pinnatipartite; inflorescence 1-flowered; petals with inconspicuous veins - - - - - - - - - 3. *mlanjense*
- Stipules divided nearly to the base:
  - Leaves divided to less than 1·5 mm. from the base:
    - Leaves softly white-tomentose below - - - - - - 4. *incanum*
    - Leaves coarsely pilose below, the hairs mostly confined to the veins 5. *vagans*
  - Leaves divided to above 2·5 mm. from the base (unless exceptionally small):
    - Cocci smooth, hirsute; leaves divided to more than three-quarters; perennial 6. *exellii*
    - Cocci with reticulate ridges, glabrous or pubescent; leaves divided to about two-thirds; annual - - - - - - - - - 7. *ocellatum*

1. **Geranium aculeolatum** Oliv., F.T.A. **1**: 291 (1868).—Knuth in Engl., Pflanzenr. IV, 129: 203, t. 26 fig. C–F (1912).—Milne-Redh. in Mem. N.Y. Bot. Gard. **8, 3**: 230 (1953).—Petit, F.C.B. **7**: 23 (1958). TAB. **20** fig. E. Syntypes from Ethiopia.

Perennial herb, climbing or trailing and rooting at the nodes; stems longitudinally furrowed, covered with sharp reflexed prickles, and with glandular hairs on the younger parts. Leaf-lamina 5-lobed, divided to about $\frac{3}{4}$; lobes 1–10 mm. across at base, rhombic in outline, pinnatifid, stiffly pilose on both the upper and lower surfaces; petiole 1–8 cm. long, longitudinally furrowed, covered with sharp reflexed prickles and often also glandular hairs; stipules 3–8 mm. long, lacerated nearly to the base into lanceolate to linear segments, often glandular-hairy towards the margin, which is often hyaline. Flowers in 2-flowered inflorescences; peduncle 2–10 cm. long, furrowed or terete, sometimes with prickles, usually densely glandular; pedicels 5–25 mm. long, furrowed or terete, sometimes with prickles, usually densely glandular; bracts 2–5 mm., linear-lanceolate, hyaline at the edge, often stiffly pilose. Sepals 7–8 × 1–4 mm., lanceolate, mucronate, 3-nerved, hyaline at the edge, usually stiffly pilose. Petals c. 10 × 5 mm., white or pale pink or pale mauve. Stamens 10, filaments about 4–5 mm. long, narrowly lanceolate.

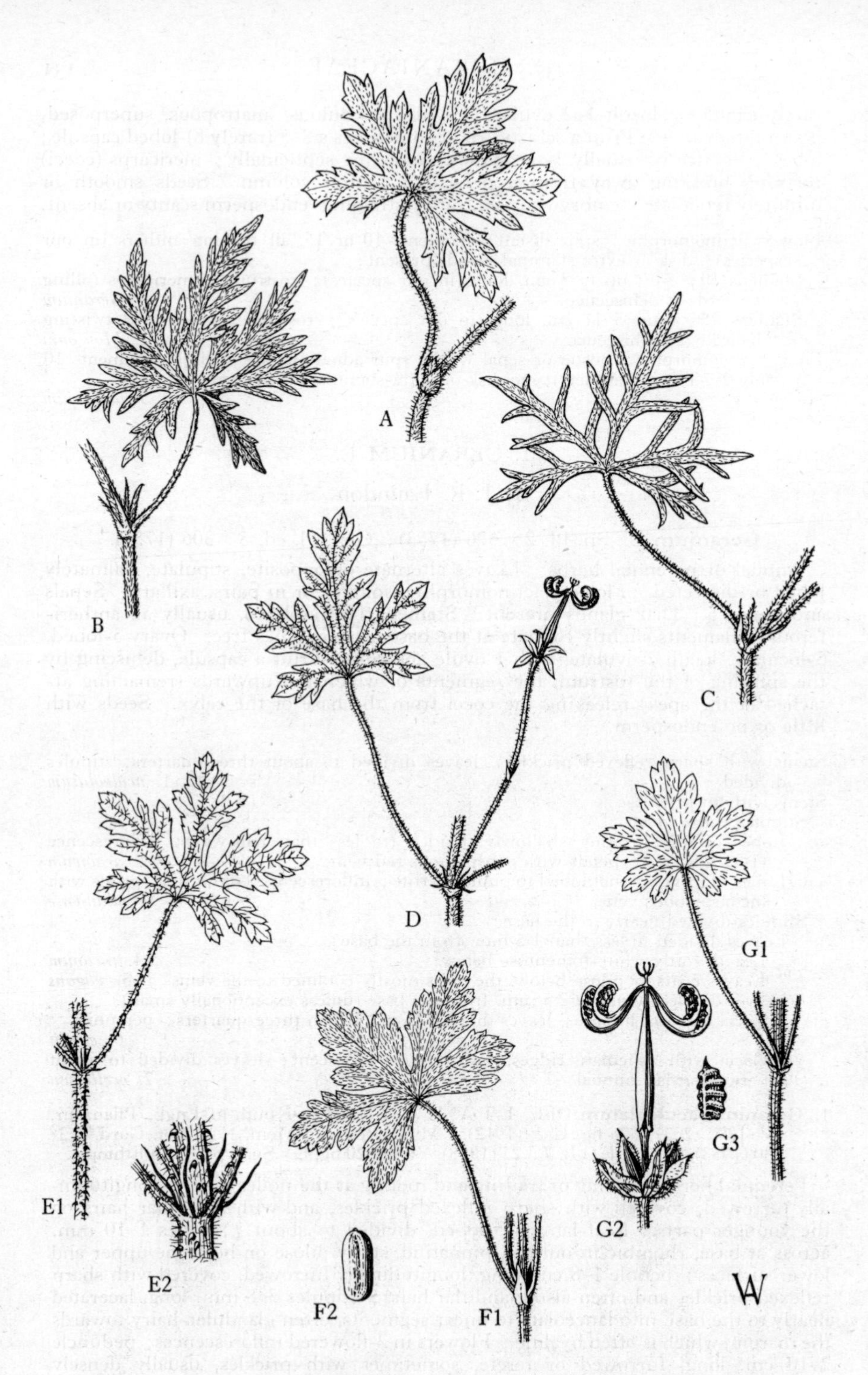

Tab. 20. A.—GERANIUM MLANJENSE, leaf (× ⅔) *Chapman* 467. B.—GERANIUM INCANUM SUBSP. NYASSENSE, leaf (× ⅔) *Brass* 17189. C.—GERANIUM VAGANS SUBSP. VAGANS, leaf (× ⅔) *Chapman* 382. D.—GERANIUM EXELLII, leaf and fruit (× ⅔) *E.M. & W.* 109. E.—GERANIUM ACULEOLATUM. E1, leaf (× ⅔) *Chapman* 297 ; E2, stem and stipules (× 4) *Chapman* 297. F.—GERANIUM ARABICUM. F1, leaf (× ⅔) *Newman & Whitmore* 195 ; F2, seed (× 4) *Newman & Whitmore* 195. G.—GERANIUM OCELLATUM. G1, leaf (× ⅔) *Chase* 4910 ; G2, fruit (× 2) *Chase* 4910 ; G3, cocci (× 4) *Chase* 4910.

Ovary pubescent; stigmas 5. Cocci smooth, pilose; mature rostrum 15–25 × 1·5 mm., shortly pubescent; stigmas 1·5–2·5 mm. long. Seeds smooth, glabrous.

Nyasaland. N: Rumpi Distr., Nyika Plateau, track to Rukuru Falls, 1800 m., fl. 6.i.1959, *Richards* 10527 (K). C: Dedza Mt., 1650 m., fl. 27.iii.1950, *Wiehe* N/466 (K; SRGH). S: Zomba Plateau, 1450 m., fl. & fr. 3.vi.1946, *Brass* 16186 (K; SRGH). **Mozambique.** N: Maniamba, S. Géci, fl. & fr. 29.v.1948, *Pedro & Pedrógão* 4060 (EA; LISC).

Also in Sudan, Ethiopia, Kenya, Uganda, Tanganyika and Congo. Montane forest, especially in clearings, 1450–2130 m.

2. **Geranium arabicum** Forsk., Fl. Aegypt. Arab.: 124 (1775).—Laundon in Bol. Soc. Brot., Sér. 2, **35**: 59, t. 1 (1961). TAB. **20** fig. F. Type from Arabia (Yemen).

Geranium simense Hochst. ex A. Rich., Tent. Fl. Abyss. **1**: 116 (1847).—Knuth in Engl., Pflanzenr. IV, 129: 203, t. 26 fig. A–B (1912).—Milne-Redh. in Mem. N.Y. Bot. Gard. **8,** 3: 230 (1953).—Hedberg in Symb. Bot. Upsal. **15,** 1: 125, 293 (1957).—E. Petit, F.C.B. **7**: 24 (1958). Type from Ethiopia.

Geranium frigidum Hochst. ex A. Rich., Tent. Fl. Abyss. **1**: 116 (1847) *in synon.* —Briq. in Ann. Cons. Jard. Bot. Genève, **11 & 12**: 184 (1908).—Knuth in Engl., Pflanzenr. IV, 129: 205 (1912). Type from Ethiopia.

Geranium simense var. *repens* Oliv., F.T.A. **1**: 291 (1868). Type from Fernando Po.

Geranium simense var. *glabrius* Oliv., F.T.A. **1**: 291 (1868). Syntypes from Cameroons.

Geranium keniense Standl. in Smithsonian Misc. Coll. **68,** 5: 7 (1917). Type from Kenya.

Perennial herb; stems stoloniferous, creeping and scrambling, longitudinally furrowed, pilose or hispid. Leaf-lamina 5-lobed, divided to ¾ or more; lobes 3–10 mm. across at base, broadly elliptic to rhombic in outline, pinnatifid to pinnatilobed, stiffly pubescent or pilose on both the upper and lower surfaces; petiole 1–16 cm. long, pilose; stipules 5–12 × 3–5 mm., entire, ovate-lanceolate, acute to acuminate, membranous, pilose or pubescent. Flowers in 2-flowered inflorescences, or, rarely, 1-flowered by reduction; peduncle 3–10 cm. long, furrowed or terete, pubescent or pilose; pedicels 1–4 cm. long, furrowed or terete, pubescent or pilose; bracts 3–5 mm., lanceolate, hyaline at the edge. Sepals 7–10 × 2–3 mm., broadly lanceolate, cuspidate or mucronate, 3-nerved, pilose and pubescent. Petals about 10 × 5–7 mm., obovate, retuse at the apex, white or pink (rarely pale purple) with conspicuous red veins. Stamens 10, filaments about 3 mm. long, narrowly lanceolate. Ovary pubescent; stigmas 5. Cocci smooth, pilose; mature rostrum 12–20 × 2 mm., pubescent; stigmas 1–2 mm. long. Seeds glabrous; testa covered with a very shallow reticulate pattern.

N. Rhodesia. N: Abercorn Distr., Kawimbe area, Lumi R. near Bridge road, 1500 m., fl. & fr. 30.viii.1956, *Richards* 6032 (K). **S. Rhodesia.** E: Umtali Distr., Engwa, 1980 m., fl. & fr. 2.ii.1955, *E.M. & W.* 91 (BM; LISC; SRGH). **Nyasaland.** N: Nyika Plateau, Lake Kaulime, 2150 m., fl. & fr. 24.x.1958, *Robson* 339 (BM; K; LISC; SRGH). S: Zomba Mt., Mulunguzi R., 1670 m., fl. & fr. 9.iii.1955, *E.M. & W.* 770 (BM; LISC; SRGH).

Widespread on the mountains of tropical Africa, occurring in Nigeria, Cameroons and Fernando Po in western Africa and extending from Eritrea, Sudan and Ethiopia to S. Rhodesia in eastern Africa. Also in Arabia (Yemen) and Madagascar. Chiefly in damp shady habitats, especially in woodland, from 1100–2130 m.

3. **Geranium mlanjense** Laundon in Bol. Soc. Brot., Sér. 2, **35**: 70, t. 4 (1961). TAB. **20** fig. A. Type: Nyasaland, Mt. Mlanje, Tuchila Plateau, *Newman & Whitmore* 132 (BM, holotype; SRGH).

Geranium latistipulatum sensu Milne-Redh. in Mem. N.Y. Bot. Gard. **8,** 3: 231 (1953).

Perennial herb; stems decumbent, pilose or pubescent. Leaf-lamina 5-lobed, divided to less than ¾; lobes 2–6 mm. across at base, obovate in outline, pinnatilobed to pinnatipartite, coarsely pubescent on both the upper and lower surfaces; petiole 0·5–16 cm. long, longitudinally furrowed, pilose or pubescent; stipules 4–7 mm. long, entire, ovate-lanceolate, acuminate, pubescent. Inflorescence 1-flowered; peduncle 3–4 cm. long, pubescent or pilose; pedicels 1–3 cm. long, pubescent; bracts 3–5 mm. long, narrowly lanceolate, acute, pubescent. Sepals 6–8 × 2–3 mm., ovate to elliptic, mucronate, pubescent. Petals about 10 × 5 mm., obovate, retuse at the apex, pale pink. Stamens 10; filaments about 5 mm. long, narrowly lanceolate, pubescent towards the base; anthers dark purple. Ovary tomentose; stigmas 5, pink. Cocci smooth, pilose; mature rostrum 15–17 × 1·5 mm., pubescent; stigmas 2–2·5 mm. long. Seeds glabrous; testa covered with a shallow reticulate pattern.

Nyasaland. S: Mlanje Mt., path from Little Ruo down to Naiwani, fl. & fr. 29.vii.1957, *Chapman* 467 (BM; LISC; SRGH).

Known only from Mt. Mlanje. Borders of woodland, amongst scrub and in grassland. 1830–2100 m.

Closely related to the widespread *Geranium arabicum* Forsk. It should be noted that the latter species occurs also on Mt. Mlanje.

4. **Geranium incanum** Burm. f., Spec. Bot. Geraniis: 28, t. 1, n. 26 (1759).—L., Sp. Pl. ed. 2, **2**: 957 (1763).—Knuth in Engl., Pflanzenr. IV, 129: 162, t. 23 fig. F–H (1912). Syntypes from S. Africa.

Geranium multisectum N.E.Br. in Kew Bull. **1901**: 120 (1901). Type from Basutoland.

Perennial herb, creeping and scrambling; stems longitudinally furrowed, pilose, but sometimes becoming nearly tomentose or densely glandular on the upper parts of the stems. Leaf-lamina 3–7-lobed, divided nearly to the base; lobes 0·5–2 mm. across at base, ovate to elliptic in outline, deeply divided into linear to oblong-lanceolate segments, pilose above and tomentose below; petiole 0·5–25 cm. long, longitudinally furrowed, pilose; stipules 3–15 mm. long, lacerated nearly to the base into linear-lanceolate segments, pilose. Inflorescence 2-flowered; peduncle 1–13 cm. long, pilose, tomentose or glandular; pedicels 1–4 cm. long, pilose, tomentose or glandular; bracts 2–7 mm. long, linear-lanceolate, hyaline at the edge, pilose or tomentose. Sepals 5–10 × 2–3 mm., ovate-lanceolate, cuspidate, 3-nerved and pilose. Petals 9–15 × 4–8 mm., obovate, retuse at the apex, white, pink or mauve with dark veins. Stamens 10, filaments about 6 mm. long, narrowly lanceolate. Ovary pubescent; stigmas 5. Cocci smooth, pilose; mature rostrum 13–27 × 1–2 mm., pubescent or glandular; stigmas 1–4 mm. long. Seeds glabrous; testa covered with a shallow reticulate pattern.

S. Africa and on mountains in east Africa in Mozambique, S. Rhodesia, Nyasaland and Tanganyika. Open habitats, especially grassland, 1500–2560 m.

Subsp. *incanum* is restricted to South Africa and Tanganyika.

Subsp. **nyassense** (Knuth) Laundon in Bol. Soc. Brot., Sér. 2, **35**: 63, t. 3 (1961). TAB. **20** fig. B. Type from Tanganyika.

Geranium nyassense Knuth in Fedde, Repert. **18**: 289 (1922).—Milne-Redh. in Mem. N.Y. Bot. Gard. **8,** 3: 231 (1953).—Martineau, Rhod. Wild Fl.: 40, t. 13 n. 2 (1954). Type as above.

Geranium ukingense Knuth in Fedde, Repert. **18**: 292 (1922). Type from Tanganyika.

Differs from subsp. *incanum* in the oblong-lanceolate leaf-segments as opposed to the linear leaf-segments of subsp. *incanum.* Subsp. *nyassense* generally has shorter petals and a shorter and narrower rostrum than subsp. *incanum* but these characters are rather variable in the latter.

S. Rhodesia. E: Inyanga Distr., Inyangani, fl. & fr. 26.xi.1949, *Chase* 1832 (BM; SRGH). **Nyasaland.** N: Nyika Plateau, 2340 m., fl. 12.viii.1946, *Brass* 17189 (BM; K; SRGH). **Mozambique.** MS: Tsetsera, 2140 m., fl. & fr. 7.ii.1955, *E.M. & W.* 248 (BM; COI: LISC: SRGH).

Also in Tanganyika.

5. **Geranium vagans** Bak. in Kew Bull. **1897**: 246 (1897).—Knuth in Engl., Pflanzenr. IV, 129: 208 (1912).—Milne-Redh. in Mem. N.Y.Bot. Gard. **8,** 3: 231 (1953).—Hedberg in Symb. Bot. Upsal. **15,** 1: 123 & 293 (1957).—Petit, F.C.B. **7**: 23 (1958). Type: Nyasaland, Nyika Plateau, *Whyte* (K, holotype).

Geranium angustisectum Knuth in Engl., Pflanzenr. IV, 129: 207 (1912). Type from Tanganyika.

Geranium schliebenii Knuth in Notizbl. Bot. Gart. Berl. **11**: 1067 (1934). Type from Tanganyika.

Perennial herb, semi-procumbent, creeping and climbing; rootstock woody; stems longitudinally furrowed, pilose on the older parts but usually densely glandular on the younger parts. Leaf-lamina 5-lobed, divided nearly to the base; lobes 1–2 mm. across at base, ovate in outline, deeply divided into linear to oblong-lanceolate segments, stiffly pilose above and stiffly pilose on the veins and margin below, otherwise glabrous on the under surface; petiole 5–35 mm. long, longitudinally furrowed, pilose or glandular; stipules 3–15 mm. long, lacerated nearly to the base into lanceolate segments, pilose or glandular. Inflorescence 2-flowered; peduncle 1–25 cm. long, pilose or glandular, rarely glabrous; pedicels 1–8 cm. long, pilose or glandular; bracts 3–8 mm. long, narrowly lanceolate, pubescent. Sepals 6–8 mm. long, 2–3 mm. broad, elliptic-lanceolate, cuspidate, 3-nerved, pubescent, coarsely pilose or glandular. Petals about 10–12 × 5–7 mm., pink or mauve, rarely white, nerved. Stamens 10, filaments about 6 mm. long, narrowly lanceolate. Ovary glandular or pilose; stigmas 5. Cocci smooth, pilose; mature rostrum 15–20 × 1–1·5 mm., glandular or pubescent; stigmas 2–4 mm. long. Seeds glabrous, the testa covered with a shallow reticulate network.

Mountains in the Congo, Uganda, Kenya, Tanganyika and Nyasaland. Grassland (1370–) 1830–2300 m.

Subsp. **vagans.** Tab. **20** fig. C.

Rostrum covered with long glands.

Nyasaland. N: Nyika Plateau, Kaulime pond, fl. & fr. 17.ii.1956, *Chapman* 382 (BM; LISC).

Subsp. **whytei** (Bak.) Laundon in Bol. Soc. Brot., Sér. 2, **35**: 71, t. 5 (1961). Syntypes: Nyasaland, Zomba, *Whyte* (K, syntype); Mt. Malosa, *Whyte* (K, syntype).

Geranium whytei Bak. in Kew Bull. **1898**: 302 (1898).—Knuth in Engl. Pflanzenr. IV, 129: 208 (1912). Type as above.

Geranium linearilobum Knuth in Jahres-Ber. Schles. Ges. Bresl., **81,** IIb, Zool. Bot. Sekt.: 17 (1904) pro parte quoad specim. Whyte. Lectotype: Nyasaland, Mt. Zomba, *Whyte* (K).

Rostrum densely hairy, without glands.

Nyasaland. S: Zomba Mountain, fl. & fr. 5.iii.1956, *Banda* 206 (BM; LISC).
Known only from Mt. Malosa and Zomba Mountain.

Knuth (tom. cit.: 202 (1912)) states that *G. whytei* is distinguished from *G. vagans* by the presence of an elongated middle segment of each lobe of the leaf. A comparison of a large amount of material shows that this character cannot be used as a means of differentiation since plants from the Nyika and Tanganyika show much variation in the length of the middle segment and specimens with segments intermediate in length occur. However, the different type of indumentum on the rostrum of the plants from Malosa and Zomba indicates that geographical differentiation has taken place, and *G. whytei* can be regarded as a subspecies.

6. **Geranium exellii** Laundon in Bol. Soc. Brot., Sér. 2, **35**: 62, t. 2 (1961). Tab. **20** fig. D. Type: S. Rhodesia, Umtali Distr., Engwa, *E.M. & W.* 109 (BM, holotype; LISC; SRGH).

Perennial herb, prostrate and straggling; stems longitudinally furrowed, pilose with spreading glandular hairs. Leaf-lamina 5-lobed, divided to about $\frac{4}{5}$; lobes 2–6 mm. across at base, elliptic in outline, pinnatilobed, pilose on both surfaces; petiole 5–80 mm. long, longitudinally furrowed, with spreading glandular hairs; stipules 4–5 mm. long, lacerated nearly to the base into linear-lanceolate segments, ± pilose. Inflorescence 2-flowered (rarely 1-flowered by reduction); peduncle 1–9 cm. long, pilose with small glands on the ends of the hairs; pedicels 1–4 cm. long, pilose with small glands on the ends of the hairs; bracts 3–4 mm. long, very narrowly lanceolate, pilose. Sepals 5–6 × 2–2·5 mm., elliptic-lanceolate, cuspidate (the tip 1 mm. long), 3-nerved, glandular. Petals about 8–9 × 4 mm., obovate, white or nearly white with pink veins. Stamens 10, filaments 5 mm. long, narrowly lanceolate. Ovary pubescent; stigmas 5. Cocci smooth, pilose; mature rostrum 14–16 × 1 mm., glandular; stigmas 2 mm. long. Seeds glabrous; testa covered with a shallow reticulate network.

S. Rhodesia. E: Umtali Distr., Himalayas, Engwa, 2130 m., fl. & fr. 2.iii.1954, *Wild* 4462 (K; SRGH). **Mozambique.** MS: Tsetsera, 1980–2140 m., fl. & fr. 10.ii.1955, *E.M. & W.* 337 (BM; LISC; SRGH).
Known only from S. Rhodesia and Mozambique. Upland grassland. 1980–2140 m.

7. **Geranium ocellatum** Jacquem. ex Cambess. in Jacquem., Voy. Ind. **4**, Bot.: 33 (1844); op. cit. Atlas **2**: t. 38 (1844).—Knuth in Engl., Pflanzenr. IV, 129: 62 (1912). TAB. **20** fig. G. Type from Pakistan.

Geranium ocellatum var. *himalaicum* Knuth in Engl., loc. cit. Type as above.

Geranium ocellatum var. *yunnanense* Knuth in Engl., loc. cit. Type from China (Yunnan).

Geranium ocellatum var. *africanum* Knuth in Engl., loc. cit. pro parte.—Petit in F.C.B. **7**: 22 (1958). Syntypes from East Africa, Arabia and Iran.

Geranium ocellatum var. *camerunense* Knuth in Engl., tom. cit.: 63 (1912). Type from the Cameroons.

Geranium favosum var. *sublaeve* Oliv., F.T.A. **1**: 292 (1868).—Knuth in Engl., tom. cit.: 62 (1912). Type from the Cameroons.

Geranium brevipes Hutch. & Dalz., F.W.T.A. **1**: 138 (1927); Kew Bull. **1928**: 30 (1928). Type as for *G. favosum* var. *sublaeve.*

Geranium ocellatum var. *sublaeve* (Oliv.) Milne-Redh. in Kew Bull. **1948**: 453 (1948). Type as for *G. favosum* var. *sublaeve.*

Geranium simense sensu Suesseng. in Proc. & Trans. Rhod. Sci. Ass. **43**: 108 (1951).

Annual herb; stems prostrate-ascendent, longitudinally furrowed, mostly covered with spreading hairs or glands up to 2 mm. in length. Leaf-lamina 5-lobed, divided to about $\frac{2}{3}$; lobes 3–10 mm. across at base, obovate in outline, pinnatifid, pubescent both above and below, usually hispid with scattered hairs on the upper surface also; petioles 5–90 mm. long, longitudinally furrowed, covered with spreading hairs or glands; stipules 2–4 mm. long, lacerated nearly to the base into lanceolate segments, pubescent or pilose. Inflorescence 2-flowered; peduncle 0–5 cm. long, glandular; pedicels 2–20 mm. long, glandular; bracts 2–4 mm. long, linear-lanceolate, hyaline at the edge, pubescent; sepals 5 × 2–3 mm., lanceolate, mucronate, 3-nerved, hyaline at the edge, pubescent, pilose or glandular. Petals about 7 × 5 mm., obovate, purple-pink with distinct dark purple centre. Stamens 10, filaments about 3 mm. long, narrowly lanceolate. Ovary pubescent; stigmas 5. Cocci with shallow reticulate ridges, glabrous or shortly pubescent; mature rostrum 8–14 mm. long, pubescent; upper portion of the styles 0–2 mm. long, stigmas 0·3–1 mm. long. Seeds glabrous; testa minutely foveolate.

N. Rhodesia. N: Abercorn Distr., Itembwe Gorge, 1520 m., fr. 24.iv.1959, *Richards* 11343 (K; SRGH). **S. Rhodesia.** W: Matobo Distr., Besna Kobila, 1460 m., fr. iii.1958, *Miller* 5175 (SRGH). C: Marandellas Distr., Cave Farm, 1370 m., fl. & fr. 6.iv.1950, *Wild* 3321 (K; SRGH). E: Inyanga Distr., north east edge of Hill Fort, Rhodes Estate, 1830 m., fl. & fr. 19.iv.1953, *Chase* 4910 (BM; K; LISC; SRGH). **Nyasaland.** S: Namassi, Mount Malosa, fl. & fr. 1897, *Cameron* 6 (K).

In West Africa on Cameroons Mt. and the Bamenda Plateau and in east Africa from the Sudan, Eritrea and Somalia to Southern Rhodesia; also in Arabia (Yemen), Iran, China and the Himalayas. Generally in densely shaded habitats such as under rocks or in caves, 1370–1830 m.

Knuth (1912, loc. cit.) distinguishes four varieties of this species, each variety being confined to a distinct geographical area. An examination of the material now available shows that there is no justification for this subdivision, for, although the species has a disjunct distribution, the isolated populations have undergone little or no differentiation.

2. MONSONIA L.

(with Lorna F. Bowden)

Monsonia L., Syst. Nat. ed. 12, **2**: 508 et Mant. Pl.: 14 (1767).

Annual or perennial herbs (sometimes with annual shoots arising from a woody basal region) or suffrutices; erect or decumbent, rarely acaulescent; variously pubescent, usually glandular. Leaves petiolate, stipulate, opposite or alternate, serrate, dentate or crenate, sometimes lobed or dissected; stipules filiform to subulate or rarely spinescent. Inflorescence a 2–several flowered pseudumbel (or flowers solitary). Pedicels often sharply bent when in fruit; bracts present. Flowers actinomorphic, 5-merous. Sepals imbricate, apiculate with membranous margin. Petals obovate, often with a truncate or lobed apex. Stamens 15, all fertile, connate at the base, in 5 bundles of 3 filaments each, the central filament of each triad being longer; extrastaminal glands present, 5, adnate to the base of the longer central filaments, sometimes inconspicuous; anthers versatile. Ovary 5-lobed, 5-locular, rostrate; loculi 2-ovulate; style absent; stigmas 5, clavate or filiform. Fruit a rostrate schizocarp; mericarps rostrate, 1-seeded, obliquely truncate at the apex of the basal portion and tapering towards the base, hirsute; rostrum persistent, helically twisting when ripe, plumose on the inside. Seeds ± oblong-obovoid; endosperm absent; embryo curved.

Plant perennial; petals (11) 14–24 mm. long; anthers 2·0–2·8 mm. long:
Petiole up to 1 (1·4) cm. long, up to half the length of the lamina, but usually shorter; lamina lorate to narrowly oblong or very narrowly elliptic to elliptic; base of lamina cuneate; extrastaminal glands clearly visible, with a ligulate apical expansion 1. *biflora*
Petiole (1) 2–4·5 (5) cm. long, half as long as to longer than the lamina; lamina lanceolate to narrowly ovate with a rounded to cordate base; extrastaminal glands merely inconspicuous cavities - - - - - - - - 2. *ovata* subsp. *glauca*
Plant annual; petals 9–12 (15) mm. long; anthers 0·8–1·2 mm. long:
Petiole as long as or longer than the lamina; lamina ovate to broadly ovate; fruit 8–9·5 cm. long - - - - - - - - - - 3. *senegalensis*
Petiole shorter than the lamina; lamina lorate to narrowly oblong, or narrowly elliptic to elliptic; fruit 5–6·5 (8) cm. long - - - - - - - 4. *angustifolia*

1. **Monsonia biflora** DC., Prodr. **1**: 638 (1824).—Burtt Davy, F.P.F.T. **1**: 193 (1926). TAB **21.** Type from S. Africa.
Monsonia burkeana Planch. ex. Harv. in Harv. & Sond., F.C. **1**: 255 (1860).—Knuth in Engl. Pflanzenr. IV, 129: 299 (1912).—Engl., Pflanzenw. Afr. **3,** 1: 705 (1915).—Eyles in Trans. Roy. Soc. S. Afr. **5**: 385 (1916).—Burtt Davy, F.P.F.T. **1**: 193 (1926).—R. A. Dyer in Fl. Pl. S. Afr. **20**: t. 794 (1940).—Exell & Mendonça C.F.A. **1,** 2: 258 (1951).—Martineau, Rhod. Wild Fl.: 41, t. 13 fig. 1 (1954). Type from S. Africa.
Monsonia malviflora Schinz in Bull. Herb. Boiss., Sér. 2, **3**: 821 (1903). Type from SW. Africa.
Monsonia glandulosissima Schinz, tom. cit.: 822 (1903). Type from SW. Africa.
Monsonia sp.—Eyles, loc. cit. quoad specim. *Eyles* 1193 et *Rand* 95.
Pelargonium sp.—Eyles, loc. cit. quoad specim. *Gardner* 64.

Erect perennial herb or suffrutex, up to 60 cm. tall, often with annual shoots arising from a woody basal region; stems branching freely; internodes 1–3 (4·5) cm. long; vegetative parts, peduncles, pedicels and sepals ± densely pubescent, covered with short recurved or patent and then often gland-tipped hairs, sometimes scattered very much longer patent hairs also present especially on the stem and calyx. Leaf-lamina 1·5–4 × 0·2–1 cm., lorate to narrowly oblong or very narrowly elliptic to narrowly elliptic, obtuse to subacute at the apex, margin irregularly serrate, base cuneate; petiole 5–10 (14) mm. long; stipules 3–6 (10) mm. long, acicular to subulate. Inflorescence a pseudumbel of 1–3 flowers; peduncle (1·5) 2·5–4·5 (7·5) cm. long, leaf-opposed or in the axil of the smaller of two apparently opposite leaves; pedicel 1·8–3·4 (4) cm. long; bracts 1–4, 3–8 mm. long, linear. Sepals 9–11 × 2·5–4 mm., narrowly oblong to lanceolate with an acicular apical appendage 1–2 mm. long. Petals 18–24 × 8–14 mm., yellowish-white to pale blue, sometimes pink to mauve, veins darker, obovate, apex truncate. Extrastaminal glands ovate with a ligulate apical appendage. Filaments 8–11·5 mm. long, pubescent; anthers 2–2·5 × 1–1·4 mm. Ovary tomentose; stigmas (2) 2·5–3·6 mm. long, filiform. Fruit 6–8 (9) cm. long; cocci 12·5–14 × 2·3–2·6 mm. Seed 5·1–5·3 × 2–2·4 mm., pale brown, minutely reticulate.

Bechuanaland Prot. SE: Gaberones, fl. & fr. 15.iii.1930, *van Son* in Herb. Transv. Mus. 28847 (BM; K; PRE). **S. Rhodesia.** W: Matobo, Farm Besna Kobila, fl. & fr. ii.1954, *Miller* 2185 (LISC; K; SRGH). C: Salisbury, south of Darwendale near railway line, fl. 18.ix.1955, *Boughey* 225 (LISC; SRGH). E: Inyanga, fl. & fr. 31.xii.1942, *Hopkins* in G.H.S. 9519 (K; SRGH).
Also in the Transvaal, N. Cape Prov., SW. Africa and Angola. Grassland 1340–2000 m.

2. **Monsonia ovata** Cav., Diss. **4**: 193, t. 113 fig. 1 (1787).—Harv. in Harv. & Sond., F.C. **1**: 255 (1860).—Eyles in Trans. Roy. Soc. S. Afr. **5**: 385 (1916). Type from S. Africa.

Subsp. **glauca** (Knuth) Bowden & T. Müll.
Monsonia glauca Knuth in Engl. Bot. Jahrb. **40**: 64 (1907).—Knuth in Engl., Pflanzenr. IV, 129: 300 (1912).—Engl., Pflanzenw. Afr. **3,** 1: 705 (1915).—Eyles in Trans. Roy. Soc. S. Afr. **5**: 385 (1916).—Burtt Davy, F.P.F.T. **1**: 193 (1926). Lectotype from Tanganyika, Kilimanjaro, *Volkens* 2128 (BM, holotype; K).

Erect perennial herb up to 40 cm. tall with ± woody basal region; stems

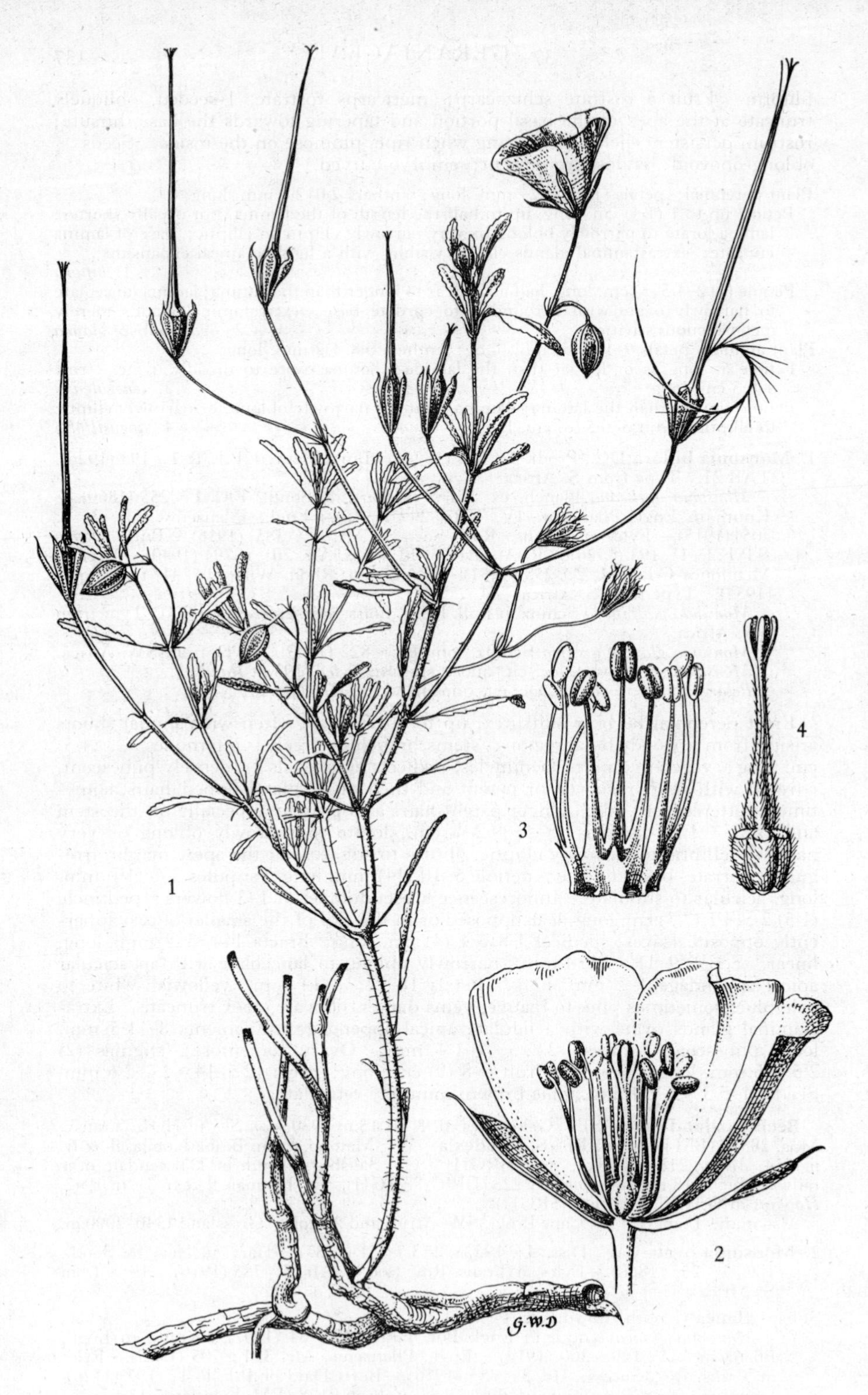

Tab. 21. MONSONIA BIFLORA. 1, habit (× $\frac{2}{3}$) ; 2, flower, front sepals and petals removed (× 2) ; 3, part of androecium (× 3) ; 4, gynoecium (× 3), all from *Drummond* 4875.

branching; internodes 0·1–3 (5) cm. long; vegetative parts, peduncles, pedicels and sepals pubescent, ± covered with short recurved or patent and then often gland-tipped hairs and with long patent hairs also present. Leaf-lamina 2–4·5 × 0·4–2 (2·2) cm., narrowly lanceolate to ovate, apex acute to subacute, margin serrate to crenate, base cordate to subcordate; petiole (1) 2–4·5 (5) cm. long; stipules (5) 7–12 (14) mm. long, acicular to spiny. Inflorescence a pseudumbel of usually 2 flowers; peduncle (1·5) 2·5–8 (9) cm. long, leaf-opposed or in the axil of the smaller of two apparently opposite leaves; pedicel (1) 2·5–4·5 (7) cm. long; bracts 2–6, 5–12 mm. long, filiform to linear. Sepals 7–12 × 3–4 mm., narrowly oblong to lanceolate with an acicular apical appendage 1–2 (2·5) mm. long. Petals (10·5) 14–20 (22·5) × (6·5) 8–14 (17) mm., white, cream or yellowish, obovate, apex truncate. Extrastaminal glands inconspicuous ovate cavities with sometimes only the two lateral rims visible. Filaments 7–9·5 mm. long, pubescent; anthers 2–2·5 × 1–1·5 mm. Ovary tomentose; stigmas 1·5–2 (2·5) mm. long, clavate. Fruit 6·5–10 cm. long; cocci 12·5–13·5 × 2·3–2·6 mm. Seeds 6–6·5 × 2·2–2·4 mm., pale brown, minutely reticulate.

Bechuanaland Prot. N: Ngamiland, Kwebe Hills, fl. & fr. 26.xii.1897, *Lugard* 69 (K). **S. Rhodesia.** W: Matopos, fl. & fr. iii.1918, *Eyles* 986 (BM; K; SRGH). S: Gwanda, fl. & fr. v.1955, *Davies* 1290 (SRGH).

Also in the Transvaal, SW. Africa, Tanganyika and Kenya. On sandy or stony soil, scrubland, grassland or mopane woodland, 600–1400 m. (outside our area 100–1500 m.).

Comparing typical specimens, *M. ovata* subsp. *ovata* differs from subsp. *glauca* in having shorter petioles (1–2 cm. long), shorter and broader leaves (ratio 2 : 1–1 : 1 instead of 5 : 1–2 : 1), longer appendages on the sepals (3–5 mm. long), slightly shorter fruits (5·5–7·5 cm. long), usually 1-flowered peduncles and extrastaminal glands which are pulvinate and conspicuous. In their typical forms they thus appear to be specifically distinct. There is, however, a certain amount of overlapping of the characters, so that the subspecific concept seems to be the most appropriate. One specimen, from Bechuanaland Prot. (*Rogers* 6067), is difficult to determine as either subspecies. It grows in the area of subsp. *glauca* and has the narrower leaves of that subspecies, but its other characters are those of subsp. *ovata*.

3. **Monsonia senegalensis** Guill. & Perr. in Guill., Perr. & Rich., Fl. Senegamb. Tent. **1**: 131 (1832).—Knuth in Engl., Pflanzenr. IV, 129: 301 (1912).—Engl., Pflanzenw. Afr. **3**, 1: 705 (1915).—Burtt Davy, F.P.F.T. **1**: 193 (1926).—Exell & Mendonça, C.F.A. **1**, 2: 258 (1951).—Keay, F.W.T.A. ed. 2, **1**, 1: 157 (1954). Type from Senegal.

Prostrate to decumbent or sometimes ± erect annual herb, branching from the often woody base; stems up to 40 cm. long, but often shorter, frequently with short lateral shoots in one of the axils of apparently opposite leaves; internodes up to 3 cm. long; vegetative parts, peduncles and pedicels ± densely covered with short patent or ± recurved, often gland-tipped hairs. Leaf-lamina 0·9–3 (4) × 0·4–2·5 (3·5) cm., ovate, apex acute, margin serrate or dentate to repand-dentate, base cordate; petiole 1·2–4·5 cm. long; stipules (3) 5–10 (12) mm. long, acicular to subulate. Peduncle 0·5–1·5 (2) cm. long, leaf-opposed or in the axil of the smaller of 2 apparently opposite leaves, often on stunted axillary branches, usually 1-flowered; pedicel 1·8–4 cm. long; bracts 1–2, 5–12 (15) × 0·5–1 mm., linear. Sepals (6) 7–12 × 1·5–3·2 mm., narrowly oblong-elliptic with an acicular apical appendage 1–2 mm. long, densely silky-pubescent with longer glandless patent hairs mixed with the normal glandular ones. Petals 9–12 × 5·5–7·8 mm., pink with darker veins, obovate, apex truncate or sometimes shallowly lobed. Extrastaminal glands ovate with sometimes only the two lateral rims clearly visible. Filaments 4·8–6·5 mm. long, pubescent; anthers 0·8–1 × 0·7–0·8 mm. Ovary tomentose; stigmas 0·8–1·2 mm. long, clavate. Fruit 7–10·5 cm. long; cocci 8·5–10 × 2·2–2·5 mm. Seeds 5–5·5 × (1·6) 2·1–2·4 mm., pale brown, minutely reticulate.

Bechuanaland Prot. N: 120 km. from Francistown, 1020 m., fl. & fr. 8.iii.1961, *Richards* 14601 (K). **S. Rhodesia.** W: Victoria Falls, fl. iii.1945, *Martineau* 724 (SRGH). S: Beitbridge fl. & fr 16.ii.1955, *E. M. & W.* 461 (BM; LISC; SRGH).

In scattered localities from Senegal to Egypt, Ethiopia and Kenya; from SW. Africa to Angola; and in N. Transvaal. In Asia from Arabia to India. On dry sandy soil,

Colophospermum mopane scrub or woodland (in our area), 300–1000 m. (outside our area 0–2100 m.).

4. **Monsonia angustifolia** E. Mey. [in Drège, Zwei Pflanz.-Docum.: 203 (1843) *nom. nud.*] ex A. Rich., Tent. Fl. Abyss. **1**: 115 (1847). Type from S. Africa.
Monsonia biflora sensu Harv. in Harv. & Sond., F.C. **1**: 255 (1860).—Oliv., F.T.A. **1**: 290 (1868).—Engl., Hochgeb. Trop. Afr.: 275 (1892).—Eyles in Trans. Roy. Soc. S. Afr. **5**: 386 (1916).—Exell & Mendonça, C.F.A. **1**, 2: 258 (1951).—Suesseng. & Merxm. in Proc. & Trans. Rhod. Sci. Ass. **43**: 108 (1951).
Monsonia biflora var. *angustifolia* Burtt Davy, F.P.F.T. **1**: 193, fig. 22 (1926). Type from S. Africa.

Prostrate to decumbent or sometimes ± erect annual herb, branching from the sometimes semi-woody base; stems up to 50 cm. long, but often shorter, frequently with short lateral shoots in the leaf-axils; vegetative parts, peduncles, pedicels and sepals pubescent, ± densely covered with extremely short recurved or patent and then often gland-tipped hairs, and also with scattered patent very much longer ones. Leaf-lamina 1·5–3 (3·5) × 0·2–0·8 (1·5) cm., lorate to narrowly oblong or very narrowly elliptic to elliptic, apex obtuse to subacute, margin irregularly serrate, base cuneate; petiole (0·3) 0·5–1·5 (2) cm. long; stipules 3–7 (10) mm. long, acicular to subulate. Inflorescence a pseudumbel of (1) 2–3 flowers; peduncle 0·2–2 (4) cm. long, leaf-opposed or in the axil of the smaller of 2 apparently opposite leaves, sometimes on stunted axillary branches; pedicel 1·4–4·1 cm. long; bracts 1–5, 4–8 (14) mm. long, filiform to linear. Sepals 6–9 (10) × 2·8–3·6 mm., narrowly oblong to lanceolate with an acicular apical appendage 1–2 mm. long. Petals 9–12 (15) × 4–6 mm., normally mauve to blue, sometimes white or yellowish, obovate, apex truncate or shallowly lobed. Extrastaminal glands inconspicuous ovate cavities with sometimes only the two lateral rims visible. Filaments 6–7·2 mm. long, pubescent; anthers 0·8–1·2 × 0·5–0·8 mm. Ovary tomentose; stigmas 1·2–2 mm. long, clavate. Fruit 5–6·5 (9) cm. long; cocci 8·5–10 × 1·7–2·2 mm. Seeds 4·8–5·5 × 1·3–1·7 mm., pale brown, minutely reticulate.

Bechuanaland Prot. N: near Tsau, Okovango, 945 m., fl. & fr. 19.iii.1961, *Richards* 14789 (K). SW: 13 km. S. of Kanga, fl. & fr. 18.ii.1960, *Wild* 5019 (BM; SRGH). **S. Rhodesia.** W: Near World's View, fl. & fr. 14.iv.1955, *E.M. & W.* 1515 (BM; LISC; SRGH). E: Inyanga, fl. & fr. 18.i.1948, *Chase* 695 (BM; K; SRGH). **Mozambique.** LM: Maputo, Bela Vista, fl. & fr. 20.xi.1940, *Torre* 2098 (LISC).
Also in S. Africa, Kenya, Uganda, Congo, Tanganyika, Eritrea, Ethiopia, and Somaliland. On grasslands and sandy soils, open hillsides and beside streams and roads. Occasionally in wet grasslands and valleys or open woodlands, 1350–2150 m. (outside our area 390–2290 m.).

Insufficiently Known Species

Monsonia betschuanica Knuth in Engl., Pflanzenr. IV, 129: 298 (1912). Type: Bechuanaland Prot., Sogosse, *Seiner* II 57 (B, holotype †).

"Herb, 22 cm. tall, with short erect branches; plant covered with short gland-tipped hairs and sparsely distributed long patent hairs. Leaf-lamina 1 × 0·6–0·7 cm., ovate, apex obtuse, margin irregularly crenate, base rounded; petiole 2–5 mm. long; stipules 3 mm. long, setose. Inflorescence (1) 2-flowered; peduncle up to 2 cm. long; pedicel 1 cm. long; bracts 3 mm. long. Sepals 8 mm. long, lanceolate. Petals 12–14 mm. long. Filaments up to 8 mm. long; anthers oblong. Fruit 5·6 cm. long."

Bechuanaland Prot.: "Sogosse", fl. & fr. ii.1906, *Seiner* II 57 (B†).

As the type has been destroyed and I have seen no specimen that matches the description, it is difficult to assess whether this is a good species or only an extreme form of *Monsonia angustifolia.*

3. PELARGONIUM L'Hérit.

Pelargonium L'Hérit. in Ait., Hort. Kew. **2**: 417 (1789).

Annual or perennial herbs, shrublets or shrubs, sometimes with underground tubers; erect or decumbent, sometimes acaulescent or with subsucculent stems; often viscid and aromatic, variously hairy, often glandular. Leaves usually petiolate, stipulate, opposite or alternate, serrate, crenate, dentate, lobed, variously dissected or compound, rarely entire; stipules membranous, sometimes setaceous, rarely forming spines. Inflorescence a 2–many-flowered pseudumbel (flowers rarely solitary); bracts present. Flowers zygomorphic, 5-merous. Sepals imbricate, often with a membranous margin, posterior sepal produced at the base into a spur which is decurrent along the pedicel and adnate to it. Petals 5 (rarely 0–4 by abortion), usually unequal, imbricate, unguiculate or sessile. Disk or extra-staminal glands absent. Stamens 10, connate at the base; 2–7 filaments bearing anthers, the remaining ones often vestigial (staminodes); anthers dorsifixed. Ovary 5-lobed, 5-locular (rarely 3–4-locular by abortion), rostrate; loculi 2-ovulate; style present, short or long; stigmas 5, usually filiform. Fruit a rostrate schizocarp; mericarps rostrate, 1-seeded, tapering from the apex to the base and ending in a point, hirsute; rostrum persistent, helically twisting when ripe, plumose on the inside. Seeds ± oblong-obovoid; endosperm absent; embryo curved.

Plant acaulescent; stipules 15–40 mm. long, linear to narrowly triangular; spur 3–6 cm. long - - - - - - - - - - - - 1. *luridum*

Plant with a long or short stem; stipules 5–10 mm. long, or if up to 15 mm. long then ovate; spur 0·1–4 cm. long, if as much as 3–4 cm. long, then hairs on spur, pedicel and underside of bracts retrorse:

Plant herbaceous, annual or perennial, sometimes with a woody base; leaf-lamina pinnate or 3-partite to 3-foliolate with the terminal lobe or leaflet larger, or almost entire to 3- or 5 (7)-palmatilobed; leaf margin not revolute:

Leaves pinnate or 3-partite to 3-foliolate with the terminal lobe or leaflet larger; sepals with conspicuous veins; style 3–8 mm. long:

Plant an erect herb with short stems and a long leafy flowering shoot; inflorescence a terminal compound cyme; bracts ovate; sepals 5–9 mm. long, obtuse or the two narrow lateral ones sometimes acute - - - 2. *dolomiticum*

Plant a decumbent herb; inflorescences leaf-opposed, or in the axil of the smaller of two apparently opposite leaves; bracts lanceolate; sepals 9–14 mm. long, all acute to acuminate - - - - - - - - 3. *whytei*

Leaves almost entire to 3–5 (7)-palmatilobed with ± equal lobes; veins on sepals not conspicuous; style 0·1–3·0 mm. long:

Hairs on spur, pedicel and underside of bracts patent or ± appressed, not distinctly retrorse; spur up to 10 mm. long, or if up to 15 mm. long then ± equal in length to the free part of the pedicel; rostrum of ovary distinct at the time of flowering:

Plant perennial; spur 6–11 mm. long, ± the same length as the free part of pedicel; petals up to 11 mm. long - - - - - 4. *mossambicense*

Plant annual; spur 1–5 mm. long, free part of pedicel longer than the spur (in our area); petals up to 6 mm. long:

Stem sparsely pubescent to glabrous, hairs up to 0·2 mm. long; peduncles 5–13 cm. long; flowers 8–30 per pseudumbel (in S. Africa 1–50); sepals cuspidate - - - - - - 6. *grossularioide*

Stem pubescent, hairs up to 1·6 mm. long; 1–2 (3) flowers per pseudumbel; sepals acute - - - - - - - - 5. *apetalum*

Hairs on spur, pedicel and underside of bracts distinctly retrorse; spur 12–30 mm. long; free part of pedicel 1–7 mm. long, always shorter than the spur; rostrum of ovary not distinct at the time of flowering - - 7. *alchemilloides*

Plant suffruticose; leaf-lamina pinnatipartite with the two basal segments much larger than the rest and often further subdivided; leaf margin revolute 8. *graveolens*

1. **Pelargonium luridum** (Andr.) Sweet in Colv. Cat. ed. 2: 2 (1822?); Geran. **3**: t. 281 (1825).—Exell & Mendonça, C.F.A. **1**, 2: 259 (1951).—Petit, F.C.B. **7**: 26 (1958). Type a plant from S. Africa cultivated in Britain (of which no specimen was kept) represented by the illustration cited.

Geranium luridum Andr., Geran. **2**: t. 34 (1813?). Type as above.

Pelargonium hurifolium Sweet in Colv. Cat.: 21 (1821) *nom. nud.* Type from S. Africa.

Polyactium aconitiphyllum Eckl. & Zeyh., Enum. Pl. Afr. Austr. Extratrop.: 67 (1835). Type from S. Africa.

Pelargonium aconitiphyllum (Eckl. & Zeyh.) Steud., Nom. Bot. **2**: 283 (1841).—Harv. in Harv. & Sond., F.C. **1**: 276 (1860).—Engl., Pflanzenw. Ost-Afr. **C**: 225 (1895).—Bak. f. in Journ. Linn. Soc., Bot. **40**: 35 (1911).—Knuth in Engl., Pflaneznr. IV, 129: 361 (1912).—Eyles in Trans. Roy. Soc. S. Afr. **5**: 386 (1916).—Burtt Davy, F.P.F.T. **1**: 90 (1926). Type as for *Polyactium aconitiphyllum*.

Pelargonium polymorphum E. Mey. in Drège, Zwei Pflanz.-Docum.: 209 (1843) *nom. nud.*

Pelargonium zeyheri Harv. in Harv. & Sond., F.C. **1**: 276 (1860).—Knuth, tom. cit.: 364 (1912). Type from S. Africa.

Pelargonium flabellifolium var. *benguellense* Welw. ex. Oliv., F.T.A. **1**: 294 (1868). —Engl., Hochgebirgsfl. Trop. Afr.: 276 (1892). Type from Angola.

Pelargonium angulosum Szyszyl., Polypet. Thalam. Rehmann.: 14 (1888). Type from S. Africa.

Pelargonium rehmannii Szyszyl., tom. cit.: 9.—Knuth, tom. cit.: 366 (1912). Type from S. Africa.

Geraniospermum aconitiphyllum (Eckl. & Zeyh.) Kuntze, Rev. Gen. Pl. **1**: 94 1891). Type as for *Polyactium aconitiphyllum*.

Geraniospermum zeyheri (Harv.) Kuntze, tom. cit.: 95 (1891). Type as for *Pelargonium zeyheri*.

Pelargonium heckmannianum Engl., Bot. Jahrb. **30**: 335 (1902); Pflanzenw. Afr. **1**, 1: 377, fig. 317 (1910).—Knuth, tom. cit.: 363, fig. 45 (1912).—R.E.Fr., Schwed. Rhod.-Kongo-Exped. **1**: 106 (1914).—Engl., Pflanzenw. Afr. **3**, 1: 709, fig. 324 (1915). Type from Tanganyika.

Pelargonium benguellense (Welw. ex Oliv.) Engl. in Warb., Kunene-Samb.-Exped. Baum: 268 (1903).—Knuth, tom. cit.: 365, fig. 65 (1912).—Engl., Pflanzenw. Afr. **3**, 1: 710 (1915). Type as for *Pelargonium flabellifolium* var. *benguellense*.

Pelargonium aconitiphyllum var. *angustisectum* Knuth, tom. cit.: 362 (1912). Type from S. Africa.

Pelargonium aconitiphyllum var. *medium* Knuth, loc. cit. Type from S. Africa.

Pelargonium aconitiphyllum var. *latisectum* Knuth, loc. cit. Type from S. Africa.

Pelargonium longiscapum Schlechter ex Knuth, tom. cit.: 365 (1912). Type from S. Africa.

Erect perennial herb up to 70 cm. tall, with a tuberous woody rootstock; acaulescent; vegetative parts, peduncles and pedicels glandular and pubescent, covered with both long (± 4 mm.) patent and much shorter ± appressed hairs, as well as sessile glands. Leaves all radical; lamina (3) 7–15 (24) × (4) 7–14 (19) cm., ovate to broadly ovate, base cuneate to cordate, extremely variously dissected, shallowly lobed to pinnatisect or bipinnatisect, later leaves often more dissected; ultimate segments filiform (in the Transvaal) to oblong or ovate, entire to serrate or crenate towards the apex, when entire apex acute to rounded or mucronate, densely tomentose (in the Transvaal on specimens with undivided leaves) to ± glabrous, with the long patent hairs absent or sparse on many leaves and always restricted to the margin and veins; petiole (6) 8–20 (30) cm. long; stipules 15–40 × 1·5–4 (5) mm., linear to narrowly triangular, acuminate. Inflorescence a terminal pseudumbel of (5) 7–30 (50) flowers; peduncles 1–3, 14–65 cm. long, unbranched; bracts (8) 11–13 (16) × 1·5–3 (4) mm., few to numerous, linear to lorate or lanceolate, acuminate, membranous; free part of pedicel 2–35 (50) mm. long. Spur 30–60 (70) mm. long. Sepals (8·5) 9–12 (15) × (1·5) 2–4 (5) mm., lanceolate to narrowly ovate, glandular and densely pubescent. Petals 5, white to pale yellow with pink venation, or to pink (or dark red in southern Mozambique and Angola), (12) 14–24 (28) × 5–12 mm.; oblanceolate to narrowly obovate, the three anterior ones unguiculate. Stamens with fertile filaments usually 7, (4) 5–7 (8) mm. long, sterile filaments (staminodes) 3, 3–5·5 mm. long, all 10 connate at the base for 1·8–3·5 mm.; anthers 1·6–2·8 (3·2) × 1·2–1·7 mm. Ovary with basal part tomentose; rostrum pubescent; style 0·1–1·2 mm. long; stigma 1·6–2·8 mm. long. Fruit 4–5 (6·5) cm. long; cocci 9–15 × 1·5–3 mm. Seeds 4·8–6 × 1·9–2·2 mm., pale brown, minutely reticulate (at × 50).

N. Rhodesia. N: Abercorn-Mpulungu road, 5 km. from Abercorn, Simanwe Farm, fl. 24.viii.1956, *Richards* 6010 (K; SRGH). W: Solwezi, on road to Kansanshi Mine, fl. 14.xi.1952, *Angus* 454 (K). E: Nyika Plateau, near edge of evergreen forest 2·5 km. from Rest House, fl. 24.ix.1956, *Benson* NR 190 (BM). **S. Rhodesia.** W: Matopo Hills, near Isotje, fl. xii.1905, *Gibbs* 245 (BM). C: Salisbury, Cleveland Dam, fl. xii.1918, *Eyles* 1393 (SRGH). E: Umtali, Engwa, fl. 1.ii.1955, *E.M. & W.* 56 (BM; LISC; SRGH). **Nyasaland.** N: Nyika Plateau, Lake Kaulime, fl. & fr. 24.ix.1958, *Robson* 324 (BM; K; LISC; PRE; SRGH). C: foot of Mt. Dedza, fl. 16.x.1937, *Longfield* 45 (BM). **Mozambique.** T: Between Furancungo and Vila Coutinho, 75 km. from Furancungo, fl. & fr. 15.vii.1949, *Barbosa & Carvalho* 3634 (LISC). MS: Manica, Moribane, fl. 17.ii.1942, *Salbany* 63 (LISC). LM: Libombos, near Namaacha, Mt. Mponduim, fl. 22.ii.1955, *E.M. & W.* 493 (BM; LISC; SRGH).

Also in Angola, Congo, Tanganyika and north-eastern parts of S. Africa. Open bush and grassland, sometimes on burnt ground, 800–2300 m. (outside our area up to 2440 m.).

P. luridum is a very variable species and the plants which resemble the type of the synonym *P. heckmannianum* (all specimens collected in Angola, N. Rhodesia, Nyasaland and some from the Inyanga district) may represent a good subspecies. These have leaves of a rougher texture, prominent venules, shorter peduncles (10–35 cm.), fewer flowers per pseudumbel (5) 7–10 (20), flowers always yellow (or red in Angola), and in most cases the flowers develop before the leaves. I have not, however, recognized *P. heckmannianum* as a subspecies since the whole species aggregate presents further problems in S. Africa, which cannot be dealt with in the scope of this Flora. I have added *P. zeyheri* and *P. rehmannii* as synonyms as they both show gradation to the typical *P. luridum*, although the types look quite distinct.

2. **Pelargonium dolomiticum** Knuth [ex Engl. in Sitzungsber. Königl. Pr. Akad. Wissensch. **11**: 877 (1906) *nom. nud.*] in Engl., Bot. Jahrb. **40**: 71 (1907); in Engl., Pflanzenr. IV, 129: 386 (1912); in Engl., Pflanzenw. Afr. **3,** 1: 711 (1915). Type from the Transvaal.
Pelargonium bechuanicum Burtt Davy, F.P.F.T. **1**: 48, 189 (1926). Type from the Transvaal.

Erect perennial herb up to 50 cm. high, with a tuberous woody rootstock and 1–several short lateral stems (each 1–5 cm. long) branching from an extremely short woody basal region; vegetative parts, peduncles and pedicels glandular and pubescent, covered with short stiff patent or ± appressed hairs and sessile glands, giving the whole plant a glaucous-canescent appearance. Leaves densely crowded on the short lateral stems; lamina 2·5–12 (15) × 1·5–8 (12) cm., narrowly ovate to ovate, 2–3-pinnate, ultimate segments 1–10 × 0·5–2 (3) mm., linear to oblong, apex rounded to acute; petiole up to 8 (18) cm. long; stipules 4–6 × 2·5–3·5 mm., oblong; to ovate, acute, membranous. 1–2 sympodial leafy, sometimes two-branched flowering shoots arising from each lateral stem, terminating in a compound, cincinnately cymose inflorescence; peduncle 3–10 (15) cm. long, opposed to a very much reduced leaf, inserted at a wide angle, each one shorter than the one below and each bearing a terminal pseudumbel of 2–5 (7) flowers; bracts 3–5 (7), 5–6·5 × 3·5–5 mm., ovate, acute, membranous; free part of pedicel 1–4 (8) mm. Spur 8–12 mm. Sepals 5–8 (9) × 1–3 mm., lorate to narrowly oblong, obtuse or sometimes acute, with conspicuous veins, studded with sessile bottle-shaped glands and a few short stiff hairs. Petals 4 (5), white to mauve with darker veins; 2 posterior 13–16 × 2–3·2 (4) mm., lorate, with a claw 5–7 mm. long, obtuse to deeply emarginate, slightly expanded laterally above the claw, expansion often inrolled; 2 (3) anterior 7–9 (10) × 3–4 (5) mm., oblong or sometimes ± elliptic, with a claw 3–5 mm. long, often with obliquely truncate base and apex. Stamens with fertile filaments 7, 4·5–10 mm. long, sterile filaments (staminodes) 3, 3–4 mm. long, all 10 connate at the base for 2–3·5 mm.; anthers 2–3 (3·5) × 1 (1·5) mm. Ovary with basal part tomentose; rostrum pubescent; style 3·5–7 mm. long; stigma 1·7–2·4 mm. long. Fruit 3·5–4·5 cm. long; cocci 8·0 × 1·6–2·0 mm. Seeds 4·8–5·5 × 1·3–1·6 mm., pale brown, very minutely reticulate (× 50).

Bechuanaland Prot. SW: Ghanzi, fl. & fr. 28.vii.1955, *Story* 5063 (PRE).
Also in the Transvaal, Orange Free State, Griqualand West and SW. Africa. On light sandy soil in open bushland, 1200–1400 m.

The larger measurements in brackets are taken from specimens collected in the Transvaal. *P. dolomiticum* is closely related to *P. senecioides* L'Hérit. from the western parts of S. Africa, but the latter can be distinguished by the shape of the petals and the shortness of the style (1 mm. or shorter). The type of *P. dolomiticum* has been destroyed, but Knuth's description agrees in most points with the material (named *P. bechuanicum*) which I have seen. It differs only in the smaller size of the leaves (3 cm. long incl. petiole) and height of the plant (17 cm.), both variable characters.

3. **Pelargonium whytei** Bak. in Kew Bull. **1897**: 246 (1897).—Knuth in Engl., Pflanzenr. IV, 129: 394 (1912).—Engl., Pflanzenw. Afr. **3,** 1: 711 (1915).— Brenan in Mem. N.Y.Bot. Gard. **8,** 3: 231 (1953).—Petit, F.C.B. **7**: 28 (1958). Type: Nyasaland, Nyika Plateau, *Whyte* (K, holotype).
Pelargonium goetzeanum Engl. in Engl., Bot. Jahrb. **30**: 334 (1901).—Knuth in Engl., Pflanzenr. IV, 129: 393 (1912). Type from Tanganyika.

Decumbent perennial herb; stems up to 1·3 m. long, branching, straggling, often reddish, arising from a woody basal region; internodes up to 13 cm. long; vegetative parts, peduncles and pedicels glandular and ± pubescent to almost glabrous, hairs ± patent or ± adpressed on the leaves; glands sessile. Leaf-lamina 1·5–5·5 × 1·5–5 cm., narrowly ovate to ovate, cordate, 3-partite to 3-foliolate, with a larger middle section or leaflet, sometimes only shallowly 3-lobed or pinnatifid to pinnatisect; segments or pinnae ± rhombic, almost unlobed to pinnatipartite, margin crenate, hairs scattered, often restricted to the margin and veins; petiole 2–7 (12) cm. long; stipules 5–8 (10) × 2–5 (7) mm., narrowly ovate to very broadly ovate, acute to apiculate, membranous. Inflorescence a terminal pseudumbel of 1–4 (5) flowers; peduncle 5–9 (13) cm. long, leaf-opposed or in the axil of the smaller of two apparently opposite leaves; bracts 4–8, 5–8 × 1·5–3 mm., lanceolate, acute, membranous; free part of pedicel 1–3 mm. long. Spur (5) 7–15 (20) mm. long. Sepals 9–14 × 1–4 mm., narrowly lanceolate to lanceolate, acuminate, glandular and sparsely pubescent, with conspicuous red veins. Petals 4, pink with red veins; 2 posterior ones 9–20 × 3–5·5 mm. and 2 anterior ones 9–15 × 2–4 mm., all oblanceolate and unguiculate to ± spathulate; 2 posterior ones normally but not always longer than the anterior ones. Stamens with fertile filaments 7, 5–12 (16) mm. long, sterile filaments (staminodes) 3, 2·5–5 (6·5) mm. long, all 10 connate at the base for 1–2·5 (5) mm.; anthers 2–2·5 × 1 mm. Ovary with basal part tomentose; rostrum pubescent; style 3–5 (8) mm. long; stigma 2–2·5 (3) mm. long. Fruit 3–4 cm. long; cocci 5–7 × 1·5–2 mm. Seeds 3·5–4 (5) × 1·5–1·7 mm., pale brown, minutely reticulate.

N. Rhodesia. E: Nyika Plateau, below Rest House on N. Rukuru Road, fl. & fr. 27.x.1958, *Robson & Angus* 409 (BM; K; LISC; PRE; SRGH). **Nyasaland.** N: Nyika Plateau, Nchena-chena Spur, fl. & fr. 20.viii.1946, *Brass* 17364 (K; SRGH). C: Kantorongonda and Kasungu Mts., NW. Lake Nyasa, fl. & fr. x. 1893, *Crawshay* (BM).

Also in the eastern Congo, Kenya, Tanganyika, and Uganda. Open bush and grassland; 1500–2440 m.

The type of the synonym *P. goetzeanum* is more glabrous, and most parts of the plant are slightly larger as indicated by the measurements in brackets (except the ones given for the stipules); but it agrees otherwise with typical *P. whytei* and, as pubescence and size vary considerably in the plants I have studied, I do not think this difference is sufficient to maintain it as a distinct species. *P. multicaule* Jacq., especially the specimens collected in the Transvaal, is very closely related to *P. whytei*, only differing in the form of leaf dissection. However, since the former shows certain connections with further plants from S. Africa of uncertain relationships, I prefer to keep *P. whytei* separate from *P. multicaule*, at least for the present.

4. **Pelargonium mossambicense** Engl., Pflanzenw. Ost-Afr. **C**: 225 (1895).—Knuth in Engl., Pflanzenr. IV, 129: 406 (1912).—Engl., Pflanzenw. Afr. **3**, 1: 712 (1915).—Martineau, Rhod. Wild Fl.: 41, t. 13 fig. 3 (1954). Type: Mozambique, Gorungosa, *Carvalho* (B, holotype †; COI).

Decumbent perennial herb, scented; stems up to 1 m. long, with very short lateral branches in one of the axils of apparently opposite leaves, often reddish, older parts woody; internodes (1) 4–7 (16) cm.; vegetative parts, peduncles and pedicels glandular and pubescent; hairs patent, or ± appressed on the leaves; glands often short-stalked. Leaf-lamina 1·5–6 × 2–7 cm., very broadly ovate to depressed-ovate, base cordate, 3–5-palmatifid or -lobed; segments with 2–3 shallow obtuse to subacute lobes with crenate margin; petiole 1–7 (16) cm. long; stipules 4–11 × 3–7 mm., lanceolate to ovate, acute to acuminate, sometimes bifid, membranous. Inflorescence a terminal pseudumbel of 2–5 flowers; peduncle 1–6 (12) cm. long, leaf-opposed or in the axil of the smaller of two apparently opposite leaves; bracts 5–8, 4–6 × 1–2 mm., lanceolate to narrowly ovate, acute, membranous; free part of pedicel 6–17 (35) mm. long. Spur 6–11 (15) mm. long. Sepals 6–7 × 1–4 mm., lorate to oblong, acute to apiculate, pubescent and glandular. Petals 5, white to pink with purple veins, 7–11 (15) × 3–4 mm.; 2 posterior ones symmetrically to asymmetrically spathulate; 3 anterior ones narrowly obovate to obovate, unguiculate. Stamens with fertile filaments 7, 3–5 (6) mm. long, sterile filaments (staminodes) 3, 1–2 mm. long, all 10 connate at the base for 0·5–1·5 mm.; anthers 1–1·5 × 0·8–1 mm. Ovary with basal part tomentose; rostrum pubescent; style 1–3 mm. long; stigma 1–1·5 mm. long. Mature fruits not seen.

S. Rhodesia. E: Umtali Distr., Banti Forest, fl. 4.ii.1955, *E.M. & W.* 192 (BM; LISC; SRGH). **Mozambique.** MS: Gorungosa, fl. 1884–85, *Carvalho* (COI).

So far only known from the mountainous parts of the eastern districts of S. Rhodesia and by the type collection from Mozambique. Among rocks in grassland or along forest edges, 1800–2400 m.

On the rocky slopes of the Inyangani west face at an altitude of 2400 m. specimens have been collected (*Goodier & Phipps* 66) which are of very stunted growth (10 cm. in height, internodes 1 cm. long), but apart from this they agree with all the characters of *P. mossambicense.* Knuth (tom. cit.: 407) states that the type in Berlin was only a side branch, which suggests that it was probably part of the specimen in the Coimbra herbarium.

5. **Pelargonium apetalum** P. Tayl. in Hook., Ic. Pl. **36**: t. 3579 (1962). Type from Tanganyika.

Decumbent annual herb; primary stem ± ascending, branched; branches up to 1 m. long, weak and straggling; internodes 2–6 (10) cm. long; vegetative parts, peduncles and pedicels glandular and pubescent; hairs patent or ± appressed on the leaves; glands short-stalked to sessile. Leaf-lamina 1·5–4·5 (6) × 1·5–4·5 (6) cm., deltoid to broadly or very broadly ovate, apex subacute to obtuse, margin crenate, base cordate, almost unlobed to shallowly 3- or 5 (or 7)-lobed, membranous; petiole 1–3 cm. long; stipules 2–4 (7) × 1–3 (4) mm., lanceolate to ovate or deltoid, acute, often 2-fid, membranous; petioles 1–3 cm. long, 0·5–2·5 cm. long. Inflorescence a 1–2 (3)-flowered pseudumbel; peduncle leaf-opposed or in the axil of the smaller of 2 apparently opposite leaves, bearing a terminal pseudumbel of 1–2 (3) flowers; bracts 3–6, 1·5–2 × 0·5–0·8 mm., lanceolate or narrowly deltate, acute, membranous; free part of pedicel (3) 5–12 mm., long. Spur 1–2 (3) mm. long. Sepals 2·5–4 × 0·5–2·5 mm., narrowly oblong to lanceolate, acute, ± patent, pilose and glandular. Petals 5 (4) or absent (in N. Nyasaland and Tanganyika), pink, 1·7–2 (2·5) × 0·5–0·7 mm., 2 posterior ones asymmetrically spathulate; 3 (2) anterior ones narrowly obovate, unguiculate, fertile filaments 2–3 (5), 1·8–2 mm. long. Stamens with sterile filaments (staminodes) (5) 7–8, 0·9–1·2 mm. long, all 9 or 10 connate at the base for 0·5 mm.; anthers 0·3–0·5 × 0·3–0·5 mm. Ovary often only 3–4-locular by abortion, with basal part tomentose; rostrum pubescent; style 0·2 mm. long; stigma 0·5 mm. long. Fruit 0·7–1·2 cm. long; cocci 2·8 × 1·2 mm. Seeds 1·8–2·4 × 0·8–1·2 mm., pale brown.

S. Rhodesia. W: Matobo, Farm Besna Kobila, fl. & fr. iv.1955, *Miller* 2735 (SRGH). C: Domboshawa, fl. & fr. 7.iii.1946, *Wild* 106 (SRGH). E: Inyanga, Punch Rock, fl. & fr., iv.1957, *Martin* in GHS. 76803 (SRGH). **Nyasaland.** N: Nyika Plateau, fl. & fr. vii.1896, *Whyte* (K).

Also in Tanganyika. Open bush among rocks, also found as a weed of cultivations, 1500–2200 m.

6. **Pelargonium grossularioides** (L.) Ait., Hort. Kew. **2**: 420 (1789).—Harv. in Harv. & Sond., F.C. **1**: 289 (1860).—Knuth in Engl., Pflanzenr. IV, 129: 410 (1912). Type a plant from S. Africa cultivated in Europe.

Geranium grossularioides L., Sp. Pl. **2**: 679 (1753). Type as above.

Pelargonium anceps L'Hérit. in Ait., loc. cit. Type a plant from S. Africa cultivated in Europe.

Geranium parviflorum Andr., Geran. **2**: t. 43 (1805?). Type a plant from S. Africa cultivated in Britain.

Geranium anceps Poir., Encycl. Méth. Suppl. **2**: 748 (1811). Type as for *Pelargonium anceps.*

Pelargonium acugnaticum Thou., Fl. Trist. d'Acugna: 44, t. 13 (1811). Type from Tristan da Cunha.

Peristera anceps Eckl. & Zeyh., Enum. Pl. Afr. Austr. Extratrop.: 72 (1836). Type from S. Africa.

Peristera nummularifolia Eckl. & Zeyh., loc. cit. Type from S. Africa.

Pelargonium micropetalum E. Mey. in Drège, Zwei Pflanz.-Docum.: 208 (1843) *nom. nud.*

Pelargonium grossularioides var. *anceps* Harv. in Harv. & Sond., F.C. **1**: 289 (1860). Type from S. Africa.

Geraniospermum grossularioides (L.) Kuntze, Rev. Gen. Plant. **1**: 95 (1891). Type as for *Pelargonium grossularioides.*

Pelargonium filicaule Knuth, tom. cit.: 408 (1912). Type from S. Africa.

Decumbent annual herb, branching from the base; stems scarcely branched, angular and furrowed, up to 50 cm. long; internodes up to 13 cm. long; vegetative parts, peduncles and pedicels ± pubescent to almost glabrous; hairs very short, patent to ± appressed; glands sessile to short-stalked. Leaf-lamina 1·5–4 × 2·5–4·5 cm., circular to reniform in outline, almost unlobed with margin crenate to serrate, or palmatisect, base cordate; in upper leaves usually smaller and more deeply dissected with obtuse lobes; petiole up to 14 cm. long; stipules 3–7 × 2–4 mm., deltate, acute, sometimes bifid, membranous. Inflorescence a terminal pseudumbel of 3–30 (50) flowers crowded into a compact head; peduncle 4–15 cm. long, leaf-opposed or in the axil of the smaller of two apparently opposite leaves; bracts 3–5·5 × 1–1·5 mm., lanceolate, acute, membranous; free part of pedicel 4–8 mm. long. Spur 1·5–3 (4) mm. long, funnel-shaped. Sepals 4–6 × 1·5–3·5 mm., lanceolate to ovate, cuspidate to acuminate, puberulous and glandular. Petals 5, pink to purple, 4–6·5 × 1–1·3 mm.; 2 posterior ones irregularly narrowly obovate-oblong; 3 anterior ones oblanceolate, unguiculate. Stamens with fertile filaments 7, 2–4 mm. long; sterile filaments (staminodes) 1·5 mm. long, all 10 connate at the base for 1–1·5 mm.; anthers 0·8–0·9 × 0·6–0·7 mm. Ovary with basal part tomentose; rostrum pubescent; style 0·3–0·8 mm. long; stigma 0·5–0·8 mm. long. Fruit 1·2–1·5 cm. long; cocci 2·4–3 × 1–1·3 mm. Seeds 1·6–2·5 × 0·8–1·2 mm., dark brown.

Mozambique. LM: Inhaca I., fl. & fr. 15.vii.1957, *Balsinhas* 118 (BM; LMJ).

Coastal regions from Inhaca I. to the Cape and Tristan da Cunha (introduced?). In damp or shady places, 0–150 m. Also recorded from California, India and Kenya as an alien.

My description refers to the plants of our area and similar plants have been found in S. Africa as far south along the east coast as the eastern Cape. In S. Africa *P. grossularioides* varies greatly in length of pedicel and spur, number of flowers per pseudumbel and degree of dissection and size of the leaves. Moreover, there seems to be a continuous gradation from the plants which occur on Inhaca I. to plants with much longer spurs (± 10 mm.) and fewer flowers per pseudumbel (3–10) which agree with the type of the synonym *P. anceps*; and also to plants mainly found at the Cape which are smaller in all parts, have a similarly short spur (1–3·2 mm.) and only 1–3 flowers per pseudumbel. The latter agree with the type of *P. grossularioides*. *P. grossularioides* is closely allied to *P. inodorum* Willd. of Australia and New Zealand.

7. **Pelargonium alchemilloides** (L.) L'Hérit in Ait., Hort. Kew. **2**: 419 (1789).—Harv. in Harv. & Sond., F.C. **1**: 295 (1860).—Knuth in Engl., Pflanzenr. IV, 129: 428 (1912).—Engl., Pflanzenw. Afr. **3**, 1: 708 (1915).—Burtt Davy, F.P.F.T. **1**: 190 (1926). Type a plant from S. Africa cultivated in Europe.

Geranium alchemilloides L., Sp. Pl. **2**: 678 (1753). Type as above.

Pelargonium alchemillifolium Salisb., Prodr.: 312 (1796). Type as above.

Pelargonium malvifolium Jacq., Eclog. Pl. Rar. **1**: 145, t. 97 (1815). Type a plant from S. Africa cultivated in Germany.

Geranium aphanoides Thunb., Fl. Cap.: 514 (1823). Type from S. Africa.

Pelargonium aphanoides (Thunb.) DC., Prodr. **1**: 680 (1824). Type as above.

Pelargonium heritieri Jacq. apud Spreng. in L., Syst. Veg. ed. 16, **3**: 61 (1826). Type from S. Africa.

Pelargonium multibracteatum Hochst. ex A. Rich., Tent. Fl. Abyss. **1**: 119 (1847). —Oliv., F.T.A. **1**: 293 (1868).—Knuth, tom. cit.: 433 (1912).—R. A. Dyer in Fl. Pl. Afr. **32**: t. 1278 (1958). Type from Ethiopia.

Pelargonium fissum Bak. in Saund., Refug. **3**: t. 149 (1870). Type a plant from S. Africa cultivated in Britain.

Geraniospermum alchemilloides (L.) Kuntze, Rev. Gen. Pl. **1**: 94 (1891). Type as for *Pelargonium alchemilloides*.

Pelargonium usambarense Engl. in Abh. Preuss. Akad. Wiss. **1894**: 61 (1894).—Knuth, tom. cit.: 434 (1912). Type from Tanganyika.

Decumbent perennial herb, with a woody stoloniferous to tuberous rootstock; stems up to 80 cm. long, little branched; internodes up to 12 (15) cm. long (but only up to 1 cm. long in our area); vegetative parts, peduncles and pedicels glandular and pubescent; hairs on spur, free part of the pedicel and under side of the bracts retrorse, those on the leaves ± appressed, elsewhere ± patent, often very much longer and most distinctly patent on the stem, petiole and lower part of the peduncle; glands sessile. Leaf-lamina 2·5–10 × 2·5–12 cm. (only 2·5–4·5 × 2·5–4·5 cm. in our area), broadly ovate to depressed-ovate, sometimes with a dark red or brownish zonal marking, palmatifid to palmatipartite (normally palmatilobed), cordate at the base; lobes rounded to oblong, crenate to serrate; petiole up to (3) 10 (15) cm. long; stipules 5–17 × 2–12 mm. (only 5–7 × 2–3 mm. in our area), membranous, narrowly ovate to very broadly ovate, acute or apiculate to cuspidate (narrowly oblong to deltate and acute in our area), sometimes bifid. Inflorescence a pseudumbel of 2–13 flowers (3–4 in our area); peduncle 6–25 cm. long, leaf-opposed or in the axil of the smaller of two apparently opposite leaves; bracts few to numerous, 3–7 × 0·8–2 mm., lanceolate or lorate to narrowly oblong, membranous; free part of pedicel 3–7 (11) mm. long. Spur 12–32 (40) mm. long.

Sepals 8–14 (17) × 2–4 (6) mm., lorate to narrowly oblong or narrowly lanceolate to lanceolate, acute, glandular and ± pubescent, sometimes accrescent. Petals 5, white, pink or dark red (pink in our area); 2 posterior ones 9–20 × 2·5–10 mm., spathulate; 3 anterior ones 7·5–15 × 2–8 mm., oblanceolate, unguiculate. Stamens with fertile filaments 7, 3·2–6·8 mm. long; sterile filaments (staminodes) 3, 3–5·2 mm. long; anthers 1·3–1·8 × 0·8–1·2 mm. Ovary tomentose; rostrum not distinct at time of flowering; style 0·1–0·8 (1) mm. long; stigma 1·2–2·5 mm. long. Fruit 3–4·5 (6) cm. long; cocci 5·5–6 × 1·5–1·8 mm. Seeds 4·2–4·6 × 1·4–1·7 mm., pale brown, minutely reticulate.

S. Rhodesia. E: Umtali Distr., Himalayas, Engwa, fl. 9.xi.1954, *Wild* 4628 (SRGH). **Mozambique.** LM: Namaacha, fl. & fr. 27.viii.1948, *Myre & Carvalho* 128 (LISC; LM).

Also in S. Africa, Tanganyika, Kenya, Somaliland and Ethiopia. In open grassland among rocks, 800–2000 m. (outside our area 0–2500 m.).

This species has so far been recorded only twice in our area. Both collections are somewhat atypical forms, with very short stunted stems, short internodes and petioles, and smaller, narrower stipules. The fruit also seems to be larger, but I have only seen mature fruits on two specimens, one of which is not from our area. Similar plants have been found, once in Natal (*Hardy* 30) and once in Tanganyika (*Greenway* 7682). In other respects this form agrees with *P. alchemilloides*, and, moreover, there are specimens showing intermediate characters. More material is needed in order to decide whether these apparently stunted plants are merely ecotypes, or represent a different species. I have united *P. multibracteatum* and *P. usambarense* with *P. alchemilloides* although there is a considerable gap in their distribution. *P. alchemilloides* is found in S. Africa from the Cape to the Transvaal, whilst *P. multibracteatum* is found from Tanganyika to N. Ethiopia and *P. usambarense* is restricted to N. Tanganyika and S. Kenya. *P. usambarense* differs from *P. multibracteatum* only in having deep red petals, and is probably a local colour form. *P. multibracteatum* and *P. usambarense* differ from *P. alchemilloides* in having larger petals (average size 13–18 × 6–8 mm. compared with 12–15 × 3–5 mm. in *P. alchemilloides* sensu stricto); more flowers per pseudumbel (average 7–8 compared with 4–5); shorter free part of the pedicel (average 3 mm. long compared with 6 mm.); longer spur (average 35–40 mm. long compared with 20 mm.). There is, however, a considerable amount of overlapping and I do not consider that the above differences are sufficient to maintain the two groups as separate species. Possibly the application of the subspecies concept might be appropriate, but this is difficult because the stunted form occurs amongst both populations, and is very uniform throughout. In fact the plants found in Tanganyika have the narrower petals typical of the S. African group.

8. **Pelargonium graveolens** L'Hérit. in Ait., Hort. Kew. **2**: 423 (1789); Geran.: t. 11 (1791–92).—Harv. in Harv. & Sond., F.C. **1**: 306 (1860).—Knuth in Engl., Pflanzenr. IV, 129: 428 (1912).—Burtt Davy, F.P.F.T. **1**: 190 (1926). TAB. **22.** Type a plant from S. Africa cultivated in Europe.
Geranium radula Roth, Bot. Abh.: 51, t. 10 and t. 12 fig. 9–16 (1787), *non* Cav. (1787). Type a plant from S. Africa cultivated in Europe.
Geranium terebenthinaceum Cav., Diss. Bot. **4**: 250, t. 114 fig. 1 (1787), non Murr. (1785). Type a plant from S. Africa cultivated in Europe.
Pelargonium asperum Ehrh. ex Willd. in L., Sp. Pl. ed. 4, **3**: 678 (1800). Type as for *Geranium radula*.
Geranium graveolens (L'Hérit.) Thunb., Prodr. Pl. Cap. **2**: 115 (1800). Type as for *Pelargonium graveolens*.
Geraniospermum terebinthenaceum (Cav.) Kuntze, Rev. Gen. Pl. **1**: 94 (1891). Type as for *Geranium terebenthinaceum*.

Suffrutex up to 1·3 m. tall, strongly aromatic; internodes 1–8 cm. (or 0·2–1 cm. in *E.M. & W.* 348) long; vegetative parts, peduncles and pedicels glandular and pubescent, tomentose when young, hairs patent or ± appressed on the leaves, glands short-stalked to ± sessile. Leaf-lamina 2–7 × 2·5–8 cm. (2–3 × 2·5–4 cm. in *E.M. & W.* 348), broadly ovate to depressed-ovate, pinnatipartite or sometimes almost palmatipartite, cordate at the base; two basal segments much larger than the rest and often further subdivided with segments often pinnatilobed and irregularly dentate, margin revolute; petiole 1·5–4 (8) cm. long (0·2–1·7 cm. in *E.M. & W.* 348); stipules 4–9 × 3–7 mm., deltate to broadly ovate, acute, often bifid, membranous. Inflorescence a pseudumbel of (1) 2–5 (7) flowers; peduncle (0·2) 1–4 cm. long, leaf-opposed or in the axil of the smaller of 2 apparently opposite leaves; bracts 3–8, 5–8 × 2–4 mm., ovate, membranous; free part of pedicel (2) 4–8 (10) mm. long. Spur 4–8 (10) mm. long. Sepals 7·5–11 × 2–5 mm., nar-

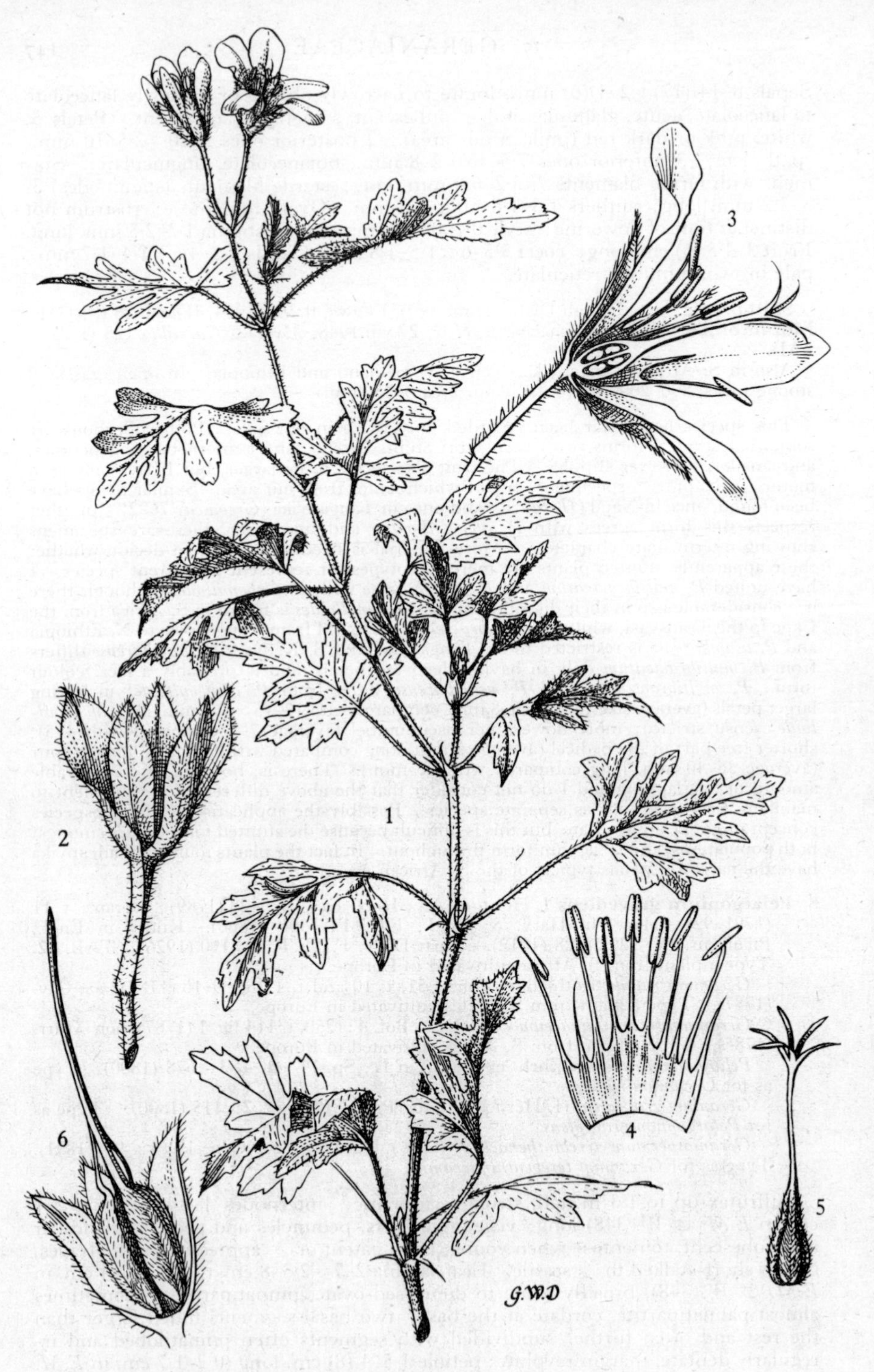

Tab. 22. PELARGONIUM GRAVEOLENS. 1, habit (× ⅔) ; 2, calyx (× 2) ; 3, vertical section of flower (× 2); 4, androecium (× 2) ; 5, gynoecium (× 2) ; 6, fruit (× 2), all from *Goodier & Phipps* 335.

rowly lanceolate to narrowly ovate or lorate to narrowly oblong, acute, patent-pubescent and glandular. Petals pink with darker veins; 2 posterior ones (14) 17–20 × 5–5·6 (6) mm., spathulate; 3 anterior ones (11) 13–15 × 2–3·5 (4) mm., oblanceolate to narrowly obovate, unguiculate. Stamens with fertile filaments 7, 10–16 mm. long; sterile filaments (staminodes) 3, 5–8 mm. long, all 10 connate at the base for 1·5–5 mm.; anthers 2–2·8 × 0·9–1·2 (1·5) mm. Ovary with basal part tomentose; rostrum pubescent; style 5–8 mm. long; stigmas 2–2·8 (3·2) mm. long. Fruit 1·8–2·4 cm. long; cocci 5–5·5 × 1·4–1·7 mm. Seeds 3·2–3·6 × 1·4–1·8 mm., obovate, pale brown, minutely reticulate.

S. Rhodesia. E: Melsetter, Chimanimani Mts., N. of upper Bundi Plain, fl. & fr. 29.xii.1959, *Goodier & Phipps* 335 (BM; COI; K; LISC; SRGH).
Also in the Transvaal and Cape Prov. Among rocks, 1500–2200 m.

Some specimens from Umtali Distr. (e.g. *E.M. & W.* 348 referred to in the description) have much shorter internodes and peduncles, and slightly smaller leaves, but they agree otherwise and are probably habitat forms. Specimens from the Transvaal have leaves with more numerous and narrower segments, and the pedicels are shorter (2–4 mm.). The few specimens I have seen from the Cape seem to agree more with the material from our area. The type of *Pelargonium graveolens* is a cultivated plant and differs from the wild specimens, besides slight differences in general appearance, in having very short filaments (up to 6 mm. long), a feature which I have noticed in all the cultivated plants I have studied, but never on a wild plant. It is possible that these cultivated plants are of hybrid origin, in which case the wild form should be given a new name. I prefer, however, at least for the present, to maintain the established name *P. graveolens* for the wild plants, as the solution of this problem is beyond the scope of this Flora. A new description of the wild material would involve the exact establishment of the somewhat obscure relationship between it and the very closely related *P. radens* H. E. Moore (and also some undescribed material).

It would be inadvisable to transfer the name *Geranium radula* Roth to *Pelargonium* as the same name was also used for *P. radens* H. E. Moore in the same year (Cav., Diss. Bot. **4**: 262, t. 101 fig. 1) and at present it is not possible to decide which publication has priority. Moore has used the name *P. asperum* Ehrh. ex Willd. for the hybrid *P. graveolens* × *P. radens* (in Baileya, **3**: 22 (1955)). I feel certain, however, that L'Héritier's plate and the one of Roth to which Willdenow refers are identical.

38. OXALIDACEAE

By A. W. Exell

Annual or perennial herbs, rarely trees or shrubs or shrublets. Leaves alternate, exstipulate (but sometimes with stipule-like expansions at the base of the petiole), digitately or pinnately 1–∞-foliolate. Flowers in axillary cymes or in pseudumbels or solitary, actinomorphic or nearly so, bisexual, 5-merous, often heterostylous, sometimes cleistogamous and reduced. Sepals 5, free, imbricate (rarely valvate). Petals 5, contorted or imbricate, free or slightly connate at the base. Stamens 10 (15), 2 (3)-seriate; anthers versatile, 2-thecous; filaments ± connate at the base. Ovary superior, 5-locular; loculi 1-∞-ovulate, with axile placentation; styles 5 (rarely 1), free; stigmas capitate, entire. Fruit a loculicidally dehiscent capsule (rarely baccate). Endosperm fleshy or absent.

Averrhoa carambola L. (Mozambique: LM: Ja Sam Farm, Umbeluzi, fl. 16.xii.1946, *Gomes e Sousa* 3475 (K; LISC; SRGH)) is occasionally cultivated in Mozambique. The genus *Averrhoa* L. is separated from the Oxalidaceae by Hutchinson (Fam. Fl. Pl. ed. 2, **1**: 356 (1959)) and placed in a unigeneric family, the Averrhoaceae.

Leaves digitately 3-foliolate (in our species) - - - - - - 1. **Oxalis**
Leaves paripinnate - - - - - - - - - - - - 2. **Biophytum**

1. OXALIS L.

Oxalis L., Sp. Pl. 1: 433 (1753); Gen. Pl. ed. 5: 198 (1754).

Caulescent or acaulescent annual or perennial herbs, usually bulbous, occasionally suffruticose. Leaves radical or cauline, alternate or often in a cluster, digitately

3-foliolate (in our area) or pinnately 3-foliolate or sometimes 1-foliolate. Flowers actinomorphic, usually heterostylous (homostylous in *O. corniculata*), solitary or subtended by bracts in pseudumbels or cymes. Petals contorted, unguiculate; claws ± coherent to form a false tube. Stamens 10, 2-seriate, the outer 5 shorter and opposite the petals, the inner 5 longer and opposite the sepals (vestiges of a third whorl sometimes present); filaments usually connate at the base and sometimes with dentate appendages; anthers introrse in bud, dehiscing by slits. Ovary with several ovules in each loculus; ovules anatropous. Capsule loculicidally dehiscent. Seeds with or without endosperm and with a testa which becomes fleshy and eventually everts explosively ejecting the contents.

In addition to the species cited below there is one record of *O. glabra* Thunb. (S. Rhodesia, Umtali, fl. 1906, *Wroughton* (K)). This is a S. African species which has never been recorded since in our area. The specimen may have been wrongly localized.

Flowers in pseudumbels:
  Flowers yellow (rarely pale pink); plant rhizomatous with branching stems; capsule cylindric exceeding the calyx - - - - - - - - 1. *corniculata*
  Flowers purple, pink or mauve (sometimes with a greenish centre and a yellowish false tube); plant with a bulb or bulbs; capsule usually ellipsoid and shorter than the calyx (broadly cylindric and slightly exceeding the sepals in the introduced *O. latifolia*):
    Plant with a vertical rhizome and with the bulb or bulbs some centimetres below the surface of the ground (Old World species):
      Leaflets entire or nearly so (not lobed or notched or conspicuously emarginate):
        Leaflets broader above the middle, subcircular to very broadly obovate, usually as broad as long or sometimes broader:
          Leaflets densely appressed–pilose - - - - - - 2. *trichophylla*
          Leaflets becoming sparsely pilose or almost glabrous - - 3. *anthelmintica*
        Leaflets usually broader below the middle, linear to lanceolate or ovate, usually 1½ to many times as long as broad (rarely only slightly longer than broad):
          Leaflets narrowly lanceolate to ovate, up to c. 6 times as long as broad 4. *oligotricha*
          Leaflets linear to linear-elliptic, 7–40 times as long as broad 5. *chapmaniae*
      Leaflets 2-lobed to 2-partite or conspicuously emarginate or notched or furcate at the apex:
        Leaflets narrowly oblong and notched or furcate at the apex - 6. *abercornensis*
        Leaflets 2-lobed or 2-partite with lobes varying from linear to broadly elliptic, or leaflets broadly obcordate - - - - - - - 7. *semiloba*
    Plant without a vertical rhizome and with bulb or bulbs at or near the surface of the ground at the base of the leaf-rosette (introduced New World species):
      Leaflets subcircular, emarginate at the apex - - - - - 8. *corymbosa*
      Leaflets obtriangular or broadly obtriangular, somewhat emarginate at the apex with a broad shallow sinus - - - - - - - - 9. *latifolia*
Flowers solitary - - - - - - - - - - 10. *obliquifolia*

1. **Oxalis corniculata** L., Sp. Pl. **1**: 435 (1753).—Sond. in Harv. & Sond., F.C. **1**: 351 (1860).—Bak. f. in Journ. Linn. Soc., Bot. **40**: 35 (1911).—Knuth in Engl., Pflanzenr. IV, 130: 146 (1930).—Salter in Journ. S. Afr. Bot., Suppl. Vol. **1**: 73 (1944). —Exell & Mendonça, C.F.A. **1, 2**: 261 (1955).—Martineau, Rhod. Wild Fl.: 42 (1954).—Wild, Common Rhod. Weeds: fig. 34 (1955).—Wilczek, F.C.B. **7**: 7 (1958). Syntypes from Italy and Sicily.

Oxalis repens Thunb., Diss. Oxal.: 16, t. 1 (1781).—Burtt Davy, F.P.F.T. **1**: 195 (1926). Syntypes from S. Africa and Ceylon.

Oxalis corniculata var. *stricta* Oliv., F.T.A. 1: 297 (1868) pro parte.—Eyles in Trans. Roy. Soc. S. Afr. **5**: 386 (1916).

A much-branched annual herb, usually creeping and often rooting at the nodes; bulb absent; stems usually procumbent, crisped-pubescent. Leaves scattered along the stem; leaflet-lamina 3–15 × 5–20 mm., obcordate-cuneate, pubescent, both surfaces minutely faveolate, apex deeply emarginate, margin entire and ciliate, base cuneate; petiole 1–5 cm. long, pubescent, broadened or winged at the base (sometimes described as stipulate). Flowers yellow (rarely pale pink), homostylous in 1–6-flowered pseudumbels; peduncle up to 5 cm. long, usually slightly exceeding the petiole, with pubescent filiform bracts up to 4 mm. long; pedicels up to 15 mm. long, pubescent. Sepals up to 5–6 × 1–2 mm., oblong-lanceolate, pubescent, without apical calli. Petals up to 10 mm. long, narrowly cuneate, glabrous. Stamens with longer filaments 4–4·5 mm. long and shorter ones 3·5–4 mm. long, edentate. Styles 5 mm. long, pubescent. Loculi ∞-ovulate. Capsule up to 25 mm. long, exserted, subcylindric, ± 5-angled, beaked at the apex and terminating in the persistent remains of the styles, pubescent with reflexed hairs. Seeds 1·5–1·8 × 1–1·2 mm., flattened-ellipsoid, rugose, glabrous.

Bechuanaland Prot. SE: Lobatsi, fl. & fr. 5.ix.1911, *Rogers* 6042 (K; SRGH). **N. Rhodesia** B: Kalabo, fr. 13.xi.1959, *Drummond & Cookson* 6417 (K; SRGH). N: Kiwimbi Distr., near Liamba Village, 1520 m., fl. & fr. 11.ii.1955, *Richards* 4463 (K). W: Ndola, fl. & fr. 17.ix.1937, *Miller* 148 (K). C: 8 km. E. of Lusaka, 1280 m., fl. & fr, 3.xii.1958, *King* 227 (K). E: Fort Jameson Distr., Kachabere, 1000 m., fl. & fr. 7.i.1959, *Robson* 1065 (BM; K; LISC; SRGH). **S. Rhodesia.** N: Miami, fl. & fr. iv.1926, *Rand* 4 (BM). W: Bulawayo, fr. v.1898, *Rand* 296 (BM). C: Salisbury, fl. ix.1919, *Eyles* 1785 (K; PRE; SRGH). E: Stapleford Forest Reserve, 1800 m., fl. & fr. 15.vi.1934, *Gilliland* 339 (BM; SRGH). S: Zimbabwe, fl. & fr. 11.viii.1929, *Rendle* 258 (BM). **Nyasaland.** N: Misuku Hills, 1520 m., fl. & fr. 13.iii.1953, *Williamson* 220 (BM). C: Dowa, Lingadzi R., fl. & fr. 31.x.1950, *Jackson* 218 (K). S: Zomba, fr. 22.v.1956, *Banda* 193 (BM; SRGH). **Mozambique.** MS: R. Curumadzi, 610 m., fl. 21.xi.1906, *Swynnerton* 2065 (BM). LM: Marracuene, fl. & fr. 3.xii.1940, *Torre* 2215 (LISC).

A cosmopolitan weed.

2. **Oxalis trichophylla** Bak. in Kew Bull. **1895**: 63 (1895).—Knuth in Engl., Pflanzenr. IV, 130: 302 (1930). Type: N. Rhodesia, Fwambo, fl. 1894, *Carson* 56 (K).

Perennial herb with vertical rhizome, usually acaulescent or stem very short. Bulb not seen. Leaves 3-foliolate, 2–10 in number, usually basal or almost so; leaflet-lamina up to 20 × 20 mm., sessile, subcircular, broadly obovate-cuneate or broadly obovate-elliptic, sometimes truncate at the apex but not conspicuously emarginate, tomentose when young, becoming appressed-pilose above and more densely so beneath, both surfaces faveolate when visible through the indumentum; petiole up to 6 cm. long, pilose. Flowers mauve or pink with a yellowish false tube, with pubescent to pilose pedicels up to 12 mm. long, in 2–5-flowered pseudumbels arising from the leaf-rosettes; peduncle up to 15 cm. long, pilose. Sepals up to 6 × 2·5 mm., lanceolate to broadly elliptic, acute, pubescent, with orange-yellow apical calli. Petals 10–12 × 3 mm., oblong-elliptic, glabrous. Stamens with 5 longer filaments pubescent and 5 shorter ones glabrous. Styles pubescent. Seeds not seen.

N. Rhodesia N: Lumi R., 1525 m., fl. 9.ii.1955, *Richards* 4397 (K).

Perhaps also in Tanganyika. Woodland and grassland.

The type is so poor that it is difficult to be certain that the species has been correctly interpreted.

3. **Oxalis anthelmintica** A. Rich., Tent. Fl. Abyss. **1**: 124 (1847).—Salter in Journ. S. Afr. Bot., Suppl. Vol. **1**: 86 (1944). Type from Ethiopia.

Oxalis purpurata var. *anthelmintica* (A. Rich.) Knuth in Engl., Pflanzenr. IV, 130: 304 (1930). Type as above.

Perennial herb with vertical rhizome up to 18 cm. long, usually acaulescent but sometimes with a stem up to 15 cm. long. Bulb usually 1–1·5 cm. long, ovoid, at the base of the rhizome. Leaves 3-foliolate, in a rosette; leaflet-lamina up to 22 × 22 mm., subcircular to very broadly obovate, rounded or shallowly emarginate at the apex, usually nearly glabrous on the upper surface; petiole up to 13 cm. long, almost glabrous or pubescent or pilose. Flowers pink, purple or bluish, with pedicels up to 20 mm. long, in 4–10-flowered pseudumbels arising from the leaf rosette or from the cauline bracts; peduncle up to over 20 cm. long, but usually much shorter; bracts linear-lanceolate or very narrowly elongate-triangular. Sepals 4–7 mm. long, narrowly lanceolate, acute, with orange-yellow apical calli. Petals up to 14 mm. long. Stamens with 5 longer filaments pubescent and 5 shorter ones glabrous. Styles pubescent. Capsule up to 5 × 2·5 mm., subglobose to ellipsoid.

N. Rhodesia. N: Abercorn Distr., Chilongowelo Farm, 1465 m., fl. 13.xii.1951, *Richards* 5 (K). **S. Rhodesia.** E: Odzani Valley, 1525 m., fl., *Gilliland* 1791A pro parte (K). **Nyasaland.** N: Nyika Plateau, fl. xi.1903, *Henderson* (BM). **Mozambique.** N: Metónia, fl. i.1934, *Gomes e Sousa* 1646 (COI).

Also in Ethiopia, Kenya, Congo and Tanganyika. In open woodland and bush.

See Salter (loc. cit.) for a note on the identity of *O. purpurata* Jacq.

4. **Oxalis oligotricha** Bak. in Kew Bull. **1895**: 64 (1895).—Knuth in Engl., Pflanzenr. IV, 130: 302 (1930). TAB. **23** fig. C. Type: N. Rhodesia, Fwambo, fl. 1890, *Carson* (K).

Perennial herb with vertical rhizome up to 10–12 cm. long, usually with a distinct stem up to 15 cm. long but often only 1 cm. long or less. Bulb up to 3 × 2 cm., brown, ovoid. Leaves 3-foliolate, usually 2–4 at the apex of the stem (appearing ± basal when the stem is very short); leaflet-lamina 8–35 × 4–20 cm., sessile, subcircular, broadly ovate, broadly elliptic, lanceolate or narrowly elliptic, not lobed or emarginate, pilose when young but ± glabrescent except on the midrib below, ciliate, ± faveolate on both surfaces; petiole up to 12 cm. long, sparsely to rather densely pilose. Flowers pale mauve with a greenish-yellow false tube, with pedicels up to 15 mm. long, in 4–10-flowered pseudumbels arising from the leaf-rosettes or sometimes subtended by cauline bracts (reduced leaves) and thus arising below the leaf-clusters; peduncle up to 15 cm. long, hairy like the petioles; bracts up to 6 mm., long, linear-lanceolate to filiform. Sepals 4–5 mm. long, ovate-lanceolate to narrowly lanceolate, acute, pubescent, with orange-yellow apical calli. Petals 9–13 × 2·5–3 mm., oblong-elliptic, glabrous. Stamens with 5 longer filaments pubescent and 5 shorter ones sparsely pubescent or glabrous. Styles pubescent. Seeds 1·5 × 1 mm., flattened-ellipsoid, glabrous.

N. Rhodesia. N: Chilongowelo, 1460 m., fl. 20.xii.1954, *Richards* 3638 (K). W: Mwinilunga Distr., W. of Matonchi Farm, fl. 2.xi.1937, *Milne-Redhead* 3057 (K). **Mozambique.** N: Planalto da Lichinga, Vila Cabral, fl. xii.1932, *Gomes e Sousa* 1067 (COI).

Also in Tanganyika. In *Brachystegia* woodland and along roadsides from c. 900 m. upwards.

5. **Oxalis chapmaniae** Exell in Bol. Soc. Brot., Sér. 2, **35**: 136 (1961). TAB. **23** fig. B. Type: Nyasaland, Nyika plateau, *Chapman* 73 (BM).

Perennial herb with vertical rhizome up to 10 cm. long, acaulescent or almost so. Bulb up to 2 × 1·5 cm., ovoid, acute. Leaves 3-foliolate, 3–5 in a basal rosette; leaflet-lamina 4–8 × 0·1–1 cm., very narrowly elliptic or linear with an apical callus, almost glabrous, faveolate beneath; midrib flattened and somewhat expanded; petiole 4–16 cm. long, sparsely pubescent or almost glabrous. Flowers purple or magenta-pink or carmine with a whitish false tube and with pubescent or nearly glabrous pedicels, in 4–7-flowered pseudumbels arising from the leaf-rosettes; peduncle 7–24 cm. long, pubescent or nearly glabrous. Sepals 5–6 × 1·5–3 mm., elliptic, pubescent or nearly glabrous with orange-coloured apical calli. Petals up to 18 mm. long, oblong-elliptic, glabrous. Stamens with 5 longer filaments 3 mm. long and 5 shorter ones 1·5 mm., long, all puberulous. Styles 4·5 mm. long, pubescent. Seeds 1 × 0·6 mm., brown, flattened-ellipsoid.

N. Rhodesia. E: Nyika Plateau, 2130 m., fl. 2.i.1959, *Robinson* 2989 (K; SRGH). **Nyasaland.** N: Nyika Plateau, 2200 m., fl. 3.v.1960, *Wright* 256 (K).

Known only from the Nyika Plateau. Upland grassland, 2130–2440 m.

6. **Oxalis abercornensis** Knuth in Fedde, Repert. **48**: 3 (1940). Type: N. Rhodesia-Tanganyika border, *Michelmore* 1049 (B, holotype†; K).

Perennial herb with vertical rhizome up to 20 cm. long (but often quite short), usually acaulescent or nearly so. Bulb 3 × 1·3 cm., ellipsoid. Leaves 2–4, 3-foliolate, usually basal or almost so; leaflet-lamina 4–7 × 0·3–0·5 cm., sessile, linear to very narrowly oblong-elliptic, notched at the apex, sparsely to rather densely pilose, not obviously faveolate; petiole up to 20 cm. long, pubescent or pilose. Flowers purple, with pubescent or pilose pedicels up to 20 mm. long, in 6–20-flowered pseudumbels arising from the leaf-rosettes; peduncle up to 25 cm. long, pubescent or pilose. Sepals 4–6 mm. long, lanceolate, acute, pubescent, with small orange-yellow apical calli. Petals up to 12 mm. long, glabrous. ("Larger stamens a little shorter than the sepals" *fide* Knuth). Styles pubescent. Seeds 1·5 × 1 mm., flattened-ellipsoid.

N. Rhodesia. N: N. Rhodesia-SW. Tanganyika border, fl. & fr. 12.xii.1936, *Michelmore* 1049 (K).

Known only from the locality cited. Along paths and as a garden weed.

Possibly merely a form of *O. oligotricha*.

7. **Oxalis semiloba** Sond. in Harv. & Sond., F.C. **1**: 350 (1860). TAB. **22** fig. A. Syntypes from S. Africa.
Acetosella semiloba (Sond.) Kuntze, Rev. Gen. Pl. **1**: 91 (1891). Type as above.

Perennial herb with vertical rhizome up to at least 15 cm. long, acaulescent or sometimes with a stem occasionally branched near the surface of the ground. Bulb 1–2 cm. long, ovoid or ellipsoid, usually solitary at the base of the rhizome (but sometimes additional laterally borne bulbs present). Leaves 3-foliolate; leaflets sessile, extremely variable in size and shape, 2-lobed to 2-partite (see key to subspp. below), nearly glabrous to densely pilose, especially beneath, often conspicuously faveolate; petiole up to 12 cm. long. Flowers 2–3 cm. in diam., pink or purplish or blue, sometimes with a yellowish centre, with pedicels up to 25 mm. long, in 4–numerous-flowered pseudumbels; peduncle 4–20 cm. long. Sepals 4–5 × 1·5–2 mm., narrowly oblong-elliptic with well-developed orange apical calli. Petals up to 15 mm. long, glabrous. Stamens with 5 longer filaments pubescent and 5 shorter ones glabrous.

Ethiopia, Kenya, Congo, Tanganyika, Angola, Bechuanaland Prot., N. & S. Rhodesia, Nyasaland, Mozambique and S. Africa. In forest woodland and grassland habitats.

Leaflets shallowly to deeply obcordate or broadly obcordate, emarginate to lobed to about half-way; lobes convex on their apical margins so that the sinus, though obtuse, has a sharp point of deepest entry (see Tab. 23 fig. A, upper row) subsp. *semiloba*
Leaflets broadly obcordate to deeply 2-lobed or 2-partite; lobes usually very narrowly oblong to oblong-lanceolate and when broader usually concave on their apical margins, so that the sinus, which varies from acute to obtuse, has a rounded point of deepest entry (see Tab. **23** fig. A, lower row) - - - subsp. *uhehensis*

Subsp. **semiloba.**—Bak. f in Journ. Linn. Soc., Bot. **40**: 35 (1911).—Eyles in Trans. Roy. Soc. S. Afr. **5**: 386 (1916).—Knuth in Engl., Pflanzenr. IV, 130: 306 (1930).—Salter in Journ. S. Afr. Bot., Suppl. Vol. **1**: 87 (1944).—Exell & Mendonça, C.F.A. **1,** 2: 262 (1951).—Wilczek, F.C.B. **7**: 10 (1958). TAB. **23** fig. A (upper row).

Bechuanaland Prot. SE: Gaberones, fl. & fr. 10.iii.1930, *van Son* in Herb. Transv. Mus. 28956 (BM). **N. Rhodesia.** S: Dundwa, 9·5 km. S. of Mapanza, 1100 m., fl. & fr. 6.iv.1953, *Robinson* 166 (K). **S. Rhodesia.** N: Mazoe, Iron Mask Hill, fl. & fr. xii.1905, *Eyles* 210 (BM; K; SRGH). W: Bulawayo, 1370 m., fl. xii.1901, *Eyles* 73 (BM; K). C: Salisbury, fl. 1914, *Craster* 215 (K). E: Vumba Mts., 1370 m., fl. & fr. 7.ii.1934, *Davies* (BM). **Nyasaland.** N: Masuku Plateau, 1980–2130 m., *Whyte* (K). C: Mua, lower Livulezi R., 600 m., fl. & fr. 20.i.1959, *Robson* 1276 (BM; K; LISC; SRGH). S: Limbe, fl. 27.iii.1948, *Goodwin* 107 (BM). **Mozambique.** N: Metónia, fl. xi.1933, *Gomes e Sousa* 1591 (COI). Z: Gúruè, fl. & fr. 28.v.1937, *Torre* 1488 (COI; LISC). MS: Chimanimani Mts., Musapa Gap, 1065 m., fl. 19.xii.1957, *Phipps* 826 (SRGH). SS: Gaza, Bilene, fl. & fr. 28.iii.1948, *Torre* 7575 (LISC). LM: Marracuene, fl. & fr. iii.1893, *Quintas* 13 (COI).

Distribution and ecology as for the species but apparently absent from the more northern parts of N. Rhodesia.

Subsp. **uhehensis** (Engl.) Exell in Bol. Soc. Brot., Sér. 2, **35**: 135 (1961). TAB. **23** fig. A (lower row). Type from Tanganyika, Uhehe Plateau, *Goetze* 499 (B†; BM).
Oxalis uhehensis Engl., Bot. Jahrb. **28**: 412 (1900); Pflanzenw. Afr. **3,** 1: 716, fig. 330 (1915).—Knuth in Engl., Pflanzenr. IV. 130: 309, fig. 20 (1930). Type as above.
Oxalis morrumbalaensis De Wild., Pl. Nov. Hort. Then.: t. 36 (1905). Type: Mozambique, Zambezia, Morrumbala, *Luja* 268 (BR).
Oxalis angustiloba R.E. Fr., Schwed. Rhod.-Kongo-Exped. 1: 107 (1914).—Wilczek, F.C.B. **7**: 10 (1958). Syntypes: N. Rhodesia, Kalambo, *Fries* 1298 (UPS) and Msisi, near Abercorn, *Fries* 1389 (UPS).

N. Rhodesia. N: Katenina Hills, fl. & fr. 27.xii.1907, *Kassner* 2163 (BM; K). W: c. 1·6 km. E. of Mwinilunga, fl. 27.xi.1937, *Milne-Redhead* 3419 (BM; K). C: Serenje, fl. & fr. 23.i.1955, *Fanshawe* 1844 (K). E: Chadiza, 850 m., fl. 28.xi.1958, *Robson* 766 (BM; K; LISC; SRGH). **Nyasaland.** C: Dedza Distr., Dzenza Forest Reserve, 1220 m., fl. & fr. 21.iii.1955, *E.M. & W.* 1100 (BM; LISC; SRGH). S: Shire Highlands, fr. 1887, *Last* (K). **Mozambique.** N: 16 km. N. of Mandimba, fl. 24.ix.1941, *Hornby* 3477 (K; PRE). Z: between Mocuba and Milange, fl. & fr. 20.v.1949, *Barbosa & Carvalho* 2748 (LISC).
Also in Tanganyika. *Brachystegia* woodland.

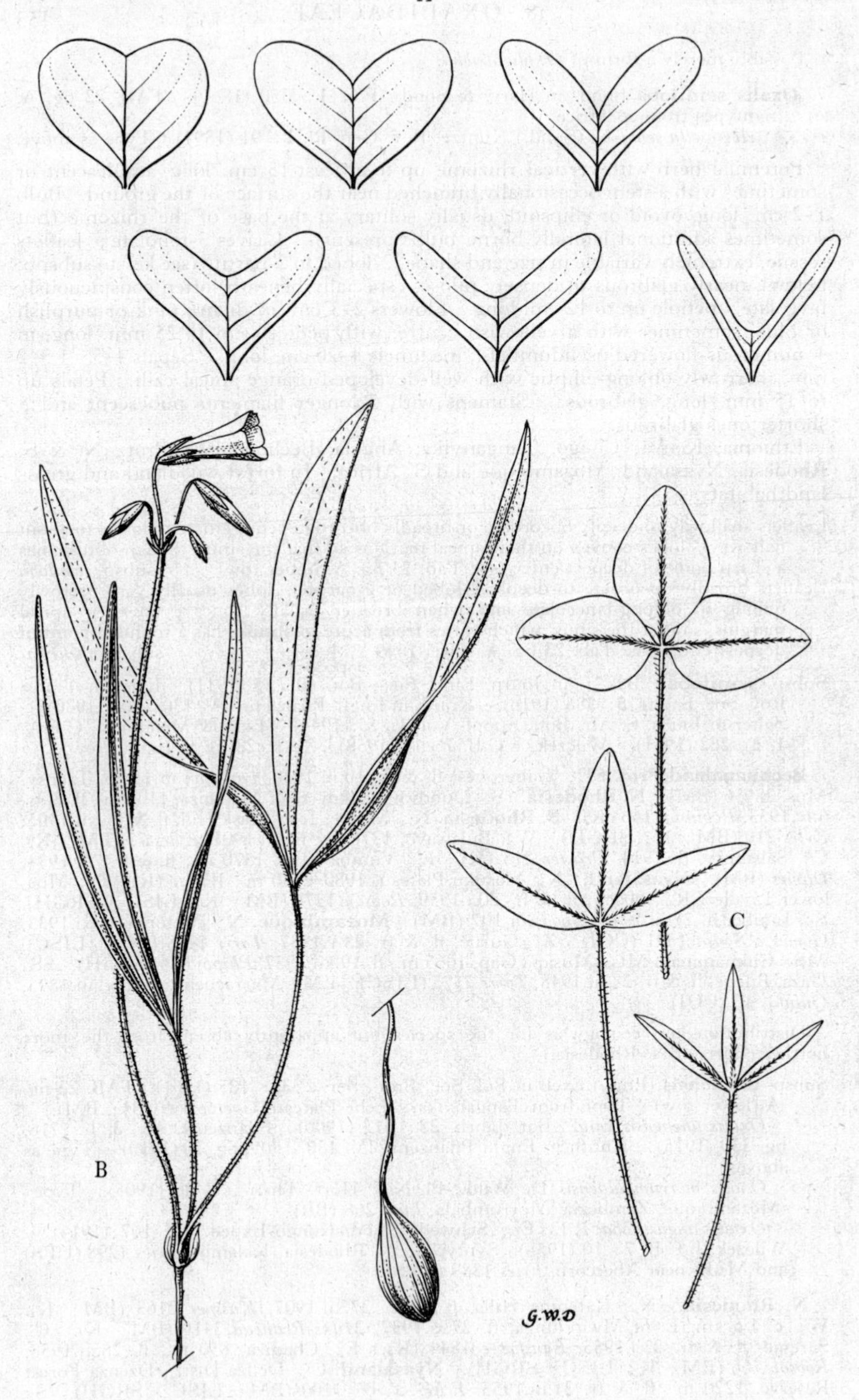

Tab. 23. A.—OXALIS SEMILOBA SUBSP. SEMILOBA (upper row) and OXALIS SEMILOBA SUBSP. UHEHENSIS (lower row) showing variation in leaflet-shape (×1). B.—OXALIS CHAPMANIAE, plant (×1) *Chapman* 73. C.—OXALIS OLIGOTRICHA. Variation in leaf-shape (×1).

O. uhuhensis is in its more extreme forms very distinct from *O. semiloba* and at first sight it appears unnecessary to reduce them to subspecies. Where they overlap, however, as in Tanganyika and Mozambique there are transitional forms and both subspecies show much variation in the lobing of the leaflets. *Kassner* 2163 (BM) shows the transition from a very narrowly lobed leaflet to a form almost indistinguishable from subsp. *semiloba* while *Torre* 4892 from the Zambezia Prov. of Mozambique shows a series ranging from the broadest-lobed form of *Kassner* 2163 to quite typical subsp. *semiloba*. The great majority of the specimens fall clearly into one or other of the two subspecies but a few cannot be satisfactorily classified.

8. **Oxalis corymbosa** DC., Prodr. **1**: 696 (1824).—Exell, Cat. Vasc. Pl. S. Tomé: 124 (1944).—Wilczek, F.C.B. **7**: 9 (1958). Syntypes from the Mascarene Is.
Oxalis martiana Zucc. in Denkschr. Akad. Muench. **9**: 144 (1825).—Knuth in Engl., Pflanzenr. IV, 130: 250 (1930). Type from Brazil.

Acaulescent perennial herb, with subspherical bulb 20–25 mm. in diam. composed of numerous bulbils. Leaves trifoliolate; leaflets up to 4·5 cm. in diam., subcircular. Flowers pink or purplish, in pseudumbels.

Mozambique. LM: Umbeluzi, near the Secretariat, fl. 9.vii.1949, *Myre* 751 (LM).
Native of tropical America; introduced. A weed of cultivation.

The specimen consists of detached leaves and an inflorescence so that it is not possible to verify whether it has a rhizome or not. Nevertheless from its appearance I strongly suspect that it is this introduced tropical American weed which is common in S. Tomé and Principe but apparently not often found elsewhere in tropical Africa though Wilczek (loc. cit.) records it from two localities in the Congo. When the bulb is collected, the New World weeds (usually *O. corymbosa* or *O. latifolia*) can at once be distinguished from native African species, some of which otherwise closely resemble them, by the fact that the bulb is directly at the base of the rosette of leaves in the American species while it is usually some centimetres below soil-level at the end of a vertical rhizome in the African species.

9. **Oxalis latifolia** Kunth, Nov. Gen. & Sp. Pl. **5**: 237, t. 467 (1828).—Knuth in Engl., Pflanzenr. IV, 130: 273 (1930).—Wilczek, F.C.B. **7**: 11 (1958). Type from Mexico.

Acaulescent perennial herb with numerous ovoid bulbs. Leaves trifoliolate; leaflets up to 7·5 × 5 cm. in outline, obtriangular to broadly obtriangular, emarginate at the apex with a broad shallow sinus. Flowers purple, in pseudumbels. Petals 10–16 mm. long. Stamens with the 5 longer filaments shortly obtusely dentate in the middle. Capsules broadly cylindric, slightly longer than the sepals.

N. Rhodesia. N: Inono valley, 1220 m., fl., *Richards* 2265a (K). C: Broken Hill, fl. & fr. 17.xiii.1907, *Kassner* 2016 (BM). **S. Rhodesia.** W: Bulawayo, 1370 m., fl. iii.1958, *Miller* 5136 (SRGH). C: Salisbury Distr., Experimental Station, fl. 5.xii.1955, *Wild* 4717 (K; SRGH). **Mozambique.** N: Metónia, fl. xi.1933, *Gomes e Sousa* 1561 (COI). LM: Lourenço Marques, garden of the Polana Hotel, fl. xi.1945, *Pimenta* 22600 (LISC; LM).
Native of tropical America; introduced. A troublesome weed of gardens and cultivated ground because its small bulbs are difficult to eradicate.

See note under *O. corymbosa*.

10. **Oxalis obliquifolia** Steud. ex A. Rich., Tent. Fl. Abyss. **1**: 123 (1847).—Knuth in Engl., Pflanzenr. IV, 130: 348 (1930).—Salter in Journ. S. Afr. Bot., Suppl. Vol. **1**: 154 (1944).—Exell & Mendonça C.F.A. **1**, 2: 262 (1951).—Brenan in Mem. N.Y. Bot. Gard. **8**, 3: 231 (1953).—Wilczek, F.C.B. **7**: 8 (1958). Type from Ethiopia.
Oxalis spp.—Eyles in Trans. Roy. Soc. S. Afr. **5**: 386 (1916).
Oxalis obliquifolia var. *transvaalensis* Knuth, loc. cit.—Suesseng. in Trans. Rhod. Sc. Assoc. **43**: 1071 (1951). Type from Transvaal.

Acaulescent perennial herb with vertical rhizome up to 10 cm. long (sometimes very short in young plants). Bulb 8–12 mm. in diam., subglobose or ellipsoid, solitary at the base of the rhizome. Leave 3-foliolate; leaflet-lamina up to 17 × 4 mm., subcircular to very broadly obovate or transversely elliptic, apex rounded to shallowly emarginate, base cuneate or slightly attenuate, glabrous or sparsely pubescent and usually distinctly faveolate; petiole 2–9 cm. long. Flowers up to 2·5 cm. in diam., pink or mauve, sometimes with a yellow centre, solitary; peduncle up to 9 cm. long. Sepals up to 4–5 × 2–2·2 mm., broadly oblong-elliptic, with or without apical calli. Petals 10–23 mm. long. Filaments with minute glandular hairs (otherwise glabrous).

N. Rhodesia. N: Abercorn Distr., 1770 m., fl. iii.1935, *Gamwell* 234 (BM). **S. Rhodesia.** N: Mazoe, Iron Mask Hill, fl. iv.1906, *Eyles* 344 (BM; SRGH). W: Besna Kobila, fl. & fr. i.1953, *Miller* 1504 (SRGH). C: Salisbury, 1520 m., fl. & fr. ii.1919, *Eyles* 1483 (BM; PRE). E: Umtali Distr., Engwa, 1670–1830 m., fl. & fr. 1.ii.1955, *E.M. & W.* 70 (BM; LISC; SRGH). S: Malangwe R., 625 m., fl. & fr. 6.v.1958, *Drummond* 5659 (SRGH). **Nyasaland.** N: Nyika Plateau, above Nchena-chena, fl. 4.v.1952, *White* 2588 (K). C: Kasungu Distr., Chipala Hill, Chamama, 1050 m., fl. 16.i.1959, *Robson* 1217 (BM; K). **Mozambique.** SS: Chibuto, st. 6.vi.1957, *Barbosa & Lemos* 7603 (LMJ).

Also in Ethiopia, Sudan, Uganda, Kenya, Tanganyika, Congo, Angola and S. Africa. Grassland and *Brachystegia* woodland from about 1,000 m. upwards.

In a note under *O. depressa* Eckl. & Zeyh. (1836), Salter (loc. cit.) says "It is closely related to *O. obliquifolia* and possibly the two species merge into one another but *O. obliquifolia* seems to be generally larger and more robust, without calli on the sepals or they are very indistinct." Salter notes (tom. cit.; 155) that the Rhodesian specimens have two rather inconspicuous apical calli on the sepals and that there are also intermediates between *O. obliquifolia* and *O. setosa* E. Mey. ex Sond., which in turn merges with *O. punctata* L.f.

2. BIOPHYTUM DC.

Biophytum DC., Prodr. **1**: 689 (1824).

Caulescent or acaulescent annual herbs or perennial herbs with woody rootstock. Leaves usually in rosettes at the apex of the stem or at ground-level on the rootstock, paripinnate (the abortive terminal leaflet represented by a bristle), ± sensitive, usually articulated just above the insertion; leaflets opposite or subopposite, very shortly petiolulate or subsessile. Flowers orange, yellow, white or pinkish, in pseudumbels with bracts forming an involucre. Sepals 5, free, imbricate. Petals 5, contorted, free at first but eventually coherent and falling off together in a mass. Stamens 10, 5 longer and 5 shorter, all slightly connate at the base. Ovary 5-locular, 5-lobed. Capsule obovoid, subglobose, or ellipsoid, loculicidally dehiscent. Seeds usually tuberculate.

Plant an annual herb with simple erect stem:
  Leaflets membranous, translucent; pedicels about twice as long as the calyx or longer; sepals 2–2·5 mm. long - - - - - - - - 1. *abyssinicum*
  Leaflets semi-rigid, ± opaque; pedicels shorter than or about equalling the calyx; sepals 4–6 mm. long:
    Leaflets 8–21 pairs; lateral nerves slender, not very prominent, forming an acute angle with the midrib; sepals 4–5 mm. long - - - - 2. *sensitivum*
    Leaflets 5–11 pairs; lateral nerves thick, rather prominent, at right angles to the midrib; sepals up to 6 mm. long - - - - - - - 3. *petersianum*
Plant a perennial, acaulescent or almost so, with a thickened woody rootstock:
  Pseudumbels pedunculate:
    Leaflets obliquely elliptic to subcircular, 2–11 mm. long:
      Peduncle ± equalling the leaves or exceeding them (rarely a little shorter); sepals not glandular - - - - - - - - 4. *crassipes*
      Peduncle $\frac{1}{3}$–$\frac{2}{3}$ as long as the leaves; sepals densely glandular-pubescent 5. *nyikense*
    Leaflets oblong, 5–11 mm. long:
      Sepals 6–6·5 mm. long; leaflets ciliate, otherwise glabrous - 6. *richardsiae*
      Sepals 4–4·5 mm. long; leaflets pubescent, rarely glabrescent - 7. *kassneri*
  Pseudumbels sessile (flowers appearing solitary) - - - - 8. *macrorrhizum*

1. **Biophytum abyssinicum** Steud. ex A. Rich., Tent. Fl. Abyss. **1**: 122 (1847).—Knuth in Engl., Pflanzenr. IV, 130: 395, fig. 23 A–C (1930).—Exell & Mendonça, C.F.A. **1**, 2: 263, t. 9 fig. A (1951).—Delhaye in Inst. Roy. Col. Belg. **23**: 855, t. 2 fig. 52–61 (1952).—Keay, F.W.T.A. ed. 2, **1**, 1: 159 (1954). TAB. **24** fig. A. Type from Ethiopia.

Oxalis abyssinica (Steud. ex A. Rich.) Walp., Ann. Bot. Syst. **2**: 241 (1851–52).—Oliv., F.T.A. **1**: 297 (1868). Type as above.

Biophytum reinwardtii (Zucc.) Klotzsch in Peters, Reise Mossamb. Bot. **1**: 85 (1861) pro parte quoad specim. mossamb.

Biophytum reinwardtii subsp. *abyssinicum* (Steud. ex A. Rich.) Steen. in Bull. Gard. Bot. Buitenz., Ser. 3, **18**, 4: 453 (1950).—Wilczek, F.C.B. **7**: 13 (1958). Type as above.

Annual herb 5–30 cm. in height with a simple slender erect appressed-pubescent stem bearing a single rosette of leaves at the apex. Leaves up to 7 cm. long;

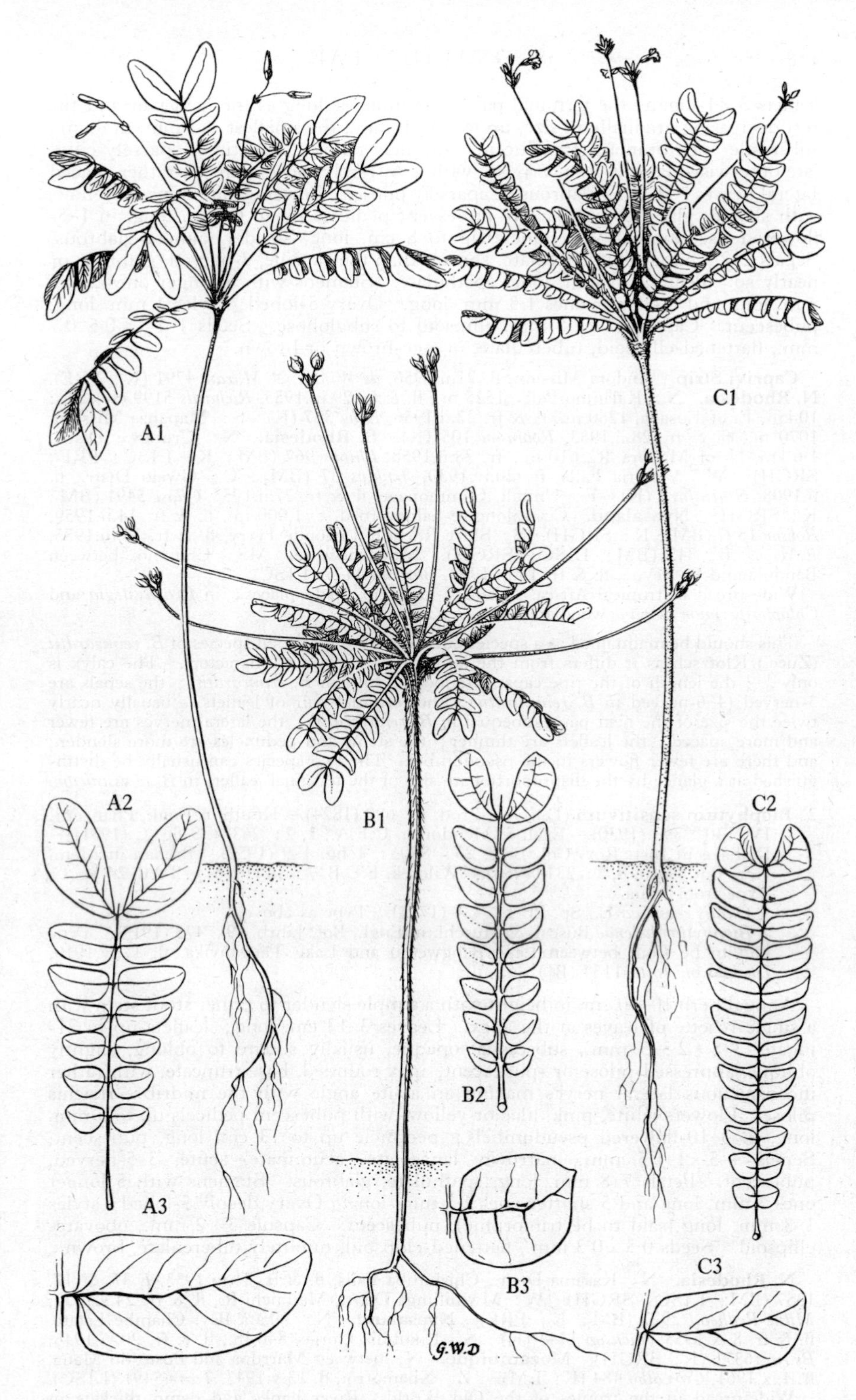

Tab. 24. A.—BIOPHYTUM ABYSSINICUM. A1, plant (×½), *Chase* 4388; A2, leaf (×1); A3, leaflet (×3), all from *Chase* 4388. B.—BIOPHYTUM PETERSIANUM. B1, plant (×½); B2, leaf (×1); B3, leaflet (×3), all from *Chase* 814. C.—BIOPHYTUM SENSITIVUM. C1, plant (×½); C2, leaf (×1); C3, leaflet (×3), all from *E.M. & W.* 1357.

leaflets 5–11-jugate, the terminal pair 1½–2 times as long as the next pair and the rest decreasing gradually in size, up to 20 × 9 mm., the smallest c. 2 mm. in diam., subsessile, membranous, translucent, obliquely elliptic, sometimes sparsely ciliolate otherwise glabrous or nearly so, with 2–6 (rarely more) pairs of rather spaced lateral nerves; rhachis glabrous or sparsely pilose. Flowers white, pink or yellow, with slender glabrous or sparsely pubescent pedicels up to 6 mm. long, in 1–5-flowered pseudumbels; peduncle up to 8 cm. long, slender, almost glabrous. Sepals 2–2·5 mm. long, lanceolate, shorter than the capsule, 3-nerved, glabrous or nearly so. Petals c. 5 mm. long, spathulate. Stamens with 5 longer ones 2–2·5 mm. long and 5 shorter ones 1·5 mm. long. Ovary 5-lobed; styles 1 mm. long, pubescent. Capsule 3 × 2·5 mm. ellipsoid to subglobose. Seeds 1–1·3 × 0·5–0·7 mm., flattened-ellipsoid, tuberculate, orange-brown or brown.

Caprivi Strip. Andora Mission, fl. 21.ii.1956, *de Winter & Marais* 4791 (K; PRE). **N. Rhodesia.** N: Kalambo Falls, 1525 m., fl. & fr. 29.iii.1955, *Richards* 5199 (K). C: 10 km. E. of Lusaka, 1280 m., fl. & fr. 22.ii.1956, *King* 317 (K). S: Mapanza Mission, 1070 m., fl. & fr. 28.ii.1953, *Robinson* 105 (K). **S. Rhodesia.** N: Urungwe Distr., 1·6 km. N. of Mauora R., 610 m., fr. 28.ii.1958, *Phipps* 967 (BM; K; LISC; PRE; SRGH). W: Victoria Falls, fr. 20.iv.1930, *Jenkins* 57 (BM). C: Gwelo Distr., fl. ii.1908, *Stuttaford* (K). E: Umtali, Commonage, fl. & fr. 27.ii.1955, *Chase* 5491 (BM; K: SRGH). **Nyasaland.** C: Lilongwe-Salima road, c. 1,000 m., fl. & fr. 14.ii.1959, *Robson* 1592 (BM; K; SRGH). S: Shire R., near Liwonde Ferry, fl. & fr. 13.iii.1955, *E.M. & W.* 845 (BM; LISC; SRGH). **Mozambique.** MS: Chimoio, between Bandula and R. Revuè, fl. & fr. 15.iii.1948, *Barbosa* 1177 (LISC).

Widespread in tropical Africa. Usually in damp shady places; in *Brachystegia* and *Colophospermum mopane* woodland.

This should be maintained as a species and not reduced to a subspecies of *B. reinwardtii* (Zucc.) Klotzsch as it differs from the latter in a number of characters. The calyx is only ½–⅔ the length of the ripe capsule (subequal to it in *B. reinwardtii*); the sepals are 3-nerved (4–6-nerved in *B. reinwardtii*); the terminal pair of leaflets is usually nearly twice the size of the next pair (subequal in *B. reinwardtii*); the lateral nerves are fewer and more spaced; the leaflets are thinner; the stems and peduncles are more slender; and there are fewer flowers in the pseudumbel. The two species can usually be distinguished at a glance by the disproportionate size of the terminal leaflets in *B. abyssinicum*.

2. **Biophytum sensitivum** (L.) DC., Prod. **1**: 690 (1824).—Knuth in Engl., Pflanzenr. IV, 130: 393 (1930).—Exell & Mendonça, C.F.A., **1,** 2: 263, t. 9 fig. C (1951).—Delhaye in Inst. Roy. Col. Belg. **23**: 850, t. 1 fig. 1–9 (1952).—Brenan in Mem. N.Y. Bot. Gard. **8,** 3: 231 (1953).—Wilczek, F.C.B. **7**: 14 (1958).—TAB. **24** fig C. Type from India.

Oxalis sensitiva L., Sp. Pl. **1**: 434 (1753). Type as above.

Biophytum helenae Buscal. & Muschl. in Engl., Bot. Jahrb. **49**: 475 (1913). Type said to be from between Lake Bangweulu and Lake Tanganyika, fl. 17.iv.1910, *Duchess of Aosta* 1154 (B†).

Annual herb 10–40 cm. in height with a simple slender to rather stout stem with a single rosette of leaves at the apex. Leaves 3–13 cm. long; leaflets up to 21-jugate, 3–8 × 2·5–5 mm., subsessile, opaque, usually square to oblong, slightly oblique, appressed-pilose or -pubescent, apex rounded, base truncate, with rather inconspicuous lateral nerves making an acute angle with the midrib; rhachis pilose. Flowers white, pink, lilac or yellow, with pubescent pedicels up to 6 mm. long, in 3–10-flowered pseudumbels; peduncle up to 13 cm. long, pubescent, Sepals 4–5 × 1–1·5 mm., narrowly lanceolate, acuminate, acute, 3–5-nerved, pubescent. Petals 7–8 mm. long, spathulate, glabrous. Stamens with 5 longer ones 3 mm. long and 5 shorter ones 1·5 mm. long. Ovary deeply 5-lobed; styles 1–3 mm. long (said to be trimorphic), pubescent. Capsule 3 × 2 mm., obovate-ellipsoid. Seeds 0·5 × 0·3 mm., flattened-ellipsoid, minutely tuberculate, brown.

N. Rhodesia. N: Kasama Distr., Chishimba Falls, fl. & fr. 31.iii.1955, *E.M. & W.* 1357 (BM; LISC; SRGH). W: Mwinilunga Distr., Matonchi R., fl. & fr. 24.x.1937, *Milne-Redhead* 2927 (BM; K; PRE). **Nyasaland.** N: Nkata Bay, Chambe Estate, fl. & fr. 8.ix.1955, *Jackson* 1744 (K). S: Likubula Gorge, 840 m., fl. & fr. 20.vi.1946, *Brass* 16371 (K; SRGH). **Mozambique.** N: between Marrupa and Posto do Maúa, fl. 4.ix.1961, *Carvalho* 484 (K; LM). Z: Nhamarroi, fl. 13.v.1943, *Torre* 5491 (LISC).

Widespread in the tropics of the Old World. River banks and damp thickets in *Brachystegia*-evergreen shrub woodland.

3. **Biophytum petersianum** Klotzsch in Peters, Reise Mossamb. Bot. **1**: 81, t. 15

(1861).—Exell & Mendonça, C.F.A. **1**, 2: 264, t. 9 fig. D (1951).—Brenan in Mem. N.Y. Bot. Gard. **8,** 3: 231 (1953).—Keay, F.W.T.A. ed. 2, **1,** 1: 159, fig. 59 (1954).—Wilczek, F.C.B. **7**: 15, t. 1 (1958). TAB. **24** fig. B. Syntypes: Mozambique, Querimba, *Peters* (B†) and from Zanzibar.

Oxalis sensitiva sensu Oliv., F.T.A. **1**: 297 (1868) pro parte.

Oxalis sessilis Buch.-Ham. [ex Wall., Numer. List: 152, n. 4344 (1831) *nom. nud.*] ex Baill. in Bull. Soc. Linn. Par. **1**: 598 (1886) *nom. illegit.* Type from India.

Biophytum sessile Knuth in Engl., Pflanzenr. IV, 130: 406 (1930).—Delhaye in Inst. Roy. Col. Belg. **23**: 852, t. 1 fig. 11–14, 16–20, 24, 25, 31, 32 (1952). Type as for *O. sessilis*.

Biophytum rotundifolium Delhaye, tom. cit.: 855, t. 1 fig. 22, 23, 28, 29 (1952). Type from Katanga.

Annual herb 1–40 cm. in height with a simple erect appressed-pilose or appressed-pubescent or tomentose stem with a simple rosette of leaves at the apex. Leaves up to 5 cm. long; leaflets 3–10-jugate, the terminal pair usually $1\frac{1}{2}$ times as long as the next pair and the rest decreasing gradually in size, up to $10 \times 6{\cdot}5$ mm., the smallest c. 3 mm. in diam., chartaceous, opaque, obliquely elliptic or oblong-elliptic to subcircular, margin sparsely ciliate, otherwise glabrous, with up to 6 (8) pairs of rather thick prominent lateral nerves at right angles to the midrib; petiolule c. 0·5 mm. long. Flowers yellow or orange, with pedicels up to 2 mm. long, in 2–5-flowered congested pseudumbels; peduncle up to 9 cm. long (sometimes very short), pubescent or pilose. Sepals up to $6 \times 1{\cdot}5$ mm., lanceolate to linear-lanceolate, 3–7-nerved, sparsely puberulous, longer than the capsule. Petals $5\text{–}6 \times 1$ mm., lorate, $\pm$ coherent. Stamens with 5 longer ones 2 mm. long, 5 shorter ones 1 mm. long. Capsule c. $4 \times 3{\cdot}5$ mm. obovoid-ellipsoid. Seeds c. 0·4 mm. long, flattened-ellipsoid, minutely cuspidate, brownish.

N. Rhodesia. B: Zambezi, above Kale, fl. 27.iv.1925, *Pocock* 146 (PRE). N: Mporokoso Distr., Lake Mweru Wantipa, 1050 m., fl. & fr. 10.iv.1957, *Richards* 9137 (K). W: Matala, fl. & fr. 1931, *Farquhar* (BM). C: 16 km. E. of Lusaka, 1220 m., fl. 19.iii.1955, *Best* 72 (K). E: Lunkwakwa valley, fl. & fr. 23.iii.1955, *E.M. & W.* 1137 (BM; LISC; SRGH). S: Livingstone, 915 m., fr. iv.1909, *Rogers* 7035 (K). **S. Rhodesia.** N: Urungwe Distr., 610 m., fl. & fr. 26.ii.1958, *Phipps* 885 (K; SRGH). W: Matobo Distr., Besna Kobila, 1400 m., fr. v.1957, *Miller* 4354 (K; PRE; SRGH). C: Salisbury Distr., Hatfield, fl. 3.ii.1957, *Whellan* 1163 (K; SRGH). E: Vumba Mts., 1520 m., fl. & fr. 3. iv.1955, *Chase* 5534 (BM; PRE; SRGH). **Nyasaland.** C: Chitala escarpment, 950 m., fl. 12.ii.1959, *Robson* 1559 (K; SRGH). S: Malosa, fl. 10.iii.1955, *E.M. & W.* 789 (BM; LISC; SRGH). **Mozambique.** N: between Ribáuè and Lalaua, fl. & fr. 24.iv.1937, *Torre* 1343 (COI). Z: Quelimane, fl. & fr. 1863, *Kirk* (K). T: between Vila Mouzinho and R. Lobuè, fl. 19.vii.1949, *Barbosa & Carvalho* 3693 (LM; LMJ).

Widespread in tropical Africa, Madagascar and tropical Asia. In *Brachystegia*-woodland and often on cultivated land especially on sandy soils.

A concoction is used in Mozambique as a remedy for snake-bite.

4. **Biophytum crassipes** Engl., Pflanzenw. Ost-Afr. **C**: 226 (1895).—Knuth in Engl., Pflanzenr. IV, 130: 408, fig. 23 F–G (1930).—Delhaye in Bull. Inst. Roy. Col. Belg. **23,** 3: 858 (1952).—Wilczek, F.C.B. **7**: 19 (1958). TAB. **25** fig. A. Type from Tanganyika.

Biophytum ringoetii De Wild. in Fedde, Repert. **11**: 510 (1913).—Knuth, tom. cit.: 409 (1930).—Delhaye, loc. cit. Type from the Congo.

Acaulescent perennial herb up to 15 cm. in height with a thickened woody rootstock usually branched just at the apex and bearing c. 1–4 rosettes of leaves. Leaves 2–13 cm. long; leaflets 4–17-jugate, $2{\cdot}5\text{–}11 \times 2\text{–}7$ mm., obliquely subcircular to obovate or subreniform, ciliate otherwise glabrous, apex obtuse to rounded, base truncate, with 3–7 pairs of prominent lateral nerves forming an acute angle with the midrib; rhachis pilose. Flowers orange (yellow or reddish), with pilose pedicels up to 4 mm. long, in 3–6-flowered pseudumbels; peduncle 5–15 cm. long, pilose or nearly glabrous. Sepals $5\text{–}6 \times 1\text{–}1{\cdot}5$ mm., linear-lanceolate, acute, 5–7-nerved, pubescent. Petals 6–9 mm. long, glabrous. Stamens with 5 longer ones 2–2·5 mm. long and 5 shorter ones 1 mm. long. Ovary deeply 5-lobed; styles 3 mm. long, pilose. Capsule 3×2 mm., obovoid-ellipsoid, glabrous. Seeds 0·6 mm. in diam., subglobose, tuberculate, brown.

N. Rhodesia. N: Abercorn Distr., Lunzua R., 1520 m., fl. 28.iii.1955, *E.M. & W.* 1245 (BM; LISC; SRGH). W: Mwinilunga Distr., Luao R., fl. 1.xi.1937, *Milne-Redhead* 3035 (K). C: Broken Hill, 1220 m., fr. 17.xii.1907, *Kassner* 2022 (BM; K).

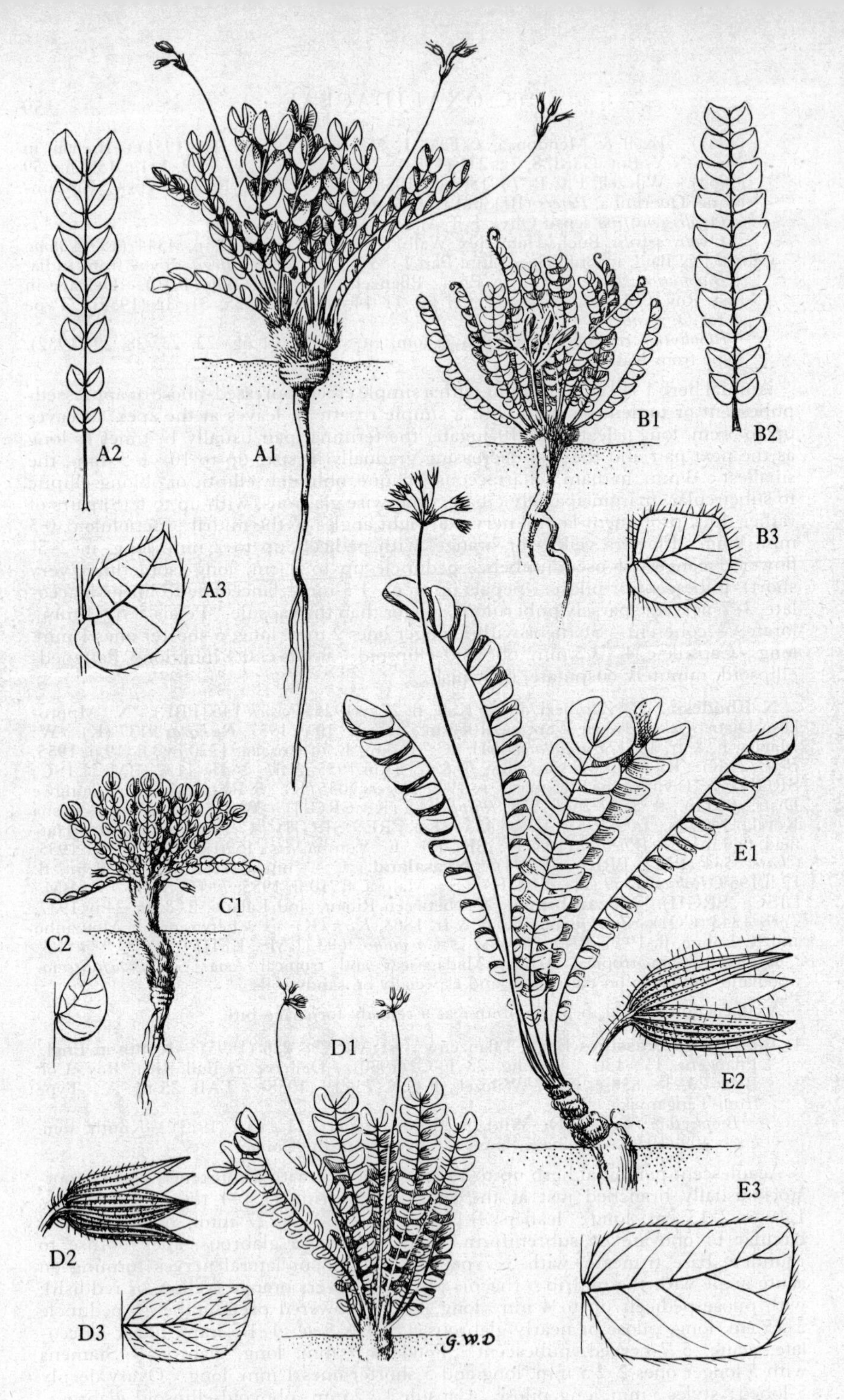

Tab. 25. A.—BIOPHYTUM CRASSIPES. A1, plant (× ½) ; A2, leaf (× 1) ; A3, leaflet (× 3), all from *Kassner* 2022. B.—BIOPHYTUM NYIKENSE. B1, plant (× ½) ; B2, leaf (× 1) ; B3, leaflet (× 3), all from *Robson* 207. C.—BIOPHYTUM MACRORRHIZUM. C1, plant (× ½) ; C2, leaflet (× 3), all from *Richards* 1116. D.—BIOPHYTUM KASSNERI. D1, plant (× ½) ; D2, calyx (× 4) ; D3, leaflet (× 3), all from *Fanshawe* 1967. E.—BIOPHYTUM RICHARDSIAE. E1, plant (× ½) ; E2, calyx, (× 4) ; E3, leaflet (× 3), all from *Richards* 11795.

E: 100 km. from Lundazi on road to Chama, 1350 m., fl. & fr. 19.x.1958, *Robson* 171 (BM; K; LISC; SRGH). **Nyasaland.** N: Mzimba Distr., base of Mt. Hora, 1370 m., fl. & fr. 11.i.1950, *Krippner & Smith* N. 15 (BM). C: near Tamanda Mission, 1200 m., fl. & fr. 8.i.1959, *Robson* 1092 (BM; COI; K; LISC; SRGH). **Mozambique.** N: Massangulo, fr. iii.1933, *Gomes e Sousa* 1334 (COI).

Also in the Congo and Tanganyika. In grassland, scrub and swamps at 1200–1550 m.

5. **Biophytum nyikense** Exell in Bol. Soc. Brot., Sér. 2, **35**: 137 (1961). TAB. **25** fig. B. Type: N. Rhodesia, Nyika Plateau, *Benson* NR 145 (BM).

Acaulescent perennial herb up to 10 cm. in height with an unbranched thickened woody rootstock with a single rosette of leaves ± at ground level. Leaves 2–8 cm. long; leaflets 4–10-jugate, 2–8 × 2–7·5 mm., obliquely subcircular, ciliate otherwise glabrous, apex rounded, base truncate; rhachis pubescent and pilose. Flowers yellow, with glandular-pubescent pedicels up to 5 mm. long, in 3–8-flowered pseudumbels; peduncle 1–3·5 cm. long, glandular-pubescent. Sepals 6–7 × 1–1·5 mm., narrowly lanceolate, acute, 6–8-nerved, glandular-pubescent. Petals 7–8 × 4–4·5 mm., spathulate, at first free, later ± coherent, glabrous. Stamens with 5 longer ones 2·5 mm. long and 5 shorter ones 1 mm. long. Ovary 5-lobed; styles 5 mm. long, pilose. Seeds 1·3 × 0·7 mm., flattened-ellipsoid, brown.

N. Rhodesia. E: Nyika Plateau, 3·2 km. SW. of Rest House, 2150 m., fl. & fr. 21.x.1958, *Robson* 207 (BM; K; LISC; SRGH). **Nyasaland.** N: Nyika Mts., 2130–2440 m., fl. xi.1932, *Sanderson* 70 (BM).

Also in Tanganyika. Upland grassland up to 2450 m.

6. **Biophytum richardsiae** Exell in Bol. Soc. Brot., Sér. 2, **35**: 138 (1961). TAB. **25** fig. E. Type: N. Rhodesia, Saisi Valley, *Richards* 11795 (BM; K, holotype).

Acaulescent perennial herb up to 20 cm. in height with an unbranched hairy thickened woody rootstock. Leaves up to 16 cm. long; leaflets 8–16-jugate, almost sessile, 5–11 × 2·5–7 mm., oblong to broadly oblong or more rarely somewhat ovate-oblong or obovate-oblong, apex rounded, base truncate, sparsely ciliate otherwise glabrous, with 5–7 pairs of lateral nerves; rhachis pubescent to pilose. Flowers dull pink with slender pilose pedicels up to 8 mm. long, in 10–16-flowered pseudumbels; peduncle up to 18 cm. long, pilose. Sepals 6–6·5 × 1·5–1·8 mm., reddish, narrowly elliptic, 5–7-nerved, pubescent. Petals c. 8 mm. long, glabrous. Stamens with 5 longer ones 4·5 mm. long and 5 shorter ones 2·5 mm. long. Ovary 5-lobed; styles 1·9 mm. long, almost glabrous. Capsule 3 × 2·5 mm., broadly ellipsoid. Seeds 0·8 × 0·6–0·7 mm., brown, subglobose to ellipsoid, tuberculate.

N. Rhodesia. N: Abercorn Distr., Chisau River Gorge, Saisi Valley, 1500 m., fl. & fr.. 18.xi.1959, *Richards* 11795 (BM; K).

Known only from the locality cited. Cliff ledges.

Nearest to *B. kassneri* Knuth, from the Congo, N. Rhodesia and Tanganyika (*Richards* 8626 (K)), but differing in having longer sepals (6–6·5 mm. compared with 4–4·5 mm.) and leaflets which are glabrous (appressed-pubescent in *B. kassneri*) except for the sparsely ciliate margins.

7. **Biophytum kassneri** Knuth in Engl., Pflanzenr. IV, 130: 409 (1930).—Delhaye in Inst. Roy. Col. Belg. **23**: 857 (1952).—Wilczek, F.C.B. **7**: 18 (1958). TAB. **25** fig. D. Type from the Congo (Katanga).

Acaulescent perennial up to 20 cm. in height with a hairy thickened woody rootstock usually branched at the apex. Leaves up to 17 cm. long; leaflets 8–16-jugate, almost sessile, 2–9 × 2–4 mm., usually oblong or obovate-oblong, apex rounded, base truncate, usually fairly densely pubescent, somewhat glabrescent when old, with 5–7 pairs of lateral nerves; rhachis pubescent to shortly pilose. Flowers white, pink or yellowish, with pubescent pedicels up to 3 mm. long, in 3–10-flowered pseudumbels; peduncle up to 11 cm. long, pubescent. Sepals 3–4·5 × 1–1·5 mm., lanceolate to oblong-lanceolate, acute, 3–5-nerved, pubescent. Petals 7 mm. long, glabrous. Stamens with 5 longer ones 3–4 mm. long and 5 shorter ones 1–1·5 mm. long. Ovary 5-lobed; styles 1·8 mm. long, almost glabrous. Capsule 2·5 × 1·2 mm., ellipsoid. Seeds 0·8 × 0·5–0·6 mm., brown, flattened-ellipsoid, tuberculate.

N. Rhodesia. N: Mpika, fl. & fr. 4.ii.1955, *Fanshawe* 1967 (K).

Tanganyika, Katanga and N. Rhodesia. On granite hills.

8. **Biophytum macrorrhizum** R.E. Fr., Schwed. Rhod.-Kongo-Exped **1**: 108, t. 6 fig. 5–6 (1914).—Knuth in Engl., Pflanzenr. IV, 130: 409 (1930). TAB. **25** fig. C. Type: N. Rhodesia, Msisi, *Fries* 1307 (UPS, holotype).

Acaulescent perennial herb up to 8 cm. in height with a branched thickened woody rootstock. Leaves up to 4 cm. long; leaflets 5–10-jugate, almost sessile, 1·5–4·5 × 1–2·8 mm., obliquely elliptic, apex obtuse or rounded, base truncate, sparsely ciliate otherwise glabrous, with 3–5 pairs of rather prominent lateral nerves; rhachis pubescent to pilose. Flowers orange, with pilose pedicels c. 7 mm. long, in sessile pseudumbels and thus appearing solitary among the leaves of the dense rosette. Sepals 5 × 1 mm. 5-nerved, pubescent. [Remaining organs not investigated owing to paucity of material available.]

N. Rhodesia. N: Abercorn Distr., Chilongowelo, 1220 m., fr. 18.iii.1952, *Richards* 1116 (K).

Also in Tanganyika. In woodland.

39. BALSAMINACEAE

By E. Launert*

Herbs, often with a suffrutescent habit, sometimes epiphytic, sometimes aquatic, glabrous or with an indumentum of simple hairs; stems herbaceous, succulent or rarely woody. Leaves simple, petiolate or sessile, verticillate or opposite or spirally arranged, pinnately veined with margins crenulate or serrate or denticulate and often with glanduliferous (tentacle-like) hairs; exstipulate or with stipular glands. Inflorescences usually axillary, more rarely of terminal racemes or pseudumbels or fascicles, or flowers solitary. Flowers bisexual, sometimes cleistogamous (often in certain species with both normal and cleistogamous ones in the same inflorescence), always zygomorphic. Sepals 3, rarely 5, deciduous, imbricate; the posterior one (apparently anterior in position by resupination) petaloid, large, funnel-shaped, obliquely navicular or bucciniform, nearly always with a nectariferous spur; the lateral ones usually small, green or coloured. Petals 5; the anterior one (apparently posterior in position in the mature flower) always large, flat or helmet-shaped, often dorsally carinate; the 4 lower ones usually connate in lateral pairs, very rarely free, usually deeply 2-lobed, rarely with the posterior lobe reduced to a small auricle. Stamens 5, alternating with the petals; filaments flattened, connate above, closely attached to the ovary; anthers 2-thecous, connate or coherent, thus forming a crown round the ovary. Ovary superior, of 5 carpels, 5-locular, with axile placentation; ovules pendulous, anatropous, 3–many in a row on each placenta; style 1, usually very short; stigmas 1–5. Fruit a 5-valved loculicidal capsule, with the valves opening elastically and coiling, rarely a berry (in the Asiatic genus *Hydrocera*). Seeds without endosperm; embryo straight.

A large family of 2 genera and more than 500 described species. Widely distributed mainly in the tropics and extending into the subtropics of the Old World, abundant in Asia and Africa, with only a few species in the temperate regions of Europe, Asia and America.

IMPATIENS L.

Impatiens L., Sp. Pl.: **2**: 937 (1753); Gen. Pl. ed. 5: 403 (1754).

Annual or perennial herbs, sometimes shrubby; stems usually succulent, rarely woody, glabrous or pilose. Leaves verticillate or opposite or spirally arranged. Inflorescences axillary and/or of terminal simple racemes or pseudumbels or fascicles, or the flowers solitary. Flowers sometimes cleistogamous (rarely both cleistogamous and normal ones in the same inflorescence), pedicellate. Sepals, petals and

* I am much indebted to my friend G. M. Schulze of the Berlin Herbarium for some advice in the taxonomy of this difficult group.

sexual organs as for the family. Fruit glabrous or pilose or sometimes tomentose. Seeds smooth or tuberculate or rugose, glabrous or pilose.

A genus of about 500 described species, of which 150 occur in Africa. Distribution as for the family.

Flowers in axillary and/or terminal racemes or, if solitary, then always with a bract about the middle of the pedicel; peduncle up to 6 cm. long or more:
 Leaves opposite; bracts and lateral sepals with 1 or 2 (rarely 3) pairs of spreading tentacle-like appendages on the margin - - - - - 1. *briartii*
 Leaves spirally arranged; bracts and lateral sepals without marginal appendages:
  Tall shrubby plant, up to 2·2 m. high or more; lateral united petals with the lobes very different in shape and size - - - - - - - 2. *prainiana*
  Juicy perennial herb, 30–50(80) cm. high or more; lateral united petals with lobes ± equal in shape and size (see t. 26 fig. 25) - - - - - 3. *wallerana*
Flowers solitary or in fascicles; peduncle scarcely exceeding 6 mm.:
 Superior sepal bucciniform (see t. 27 fig. 3; t. 26 fig. 1):
  Spur abruptly recurved (see t. 26 fig. 1), with end somewhat thickened and sometimes inconspicuously bifid; petiole with stipular glands at the very base
   4. *gomphophylla*
  Spur gradually recurved (see t. 27 fig. 3), with the end subacute, never thickened nor bifid; petiole eglandular - - - - - - - - - 5. *salpinx*
 Superior sepal shortly funnel-shaped or boat-shaped:
  Spur shorter than, rarely as long as, or very rarely slightly longer than the upper margin of the superior petal (see t. 26 figs. 2–4):
   Leaves opposite or rarely in whorls of three - - - - - 6. *assurgens*
   Leaves spirally arranged:
    Ovary and capsule tomentose - - - - - - - - 7. *eriocarpa*
    Ovary and capsule glabrous or very sparsely hairy:
     Leaves sessile or very shortly petiolate - - - - - 8. *brachycentra*
     Leaves distinctly petiolate; petiole 1–6 cm. long:
      Leaf-lamina glabrous; petiole 1–2·5(3) cm. long; seeds smooth
       9. *shirensis*
      Leaf-lamina pilose on the venation beneath; petiole (1)2·5–6 cm. long; seeds with a wrinkled surface - - - - - - - 10. *quisqualis*
  Spur always longer (usually several times) than the upper margin of the superior sepal (see t. 26 figs. 5–7):
   Ovary pilose:
    Lateral united petals 8–14 mm. long; the lobes nearly equal in size (see t. 26 fig. 17); lateral sepals usually with long hairs along the margins; spur 11–15(18) mm. long, not much longer than the whole flower 11. *sylvicola*
    Lateral united petals 18–22 mm. long; the posterior lobe ⅓ of the anterior one in size (or even smaller) (see t. 26 fig. 14); lateral sepals without long hairs along the margins (often quite glabrous); spur up to 25(30) mm. long, much longer than the rest of the flower - - - - - - - 12. *limnophila*
   Ovary glabrous:
    Posterior lobe of the lateral united petals reduced to a small auricle (sometimes nearly obsolete) (see t. 26 fig. 19), anterior lobe distally extended to a ligulate appendage at the very base - - - - - 13. *duthieae*
    Posterior lobe of the lateral united petals always well developed; anterior lobe not produced into an appendage:
     Posterior lobe of the lateral united petals apically acute and produced into a filiform appendage (see t. 26 figs. 20–23) - - - - 16. *cecilii*
     Posterior lobe of the lateral united petals rounded or sometimes subacute but never with a filiform appendage:
      Flowers small; lateral united petals 5–10(13) mm. long; anterior petal 3·5–6 mm. long:
       Leaves sessile; posterior lobe of the lateral united petals truncate, with the upper edge usually slightly retuse (see t. 26 fig. 8); seeds with a papillose surface - - - - - - - - 20. *oreocallis*
       Leaves always petiolate; posterior lobe of the lateral united petals usually entire and rounded (see t. 26 fig. 15); seeds densely covered with very short thick conical multicellular hairs - - 21. *zombensis*
      Flowers larger; lateral united petals (13)17–33 mm. long; anterior petal (6)8–15 mm. long:
       Lateral united petals with the posterior lobe broadly obcuneate and with the upper margin retuse to emarginate (see t. 26 fig. 11); leaves sessile - - - - - - - - - - 18. *psychantha*
       Lateral united petals with the posterior lobe rounded and entire; leaves sessile or petiolate:
        Anterior petal dorsally broadly saccate at the base 19. *hydrogetonoides*

Anterior petal not saccate at the base:
Posterior lobe of the lateral united petals $\frac{1}{6}$–$\frac{1}{3}$ the size of the anterior one:
Pedicels, lateral sepals and posterior sepal (including spur) pilose; posterior lobe of the lateral united petals $\frac{1}{6}$–$\frac{1}{5}$ the size of the anterior one; younger parts of the plant densely hairy 17. *psychadelphoides*
Pedicels, lateral sepals and posterior sepal (including spur) glabrous; posterior lobe of the lateral united petals c. $\frac{1}{3}$ the size of the anterior one; younger parts of the plant glabrous or very sparsely pilose - - - 19. *hydrogetonoides*
Posterior lobe of the lateral united petals equalling the anterior one or at least not much different in size:
Leaves sessile or rarely shortly petiolate; lobes of the lateral united petals always distinctly overlapping (see t. 26 fig. 13), usually not much longer than broad - - 22. *irvingii*
Leaves always distinctly petiolate; lobes of the lateral united petals separated by a sinus, touching at their margins, or rarely slightly overlapping, usually longer than broad:
Anterior petal 6–10 mm. long; lateral united petals (10)13–18 mm. long; posterior sepal longitudinally striate (always?) 21. *zombensis*
Anterior petal 11–14 mm. long; lateral united petals 17–25 mm. long:
Lateral sepals glabrous; lateral united petals 17–20 mm. long with the anterior lobe distally rounded (see t. 26 fig. 12) 15. *eryaleia*
Lateral sepals pilose, at least at the margins; lateral united petals 22–25 mm. long with the anterior lobe distally somewhat drawn out at the very base thus forming a subacute point (see t. 26 fig. 10) - - - 14. *schulziana*

1. **Impatiens briartii** De Wild. & Dur. in Bull. Soc. Roy. Bot. Belg. **38,** 2: 185 (1899).—Gilg in Engl., Bot. Jahrb. **43**: 104 (1909).—De Wild. in Ann. Soc. Sci. Brux., Sér. B. **38,** 2: 18 (1914); in Contr. Fl. Katanga: 119 (1921).—G.M. Schulze in Fedde, Repert. **39**: 21–22 (1935); in Bol. Soc. Brot., Sér. 2, **39**: 12–14 (1955); in Exell & Mendonça, C.F.A. **2**: 156 (1956).—Wilczek & Schulze, F.C.B. **9**: 405 (1960). TAB. **26** fig. 18. Type from the Congo.

Weak perennial herb; stems 0·3–0·8(1·3) m. high, erect or decumbent and rooting at the lower nodes, simple or more often laxly branched (mainly in the upper parts), glabrous or rarely with the younger parts sparsely pilose, succulent, reddish or dark green. Leaves opposite, stipulate, petiolate; lamina (3)4–8 × 1·2–3 cm., lanceolate or ovate-lanceolate to lanceolate-oblong, rarely ovate, chartaceous to papyraceous, somewhat olive-green above, glaucous or pale silvery below, when young somewhat pilose (mainly below) or glabrous, when older usually glabrous but sometimes somewhat pilose (mainly on the midrib underneath), apex acute or subobtuse, margin crenate usually with small teeth in the sinuses, base rounded or broadly cuneate or very rarely subcordate; secondary nerves 6–8 pairs, very fine, somewhat prominent underneath; petiole (0·2)0·6–1·7(2·3) cm. long, slender, canaliculate, pilose or glabrous; stipules up to 5 mm. long, triangular-subulate, very slender. Flowers in few- to many-flowered axillary and/or terminal racemes, white or pink or pale mauve, usually with 2 lateral red spots at the throat of the posterior sepal and with small white spots in front of them (the spur often bright yellow edged with red); pedicels 2–12 mm. long, very slender, usually glabrous; bracts 1·5–4 mm. long, oblong-obovate, apically obtuse and somewhat thickened (glandular tip?) with 1 or 2 (rarely 3) pairs of 1–3 mm. long spreading filiform appendages on the margin. Lateral sepal 3–5 × c. 2 mm., ovate to ovate-oblong, somewhat oblique, with the apex obtuse and slightly thickened (glandular tip?), with 1 or 2 (rarely 3) pairs of spreading filiform appendages on the margin in the lower half, glabrous; posterior sepal 7–8·5 mm. long, 2·5–4 mm. deep, abruptly contracted into the spur, distally acute, rarely subcaudate; spur 10–14(17) mm. long, slightly incurved, usually cylindric, with the end acute or bluntish. Anterior petal 4–6 mm. long, c. 5 mm. broad (when flattened), dorsally not cristate, glabrous; lateral united petals (10)12–14(17) mm. long, deeply 2-lobed, with a very long finely filiform appendage (inserted just below the sinus of the lobes) which extends into the spur of the posterior sepal at the base (see t. 26 fig. 18); anterior

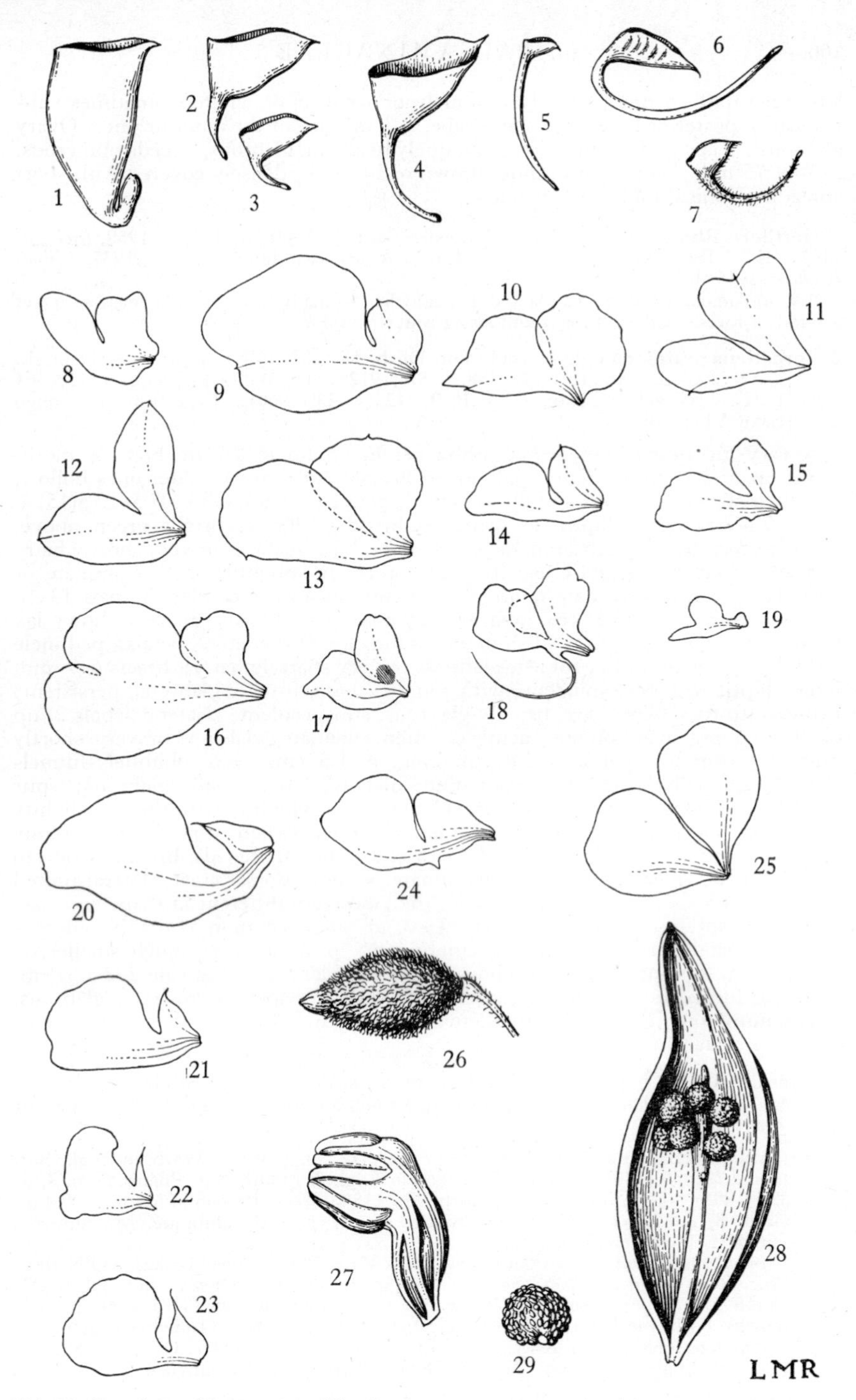

Tab. 26. Posterior sepals of IMPATIENS (all ×1): 1. I. GOMPHOPHYLLA; 2 and 3. I. ASSURGENS; 4. I. ERIOCARPA; 5. I. OREOCALLIS; 6. I. CECILII subsp. CECILII; 7. I. SYLVICOLA. Lateral petals of IMPATIENS (all ×1): 8. I. OREOCALLIS; 9. I. PSYCHADELPHOIDES; 10. I. SCHULZIANA; 11. I. PSYCHANTHA; 12. I. ERYALEIA; 13. I. IRVINGII; 14. I. LIMNOPHILA; 15. I. ZOMBENSIS; 16. I. ERIOCARPA; 17. I. SYLVICOLA; 18. I. BRIARTII; 19. I. DUTHIAE; 20. I. CECILII subsp. GRANDIFLORA; 21–23. I. CECILII subsp. CECILII; 24. I. ASSURGENS; 25. I. WALLERANA. 26. Capsule of I. ERIOCARPA (×2); 27. Androecium surrounding the gynoecium of I. GOMPHOPHYLLA (×3); 28. Open mature capsule of I. GOMPHOPHYLLA (×3); 29. Seed of I. GOMPHOPHYLLA (×3).

lobe up to 10 × 8 mm., usually obliquely obovate or obdeltate, sometimes subtruncate; posterior lobe similar in shape, ½–⅓ the size of the anterior one. Ovary glabrous. Capsule 5–10 mm. long, obliquely fusiform, glabrous. Seeds numerous, c. 3 × 1·75 mm., elliptic in outline, brown or blackish, densely covered with short spreading conical multicellular hairs.

Northern Rhodesia. N: Abercorn Distr., Saisi R., 1500 m., fl. 15.iv.1959, *Richards* 11232 (K; LISC; SRGH). W: near Lunga R. at Mwinilunga, fl. 29.xi.1937, *Milne-Redhead* 3445 (K).

Also in Angola, Congo, Uganda and Tanganyika. Usually forming large masses on wet ground in forests, gallery forests and along water courses.

2. **Impatiens prainiana** Gilg in Engl., Bot. Jahrb. **43**: 127 (1909); in Mildbr., Deutsch. Z.-Afr. Exped. 1907–1908, **2**: 488, t. 57 (1912).—De Wild., Pl. Bequaert. **1**: 364 (1922).—Wilczek & Schulze, F.C.B. **9**: 422, t. 43 (1960). Type from the Congo (Lake Albert Distr.).

A very tall much-branched shrubby perennial, up to 2·2 m. high or more; stems up to 3·8 cm. in diam., pale green flecked with brown, older ones hollow, usually glabrous. Leaves spirally arranged, petiolate; lamina (5·5)7·5–12·5(15) × (3·5)5–7·5(8·5) cm., elliptic to ovate or broadly elliptic, bright green above, greyish-green below, membranous, glabrous on both surfaces or very shortly hairy beneath when young, apex usually acute or rarely subobtuse and acuminate or cuspidate, margin crenulate-denticulate, base cuneate; secondary nerves 13–18 pairs; petiole (1)2–5(6·5) cm. long, usually glabrous. Flowers in very showy lax open racemes, white with reddish marks at the base of the lateral petals; peduncle 7–18(25) cm. long, rather robust, glabrous or very sparsely hairy; bracts 6–9 mm. long, elliptic or ovate, apiculate, with rounded base, usually glabrous, persistent; pedicels up to 3·3 cm. long, usually glabrous, subsucculent. Lateral sepals 2, up to 10 × 4 mm., ovate-oblong, acute or often apiculate, glabrous or very shortly pilose; posterior sepal 2·25–2·8 cm. long, c. 1·5 cm. deep, obliquely-funnel-shaped, gradually tapering into the spur, distally acute, usually glabrous; spur 7–10(13) cm. long, relatively slender, subconical or cylindric, with the end slightly thickened, usually incurved, sometimes recurved (when very young), rarely straight. Anterior petal 1·3–1·7 × 1·4–1·6 cm. (when flattened), broadly ovate to semi-circular, apically entire and acuminate, without dorsal crest; lateral united petals 3·5–4·5 cm. long, deeply 2-lobed, the lobes very different in shape and size, glabrous; anterior lobe large, up to 3 × 3 cm. or even more, broadly oblique-obovate, distally usually retuse to emarginate; posterior lobe much smaller, c. 7·5 × 6 mm., obovate-oblong to obovate. Ovary glabrous. Capsule 2–2·5(3) cm. long, fusiform, broader in the upper half, with the apex ± rostrate, glabrous. Seeds numerous, c. 3·5 × 1·8 mm., ovate-oblong in outline, with a wrinkled or papillose surface, reddish-brown.

Nyasaland. N: Nyika Plateau, fl. 2.i.1959, *Richards* 10395 (K; SRGH).

Congo and Nyasaland. In damp places in montane forest, beside water courses or along roads in forests.

3. **Impatiens wallerana** Hook. f. in Oliv., F.T.A. **1**: 302 (1868).—Warb. in Engl., Bot. Jahrb. **22**: 49 (1895).—Warb. & Reiche in Engl. & Prantl, Nat. Pflanzenfam. **3,** 5: 392 (1895).—Gilg in Engl., Bot. Jahrb. **43**: 118 (1909).—Brenan in Mem. N.Y. Bot. Gard. **8,** 3: 232 (1953). TAB. **26** fig. 25. Type: Mozambique, Morrumbala, viii.1864, *Waller* (K, holotype).

Impatiens sultani Hook. f. in Curt., Bot. Mag., **108**: t. 6643 (1882).—O'Neill in Proc. Roy. Geogr. Soc., N. Ser. **7**: 454 (1885).—Warb. in Engl., Bot. Jahrb. **22**: 48 (1895). Type: description based on a plant grown at Kew Gardens from seeds sent by Kirk, the latter probably collected on the Mamboia (Zanzibar coast) hills.

Impatiens holstii Engl. & Warb. in Engl., Pflanzenw. Ost-Afr. **C**: 254 (1895).—Warb. in Engl., Bot. Jahrb. **22**: 48 (1895). Type from Tanganyika.

Juicy perennial herb, 30–50(80) cm. high, usually branched to form a flat top; stems erect or straggling, succulent, thick, glabrous or rarely scattered-pilose, usually translucent, green or reddish, sometimes rooting at lower nodes. Leaves spirally arranged, very rarely apparently opposite, petiolate; lamina very variable in shape and size, (3)4·5–10(15) × (2)2·5–5(7·5) cm., broadly elliptic or ovate to elliptic-oblong or ovate-oblong, subsucculent, usually translucent, dark or pale green to reddish-green on both surfaces, glabrous on both surfaces or rarely scattered-pilose beneath (mainly on the nerves), apex acute, often acuminate or cuspidate, rarely obtuse, margin crenate-denticulate, often with tentacular hairs on the apex of the crenations towards the base, base usually narrowly (but sometimes broadly) cuneate or rarely rounded and usually with the lamina running down the petiole and thus forming a narrow wing; secondary nerves 7–12 pairs, not very prominent beneath; petiole (1)2–4·5(6·5) cm. long, succulent, usually glabrous. Flowers in few-flowered axillary and/or terminal racemes, rarely apparently 1-flowered by reduction (but then the " pedicel " always with a small bract about the middle), very variable in colour and size, dark carmine, wine-red, pink, orange, violet, rarely white; bracts up to 6 mm. long, lanceolate-triangular, acute; pedicel 1–2·5(3) cm. long, slender, succulent, usually glabrous. Lateral sepals (3)4–7 × 2·25–3 mm. (in cultivated specimen sometimes slightly larger), ovate-lanceolate or lanceolate-triangular, acute, glabrous (always?), green; posterior sepal relatively small, up to 1·5 cm. long, c. 0·5 cm. deep, shallowly navicular, bluntish-keeled, abruptly constricted into the spur, distally sometimes subcaudate, usually acute or subobtuse; spur up to 4 cm. long, slender, slightly curved, usually with the end somewhat thickened, glabrous. Anterior petal up to 12 mm. long, up to 18 mm. broad (when flattened), very broadly ovate or obcordate, with the apex usually retuse to emarginate, rarely acute, dorsally very narrowly cristate; lateral united petals very deeply 2-lobed (the lobes appearing nearly separate), with the lobes nearly equal in shape and size, 10–20 × 7–15 mm. (sometimes even larger), obliquely obovate, obovate-spathulate or obovate-cuneate, with the apex rounded or subtruncate (see t. 26 fig. 25). Ovary usually glabrous. Capsule up to 18 mm. long, broadly oblique-fusiform, glabrous. Seeds numerous, c. 1·75 × 1·5 mm., pyriform, rusty or blackish, ± densely covered with short thick hairs on papillose bases.

S. Rhodesia. E: Chirinda, fl. 28.iii.1950, *Hack* 131/50 (SRGH). **Nyasaland.** S: Chola Distr., Chola Mt., 1200 m., fl. 24.ix.1946, *Brass* 17779 (SRGH). **Mozambique.** Z: Morrumbala Mt., E. hillside, R. Inhambondua, fl. & fr. 27.x.1945, *Pedro* 485 (LMJ; PRE; SRGH). MS: Manica, Moribane, R. Furozi, fl. & fr. 30.x.1953, *Pedro* 4169 (LMJ).

Kenya, Zanzibar, Tanganyika, Nyasaland, S. Rhodesia and Mozambique. Frequent in rain-forest gullies, along water courses in forests, in damp shady places, on decaying tree trunks and on rocky river banks.

4. **Impatiens gomphophylla** Bak. in Kew Bull. **1895**: 64 (1895).—Warb. in Engl., Bot. Jahrb. **22**: 47 (1895) in annot.—Gilg in Engl., Bot. Jahrb. **43**: 104 (1909).—De Wild. in Ann. Mus. Cong. Belg., Bot. Sér. 4, **2**: 107 (1913); in Ann. Soc. Sci. Brux., Sér. 37, **2**: 83 (1913); in Contr. Fl. Katanga: 119 (1921).—Wilczek & Schulze, F.C.B. **9**: 406 (1960). TAB. **26** fig. 1, 27–29. Type from Tanganyika.

Impatiens verdickii De Wild. in Ann. Mus. Cong. Belg., Bot. Sér. 4, **1**: 84, t. 29 fig. 1–9 (1903); Contr. Fl. Katanga: 120 (1921).—Th. & H. Dur., Syll. Fl. Cong.: 80 (1909).—Gilg in Engl., Bot. Jahrb. **43**: 105 (1909). Type from Katanga.

Impatiens homblei De Wild. in Bull. Jard. Bot. Brux. **5**: 39 (1915); in Ann. Soc. Sci. Brux. **40**: 102 (1921); Contr. Fl. Katanga: 119 (1921). Type from Katanga.

Annual (always?) herb, up to 1·2 m. high; stems simple or moderately branched, erect or semi-prostrate, succulent, glabrous or sometimes somewhat pubescent mainly on younger parts. Leaves spirally arranged, very rarely pseudopposite, petiolate; lamina 3–10(14) × 1–2·5(4) cm., lanceolate-elliptic, ovate-lanceolate or very rarely obovate-lanceolate, usually dark green on the upper surface, greyish-green on the lower one, glabrous above, shortly pilose or rarely glabrous beneath, apex acute, often acuminate, margin serrate-denticulate, base cuneate; secondary nerves 6–10 pairs, ± prominent beneath; petiole 0·5–3(4) cm. long, rather slender, usually glabrous, with stipular glands at the very base. Flowers 2–6 in axillary fascicles, very rarely solitary, green-yellow tipped orange-red, scarlet shaded yellow, or orange; peduncle very short, not exceeding 4 mm.; bracts up to 4 mm. long, lanceolate-triangular, pilose or glabrous; pedicels 1·5–2·5(4) cm. long, rather slender, glabrous or sometimes pubescent. Lateral sepals 4–7 × 1–2(2·5) mm., ± asymmetrically lanceolate, triangular-lanceolate to ovate-lanceolate, usually glabrous; posterior sepal 7–9 mm. in diam. at the mouth, 1·5–2(2·4) cm. deep, bucciniform, gradually tapering into the spur; spur up to 8 mm. long, sharply recurved (see t. 26 fig. 1), medium thick, ± cylindric, the end somewhat thickened, entire or very rarely inconspicuously bifid. Anterior petal 7–12 mm. long, helmet-shaped, dorsally distinctly and broadly cristate; lateral united petals (7)8–10(12) mm. long, distinctly 2-lobed; anterior lobe c. 4·5–2·5 mm., obliquely elliptic to obovate-oblong; posterior lobe c. 7 × 4 mm., obliquely and broadly ovate. Ovary glabrous. Capsule 1·5–2 cm. long, fusiform, usually glabrous. Seeds numerous, c. 2 mm. in diam., subglobose, glabrous, rugose (see t. 26 fig. 29).

N. Rhodesia. N: Kali Dambo, fl. 17.iii.1955, *Richards* 4997 (K; SRGH). **Nyasaland.** N: Vipya Plateau, fl. iii.1948, *Benson* 1468 (BM). S: Blantyre Distr., fl. & fr. 27.i.1956, *Jackson* 1818 (K). **Mozambique.** N: Vila Cabral, fl. & fr. 1.vi.1934, *Torre* 165 (COI; LISC).

N. Rhodesia, Nyasaland, Mozambique, Tanganyika and the Congo. Growing in dambos, in dense vegetation on river banks or wet places on road-sides, often gregarious.

5. **Impatiens salpinx** Schulze & Launert in Bol. Soc. Brot., Sér. 2, **36**: 63 (1962). TAB. **27**. Type: Mozambique, Manica e Sofala, *Phipps & Goodier* 353 (SRGH, holotype).

Perennial herb, often with a shrubby habit, erect or decumbent, up to 70 cm. high or more; main stems often woody, terete or indistinctly quadrangular, with a dark brown somewhat rimose bark, glabrous, younger branches somewhat pilose but soon glabrescent, striate, yellowish-brown. Leaves spirally arranged, petiolate; lamina 2·5–5·5 × 1·6–2·8 cm., ovate, ovate-lanceolate or elliptic, usually glabrous on both surfaces, rarely the younger leaves scattered-pilose but soon glabrescent, bright green on the upper, greyish-green on the lower surface, apex acute, margins crenulate-denticulate, base broadly to narrowly cuneate herbaceous; secondary nerves 6–8 pairs; petiole (0·3)0·6–1(1·7) cm. long, rather slender, glabrous, eglandular. Flowers solitary, axillary, deep red; bracts up to 5 mm. long, lanceolate, acute, glabrous; pedicel 1·5–3·5 cm. long, very slender, glabrous. Lateral sepals up to 6 × 2·2 mm., oblong-elliptic, apically finely caudate, glabrous or rarely finely puberulous; posterior sepal c. 6 mm. in diam. at the mouth, 2·3–3 cm. long, bucciniform, tapering into the spur, usually glabrous; spur medium-thick, recurved, with the end slightly thickened and sometimes very shortly bifid. Anterior petal c. 6·5 × 7 mm. (when flattened), cucullate, apically produced into a short appendage, dorsally cristate; lateral united petals up to 10 mm. long, deeply 2-lobed, usually glabrous; anterior lobe c. 8 × 4·5 mm., broadly oblique-elliptic to ovate-elliptic, entire; posterior lobe c. 5 mm. long, 2–3 mm. high, very broadly oblique-obovate to transversely semi-elliptic. Ovary glabrous. Capsule not known in a mature state. Seeds numerous.

S. Rhodesia. E: Chimanimani Mts., fl. 12.x.1950, *Chase* 2953 (BM; SRGH). **Mozambique.** MS: Chimanimani Mts., 1550 m., fl. 30.xii.1959, *Phipps & Goodier* 353 (SRGH).

Not known from elsewhere. Growing in damp grass, among ferns, in shade along river sides, on water edges, in well-lit areas in kloofs, on wet forest ground or on forest edges.

6. **Impatiens assurgens** Bak. in Kew Bull. **1895**: 64 (1895).—Warb. in Engl., Bot. Jahrb. **22**: 47 (1895) in adnot.—Gilg in Engl., Bot. Jahrb. **43**: 99 (1909).—De Wild. in Ann. Mus. Cong. Bot., Sér. 4, **2**: 107 (1913); in Contr. Fl. Katanga: 119 (1921). —Brenan in Mem. N.Y. Bot. Gard. **8,** 3: 231 (1953).—G. M. Schulze in Notizbl. Bot. Gart. Berl. **12**: 74 (1935); in Bol. Soc. Brot., Sér. 2, **29**: 10 (1955); in Exell & Mendonça, C.F.A. **2**: 156 (1956). TAB. **26** figs. 2–3. Type from Lake Tanganyika.

Impatiens sweertioides Warb. in Engl., Bot. Jahrb. **22**: 49 (1895).—Gilg in Engl., Bot. Jahrb. **43**: 98 (1909). Type from Angola.

Impatiens gratioloides Gilg, op. cit. **30**: 350 (1901). Type from Tanganyika (Uhehe).

Impatiens katangensis De Wild., Études Fl. Katanga: 82 (1903).—Gilg, op. cit. **43**: 99 (1909). Type from Katanga.

Impatiens bussei Gilg, op. cit. **43**: 98 (1909). Type from tropical E. Africa.

Impatiens jodotricha Gilg, loc. cit. Type from Tanganyika.

Impatiens katangensioides De Wild. in Ann. Soc. Sci. Brux. **40,** 2: 103 (1921). Type from the Congo.

Annual or perennial juicy herb (5)10–60(80) cm. high or more, usually erect; stems simple or little branched, glabrous or pubescent mainly on younger parts. Leaves opposite, rarely in threes, sometimes ± densely crowded on the upper part of the stem, sessile or shortly petiolate; lamina (1·5)2·5–7(10) × (0·3)0·8–2(3) cm., linear, lanceolate, ovate or elliptic to (rarely) subcircular, the lower ones often much broader than long and shorter than the higher ones, chartaceous or thinly chartaceous, usually glabrous but sometimes pubescent on both surfaces, dark green

Tab. 27. IMPATIENS SALPINX. 1, upper part of plant with leaves and flowers ($\times \frac{2}{3}$); 2, leaf-margin (×2); 3, flower (×2); 4, anterior petal (×3); 5, lateral united petals (×3); all from *Phipps & Goodier* 353.

on the upper, greyish-green or glaucous on the lower surface, apex acute or obtuse, margins slightly crenate-denticulate, base cuneate to rounded or very rarely subcordate; petiole 0·2–0·5 cm. long, pubescent or glabrous. Flowers solitary, very rarely in pairs, axillary, usually white but sometimes white with yellow spots (raised honey guides) inside and a pinkish spur, pink or often with an orange throat; pedicel 2–5(7) cm. long, glabrous or pubescent; bracts very short, lanceolate-triangular, acute. Lateral sepals 3–5 × 0·7–2 mm., lanceolate to broad-lanceolate (rarely triangular-lanceolate) usually glabrous; posterior sepal broadly and obliquely infundibuliform, 4–10 mm. deep (excluding the spur), up to 1·3 cm. in diam. at the mouth, abruptly constricted into the spur, distally acute; spur 4–10 mm. long, slightly incurved or more rarely straight, usually tapering into a fine point (see t. 26 figs. 2–3). Anterior petal 8–12 × 8–12 mm. (when flattened), dorsally with a slender crest, glabrous or with a row of ± stiff white hairs on the crest; crest terminating in an acute or bluntish point. Lateral pair of united petals 15–20(25) mm. long, deeply 2-lobed; anterior lobe 9–15 × 6–9 mm., transversely broadly oblique-obovate, rarely obtrapeziform; posterior lobe about $\frac{1}{2}$ to $\frac{2}{3}$ the size of the anterior one, oblique-obovate to broadly obdeltate (see t. 26 fig. 24). Ovary glabrous. Capsule 0·9–1·5 cm. long, fusiform, glabrous or rarely sparsely hairy when young. Seeds numerous, c. 1·6 mm. in diam., broadly elliptic in outline, laterally slightly flattened, rusty or dark brown, with a densely papillose surface, glabrous.

N. Rhodesia. N: Abercorn Distr., Lunzua R., 1520 m., fl. & fr. 28.iii.1955, *E.M. & W.* 1249 (BM; LISC; SRGH). W: Mwinilunga Distr., S.W. of Dobeka Bridge, fl. 27.x.1937, *Milne-Redhead* 2975 (BM; K). C: Mkushi, fl. 23.i.1955, *Fanshawe* 1837 (K). E: Lundazi Distr., Nyika Plateau, 2100 m., fl. 3.i.1959, *Richards* 10417 (K). **Nyasaland.** N: Nyika Plateau, Lake Kaulime, 2150 m., fl. 15.xi.1958, *Robson* 621 (BM; K; LISC; SRGH). C: Dzalanyama Forest Reserve, near Chiungiza, 1550 m., fl. 9.ii.1959, *Robson* 1522 (BM; K; LISC; PRE; SRGH). **Mozambique.** N: Vila Cabral, Massangulo, fl. iii.1933, *Gomes e Sousa* 1269 (COI).

N. Rhodesia, Nyasaland, Mozambique, Tanganyika, Angola and Congo. In grassland, open woodland and secondary woodland in semi-swamp areas, and in dambos, on wet sandy soil, often on river banks; usually common.

7. **Impatiens eriocarpa** Launert in Bol. Soc. Brot., Sér. 2, **36**: 59, t. 7 (1962). TAB. **26** fig. 16. Type: Mozambique, Niassa, Maniamba, *Pedro & Pedrógão* 3794 (LISC).

Annual (?) herb; stem up to 60 cm. high or more, usually simple, erect, younger parts somewhat pilose, older ones glabrous. Leaves spirally arranged, very shortly petiolate or sessile; lamina 5–8 × 1·5–2·6 cm., oblanceolate, chartaceous, bright green, glabrous on both surfaces, or only younger ones very sparsely pilose, with 1 or 2 pairs of marginal circular glands near the base, apex acute, margin crenulate-denticulate, base narrowly cuneate; secondary nerves 4–7 pairs, ± prominent beneath; petiole up to 3 mm. long, canaliculate. Flowers in few-flowered axillary fascicles, rarely solitary, white; bracts very short, ovate; pedicels 1–1·5 cm. long, slender, pilose or sometimes glabrescent. Lateral sepals c. 2·5 × 1·2 mm., ovate-oblong, acute, glabrous; posterior sepal 13–17 mm. long, 0·4–0·6 mm. deep (excluding the spur), pilose, abruptly constricted into the spur, distally acute; spur 16–20 mm. long, incurved, slender, tapering towards a slightly thickened end, shortly pilose or glabrous (see t. 26 fig. 4). Anterior petal c. 10 × 10 mm. (when flattened), very narrowly cristate towards the apex, with the crest terminating in an acute point, somewhat pilose. Lateral united pair of petals 23–27 mm. long, deeply 2-lobed; anterior lobe c. 18 × 20 mm., broadly transverse-elliptic to sub-circular, distally slightly emarginate; posterior lobe $\frac{1}{2}$–$\frac{2}{3}$ the size of the anterior one, obtusely deltoid. Ovary densely tomentose. Capsule 12–17(20) mm. long, ovate-triangular to elliptic in outline, acute, densely tomentose (see t. 26 fig. 26). Seeds numerous, glabrous (?).

Mozambique. N: Maniamba, Missão de S. António de Unango, fl. & fr. i.1934, *Gomes e Sousa* 1668 (LISC).

Known only from Mozambique. In forests (?).

8. **Impatiens brachycentra** Schulze & Launert in Bol. Soc. Brot., Sér. 2, **36**: 62 (1962). Type: N. Rhodesia, Nyika Plateau, *Richards* 10423 (K, holotype).

Upright or straggling perennial (?) herb, up to 80 cm. high or more; stems glabrous or rarely scattered-pilose, pale green or brownish, somewhat succulent.

Leaves spirally arranged, sessile or very shortly petiolate; lamina 3·5–9 × 1–2·5 cm., oblong-lanceolate, rarely elliptic or oblanceolate, glabrous on both surfaces, pale green on the upper, greyish, silvery or glaucous below, chartaceous, apex acute, margin crenulate-denticulate, base cuneate (usually narrowly so); secondary nerves 8–12 pairs, somewhat prominent below. Flowers axillary, single or in pairs or in very few-flowered clusters, white; bracts up to 6 mm. long, lanceolate-triangular to nearly subulate; pedicels (1·5)2·5–4(5) cm. long, slender, usually glabrous. Lateral sepals 6–8 × 3·5–5 mm., ovate or ovate-triangular, acute and usually apiculate; posterior sepal 1·3–1·5 and 0·5–0·8 cm. deep (excluding the spur), broadly and obliquely funnel-shaped in outline, ± abruptly constricted into the spur, distally very acute; spur 6–9(11) mm. long, c. 1 mm. thick, slightly incurved, cylindric or subconical, with an obtuse end, usually glabrous. Anterior petal 9–12 × 10–14 cm. (when flattened), helmet-shaped, dorsally broadly cristate (crest up to 3 mm. broad), shortly apiculate at the apex, apparently glabrous; lateral pair of united petals 20–27 mm. long, deeply 2-lobed, glabrous; anterior lobe 10–16 × 6–9 mm., transversely semi-elliptic; posterior lobe ½ to ¾ the size of the anterior one, transversely oblique-elliptic to oblique-ovate. Ovary glabrous or very sparsely pilose. Capsule (not seen in a mature state) glabrous or somewhat pilose towards the apex. Seeds numerous.

N. Rhodesia. E: Lundazi Distr., Nyika Plateau, 2100 m., fl. 3.i.1959, *Richards* 10423 (K; SRGH).

N. Rhodesia and Tanganyika. Sides of streams and also in open grassland.

9. **Impatiens shirensis** Bak. f. in Trans. Linn. Soc., Bot. Ser. 2, **4**: 7 (1894).—Warb. in Engl., Bot. Jahrb. **22**: 48 (1895).—Gilg in Engl., Bot. Jahrb. **43**: 111 (1909).—Brenan in Mem. N.Y. Bot. Gard. **8**, 3: 232 (1953).—Launert in Bol. Soc. Brot., Sér. 2, **36**: 61 (1962). Type: Nyasaland, Mt. Mlanje, *Whyte* 55 (BM, holotype; K, isotype).

Much-branched dense straggling shrub, up to 2 m. high or more; young shoots often zigzag, bright red, finely pilose or glabrous, succulent, with swollen nodes; older shoots brownish, glabrous. Leaves spirally arranged, petiolate; lamina (4)5–9(12) × 1·5–3(4·5) cm., usually lanceolate, rarely broadly lanceolate to elliptic, sparsely pilose or glabrous when young, usually glabrous when older, pale green with marginal teeth often red, apex acute and often acuminate, margin crenulate-denticulate or serrate, base cuneate; secondary nerves more than 12 pairs; petiole 1–2·5(3) cm. long, slender, subcanaliculate, glabrous. Flowers in axillary few-flowered fascicles, white or pinkish-white, with 2 orange spots inside the lateral petals, the posterior sepal with red markings and the spur usually purple, fragrant; bracts up to 5 × 1 mm., triangular or subulate, acute; pedicels (1)1·5–2·5(3) cm. long, slender, finely pilose or glabrous. Lateral sepals up to 6 × 3 mm., ovate-oblong, often cuspidate, usually glabrous; posterior sepal 11–15 × 3–4·5 mm. (excluding the spur), broadly navicular, abruptly constricted into the spur, distally acute or subcaudate, usually glabrous; spur up to 6 mm. long, straight or slightly incurved or often only somewhat bent inwards in the lower half, cylindric or subconical with the end slightly thickened. Anterior petal up to 1·5–2 cm. long (when flattened), helmet-shaped, dorsally with a small crest, apically mucronulate, glabrous; lateral united petals 1·6–2·4 cm. long, deeply 2-lobed; anterior lobe 10–13 × 4–6 mm., broadly ovate to transversely semi-elliptic; posterior lobe c. ½–⅔ the size of the anterior one, ovate. Ovary glabrous. Capsule c. 2 cm. long, fusiform, glabrous or very rarely sparsely pilose. Seeds numerous, c. 2·75 × 1·8 mm., ovate in outline, bright reddish-brown, smooth and glabrous.

Nyasaland. S: Mt. Mlanje, Luchenya Plateau, fl. 25.vi.1946, *Brass* 16423 (BM; K; SRGH); Mt. Mlanje, Tuchila, Upper Valley, fl. 20.vii.1956, *Newman & Whitmore* 126 (BM; SRGH).

Endemic to Mt. Mlanje as far as is known. On forest edges, on banks of streams in forests or in rocky grassland, in forest undergrowth, usually on moist sandy soil; frequent.

10. **Impatiens quisqualis** Launert in Bol. Soc. Brot., Sér. 2, **36**: 60, t. 8 (1962). Type: Nyasaland, Mt. Mlanje, *Chapman* 492 (K, holotype; SRGH).

A weak upright or decumbent perennial herb; stems up to 1·7 m. or more high, succulent, glabrous or somewhat pilose mainly on younger parts. Leaves spirally arranged, petiolate; lamina (5)9–13(16) × (1)3–4·5(6) cm., ovate to ovate-oblong, membranous, usually glabrous on the upper surface, pilose on the venation of the lower surface, dark green on the upper surface, greyish-green on the lower one, apex acute, sometimes acuminate, margin serrate-denticulate, base cuneate; secondary nerves 7–9 pairs, prominent beneath; petiole (1)2·5–4·5(6) cm. long, relatively robust, glabrous or scattered-pilose. Flowers in few-flowered axillary fascicles, rarely apparently solitary, white (as far as known); bracts up to 4 mm. long, lanceolate-triangular to subulate, acute; pedicels up to 5 cm. long, succulent, glabrous or very sparsely pilose. Lateral sepals 4·5–9 × 3–5 mm., ± obliquely ovate-lanceolate to ovate, acute, apiculate, glabrous; posterior sepal 1·4–1·8 cm. long, 0·5–0·7 mm. deep, deeply and obliquely navicular or broadly funnel-shaped, ± abruptly constricted to the spur, distally acute or slightly caudate, glabrous; spur up to 7 mm. long, cylindric or subconical, rather blunt, straight or slightly incurved, glabrous, with the end somewhat thickened. Anterior petal up to 1·1 × c. 1·5 cm. (when flattened), helmet-shaped, dorsally distinctly cristate (crest up to 2 mm. broad), apically slightly retuse, mucronulate, glabrous; lateral united petals 2·5–3·5 cm. long, deeply 2-lobed, usually glabrous; anterior lobe 2–2·3 × 0·6–0·9 cm., transversely oblique-elliptic; posterior lobe about ⅓ the size of the anterior one (obovate? entire?). Ovary glabrous. Capsule up to 2·2 cm. long, fusiform, glabrous. Seeds numerous, c. 2·5 × 2 mm., obovate in outline, dark brown to blackish, with the surface wrinkled, glabrous.

Nyasaland. S: Mt. Mlanje, fl. & fr. 25.v.1957, *Goodier* 252 (SRGH).
Endemic to Mt. Mlanje, as far as is known. Common in shade in damp places by watercourses, in woodland, and among shrubs along forest margins.

11. **Impatiens sylvicola** Burtt Davy, F.P.F.T. **1**: 44, 196 (1926). TAB. **26** fig. 17. Type from the Transvaal.

Low scrambling or upright (perennial?) herb; stems succulent, pinkish or green, often rooting at the nodes, glabrous or sparsely pilose (mainly when young). Leaves spirally arranged, petiolate; lamina 2–7 × 1·8–4·5 cm., ovate to ovate-lanceolate, membranous, pilose on both surfaces (mainly on the nerves) or glabrescent or rarely quite glabrous, dark green on the upper surface, bright green or greyish-green on the lower one, apex acute or subobtuse, margin crenate or crenate-denticulate to serrate, base usually broadly cuneate, rarely ± rounded; secondary nerves 5–8 pairs, prominent beneath; petiole 0·8–3·2 cm. long, canaliculate, sometimes densely but usually sparsely pilose or often glabrous. Flowers axillary, usually solitary, rarely in fascicles of 2 or 3 (but then usually only one developed at a time), pinkish-purple or pink with deep red blotches at the base of the lateral petals; bracts very small, lanceolate-triangular, acute, usually pilose; pedicels 1·5–3(3·5) cm. long, very slender, usually pilose (sometimes densely so), rarely glabrous. Lateral sepals 4–6·5 × 0·9–1·25 mm., lanceolate-triangular, acute, ± densely pilose, with longer hairs along the margins; posterior sepal 9–11(13) mm. long, 3–4 mm. deep, shallowly and obliquely navicular, abruptly constricted into the spur, distally subcaudate, pilose (especially along the keel-line); spur 11–15(18) mm. long (or even somewhat longer), rather slender, tapering into a bluntish point, rarely with the end slightly thickened, always ± incurved. Anterior petal 7–11 × 8–13 mm. (when flattened), broadly ovate to transversely elliptic (when flattened), dorsally narrowly cristate, with the apex usually apiculate but sometimes slightly emarginate and with long hairs along the crest and denser towards the base, otherwise shortly hispid; lateral united petals 8–14 mm. long, deeply 2-lobed, usually shortly hispid outside; anterior lobe 8–10 mm. long, 5·25–7 mm. high, obliquely obovate, distally rounded or subtruncate and produced into a short ligulate appendage (1–2 × c. 1 mm.) at the lower margin; posterior lobe nearly equal in size, broadly oblique-obovate, with the apex rounded or obliquely truncate, entire or sometimes irregularly retuse. Ovary ± densely pilose. Capsule up to 1·5 cm. long, fusiform, pilose or glabrescent. Seeds numerous, not seen in a mature state (c. 1·4 × 0·9 mm. ?), dark brown, glabrous (the surface smooth?).

Nyasaland. S: Mt. Mlanje, Tuchila Plateau, 1830 m., fl. 19.vii.1956, *Newman & Whitmore* 72 (BM; SRGH). **S. Rhodesia.** E: Umtali Distr., Engwa, 1830 m., fl. 2.ii.1955, *E.M. & W.* 128 (BM; LISC; SRGH). **Mozambique.** MS: Manica, Tsetsera, 1830 m., fl. 9.ii.1955, *E.M. & W.* 320 (BM; LISC; SRGH).
Nyasaland, S. Rhodesia, Mozambique and the Transvaal. Growing in humid shady places; in forest undergrowth, in clearings and along watercourses in forests or on forest edges; sometimes common.

12. **Impatiens limnophila** Launert in Bol. Soc. Brot., Sér. 2, **36**: 50, t. 2 (1962). TAB. **26** fig. 14. Type: N. Rhodesia, Abercorn, Lunzua Falls, *Whellan* 1948 (K, holotype; SRGH).

Usually prostrate, sometimes erect perennial (?) herb; stems up to 180 cm. long or more, rooting at the lower nodes, red, pilose on younger usually glabrescent on older parts. Leaves spirally arranged, petiolate to subsessile; lamina (3·5)4–13 × (1·5)1·8–4 cm., lanceolate-oblong or ovate-lanceolate to elliptic, membranous to subchartaceous, bright to medium green on the upper surface, greyish-green to glaucous on the lower one, shortly pilose on both surfaces, more densely on the lower one with an increasing density towards the base, usually glabrescent when older, sometimes glabrous on both surfaces, apex subobtuse to acute, often acuminate, margin crenulate-denticulate to serrate, base cuneate or rarely rounded; secondary nerves 7–10 pairs, somewhat prominent beneath; petiole up to 1 cm. long, canaliculate, usually pilose or glabrescent or glabrous. Flowers axillary, solitary or in bundles of 2–3, pale mauve or pink; bracts very small, lanceolate-triangular or subulate, tapering into a fine point; pedicels 2–6 cm. long, very slender, pilose or glabrous. Lateral sepals 4–5 × 1·75–2·5 mm., broadly lanceolate-triangular, acute or finely acuminate, glabrous or sparsely pilose; posterior sepal 8–11 mm. long and c. 2·5 mm. deep (excluding the spur), shallowly obliquely navicular, abruptly constricted into the spur, distally acute or subcaudate (purplish-vittate?) somewhat pilose mainly towards the keel-line; spur up to 25(30) mm. long, slightly incurved, slender, tapering into a bluntish (rarely slightly thickened) end, usually pilose, very rarely glabrous. Anterior petal 8–10 mm. long and up to 10 mm. broad (when flattened), helmet-shaped, dorsally distinctly cristate, with the crest terminating in a sharp point, usually pilose on the crest; lateral pair of united petals 18–22 mm. long, deeply 2-lobed, usually glabrous; anterior lobe 12–17 × 7–10 mm., transversely broadly semi-elliptic, entire; posterior lobe c. $\frac{1}{3}$ the size of the anterior one (rarely smaller), obovate to oblong (see t. 26 fig. 14). Ovary ± densely pilose. Capsule not seen in a mature state, pilose or glabrescent. Seeds numerous.

N. Rhodesia. N: Abercorn Distr., Kambole Area, fl. 25.viii.1956, *Richards* 5954 (SRGH).

Known only from N. Rhodesia. In wet places, generally in swampy patches, often growing in mud on river banks or on sides of water courses, sometimes floating.

This species shows a great range of variability in its vegetative characters (see original description, loc. cit.). Gatherings from the Chishimba Falls are prostrate plants with chartaceous leaf-laminas, while the remaining specimens I have seen are erect or semi-erect plants with finely membranous leaves.

13. **Impatiens duthieae** L. Bolus in Ann. Bolus Herb. **3**: 70 (1921).—B. L. Burtt in Kew Bull. **1938**: 161 (1938).—G. M. Schulze in Bol. Soc. Brot., Sér. 2, **29**: 14 (1955).—Launert in Bol. Soc. Brot., Sér. 2, **36**: 48 (1962). TAB. **26** fig. 19. Type from S. Africa (Knysna Distr.).

Impatiens capensis Thunb., Prodr. Pl. Cap.: **41** (1794) *nom. illegit.* non *I. capensis* Meerb. (1775).—O. Hoffm. in Linnaea, **43**: 122 (1881).—Hiern, Cat. Afr. Pl. Welw. **1**: 110 (1896). Type from S. Africa (Cape Prov.).

Balsamina capensis DC., Prodr. **1**: 686 (1824). Type as above.

Impatiens marlothiana G. M. Schulze in Notizbl. Bot. Gart. Berl. **13**: 665 (1937). Type as above.

Erect or straggling perennial (always?) herb; stems 20–60(75) cm. high, usually glabrous. Leaves spirally arranged, petiolate; lamina (3)4–8·5(11) × (2)2·5–4(5) cm., ovate-lanceolate to narrowly elliptic or elliptic-lanceolate, membranous, sparsely pilose on both surfaces or glabrous; secondary nerves 8(10) pairs, somewhat prominent beneath, apex subacute or rarely obtuse, usually acuminate, margin crenulate-denticulate (the teeth usually in the sinuses), base cuneate; petiole (1)1·5–4(5) cm. long, rather slender, pilose or glabrous. Flowers axillary, solitary or in few-flowered clusters, usually very pale pink, white on the back with two spots at the base of the posterior sepal; bracts very small, lanceolate-triangular or subulate, acute; pedicels 2–3·5(4·5) cm. long, very slender, usually glabrous. Lateral sepals 2–3(4) × 0·75–1·5(1·75) mm., lanceolate to triangular, acute, usually glabrous; posterior sepal (4)5–6(7) mm. long, shallowly navicular, abruptly constricted into the spur, distally subcaudate, glabrous or very sparsely pilose; spur (8)10–18(25) mm. long, slightly incurved, rather slender, subconical, tapering into a subacute or sometimes slightly thickened end, usually glabrous, rarely sparsely pilose. Anterior petal 5–7·5 mm. high, 4·5–8 mm. broad when flattened, with a distinct dorsal crest, apically apiculate; lateral united petals 10–17 mm. long, 2-lobed but the lobes very dissimilar in shape and size; anterior lobe variable in shape and size, 8–13 mm. long (at the base) and 6–10 mm. high, broadly transverse-elliptic, broadly obovate to subcircular, distally prolonged at the very base into a ligulate appendage 2–7 × 1–2(3) mm.; posterior lobe reduced to a small auricle not exceeding 2 × 2 mm. (see t. 26 fig. 19). Ovary glabrous or rarely scattered-hispid. Capsule 6–13 mm. long, fusiform, glabrous or scattered-hispid. Seeds numerous.

Nyasaland. C: Dedza Mt., 1670–1830 m., fl. 20.iii.1955, *E.M. & W.* 1091 (BM; LISC; SRGH).

Transvaal, Natal, Nyasaland and Angola.

In damp and shady places, on wet rocks, in forests and beside watercourses.

14. **Impatiens schulziana** Launert in Bol. Soc. Brot., Sér. 2, **36**: 57, t. 6 (1962). TAB. **26** fig. 10. Type: Nyasaland, Nyika Plateau, *Brass* 17204 (BM, holotype; SRGH, isotype).

Usually much-branched perennial shrubby herb, up to 1 m. high or more; stems succulent, older ones woody (?), reddish or green, pilose on younger parts, even glabrescent. Leaves spirally arranged, petiolate; lamina (2·3)3–10 × (1)1·5–5 cm., ovate-oblong, ovate or narrowly to broadly elliptic, membranous, intense green on the upper, greyish-green on the lower surface, shortly pilose on both surfaces (more densely so on younger leaves) rarely glabrous, apex acute, usually acuminate, margin crenate-denticulate to serrate-denticulate and usually ciliate towards the base, base cuneate; petiole 0·7–2·5 cm. long, subcanaliculate, somewhat pilose (often densely so when young) or glabrous. Flowers axillary, solitary or in pairs but only one fully developed at a time (as far as known), mauve, the lateral petals (always?) with a purplish blotch, the lateral sepals and the spur usually red; bracts very small, lanceolate-triangular, acute, densely pilose; pedicels up to 3 cm. long, slender, pilose or glabrescent. Lateral sepals c. 7 × 2 mm., ovate-oblong, subacute, pilose on the margins; posterior sepal c. 3 × 7 mm., shallowly and obliquely navicular, abruptly constricted into the spur, distally acute, pilose (mainly towards the keel-line) or glabrescent; spur 23–28 mm. long, rather slender, tapering into a bluntish or slightly thickened end, incurved (often incurved in the upper half and straight in the lower one), usually pilose, rarely glabrescent. Anterior petal up to 13 × 18 mm. (when flattened), apically retuse to emarginate, distally very narrowly cristate, the crest terminating in a subacute point, glabrous (as far as known); lateral united petals up to 25 mm. long, deeply 2-lobed, glabrous or somewhat hispid outside; anterior lobe c. 15 × 8–10 mm., transversely and broadly semi-elliptic to semi-circular, distally usually somewhat drawn out at the very base and forming a subacute point; posterior lobe c. 13 mm. broad and c. 10 mm. high, broadly ovate to semi-circular (see t. 26 fig. 10). Ovary glabrous. Capsule not seen in a mature state, glabrous. Seeds numerous.

Nyasaland. N: Nyika Plateau, Nchena Waterfall, c. 2400 m., fl. xii.1952, *Chapman* 54 (BM).

Endemic to the Nyika Plateau as far as known. Growing on borders of montane forest (*Brass* 17204), or beside the lips of waterfalls (*Chapman* 54).

15. **Impatiens eryaleia** Launert in Bol. Soc. Brot., Sér. 2, **36**: 56, t. 5 (1962). TAB. **26** fig. 12. Type: Nyasaland, Misuku hills, *Richards* 10591 (K, holotype; SRGH).

Erect or semi-prostrate perennial herb, up to 70 cm. high or more; stem simple or branched, sparsely pilose on younger parts, but soon glabrescent, succulent, often with a woody base. Leaves spirally arranged, petiolate, sometimes crowded towards the apex of the branches and stems; lamina 4–9·5 × 2·5–4·5 cm., ovate to ovate-oblong or narrowly elliptic, thinly membranous, bright green on the upper surface, greyish-green on the lower one, glabrous on the upper, sometimes pilose (mainly on the nerves) on the lower surface, apex acute and often acuminate, margin crenulate-denticulate, ciliolate towards the base, base cuneate; secondary nerves 6–8 pairs, somewhat prominent beneath; petiole 1–3(4·5) cm. long, usually glabrous. Flowers axillary, solitary, colour not properly known (pale pink or mauve?); bracts very small, lanceolate-triangular, acute, glabrous or sparsely pilose; pedicels up to 6 cm. long, rather slender, glabrous. Lateral sepals up to

8 × 3·5 mm., ovate-oblong to oblong, acute and finely acuminate, glabrous; posterior sepal c. 11 mm. long, obliquely funnel-shaped, gradually constricted into the spur, distally acute to subcaudate, glabrous; spur up to 35 mm. long, strongly incurved to nearly recurved, rather slender, conical in the upper, subcylindric in the lower half, with the end slightly thickened, glabrous. Anterior petal c. 11–14 mm. high and up to 12 mm. broad (when flattened), dorsally cristate, the crest terminating in an acute point, glabrous. Lateral united petals 17–20 mm. long, deeply 2-lobed, the lobes in a characteristic position, their axes forming (? always) an angle of c. 90°, glabrous; anterior lobe up to 18 mm. long and up to 10 mm. high, transversely semi-elliptic, entire; posterior lobe similar to the anterior one but usually a little longer and somewhat narrower (see t. 26 fig. 12). Ovary glabrous. Capsule not known in a mature state, glabrous. Seeds numerous.

Nyasaland. N: Karonga Distr., Misuku Hills, fl. 10.i.1959, *Richards* 10591 (K; SRGH).

Known only from Nyasaland. On floor of rain forest.

16. **Impatiens cecilii** N.E. Br. in Kew Bull. **1906**: 101 (1906).—Warb. in Engl., Bot. Jahrb. **43**: 113 (1909).—Launert in Bol. Soc. Brot., Sér. 2, **36**: 64 (1962). Type: Mozambique, Manica, *Cecil* 169 (K).

Perennial herb, very variable in habit, erect, spreading or decumbent; usually (20)30–60 cm. (rarely up to 90 cm. or more) high; stems succulent, rarely with a woody base, often rooting at lower nodes, ± densely pilose (mainly on younger parts), glabrescent or sometimes quite glabrous, green or reddish. Leaves spirally arranged, petiolate or rarely nearly sessile; lamina extremely variable in shape, size and texture, 1–14 × 0·7–5·5 cm., lanceolate-ovate, narrowly elliptic, oblong-elliptic, elliptic, ovate or rarely oblanceolate, thinly membranous to chartaceous, ± densely pilose on both surfaces (mainly when young) to quite glabrous, rarely hispid on the upper and glabrous or glabrescent on the lower surface, bright green on both surfaces to dark green on the upper and greyish-green or glaucous on the lower surface, apex usually acute but sometimes subobtuse to obtuse and often acuminate, margin crenulate-denticulate to serrate and very often ciliolate towards the base, base narrowly to broadly cuneate; secondary nerves 5–8(9) pairs, somewhat prominent beneath; petiole 0·2–5 cm. long, fairly slender, often with the lamina decurrent nearly to the base, ± densely pilose to glabrescent or glabrous. Flowers axillary, solitary or in clusters of 2–3, bright to dark pink or mauve, with the posterior sepal distally longitudinally striate; bracts very small, lanceolate-triangular to mostly subulate, acute, glabrous or sometimes pilose; pedicels 2·5–7 cm. long, rather slender, glabrous or pilose. Lateral sepals (3)3·5–7(9) × (0·8)1·5–2·75(3·5) mm., very variable, lanceolate-triangular, triangular-subulate, ovate to ovate-oblong, acute or usually narrowly acuminate, glabrous or sparsely hairy; posterior sepal 9–15 mm. long and 2·5–5 mm. deep, shallowly obliquely navicular, abruptly constricted into the spur, distally subcaudate or acute, almost always striate as described above, glabrous or sparsely pilose; spur (13)17–32 mm. long, always incurved (in immature flowers often recurved), rather slender cylindric or somewhat conical, the end bluntish or only slightly thickened, very rarely acute, glabrous, glabrescent or pilose (see t. 26 fig. 6). Anterior petal 7–15 mm. high and 8–18 mm. broad (when flattened), helmet-shaped, dorsally narrowly cristate (broader towards the apex), with the crest terminating in an acute point. Lateral united petals 13–35 mm. long, deeply 2-lobed, the lobes very different in shape and size, usually glabrous; anterior lobe 10–27 (at the base) × 10–18 mm., rather variable in shape, transversely semi-elliptic to subcircular, broadly obliquely ovate to broadly obliquely obovate, with the upper edge rounded or oblique-truncate, distally rounded to (very rarely) subacute, always entire; posterior lobe much smaller than the anterior one, broadly ovate, ovate-triangular, rarely ovate-oblong, apically acute and nearly always produced into a filiform appendage (see t. 26 fig. 20–23). Ovary glabrous. Capsule long, fusiform, glabrous. Seeds numerous, 1·5–1·7 × 2–2·5 mm., light to dark brown or reddish-brown, ovate-oblong in outline, glabrous, ± densely covered with protuberant papilli.

A very variable species. There are two distinct subspecies, one of them is rather uniform while the other one may be divided into at least 2 groups to which, because of so many intermediate forms, taxonomic rank can scarcely be given in our present state of knowledge (see also discussion under original descriptions of the subspecies).

Subsp. **cecilii.** TAB. **26** fig. 6, 21–23.

Lateral united petals 14–20(24) mm. long, 8–13 mm. broad. Posterior sepal 7–12 mm. long, 2–3 mm. deep; spur 14–24 (very rarely up to 30) mm. long. Rather weak and delicate plant, usually glabrous or rarely somewhat pilose in younger parts.

Group A:

Leaf lamina (4)5–10(13) × 2·5–4·5(5·5) cm., membranous, usually translucent, nearly always glabrous on both surfaces; petiole (2)3–5 cm. long. Pedicels 3–7 cm. long.

S. Rhodesia. E: Umtali Distr., Engwa, 1980 m., fl. 2.ii.1955, *E.M. & W.* 90 (BM; LISC; SRGH). **Mozambique.** MS: Manica, Tsetsera, 1930 m., fl. 9.ii.1955, *E.M. & W.* 318 (BM; LISC; SRGH).

Not known from elsewhere. On damp forest ground, at foot of cliffs and in kloofs.

Group B:

Leaf-lamina (1)2–5(7) × (0·7)1·5–3·5(4·5) cm., membranous to chartaceous, usually not translucent as in the other group, glabrous or often somewhat pilose on the lower surface (mainly on younger leaves); petiole (0·2)0·6–2(3) cm. long. Pedicels 2·5–4·5 cm. long.

Subsp. **grandiflora** Launert in Bol. Soc. Brot., Sér. 2, **36**: 64 (1962). TAB. **26** fig. 20. Type: S. Rhodesia, Inyanga, *Chase* 2869 (BM, holotype; LISC; SRGH).

Lateral united petals 27–35 mm. long, 14–17 mm. broad (see t. 26 fig. 20). Posterior sepal c. 15 mm. long, c. 5 mm. deep; spur 28–40 mm. long. Rather robust plant with the younger parts of the stems, the petioles and often at least the lower surface of the leaves ± densely pilose.

S. Rhodesia. E: Inyanga, 2130 m., fl. 4.ix.1954, *Wild* 4596 (SRGH).

Endemic to this district as far as is known.

Growing in forest patches and on forest edges, beside waterfalls in shelter of rocks and trees, sometimes on river banks, usually frequent.

S. Rhodesia. C: Salisbury, fl. 27.ix.1932, *Mundy* in GHS. 6155 (SRGH) cult. E: Umtali Distr., Stapleford Forest Reserve, Inyamakwakwakwa, 1700 m., fl. 28.xi.1955, *Chase* 5896 (BM; LISC; SRGH). **Mozambique.** MS: Manica, Chimanimani Mts., east side of point 71, 1650 m., fl. 10.ii.1958, *Hall* 426 (SRGH).

Not known from elsewhere. Growing in riverine forest patches by streams, on wet shady places along roads, in river beds amongst rocks, hanging down cliffs of waterfalls and on damp forest floors.

17. **Impatiens psychadelphoides** Launert in Bol. Soc. Brot., Sér. 2, **36**: 48, t. 1 (1962). TAB. **26** fig. 9. Type: Mozambique, Zambezia, Gúruè, *Mendonça* 2092 (BM; BR; LISC, holotype; LM).

Herb, up to 1·5 m. high or more; stems erect to decumbent, succulent, branched, younger ones ± densely hairy (usually villous), older ones glabrescent. Leaves spirally arranged, petiolate; lamina (2·5)3·5–8(11) × (2)2·5–5(6) cm., ovate to narrowly elliptic or rarely ovate-lanceolate, membranous, pilose on both surfaces when young, older ones hairy on the lower surface (mainly on the nerves), glabrescent on the upper one, green on the upper, greyish-green or glaucous on the lower surface, apex acute, sometimes acuminate, margin serrate-denticulate and ciliolate towards the base, base cuneate or very rarely rounded; secondary nerves 10–16 pairs; petiole (0·6)1–2·5 cm. long, canaliculate, ± densely hairy. Flowers solitary, rarely in fascicles of 2 or 3, axillary, pinkish; bracts up to 3·5 mm. long, lanceolate-triangular or subulate, acute, usually densely hairy. Pedicels 4–6(7) cm. long, usually densely hairy. Lateral sepals 4–7 × 2–3·5 mm., ovate-lanceolate to lanceolate-triangular, ± acute, pilose, rarely glabrous; posterior sepal 9–18 × c. 3·5 mm. (excluding the spur), obliquely navicular, abruptly constricted into the spur, distally caudate, ± densely pilose all over, rarely glabrescent; spur up to 3·5 cm. long, usually incurved, rarely straight, rather slender, narrowly conical in the upper, cylindrical in the lower half, with the end bluntish or rarely slightly thickened, pilose. Anterior petal up to 15 × 20 mm. (when flattened), dorsally with a very narrow crest, apically mucronulate, pilose especially along the crest; lateral pair of united petals up to 33 mm. long, deeply 2-lobed, usually glabrous; anterior lobe large and showy, up to 15 × 25 mm., very broadly obovate to transversely semi-elliptic, with the upper edge entire, distally usually ± emarginate towards the base; posterior lobe much smaller, $\frac{1}{5}$–$\frac{1}{6}$ the size of the anterior one, semi-obovate to subcircular, with the upper edge rounded and sometimes somewhat mucronulate (see t. 26 fig. 9). Ovary glabrous. Capsule up to 2 cm. long, fusiform, glabrous. Seeds numerous, c. 2·8 × 1·75 mm., reddish-brown, with a densely papillose surface, obovate-elliptic in outline.

Mozambique. Z: Gúruè, slopes of the serra, fl. 1.x.1941, *Torre* 3555 (LISC). MS: Serra de Gorongosa, fl. & fr. 6.x.1946, *Simão* 969 (BM; LISC).
Known only from Mozambique. In humid places; usually in forest.

18. **Impatiens psychantha** Launert in Bol. Soc. Brot., Sér. 2, **36**: 55, t. 4 (1962). TAB. **26** fig. 11. Type: Mozambique, Milange, *Mendonça* 2326 (LISC, holotype).

Annual (?) herb; stems up to 50 cm. high or more, simple or sparsely branched, erect or ascending from a prostrate base, glabrous or somewhat pilose, on very young parts reddish or green. Leaves spirally arranged, sessile or very shortly petiolate; lamina (2·5)3·5–9(11) × (1)1·5–3(4) cm., obovate to broadly oblanceolate or elliptic or rarely obtusely rhombic, apex acute and usually acuminate, margin coarsely crenulate-denticulate or serrate, often sparsely long-ciliate towards the base, base usually narrowly cuneate, membranous, usually intense green on the upper, greyish-green on the lower surface, pilose (rarely glabrous) on both surfaces when young—often with an increasing density towards the base, older ones glabrescent but sometimes scattered pilose beneath (mainly on the nerves); secondary nerves 4–6 pairs. Flowers axillary, solitary or in bundles of 2 or 3, purplish (as far as known); bracts up to 5 mm. long, lanceolate-triangular or subulate, acute, usually pilose; pedicels 2·75–5 cm. long, slender, pilose (usually more densely so in the lower half), glabrous or glabrescent. Lateral sepals 5–7 × 1–1·25 mm., narrowly triangular, acute, usually long-pilose along the margins; posterior sepal c. 7 mm. long, 2·5–3·5 mm. deep, shallowly and obliquely navicular, abruptly constricted into the spur, distally acute to subcaudate, usually glabrous; spur up to 35(40) mm. long, very slender, with the end bluntish or slightly thickened, glabrous or very sparsely pilose (when young), slightly incurved (sometimes twisted?). Anterior petal up to 10 mm. high, 10–14 mm. broad (when flattened), dorsally very narrowly cristate, the crest terminating in a subacute point, glabrous(?) or with a row of long multicellular hairs along the crest. Lateral united petals 17–20 mm. long, deeply 2-lobed the symmetrical axes of the lobes forming an angle of c. 90°, usually glabrous; anterior lobe c. 17 × 8 mm., transversely semi-elliptic, distally entire rounded, with the lower edge straight or slightly curved; posterior lobe 10–15 × 13–15 mm., obcuneate, with the upper margin retuse to emarginate and with a slender mucro in the sinus. Ovary glabrous. Capsule up to 18 mm. long, fusiform, glabrous. Seeds numerous, c. 2 × 1·2 mm., obovate in outline, reddish brown, with a densely papillose surface, glabrous.

Mozambique. N: Malema, near Mutuali, fl. & fr. 28.ix.1944, *Mendonça* 2300 (BM; LISC). Z: Milange, slopes of the serra, fl. & fr. 2.x.1944, *Mendonça* 2326 (BM; BR; LISC; LM; PRE; SRGH).
Not known elsewhere. Growing in humid places.

19. **Impatiens hydrogetonoides** Launert in Bol. Soc. Brot., Sér. 2, **36**: 54 (1962). Type: N. Rhodesia, Isoka, *Angus* 816 (K, holotype).

Soft erect or prostrate perennial (?) herb; stems up to 50 cm. high, glabrous or very sparsely pilose, succulent. Leaves spirally arranged, petiolate; lamina 6–11 × 2–4·5 cm., narrowly elliptic or ovate-lanceolate, apex acute to subobtuse, usually acuminate, margin crenulate-denticulate, base cuneate with some slender tentacle-like hairs on the margins, membranous, bright to medium green on the upper surface, greyish-green to glaucous on the lower one, sparsely pilose on the upper surface, less densely so or glabrous on the lower one; secondary nerves 6–9 pairs, somewhat prominent beneath; petiole up to 5 cm. long, slender, subcanaliculate. Flowers axillary, solitary, in pairs or clusters of 3, pink; bracts up to 5 × 1·25 mm., triangular, subulate, tapering to a very fine point, glabrous; pedicels up to 6 cm. long, slender, succulent, usually glabrous. Lateral sepals 5–7 × 3·75–4·25 mm., broadly ovate, acuminate and acute, with a subcordate base, glabrous; posterior sepal up to 13 mm. long and c. 2·5 mm. deep (excluding the spur), shallowly and obliquely navicular, ± abruptly constricted into the spur, distally caudate, glabrous (always?); spur up to 40 mm. long, slender, tapering towards the slightly thickened end, glabrous. Anterior petal c. 14 mm. long and c. 12 mm. broad when flattened, helmet-shaped, dorsally very narrowly cristate with the crest terminating in a sharp point, slightly cordate and broadly saccate at the base, glabrous (always?). Lateral pair of united petals c. 20 mm. long, deeply 2-lobed; lobes with the margins slightly overlapping, glabrous; anterior lobe c. 15 × 10 mm., transversely broadly semi-elliptic, distally rounded, with a distinct small ridge at the very base; posterior lobe c. ⅓ the size of the anterior one, broadly elliptic or obliquely obovate (broadest slightly above the middle), entire. Ovary glabrous. Capsule c. 14 mm. long, fusiform. Seeds numerous, c. 2 × 1·6 mm., broadly elliptic to oblong in outline, reddish-brown, finely tuberculate, glabrous.

N. Rhodesia. N: Isoka Distr., Mafingi Mts., near Chisenga, fl. 21.xi.1952, *Angus* 816 (K).

Not known from elsewhere. Gregarious in the spray zone at foot of a waterfall in a ravine with evergreen forest, in dense shade.

20. **Impatiens oreocallis** Launert in Bol. Soc. Brot., Sér. 2, **36**: 52, t. 3 (1962). TAB. **26** fig. 5, 8. Type: Nyasaland, Dedza Distr., Kalichero Hill, 1700 m., fl. & fr. 21.i.1959, *Robson* 1287 (K, holotype; SRGH).

Impatiens zombensis var. *micrantha* Brenan in Mem. N.Y. Bot. Gard. **8,** 3: 232 (1953). Type: Nyasaland, Zomba Plateau, 1500 m., fl. & fr. vi.1946, *Brass* 16321 (K, holotype; SRGH).

Delicate weak straggling (annual?) herb; stems decumbent or suberect, rooting at the nodes, red or green tinged with red, glabrous or somewhat pilose on very young parts only. Leaves spirally arranged, sessile or with the lamina tapering into a very short petiole; lamina (0·7)1·5–3·5(5·5) × (0·4)0·7–1·8(2·3) cm., ovate to ovate-lanceolate, rarely narrowly elliptic, membranous, glabrous or sparsely pilose, dark green on the upper surface, greyish or glaucous on the lower one, apex acute, margin serrate and ciliate towards the base, base cuneate (usually narrowly so); secondary nerves 5–8 pairs, ± prominent beneath. Flowers in few-flowered axillary fascicles, pinkish (details of colours not known); bracts up to 3 mm. long, triangular-subulate, acute, usually glabrous; pedicels (0·8)1·2–2(2·5) cm. long, very slender, pilose, sometimes glabrescent. Lateral sepals 2·2–3(3·75) × 1–1·25 mm., lanceolate-triangular, acute, glabrous or pilose; posterior sepal 4·5–6(7) mm. long, 1·4–1·8 mm. deep, shallowly and obliquely navicular, abruptly constricted into the spur, distally acute to subcaudate, apparently glabrous except for a row of long hairs on the keel-line; spur up to 20 mm. long, slender, slightly incurved, tapering towards the somewhat swollen end. Anterior petal 3·5–6 mm. high and 4·5–7 mm. broad (when flattened), helmet-shaped, dorsally narrowly cristate with the crest terminating in an acute point, glabrous. Lateral united petals 5–10(13) mm. long, deeply 2-lobed, glabrous; lobes usually separated by a sinus, rarely overlapping; anterior lobe c. 0·8 × 0·6 mm., oblong-obovate; posterior lobe c. 4·8 × 5·6 mm., broadly obcuneate, always truncate and often slightly retuse at the upper margin (see t. 26 fig. 8). Ovary glabrous. Capsule up to 7 mm. long, fusiform, glabrous. Seeds numerous, c. 1·5 × 0·95 mm., obovate-oblong in outline, bright or reddish-brown, papillose.

Nyasaland. C: summit of Mt. Dedza, 2230 m., fl. & fr. 20.iii.1955, *E.M. & W.* 1081 (BM; LISC; SRGH). S: Blantyre Distr., Shire Highlands, fl. 1887, *Last* (K).

Not known from elsewhere. Growing on moist shady banks of streams in rain forest or on wet places in montane forest, often forming carpets on wet rocks.

21. **Impatiens zombensis** Bak. in Kew Bull. **1897**: 247 (1897).—Gilg in Engl., Bot. Jahrb. **43**: 114 (1909).—Brenan in Mem. N.Y. Bot. Gard. **8,** 3: 232 (1953). TAB. **26** fig. 15. Type: Nyasaland: Mt. Mlanje, *Whyte* (BM).

Shrubby perennial herb, usually much branched, 25–60(100) cm. high; stems erect or spreading, succulent, with swollen nodes, light green tinged with red or red, pilose (mainly on younger parts), glabrous or glabrescent. Leaves spirally arranged, petiolate; lamina (2·5)3·5–6(7·5) × (2)2·5–3·5(4) cm., usually ovate, rarely ovate-lanceolate or ovate-oblong, apex acute and usually acuminate, margin crenulate-denticulate to serrate, base cuneate (usually broadly so) rarely rounded, membranous, mid-green or yellowish-green and shining above, paler green or greyish-green below, the marginal teeth often red, usually sparsely pilose on the upper surface, glabrous on the lower one, but sometimes also pilose below (mainly on the nerves), rarely quite glabrous on both surfaces; secondary nerves 6–8 pairs; petiole 0·5–1·8(2·5) cm. long, subcanaliculate, usually pilose, rarely glabrous. Flowers axillary, usually solitary, rarely in pairs or fascicles of three, pinkish, lilac, pink-magenta, or purplish; bracts up to 3·5 mm. long, subulate; pedicels (2)3–6(7) cm., slender, usually pilose, rarely glabrous, sometimes glabrescent. Lateral sepals 2·9–6 × 0·8–2 mm., lanceolate, lanceolate-triangular or rarely ovate-oblong, often acuminate, pilose (sometimes glabrescent) or glabrous. Posterior sepal c. 7·5 × 2·5–3 mm., shallowly and obliquely navicular, abruptly constricted into the spur, distally acute or subcaudate, sparsely (rarely ± densely) pilose, glabrescent or glabrous, longitudinally striped (always?); spur very variable in length, 14–32 mm. long, slender, usually sharply incurved in the upper third, slightly incurved or nearly straight in the lower part, with the end bluntish or slightly thickened, usually glabrous, rarely pilose but then soon glabrescent. Anterior petal 6–10 mm. high, 8–13 mm. broad (when flattened), dorsally cristate with the crest terminating in an acute point, somewhat pilose or glabrous. Lateral united petals (10)13–18 mm. long, deeply 2-lobed, usually glabrous, rarely shortly and sparsely hispid outside; lobes usually distinctly separated by a sinus, rarely overlapping; anterior lobe 8–13 × 4–7 mm., transversely oblique-elliptic, entire, the lower margin straight or slightly concave; posterior lobe a little smaller than the anterior one, obovate-oblong or obliquely elliptic, distally rounded, entire (see t. 26 fig. 15). Ovary glabrous. Capsule 0·8–1·6 cm. long, fusiform, glabrous. Seeds numerous, c. 2·25 × 1·5 mm., obovate in outline, light or reddish-brown, densely covered with very short thick conical multicellular hairs.

Nyasaland. N: Nyika Plateau, 2300 m., fl. xi.1932, *Sanderson* 17 (BM). S: Zomba, 1830 m., fl. 9.iii.1955, *E.M. & W.* 736 (BM; LISC; SRGH); Mt. Mlanje, 1750 m., fl. 17.viii.1956, *Newman & Whitmore* 483 (BM). **Mozambique.** Z: Milange, Serra de Tumbrina, fl. 12.xi.1942, *Mendonça* 1376 (BM; LISC).

Not known from elsewhere. Growing on wet rocks, on moist shady banks of streams, on forest edges or in open secondary scrub; often common and appearing in clumps.

22. **Impatiens irvingii** Hook. f. ex Oliv. F.T.A. **1**: 300 (1868).—Warb. in Engl., Bot. Jahrb. **22**: 48 (1895).—Gilg in Engl., Bot. Jahrb. **43**: 111 (1909).—De Wild., Plant. Bequaert. **1**: 360 (1922).—Keay, F.W.T.A. ed. 2, **1**: 161 (1954).—G. M. Schulze in Exell & Mendonça, C.F.A. **2**, 2: 160 (1956).—Wilczek & Schulze, F.C.B. **9**: 415 (1960). TAB. **26** fig. 13. Type from Nigeria.

Impatiens kirkii Hook. f. ex Oliv., F.T.A. **1**: 300 (1868).—Warb. in Engl., Bot. Jahrb. **23**: 48 (1895). Type: Nyasaland, W. side of Lake Nyasa, *Kirk* (K, holotype).

Impatiens kirkii var. *hypoleuca* Welw. ex. Hiern, Cat. Afr. Pl. Welw. **1**: 110 (1896). Type from Angola.

Impatiens thonneri De Wild. & Dur., Pl. Thonn. Congol.: **24,** t. 11 (1900).—T. & H. Dur., Syll. Fl. Cong.: **80** (1901).—De Wild., Études Fl. Bang. Ubang.: 342 (1911). Type from the Congo.

Perennial herb usually much branched; stems up to 1·5 m. long or even more, succulent, prostrate or ascending, rarely erect, red or green tinged with red, usually somewhat swollen at the nodes, glabrous or hairy (sometimes all over) with a very dense velvety indumentum. Leaves spirally arranged, sessile or petiolate; lamina (4)6–13(16) × (1)1·5–3·5(4·5) cm., lanceolate, oblong-lanceolate, ovate-lanceolate, oblong-elliptic or rarely elliptic, chartaceous, usually olive or dark green on the upper surface, glaucous, silvery or greyish-green on the lower one, glabrous on both surfaces or pubescent on the lower and glabrous or sparsely pilose on the upper one, or often with a very dense velvety indumentum below, apex acute or very rarely subobtuse and usually acuminate, margin crenate-denticulate or rarely serrate and sometimes ciliate towards the base, base cuneate or rarely subrounded; secondary nerves 8–13 pairs; petiole up to 3 cm. (in our region leaves subsessile in only specimen known), glabrous or more often pilose or pubescent. Flowers axillary, solitary or in clusters of 2 or 3, pale purple or violet; bracts up to 4 mm. long, lanceolate-triangular or subulate, acute, usually densely pilose, rarely glabrous; pedicels up to 9 cm. long, slender, pilose or glabrous. Lateral sepals 4–7 × 1·5–2·5(3) mm., lanceolate to ovate-lanceolate, rarely oblong-lanceolate, acuminate, usually ± densely pilose, sometimes glabrous. Posterior sepal 9–14 mm. long, c. 2·5 mm. deep, shallowly and obliquely navicular, abruptly constricted into the spur, distally caudate, usually ± densely pilose, rarely glabrous; spur up to 5·5 cm. long, rather slender, subconical in the upper half, cylindric in the lower one, with a subacute end, ± densely pilose, rarely glabrous, strongly incurved when young, slightly incurved or often nearly straight when older. Anterior petal 8–15 mm. high and up to 17 mm. broad (when flattened), dorsally very narrowly cristate with crest often rim-like, apex usually entire, mucronate. Lateral united petals very variable in size, (13)17–25(30) mm. long, deeply 2-lobed; lobes nearly always ± overlapping, usually glabrous, rarely pubescent outside; anterior lobe 10–18 × 7–15 mm., obliquely semi-elliptic, semi-circular or rarely obovate, entire, with the lower margin usually ± straight; posterior lobe usually somewhat smaller (very rarely a little larger) than the anterior one, usually obovate, rarely subcircular, entire (see t. 26 fig. 13). Ovary glabrous. Capsule up to 1·8 cm. long, fusiform, glabrous. Seeds numerous, c. 2·5 × 1·7 mm., obovate in outline, brownish, with a papillose surface, glabrous.

Nyasaland. N: Nkata Bay, 600 m., fl. 21.ii.1961, *Richards* 14440 (K; SRGH).
Nyasaland, Angola, Congo, Cameroons, Uganda, Guinea, Tanganyika.

In moist places in forests, beside watercourses and in swamps. A very variable species in accordance with its wide distribution.

40. RUTACEAE

By F. A. Mendonça

Trees or shrubs, rarely suffrutices or perennial herbs, with odoriferous oil-glands, sometimes armed with prickles or spines. Leaves simple or pinnate or (1)3-foliolate; lamina dotted with pellucid glands all over the surface or on the margin only. Inflorescences of panicles, racemes, cymose clusters or glomerules, terminal, axillary, or terminal and axillary, sometimes also on older leafless branches. Flowers (2–3)4–5-merous, actinomorphic, bisexual or unisexual by abortion and dioecious. Sepals free or united, usually imbricate. Petals free, imbricate or rarely valvate. Stamens as many as or twice as many as the petals, free or rarely numerous and in phalanges, inserted at the base of a disk; anthers 2-lobed at the base, sometimes with an apical gland; staminodes well-developed or vestigial or absent. Ovary vestigial in the male flowers, in the female flowers of a single carpel which is sessile, subsessile or stipitate, 1-locular and 2-ovulate, or of 2–5(7) carpels united for their full length and sessile or on a short or long gynophore, or slightly united at the base and by the stigma (otherwise free), each carpel 1-locular and 2-ovulate, or rarely 1–∞ ovulate; style long or short or absent, terminal or lateral; stigma capitate or discoid, ± deeply lobed. Fruit baccate or drupaceous, 1–4-locular (1–3 loculi abortive), with 1-seeded loculi, or a 5-locular capsule with 2-seeded loculi, or 2-locular with 5–6-seeded loculi, or rarely a 1-seeded follicle. Seeds oblong, reniform or rounded; endosperm present or absent.

As well as the cultivated species of *Citrus* (see p. 209), *Murraya exotica* L., an ornamental shrub, and *Casimiroa edulis* Llave, a tree with edible fruits, are occasionally grown in gardens.

Leaves simple; flowers bisexual:
  Flowers 5-merous; stamens 5, alternating with 5 petaloid staminodes; fruit a septicidal 5-locular capsule; trees with broad opposite leaves - - **1. Calodendrum**
  Flowers 4-merous; stamens 8; fruit a loculicidal 2-locular capsule; shrublets with alternate linear deeply divided leaves - - - - - **2. Thamnosma**
Leaves compound, pinnate or (1)3-foliolate; flowers unisexual or bisexual:
  Leaves pinnate, alternate or rarely opposite; rhachis wingless or winged:
    Rhachis of leaf wingless:
      Trees or shrubs usually with the stems, leaves and sometimes the panicles aculeate; flowers unisexual, dioecious (in our area); fruit dehiscent, 1-seeded **3. Fagara**

Trees or shrubs not aculeate; flowers unisexual or bisexual; fruit indehiscent:
Deciduous trees; flowers unisexual, in panicles borne at the apex of leafless branches or a few occasionally with the young branches; fruit 4-seeded **4. Fagaropsis**
Evergreen shrubs or trees; flowers bisexual, in panicles borne in the axils of alternate leaves; fruit 1–2-seeded - - - - - - **5. Clausena**
Rhachis of leaf winged; branches with (0)1–2 straight axillary spines; flowers 4-merous, bisexual; fruit a globose berry - - - - - **6. Citropsis**
Leaves (1)3-foliolate:
Carpels (2)4, united at the base and by the stigma, otherwise free:
Stamens 4; ovary glabrous; carpels (2)4 - - - - - **7. Oricia**
Stamens (6–7)8; ovary densely pilose; carpels 2 - - - **8. Diphasiopsis**
Carpels 1–7, united for their full length:
Scrambling shrubs or lianes with stems, leaves and inflorescences retrorse-aculeate; flowers 5-merous; fruit 5–7-seeded - - - - - - **9. Toddalia**
Shrubs or trees not aculeate; flowers (2)4–5-merous; fruit 1–4(∞)-seeded:
Flowers unisexual, 4-merous; stamens (6–7)8:
Ovary (2)4-locular (or 1-locular by abortion), smooth; fruit smooth, (1)4-seeded - - - - - - - - - - - - - **10. Vepris**
Ovary 4-locular, verrucose; fruit verrucose, 4-seeded **11. Toddaliopsis**
Flowers unisexual or bisexual; stamens 4–5 or numerous:
Flowers unisexual (rarely bisexual), 4(5)-merous; ovary 1-locular; fruit 1-seeded - - - - - - - - - - - - **12. Teclea**
Flowers bisexual, 5-merous; stamens numerous, in phalanges; ovary (4)5–many-locular; fruit ∞-seeded (introduced, sometimes subspontaneous) **13. Citrus**

1. CALODENDRUM Thunb.

Calodendrum Thunb., Nov. Gen. **2**: 41 (1782) *nom. conserv.*

Tall trees with opposite branches. Leaves opposite, simple, exstipulate, petiolate, dotted with pellucid glands. Inflorescence a terminal panicle. Flowers bisexual, large. Sepals 5, small, ovate. Petals 5, imbricate. Stamens 5, opposite the sepals; anthers with an apical gland; staminodes 5, petaloid, opposite the petals, inserted at the base of a cupuliform disk. Ovary on an elongated gynophore, verrucose-glandular, 5-lobed, 5-locular; loculi 2-ovulate; style filiform, long; stigma small. Fruit a globose woody verrucose stipitate 5-locular capsule; loculi 2-seeded, dehiscence valvate. Seeds large, angular, black, shiny.

Calodendrum capense (L.f.) Thunb., Nov. Gen. Pl. **2**: 41, 43 (1782).—Sond. in Harv. & Sond., F.C. **1**: 371 (1860).—Burtt Davy, F.P.F.T. **2**: 479 (1932).—Steedman, Trees etc. S. Rhod.: 28 (1933).—Palgrave, Trees Centr. Afr.: 396 cum tab. et phot. (1957).—Dale & Greenway, Kenya Trees & Shrubs: 481, fig. 87 & phot. 66 (1961). TAB. **28**. Type from S. Africa (Cape Prov.).
Dictamnus capensis L.f., Suppl. Pl.: 232 (1781). Type as above.

Tree up to 20 m. tall; young branches, leaves and inflorescences densely tomentose with mixed simple, stellate and branched hairs, soon becoming glabrous. Leaf-lamina 6–13(18) × 3·5–7·5(11) cm., elliptic to broadly elliptic, acute or slightly apiculate or rounded at the apex, cuneate or obtuse or slightly cordate at the base; petiole 2–10 mm. long. Inflorescence up to 15 cm. long, with opposite cymose branches. Flowers erect, mauve; bracteoles 1·3–1·5 mm. long, linear, caducous; pedicels up to 1·5 cm. long, stout. Sepals 3·2 × 1·8 mm., free or slightly united at the base, ovate, minutely tomentose with branched hairs on the outer surface and simple sericeous hairs within. Petals 3–3·5 × 0·6–0·7 cm., narrowly oblanceolate, straight or somewhat recurved, whitish or mauve, tomentose with simple or branched hairs on the outside and simple hairs directed downwards inside. Stamens as long as the petals, inserted at the base of an annular disk; filaments filiform; anthers 1·3–1·5 mm. long, dorsifixed, sagittate; staminodes longer than the petals, very narrowly oblanceolate, caudate, dotted with bright crimson glands. Gynophore 5 mm. long, glandular. Ovary 2–3 mm. in diam., globose, apiculate, papillose; style 18–20 mm. long; stigma capitate. Fruit a 5-lobed strongly rugose warty septicidal capsule 3·5 cm. in diam., dehiscing from below with the 5 valves remaining attached to the top of a central column.

S. Rhodesia. W: Matobo Distr., Besna Kobila, fr. i.1954, *Miller* 2073 (EA; K; LISC; SRGH). E: Umtali, Commonage, fl. 10.xi.1948, *Chase* 1324 (BM; COI;

Tab. 28. CALODENDRUM CAPENSE. 1, branch with flowers and leaves ($\times\frac{2}{3}$); 2, pendent dehisced fruit ($\times\frac{2}{3}$); 3, seed ($\times\frac{2}{3}$); 4, cross section of ovary ($\times 6$); 5, stamen and staminode ($\times 1\frac{1}{2}$); 6, gynoecium, gynophore and disk ($\times 2$); 7, longitudinal section of ovary ($\times 8$). All from *Chase* 1930, except 3 from *Palgrave & Trimen* s.n.

LISC; SRGH). S: Bikita, fl. 16.i.1957, *Cleghorn* 225 (SRGH). **Nyasaland.** S: Port Herald, Malawe Hills, fl. 15.ii.1933, *Lawrence* 131 (K).
Also in Cape Prov., Natal, Transvaal, Tanganyika and Kenya. In evergreen riverine and higher-altitude forests.

The Cape Chestnut. An ornamental shade-tree with sweet-scented showy mauve flowers.

2. THAMNOSMA Torr. & Frém.

Thamnosma Torr. & Frém. in Frém., Rep. Expl. Exped.: 313 (1845).

Shrublets or perennial herbs 0·2–0·6 m. high, with pellucid gland-dots on all parts. Leaves alternate, entire or deeply divided, sessile or petiolate, linear to narrowly oblanceolate. Inflorescence of terminal or axillary racemes or panicles. Flowers bisexual, 4-merous, pedicellate. Stamens 8, inserted at the base of an annular disk. Ovary of 2 united carpels, sessile, 2-locular and prominently 2-lobed at the apex; loculi 4–6-ovulate; style filiform, arising between the lobes of the ovary; stigma capitate. Fruit a capsule dehiscing ventrally by a slit. Seeds echinate.
The African species belong to Subgen. *Palaeothamnosma* Engl.; New World species belong to Subgen. *Thamnosma*.

Thamnosma rhodesica (Bak. f.) Mendonça, stat. nov. TAB. **29** fig. A. Lectotype: S. Rhodesia, Bulawayo, *Rand* 83 (BM).
Thamnosma africanum var. *rhodesica* Bak. f. in Journ. of Bot. **37**: 426 (1899).— Eyles in Trans. Roy. Soc. S. Afr. **5,** 4: 387 (1916). Syntypes: S. Rhodesia, Bulawayo, *Rand* 83 (BM), 297 (BM).

Shrublet or perennial herb, corky at the base, 0·2–0·6 m. tall, with conspicuous glandular dots in all parts, aromatic. Leaves simple, alternate; lamina 8–32 × 0·9–1·4 mm., linear, blunt at the apex, entire or 3(5)-fid with the entire leaves localized at the lower part of the stem and extremities of the branches; petiole 0–8 mm. long. Inflorescence of racemes or axillary and terminal panicles. Flowers yellow; pedicels 3–5 mm. long (elongating to 15 mm. in fruit). Sepals 4, 1·3–1·6 mm. long, ovate, slightly united at the base. Petals 5·3–5·7 × 3·3 mm., free, imbricate, elliptic. Stamens as long as the petals; anthers small, ovate, with an apical gland. Ovary 1·5 mm. long, subglobose; style 3 mm. long. Fruit 8–11 × 5–6 mm., coriaceous, 2-lobed, somewhat compressed; loculi 4–6-seeded. Seeds 1 mm. in diam., subreniform.

Bechuanaland Prot. SE: Serowe, fl. & fr. 13.i.1955, *Mogg* 24562 (K; PRE). **S. Rhodesia.** W: Matobo Distr., fl. & fr. 29.xii.1948, *West* 2828 (SRGH). S: Essexvale, fl. & fr. 28.ix.1942, *Hopkins* (K; SRGH).
In southern and western S. Rhodesia and SE. Bechuanaland. Dry bush and open woodland.

This species is intermediate, both in morphology and distribution, between *T. africana* Engl. from southern Angola and SW. Africa and *T. densa* N.E.Br. (*T. africana* var. *crenata* Engl.) from the Transvaal.

3. FAGARA L.

Fagara L., Syst. Nat. ed. 10, **2**: 897, 1362 (1759) *nom. conserv.*

Deciduous or evergreen aculeate trees or shrubs; trunk usually with bosses. Leaves alternate, imparipinnate; leaflets (1)2–12 (or more)-jugate, opposite or alternate, sessile or petiolulate, usually ± asymmetric; lamina crenulate, serrate or entire, dotted with pellucid glands (sometimes only in the marginal sinuses). Inflorescence of terminal or axillary and terminal panicles, or of racemes borne at the base of new branches below the leaves. Flowers unisexual by abortion, dioecious (in our area), 4–5-merous. Sepals very small, usually persistent. Petals valvate or slightly imbricate, much longer than the sepals. Male flowers: stamens opposite the sepals; gynophore short; ovary vestigial. Female flowers: staminodes vestigial, opposite the sepals (in our area); gynophore short; ovary of 1 carpel, or rarely of 2 carpels of which 1 aborts (in our area), 1-locular, with 2 subapical collateral pendulous ovules (1 aborted); style terete, shorter than the ovary, oblique, incurved; stigma discoid or subcapitate. Fruit a globose or subglobose

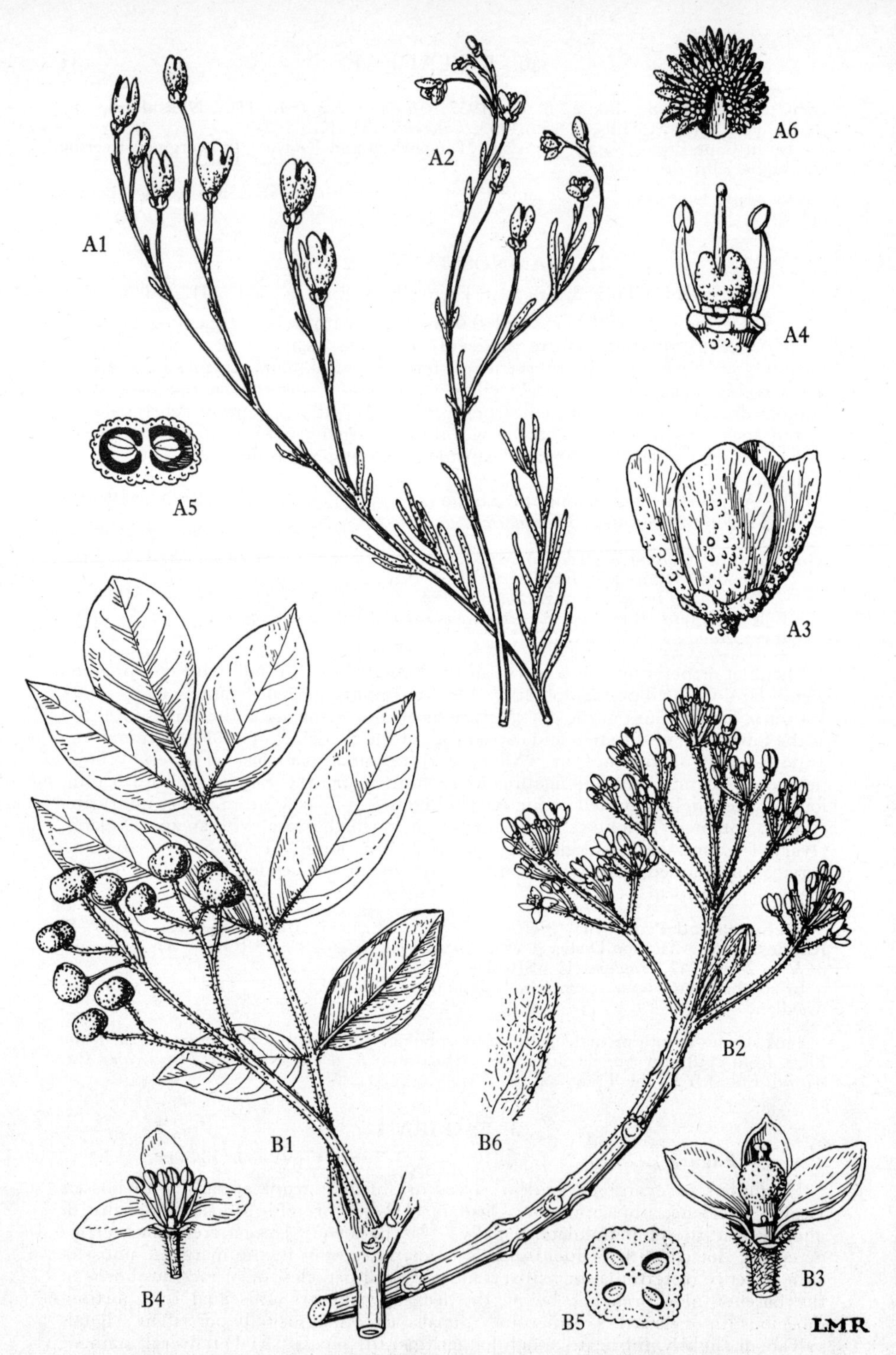

Tab. 29. A.—THAMNOSMA RHODESICA. A1, branchlet with fruits ($\times \frac{2}{3}$); A2, branchlet with flowers ($\times \frac{2}{3}$); A3, flower ($\times 4$); A4, gynoecium, disk and 2 of the 8 stamens ($\times 4$); A5, cross section of ovary ($\times 8$); A6, seed ($\times 9$). 1 and 6 from *Rand* 297, 2–5 from *Rand* 83. B.—FAGAROPSIS ANGOLENSIS VAR. MOLLIS. B1, apex of stem with fruit and mature leaves ($\times \frac{2}{3}$) *Chase* 4289; B2, apex of stem with female flowers ($\times \frac{2}{3}$) *Chase* 953; B3, single female flower with one petal and one sepal removed ($\times 3$) *Chase* 953; B4, single male flower with three sepals, one petal and two stamens removed ($\times 3$) *Gossweiler* 10564; B5, cross section of ovary ($\times 8$) *Chase* 953; B6, leaf margin showing glands ($\times 6$) *Chase* 4289.

dehiscent 1-seeded follicle, stipitate or subsessile, glandular-foveolate, usually with the seed hanging from the detached placenta during dehiscence. Seed ± globose, black, shiny.

Leaflets on flowering shoots with pellucid glands confined to the margins:
  Flowers 4-merous, in terminal panicles; rhachis of leaves on sterile (and frequently on fertile) shoots glabrous, with spines (when present) directed upwards or almost straight:
    Leaflets usually obtuse or rounded at the apex, ovate to elliptic or obovate, with 4–8 pairs of primary lateral nerves - - - - - - - 1. *capensis*
    Leaflets acute or acuminate at the apex, lanceolate to oblong or narrowly elliptic, with 16–20 or more pairs of primary lateral nerves - - - - - 2. *davyi*
  Flowers 5(6)-merous, in racemes borne at the base of the new branches below the leaves; rhachis of leaves on sterile and fertile shoots pubescent, with spines (when present) down-curved - - - - - - - - - - 3. *humilis*
Leaflets on flowering shoots with pellucid glands scattered throughout the lamina:
  Deciduous shrubs or trees with racemose or paniculate inflorescences borne at the base of the new branches below the leaves; spines and thorns down-curved 4. *chalybea*
  Deciduous shrubs or trees with elongate terminal panicles:
    Leaflets papyraceous, 3–4-jugate, usually pubescent beneath:
      Lateral leaflets with petiolules (1·5) 2–4 mm. long; lamina-base broadly to narrowly cuneate; inflorescence glabrous or very sparsely pubescent 5. *holtziana*
      Lateral leaflets sessile or with petiolules up to only 1(1·5) mm. long; lamina-base broadly cuneate to shallowly cordate; inflorescence persistently ± densely pubescent, rarely glabrescent - - - - - - - 6. *trijuga*
    Leaflets chartaceous to coriaceous, (1)3–8-jugate, glabrous beneath:
      Leaflets 2–6(7) cm. long; spines on rhachis down-curved; shrub 7. *schlechteri*
      Leaflets (5)7–17 cm. long; spines on rhachis straight (or absent); inflorescence pyramidal; trees:
        Leaflets slightly unequal at the base, chartaceous, up to 10 cm. long, with veins impressed on the upper surface - - - - - - 8. aff. *braunii*
        Leaflets (at least on flowering shoots) strongly asymmetric at the base, coriaceous, up to 17 cm. long, with veins not impressed on the upper surface, usually slightly prominent - - - - - - - - - 9. *macrophylla*

1. **Fagara capensis** Thunb., Prodr. Pl. Cap. **1**: 28 (1794).—Verdoorn in Journ. of Bot. **57**: 204 (1919) pro parte excl. ref. Sim sub syn. *Xanthoxylon capense*.—Engl. in Engl. & Prantl, Nat. Pflanzenfam. ed. 2, **19a**: 223 (1931).—Steedman, Trees etc. S. Rhod.: 28 (1933). TAB. **30** fig. A. Type from the Cape.

Fagara armata Thunb., loc. cit. Type from the Cape.

Zanthoxylum capense (Thunb.) Harv. in Harv. & Sond., F.C. **1**: 446 (1860) (" Xanthoxylon "). Type as for *F. capensis.*

Fagara multifoliolata Engl., Bot. Jahrb. **23**: 149 (1896). Type from S. Africa (Cape Prov.).

Fagara magalismontana Engl., op. cit. **46**: 408 (1911).—Burtt Davy, F.P.F.T. **2**: 478 (1932). Type from the Transvaal.

Tree up to 10 m. tall; trunk usually armed with aculeate corky bosses; young branches and inflorescences minutely pubescent or glabrous; stems and the rhachis of the leaves sometimes aculeate; aculei straight, greyish, up to 8 mm. long. Leaves 4–12 cm. long; rhachis together with the petiole 4–8 cm. long, deeply grooved above; leaflets sessile, opposite or alternate, (2)3–8-jugate, becoming progressively larger from the basal pair upwards to the apical one; lamina 1–4 × 1–2·3 cm., elliptic to broadly elliptic or obovate, obtuse or rounded or rarely acute at the apex, margin slightly crenulate or serrulate with pellucid glands in the sinuses only, obtuse and usually with small callose auricles at the base, ± asymmetric, the terminal leaflet frequently atrophied or aborted; nerves 4–8 pairs. Inflorescences of terminal panicles, 2–6 cm. long; bracts 0·6 mm. long, ovate, ciliolate. Flowers 4-merous, subsessile or with pedicels up to 1·5 mm. long. Sepals 4, free, 0·5 mm. long, ovate, persistent on the fruit. Petals 4, imbricate, 2·7–3·1 × 1–1·3 mm., narrowly elliptic. Male flowers: stamens 4; vestigial ovary very small. Female flowers: staminodes 4, reduced to the aborted anthers, inserted at the base of a stout gynophore 0·3 mm. long; ovary glabrous, ovoid, conspicuously dotted with glands, 1-locular, 2-ovulate; style c. 0·8 mm. long; stigma discoid, black. Fruit 4–4·5 mm. in diam., subsessile or with a short stipe up to 2 mm. long, subglobose, gland-dotted. Seed black, shiny.

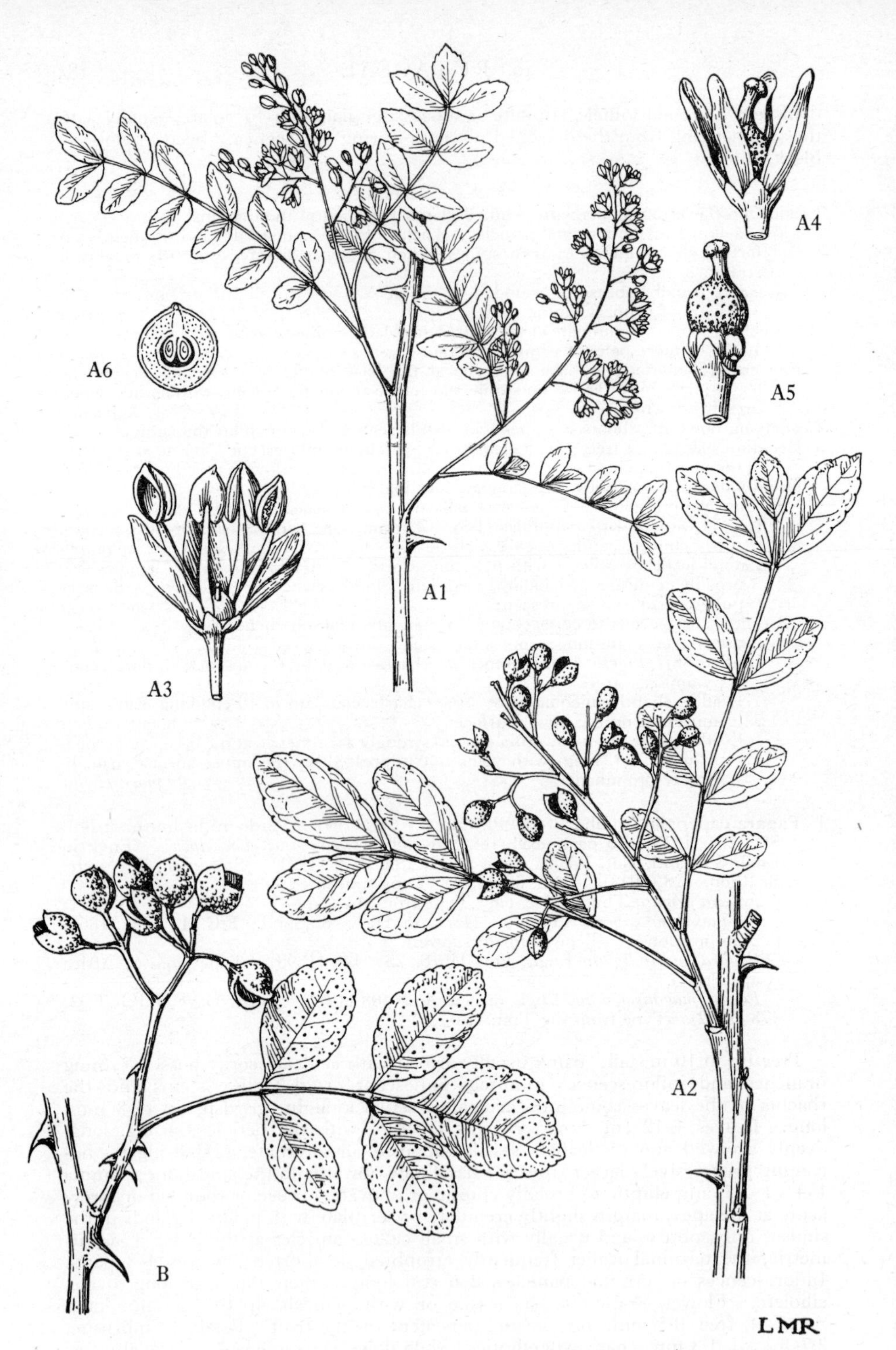

Tab. 30. A.—FAGARA CAPENSIS. A1, branchlet with leaves and male inflorescences (× ⅔) *Mendonça* 2853; A2, branchlet with leaves and fruits (× ⅔) *Torre* 2563; A3, male flower with 1 petal removed (× 6) *Mendonça* 2853; A4, female flower (× 6) *Monro* 791; A5, female flower with 1 sepal and petals removed (× 6) *Monro* 791; A6, cross section of ovary (× 12) *Monro* 791. B.—FAGARA SCHLECHTERI, branchlet with fruits (× ⅔) *E.M. & W.* 701.

S. Rhodesia. W: Matopos, World's View, fr. 13.iv.1955, *E.M. & W.* 1487 (BM; LISC; SRGH). C: Chilimanzi, fr. 24.i.1951, *McGregor* 10/51 (K; SRGH). E: Inyanga, 1740 m., st. 28.v.1954, *Chase* 5257 (BM; LISC; SRGH). S: Victoria Distr., ♀ fl. 1909, *Monro* 791 (BM; BOL; SRGH). **Mozambique.** SS: Massinga, Inhachengo, fr. 26.ii.1955, *E.M. & W.* 617 (BM; LISC; SRGH). LM: Maputo, Bela Vista, ♂ fl. 14.xi.1944, *Mendonça* 2853 (K; LISC; LM; SRGH).

In S. Africa from the Cape to the east and north, S. Rhodesia and southern Mozambique. Dry woodland or bush on sandy soils and in rocky places.

Knobwood.

Fagara capensis Thunb. has been frequently and variously misinterpreted. In the Uppsala Herbarium there are four Thunberg specimens lettered α, β, γ and δ respectively, of which only specimen β, with ripe fruit, is a *Fagara* and this must be considered to be the type of the species. The other three specimens are *Clausena anisata*. In the Lund Herbarium there are two Thunberg specimens, identified as *F. capensis* in his own handwriting, one with flower and the other with immature fruit; both are *C. anisata*. A third (sterile) specimen collected by Sparmann is also *C. anisata*. Many subsequent authors have neglected Thunberg's name and created numerous new epithets, specific and varietal, which Verdoorn (loc. cit.) later reduced to synonymy or left in doubt (e.g. *Fagara multifoliolata* Engl.). The great variability found in juvenile shoots, suckers and coppiced specimens has also led to confusion. Sketches made by Exell of the types (now destroyed) of *F. magalismontana* and *F. multifoliolata* in the Berlin Herbarium, when compared with coppice shoots from our area, leave no doubt that they are both *F. capensis*.

2. **Fagara davyi** Verdoorn in Journ. of Bot. **57**: 203 (1919).—Burtt Davy, F.P.F.T. **2**: 478 (1932) pro parte.—Engl. in Engl. & Prantl, Nat. Pflanzenfam. ed. 2, **19a**: 222 (1931). Syntypes from S. Africa.

Xanthoxylon thunbergii var. *grandifolia* Harv. in Harv. & Sond., F.C. **1**: 446 (1860). Type from S. Africa.

Tree 8–30 m. tall; trunk armed with thorny bosses; branches glabrous, aculeate with aculei 1–5 mm. long, straight or slightly curved upwards. Leaves 8–25 cm. long; rhachis grooved above. Leaflets sessile, opposite or subopposite, (2)3–6-jugate, becoming progressively larger from the basal pair upwards to the apical one, terminal leaflet sometimes petiolulate or sometimes aborted; lamina 2–7 × 0·8–2·5 cm., lanceolate to narrowly elliptic and ± asymmetric, dotted with pellucid glands in the marginal sinuses only (in our area), acuminate and slightly retuse at the apex, margin serrate or crenulate-serrate, cuneate or obtuse and slightly callose-auriculate at the base; nerves very numerous and rather closely spaced. Inflorescence of terminal panicles 4–6 cm. long; bracts and bracteoles very small, caducous. Flowers 4-merous. Male flowers: pedicellate, with pedicels 1–1·5 mm. long, usually in clusters at the ends of the panicle-branches; sepals 4, slightly united at the base, ovate, 0·6 mm. long; petals 2·7 × 1·3 mm., imbricate, elliptic; stamens 4; filaments 3·5 mm. long, linear; anthers dorsifixed, ovate, 1 mm. long; gynophore very short, surmounted by the vestigial ovary. Female flowers: not seen from our area. Fruit 5 mm. in diam., subsessile, gland-dotted. Seed globose, black, shiny.

S. Rhodesia. E: Umtali, ♂ fl. 17.i.1950, *Chase* 1926 (BM; COI; K; SRGH). S: Victoria Distr., fr. iv.1921, *Eyles* 6596 (SRGH).

Also in the Transvaal, Swaziland and Cape Prov. (Transkei). In forests or forest patches.

3. **Fagara humilis** E. A. Bruce in Bothalia, **6**: 234 (1951). Type from the Transvaal (Kruger National Park).

Deciduous shrub up to 2 m. high; branches and rhachis of the leaves cinereous-tomentose or pubescent or glabrescent, ± densely aculeate, aculei on the stems straight or ± recurved, 2–8 mm. long, greyish or reddish, those of the rhachis recurved and up to 2 mm. long; branches with terminal buds protected by cottony scales. Leaves 4–16 cm. long; rhachis terete, rarely slightly grooved above. Leaflets chartaceous, sessile, (4)6–12(14)-jugate, opposite or alternate, the distal ones smaller, the terminal one usually present (rarely aborted), 0·8–2·5 × 0·5–1·2 cm., elliptic to broadly elliptic or ovate, slightly asymmetric, with pellucid gland-dots in the marginal sinuses only, obtuse or rounded at the apex, margin crenulate, rounded or almost cuneate at the base; lateral nerves 3–4 pairs. Inflorescence of racemes 1–3 cm. long, borne on branches of the previous year before or with the leaves but not or very rarely in the axils of the latter. Flowers 5-merous. Male flowers: pedicels 1–1·5 mm. long; sepals 5, slightly united at the base, 0·6 mm. long, persistent on the fruit; petals 5, 3–4 × 1–1·3 mm., imbricate, narrowly elliptic; stamens 5, with linear filaments c. 3 mm. long; anthers 1·5 mm. long, narrowly oblong, basifixed, deeply 2-lobed at the base and laterally dehiscent; ovary vestigial, on a short gynophore. Female flowers: sessile or subsessile; staminodes 5, reduced to the abortive anthers at the base of a short gynophore; ovary c. 1 mm. long, ovoid, oblique, 1-locular, 2-ovulate; style 1 mm. long, lateral, incurved; stigma subglobose. Fruit 4–5 mm. in diam., subglobose, gland-dotted. Seed black, shiny.

S. Rhodesia. S: Nuanetzi, ♀ fl. 3.xi.1955, *Wild* 4706 (K; LISC; SRGH). **Mozambique.** SS: Limpopo, Mapai, fr. 8.v.1944, *Torre* (K; LISC; PRE; SRGH). LM: Magude, between Moine and Uanetze, ♂ fl. 29.xi.1944, *Mendonça* 3144 (K; LISC; SRGH).

Also in the Transvaal. In *Colophospermum mopane* woodland.

4. **Fagara chalybea** (Engl.) Engl. in Engl. & Prantl, Nat. Pflanzenfam. **3,** 4: 118 (1896).—White, F.F.N.R.: 170 (1962). Type from Tanganyika.
Zanthoxylon chalybeum Engl., Pflanzenw. Ost-Afr. **C**: 227 (1895). Type as above.
Zanthoxylon olitorium Engl., loc. cit. Type from Tanganyika.
Fagara olitoria (Engl.) Engl. in Engl. & Prantl, Nat. Pflanzenfam. **3**, 4: 118 (1896). Type as above.
Fagara merkeri Engl., Bot. Jahrb. **36**: 242 (1905). Type from Tanganyika.
Fagara mpwapwensis Engl., op. cit. **46**: 410 (1911). Type from Tanganyika.

Deciduous shrub or small tree up to 6 m. tall; branches glabrous with terminal buds protected by dark scales, aculeate, aculei 2–10 mm. long, ± recurved, reddish becoming greyish. Leaves 6–20 cm. long; petiole 1–5 cm. long, somewhat flattened above at the base; rhachis subterete, slightly grooved above, usually aculeate below; leaflets papyraceous to subchartaceous, (1)2–5-jugate, opposite or occasionally subopposite, sessile or the terminal one sometimes petiolulate with petiolule up to 15 mm. long; lamina 2·5–7 × 1–2·5 cm., ovate-oblong to elliptic, acute or bluntly rounded at the apex, margin slightly crenulate or subentire, rounded or obtuse at the base, sparsely dotted with pellucid glands; lateral nerves 6–9 pairs. Inflorescence glabrous, of racemes or panicles up to 9 cm. long, borne at the base of the new branches below the first leaves (rarely also in the axils); rhachis flexuous and ± pendulous in male and straight in female plants. Flowers 4-merous. Male flowers: pedicellate, usually in clusters, with slender pedicels 1·5–2 mm. long; sepals 4, 0·4 mm. long, united at the base; petals 2·5 × 1 mm., imbricate, elliptic; stamens 4 with filaments as long as the petals; anthers basifixed, deeply 2-lobed at the base; gynophore very short; vestigial ovary 1·5 mm. long, ellipsoid. Female flowers: subsessile, usually glomerate on pulvinoid nodes of the racemes; sepals 4; staminodes vestigial, reduced to the aborted anthers; ovary 1·5 mm. long, very oblique, 1-locular, 2-ovulate; ovules subapical, collateral, pendulous, 1 aborted; style short, incurved; stigma broadly discoid, peltate. Fruit c. 6 mm. in diam., a subsessile or stipitate somewhat oblique subglobose follicle, glandular-foveolate, with stipe up to 1·5 mm. long. Seed black, shiny.

N. Rhodesia. W: Mufulira, fr. 29.iv.1934, *Eyles* 8149 (K; SRGH). C: Mazabuka, st. 1930, *Stevenson* 23/30 (K). **S. Rhodesia.** N: Urungwe, st. 14.vii.1953, *Lovemore* 368 (K; SRGH). C: Hartley, 16.xi.1932, *Hornby* 3266 (K; SRGH). **Nyasaland.** N: Mzimba R., 1370 m., fr. 26. ii 1959, *Robson* 1725 (BM; K; LISC; PRE; SRGH).

From Somaliland and Ethiopia to S. Rhodesia. In dry woodland on termite mounds.

5. **Fagara holtziana** Engl., Bot. Jahrb. **36**: 242 (1905).—Brenan, T.T.C.L.: 541 (1949). Type from Tanganyika (Dar es Salaam).

Shrub or small tree (2)3–15 m. high, sometimes scandent; trunk armed with conical corky bosses c. 5 cm. in diam.; branches, leaves and inflorescences sometimes shortly rusty- or brown-pubescent; stems ± sparsely aculeate with straight or slightly recurved aculei 2–7 mm. long; rhachis of leaves usually unarmed or rarely with 1 or 2 short aculei. Leaves 15–37 cm. long; leaflets papyraceous, 3–4-jugate, opposite, with petiolule (1·5)2–4 mm. long, the terminal one with petiolule up to 30 mm. long; lamina 7–15 × 3–6·5 cm., ovate to elliptic or oblong-elliptic, densely dotted with pellucid glands, obtuse to acute or shortly acuminate at the apex, margin crenulate to minutely and obtusely serrulate, acute or obtuse and sometimes ± asymmetric at the base; lateral nerves 8–12 pairs. Inflorescence a terminal panicle 8–15 cm. long. Flowers in cymose clusters (rarely solitary) with pedicels (1)2–3 mm. long; buds 2 mm. long, ellipsoid-cylindric. Sepals 4, 0·5 mm. long, slightly united at the base, ovate to subcircular, imbricate, persistent, margin ciliolate. Petals 4, 2·5 × 1 mm., white, imbricate. Male flowers: stamens 4; filaments linear, bent inwards near the apex in bud; anthers c. 0·8 mm. long, dorsifixed, deeply 2-lobed at the base, laterally dehiscent; disk or gynophore short, glabrous; vestigial ovary ovoid, obsoletely lobed. Female flowers: petals c. 3 × 1 mm.; staminodes 4, vestigial, reduced to the aborted anthers; gynophore c. 0·4 mm. long; ovary of 1 carpel, c. 1·5 mm. long, oblique; style lateral, c. 1 mm. long, incurved; stigma discoid, black. Fruit 5 mm. in diam., red, globose, densely gland-dotted. Seed compressed-globose, black.

Mozambique. N: between Diaca and Mueda, fr. 23.iii.1961, *Gomes e Sousa* 4662 (K).
Also in Kenya and Tanganyika. In forest or woodland or in thickets on coral rock, never far from the coast, 0–240 m.

6. **Fagara trijuga** Dunkley in Kew Bull. **1934**: 186 (1934).—Brenan, T.T.C.L.: 542 (1949).—White, F.F.N.R.: 170 (1962). Type: N. Rhodesia, Bombwe, *Martin* 137 (K).

Shrub or small tree; branches, leaves and inflorescences rusty-pubescent or greyish-tomentellous, glabrescent; stems and leaf-rhachis ± sparsely aculeate with straight or slightly recurved aculei 2–8 mm. long, rhachis of leaves sometimes unarmed. Leaves 15–35 cm. long; leaflets papyraceous, 3–4-jugate, opposite or subopposite, sessile or petiolulate with petiolules up to 1·5 mm. long, the terminal one with petiolule up to 30 mm. long, obtuse or rounded to shallowly cordate at the base; lamina 4–10 × 3–5·5 cm., ovate to elliptic, densely dotted with pellucid glands, acute at the apex, margin crenulate to minutely serrate, obtuse or rounded at the base; lateral nerves 6–10 pairs. Inflorescence a terminal panicle 12–18 cm. long. Flowers usually in cymose clusters with pedicels up to 1 mm. long; buds 3 mm. long, ellipsoid. Sepals 4, 1 mm. long, slightly united at the base, ovate to subcircular, imbricate, persistent, margin ciliolate. Petals 4, 3·5 × 1·3 mm., imbricate. Male flowers: stamens 4; filaments linear, bent inwards near the apex in bud; anthers dorsifixed, deeply 2-lobed at the base, laterally dehiscent; disk or gynophore pulviniform, glabrous; vestigial ovary obsoletely lobed. Female flowers: not seen. Fruit 5 mm. in diam., globose, densely gland-dotted. Seed subglobose, black.

N. Rhodesia. B: Sesheke Distr., Sisisi Forest, near Masese, ♂ fl. 23.xii.1952, *Angus* 1008 (BM; FHO; K). S: Bombwe, fr. 1933, *Martin* 571 (EA; BM; K).
In south-western N. Rhodesia and central and south-western Tanganyika. Woodland and thickets in dry regions, mainly in *Baikiaea* woodland in our area.

7. **Fagara schlechteri** Engl., Bot. Jahrb. **46**: 409 (1911); Pflanzenw. Afr. **3,** 1: 750 (1915); in Engl. & Prantl, Nat. Pflanzenfam. ed. 2. **19a**: 223 (1931). Type: Mozambique, Lourenço Marques, *Schlechter* 12005 (B†; BM). TAB. **30** fig. B.
Xanthoxylum capense sensu Sim, For. Fl. Port. E. Afr.: 23, t. 17A (1909). The plate is very inaccurate.
Fagara capensis sensu Verdoorn in Journ. of Bot. **47**: 204 (1919) pro parte quoad syn. *Xanthoxylon capense*.

Shrub or small tree up to 4 m. tall; young branches glabrous, reddish, sparsely aculeate; aculei recurved, 4–10 mm. long. Leaves 5–18 cm. long; petiole and rhachis terete, narrowly grooved above, sometimes aculeate. Leaflets (1)2–4-jugate, chartaceous or subcoriaceous, opposite or subopposite, subsessile or shortly petiolulate, terminal one usually with a petiolule (or segment of the rhachis) 0·5–2·5 cm. long; lamina (2)3–7 × (1)2–3·5 cm., elliptic to broadly elliptic, sparsely dotted with very conspicuous pellucid glands, rounded or obtuse at the apex, margin crenulate, abruptly and narrowly cuneate at the base; midrib impressed on the upper surface, prominent below; lateral nerves 5–10 pairs. Inflorescence 2–10 cm. long, of terminal and sometimes also axillary subcorymbose panicles, glabrous; bracts and bracteoles very small; pedicels 1–2 mm. long. Flowers 4-merous, usually in cymose clusters at the ends of the panicle-branches. Male flowers: sepals 4, 0·4–0·6 mm. long, united at the base, obtuse or rounded; petals 4, 3·5 × 1·5 mm., imbricate; stamens 4, inserted at the base of the gynophore; filaments flattened, subulate, longer than the petals; anthers 1·3 mm. long, basifixed, deeply lobed at the base; gynophore 0·6 mm. long; vestigial ovary c. 0·8 mm. long. Female flowers. petals 4·5 × 1·7 mm.; staminodes 4, vestigial, reduced to the aborted anthers; gynophore very short; ovary of 1 carpel, 1·5 mm. long, oblique, 1-locular, 2-ovulate; style lateral, short, incurved; stigma broadly discoid, black. Fruit c. 8 mm. in diam., with stipe c. 1 mm. long, globose, densely glandular-foveolate, orange-red. Seed c. 6 mm. in diam., subglobose, black, shiny.

Mozambique. SS: Zavala, Quissico dunes, fr. 28.ii.1955. *E.M. & W.* 701 (BM; LISC; SRGH). LM: Lourenço Marques, dunes of the Costa do Sol, ♀ fl. 15.xi.1944, *Mendonça* (LISC; K; LM; SRGH).

Known only from the coastal dunes of southern Mozambique.

8. **Fagara** sp. aff. **braunii** Engl., Bot. Jahrb. **46**: 407 (1911).—Brenan, T.T.C.L.: 541 (1949). Type from Tanganyika (Usambara).

Tree c. 10 m. tall; trunk with aculeate bosses; branches glabrous, aculeate, aculei 2–4 mm. long, conical, reddish-brown. Leaves c. 30 cm. long; petiole 5–6 mm. long, terete, with a dorsal line of pubescence; rhachis 2–3 mm. in diam., terete, dorsally pubescent in the lower part. Leaflets chartaceous, glabrous, with petiolules c. 1 mm. long, subopposite; lamina 7–10 × 2·5–3·4 cm., oblong to elliptic-oblong, abruptly acuminate to ± attenuate and subacute at the apex, with margin rather remotely crenulate-denticulate, reflexed, broadly cuneate to rounded and ± asymmetric at the base; lateral nerves 18–24 pairs, impressed above; glandular dots conspicuous, those beside the marginal teeth rather larger with darker contents. Inflorescence a terminal panicle 15 cm. long, pubescent; bracts and bracteoles very small; pedicels 2–4 mm. long. Flowers 4-merous, borne in clusters along the panicle branches. Female flowers: sepals c. 0·5 mm. long, united at the base, lobes rounded; petals 2·5 (3) × 1·5 mm., imbricate, reflexed; staminodes vestigial, reduced to the aborted anthers, opposite the sepals; gynophore short; ovary of a single carpel, oblique, 1-locular, 2-ovulate; style lateral, short, incurved; stigma discoid. Fruit immature.

Nyasaland. N: Vipya Mts., near Chikangawa, Chamambo rain forest, fr. 22.i.1956, *Chapman* 268 (BM; FHO; K).

"In a slightly open patch in the rain forest."

This specimen may belong to *F. braunii* Engl., hitherto known only from south and east Tanganyika. *F. braunii* differs, however, in having leaflets softly pubescent (at least on the nerves below) and with plane margins and denser teeth and crenulations. It seems advisable, therefore, to refrain from recording *F. braunii* from Nyasaland, at least until more material is available.

9. **Fagara macrophylla** (Oliv.) Engl. in Engl. & Prantl, Nat. Pflanzenfam. **3**, 4: 118 (1896).—White, F.F.N.R.: 170 (1962). Type from Principe.

Zanthoxylum? macrophyllum Oliv., F.T.A. **1**: 304 (1868). Type as above.

Fagara inaequalis Engl., Bot. Jahrb. **54**: 303 (1917).—Gilbert, F.C.B. **7**: 88 (1958). Syntypes from southern Cameroons.

Tree 10–30 m. tall; trunk usually with aculeate bosses; branches glabrous, aculeate; aculei 2–5 mm. long, conical, reddish. Leaves 20–60 cm. long; petiole 3–15 cm. long, broadly channelled or flat above at the base; rhachis 2–4 mm. in diam., terete; leaflets coriaceous, with petiolules 2–7 mm. long, alternate or subopposite; lamina (5·5)8–17 × 3·5–5·5 cm., broadly elliptic to elliptic, gland-dots almost inconspicuous, attenuate or abruptly acuminate and obtuse at the apex, margin entire or slightly crenulate, very asymmetric at the base; lateral nerves 8–14 pairs. Flowers not seen from our area. Infructescence of rather stout terminal and sometimes also axillary panicles, up to 20 × 14 cm. Fruit 3·5 mm. in diam., sessile or subsessile or shortly stipitate, subglobose, glandular-foveolate, with a very small 5-lobed persistent calyx. Seed 2 mm. in diam., globose, black, shiny.

N. Rhodesia. N: Fort Rosebery Distr., near Samfya, st. 31.viii.1952, *Angus* 354 (BM; FHO; K). W: Mwinilunga, fr. 15.ii.1938, *Milne-Redhead* 4586 (BM; K). **S. Rhodesia.** E: Chirinda Forest, st. 24.x.1947, *Wild* 2197 (COI; K; PRE; SRGH).

Also in the Cameroons, Congo and north of Angola. In evergreen, riverine and swamp forests.

4. FAGAROPSIS Mildbr. ex Siebenlist

Fagaropsis Mildbr. ex Siebenlist, Forstw. Deutsch-Ostafr.: 90 (1914).

Deciduous trees or shrubs, usually with buttresses. Leaves opposite, imparipinnate. Inflorescence of terminal panicles arising from buds at the end of the previous year's branches, before or together with the new leafy branches. Flowers unisexual by abortion, pedicellate, arranged in cymose clusters in branched panicles and in sessile clusters on the main rhachis. Sepals 4, united at the base. Petals 4 (sometimes 5–6 in the female flowers), imbricate. Male flowers: (6)8 stamens*; ovary vestigial, glabrous. Female flowers: (staminodes not seen); disk annular; ovary globose, 4–5-locular with 1 pendulous ovule in each loculus; style short; stigma 4–5-lobed. Fruit baccate, subglobose, somewhat depressed, glandular-foveolate. Seed ovoid, pendulous; endosperm present.

Fagaropsis angolensis (Engl.) Dale in H.M.Gardn., Trees & Shrubs Kenya: 99 (1936).—Exell & Mendonça, C.F.A. **1,** 2: 272 (1951). Type from Angola (Cuanza Norte).
Vepris angolensis Engl. in Engl. & Prantl, Nat. Pflanzenfam. **3,** 4: 178 (1896). Type as above.

Var. **mollis** (Suesseng.) Mendonça, comb. nov. TAB. **29** fig. B. Syntypes: S. Rhodesia, Marandellas, *Dehn* 659 (M), 660 (M).
Fagaropsis sp. nov.—Burtt Davy & Hoyle, N.C.L.: 68 (1936).
Clausenopsis angolensis var. *mollis* Suesseng. in Proc. & Trans. Rhod. Sci. Ass. **43**: 108 (1951). Syntypes as above.

Deciduous tree 7–15 m. tall; young branches, leaves and inflorescences cinereous-pubescent, older branches glabrous. Leaves 12–30(40) cm. long; petiole 3·5–7(9) cm. long; leaflets opposite, 2–4(6)-jugate, subsessile or petiolulate; petiolules 1–2 mm. long, that of the terminal leaflet sometimes up to 20 mm.; lamina 4–9(11) × 2–4(5) cm., papyraceous, ovate or oblong-ovate to elliptic, acuminate or acute at the apex, margin entire with a row of rather closely spaced pellucid gland-dots, asymmetric at the base, softly pubescent on both surfaces. Inflorescence of panicles arising from the buds at the end of the previous year's branches and occasionally between two new branches. Flowers 4-merous; pedicels 6–10 mm. long. Sepals 4, 0·9–1·3 mm. long, united at the base, densely whitish-pubescent, ovate, acute. Petals 4, 5–6 mm. long, imbricate in bud, oblanceolate. Male flowers: stamens 8, shorter than the petals; anthers 1 mm. long, basifixed, 2-lobed at the base; vestigial ovary glabrous. Female flowers: staminodes present; disk annular; ovary 4-locular, slightly 4-lobed, with 1 apical ovule in each loculus; style very short; stigma slightly 4-lobed. Fruit 6–7 mm. in diam.

S. Rhodesia. C: Marandellas, Cave Farm, fr. 5.iv.1950, *Wild* 2364 (K; SRGH). E: Umtali, Commonage, fl. 3.ii.1948, *Chase* 952 (BM; K; SRGH). **Nyasaland.** N? Nkope, fr. 1895, *Buchanan* 285 (BM). C: Dedza, fl. 2.xii.1938, *McGregor* 48/38 (FHO). S: Zomba, fr. iii.1934, *Clements* 426 (BM; FHO).

Known only from S. Rhodesia and Nyasaland. *F. angolensis* var. *angolensis* occurs in tropical East Africa and Angola. Slopes of mountains in rocky places, 1000–1400 m.

5. CLAUSENA Burm. f.

Clausena Burm. f., Fl. Ind.: 87, 243 et sub t. 29 (1768).

Unarmed trees or shrubs. Leaves alternate, pinnate, densely dotted with pellucid glands. Inflorescence of axillary cymose panicles. Flowers small, 4-merous, bisexual. Sepals 4–5, very small, united at the base. Petals 4–5, imbricate in bud. Stamens 8–10, inserted at the base of the disk; filaments subulate, flattened towards the base; anthers elliptic, dorsifixed, introrse. Ovary (2–3)4–5-locular, with 2 ovules in each loculus; style short, caducous; stigma broad, obtuse, 2–5-lobed. Fruit baccate, 1–2-seeded.

Clausena anisata (Willd.) Hook. f. ex Benth. in Hook., Niger Fl.: 256 (1849) (" Claussena ").—Oliv. in Journ. Linn. Soc., Bot. **5**, Suppl. 2: 34 (1861); F.T.A. **1**: 308 (1868).—Engl., Pflanzenw. Afr. **3,** 1: 758 (1915); in Engl. & Prantl, Nat. Pflanzen-

* Engler (in Engl. & Prantl, Nat. Pflanzenfam. ed. 2. **19a**: 298 (1931)) states (under *Clausenopsis*) that there are 6 stamens but I have found 8 (even in the type). See also Suessenguth's statement (in Proc. & Trans. Rhod. Sc. Ass. **43**: 168 (1951)) under *Clausenopsis angolensis* var. *mollis*.

Tab. 31. CLAUSENA ANISATA. 1, branch with inflorescences and leaves ($\times\frac{2}{3}$); 2, infructescence ($\times\frac{2}{3}$); 3, portion of under surface of leaf to show glands ($\times 4$); 4, flower ($\times 6$); 5, flower with 2 sepals, petals and stamens removed ($\times 6$); 6, cross section of ovary ($\times 12$); 7, stamen (from bud) ($\times 8$). All from *Chase* 1607, except 2 from *Swynnerton* 122.

fam. ed. 2, **19a**: 322 (1931).—Burtt Davy & Hoyle, N.C.L.: 68 (1936).—Brenan, T.T.C.L.: 540 (1949).—Exell & Mendonca, C.F.A. **1**, 2: 273 (1951).—Brenan in Mem. N.Y. Bot. Gard. **8**, 3: 233 (1953).—Keay, F.W.T.A. ed. 2, **1**, 2: 686, fig. 191 (1958).—Gilbert, F.C.B. **7**: 94 (1958).—Dale & Greenway, Kenya Trees & Shrubs: 481 (1961).—White, F.F.N.R.: 169 (1962). TAB. **31**. Type from Guinea.

Amyris anisata Willd. in L., Sp. Pl. ed. 4, **2**: 337 (1779). Type as above.

Elaphrium? inaequale DC., Prodr. **1**: 724 (1824). Type from S. Africa.

Amyris inaequalis (DC.) Spreng. in L., Syst. Veg. ed. 16, **2**: 218 (1825). Type as above.

Clausena inaequalis (DC.) Benth. in Hook., Niger Fl.: 257 (1849).—Engl., Pflanzenw. Ost-Afr. **C**: 229 (1895).—Bak. f. in Journ. Linn. Soc., Bot. **40**: 36 (1911).—Burtt Davy, F.P.F.T. **2**: 477 (1932).—Steedman, Trees etc. S. Rhod.: 28, t. 24 (1933).—Burtt Davy & Hoyle, N.C.L.: 68 (1936). Type as above.

Clausena anisata var. *mollis* Engl., Pflanzenw. Ost-Afr. **C**: 228 (1895); in Engl. & Prantl, Nat. Pflanzenfam. **3**, 4: 189 (1896); op. cit. ed. 2, **19a**: 322 (1931).—Brenan, T.T.C.L.: 540 (1949). Syntypes from Zanzibar, Kilimanjaro and Lakes region.

Clausena inaequalis var. *abyssinica* Engl., Pflanzenw. Ost-Afr. **C**: 229 (1895); in Engl. & Prantl, Nat. Pflanzenfam. **3**, 4: 189 (1896).—Burtt Davy, loc. cit. Syntypes from Abyssinia and Kilimanjaro.

Clausena abyssinica (Engl.) Engl., Pflanzenw. Afr. **3**, 1: 757 (1915). Syntypes as above.

Shrub or small tree up to 10 m. tall; bark grey, mottled; young branches and leaves pubescent becoming glabrous. Leaves 6–20 (28) cm. long, imparipinnate; petiole 1–2·5 cm. long, terete or slightly grooved at the base; leaflets alternate, (4)5–8(9–10) on each side of the rhachis; petiolules slender, 1–3·5 mm. long; lamina 1–6(7) × 0·6–2·5(3·5) cm., rather smaller in the proximal than in the distal leaflets, ± asymmetric, ovate or ovate-oblong or elliptic to narrowly elliptic, pubescent or glabrous, densely dotted with pellucid glands, acute or slightly emarginate at the apex, margin crenulate or entire, cuneate or rounded at the base; lateral nerves 5–8 pairs. Inflorescence of elongate puberulous panicles, as long as the subtending leaves or a little shorter; peduncle long; bracts and bracteoles scarious, very small, caducous. Flower-buds ± 3 mm. long, obovoid; pedicels very slender, 3–4(5) mm. long, puberulous. Sepals 4, 0·8 mm. long, acute, slightly joined at the base, pubescent. Petals 4, 4–5 mm. long, whitish, imbricate in bud, elliptic. Stamens 8, 4 mm. long, inserted at the base of the disk; filaments subulate, flattened towards the base; anthers basifixed, 1 mm. long, introrse, elliptic, 2-lobed at the base. Ovary 1·3 mm. long, subellipsoid, 4(5)-locular with 2 ovules in each loculus; style stout, stigma broadly obtuse.

N. Rhodesia. N: Lake Mweru, fl. 11.xi.1957, *Fanshawe* 3918 (K). W: Mwinilunga, Kakema R., fl. 24.viii.1930, *Milne-Redhead* 961 (K). E: Nyika Plateau, c. 3·2 km. SW. of Rest House, 2150 m., fl. 21.x.1958, *Robson* 216 (BM; K; LISC; PRE; SRGH). **S. Rhodesia.** E: Vumba Mts., Cloudlands, fl. 28.xi.1948, *Chase* 1607 (BM; COI; LISC; SRGH). S: Belingwe, Mt. Buhwe, fl. & fr. 10.xii.1953, *Wild* 4329 (K; LISC; SRGH). **Nyasaland.** N: Nyika Plateau, fl. 27.x.1958, *Robson* 420A (BM; K; LISC; PRE; SRGH). C: Chenga Hill, fl. 9.ix.1946, *Brass* 17609 (BM; K; SRGH). S: Mt. Mlanje, fl. x.1891, *Whyte* 49 (BM; K). **Mozambique.** MS: Báruè, Vila Gouveia, Serra de Choa, fl. 8.ix.1943, *Torre* 5888 (LISC). SS: Gaza, Chipenhe, fr. 3.x.1957, *Barbosa & Lemos* 8029 (COI; K; LISC; LMJ). LM: Maputo, Rio Masiminhana, fl. 19.xi.1944, *Mendonça* 2962 (COI; LISC; LM; SRGH).

From Sierra Leone to Ethiopia and the Sudan, and southward to Cape Prov. Evergreen and fringing forest, from sea-level to 2150 m.

Swynnerton notes that the flowers are sweet-scented and that the leaves have a strong and not unpleasant smell which, however, scarcely resembles aniseed.

6. CITROPSIS (Engl.) Swingle & Kellerm.

Citropsis (Engl.) Swingle & Kellerm. in Journ. Agric. Res. **1**: 419, 421 (1914).

Small trees or shrubs; branches with (0)1–2 axillary spines. Leaves alternate, imparipinnate, (1–3)5–9-foliolate; petiole and rhachis winged (in our area); leaflets ovate, elliptic or lanceolate, dotted with pellucid glands, margin slightly crenate. Inflorescence of short axillary racemes or cymose clusters. Flowers 4(5)-merous, bisexual. Sepals united in a 4(5)-lobed cupuliform calyx. Petals 4(5). Stamens 8(10). Disk annular. Ovary 4(5)-locular, loculi 1-ovulate; style long; stigma

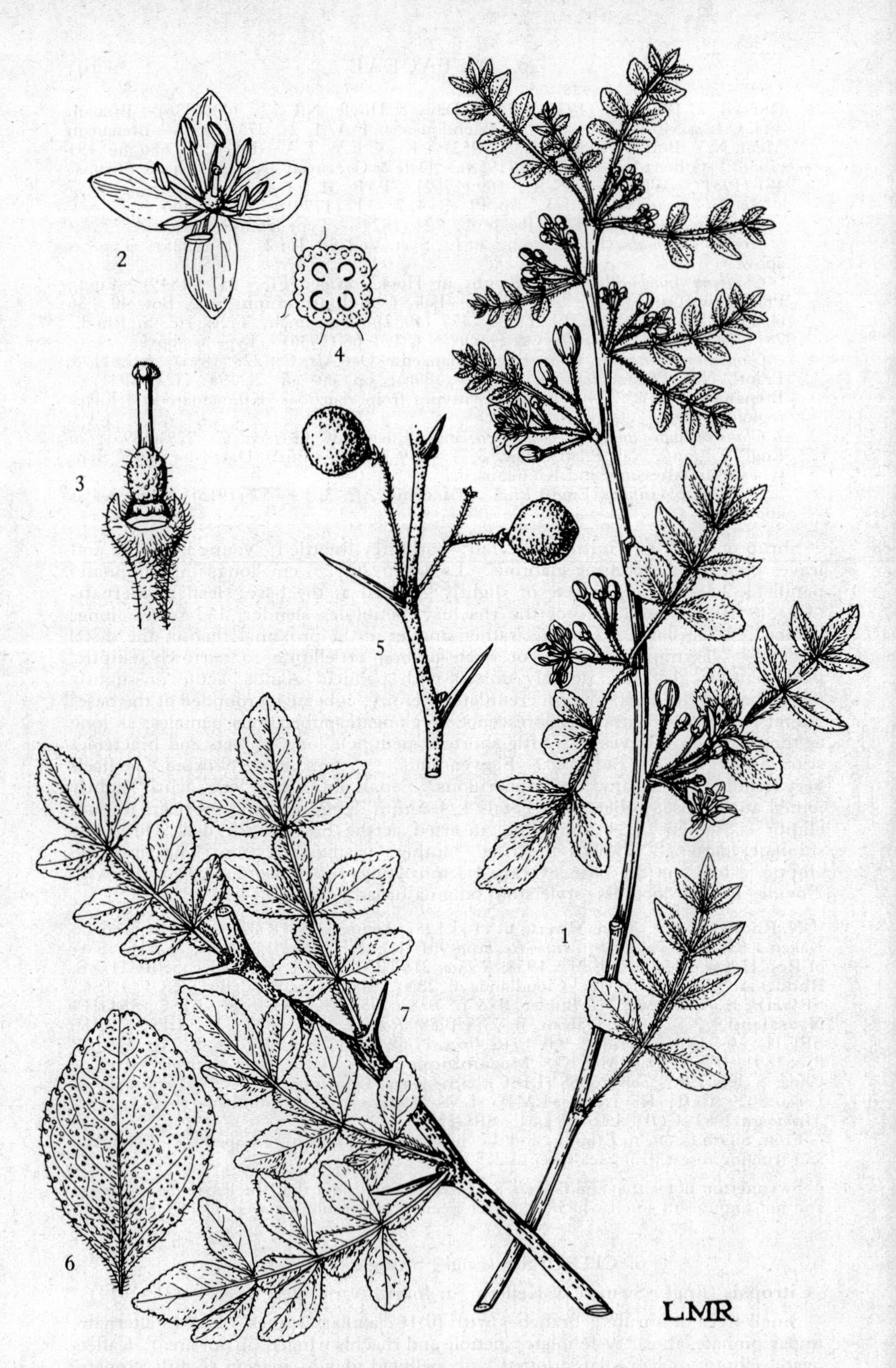

Tab. 32. CITROPSIS DAWEANA. 1, branchlet with leaves and inflorescences ($\times \frac{2}{3}$) *Lovemore* 371; 2, flower ($\times 2$) *Lovemore* 371; 3, gynoecium ($\times 4$) *Lovemore* 371; 4, cross section of ovary ($\times 8$) *Lovemore* 371; 5, branchlet with fruit ($\times \frac{2}{3}$) *Lovemore* in GHS 40313 and *Fanshawe* 4549; 6, leaflet ($\times 1\frac{1}{2}$) *Lovemore* 242; 7, branchlet with spines and leaves ($\times \frac{2}{3}$) *Lovemore* in GHS 40313 and 44064.

subglobose, 4(5)-lobed. Fruit baccate, globose, punctate with oil-glands; endocarp fleshy. Seeds with fleshy cotyledons; endosperm absent.

Citropsis daweana Swingle & Kellerm. [in Journ. Wash. Acad. Sci. **28**: 533 (1938) *nom. nud.*] ex Swingle in Journ. Arn. Arb. **21**: 123 (1940).—White, F.F.N.R.: 169 (1962). TAB. **32**. Type: Mozambique, Mossurize, Madanda Forest, *Dawe* 443 (K).
Hesperethusa villosa Tanaka, Citrus Studies: 77, t. 70, fig. dextr. (1933) *nom. nud.*—O. B. Mill. in Journ. S. Afr. Bot. **18**: 37 (1952) *sine descr. lat.*

Shrub or small tree up to 6 m. tall; young branches, leaves, spines and inflorescences pubescent or puberulous. Leaves 6–13 cm. long, alternate, imparipinnate; petiole 0·8–2·5 cm. long, winged or sometimes wingless; rhachis winged with wings 1–4(6) mm. broad on each side at the apex and gradually narrowing from the distal extremity downwards; leaflets opposite, 2–4-jugate, sessile or subsessile; lamina 2–4·5(6·5) × 0·7–2·5 cm., narrowly elliptic to elliptic, ovate or obovate, greyish-pubescent on both surfaces, glabrescent, densely dotted with pellucid glands, obtuse or rounded at the apex, margin crenulate, cuneate at the base; lateral nerves 5–7 pairs; spines (0)1–2, 1–3 cm. long, borne in the axils of the leaves, straight, robust. Inflorescence of small racemes or clusters borne on short lateral branches or from sessile buds on leafless branches of the previous year. Flowers 4-merous, cream-coloured; pedicels 4–8 mm. long, slender, puberulous. Sepals 4, 1–1·5 mm. long, ovate, puberulous, slightly united at the base. Petals 6 × 4 mm., free, imbricate. Stamens 8, free, inserted at the base of an annular disk, flattened towards the base, subulate, shorter than the petals; anthers dorsifixed, 2-lobed at the base, laterally dehiscent. Ovary 1·5 mm. long, ellipsoid; style 2·5 mm. long; stigma subcapitate, 4-lobed. Fruit 1·4 cm. in diam., punctate with oil-glands; endocarp fleshy. Ripe seeds not seen.

Caprivi Strip. Banoni Camp, st. 1956, *De Winter* 4352 (K). **Bechuanaland Prot.** N: Kabula-bula, Chobe R., st. vii.1930, *van Son* in Herb. Transv. Mus. 28984 (BM; PRE). **N. Rhodesia.** B: Nangweshi, st. 21.vii.1953, *Codd* 7135 (BM; K; PRE; SRGH). C: Kafue R., fr. 7.vi.1958, *Fanshawe* 4549 (K; SRGH). S: Bombwe Forest, st. 3.i.1953, *Angus* 1010B (FHO; K). **S. Rhodesia.** N: Urungwe, Zambezi Valley, Rifa R., fl. 24.ix.1953, *Lovemore* 371 (BM, fragm.; K; LISC; PRE; SRGH). W: Wankie, st. 16.ii.1952, *Lovemore* 242 (BM, fragm.; K; SRGH). **Mozambique.** MS: Mossurize, Machase, Madanda Forest, st. 31.ii.1943, *Pedro & Pedrógão* 7878 (BM, fragm.; SRGH).

Known only from our area. In woodland, bush and thickets in the hotter and drier regions.

7. ORICIA Pierre

Oricia Pierre in Bull. Soc. Linn. Par. **2**: 1288 (1897).

Trees with alternate or opposite 3–5-foliolate leaves. Inflorescence of terminal and axillary panicles. Flowers 4-merous, unisexual by abortion. Sepals 4. Petals 4, valvate or slightly imbricate. Male flowers: stamens 4, opposite the sepals; filaments short; anthers introrse; carpels vestigial. Female flowers: staminodes 4; carpels 2–4 (1–3 aborting), slightly united at the base and by the stigma (otherwise free) into a 2–4-locular ovary with 2 pendulous apical ovules in each loculus; style short; stigma discoid, 2–4-lobed. Fruit of 1–4 1-seeded drupes. Seed with a thin testa; endosperm absent.

Oricia swynnertonii (Bak. f.) Verdoorn in Kew Bull. **1926**: 413 (1926).—Brenan in Mem. N.Y. Bot. Gard. **8,** 3: 233 (1953). TAB. **33**. Type: S. Rhodesia, Chirinda Forest, *Swynnerton* 12 (BM, holotype).
Teclea swynnertonii Bak. f. in Journ. Linn. Soc., Bot. **40**: 35, t. 2 fig. 1–5 (1911).—Steedman, Trees etc. S. Rhod.: 29 (1933). Type as above.
Teclea welwitschii sensu Hutch., Botanist in S. Afr.: 469 (1946).

Evergreen tree up to 15 m. tall; branches glabrous, grey or brown, with large lenticels. Leaves opposite or subopposite, 3-foliolate; petiole 1–7 cm. long, terete or somewhat grooved from the middle upwards; leaflets with petiolule up to 1 cm. long; lamina 5–16(18) × 2·5–6 cm., coriaceous, elliptic to oblong-elliptic, densely dotted with pellucid glands, attenuate or obtuse at the apex, cuneate at the base; lateral nerves numerous, up to 30 or more pairs. Inflorescence of axillary and terminal panicles, ferruginous-pubescent; peduncle and rhachis somewhat flattened; bracteoles 0·6 mm. long, caducous. Flowers creamy-white, unisexual

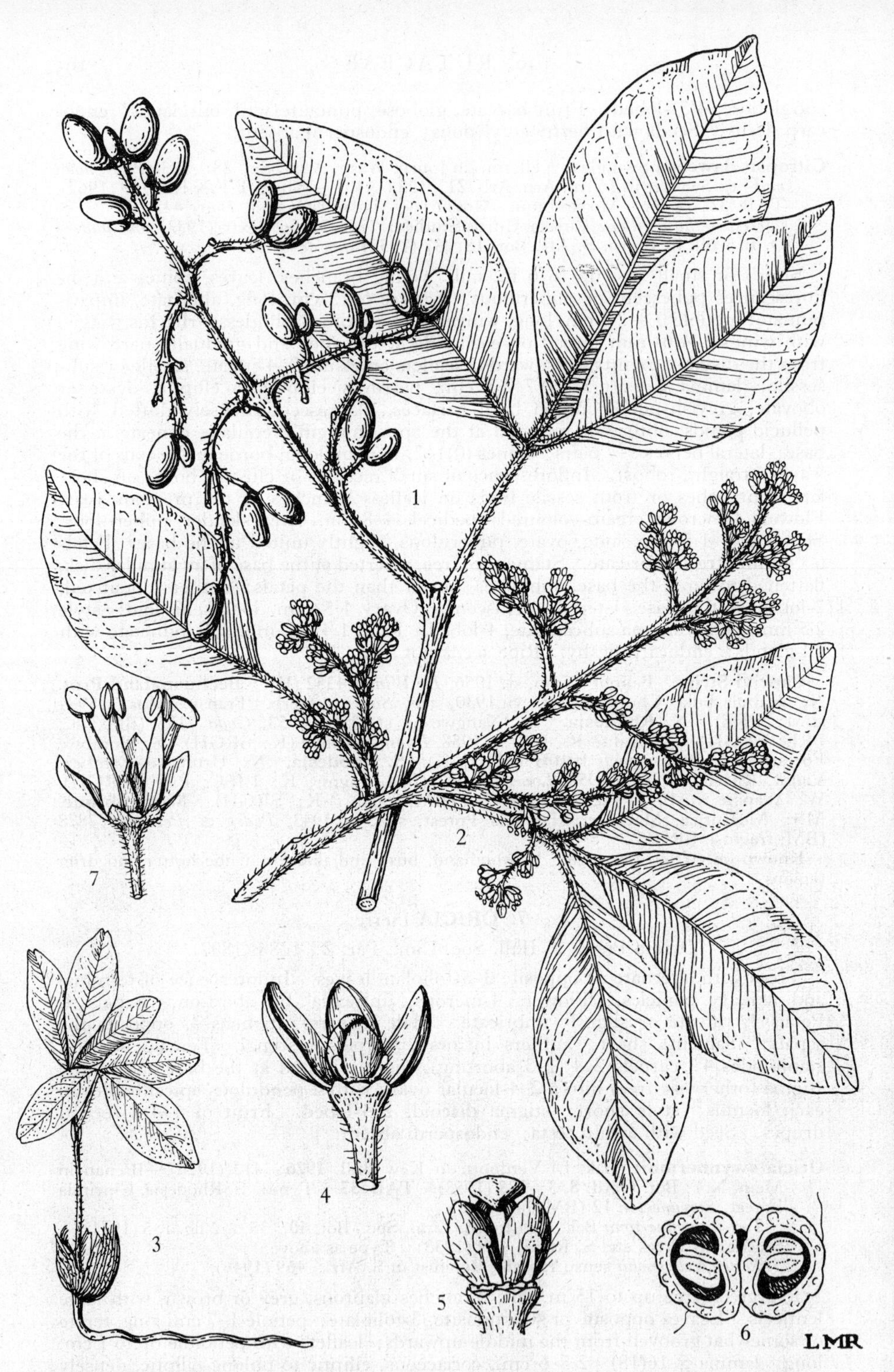

Tab. 33. ORICIA SWYNNERTONII. 1, branchlet with leaves and fruits (× ⅔) *Swynnerton* s.n.; 2, branchlet with male inflorescence (× ⅔); 3, seedling (× ⅔); 4, female flower with 1 petal removed (× 6); 5, gynoecium and staminodes (× 8); 6, cross section of gynoecium from older flower (× 8); 7, male flower with 1 petal removed (× 4). All from *Swynnerton* 12.

by abortion, dioecious, in clusters; pedicels 1–3·5 mm. long, stout in female slender in male flowers; sepals 4, not exceeding 0·4 mm. long, thin, ciliate; petals 4, 3 × 2 mm., imbricate at the base, valvate at the apex. Male flowers: stamens 4, 3–4 mm. long, flattened towards the base, subulate; vestigial ovary of 2 carpels, pilose at the base. Female flowers: staminodes 4; ovary of 2(4) carpels, united at the base and by the stigma; styles very short, concrescent to form a broad 2-lobed stigma. Fruit of 1–4 druplets, 15 × 8 mm., orange-coloured, ellipsoid, 1-seeded (1–3 carpels aborted). Seeds with membranous testa; cotyledons 9 × 5 mm.

N. Rhodesia. E: Fort Jameson, fl. 1.vi.1958, *Fanshawe* 4491 (K). **S. Rhodesia.** E: Chirinda Forest, 1130–1220 m., fl. & fr. viii.1905, *Swynnerton* 12 (BM; K). S: Victoria Distr., Zimbabwe, fl. 10.x.1949, *Wild* 3012 (K; SRGH). **Nyasaland.** S: Cholo Mt., fl. 20.ix.1946, *Brass* 17663 (BM; K; SRGH). **Mozambique.** T: Tete, Mt. Zóbuè, fl. 3.x.1942, *Mendonça* 586 (LISC). MS: Báruè, Serra de Choa, fr. 31.x.1943, *Torre* 6110 (LISC).

Eastern N. Rhodesia, Mozambique, Nyasaland and S. Rhodesia. Montane mist-forests.

Fanshawe 4491, cited above, seems to be a hairy form of this species but may prove to be a new taxon.

8. DIPHASIOPSIS Mendonça

Diphasiopsis Mendonça in Mem. Junt. Invest. Ultram., Sér. 2, **28**: 81 (1961).

Shrubs with pilose branches, petioles and inflorescences. Leaves alternate, 3-foliolate; petiole subterete; leaflets sessile, elliptic or obovate, densely dotted with minute pellucid glands; nerves few, remote from each other and anastomosing in arches near the margin. Inflorescence of terminal and axillary panicles; bracts and bracteoles small, caducous. Flowers 4-merous, unisexual by abortion, dioecious. Sepals 4, slightly united at the base or free. Petals 4, valvate. Male flowers: stamens usually 8, inserted at the base of the disk; vestigial ovary of 2 free carpels with atrophied styles. Female flowers: staminodes 8, short or vestigial; ovary of 2(3) free 1-locular carpels, each with 2 pendulous apical ovules; styles united at the apex in a discoid peltate stigma. Fruit unknown.

Diphasiopsis whitei Mendonça in Mem. Junt. Invest. Ultram., Sér. 2, **28**: 81 (1961). TAB. **34** fig. A. Type: N. Rhodesia, Kawambwa, *White* 3536 (BM; FHO; K, holotype).
Oricia sp.—White, F.F.N.R.: 170 (1962).

Shrub 1–1·5 m. high; young branches, petioles and inflorescences ± densely pubescent. Leaves with subterete petioles 1–8 cm. long; leaflets sessile; lamina 2–7(11·5) × 1–3(4) cm., elliptic to obovate, glabrous except on the midrib or pubescent-glabrescent with long sparse appressed hairs, acute or obtuse at the apex, obtuse or rounded at the base; lateral nerves 5–10 pairs; midrib and nerves somewhat impressed or flat on the upper surface and raised below and densely dotted with very minute pellucid glands, which are inconspicuous on both younger and older leaflets. Inflorescence of terminal and axillary panicles up to 7 cm. long; bracts 1·1 mm. long, linear, pubescent; bracteoles 0·5 mm. long, linear; pedicels 1–1·5 mm. long. Flowers yellow. Sepals 4, 0·8 mm. long, ovate, acute. Petals 4, 4·5–6 × 1·8–2·3 mm., valvate, elliptic to oblong-elliptic. Male flowers: stamens (6–7)8, inserted at the base of the pilose disk; filaments alternately unequal, shorter than the petals; anthers basifixed, 2-lobed at the base, laterally dehiscent; vestigial carpels pilose with short atrophied styles. Female flowers: staminodes at the base of a pilose disk; carpels densely pilose; styles 0·7–1·4 mm. long, pilose. Fruit not seen.

N. Rhodesia. N: Kawambwa, ♂ fl. 16.xi.1957, *Fanshawe* 4053 (FHO; K); Abercorn Distr., Lunzua Valley, 1200 m., ♂ fl. 19.xi,1956, *Richards* 7021 (K).

Known only from the Northern Distr. of N. Rhodesia. Fringing forest and river banks.

According to Mrs. Richards the flowers are lemon-yellow and the plant has an unpleasant smell. Fanshawe describes the flowers as creamy.

9. TODDALIA Juss.

Toddalia Juss., Gen. Pl.: 371 (1789).

Sarmentose or climbing shrubs; branches, leaves and inflorescences aculeate.

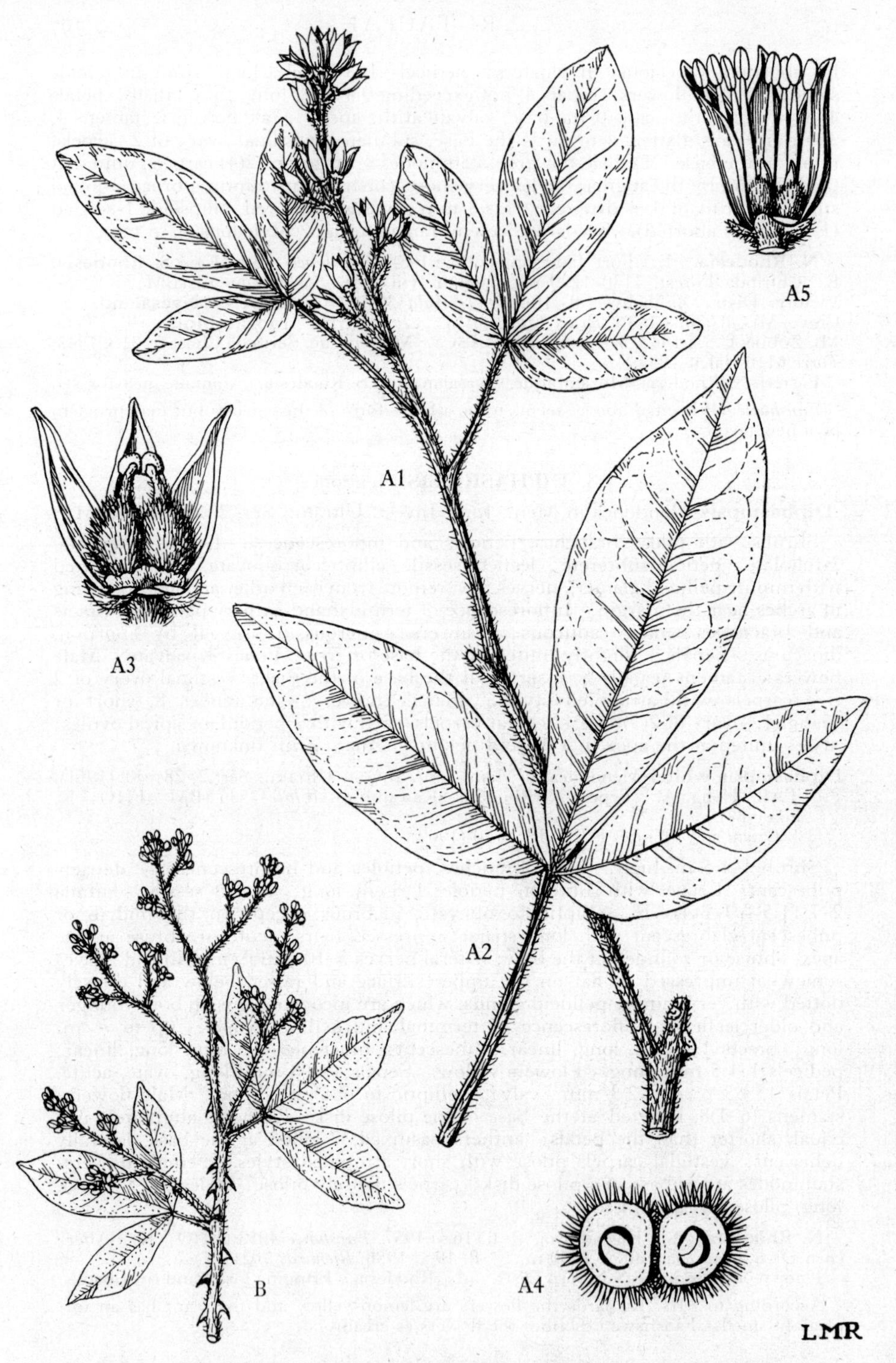

Tab. 34. A.—DIPHASIOPSIS WHITEI. A1, branchlet with inflorescence (×⅔); A2, leaf (×⅔); A3, flower with 1 sepal and 1 petal removed (×4); A4, cross section of ovary (×6); A5, male flower with 1 petal removed (×4), all from *White* 3536. B.—TODDALIA ASIATICA, branchlet with inflorescences (×⅔) *Chase* 5144.

Leaves alternate, 3-foliolate. Inflorescence of axillary and terminal panicles. Flowers small, 5-merous, unisexual by abortion. Sepals united to form a cupuliform 5-dentate calyx. Petals 5, free, imbricate. Male flowers: stamens 5, inserted at the base of a short gynophore, opposite the sepals; filaments filiform; anthers introrse; ovary vestigial, conical. Female flowers: staminodes filiform; ovary on a short gynophore, 5–7-locular, with 2 ovules in each loculus; stigma disk-shaped, 5-lobed. Fruit globose, drupaceous with woody endocarp, 3–7-locular; loculi 1-seeded. Seeds reniform, oblong; testa thick; cotyledons linear.

Toddalia asiatica (L.) Lam., Tabl. Encycl. Méth. Bot. **2**: 116 (1797).—Verdoorn in Kew Bull. **1926**: 400 (1926).—Steedman, Trees etc. S. Rhod.: 29, t. 25 (1933).—Brenan, T.T.C.L.: 545 (1949); in Mem. N.Y. Bot. Gard. **8,** 3: 233 (1953).—Gilbert, F.C.B. **7**: 99, t. 13 (1958).—Dale & Greenway, Kenya Trees & Shrubs: 491 (1961).—White, F.F.N.R.: 171 (1962). TAB. **34** fig. B. Syntypes from India.
Paullinia asiatica L., Sp. Pl. **1**: 365 (1753). Syntypes as above.
Scopolia aculeata Sm., Pl. Ic. Hact. Ined. **1,** 2: sub. t. 34 (1790) *nom. illegit.* Syntypes as above.
Toddalia aculeata Pers., Syn. Pl. **1**: 249 (1805) *nom. illegit.*—Bak. f. in Journ. Linn. Soc., Bot. **40**: 35 (1911).—Engl., Pflanzenw. Afr. **3,** 1: 754, fig. 351 A–K (1915).—Burtt Davy, F.P.F.T. **2**: 478 (1932).—Burtt Davy & Hoyle, N.C.L.: 68 (1936). Syntypes as above.

Sarmentose or climbing shrubs, retrorsely aculeolate; young branches and leaves ferruginous-pubescent, glabrescent. Leaves with petiole 1–4 cm. long, grooved above and persistently pubescent in the groove, sometimes aculeolate; leaflets sessile; lamina 3–7(8) × 1·2–3 cm., elliptic or obovate or oblanceolate, that of the lateral leaflets asymmetric, chartaceous, conspicuously glandular on both surfaces, apex shortly acuminate or obtuse, margin entire or slightly crenulate; midrib prominent on the lower surface and occasionally aculeolate; lateral nerves and veins numerous, not raised. Inflorescence shorter or as long as the subtending leaf, ferruginous-pubescent; bracts 0·8–1·3 mm. long, linear, caducous. Flowers 5-merous, unisexual by abortion, usually in cymose clusters on the panicle-branches; pedicels 1–3 mm. long, acute, caducous. Petals 2·5 mm. long, valvate. Male flowers: stamens as long as the petals; anthers 1 mm. long, 2-lobed, ovate. Female flowers not seen. Fruit 7–8 mm. in diam., drupaceous, 3–5(7)-locular. Seeds 4 mm. long, reniform; testa dark brown, hard, smooth; endosperm fleshy.

N. Rhodesia. N: Mpika, ♂ fl. 4.ii.1956, *Fanshawe* 1979 (K). W: Mwinilunga, Kakoma, Muzera R., fr. 29.ix.1952, *White* 3414 (BM; FHO; K). E: Nyika Plateau, fr. 30.x.1958, *Robson* 487 (BM; K; LISC; PRE; SRGH). **S. Rhodesia.** E: Umtali Distr., Engwa, 1600 m., fr. 1.ii.1955, *E.M. & W.* 5 (BM; LISC; SRGH). S: Belingwe, Mt. Buhwe, fl. 10.xii.1953, *Wild* 4334 (SRGH). **Nyasaland.** N: Karonga Distr., Misuku Hills, ♂ fl. 12.i.1959, *Richards* 10617 (K; SRGH). S: Mt. Mlanje, Swazi Estate, fl. 24.i.1949, *Faulkner* 366 (COI; K). **Mozambique.** Z: Milange, fl. 23.ii.1943, *Torre* 4806 (LISC). MS: Manica, Mavita, Mt. Xiroso, fl. 26.x.1944, *Mendonça* 2641 (COI; K; LISC; LM; SRGH).

Eastern and central Africa from the Sudan southwards, Transvaal, Madagascar, Mascarene Is. and India. In forest, woodland and grassland, 0–1600 m.

Walking-stick Climber. Africans use the stems for walking-sticks.

10. VEPRIS Comm. ex A. Juss.

Vepris Comm. ex A. Juss. in Mém. Mus. Hist. Nat. Par. **12**: 509 (1825).

Trees or shrubs, sometimes subscandent. Leaves alternate or rarely opposite, 1- or 3-foliolate; leaflets dotted with pellucid glands. Inflorescence of axillary and terminal racemes or panicles or clustered cymes, sometimes on older leafless branches. Flowers (2–3)4-merous, unisexual by abortion. Sepals united to form a cupuliform (3)4-lobed calyx. Petals (3)4, oblong, imbricate in bud. Male flowers: stamens (6–7)8, inserted at the base of the disk, those opposite the petals shorter than those opposite the sepals; filaments flattened at the base, subulate; anthers dorsifixed, 2-lobed at the base; ovary vestigial with 2–4 styles. Female flowers: staminodes 8 or fewer; ovary (1)2–4-locular; loculi 2-ovulate; ovules pendulous; stigma sessile, 2–4-lobed. Fruit drupaceous, (1)2–4-locular; loculi 1-seeded. Endosperm present.

Ovary 3–4-locular; leaves 3-foliolate:
Panicle terminal, much branched, broader than long, densely flowered; fruit slightly 4-lobed, somewhat depressed, 4-seeded, 4–5 mm. in diam. - - 1. *undulata*
Panicles or racemes axillary and terminal, sometimes also on defoliate branches; fruit 10–15 mm. in diam.:
Flowers and fruits pedicellate or rarely subsessile; inflorescences paniculate or more rarely simple by reduction; leaf-venation more prominent than the inconspicuous glands - - - - - - - - - - 2. *stolzii*
Flowers and fruits sessile, in a simple raceme or pedunculate cluster (rarely solitary); leaf-venation as prominent as the ± protruding glands - 3. *whitei*
Ovary 2-locular (sometimes 1-locular by abortion); leaves 1- or 3-foliolate:
Leaves 3-foliolate; shrubs or small trees up to 3 m. tall:
Petioles winged; wings narrowing from the apex towards the base:
Young branches, leaves and inflorescences greyish-tomentose - 4. *zambesiaca*
Young branches, leaves and inflorescences glabrous; flowers in small terminal racemes or panicles on short lateral branches - - - 5. *carringtoniana*
Petioles not or only slightly winged, sometimes flattened and broadening towards the apex:
Branches, petioles and inflorescences pubescent; flowers in axillary racemes or panicles shorter than the petiole of the subtending leaf; pedicels up to 2 mm. long 6. *termitaria*
Branches, petioles and inflorescences glabrous:
Flowers sessile, usually glomerate on a short axillary rhachis; petals 4·5 × 0·8 mm., oblong, reflexed - - - - - - - 7. *reflexa*
Flowers pedicellate, in racemes or panicles:
Flowers in slender racemes or panicles up to 1·5 cm. long on leafless stems; pedicels filiform, up to 5 mm. long; calyx 0·2 mm. long; leaves opposite 8. *allenii*
Flowers in stout axillary panicles up to 5 cm. long; leaves alternate, with drooping leaflets - - - - - - - - - 9. *fanshawei*
Leaves 1-foliolate; young branches puberulous; vestigial ovary pilose with 2 setigerous styles - - - - - - - - - - - 10. *drummondii*

1. **Vepris undulata** (Thunb.) Verdoorn & C.A.Sm. in Journ. S. Afr. For. Ass. no. **20**: 35, 50 (1951) in obs. Type from S. Africa (Cape).
Boscia undulata Thunb., Prodr. Pl. Cap.: 32 (1794). Type as above.
Toddalia lanceolata Lam., Tabl. Encycl. Méth. Bot. **2**: 117 (1797).—Oliv., F.T.A. **1**: 307 (1868).—Engl., Pflanzenw. Ost-Afr. **C**: 228 (1895).—Sim, For. Fl. Port. E. Afr.: 24, t. 25 (1909). Type from Mauritius.
Vepris lanceolata (Lam.) G. Don, Gen. Syst. **1**: 806 (1831).—Engl., tom. cit.: 433 (1895); in Engl. & Prantl, Nat. Pflanzenfam. **3,** 4: 174, fig. 101 L–V (1896); op. cit. ed. 2, **19a**: 306, fig. 136 L–V (1931).—Bak. f. in Journ. Linn. Soc., Bot. **40**: 35 (1911).—Verdoorn in Kew Bull. **1926**: 395 (1926).—Burtt Davy, F.P.F.T. **2**: 479 (1932).—Burtt Davy & Hoyle, N.C.L.: 68 (1936).—Brenan, T.T.C.L.: 546 (1949). Type as above.
Vepris querimbensis Klotzsch in Peters, Reise Mossamb. Bot. **1**: 83 (1961). Type: Mozambique, Querimba I., *Peters* (B, holotype †; K, isotype).

Evergreen shrub or small tree up to 5 m. tall; branches, inflorescences and leaves glabrous. Leaves alternate, 3-foliolate; petiole 1–5(7) cm. long, rather slender, narrowly grooved above. Leaflets sessile; lamina 5–12 × 1·5–3·2 cm., narrowly elliptic, densely dotted with minute pellucid glands, acute or obtuse or blunt at the apex, margin somewhat undulate, cuneate at the base. Inflorescence of terminal dense much-branched panicles; bracteoles very small, glabrous or slightly puberulous, caducous. Flowers 4-merous, unisexual by abortion; pedicels 1–3(5) mm. long. Sepals 4, united at the base into a very short 4-lobed calyx 0·5 mm. long, ciliolate on the margin. Petals 4, 2 × 1 mm., slightly imbricate in bud, obovate or elliptic. Male flowers: stamens 8, inserted at the base of a narrow disk, with subulate filaments flattened at the base and shorter than the petals; anthers basifixed, 2-lobed at the base, laterally dehiscent. Female flowers: vestigial staminodes 8; ovary subglobose, 4-locular; stigma subsessile, broadly discoid. Fruit 5 mm. in diam., black when ripe, somewhat depressed, 4-locular, slightly 4-lobed; loculi 1-seeded, 1 or 2 seeds sometimes aborting. Seeds 3 mm. long, black, subtrigonal; testa hard with subapical hilum; endosperm scanty.

Mozambique. Z: Chinde, Luabo, 17.v.1858, *Kirk* (K). MS: Beira, fl. & fr. 25.ii.1912, *Rogers* 4568 (BM; BOL; K; SRGH). SS: Zavala, on dunes, fl. 10.xii.1944, *Mendonça* 3376 (LISC). LM: Lourenço Marques, fr. 8.v.1946, *Gomes e Sousa* 3430 (LISC; K; SRGH).

Also in S. Africa, Tanganyika, Kenya, Mauritius and Réunion. In littoral evergreen thickets on sandy soils and on dunes.

2. **Vepris stolzii** Verdoorn in Kew Bull. **1926**: 396 (1926).—Brenan, T.T.C.L.: 546 (1949).—Letouzey in Adansonia, **2,** 1: 135, fig. 1 (1962). TAB. **35** fig. C. Type from Tanganyika.
   Vepris sp. 1.—White, F.F.N.R.: 171 (1962).

Evergreen tree, 6–15 m. tall; branches, leaves and infructescences glabrous. Leaves alternate, 3-foliolate; petiole 1–3·5 cm. long, subterete, narrowly grooved above; leaflets with grooved petiolules 3–15 mm. long; lamina 8–13 × 2·5–5(6) cm., entire, narrowly elliptic to elliptic, shortly acuminate, blunt at the apex, cuneate at the base; glandular dots usually inconspicuous. Flowers not seen. Infructescence of panicles up to 6 cm. long, axillary and terminal, also on the older branches. Fruit up to c. 20 × 20 mm., drupaceous, on a stipe 3–7 mm. long, subtetragonal, 4-locular; endocarp rather hard. Seeds 10 × 4 mm.; testa thin, yellowish, with a large brown ventral blotch; cotyledons 10 mm. long, thick, planoconvex; endosperm absent.

N. Rhodesia. N: Mporokoso, fr. 11.x.1958, *Fanshawe* 4931 (EA; FHO; K; SRGH). W: Chingola, fr. 5.viii.1955, *Fanshawe* 2410 (EA; K; LISC; SRGH). E: Nyika Plateau, fr. 27.x.1958, *Robson* 419 (BM; K; LISC; PRE; SRGH). **Mozambique.** Z: Milange, Mt. Tumbine, fr. 13.x.1942, *Torre* 4602 (LISC).

N. Rhodesia, Tanganyika and Mozambique. In evergreen forests from 1200 to 2150 m.

3. **Vepris whitei** Mendonça in Mem. Junt. Invest. Ultram., Sér. 2, **28**: 82 (1961). Type: N. Rhodesia, Mporokoso, Lake Mweru, *White* 3620 (FHO; K, holotype).
   Vepris sp. 2.—White, F.F.N.R.: 171 (1962).

Evergreen shrub or tree up to 12 m. tall; branches, leaves and inflorescences glabrous. Leaves alternate, 3-foliolate; petiole 1–5 cm. long, terete or slightly grooved above at the apex; leaflets sessile or petiolulate; lamina 6–13 × 1·5–3(4) cm., narrowly elliptic, chartaceous, densely dotted with pellucid glands, acuminate or acute or blunt at the apex, cuneate at the base; nerves and veins numerous, flat on the lower surface. Inflorescence of axillary pseudoracemes. Flowers 4-merous, sessile, unisexual by abortion. Sepals 4, united into a cupuliform calyx 1·4 mm. long; lobes very short, slightly imbricate, with scarious ciliolate margin. Petals 4, 3·5 × 1·3 mm., imbricate, elliptic to ovate. Male flowers: stamens (7)8, inserted at the base of a narrow disk; filaments flattened at the base, those opposite the petals shorter; anthers ovate, dorsifixed, 2-lobed at the base, laterally dehiscent; vestigial ovary glabrous with 2 stout styles. Female flowers: staminodes absent; ovary globose, sessile, glabrous, 3-locular; stigma sessile, discoid. Fruit c. 1·4 cm. in diam., globose, sessile, 3-locular; endocarp woody. Seeds 9 mm. long; testa membranaceous; cotyledons plano-convex.

N. Rhodesia. N: Mpika, base of Muchinga escarpment, 61 km. S. of Shiwa Ngandu, fr. 28.xi.1952, *Angus* 858 (FHO; K).

Known only from N. Rhodesia. Evergreen riverine forest.

According to Angus the fruits are " orange-yellow and sweet to the taste ".

4. **Vepris zambesiaca** S. Moore in Journ. of Bot. **57**: 86 (1919).—Verdoorn in Kew Bull. **1926**: 397 (1926).—Engl. in Engl. & Prantl, Nat. Pflanzenfam. ed. 2, **19a**: 306 (1931).—White, F.F.N.R.: 171 (1962). Type: N. Rhodesia, Livingstone, banks of Zambezi R., *Rogers* 7486 (BM, holotype).

Deciduous shrub c. 1·5 m. tall or small tree; young branches, leaves and panicles greyish-tomentose. Leaves alternate, 3-foliolate; petiole 1–3 cm. long, narrowly winged; leaflets sessile; lamina 1·8–4·5 × 0·8–1·5 cm., papyraceous, narrowly elliptic to elliptic, sparsely dotted with minute pellucid glands, softly velutinous on both surfaces, blunt or sometimes emarginate at the apex, margin slightly crenulate, cuneate to obtuse at the base. Inflorescence a terminal panicle 2–4 cm. long and 2–3 cm. broad at the base; bracteoles 0·3 mm. long, linear, pilose, caducous; pedicels 1–2 mm. long. Sepals 4, united into a cupuliform shortly lobed calyx 0·8–1·1 mm. long. Petals 4, 4·5–5 × 1·7–2 mm., imbricate, narrowly elliptic. Male flowers: stamens (6–7)8, inserted at the base of a narrow disk; filaments flattened at the base, those opposite the petals a little shorter; anthers 1·5 mm. long, basifixed, ovate, 2-lobed at the base, with lateral dehiscence; vestigial ovary with styles united for their full length. Female flowers and fruits unknown.

N. Rhodesia. S: Livingstone, north bank of the Zambezi R., *Rogers* 7486 (BM). **S. Rhodesia.** N: Urungwe, st. v.1959, *Griffiths* in GHS 96945 (SRGH). W: Bubi Distr., Lonely Mine, fl. xii.1955, *Goldsmith* 84/56 (LISC; SRGH).
Known only from N. and S. Rhodesia. Dry scrub belt, not common.

Related to *V. zambesiaca* but possibly either specifically or varietally distinct are two sterile gatherings from Bechuanaland Prot., Tati Distr., *Miller* B/1147 (PRE) and *Miller* B/2974 (PRE).

5. **Vepris carringtoniana** Mendonça in Mem. Junt. Invest. Ultram., Sér. 2, **28**: 83 (1961). TAB. **35** fig. B. Type: Mozambique, Lourenço Marques Distr., Magude, *Mendonça* 3156 (LISC, holotype).

Deciduous shrub up to 3 m. tall; young branches glabrous. Leaves alternate, 3-foliolate; petiole 0·5–2·3 cm. long, with wings 1–4 mm. broad on each side at the apex and narrowing downwards; leaflets sessile; lamina 1·5–4(5) × 0·6–1·5(2) cm., narrowly elliptic to elliptic or obovate, papyraceous, densely dotted with pellucid glands, obtuse or rounded at the apex, margin crenulate, cuneate at the base; nerves and veins conspicuous on both surfaces. Inflorescences up to 2 cm. long, of terminal racemes or panicles, glabrous or rarely slightly puberulous; bracteoles up to 0·5 mm. long, ovate, acute, deciduous; pedicels 0·5–2 mm. long. Flowers yellow, 4-merous, unisexual by abortion. Sepals 4, united into a short-lobed cupuliform calyx. Petals 4, 4 × 1·6 mm., imbricate in bud, elliptic. Male flowers: stamens 8, inserted at the base of a very narrow disk; filaments flattened at the base, those opposite the petals shorter and all shorter than the petals; anthers basifixed, ovate, 2-lobed at the base; vestigial ovary glabrous with the styles united almost to the apex. Female flowers not seen. Fruit c. 1 cm. long, ellipsoid, 1-locular; endocarp woody. Seed immature.

Mozambique. LM: between Matola and Umbeluzi, fr. 2.xii.1947, *Barbosa* 626 (LISC; LMJ; SRGH).
At present known only from the Lourenço Marques Distr. of Mozambique but likely to occur in the Transvaal and Swaziland. Deciduous thickets with *Acacia* on dry alluvial or sandy soils in low-lying sublittoral regions.
A specimen without flowers or fruit collected from a shrub 1·5 m. tall, with longer petioles and longer and narrower leaflets, from Govuro between Mabote and Zimane (about 259 km. N. of the type-locality of *V. carringtoniana*), at 50 m. alt., in thickets on dry sandy soil, 2.ix.1944, *Mendonça* 1959 (COI; LISC; LM), appears to be related to this species.

6. **Vepris termitaria** Mendonça in Mem. Junt. Invest. Ultram., Sér. 2, **28**: 83 (1961). Type: N. Rhodesia, Kitwe, *Fanshawe* 4336 (FHO; K, holotype; SRGH).
Vepris sp. 4.—White, F.F.N.R.: 171 (1962).

Evergreen shrub or small tree 3 m. tall; young branches, petioles and inflorescences minutely puberulous, older branches glabrous, lenticellate. Leaves alternate, 3-foliolate; petiole 1–4 cm. long, broadly grooved above, sometimes flattened and slightly winged towards the apex; leaflets sessile; lamina 3–6 × 1·3–3·5 cm., chartaceous, elliptic to obovate, dotted with pellucid glands, obtuse or rounded at the apex, cuneate at the base; nerves and veins conspicuous on both surfaces. Inflorescence of small axillary racemes or panicles, shorter than the petiole of the subtending leaf; bracteoles very small, caducous; pedicels up to 2·5 mm. long. Flowers 4-merous, unisexual by abortion. Sepals 4, united into a shallowly lobed cupuliform calyx 0·8–1·2 mm. long. Petals 4, 3·5 × 1·3 mm., imbricate in bud, elliptic. Male flowers : stamens 8, inserted at the base of a narrow disk; filaments subulate, flattened at the base, those opposite the petals shorter and all shorter than the petals; anthers 0·6 mm. long, basifixed, ovate, 2-lobed at the base. Female flowers not seen. Fruit (immature) glabrous, 2-locular with only 1 loculus fertile.

N. Rhodesia. W: Mufulira, fr. 24.iv.1934, *Eyles* 8148 (BM; K; SRGH); Ndola, fr. 7.xii.1953, *Fanshawe* 538 (K).
Known only from the Western Province of N. Rhodesia. On termite mounds in woodland.

7. **Vepris reflexa** Verdoorn in Kew Bull. **1926**: 398 (1926).—Engl. in Engl. & Prantl, Nat. Pflanzenfam. ed. 2, **19a**: 306 (1931).—Burtt Davy, F.P.F.T. **2**: 479 (1932). TAB. **35** fig. A. Type from the Transvaal (Babiaanspoort).

Shrub or small tree up to 6 m. tall; branches, leaves and inflorescences glabrous. Leaves alternate, 3-foliolate; petiole 1–3 cm. long, terete or somewhat grooved at

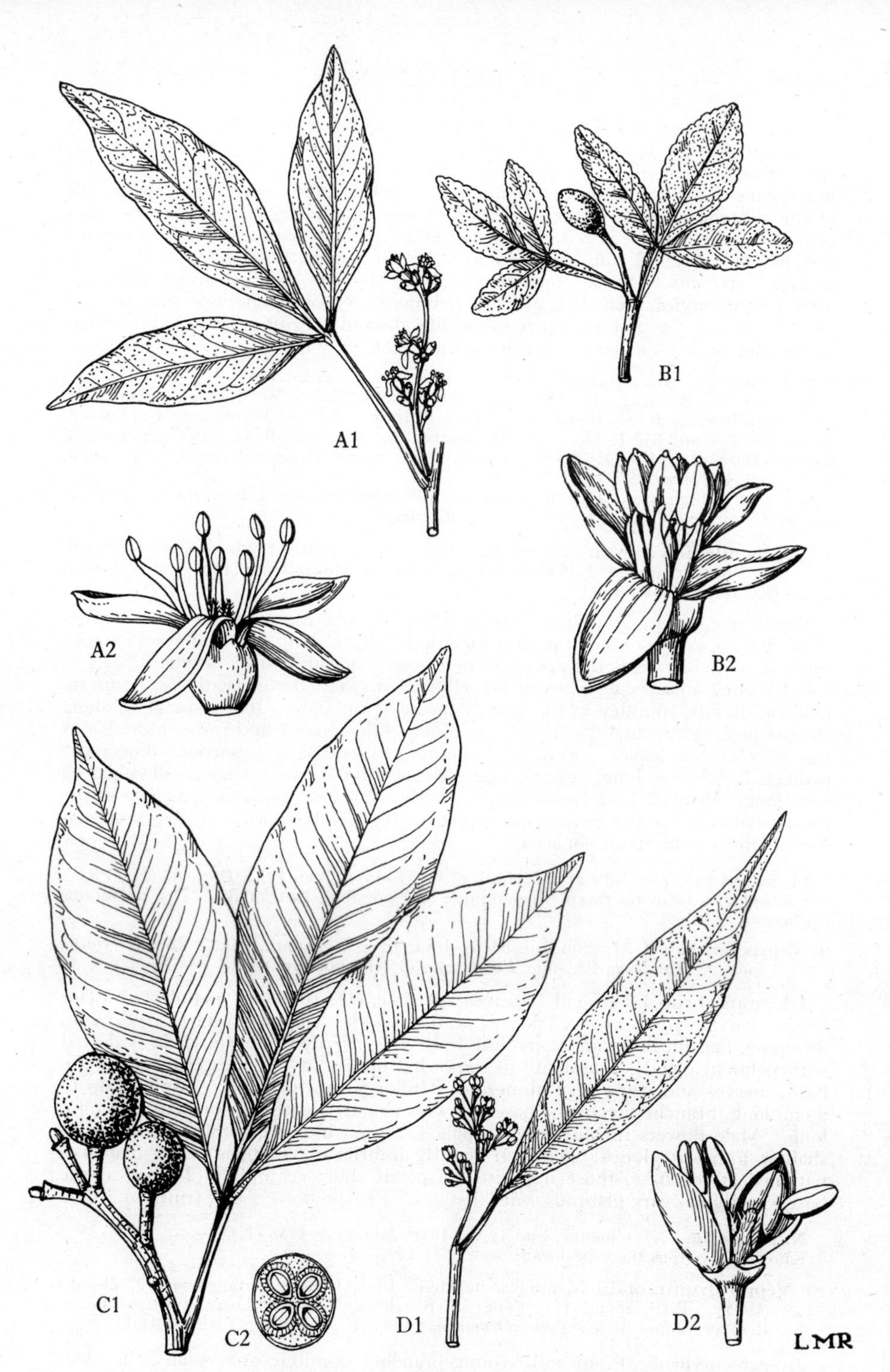

Tab. 35. A.—VEPRIS REFLEXA. A1, branchlet with leaf and inflorescence (× ⅔); A2, male flower (× 6), both from *Torre* 6651. B.—VEPRIS CARRINGTONIANA. B1, branchlet with leaves and fruit (× ⅔), *Barbosa* 625; B2, male flower (× 6) *Mendonça* 3156. C.—VEPRIS STOLZII. C1, branchlet with leaf and fruits (× ⅔) *Fanshawe* 2410; C2, cross section of fruit (× ⅔) *Robson* 419. D.—VEPRIS DRUMMONDII. D1, branchlet with leaf and inflorescence (× ⅔); D2, male flower with 1 petal and 1 stamen removed (× 6), both from *Drummond* 4995.

the apex; leaflets usually sessile; lamina 3–8 × 1–4 cm., coriaceous, narrowly elliptic to elliptic, densely dotted with pellucid glands, obtuse or ± acute at the apex, cuneate at the base; nerves and veins conspicuous on the upper surface. Inflorescence of axillary racemes or panicles, shorter or a little longer than the petiole of the subtending leaf. Flowers sessile, 4-merous, unisexual by abortion, dioecious, usually clustered on the rhachis. Sepals ciliolate, united into a shortly lobed cupuliform calyx 1·2 mm. long. Petals 4·5 × 0·8 mm., oblong, reflexed. Male flowers: stamens 8, those opposite the petals shorter; vestigial ovary glabrous, with 2 short styles. Female flowers: staminodes 8; ovary globose, glabrous, 2-locular; loculi 2-ovulate; stigma subsessile, discoid. Fruit 1·2 cm. long, drupaceous, ellipsoid, 1-locular. Mature seed not seen.

S. Rhodesia. W: Matobo Distr., ♂ fl. xi.1953, *Miller* 1969 (K; SRGH). E: Chipinga, Sabi R., ♂ fl. 16.xi.1959, *Goodier* 666 (SRGH). S: Nuanetsi, immat. fr. 5.v.1958, *Drummond* 5583 (SRGH). **Mozambique.** MS: Cheringoma, Conduè, fl. 24.viii.1946, *Simão* 812 (LM). SS: Inharrime, Ponta Závora, fl. 16.x.1957, *Barbosa & Lemos* in *Barbosa* 8081 (COI; LISC; LMJ). LM: Sábiè, between Boane and Namaacha, fl. 30.vi.1944, *Torre* 6651 (LISC).

Also in the Transvaal. Dry deciduous woodland and grassland, from the low-altitude belt in Mozambique up to c. 800 m. in S. Rhodesia.

8. **Vepris allenii** Verdoorn in Kew Bull. **1926**: 398 (1926).—Engl. in Engl. & Prantl, Nat. Pflanzenfam. ed. 2, **19a**: 306 (1931). Type: Mozambique, mouth of R. Msalo, *Allen* 68 (K, holotype).

Small tree up to 3 m. tall; young branches glabrous, lenticellate. Leaves 3-foliolate, opposite or subopposite; petiole 1–2·5 cm. long, grooved above, sometimes flattened towards the apex; leaflets sessile or subsessile; lamina 2·5–7·5 × 1·4–3·5 cm., coriaceous, obovate or elliptic, densely dotted with very minute pellucid glands, rounded at the apex, cuneate at the base. Inflorescence of short slender racemes up to 1·3 cm. long on older leafless stems and sometimes also in the axils of the leaves. Flowers 4-merous, unisexual by abortion, dioecious; pedicels 2–4·5 mm. long, very slender. Sepals united into a very small calyx 0·2 mm. long. Petals 2·4 × 1·1 mm., oblong. Male flowers: stamens 8, as long as the petals, flattened at the base, those opposite the petals shorter; ovary vestigial. Female flowers and fruit not seen.

Mozambique. N: hills at the mouth of R. Msalo, fl. 7.xii.1912, *Dawe* 68 (K; PRE).

Known only from northern Mozambique and possibly Tanganyika. Dry deciduous thickets.

9. **Vepris fanshawei** Mendonça in Mem. Junt. Invest. Ultram., Sér. 2, **28**: 84 (1961). Type: N. Rhodesia, Chienge, *Fanshawe* 4735 (K, holotype).

Glabrous shrub 1·5 m. tall. Leaves alternate, (1)3-foliolate; petiole 1–4·5 cm. long, terete, narrowly grooved above; leaflets sessile; lamina 6·5–11·5 × 1·9–3 cm., drooping, flaccid, becoming convolute in drying, narrowly elliptic, densely dotted with pellucid glands, acuminate to somewhat obtuse at the apex, cuneate at the base; nerves and veins very numerous. Inflorescence of axillary panicles up to 5 cm. long, branched from the base; flowers usually clustered; pedicels 1–4 mm. long. Male flowers in bud only: sepals 4, united into a cupuliform calyx with shallow imbricate lobes; petals 4, slightly imbricate; stamens (7)8; filaments flattened at the base, those opposite the petals shorter; anthers 2-lobed at the base; vestigial ovary glabrous, with 2 styles. Female flowers and fruit not seen.

N. Rhodesia. N: Chienge, bud 12.viii.1958, *Fanshawe* 4735 (K).

Known only from the type-locality.

10. **Vepris drummondii** Mendonça in Mem. Junt. Invest. Ultram., Sér. 2, **28**: 84 (1961). TAB. **35** fig. D. Type: S. Rhodesia, Melsetter Distr., Glencoe Forest Reserve, slopes of Mt. Pene, *Drummond* 4995 (K, holotype; PRE; SRGH).

Evergreen shrub, 0·7 m. tall; young branches minutely puberulous, the older ones glabrous. Leaves alternate, 1-foliolate; petiole 3–10 mm. long, grooved above; leaflet-lamina 4–12·5 × 1–3·2 cm., narrowly elliptic, ± papyraceous, acuminate, dotted with very minute pellucid glands, acute or blunt at the apex, cuneate at the base; nerves and veins rather fine. Inflorescence of axillary panicles up to 3 cm. long; bracts and bracteoles very small, caducous; flowers 4-merous, unisexual, dioecious, pale yellowish; pedicels 2·5 mm. long, very slender. Sepals 4, united into a short-lobed cupuliform calyx 0·5–0·7 mm. long. Petals 4, 2·6 × 1·3 mm., slightly imbricate, elliptic. Male flowers: stamens 8; filaments flattened at the base, those opposite the petals shorter; anthers 0·5 mm. long, 2-lobed at the base; disk inconspicuous; vestigial ovary pilose with 2 densely setigerous styles. Female flowers and fruits not seen.

S. Rhodesia. E: Melsetter Distr., Glencoe Forest Reserve, slopes of Mt. Pene, c. 1200 m., fl. 24.xi.1955, *Drummond* 4995 (K; PRE; SRGH). Known only from the type locality.

Note

Vepris glomerata (F. Hoffm.) Engl. (*Toddalia glomerata* F. Hoffm.; *Teclea glomerata* (F. Hoffm.) Verdoorn) was wrongly localized by Verdoorn (in Kew Bull. **1926**: 406 (1926)) as coming from Kakoma, Nyasaland. The type was from Kakoma in Tanganyika and the species has not been found in our area.

11. TODDALIOPSIS Engl.

Toddaliopsis Engl., Pflanzenw. Ost-Afr. **C**: 443 (1895).

Shrubs or small trees. Leaves alternate, 3-foliolate; leaflets densely dotted with pellucid glands. Inflorescence of short racemes or panicles, axillary and terminal. Flowers unisexual by abortion. Sepals 4, ovate, united at the base. Male flowers: stamens 8, inserted at the base of the disk; filaments subulate; anthers cordate; ovary vestigial. Female flowers: staminodes 8, filiform, with atrophied anthers; ovary slightly 4-lobed with thick verrucose walls, 4-locular; loculi 2-ovulate; stigma sessile, 4-lobed. Fruit very woody, globose, subtetragonal, 4-locular; loculi 1-seeded. Seed oblong; endosperm absent.

Toddaliopsis bremekampii Verdoorn in Kew Bull. **1935**: 204 (1935). TAB. **36** fig. B. Type from the Transvaal (Soutpansberg).

Shrub or small evergreen tree up to 6 m. tall; young branches reddish, glabrous, older ones dark grey. Leaves petiolate; petiole 1–4 cm. long, grooved above; leaflets sessile or the central one sometimes petiolulate; lamina 3·5–8(9) × 1·2–3·5 cm., elliptic or oblong-elliptic, shining above and light yellowish below, dotted with pellucid glands, acuminate or obtuse at the apex, margin entire and slightly revolute, narrowly cuneate at the base; secondary nerves slightly raised above. Inflorescence of axillary and terminal racemes as long as or shorter than the petiole of the subtending leaf; peduncles and pedicels sparsely and minutely puberulous; bracteoles 0·5 × 0·5 mm., caducous. Male flowers: pedicel 1·5–2 mm. long; sepals 1·5 mm. long, broadly ovate, imbricate; petals 3·5 × 1·5 mm., elliptic, imbricate; stamens somewhat flattened at the base, as long as the petals or slightly longer; anthers basifixed, 0·9 mm. long, deeply 2-lobed at the base; ovary vestigial. Female flowers not seen. Fruit 1·4 cm. in diam., densely verrucose, reddish-brown. Mature seeds not seen.

Mozambique. SS: Muchopes, Manjacaze, fr. 26.i.1941, *Torre* 2543 (LISC). LM: Maputo, Porto Henrique, ♂ fl. & fr. 14.xi.1944, *Mendonça* 2833 (LISC).
Also in the Transvaal. In dry woodland or bush on sandy soils.

12. TECLEA Del.

Teclea Del. in Ann. Sci. Nat., Sér. 2, Bot., **20**: 90 (1843).

Trees or shrubs. Leaves alternate, (1)3-foliolate; leaflets densely dotted with pellucid glands. Inflorescence of axillary and terminal panicles, racemes or cymose clusters. Flowers 4–5-merous, unisexual by abortion, rarely bisexual. Sepals united into a 4–5-lobed cupuliform calyx; lobes short, imbricate. Petals longer than the calyx, imbricate in bud. Male flowers: stamens 4–5 inserted at the base of the disk; filaments linear, subulate; anthers dorsifixed, cordate, deeply 2-lobed at the base; ovary vestigial. Female flowers: staminodes 4–5 or fewer, opposite the sepals; ovary 1-locular with 2 pendulous ovules; style short; stigma broadly peltate. Fruit drupaceous, globose or ovoid or ellipsoid, 1-seeded; endocarp woody. Seed ovoid, testa thin; endosperm absent.

Flowers bisexual in axillary and terminal panicles - - - - - 1. *grandifolia*
Flowers unisexual by abortion:
Flowers in axillary and terminal panicles; petioles terete or winged:
Petioles terete; leaflets elliptic, entire; flowers sessile or shortly petiolate, in axillary and terminal panicles - - - - - - - - - - 2. *nobilis*
Petioles winged; leaflets obovate-oblong, crenulate; flowers in terminal panicles 3. *crenulata*
Flowers in short racemes or glomerate in the axils of the leaves; petiole broadly grooved:
Ovary densely pilose; fruit glabrescent; flowers usually pedicellate, in short axillary racemes, rarely in pedunculate clusters - - - - 4. *gerrardii*
Ovary and fruit glabrous; flowers sessile on a peduncle or clustered:
Petioles and young shoots glabrous; terminal leaflet (4)5–10 cm. long 5. *rogersii*
Petioles and young shoots pubescent; terminal leaflet 2–5 cm. long 6. *fischeri*

1. **Teclea grandifolia** Engl., Bot. Jahrb. **23**: 153 (Sept. 1896)—Exell & Mendonça, C.F.A. **1,** 2: 270 (1951). Syntypes from Angola (Cuanza Norte).
Zanthoxylum welwitschii Hiern, Cat. Afr. Pl. Welw. **1**: 114 (Dec. 1896). Type from Angola (Cuanza Norte).
Teclea welwitschii (Hiern) Verdoorn in Kew Bull. **1926**: 408 (1926) pro parte excl. specim. *Swynnerton* 1322.

Small evergreen tree up to 6 m. tall; branches and leaves glabrous. Leaves 3-foliolate; petiole 1·5–5(6·5) cm. long, terete, sometimes slightly grooved near the apex; leaflets subsessile or with a petiolule up to 5 mm. long; lamina 7–18 × 2–7 cm., narrowly elliptic to elliptic, acuminate or acute or blunt at the apex, cuneate at the base; lateral nerves and veins numerous. Inflorescence of axillary and terminal panicles; peduncle and rhachis reddish-brown, minutely puberulous. Flowers bisexual, 4-merous, sessile, usually glomerate on the branches of the panicle; bracteoles very small, scarious, ciliolate. Sepals forming a cupuliform calyx 1·1 mm. long, including lobes 0·3 mm. long. Petals 3 × 1·1 mm., oblong, slightly imbricate in bud, reflexed. Stamens opposite the sepals, inserted at the base of a very narrow disk; filaments linear, flattened at the base, as long as the petals; anthers basifixed, 2-lobed at the base; ovary ellipsoid; style short; stigma capitate, 2-lobed. Fruit 7 × 5 mm., ovoid. Mature seeds not seen.

N. Rhodesia. W: Mufulira, fr. x.1954, *Fanshawe* 1740 (K); Mwinilunga, boma, fl. 20.x.1955, *Holmes* 1275 (K).
Also in Angola, Uganda and perhaps Tanganyika. Mist and riverine evergreen forest, 800–1700 m.

2. **Teclea nobilis** Del. in Ann. Sci. Nat., Sér. 2, Bot. **20**: 90, t. 1 fig. 1 (1843).—Engl., Pflanzenw. Afr. **3,** 1: 756, fig. 353 A–D (1915); in Engl. & Prantl, Nat. Pflanzenfam. ed. 2, **19a**: 314, fig. 143 A–D (1931).—Verdoorn in Kew Bull. **1926**: 408 (1926).—Steedman, Trees etc. S. Rhod.: 29 (1933).—Burtt Davy & Hoyle, N.C.L.: 68 (1936).—Gilbert, F.C.B. **7**: 100 (1958).—Dale & Greenway, Kenya Trees & Shrubs: 488, fig. 89 (1961).—White, F.F.N.R.: 170 (1962). TAB. **36** fig. A. Type from Ethiopia.
Toddalia nobilis (Del.) Hook. f. in Oliv., F.T.A. **1**: 306 (1868). Type as above.
Teclea natalensis sensu Bak. f. in Journ. Linn. Soc., Bot. **40**: 35 (1911) non *Toddalia natalensis* Sond.
Teclea welwitschii (Hiern) Verdoorn in Kew Bull. **1926**: 408 (1926) pro parte quoad specim. *Swynnerton* 1322.

Evergreen shrub or tree up to 10 m. tall; branches, leaves and inflorescences glabrous. Leaves 3-foliolate; petiole 1·5–6(8) cm. long, terete or sometimes slightly grooved at the apex; leaflets subsessile or petiolulate; petiolule up to 10 mm. long; lamina 5–15(18) × 1·5–4(5·5) cm., narrowly elliptic to elliptic, acuminate or broadly attenuate, blunt at the apex, narrowly cuneate at the base; lateral nerves and veins numerous. Inflorescence of axillary and terminal panicles. Flowers unisexual by abortion, sessile or with pedicels up to 2 mm. long, usually in clusters on the branches of the panicle. Flower-buds 2–2·5 mm. long, ellipsoid or obovoid. Sepals 4, united into a cupuliform calyx 0·6–0·8 mm. long with very small ovate lobes. Petals 4, 4 × 1·7 mm., slightly imbricate in bud, narrowly elliptic. Male flowers: stamens 4, alternating with the petals, inserted at the base of the disk; filaments 5 mm. long, linear; anthers basifixed, 2-lobed

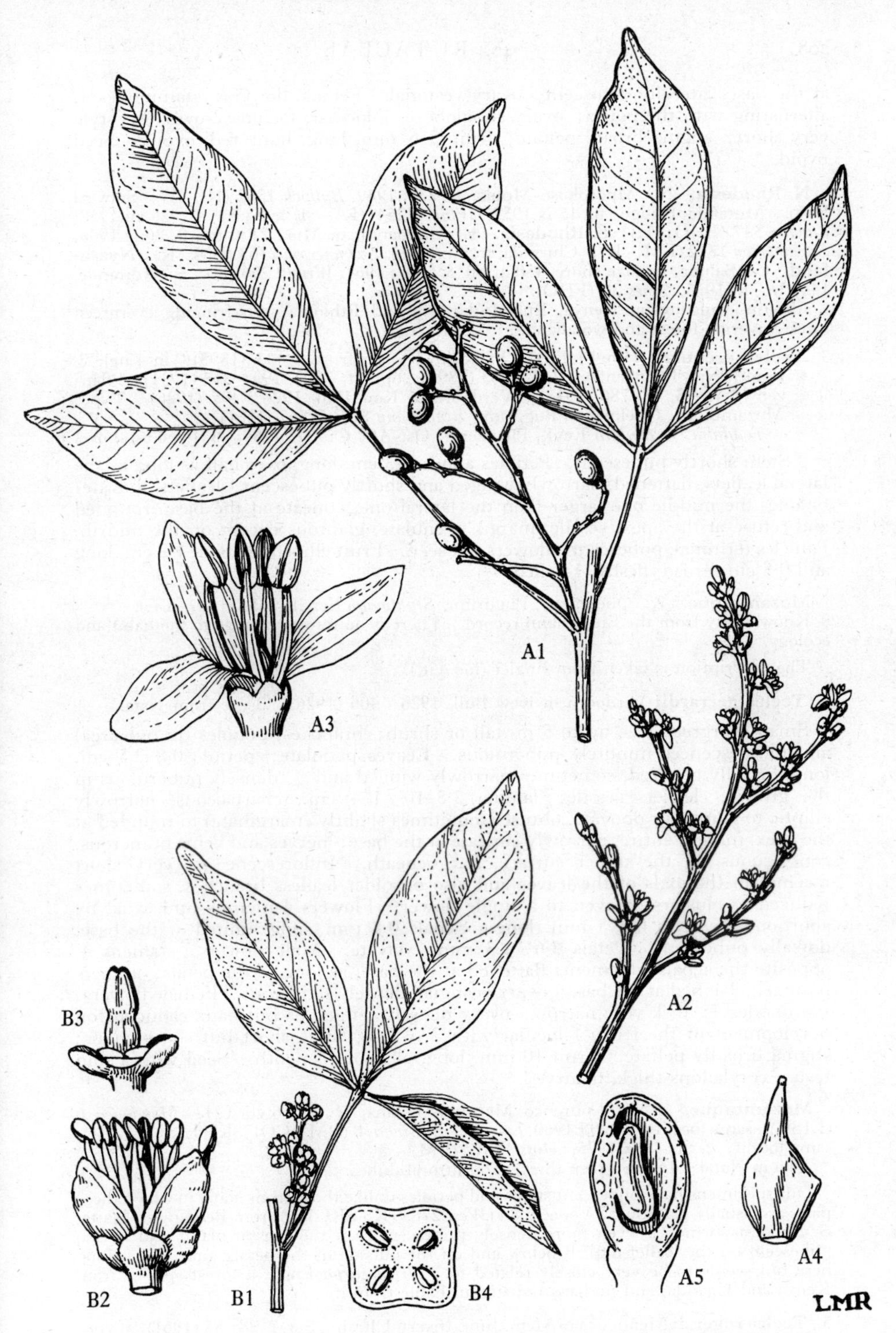

Tab. 36. A.—TECLEA NOBILIS. A1, branchlet with leaves and fruits ($\times\frac{2}{3}$) *Chase* 4278; A2, male inflorescence ($\times\frac{2}{3}$); A3, male flower with 1 petal removed ($\times 6$); A4, vestigial ovary ($\times 12$); A5, longitudinal section of young fruit ($\times 8$), all from *Chase* 5297. B.—TODDALIOPSIS BREMEKAMPII. B1, branchlet with leaf and inflorescence ($\times\frac{2}{3}$); B2, male flower ($\times 4$); B3, vestigial gynoecium ($\times 4$); B4, cross section of ovary ($\times 12$), all from *Mendonça* 2833.

at the base, laterally dehiscent; ovary vestigial. Female flowers: staminodes 4, alternating with the petals; ovary subglobose, 1-locular, loculus 2-ovulate; style very short; stigma broad, peltate; fruit 5–6 mm. long, light red, ovoid. Seed ovoid.

N. Rhodesia. N: Mporokoso-Mpuka, fr. 25.x.1949, *Bullock* 1362 (K). W: Solwezi Distr., Mutanda Bridge, fl. 15.ix.1952, *White* 3254 (K). E: Nyika, fr. 26.xii.1958, *Lawton* 547 (FHO; K). **S. Rhodesia.** E: Chimanimani Mts., 2130 m., fl. 26.ix.1906, *Swynnerton* 1322 (BM; K); Chirinda, fr. 9.ix.1906, *Swynnerton* 2163 (BM; K). **Nyasaland.** C: Salima, on lake shore, 450 m., fl. & fr. 8.i.1961, *Wright* 296 (K). S: Ndirande, 1460 m., fl. 1955, *Jones* 39 (FHO).

Also in Tanganyika, Kenya, Ruanda-Urundi and Ethiopia. In fringing evergreen forest, riverine forest and woodland.

3. **Teclea crenulata** (Engl.) Engl., Pflanzenw. Ost-Afr. **C**: 433 (1895); in Engl. & Prantl, Nat. Pflanzenfam. **3,** 4: 183 (1896); op. cit. ed. 2, **19a**: 315 (1931); Pflanzenw. Afr. **3,** 1: 786 (1915).—Verdoorn in Kew Bull. **1926**: 407 (1926). Type: Mozambique, Quelimane, Pugurimi, *Stuhlmann* Ser. I, 562 (HBG).
Toddalia ? crenulata Engl., Pflanzenw. Ost-Afr. **C**: 228 (1895). Type as above.

" Stem shortly pubescent. Petioles about 2–3 cm. long, over half as long as the lateral leaflets, flattened, narrowly winged and shortly pubescent; leaflets obovate-oblong, the middle one larger than the lateral one, cuneate at the base, rounded and retuse at the apex, sessile, margin crenulate, glabrous except on the midrib. Panicles terminal, pubescent; flowers not seen. Fruit ellipsoid, about 1·2 cm. long and 0·8 cm. broad, fleshy, 1-celled."

Mozambique. Z: Quelimane, Pugurimi, *Stuhlmann* Ser. I, 562 (HBG, n.v.).

Known only from the Stuhlmann record. There is no information as to the habit and ecology.

The description is taken from Engler (loc. cit.).

4. **Teclea gerrardii** Verdoorn in Kew Bull. **1926**: 406 (1926). Type from Natal.

Small evergreen tree up to 5 m. tall or shrub; branches, petioles (in our area) and inflorescences minutely puberulous. Leaves petiolate; petiole 0·8–2·5 cm. long, broadly grooved, sometimes narrowly winged and ± densely puberulous in the groove; leaflets sessile; lamina 3·5–10 × 1–4 cm., chartaceous, narrowly elliptic to elliptic or obovate, obtuse (sometimes slightly emarginate) to rounded at the apex, margin entire, narrowly cuneate at the base; nerves and veins numerous, conspicuous on the upper surface, flat beneath. Inflorescence of very short racemes in the axils of the leaves and also on older leafless branches, sometimes reduced to clusters or even to a single flower. Flowers 4-merous, unisexual by abortion; pedicels 0·8–3 mm. long. Sepals 0·8 mm. long, united at the base, dorsally pubescent. Petals 3·6 × 1·3 mm., valvate. Male flowers: stamens 4, opposite the sepals; filaments flattened at the base, as long as the petals; anthers basifixed, 2-lobed at the base; ovary vestigial, densely setigerous. Female flowers: staminodes 4; disk very narrow; ovary densely setigerous but hairs caducous on development of the fruit, 1-locular with 2 ovules; style short but conspicuous; stigma broadly peltate. Fruit 10 mm. long, ellipsoid, smooth. Seed with a thin testa; cotyledons thick, incurved.

Mozambique. LM: Lourenço Marques, Polana, ♂ fl. 16.viii.1942, *Mendonça* 6 (LISC); same locality, fr. 20.i.1960, *Lemos & Balsinhas* 8 (BM; COI; K; LISC; LMJ); same locality, ♀ fl. 30.ix.1925, *Moss* 11812 (BM).

Also in Natal. In evergreen thickets on littoral dunes.

Our specimens have pubescent stems and petioles unlike those from Natal in which these parts are usually glabrous. A specimen (*West* 2783 (K; SRGH)) from Belingwe, Makita, S. Rhodesia, which is even more densely pubescent than the coastal plants and is also pubescent on the leaflet-midrib below and on the outside of the petals, may also belong here but seems to be very closely related to *T. trichocarpa* Engl., a forest species from Kenya and Uganda, and perhaps indistinguishable from it.

5. **Teclea rogersii** Mendonça in Mem. Junt. Invest. Ultram., Sér. 2, **28**: 85 (1961). Type: N. Rhodesia, Kalomo, *Rogers* 26015 (K, holotype).
Teclea fischeri sensu White, F.F.N.R.: 170 (1962) pro parte quoad specim. *White* 2473.

Evergreen shrub up to 3 m. tall; glabrous in all parts. Leaves petiolate, 3-

foliolate; petiole 0·7–3 cm. long, broadly grooved; leaflets sessile; lamina 2–8·5(12) × 1–2·6(4) cm., coriaceous, elliptic, apex obtuse or rounded or sometimes shortly and broadly acuminate or sometimes emarginate, cuneate at the base; nerves and veins numerous, quite conspicuous on the upper surface, flat beneath. Inflorescence of axillary sessile glomerules, sometimes also on older leafless stems. Flowers 4-merous; buds 4·5 mm. long, narrow. Sepals 4, united into a cupuliform calyx 1 mm. long with short truncate lobes. Male flowers: petals 4, 4·5 × 0·8 mm., very narrow, slightly imbricate; stamens 4, opposite the sepals; filaments filiform, as long as the petals; anthers basifixed, 2-lobed at the base; disk obsolete, ovary vestigial, conical, glabrous; style slender. Female flowers: not seen. Fruit 11 × 7 mm., scarlet (*fide* Miller), black when dry, smooth; endocarp rather thin. Seed ellipsoid; testa membranous; cotyledons thick; embryo very small.

N. Rhodesia. E: Fort Jameson to Lundazi, km. 48, bud 26.iv.1952, *White* 2473 (FHO; K). S: Siamambo Forest Reserve, 6 km. SE. of Choma, bud 9.ix.1957, *Angus* 1700 (FHO; K). **S. Rhodesia.** N: Lomagundi, Hunyani R., near Sinoia, st. 16.v.1952, *Pardy* in GHS 5926 (K; SRGH). W: Victoria Falls, fl. 18.xi.1949, *Wild* 3085 (K; LISC; SRGH). S: Ndanga, north and west of Sabi-Lundi Junction, bud iv.1955, *Middleton-Stokes* 16 (K; SRGH). **Mozambique.** T: Lupata Gorge, fl. x.1868, *Kirk* (K).

Known only from our area. In dry rocky places in the *Colophospermum mopane* belt of the Zambezi and Sabi basins.

6. **Teclea fischeri** (Engl.) Engl., Pflanzenw. Ost-Afr. **C**: 433 (1895); in Engl. & Prantl, Nat. Pflanzenfam. ed. 1, **3,** 4: 183 (1896), ed. 2, **19a**: 315 (1931).—Verdoorn in Kew Bull. **1926**: 404 (1926).—White, F.F.N.R.: 170 (1962) pro parte excl. specim. *White* 2473. Type from Tanganyika.
Toddalia fischeri Engl. tom. cit.: 228 (1895). Type as above.

Shrub up to 2·5 m. high. Leafy branches, petioles and leaflet midrib ± densely and minutely puberulous with straight hairs. Leaves alternate, 3-foliolate; petiole 0·5–1·5 cm. long, terete at the base, grooved above to the top. Leaflets coriaceous, sessile; lamina 1·5–4 × 1–2 cm., glossy above, elliptic, obovate or oblong-obovate, densely dotted with glands below, inconspicuous on the upper surface, emarginate or rounded at the apex, margin usually crenulate from the middle upwards, ± narrowly cuneate at the base; nerves and veins raised on the upper surface. Flowers (young buds) 4-merous, sessile, in axillary clusters or in short spikes. Sepals united into a cupuliform 4-lobed calyx 1·6 mm. long, lobes rounded and 0·2 mm. long, with the margin scarious, ciliolate. Petals imbricate. Male flowers: young buds with 4 stamens; ovary vestige glabrous. Female flowers and fruit not seen from our area.

N. Rhodesia. N: Samfya, Lake Bangweulu, bud 1.ix.1952, *White* 3184A (FHO; K). C: between Rufunsa and Lusaka, bud 26.iii.1955, *E.M. & W.* 1209 (BM; LISC; SRGH). S: Mumbwa, Mambala Mt., bud 18.ix.1947, *Brenan* 7876 (K).
Also in Tanganyika. In thickets, sandy soils, rocky slopes and stony ground.

13. CITRUS L.

Citrus L., Sp. Pl. 2: 782 (1753); Gen. Pl. ed. 5: 341 (1754).

Small trees or shrubs. Leaves 1-foliolate, with winged rhachis. Flowers bisexual, (4)5-merous. Stamens numerous, in phalanges. Ovary (4)5–many-locular; loculi 4–8-ovulate. Fruit a large globose or ovoid or obovoid hesperidium, many-seeded and usually composed of numerous carpels.

No indigenous species of *Citrus* are found in our area but some of the cultivated species (especially *C. limon* (L.) Burm. f. and *C. aurantium* L.) may become naturalized. Dr. G. R. Bates has kindly provided the appended note on the cultivation of *Citrus* in the Federation.

1. *Citrus in the Federation.*

Citrus is the most widely grown subtropical fruit crop in the Federation. The majority of commercial orchards are established in areas ranging from 1000–1350 m. where suitable soil and irrigation facilities exist. Recently substantial plantings have also been made in S. Rhodesia along the Lundi R. (450–500 m.) By far the most extensive plantings are in the Mazoe Valley (1300 m.) where the so-called Mazoe Rough Lemon, generally used as a citrus rootstock throughout southern

Africa, is spontaneous in scattered thickets along the banks of the river. Oranges constitute the bulk of commercial plantings, the Valencia Late variety being the most favoured, although appreciable quantities of mid-season varieties and the early Washington Navel variety are also grown. Grapefruit is best suited to the lower altitudes. Increasing numbers of lemon trees are being planted, mainly to provide fruit for processing purposes. The development of the citrus industry in recent years foreshadows a considerable expansion of production in the near future. [G.R.B.]

2. *Citrus in Mozambique.*

In Mozambique the Sweet Orange (*C. sinensis* (L.) Osbeck) is cultivated from sea-level to an altitude of 1000 m. The Lemon (*C. limon* (L.) Burm. f.) is cultivated throughout the province and in favourable localities, such as the banks of the tributaries of the Zambeze, it often becomes naturalized, as in the " floresta dos limoeiros " near Inhaminga. The Mandarin or Tangerine Orange (*C. reticulata* Blanco) and the Grapefruit (*C. paradisi* Macf.) are cultivated here and there on a small scale. The Lime (*C. aurantifolia* (Christm.) Swingle) and the Sour or Seville Orange (*C. aurantium* L.) are infrequent.

41. SIMAROUBACEAE

By H. Wild and J. B. Phipps

Trees, shrubs or shrublets, occasionally ± sarmentose, sometimes spiny and with pubescence of simple, sometimes glandular hairs. Leaves alternate (at least in our area), pinnate or sometimes simple, exstipulate. Inflorescences paniculate, bracteate (in our area). Flowers bisexual, unisexual or polygamous. Calyx 2–5-lobed, often very deeply so. Petals (3) 4–5 (9), imbricate or valvate, very rarely united in a tube (not in our area), very rarely absent (not in our area). Disk annular, cupular, or sometimes fused with the ovary, rarely absent. Stamens 4–10 (18), rarely ∞ (not in our area), free, sometimes with a scale adnate at the base of the filament. Ovary subentire or rarely (2) 4–5-lobed, (1) 2–5-locular; loculi 1-ovulate or rarely 2-ovulate, or sometimes with 2–5 free carpels; styles (2) 4–5, free or variously united. Fruit a berry, or of separate drupaceous mericarps or dry angled mericarps suspended from a central carpophore.

Leaves simple; carpels 2-ovulate; maritime shrub - - - - - 1. **Suriana**
Leaves pinnate; carpels 1-ovulate:
  Stamens (3) 4 (5), without appendages at base:
    Fruit of 1–4 drupaceous mericarps; indumentum ± golden-brown 2. **Brucea**
    Fruit of 4 ± dry triquetrous mericarps suspended from a central carpophore; indumentum lacking or greyish - - - - - - - 3. **Kirkia**
  Stamens 8–10 or more, with hairy appendages at the base of the filaments:
    Carpels free in flower and fruit; branchlets unarmed - - - - 4. **Hannoa**
    Carpels united forming a 4–5-locular ovary; branchlets with short, often paired, prickles - - - - - - - - - - - - 5. **Harrisonia**

1. SURIANA L.

Suriana L., Sp. Pl. **1**: 284 (1753); Gen. Pl. ed. 5: 137 (1754).

A maritime shrub. Leaves alternate, crowded along the branches, simple. Inflorescences of few-flowered panicles or solitary and axillary near the ends of the branches. Flowers bisexual. Calyx deeply 5-lobed; lobes imbricate, persistent. Petals 5, imbricate. Stamens 10 (18); filaments densely pilose towards the base; alternate filaments somewhat shorter, and sometimes 5 filaments without anthers. Ovary of 5 free, densely pilose, 2-ovulate carpels; styles 5, free, lateral; ovules collateral, ascending. Fruiting carpels dry, 1-seeded. Seeds without endosperm; embryo conduplicate.

Gutzwiller (in Engl., Bot. Jahrb. **81**: 1–49 (1961)) considers that *Suriana* should be placed in a separate family, the *Surianaceae*, close to *Sapindaceae* and *Connaraceae*.

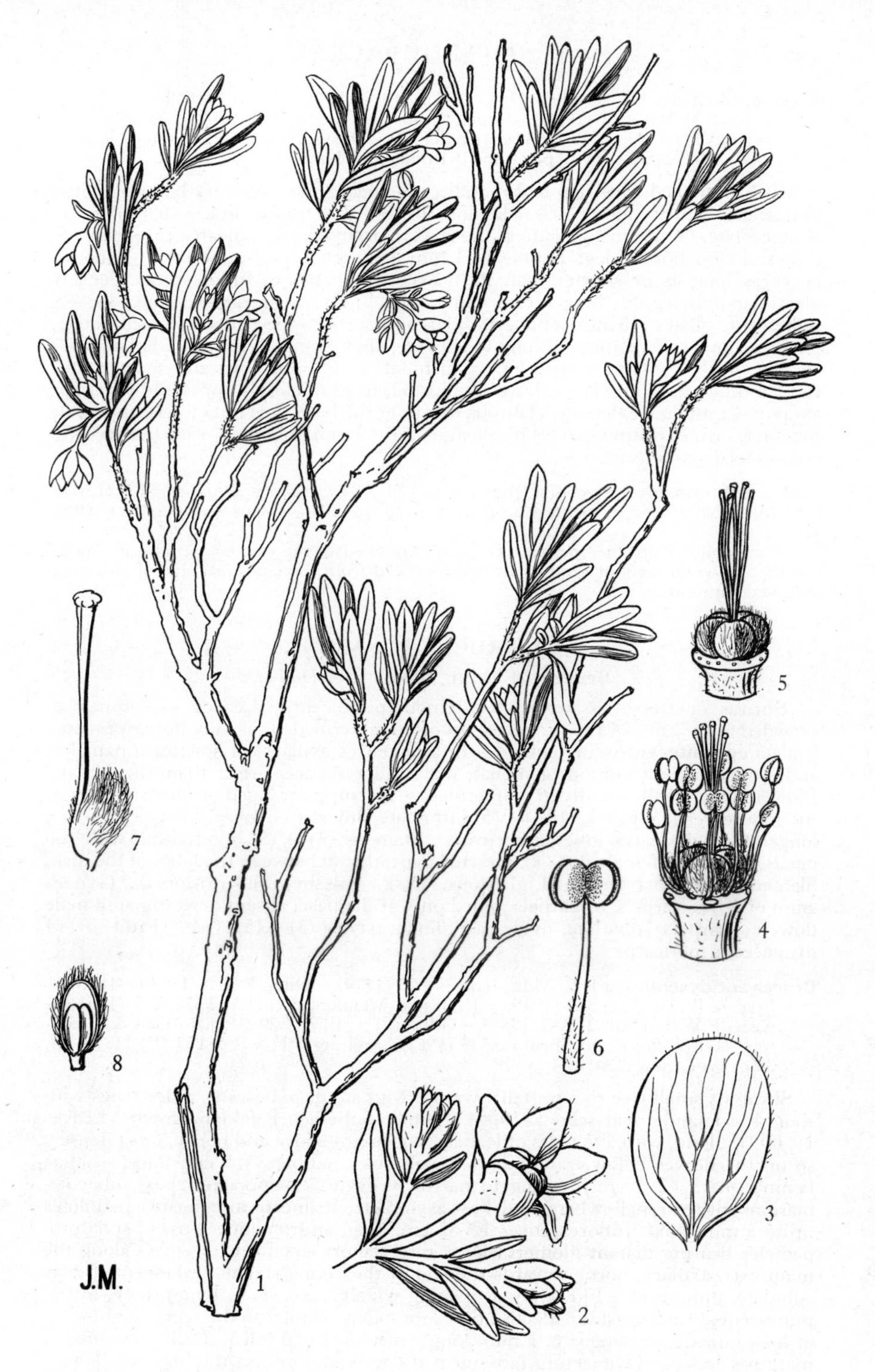

Tab. 37. SURIANA MARITIMA. 1, flowering branch (× ⅔); 2, inflorescence (× 2); 3, petal (× 4); 4, flower with calyx and petals removed (× 4); 5, gynoecium (× 4); 6, anther (× 6); 7, single carpel with style (× 10); 8, vertical section of carpel with 2 ovules (× 10) all from *Drummond & Hemsley* 1044.

Suriana maritima L., Sp. Pl. **1**: 284 (1753).—Oliv., F.T.A. **1**: 313 (1868).—Sim, For. Fl. Port. E.Afr.: 24 (1909).—Engl., Pflanzenw. Afr. **3**, 1: 765, fig. 356 (1915).—Brenan, T.T.C.L.: 573 (1949).—Perrier, Fl. Madag. Simarub.: 7, tab. 2 fig. 3–6 (1950). TAB. **37**. Syntypes from Bermuda and Jamaica.

Much-branched shrub up to 2 m. tall; branches greyish-velvety-hairy or thinly pubescent when older. Leaf-lamina 10–40 × 3–4 mm., linear-spathulate to oblanceolate, obtuse to subacute at the apex, margin entire, cuneate at the base, ± pubescent on both sides; petiole c. 1 mm. long, glandular-pubescent. Inflorescence as long as or shorter than the leaves; branches of inflorescence densely glandular-pubescent; bracts c. 4 mm. long, linear or linear-subulate, densely pubescent. Calyx c. 8 mm. long, glandular-pubescent; lobes lanceolate-acuminate. Petals yellow, c. 7 × 5 mm., oblong to obovate, apex rounded, margin ciliolate, base shortly clawed. Stamens with longer filaments c. 4·5 mm. long and shorter ones c. 3·5 mm. long. Carpels ± obovoid-ellipsoid, free but closely appressed together; styles c. 4 mm. long, slender, glabrous except at the base, free but closely appressed together. Ripe fruiting carpels blackish, c. 3·5 × 3 mm., obovoid, with two faces ± compressed, pubescent.

Mozambique. N: Macomia, Ingoane, fl. 12.ix.1948, *Pedro & Pedrógão* 5164 (LMJ). SS: Inhambane, Ilha de Santa Carolina, fl. & fr. viii.1936, *Gomes e Sousa* 1883 (COI; K).

A pantropical maritime species. In Africa confined to the east coast; also in Madagascar. On coral reefs or sandy soils and recorded from the landward side of *Avicennia marina* communities.

2. BRUCEA J.F. Mill.

Brucea J.F. Mill., Icon.: t. 25 (1779).

Shrubs or trees, ± velvety-ferruginous-pubescent. Leaves sometimes ± crowded at the ends of the branches but often not crowded, usually imparipinnate; leaflets opposite, entire or toothed. Inflorescences axillary, of spiciform panicles, andromonoecious (? or polygamous), male inflorescence larger than the female. Flowers very small, usually in glomerules ± distantly arranged on the branches of the inflorescence. Calyx of (3) 4 (5) imbricate almost free lobes. Petals (3) 4 (5), longer than the calyx-lobes, imbricate. Stamens (3) 4 (5), alternating with the petals, smaller in female flowers, inserted beneath and between the lobes of the disk, filaments free, unappendaged, glabrous. Disk ± fleshy, (3) 4 (5)-lobed. Gynoecium of 4 free carpels, or carpels joined only at the base, absent or vestigial in male flowers; carpels 1-ovulate, ovule pendulous; styles (3) 4 (5), free. Fruit of 1–4 drupaceous mericarps.

Brucea antidysenterica J. F. Mill., Icon.: t. 25 (1779).—Oliv., F.T.A. **1**: 309 (1868).—Brenan, T.T.C.L.: 572 (1949).—Exell and Mendonça, C.F.A. **1**, 2: 277 (1951).—Keay, F.W.T.A. ed. 2, **1**, 2 (1958).—Gilbert, F.C.B. **7**: 129 (1958).—Dale & Greenway, Kenya Trees and Shrubs: 535 (1961).—White, F.F.N.R.: 172 (1962). TAB. **38**. Type from " Africa ".

Shrub or small tree to 5 m. tall; young twigs stout, pubescent, older twigs with roughly triangular leaf-scars 5–10 × 4–9 mm.; pubescence golden-brown. Leaves 15–60 cm. long, with 3–5 pairs of leaflets, pubescent above and below, most densely so on the nerves below, rarely almost glabrous; petiole 8–15 cm. long; leaflet-laminas 4–13 × 2–6 cm., oblong to narrowly ovate, ± opposite, apex subacute, margins entire to shallowly repand, base asymmetric truncate to cuneate; petiolules up to 5 mm. long. Inflorescences 12–35 cm. long, andromonoecious, of spiciform panicles bearing distant glomerules (or rarely short spiciform racemes) along the main axis, axillary, borne towards the end of the young shoots, pubescent; bracts subulate, pubescent. Flowers not opening widely. Sepals c. 1 mm. long, ovate, pubescent at least outside. Petals up to 2 mm. long, similar to the sepals, pubescent at least outside. Stamens c. 1 mm. long; filaments glabrous. Disk pulviniform, margin 4-lobed. Gynoecium (absent in the male flowers) in the bisexual flowers consisting of ± free, ovoid, usually glabrous carpels pressed together at first with the stigmas bent outwards in a cruciform arrangement. Fruit developing from 1–4 of the carpels in each flower and consisting of red ovoid drupaceous mericarps c. 10 × 6 mm. acute at the apex.

Tab. 38. BRUCEA ANTIDYSENTERICA. 1, flowering branch ($\times \frac{2}{3}$) *Bally* 1004; 2, male flower ($\times 12$) *Bally* 1004; 3, bisexual flower (after the falling off of the anthers) ($\times 12$) *Dale* 1930; 4, portion of infructescence ($\times \frac{2}{3}$) *Dale* 1930.

N. Rhodesia. N. Abercorn Distr., Mushambishi Source, fl. 20.ii.1933, *Michelmore* 481 (K). W: Mwinilunga, 16 km. E. of the Boma, fl. 9.x.1930, *Milne-Redhead* 1073 (K) **Nyasaland.** N: Mugesse Forest Reserve, fl. 9.xi.1952, *Chapman* 35 (FHO).

Also in French Guinea, Nigeria, Cameroons, Congo, Angola and in eastern Africa from Ethiopia to Tanganyika and N. Rhodesia. In the lower storeys of evergreen forest and often as secondary growth in deforested areas or at forest edges.

3. KIRKIA Oliv.

Kirkia Oliv., F.T.A. **1**: 310 (1868); in Hook., Ic. Pl.: t. 1036 (1868).

Trees. Leaves crowded near the ends of the branches in close spirals, imparipinnate, with numerous opposite leaflets. Inflorescences axillary, near the ends of the branches, of paniculate cymes. Flowers apparently bisexual but many functionally male with longer stamens and abortive ovaries; others with smaller stamens and normal ovaries. Calyx-lobes 4, free almost to the base. Petals 4. Stamens 4, opposite the sepals, inserted beneath the disk; filaments filiform or broadening slightly towards the base, without scales. Disk annular or cushion-like, rather fleshy. Gynoecium composed of 4 united carpels forming a 4-locular ovary with 1 ovule in each loculus; style slender, with a capitate stigma (the style is formed by the union of 4 original styles and after anthesis the style splits towards the base into its 4 constituent parts and the upper portion falls off, leaving behind the persistent basal portions which simulate 4 free short styles*). Fruit dry, 4-angled, 4-locular, when ripe consisting of 4 triquetrous mericarps suspended from a central carpophore.

Kirkia acuminata Oliv., F.T.A. **1**: 311 (1868); in Hook., Ic. Pl.: t. 1036 (1868).—Sim, For. Fl. Port. E.Afr.: 24 (1909).—Bak. f. in Journ. Linn. Soc., Bot. **40**: 36 (1911).—Eyles in Trans. Roy. Soc. S. Afr. **5**: 388 (1916).—Burtt Davy, F.P.F.T. **2**: 481 (1932).—Steedman, Trees etc. S. Rhod.: 29, t. 26 (1933).—Exell and Mendonça, C.F.A. **1**, 2: 276 (1951).—Wild, Guide Fl. Vict. Falls: 151 (1952).—O.B. Mill. in Journ. S. Afr. Bot. **18**: 37 (1952).—Palgrave, Trees Centr. Afr.: 411 cum tab. et photo (1956).—White, F.F.N.R.: 172 (1962). TAB. **39**. Syntypes: Mozambique, Lupata, *Kirk* (K); Sena, *Kirk* (K).

Medium-sized deciduous tree 6–16 m. tall with clean bole, well branched above with straightish branches; bark grey, smooth when young, flaking later, corrugated in old age; young twigs with prominent ± rhombic leaf scars 2–5 × 3–7 mm. Leaves 10–45 cm. long, viscid when young, densely pubescent (then ± glabrescent) to glabrous, hairs proportionately more numerous on the rhachis and on the midrib of the leaflet above and below than on the lamina; petiole 3–10 cm. long; leaflets sessile or subsessile, 6–10-jugate, somewhat larger towards the distal end; lamina 2–8 × 1–2·5 cm., narrowly ovate to lanceolate, apex acuminate to long-acuminate, margin crenulate, base ± asymmetric. Inflorescences much branched, with minute bracts, pubescent (then ± glabrescent) to glabrous, viscid, with numerous flowers. Flowers pale green to cream, pubescent to glabrous, dimorphic (some functionally ♂, some functionally ♀). Sepals 1·5–2 × 1·5–2 mm., triangular to circular, densely pubescent to glabrous outside, glabrous within. Petals spreading, c. 4 mm. long, narrowly triangular with subacute apex and minute globose loosely attached glands at the base within, glabrous within, sometimes pubescent along the middle outside. Stamens in male flowers with filaments 3–4 mm. long, tapering towards the apex, glabrous; anthers 0·7 mm. long; in female flowers stamens about half the size of those in the male flowers and apparently non-functional. Gynoecium in male flowers a minute vestige lying centrally on the disk; in female flowers glabrous to pubescent consisting of (before anthesis) four erect narrowly flask-shaped carpels fused at the base but diverging above into the styles which are free at the base but united above, stigmas united, capitate; after anthesis styles deciduous above the base. Fruit 0·8–2 × 0·5–1 cm., woody, oblong-ellipsoid, tetragonal with very acute angles, glabrous, with what are apparently recurved style-bases at the top. Seed almost as large as, and apparently fused with the mericarp.

* Because of the falling off of the upper portion of the style after anthesis as described above, Oliver (loc. cit.) and some later authors have mistakenly described the persistent style-bases as free styles.

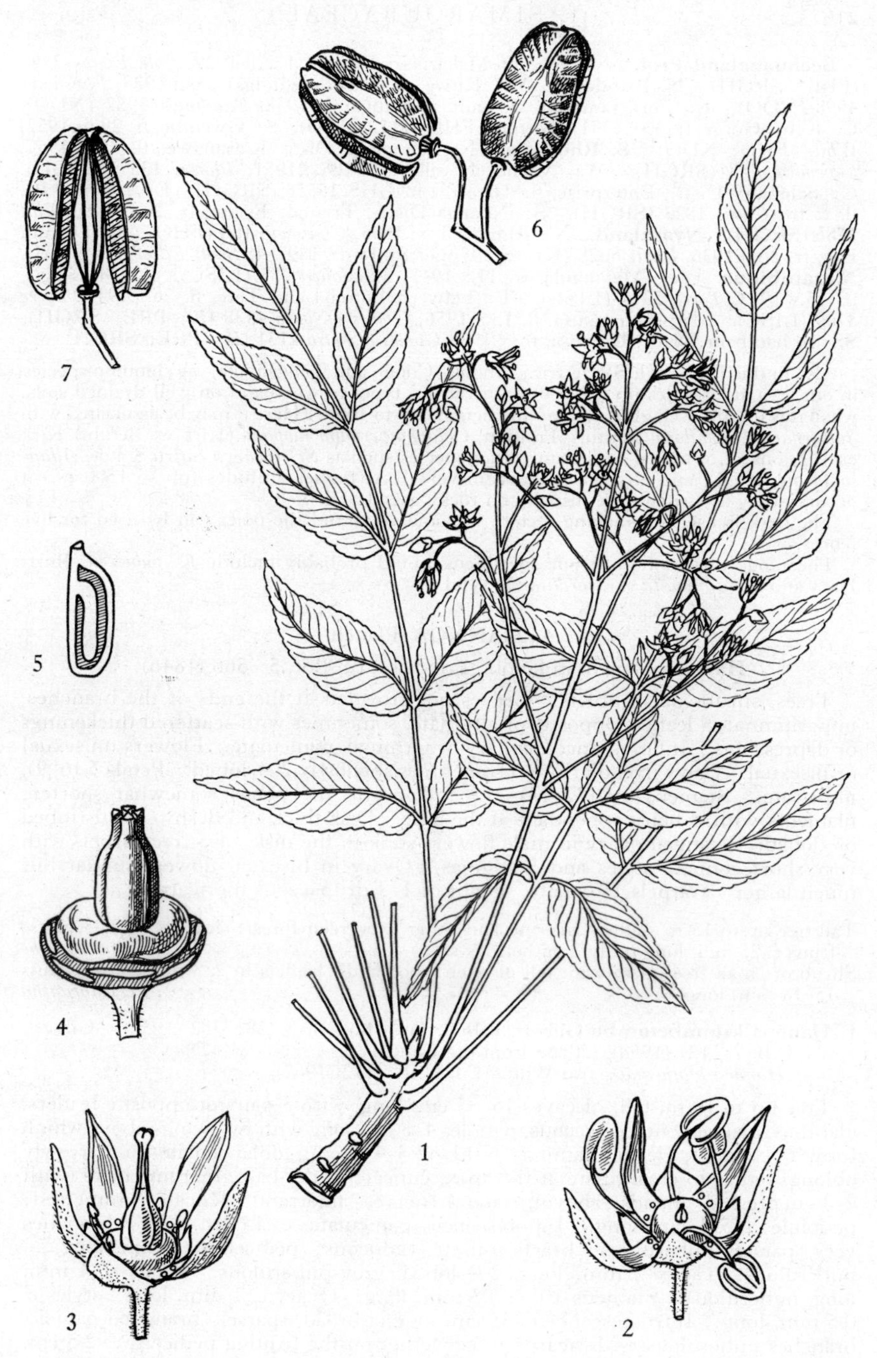

Tab. 39. KIRKIA ACUMINATA. 1, flowering branch ($\times \frac{2}{3}$) *Chase* 1823; 2, male flower ($\times 4$) *Zealley* 132; 3, female flower ($\times 4$) *Eyles* 197; 4, gynoecium after dehiscence of style ($\times 8$); 5, vertical section of one carpel ($\times 8$) *Chase* 1822; 6, fruits ($\times 1$) *Denny* 49; 7, fruit showing suspension of mericarps from a central carpophore ($\times 1$) *Denny* 49.

Bechuanaland Prot. N: NE. of Makarikari Pan, fl. 12.xii.1929, *Pole Evans* 2590 (PRE: SRGH). **N. Rhodesia.** W: Kitwe, st. (cult. seedlings) 25.xi.1958, *Fanshawe* 4998 (NDO). E: Fort Jameson, st. (cult. seedlings) ix.1958, *Fanshawe* 4952 (NDO). C: Kafue Gorge, fr. 5.vi.1931, *Martin* in FHO 31112 (FHO). S: Gwembe, fr. 29.iii.1952, *White* (FHO; NDO). **S. Rhodesia.** N. Sebungwe Distr. Kariangwe, fl. 19.xi.1951, *Lovemore* 188 (SRGH). W: Mtshekeli valley, fl. 29.xi.1951, *Plowes* 1357 (SRGH). C: Salisbury Distr., Enterprise, fl., *Greatrex* in GHS 15556 (SRGH). E: Impodsi R., fl. & fr., *Chase* 1823 (SRGH). S: Ndanga Distr., Triangle Estate, fr. 26.i.1949, *Wild* 2786 (SRGH). **Nyasaland.** N: Rumpi, fl. xi.1953, *Chapman* 180 (FHO). C: Domira Bay, fr. 2.vii.1936, *Burtt* 6028 (K). S: Upper Shire, fl. 1893–4, *Scott Elliot* 8430 (BM). **Mozambique.** N: Metangula, fr. 11.x.1942, *Mendonça* 755 (LISC): Z: Massingir, fr. 15.v.1943, *Torre* 5313 (LISC). T: between Boroma and Tete, fl. 26.x.1941, *Torre* 3709 (LISC). MS: Gorongosa, fl. 18.x.1956, *Gomes e Sousa* 4338 (K; PRE; SRGH). SS: Inhambane Distr., Umabsa, fr. x.1938 *Gomes e Sousa* 2131 (K; PRE; SRGH).

Also in the Transvaal, SW. Africa, Angola, Congo and Tanganyika. A common species in our area at altitudes up to 1200 m. in various types of woodland on well-drained soils, when it is usually well grown, attaining heights up to 16 m. Here it may be associated with *Julbernardia globiflora* (Benth.) Troupin, *Colophospermum mopane* (Kirk ex Benth.) Kirk ex J. Léonard, or mixed with many other species such as *Sclerocarya caffra* Sond., *Adansonia digitata* L. and *Afzelia quanzensis* Welw. At higher altitudes (up to 1500 m.) it occurs rarely as a small tree, usually on rocky granite hills.

The bark is used for making a cloth in some areas and the poles can be used for live fences.

The range of variation in our specimens would probably include *K. pubescens* Burtt Davy and *K. acuminata* var. *cordata* De Wild.

4. HANNOA Planch.

Hannoa Planch. in Hook., Lond. Journ. Bot. **5**: 566 (1846).

Trees, shrubs or shrublets. Leaves not crowded at the ends of the branches, imparipinnate; leaflets opposite or alternate, sometimes with scattered thickenings or depressions. Inflorescences axillary or terminal, paniculate. Flowers unisexual or bisexual. Calyx irregularly and sometimes shallowly 2–4-lobed. Petals 5 (6–9), imbricate. Stamens 10 (12–14), the 5 opposite the petals somewhat shorter; filaments with a hairy appendage at the base. Disk thick and fleshy, ± 10-ribbed or -lobed. Vestigial ovary in male flowers sunk in the disk, of 5 free carpels with very short connate styles and 5 stigmas. Ovary in bisexual flowers similar but much larger. Carpels 1-ovulate. Fruit of 1–3 drupaceous mericarps.

Tall tree up to 15 m. tall of riverine and other evergreen forest; leaflets in c. 5 pairs; fruits c. 25 mm. long (only 2 gatherings seen) - - - 1. *H. kitombetombe*
Shrub or small tree up to 6 m. tall of open woodland; leaflets in 2–3 pairs; ripe fruits 15–18 mm. long - - - - - - - - - - - 2. *H. chlorantha*

1. **Hannoa kitombetombe** Gilbert in Bull. Jard. Bot. Brux. **28**: 382 (1958).—Gilbert, F.C.B. **7**: 124 (1958). Type from the Congo.
Hannoa klaineana sensu White, F.F.N.R.: 172 (1962).

Tree up to 16 m. tall. Leaves 15–35 cm. long, with 5 pairs of opposite leaflets, glabrous, glaucescent, coriaceous, petioles 4–8 cm. long with two wings above which form a channel; leaflet laminas 6–12 × 2·5–4·5 cm., oblanceolate to narrowly oblong, retuse to emarginate at the apex, cuneate at the base, margin entire (with 2–3 depressed glands above), lateral nerves fine and delicate, immersed; petiolules 0·5–1 cm. long. Inflorescences paniculate, c. 15 cm. long; branches very sparsely puberulous; bracts minute, caducous; pedicels c. 1 mm. long, ± puberulous. Calyx 2 mm. long, 2–4-lobed, grey-puberulous. Petals c. 4 mm. long, puberulous. Stamens 10, c. 3·5 mm. long. Ovary c. 1 mm. long; styles c. 0·5 mm. long. Infructescence c. 25 cm. long, a broad, sparsely branched panicle, branches puberulous. Mericarps 1–3-together on the fruiting pedicels, c. 2·5 cm. long, ± ellipsoid; pericarp smooth.

N. Rhodesia. N: Kawambwa Distr., Lusengu, fr. 27.viii.1957, *Fanshawe* 3671 (K; NDO; SRGH). So far recorded only from evergreen forest (mushitu) in the north of Rhodesia and Upper Katanga.

Flowering material has not yet been seen from our area; the flower descriptions are based on the type specimen from the Congo. It is possible that *H. kitombetombe* may eventually prove to be a form of *H. klaineana* Pierre ex Engl. but in the limited material

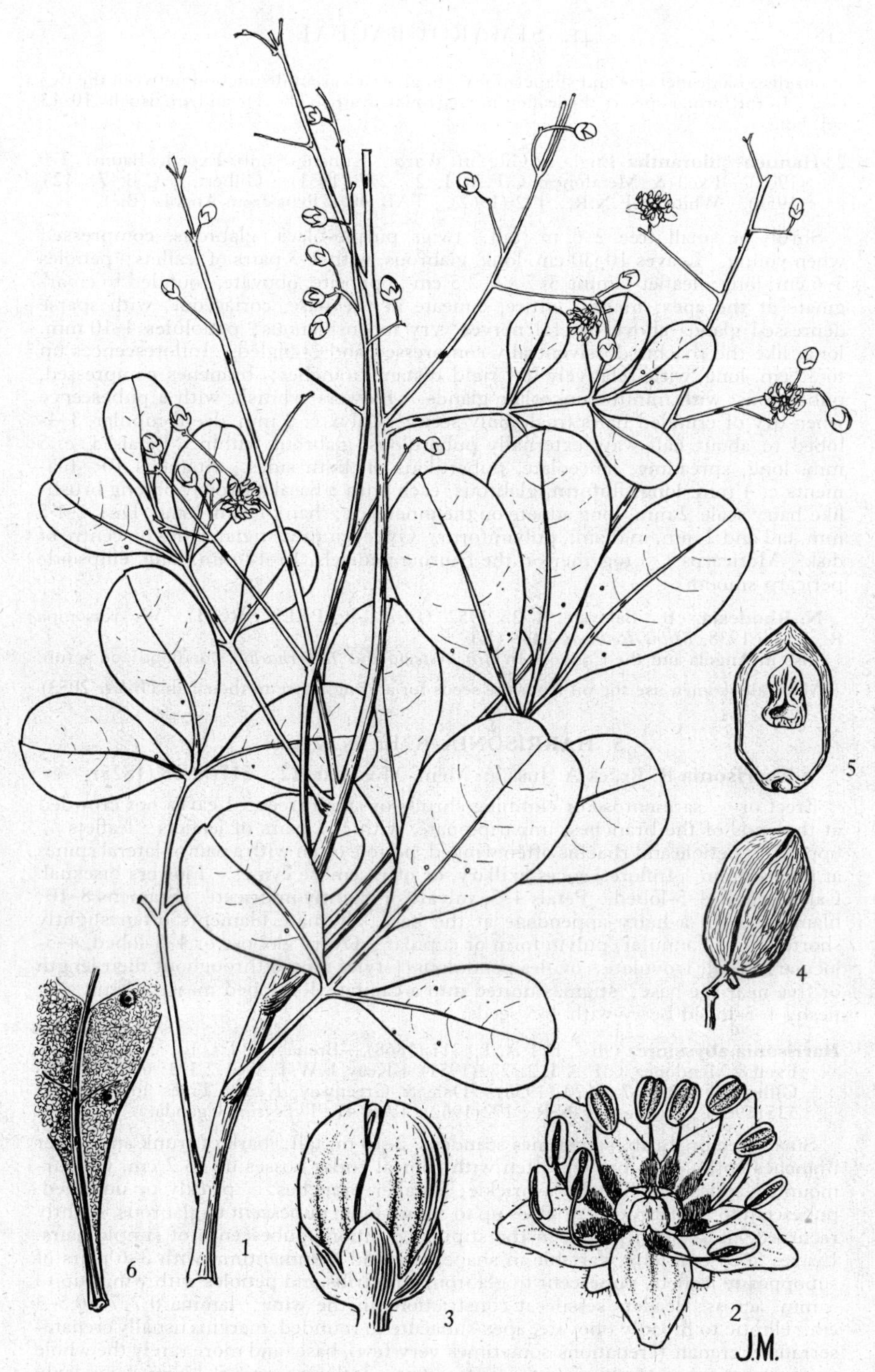

Tab. 40. HANNOA CHLORANTHA 1, flowering branch (× ⅔) *White* 2053; 2, male flower (× 4) *White* 2053; 3, flower-bud (× 6) *White* 2053; 4, fruit (× 1⅓) *Gilges* 255; 5, vertical section of fruit showing seed (× 1⅓) *Gilges* 255; 6, upper side of leaflet with depressed glands and showing roughened surface of epidermis (× 3) *White* 2053.

at our disposal leaflet size and shape appears to give a clearcut distinction between the two taxa. In the former species the leaflets are narrowly ovate to obovate and are usually 10–13 cm. long.

2. **Hannoa chlorantha** Engl. & Gilg in Warb., Kunene-Samb.-Exped. Baum: 270 (1903).—Exell & Mendonça, C.F.A. **1,** 2: 277 (1951).—Gilbert, F.C.B. **7**: 124 (1958).—White, F.F.N.R.: 172 (1962). TAB. **40.** Type from Angola (Bié).

Shrub or small tree 2–6 m. tall; twigs purple-black, glabrous, compressed when young. Leaves 10–30 cm. long, glabrous, with 2–3 pairs of leaflets; petioles 3–6 cm. long; leaflet-lamina 3–7 × 1–2·5 cm., opposite, obovate, rounded to emarginate at the apex, margin entire, cuneate at the base, coriaceous, with sparse depressed glands above, lateral nerves very inconspicuous; petiolules 4–10 mm. long, like the rhachis dorsiventrally compressed and 2-angled. Inflorescences up to 35 cm. long, with relatively few rigid distant branches; branches compressed, puberulous, with minute lanceolate glands. Flowers whitish, with a pubescence when dry of crinkled hairs (male only seen). Calyx c. 2 mm. deep, cupular, 3–4-lobed to about half-way, externally puberulous, glabrous within. Petals 5, c. 5 mm. long, spreading, lanceolate, puberulous on both sides. Stamens 10; filaments c. 4 mm. long, filiform, glabrous, each with a basal narrowly oblong brush-like hairy scale 2 mm. long adnate on the inner side; hairs 1 mm. long. Disk 1–1·5 mm. tall and 2 mm. in diam., pulviniform. Gyneocium vestigial, lying on centre of disk. Mericarps 1–3 together on the fruiting pedicels, 15–18 mm. long, ellipsoid; pericarp smooth.

N. Rhodesia. B: Balovale, fl. 2.x.1952, *Gilges* 150 (PRE; SRGH). W: Kasompa R., fl. 1.ii.1938, *Milne-Redhead* 4444 (K).

Also in Angola and the Congo. In *Brachystegia* and *Julbernardia* woodland and scrub.

Malovale women use the oil from the seeds for adding lustre to their hair (*White* 2053)

5. HARRISONIA R. Br. ex A. Juss.

Harrisonia R. Br. ex A. Juss. in Mém. Mus. Par. **12**: 517, t. 28 (1825).

Erect or ± sarmentose or climbing shrubs or small trees. Leaves not crowded at the ends of the branches, imparipinnate, with 1–7 pairs of leaflets; leaflets ± opposite; petiole and rhachis often winged, petiole often with a pair of lateral spines at its insertion. Inflorescences axillary, or of racemose cymes. Flowers bisexual. Calyx deeply 4–5-lobed. Petals 4–5, valvate or slightly imbricate. Stamens 8–10; filaments with a hairy appendage at the base, alternate filaments often slightly shorter. Disk annular, pulviniform or cupular. Ovary globose or 4–5-lobed, 4–5-locular; loculi 1-ovulate; ovules pendulous; styles united throughout their length or free near the base; stigmas united into a capitate 4–5-lobed mass. Fruit a ± fleshy 4–5-lobed berry with 4–5 seeds.

Harrisonia abyssinica Oliv., F.T.A. **1**: 311 (1868).—Brenan, T.T.C.L.: 572 (1949).—Exell & Mendonça, C.F.A. **1,** 2: 279 (1951).—Keay, F.W.T.A. ed. 2, **1,** 2: 690 (1958).—Gilbert, F.C.B.: **7**: 120 (1958).—Dale & Greenway, Kenya Trees and Shrubs: 535 (1961).—White, F.F.N.R.: 172 (1962). TAB. **41.** Type from Uganda.

Small tree or shrub, sometimes scandent, 2–10 m. tall; bark of trunk and larger branches corky, pale brown, often with conical corky bosses up to 2 cm. tall surmounted at least at first by a prickle; smaller branches ± prickly or unarmed, pubescent to ± glabrous; prickles up to 5 mm. long, pubescent to glabrous, slightly recurved, most often paired in the stipular position; pubescence of simple hairs. Leaves 5–25 cm. long, variable in shape, size and indumentum, with 3–6 pairs of subopposite leaflets, pubescent to glabrous; midribs and petioles with wings up to 3 mm. across; leaflets sessile at constrictions in the wing; lamina 0·7–7 × 0·5–3 cm., elliptic to broadly obovate, apex subacute to rounded, margins usually crenate-serrate to crenate (crenations sometimes very few), base (and more rarely the whole margin) entire, narrowly cuneate, asymmetric. Inflorescences 5–15 cm. long, laxly paniculate, pubescent, often leafy in the lower part, the leaves rapidly grading upwards into oblong foliose bracts and then to minute caducous bracts. Calyx-lobes c. 1 mm. long, spreading, ovate, pubescent outside, glabrous within. Petals white to yellow, 4–5 mm. long, spreading, pubescent outside, glabrous within, lanceolate. Stamens c. 3 mm. long, inserted at the base of the disk; filaments 2–2·5 mm. long,

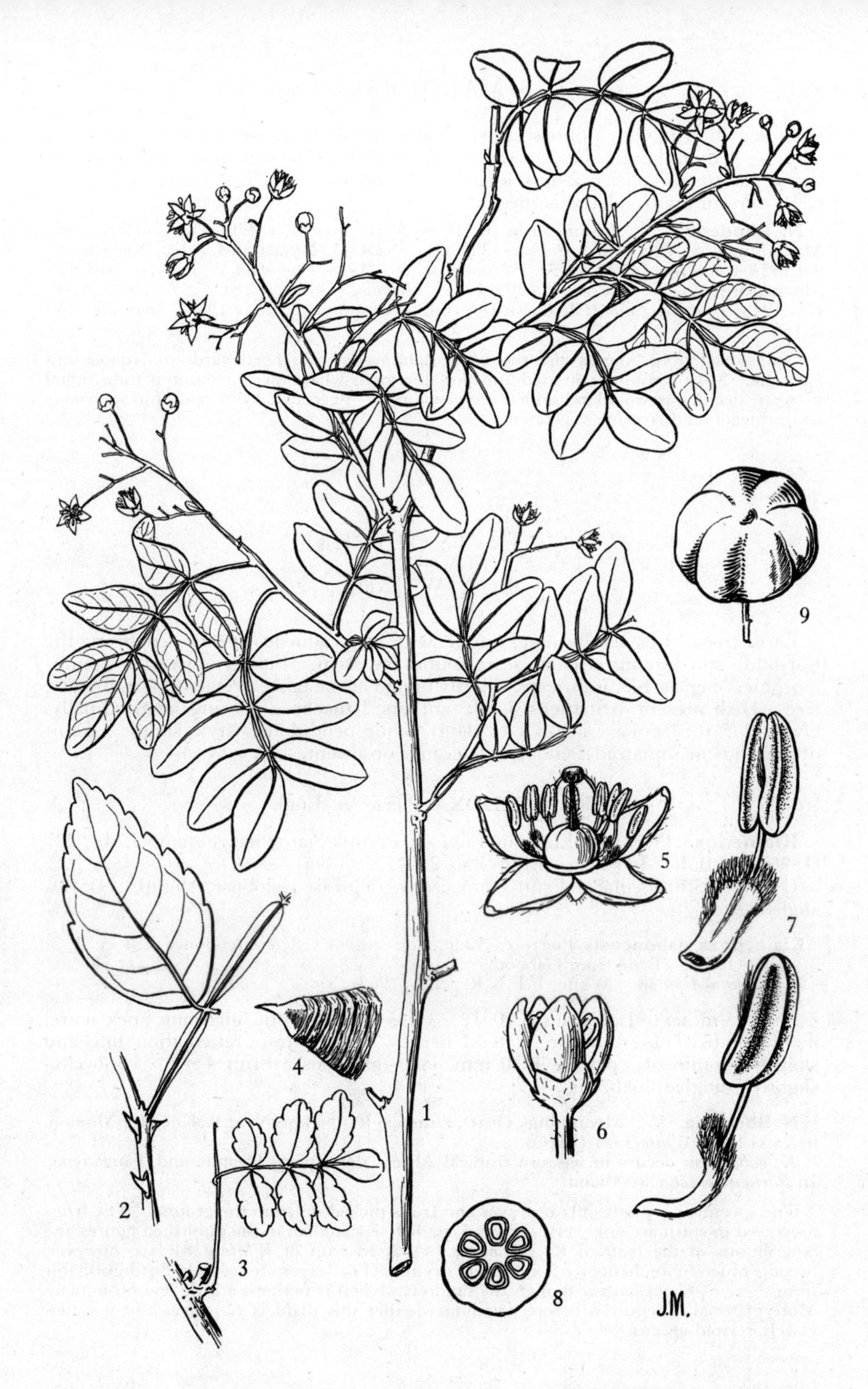

Tab. 41. HARRISONIA ABYSSINICA. 1, flowering branch (× ⅔) *Barbosa & Carvalho* 4313; 2, portion of leaf and twig showing prickles (× ⅔) *Chapman* 582; 3, leaf illustrating variation in leaflet shape and margins (× ⅔) *Germain* 6394; 4, corky boss with prickle from main stem (× ⅔) *Chapman* 582; 5, flower with 2 stamens removed (× 3) *Barbosa & Carvalho* 4313; 6, flower-bud (× 3) *Barbosa & Carvalho* 4313; 7, two stamens each with scale (× 14) *Barbosa & Carvalho* 4313; 8, cross-section of ovary (× 8) *Barbosa & Carvalho* 4313; 9, fruit (× 4) *Germain* 5196.

glabrous, filiform, with a pubescent scale attached by the base to their inner faces; anthers 1 mm. long. Disk c. 1·5 mm. in diam., inconspicuous, pulviniform. Ovary glabrous; styles c. 2 mm. long, usually pubescent. Ripe fruits bright red, c. 10 mm. in diam., glabrous, fleshy.

N. Rhodesia. N: Kalungwishi, fl. 16.viii.1958, *Fanshawe* 4705 (K; NDO). W: Mwinilunga, fl. 20.x.1955, *Holmes* 1282 (K; NDO). **Nyasaland.** N: Njakwa, fr. 9.ii.1954, *Chapman* 183 (K). S: Palombe, fr. 29.v.1958, *Chapman* W/582 (K; SRGH). **Mozambique.** N: Lúrio, fl. 29.x.1942, *Mendonça* 1118 (LISC). Z: Nacala, fl. 16.x.1948, *Barbosa* 2448 (LISC; LM). MS: Chimoio, fl. 23.iv.1948, *Andrada* 1155 (LISC).

Widespread from Angola and the Flora Zambesiaca area northwards to Guineé and Ethiopia. Most commonly in medium-rainfall regions where the vegetation is transitional between deciduous woodland and evergreen forest. Recorded from open forest, woodland, thicket, and riverine vegetation.

42. IRVINGIACEAE

By A. W. Exell

Large trees. Leaves alternate, simple, with long caducous stipules enclosing the leaf-buds and leaving circular scars round the stem. Flowers bisexual, actinomorphic. Sepals 4–5, imbricate, slightly united at the base. Petals 4–5, imbricate, free. Disk present. Stamens 8–10; anthers 2-thecous, opening longitudinally. Ovary 2–5 (6)-locular; loculi 1-ovulate; ovule pendulous; style simple. Fruit drupaceous or samaroid; endosperm scanty or absent.

KLAINEDOXA Pierre ex Engl.

Klainedoxa Pierre ex Engl. in Engl. & Prantl, Nat. Pflanzenfam. **3,** 4: 227 (1896); in Bull. Mens. Soc. Linn. Par. **2**: 1235 (1896).

Ovary (4)5(6)-locular. Fruit drupaceous, depressed-globose, slightly (4)5(6)-angled.

? Klainedoxa gabonensis Pierre ex Engl., loc. cit.—Exell & Mendonça, C.F.A. **1,** 2: 279 (1951). Type from Gaboon.
Klainedoxa sp.—White, F.F.N.R.: 171 (1962).

Tree 25 m. tall. Leaf-lamina 10–12 × 3·8–5·5 cm., elliptic, glabrous, apex acute, margin entire, base cuneate; lateral nerves 17–20 pairs; reticulation fine and slightly prominent; petiole 4–10 mm. long, glabrous. Fruit 4–4·5 × 5·5–6 cm., slightly 5-angled, glabrous.

N. Rhodesia. W: Mwinilunga Distr., Zambezi R., 6·4 km. N. of Kalene Hill Mission, fr. 26.xi.1952, *White* 3381 (FHO).

K. gabonensis occurs in western tropical Africa, the Sudan, Uganda and Tanganyika. In *Parinari excelsa* woodland.

The specimen consists only of leaves and fruits picked up from the ground. The fruits preserved in spirit are only very slightly 5-angled, less so than in the published figures and descriptions of the fruit of *K. gabonensis*. A dried fruit of *White* 3381 has, however, become much more distinctly 5-angled on drying. The leaves are scarcely distinguishable from those of *K. gabonensis* though the fine reticulation is perhaps a little less prominent. More material is required before deciding whether this plant is *K. gabonensis* or a new closely related species.

43. BALANITACEAE

By E. Launert

Small trees or shrubs, usually spiny; spines simple or forked. Leaves spirally arranged, 2-foliolate, petiolate or subsessile; leaflets coriaceous or subsucculent, entire. Flowers bisexual, actinomorphic, usually scented. Sepals 5, free, imbricate, usually caducous. Petals 5, imbricate, free, glabrous or variously hairy. Stamens 10; anthers 2-thecous and dorsifixed; filaments free, filiform, somewhat broadened towards the base, without appendages. Disk succulent, longitudinally 10-grooved, conical or invertedly cup-shaped or cupular, surrounding the ovary. Ovary superior, semi-globose, 5-locular with 1 pendulous ovule in each loculus. Fruit a 1-seeded drupe; endocarp often woody; mesocarp spongy or fibrous; exocarp crustaceous or coriaceous, usually oily. Seed without endosperm; cotyledons plano-convex, thick.

1 genus occurring in Africa and Asia.

BALANITES Del.

Balanites Del. in Descr. Égypt., Hist. Nat. **2**: 221 (1813) *nom. conserv.*

Characters as for the family.

A genus of about 20 described species.

Spines usually forked, rarely simple; petals ± densely hairy on the adaxial surface - 1. *maughamii*

Spines always simple; petals glabrous on both surfaces:

Petiole 1·5–4(6) mm. long; leaflets sessile, obcuneate or obovate to broadly obovate; petals narrowly elliptic, (2·5)2·8–3·4 mm. broad; drupe subglobose or broadly ellipsoid - - - - - - - - - - - - 3. *pedicellaris*

Petiole (5)8–20 mm. long; leaflets always shortly but distinctly petiolulate, elliptic to elliptic-obovate; petals elliptic-oblong, rarely lanceolate-oblong, 2–2·4(2·8) mm. broad - - - - - - - - - - - - 2. *aegyptiaca*

1. **Balanites maughamii** Sprague in Kew Bull. **1913**: 138 (1913).—Mildbr. & Schlechter in Engl., Bot. Jahrb. **51**: 157 (1913).—Engl. in Engl. & Prantl, Nat. Pflanzenfam. ed. 2, **19a**: 182 (1931).—Burtt Davy, F.P.F.T. **2**: 482 (1932).—Brenan, T.T.C.L.: 571 (1949); in Mem. N.Y. Bot. Gard. **8**, 3: 234 (1953). TAB. **42**. Syntypes: Mozambique, Libombo Mts., and by the R. Umbeluzi, *Maugham* (K); by the R. Rovuma, *Kirk* (K); Madanda Forest, *Dawe* 428 (BM; K).

Balanites dawei Sprague, tom. cit.: 140 (1913).—Mildbr. & Schlechter, loc. cit.—Engl., loc. cit.—Brenan, loc. cit. Type: Mozambique, Madanda Forest, *Dawe* 435 (K).

Balanites sp. 1.—White, F.F.N.R.: 168 (1962).

Tree up to 20 m. high, with a fluted trunk up to 0·75 m. in diam.; branches of two kinds, some sterile and mostly spiny, others flower-bearing and unarmed or with only tiny spines; younger branches green and pubescent, older ones glabrescent, often lenticellate; spines of the sterile shoots (1)3–6 cm. long, stout, usually ± regularly forked, rarely simple or with one of the branches shorter than the other. Leaves petiolate; leaflets of the flower-bearing shoots ± asymmetric, 2·5–6 × 2·25–5·7 cm., lanceolate-elliptic, ovate or broadly ovate to circular, apex rounded or obtuse, rarely shortly acuminate, base broadly cuneate or rounded, coriaceous, densely pubescent to subtomentose on the lower surface, pubescent or rarely glabrous on the upper one; petiole 1–2·8 cm. long, stout, pubescent or glabrous; petiolules 0·4–1 cm. long, ± densely pubescent; leaflets of the sterile shoots, 6·5–8 × 5–7 cm. ± asymmetric, broadly ovate to subcircular, apex obtuse and usually shortly acuminate, base ± broadly cuneate, rounded or very rarely subcordate, coriaceous, ± densely pubescent on the lower surface and glabrous on the upper one or glabrous on both; petiole 2·3–3 cm. long, stout, pubescent or glabrous; petiolules 0·6–1·2 cm. long, ± densely pubescent at first, later usually glabrescent. Inflorescences of dense 3–7-flowered umbel-like cymes, pedunculate

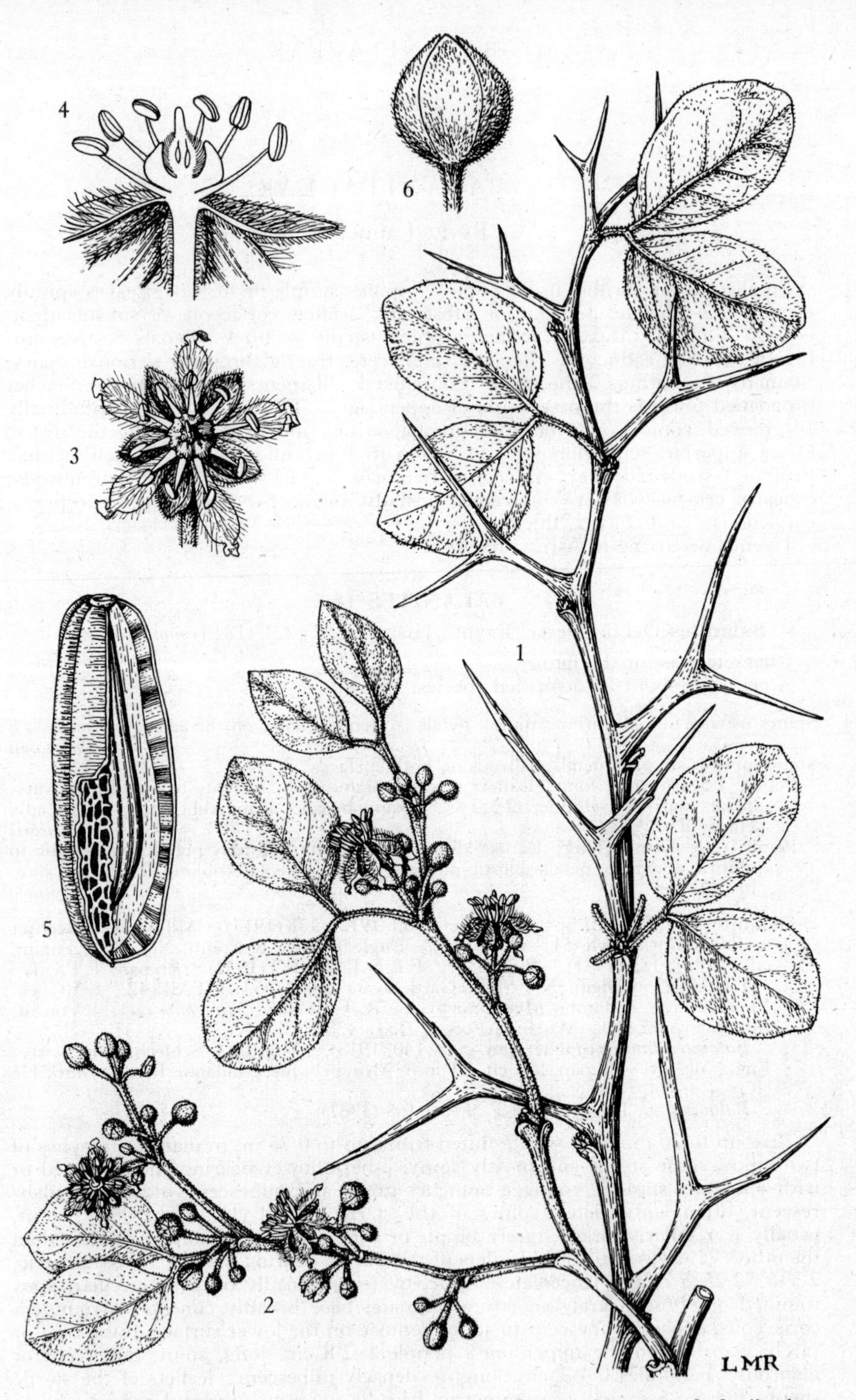

Tab. 42. BALANITES MAUGHAMII. 1, sterile shoot ($\times \frac{2}{3}$) *Maugham* s.n.; 2, fertile shoot ($\times \frac{2}{3}$) *Gomes e Sousa* 3457; 3, flower from above ($\times 4$) *Gomes e Sousa* 3457; 4, longitudinal section of flower ($\times 3$) *Gomes e Sousa* 3457; 5, fruit, partly in longitudinal section, showing exocarp, mesocarp, endocarp and part of the seed with the exocarp partly cut away to show the mesocarp in surface view ($\times \frac{2}{3}$) *Dawe* 435; 6, flower bud ($\times 4$) *Maugham* s.n.

or often nearly sessile, ± densely yellowish-tomentose all over. Flowers 1·5–2 cm. in diam. Sepals (4)4·5–5·6 × (2)2·4–3(3·3) mm., coriaceous, densely yellowish- or greenish-pubescent to subtomentose outside, whitish-sericeous inside, caducous. Petals greenish or yellowish, (6·5)7–8(10) × 1·9–2·6 mm., oblong-lanceolate or lanceolate, ± densely villous on the upper surface, glabrous on the lower one, somewhat crumpled at the apex, scented. Stamens with anthers 1·2–1·6 mm. long; filaments (2·5)3–3·2(3·8) mm. long. Ovary densely covered with long ± stiff white hairs, very rarely glabrescent. Drupe (3)4–6(7·5) × (2)2·5–3(3·5) cm., cylindric-ellipsoid, rarely subclavate, longitudinally 5-grooved (more distinctly so in the upper half), with a deep basal depression and a smaller apical one left by the pedicel and the style respectively, with a rather thick spongy mesocarp, endocarp woody (3–6 mm.) thick, exocarp crustaceous.

N. Rhodesia. S: Zambezi Valley, Sinazongwe Chieftaincy, Gwembe Valley, fr. 12.vii.1961, *Bainbridge* 519 (BM; SRGH). **S. Rhodesia.** N: Sebungwe, 24.ix.1951, *Lovemore* 114 (SRGH). W: Wankie Distr., near Kabira, fr. 9.vii.1954, *Orpen* 39/54 (SRGH). S: Belingwe, fr. v.1956, *Judge* 7/56 (SRGH). **Nyasaland.** S: Chikwawa, fl. 3.x.1946, *Brass* 17924 (K; SRGH). **Mozambique.** N: between Mocímboa da Praia and Palma, fl. 15.ix.1948, *Barbosa* 2129 (LISC). Z: between Mulevala and Nampevo at 10 km. from Mulevala, fr. 31.v.1949, *Barbosa & Carvalho* 2932 (LISC). T: Boroma, Disitso, 275 m., 17.vii.1950, *Chase* 2748 (BM; SRGH). MS: Nhamacolongo, Chemba, 6.vii.1947, *Simão* 1383 (LISC). SS: Gaza, Guijá region, 10.v.1944, *Torre* 6605 (LISC). LM: Bela Vista to Catuane road, fl. 6.ix.1948, *Gomes e Sousa* 3825 (LISC; LM; PRE).

In Tanganyika, Nyasaland, N. Rhodesia, S. Rhodesia, Mozambique, Natal and N. Transvaal.

In forests, in open woodland, along rivers, near springs, and in pans.

2. **Balanites aegyptiaca** (L.) Del. in Descr. Égypt., Hist. Nat. **2**: 221, t. 28 fig. 1 (1813).—DC., Prodr. **1**: 708 (1824).—Oliv., F.T.A. **1**: 315 (1868).—Engl. in Engl. & Prantl, Nat. Pflanzenfam. ed. 2, **19a**: 180 (1931).—Brenan, T.T.C.L.: 571 (1949).—Exell & Mendonça, C.F.A. **1**, 2: 282 (1951).—Keay, F.W.T.A. ed. 2, **1**, 2: 364 (1958).—Gilbert, F.C.B. **7**: 66 (1958).—White, F.F.N.R.: 168 (1962). Type from Egypt.

Ximenia aegyptiaca L., Sp. Pl. **2**: 1194 (1753). Type as above.

Small evergreen tree (rarely a shrub) up to 15 m. high, ± spiny, with a flat or rounded crown; stem with a reticulate dark brown or grey (rarely green) bark; branches green or greyish, stiff and brittle, always armed with stout simple green or yellowish spines. Leaves petiolate; leaflets shortly petiolulate; lamina 2·5–6 × 1·5–4 cm., slightly asymmetric, elliptic to elliptic-obovate, apex subacute to obtuse, sometimes slightly retuse, base cuneate or rarely rounded, coriaceous, puberulous when young, later glabrescent or sometimes remaining puberulous on the lower surface; secondary nerves 4–6 pairs, ± prominent beneath; petiole (0·5)0·8–2 cm. long, puberulous or glabrescent, canaliculate. Flowers in usually few-flowered sessile or shortly pedunculate fascicles; pedicels up to 1·5 cm. long, ± densely greyish-pubescent. Flowers c. 1·4 cm. in diam. Sepals 5·2–7 × 2·7–3·25 mm., ovate or ovate-lanceolate, coriaceous, caducous, densely pubescent outside, with long silky whitish hairs inside. Petals 7·2–9·5 × 2–2·4(2·8) mm., narrowly elliptic or elliptic-oblong, rarely lanceolate-oblong, glabrous on both surfaces. Stamens with the anthers 1·7–2·2 mm. long, ovate or ovate-oblong, glabrous; filaments c. 3·75 mm. long. Ovary densely covered with long silky hairs. Drupe yellowish or green, up to 5 × 2·5 cm., usually subcylindric, more rarely narrowly ellipsoid or subobclavate, finely puberulous, sometimes glabrescent.

N. Rhodesia. N: Kaputa-Kabwe, fr. 17.x.1949, *Bullock* 1308 (K; SRGH). W: Ndola, st. 16.vii.1954, *Fanshawe* 1382 (K). S: Kafue R., *Hutchinson & Gillett* 3580 (K); Livingstone, 880 m., fl. 18.x.1955, *Gilges* 479 (PRE). **S. Rhodesia.** N: Mazoe Distr., Chipoli, 825 m., fl. 1.ix.1958, *Mowbray* 54 (SRGH). W: Bubi Distr., Shangani, Mateme R., fl. x.1920, *Eyles* 3553 (SRGH).

3. **Balanites pedicellaris** Mildbr. & Schlechter in Engl., Bot. Jahrb. **51**: 162 (1913).—Engl. in Engl. & Prantl, Nat. Pflanzenfam. ed. 2, **19a**: 182 (1931).—Brenan, T.T.C.L.: 571 (1949); in Mem. N.Y. Bot. Gard. **8,** 3: 233 (1953). Type from Kenya.

Balanites australis Bremek. in Ann. Transv. Mus. **15**: 244 (1933).—Brenan, tom. cit.: 234 (1953). Type from the Transvaal.

Much-branched shrub or small tree with pendulous branches, up to 6 m. or more high, variable in shape and size. Branches yellowish or greyish-green or grey, rather stiff, usually spiny, younger parts densely puberulous, glabrous when older; spines rather stout, simple. Leaves shortly petiolate; leaflets sessile, 1–3 (4) × 0·5–2·3(3) cm., obovate, broadly obovate, obcuneate or rarely obovate-elliptic, apex rounded, rarely subtruncate, entire or slightly retuse, base always narrowly cuneate, subsucculent, very shortly and densely puberulous to subtomentose when young, later glabrescent; secondary nerves 3–5 pairs, not very prominent beneath; petiole 1·5–4(6) mm. long, rather stout, ± densely pubescent at first, later glabrescent. Flowers c. 1·4 cm. in diam., in few- to many-flowered sessile or rarely shortly pedunculate fascicles; pedicels up to 2 cm. long, fairly stout, ± densely puberulous, very rarely glabrous. Sepals 6·5–7·5 × 3·2–5 mm., ovate to ovate-lanceolate, coriaceous, densely tomentellous outside, with long whitish silky hairs inside, acute, caducous. Petals 6·7–9 × (2·5)2·8–3·4 mm., narrowly elliptic with an acutish apex, glabrous on both surfaces. Stamens with oblong anthers 2–2·8 mm. long; filaments c. 3·75 mm. long. Ovary densely covered with whitish silky hairs. Drupe orange, subglobose or broadly ellipsoid, usually somewhat flattened on both ends, 1·2–2·5 × 1·5–2 cm., finely puberulous when young, later usually glabrescent.

S. Rhodesia. S: Beitbridge, fl. 22.iii.1959, *Drummond* 5937 (BM; K; PRE; SRGH). **Nyasaland.** S: Chikwawa, 200 m., 2.x.1946, *Brass* 17906 (K; PRE; SRGH). **Mozambique.** SS: Musamane, fl. 25.vi.1947, *Pedro & Pedrógão* 2061 (PRE). LM: Maputo, Catuane, fl. 29.iv.1948, *Torre* 7744 (LISC).

Kenya, Tanganyika, Nyasaland, S. Rhodesia, Mozambique, Natal and N. Transvaal. In dry bushy or open woodland, on alluvial flood plains, amongst scattered trees.

44. OCHNACEAE

By N. K. B. Robson

Trees, shrubs or shrublets, or more rarely perennial or annual herbs, glabrous or rarely shortly pubescent; bark rough or smooth, sometimes flaking, occasionally pigmented beneath; stem with cortical vascular bundles. Leaves alternate, simple or very rarely pinnately compound, entire or more often with undulate to serrate or ciliate margins, penninerved (often densely so), membranous to coriaceous, with stipules entire to fringed or ± dissected, free or ± united intrapetiolarly, caducous or ± persistent. Flowers bisexual, actinomorphic or rarely ± zygomorphic, sometimes fragrant, in terminal or axillary paniculate to umbellate or fascicled cymes or in racemes or solitary; pedicels usually articulated. Sepals (3–4)5(6–10), imbricate (usually quincuncial) in bud, free or rarely shortly united at the base, persistent or deciduous, sometimes enlarging in fruit. Petals (4)5(6–12), free, contorted in bud, deciduous. Stamens (1)5–∞, apparently whorled or diplostemonous or antisepalous, more rarely excentrically grouped, free, inserted on the receptacle; staminodes sometimes present outside the stamens, subulate or petaloid, sometimes connate into a tube; anthers usually ± elongated, extrorse, basifixed, often deciduous, dehiscing by longitudinal slits or apical pores; connective occasionally with a slender prolongation; filaments persistent. Ovary free, sessile, syncarpous, either entire to shallowly lobed with apical style and 2–5 parietal (more rarely axile or basal) placentas each bearing 1–∞ ovules in two rows, or with (3)5(6–15) 1-ovulate lobes and gynobasic style; styles as many as the placentas or ovary-lobes, free towards the apex or completely united. Fruit either a septicidal 2–5-valved 1–∞-seeded capsule, or of (3)4–12 1-seeded drupelets with fleshy or coriaceous mesocarp borne on a ± swollen receptacle. Seeds exarillate, with or without endosperm; embryo straight or curved, incumbent or accumbent, isocotylous or heterocotylous.

A family of over 30 genera distributed over the tropical regions of Asia, Africa and America, but almost absent from Australasia.

Fruit (in our genera) comprising 1 or more drupelets (frequently black) on an enlarged red or purplish receptacle; ovary lobed, with (3)5–15 1-ovulate loculi; style gyno-

basic; staminodes absent; seeds without endosperm; trees, shrubs or shrublets (*Exalbuminosae*):
Stamens 13 or more or, if 8–10, then anthers dehiscing longitudinally and equalling the filaments; petals yellow to orange, pink or white:
Stipules entire or fringed or bifid, not striate, deciduous; petals pale yellow to orange (very rarely white); drupelets without internal projection of endocarp; pigment under bark absent - - - - - - - - - - - 1. **Ochna**
Stipules laciniate or deeply divided into linear segments, markedly longitudinally striate, persistent on 1st-year shoots; petals white to pink; drupelets with internal projection of endocarp; yellow pigment present under the bark
2. **Brackenridgea**
Stamens 10, with anthers much longer than the filaments and dehiscing by apical pores; petals yellow - - - - - - - - - - - 3. **Ouratea**
Fruit (in our genera) a 3-valved septicidal capsule; ovary 1-locular (or 3-locular at the base) with 3 parietal placentas each bearing numerous ovules in 2 rows; style terminal; staminodes present; seeds with abundant endosperm; perennial or annual herbs (in our species) (*Albuminosae*):
Petaloid staminodes free, not united with the base of the stamen filaments; outer filiform staminodes present (in our species); flowers solitary or rarely paired in leaf axils (in our species) - - - - - - - - - - 4. **Sauvagesia**
Petaloid staminodes connate at the base, each united with the base of a stamen filament; outer filiform staminodes absent; flowers in monochasial cymes (rarely solitary) in leaf axils - - - - - - - - - - - 5. **Vausagesia**

1. OCHNA L.

Ochna L., Sp. Pl. **1**: 513 (1753); Gen. Pl. ed. 5: 229 (1754).—Schreb. in L., Gen. Pl. ed. 8, **1**: 354 (1789).
Diporidium Bartl. & Wendl., Beitr. Bot. **2**: 24 (1825).

Trees, shrubs or shrublets, usually completely glabrous. Leaves petiolate; lamina with margin serrate to ciliate or rarely entire; stipules entire or with ciliate or fimbriate margins or ± deeply bifid, not striate, free, deciduous or caducous. Inflorescence paniculate to racemose or umbellate or reduced to a single flower, terminal or terminating short axillary shoots; bracts scale-like, caducous; pedicels articulated at or above the base. Sepals (4)5, quincuncial in bud, green or yellow in flower, persistent, enlarging and becoming varying shades of red and coriaceous in fruit. Petals 5 (rarely up to 12), yellow or rarely orange or white, often unguiculate, deciduous. Stamens (14)20–∞, free; anthers yellow, dehiscing by longitudinal slits or terminal pores, deciduous; filaments ± slender, longer or shorter than the anthers, persistent. Carpels (3)5–15, apparently free at the base, 1-ovulate; styles slender, gynobasic, united except sometimes towards the apex; stigmas terminal, enlarged. Fruit 1 to several free black 1-seeded drupelets with fleshy mesocarp, inserted on the enlarged red receptacle. Seeds straight or ± curved or reniform, cylindric or somewhat flattened but not angular, without endosperm or an internal projection of the endocarp; embryo straight or curved, incumbent or accumbent, isocotylous or heterocotylous.

A genus of c. 85 species in the Old World tropics and subtropics, mostly in Africa, Madagascar and the Mascarene Islands but c. 10–12 species occur in Asia from India and Ceylon to N. Malaya, Indo-China and Hainan, and one is recorded from Timor.

Carpels and drupelets reniform, with point of attachment near centre of long side; inflorescence ± elongate, racemose, simple or compound; styles free at the apex; bark smooth, flaking in patches (Sect. *Renicarpus*):
Anther dehiscence longitudinal; raceme simple; carpels 5 - - 1. *multiflora*
Anther dehiscence porose; raceme simple or ± compound; carpels 5–8:
Carpels 7–8; 1st year shoots creamy white, rather stout, usually widely spreading:
Plant a shrub or small tree; raceme simple or more rarely compound at the base; pedicels articulated at or above the base - - - - 2. *pulchra*
Plant a rhizomatous shrublet up to 60 cm. high; raceme simple; pedicels articulated above the base - - - - - - - - 3. *manikensis*
Carpels 5(6); 1st year shoots dark reddish-purple, rather slender, usually ± ascending:
Raceme usually ± compound, (10)15–85-flowered, 2 cm. long or more in fruit
4. *oconnorii*
Raceme almost always simple, 3–7(11)-flowered, 0·5–1·5 cm. long in fruit
5. *arborea*

Carpels and drupelets cylindric to globose with point of attachment near base of long side (rarely subreniform in *O. afzelii*, but then styles completely united); inflorescence elongate or condensed, racemose or paniculate or 1-flowered; styles free at the apex or not; bark smooth or rough but not flaking in patches:

Anther dehiscence porose*; inflorescence paniculate or fascicled or umbellate to 1-flowered (more rarely shortly racemose); styles usually free at the apex (Sect. *Ochna*):

Carpels and style-branches (6)8–13; inflorescence paniculate to fasciculate (rarely subumbellate in *O. natalitia*):

Anthers 4–7 mm. long; sepals in fruit 12–26 mm. long; leaves usually obovate to oblanceolate:

Leaf-margin entire or ± undulate or ciliate (frequently only towards the base); anthers not longer than the filaments, frequently curved - - 6. *kirkii*

Leaf-margin serrulate or spinulose-serrulate; anthers not shorter than the filaments, straight - - - - - - 7. *mossambicensis*

Anthers 1·5–3 mm. long; sepals in fruit 7–12(14) mm. long; leaves usually oblong to elliptic-oblong - - - - - - - - 8. *natalitia*

Carpels and style-branches 5(7)†; inflorescence shortly racemose to umbellate or 1-flowered (but usually fasciculate or paniculate in *O. macrocalyx*):

Sepals in fruit 12–30 mm. long or, if 10–11 mm. long, then convex; petals 7·5–28 mm. long:

Sepals in fruit with plane apex, neither incurved round the maturing drupelets nor becoming navicular; petals 16–28 mm. long:

Flowers (2)3–14 in each inflorescence; anthers 4–7 mm. long; leaves 6·8–24·6 cm. long - - - - - - - - - 9. *macrocalyx*

Flowers solitary; anthers c. 2·5 mm. long; leaves 1·8–4·3(6·5) cm. long 10. *cinnabarina*

Sepals in fruit convex, incurved round the maturing drupelets, afterwards spreading and becoming navicular; petals (where known) 7·5–13 mm. long:

Pedicels articulated (1)2–4 mm. above the base; flowers 1–5:

Flowers 3–5 in an umbellate inflorescence; leaves narrowly cuneate to angustate at the base - - - - - - 11. *beirensis*

Flowers solitary (rarely paired) in leaf axils; leaves subcordate to truncate or broadly cuneate at the base - - - - 12. *inermis*

Pedicels articulated at the base; flowers solitary (rarely paired):

Sepals in fruit 21–30 mm. long; leaves ± broadly elliptic 13. *rovumensis*

Sepals in fruit 12–18 mm. long; leaves narrowly elliptic to oblong or oblanceolate - - - - - - - - 14. *barbosae*

Sepals in fruit 7–11 mm. long, spreading or reflexed, not convex; petals 8–12 mm. long:

Flowers 3–5 in an umbellate inflorescence; branches brownish-white; leaves not glaucous; pedicels usually articulated above the base - 15. *angustata*

Flowers solitary in leaf axils; branches glaucous, purple-grey; leaves glaucous; pedicels articulated at the base - - - - - 16. *glauca*

Anther dehiscence longitudinal; inflorescence racemose (simple or very rarely branched at the base) to umbellate or 1-flowered; styles usually completely united (Sect. *Schizanthera*):

Pedicels articulated at least 2 mm. above the base or, if less, then either leaves and young stems (and frequently sepals and stamens) tinged blue-green when dry or young shoots and pedicels pubescent or puberulous:

Pedicels articulated at least 2 mm. above the base or, if less, then either leaves ± rounded at the apex and with margin entire or remotely spinulose-serrulate, or drupelets cylindric, straight, surrounded by flat spreading sepals:

Sepals in fruit spreading, not enclosing developing drupelets, flat or, if somewhat convex, then inflorescence elongate-racemose, not subumbellate:

Leaves rhombic to elliptic or oblanceolate, usually ± acute or acuminate, neither markedly copper-tinged nor glaucous; young shoots ± striate or angular, obviously lenticellate, with bark persisting or fissuring and flaking but the epidermis not peeling in strips:

Carpels 5, rarely 6 but then sepals in fruit 9–18 mm. long and plant a tree of evergreen forest:

Flowers (5)7–14(20), usually in ± elongated racemes with rhachis up to 20 mm. long; pedicels usually articulated more than 1 mm. above the base; tree (rarely shrub) (2)3–20 m. high of evergreen forest and rocky places - - - - - - - - 17. *holstii*

Flowers solitary or paired (more rarely in fascicles or condensed racemes of 3–5(7)) with rhachis up to 2 mm. (very rarely to 8 mm.) long;

* *O. beirensis* assumed to have porose dehiscence.

† *O. macrocalyx* sometimes has 6–8 carpels in Tanganyika.

pedicels articulated within 1 mm. of the base; small trees, shrubs or shrublets up to 8 m. high:

Flowers erect or spreading; leaves sharply acuminate; branchlets ± slender; small tree (1)2–8 m. tall in upland rain-forest 18. *oxyphylla*

Flowers becoming pendulous; leaves subacute to obtuse or rounded; branchlets much branched, stouter; shrub or shrublet up to 1·2(2) m. tall in upland grassland or forest relicts - - 19. *stolzii*

Carpels 6–8, rarely 5 but then sepals in fruit 7–9 mm. long; tree or shrub of deciduous woodland - - - - - - - - 20. *afzelioides*

Leaves obovate to oblong-elliptic or oblanceolate, usually rounded or obtuse, usually markedly copper-tinged or glaucous; young shoots usually terete with lenticels very small or absent and epidermis often peeling in strips:

Leaves with lamina rounded or truncate at the base and margin densely curved-serrulate; leaf-lamina and fruiting sepals markedly copper-tinged; inflorescence-rhachis 3–25 mm. long; pedicels articulated at least 1 mm. above the base; young stems reddish-brown, frequently pubescent or puberulous - - - - - - - - - 21. *polyneura*

Leaves with lamina attenuate at the base and margin entire or remotely spinulose-serrulate; leaf-lamina ± glaucous; inflorescence-rhachis absent or very short; pedicels articulated at the base; young stems whitish, glabrous - - - - - - - 22. *leptoclada*

Sepals in fruit enclosing developing drupelets like a bud, eventually spreading, broad and convex; inflorescence ± subumbellate; young shoots and inflorescence papillose-puberulous - - - - 23. *puberula*

Pedicels articulated at or within 1 mm. of the base; leaves acute at the apex and/or with ± densely serrulate margin; drupelets curved-obovoid or subglobose, surrounded by usually ± convex incurving sepals:

Plant a small tree 3·5–6 m. high; carpels 6–8; anthers 1–1·5 mm. long; leaves tinged bluish-green when dry - - - - - - 24. *cyanophylla*

Plant a shrublet up to 0·6 m. high; carpels 5; anthers (1·75)2–3 mm. long; leaves rarely tinged bluish-green when dry:

Shoots glabrous, numerous from polycephalous rootstock, forming cushions up to 15 cm. high - - - - - - - - 25. *katangensis*

Shoots papillose-puberulous when young, erect, slender, little branched, up to 60 cm. high - - - - - - - - - 26. *richardsiae*

Pedicels articulated not more than 1 mm. above the base; leaves, young stems, etc. not tinged blue-green when dry; young shoots and pedicels glabrous:

Branches when young white to yellow-brown, smooth or ridged, usually without lenticels but with bark peeling in papery strips; petals and anthers bright yellow:

Leaves with margin densely serrulate and lamina flat or incurved (concave) towards the petiole; petiole (3)5–20 mm. long:

Sepals in fruit 16–25 mm. long, convex; drupelets 10–13 mm. long; inflorescence rhachis up to 13 mm. long; leaves ± glaucous 27. *gambleoides*

Sepals in fruit 10–15 mm. long, flat; drupelets 7–9 mm. long; inflorescence rhachis absent or up to 8 mm. long; leaves not glaucous 28. *schweinfurthiana*

Leaves with margin entire or remotely spinulose-serrulate and lamina base recurved (convex) towards the petiole; petiole 2·5–4 mm. long 22. *leptoclada*

Branches when young purple to reddish-brown or whitish, ridged or angular, with numerous prominent lenticels but usually without peeling bark; petals white to orange-yellow; anthers orange to bright yellow:

Plant a shrub or tree (1·5)3–12 m. high; petals white to pale or rarely bright yellow; carpels (5)6–8 - - - - - - - - 29. *afzelii*

Plant a shrublet or shrub up to 1(2) m. high; petals bright to orange-yellow; carpels 5:

Anthers 1–1·5 mm. long; petals 7–10 mm. long, bright yellow; leaf-lamina elliptic to oblong-elliptic or oblanceolate; stems erect, ± clustered:

Sepals in fruit 8–11(13) mm. long; petiole (1)2–5 mm. long; leaf-lamina elliptic to oblanceolate, narrowed to an acute or subacute apex 30. *confusa*

Sepals in fruit (11)13–16 mm. long; petiole 0·5–1 mm. long; leaf-lamina usually oblanceolate, rounded at the apex - - - 31. *pygmaea*

Anthers (1·75)2–3 mm. long; petals 12–20 mm. long, bright to orange-yellow; leaf-lamina narrowly elliptic or narrowly oblanceolate to linear-oblong; stems much branched, forming low cushions up to c. 15 cm. high - - - - - - - - - - - 25. *katangensis*

1. **Ochna multiflora** DC. in Ann. Mus. Par. **17**: 412, t. 3 (1811); Prodr. **1**: 735 (1824).—Oliv., F.T.A. **1**: 317 (1868).—Van Tiegh. in Journ. de Bot. **16**: 119 (1902); in Ann. Sci. Nat., Sér. 8, Bot. **16**: 385 (1902).—Gilg in Engl., Bot. Jahrb. **33**: 233 (1903).—Keay, F.W.T.A. ed. 2, **1**, 1: 223 (1954).—White, F.F.N.R.: 251, fig. 43E (1962). Type from Sierra Leone.

Ochna mannii Van Tiegh. in Journ. de Bot. **16**: 121 (1902); in Ann. Sci. Nat., Sér. 8, Bot. **16**; 385 (1902).—Exell & Mendonça, C.F.A. **1**, 2: 289 (1951).—Keay, loc. cit. Type from Nigeria.

Shrub or small tree up to c. 10·5 m. high, with bark very smooth, mottled pale to dark grey and purplish-brown and flaking in thin sheets to reveal greenish-yellow patches; branches slender, spreading, reddish-purple at first, becoming dark grey. Leaves petiolate; lamina (7)8·5–15 × 3–6·4 cm., elliptic to oblong or obovate, acute or shortly acuminate at the apex, with margin remotely and very shallowly appressed-spinulose-serrulate, cuneate to rounded at the base, subcoriaceous, with few (c. 12) lateral nerves and densely reticulate tertiary venation more prominent below than above; petiole 3–4 mm. long, slender; stipules subulate, c. 1·3 cm. long. Flowers (5)12–20, in simple terminal erect (? or pendulous) racemes (1·5)2·5–5(6) cm. long in fruit; pedicels 1·5–2·3 cm. long in fruit, articulated at the base or in the lower ½. Sepals 5–6 mm. long in flower, elliptic, rounded, becoming pink to red, 9–12 mm. long, slightly convex and reflexed in fruit. Petals bright yellow, 8–10 × 4–5 mm., obovate to elliptic. Stamens with anthers 1·5–2 mm. long, about ½ as long as the filaments, straight, dehiscing by longitudinal slits. Carpels 5, with styles almost completely united; stigmas small. Drupelets 6–9 × 4–6 mm., reniform, inserted centrally; embryo curved.

N. Rhodesia. W: bank of Kabompo R. just above crossing with Solwezi-Mwinilunga road, fr. 17.ix.1952, *Angus* 472 (BM; COI; FHO; K).

From Gambia and Sierra Leone to the Congo, Angola (Lunda) and N. Rhodesia. In evergreen fringing forest in our area, but also in drier habitats in W. Africa.

The position of the pedicel articulation is variable and cannot be used to separate *O. mannii* from *O. multiflora*. Specimens from southern parts of the range of *O. multiflora* are said to have sepals which turn pink in fruit rather than wine-red; but more material would be required in order to decide whether this variation merits taxonomic recognition.

2. **Ochna pulchra** Hook., Ic. Pl. **6**: t. 588 (" pulchrum ") (1843).—Harv. in Harv. & Sond., F.C. **1**: 449 (1860).—Oliv., F.T.A. **1**: 317 (1868).—Gilg. in Engl., Bot. Jahrb. **33**: 234 (1903).—Bak. f. in Journ. Linn. Soc., Bot. **40**: 37 (1911).—Eyles in Trans. Roy. Soc. S. Afr. **5**: 419 (1916).—Phillips in Bothalia, **1**: 91 (1922).—Burtt Davy, F.P.F.T. **1**: 238 (1926).—Burtt Davy & Hoyle, N.C.L.: 56 (1936).—Exell & Mendonça, C.F.A. **1**, 2: 291, t. XIIB (1951).—Suesseng. in Proc. & Trans. Rhod. Sci. Ass. **43**: 88 (1951).—O.B. Mill. in Journ. S. Afr. Bot. **18**: 58 (1952).—Verdoorn in Fl. Pl. Afr. **29**: t. 1139 (1952).—Pardy in Rhod. Agr. Journ. **49**: 10, tt. (1952).—Palgrave, Trees of Central Afr.: 298, tt. (1956).—Story, Bot. Survey of S. Africa, Mem. 30: 35, photo. 33 (1958).—White, F.F.N.R.: 250, fig. 43C (1962). Type from the Transvaal.

Ochna rehmannii Szyszyl., Polypet. Disc. Rehm.: 28 (1888). Type from the Transvaal.

Ochna aschersoniana Schinz in Verh. Bot. Verein. Brand. **29**: 61 (1888).—Gilg, tom. cit.: 235 (1903). Type from SW. Africa.

Ochna quangensis Büttn. in Verh. Bot. Verein. Brand. **32**: 49 (1890).—Gilg, tom. cit.: 235 (1903).—De Wild. in Rev. Zool. Afr., Suppl. Bot. **7**: 38 (1919). Type from the Congo.

Polythecium pulchrum (Hook.) Van Tiegh. in Ann. Sci. Nat., Sér. 8, Bot. **16**: 368 (1902). Type as for *Ochna pulchra*.

Porochna huillensis Van Tiegh., tom. cit.: 387 (1902). Type from Angola.

Porochna antunesii Van Tiegh., loc. cit. Type from Angola.

Porochna brunnescens Van Tiegh., tom. cit.: 388 (1902). Type from Angola.

Porochna davilliflora Van Tiegh., loc. cit. Type from Angola.

Porochna bifolia Van Tiegh., tom. cit.: 389 (1902). Type from " upper Zambezi ".

Ochna antunesii [Engl. ex Van Tiegh., tom. cit.: 388 (1902) *in synon.*] Engl. & Gilg. in Warb., Kunene-Samb.-Exped. Baum: 304 (1903).—Gilg, tom. cit.: 235 (1903).—R.E.Fr., Wiss. Ergebn. Schwed. Rhod.-Kongo Exped. **1**: 150 (1914).—Eyles, loc. cit. Type as for *Porochna antunesii*.

Ochna brunnescens [Gilg ex Van Tiegh., tom. cit.: 388 (1902) *in synon.*] Engl. & Gilg, tom. cit.: 302 (1903).—Gilg, tom. cit.: 235 (1903). Type as for *Porochna brunnescens*.

Polythecium rehmannii (Szyszyl.) Van Tiegh., op. cit. **18**: 55 (1903). Type as for *Ochna rehmannii.*

Ochna pulchra forma *integra* Suesseng. in Proc. & Trans. Rhod. Sci. Ass. **43**: 88 (1951). Type: S. Rhodesia, Marandellas Distr., *Dehn* 474 (M, holotype; SRGH).

Ochna fuscescens Heine in Mitt. Bot. Staatssamml. Münch. **1**: 340 (1953). Type from SW. Africa.

Shrub or small tree up to 8 (10·5) m. high, with bark smooth, grey to bluish-grey, flaking in thin sheets to reveal cream to yellow-brown or pale grey patches; branches slender or rather stout, spreading widely, creamy white at first, becoming dark grey. Leaves petiolate; lamina 3·5–18·5 × 1·3–5·7 cm., elliptic to oblong or oblanceolate (rarely obovate), acute or obtuse to rounded (often apiculate) at the apex, with margin entire or densely appressed-spinulose-serrulate towards the apex, narrowly cuneate or attenuate to rounded (rarely truncate to subcordate) at the base, subcoriaceous, with numerous (c. 20–30) lateral nerves and densely reticulate tertiary venation more prominent above than below; petiole 2–6(8) mm. long, triangular in section or ± flattened, stout, often ± swollen; stipules oblanceolate or elliptic to subulate, 1–2 cm. long. Flowers (20)25–c. 50, in a simple (rarely compound towards the base) terminal erect or pendulous raceme (1·5)2·5–10·5(15) cm. long in fruit; pedicels 1·2–2·7 cm. in fruit, articulated at the base or in the lower ⅓. Sepals 5–9 mm. long in flower, elliptic to obovate, rounded, becoming red or pink, 10–19 mm. long, convex and eventually reflexed in fruit. Petals golden to pale yellow, 9–11(13) × 4–6·5 mm., obovate to oblanceolate. Stamens with anthers 1–2 mm. long, about ½ as long as the filaments, straight, dehiscing by apical pores. Carpels 7 (rarely 8), with styles free at the apex. Drupelets reniform, inserted centrally, 11–14 × 7–10 mm.: embryo curved.

From N. Rhodesia, Angola and the Congo south to SW. Africa, Bechuanaland Prot. and the Transvaal and east to western Mozambique. In open deciduous woodland or scrub, frequently on sand or rocky terrain, 900–1700 m. in our area.

O. pulchra is recorded from Nyasaland by Burtt Davy and Hoyle, Pardy and Palgrave, but I have seen no specimens from that region and these records may be based on misidentifications.

O. pulchra is very variable in leaf-size and shape and several " species " have been split off from it. Only one of these variations appears to merit taxonomic recognition, on account of its correlation with an inflorescence character and its geographical distribution. Since the correlation is incomplete, however, it can most conveniently be treated as a subspecies.

Subsp. **pulchra**

Leaf-lamina 3·5–10(11) long. Raceme nearly always simple. Pedicels articulated at the base or up to 1·5 mm. from it. Sepals in fruit 10–13 mm. long.

Caprivi Strip. Bwabwata Rest Camp, st. 1.x.1957, *Watt* 18 (K; PRE). **Bechuanaland Prot.** N: Mabele-ea-pudi, c. 123 km. SW. of Maun, fl. 19.ix.1954, *Story* 4722 (K; PRE). SE: 5 km. N. of Kanye, 1100 m., fl. x.1940, *Miller* B/222 (PRE; SRGH). **N. Rhodesia.** Balovale, near Chavuma, fr. 13.x.1952, *White* 3486 (BM; FHO; K). N: L. Bangweulu, N. of Samfya Mission, fl. 6.x.1947, *Brenan* 8034 (FHO; K). W: Ndola Distr., Ichimpi Forest Reserve, fl. 24.ix.1951, *Holmes* 179 (FHO). C: Broken Hill Forest Reserve, fl. 21.ix.1947, *Brenan* 7902 (FHO; K). E: Petauke Distr., Sivwa Camp, st. 28.viii.1929, *Burtt Davy* 20981 (FHO). S: Livingstone, N. bank of Zambezi R., fl. 15.x.1911, *Rogers* 7459 (K; SRGH). **S. Rhodesia.** N: Sebungwe Distr., between the Chikomba and Dett Rivers, fr. 15.x.1956, *Lovemore* 468 (K; SRGH). W: Wankie Distr., Victoria Falls, 1800 m., fl. 1.ix.1955, *Chase* (BM; COI; K; SRGH). C: Salisbury, Prince Edward Dam, fl. 21.x.1948, *Armitage* 15/48 (K; SRGH). E: Chipinga Distr., E. side of Sabi R. at Dott's Drift, fr. 16.xii.1959, *Goodier* 662 (K; SRGH). **Mozambique.** MS: Mossurize, between Espungabera and Chibabava, fr. 10.xi.1943, *Torre* 6139 (LISC).

Throughout the range of the species except in north-western N. Rhodesia and adjacent parts of the Congo and Angola, where it is replaced by subsp. *hoffmanni-ottonis.*

Subsp. **hoffmanni-ottonis** (Engl.) N. Robson in Bol. Soc. Brot., Sér. 2, **36**: 16 (1962). Type from Angola.

Ochna pulchra sensu O. Hoffm. in Linnaea, **43**: 122 (1881).

Ochna hoffmanni-ottonis Engl., Bot. Jahrb. **17**: 78 (1893).—Gilg in Engl., Bot. Jahrb. **33**: 235 (1903).—R.E.Fr., Wiss. Ergebn. Schwed. Rhod.-Kongo Exped. **1**: 150 (1914).—Exell & Mendonça, C.F.A. **1**, 2: 291, t. 12 fig. F (1951). Type as above.

Porochna hoffmanni-ottonis (Engl.) Van Tiegh. in Ann. Sci. Nat., Sér. 8, Bot. **16**: 387 (1902). Type as above.

Leaf-lamina (9)11–18·5 cm. long. Raceme sometimes compound towards the base. Pedicels (at least the lower ones) usually articulated 2–5 mm. from the base. Sepals in fruit 13–15 mm. long.

N. Rhodesia. N: Kawambwa Distr., Nchelengi road, 21 km. from Kawambwa, 1290 m., fr. 29.xi.1961, *Richards* 15435 (K). W: Solwezi Distr., Mutanda R., W. of Solwezi, fl. 17.ix.1930, *Milne-Redhead* 1144 (K).

In western N. Rhodesia and adjacent parts of the Congo and Angola, where it replaces the typical subspecies.

Intermediate forms between the two subspecies occur where their areas of distribution overlap i.e. in northern and central N. Rhodesia, and in Barotseland; but they do not seem to constitute a large proportion of the population in these areas.

3. **Ochna manikensis** De Wild. in Rev. Zool. Afr., **7**, Suppl. Bot.: B 36 (1919). Type from the Congo (Manika).
Ochna sapinii De Wild., tom. cit.: B 38 (1919). Type from the Congo (Bienge).
Ochna angolensis I. M. Johnston in Contrib. Gray Herb. N.S. **73**: 38 (1924).—Exell & Mendonça, C.F.A. **1,** 2: 290, t. 12 fig. D (1951).—White, F.F.N.R.: 250, fig. 43B (1962). Type from Angola (Benguela).

Closely related to *O. pulchra* and perhaps only a fire-induced form of subsp. *hoffmanni-ottonis*, but differing from the latter as follows:

Shrublet 5–30(60) cm. high, with branched rhizomes arising from stout woody rootstock. Leaf-lamina (7)8·5–14·5 × (1·6)2–4 cm., oblanceolate to oblong, always attenuate at the base. Pedicels articulated 2–18 mm. from the base (in the lower $\frac{2}{3}$). Sepals in fruit 12–15 mm. long. Petals up to 9 mm. broad.

N. Rhodesia. B: road from Kabompo to Mankoya, fl. 7.x.1957, *West* 3544 (K; SRGH). N: Puta, fl. 18.viii.1958, *Fanshawe* 4721 (K). W: 48 km. from Mwinilunga on the Solwezi road, fr. 18.ix.1952, *Angus* 486 (FHO; K).

In western N. Rhodesia, Angola and the Congo. On Kalahari sand in areas subjected to burning, often on margins of seasonal swamps.

4. **Ochna oconnorii** Phillips in Bothalia, **1**: 92 (1922).—Burtt Davy, F.P.F.T. **1**: 239 (1926). TAB. **43**. Syntypes from the Transvaal.

Shrub or small tree (1·5)2·5–6 m. high, with bark smooth, greyish (?) and flaking to reveal beetroot-red patches; branches slender, ascending, reddish-purple to dark brown, not peeling. Leaves petiolate; lamina (5)6·8–14·6 × (1·8)2·2–5 cm., oblong or elliptic, acute or very shortly and broadly acuminate to obtuse at the apex, with margin densely but shallowly crenate-serrate, rounded to cuneate at the base, chartaceous, with numerous (c. 20–40) lateral nerves and densely reticulate tertiary venation prominent on both sides; petiole 2–3 mm. long, slender. Flowers (10)15–50(85), in a compound terminal erect raceme (2)4–8·5 cm. long; pedicels (1·2)1·5–2(2·6) cm. long in fruit, articulated in the lower third. Sepals (4)5–6(7) mm. long in flower, elliptic-oblong, rounded, becoming red, 8–10 mm. long, ± convex and spreading in fruit. Petals bright yellow, 9–12 × 5–6(7) mm., obovate. Stamens with anthers 1·5–2 mm. long, $\frac{3}{4}$ as long to as long as the filaments, straight, dehiscing by apical pores. Carpels 5(6), with styles free and recurved at the apex; stigmas small. Drupelets (10)11–13 × 7–9 mm., reniform or flattened-cylindric, inserted eccentrically; embryo almost straight.

S. Rhodesia. E: Chirinda Distr., Chipete, 1160 m., fr. 22.x.1947, *Wild* 2137 (K; SRGH). **Mozambique.** MS: Mossurize, Macuiana, fl. & fr. 30.x.1944, *Mendonça* 2690 (LISC).

Also in NE. Transvaal. In forest and along stream-banks in higher-rainfall regions, 900–1500 m.

Related to *O. membranacea* Oliv., from W. Africa and the Congo basin, but differing from it *inter alia* in flower size, pedicel length and leaf shape and texture.

5. **Ochna arborea** Burch. ex DC., Prodr. **1**: 736 (1824).—Harv. in Harv. & Sond., F.C. **1**: 449 (1860) pro parte excl. specim. Forbes.—Gilg in Engl., Bot. Jahrb. **33**: 235 (1903).—Sim, For. Fl. Port. E. Afr.: 28 (1909) pro parte quoad descr. pars.—Phillips in Bothalia, **1**: 92 (1922).—Burtt Davy, F.P.F.T. **1**: 238 (1926). Type from Cape Prov.
Diporidium arboreum (Burch. ex DC.) Wendl. in Bartl. & Wendl., Beitr. Bot. **2**:

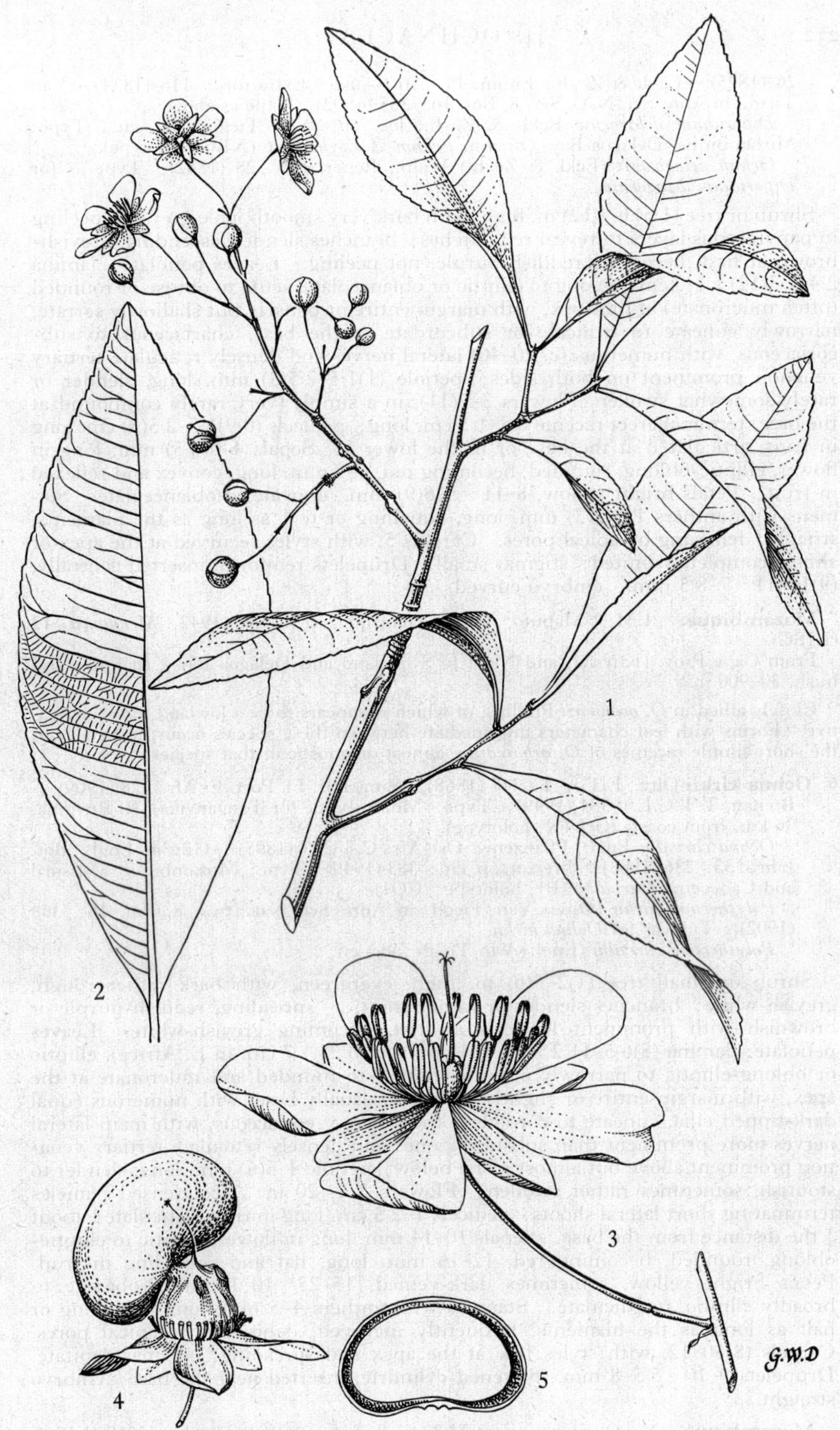

Tab. 43. OCHNA OCONNORII. 1, flowering branch (×1), *Crook* 575; 2, mature leaf (×1), *Swynnerton* 193; 3, flower (×3), *Crook* 575; 4, fruit (×2); 5, vertical section of drupelet (×2), both from *Pardy* in GHS 56660.

26 (1825).—Eckl. & Zeyh., Enum. Pl. Afr. Austr. Extratrop.: 118 (1834).—Van Tiegh. in Ann. Sci. Nat., Sér. 8, Bot. **16**: 355 (1902). Type as above.
Diporidium delagoense Eckl. & Zeyh., loc. cit.—Van Tiegh., loc. cit. Type: Mozambique, Delagoa Bay, *Owen* in *Ecklon & Zeyher* 926 (NBG, holotype).
Ochnn delagoensis (Eckl. & Zeyh.) Walp., Repert. **1**: 528 (1842). Type as for *Diporidium delagoense.*

Shrub or tree (1·5)3–9(12) m. high, with bark very smooth, pale grey and peeling in papyraceous layers to reveal red patches; branches slender, ascending, greyish-brown at first, becoming reddish-purple, not peeling. Leaves petiolate; lamina 2·4–7·5(12) × 1–3 cm., oblong to elliptic or oblanceolate, acute to obtuse or rounded (often mucronate) at the apex, with margin entire or densely but shallowly serrate, narrowly cuneate to truncate or subcordate at the base, chartaceous to sub-coriaceous, with numerous (c. 20–40) lateral nerves and densely reticulate tertiary venation prominent on both sides; petiole (1)1·5–2·5(3) mm. long, slender or rarely somewhat swollen. Flowers 3–7(11), in a simple (very rarely compound at the base) terminal erect raceme 0·5–1·5 cm. long; pedicels (0·9)1·5–2·5(3) cm. long in fruit, articulated at the base or in the lower ¼. Sepals 4-6(6·5) mm. long in flower, elliptic-oblong, rounded, becoming red, 6–9 mm. long, convex and reflexed in fruit. Petals bright yellow, 8–11 × 5–6(9) mm., obovate to oblanceolate. Stamens with anthers 1·5–2(3) mm. long, equalling or to ⅓ as long as the filaments, straight, dehiscing by apical pores. Carpels 5, with styles recurved at the apex or almost completely united; stigmas small. Drupelets reniform, inserted centrally, (9)10–11 × 7–8·5 mm.; embryo curved.

Mozambique. LM: Maputo, Porto Henrique, fr. 19.viii.1942, *Mendonça* 12 (LISC).
From Cape Prov. (Knysna) and Natal to Swaziland and Delagoa Bay. In forest and bush, 30–900 m.

Closely allied to *O. oconnorii* Phillips, of which it appears to be a lowland forest derivative. Forms with leaf characters intermediate between these species occur in Natal, but the short simple racemes of *O. arborea* are almost diagnostic of that species.

6. **Ochna kirkii** Oliv., F.T.A. **1**: 317 (1868).—Sim, For. Fl. Port. E. Afr.: 28 (1909).—Brenan, T.T.C.L.: 384 (1949). Type: Mozambique (or Tanganyika), R. Rovuma, 36 km. from coast, *Kirk* (K, holotype).
Ochna carvalhi Engl., Pflanzenw. Ost-Afr. **C.**: 273 (1895).—Gilg in Engl., Bot. Jahrb. **33**: 236 (1903).—Brenan, op. cit.: 383 (1949). Type: Mozambique, Mussoril and Cabeceira, *Carvalho* (B†, holotype; COI).
Polythecium kirkii (Oliv.) Van Tiegh. in Ann. Sci. Nat., Sér. 8, Bot. **16**: 368 (1902). Type as for *Ochna kirkii*.
Polythecium carvalhi (Engl.) Van Tiegh., loc. cit.

Shrub or small tree (1)2–5(6) m. high, evergreen, with bark rather rough, greyish-white; branches slender or rather stout, ± spreading, reddish-purple or brownish with prominent lenticels at first, becoming greyish-white. Leaves petiolate; lamina (5)6·5–11·2 × 2·6–5·1 cm. (up to 21 × 7 cm. in E. Africa), elliptic or oblong-elliptic to narrowly obovate, obtuse to rounded and mucronate at the apex, with margin entire or slightly undulate, usually beset with numerous equal dark-tipped cilia, cuneate to cordate at the base, ± coriaceous, with main lateral nerves more prominent than subsidiary ones and densely reticulate tertiary venation prominent above but almost plane below; petiole 4–5(6) mm. long, slender to stoutish, sometimes rather swollen. Flowers 3–c. 20 in ± condensed panicles terminating short lateral shoots; pedicels 1–2·5 cm. long in fruit, articulated about ¼ the distance from the base. Sepals 10–14 mm. long in flower, elliptic to elliptic-oblong, rounded, becoming red, 12–16 mm. long, flat and spreading in fruit. Petals bright yellow, sometimes dark-veined, 15–25 × 10–18 mm., obovate to broadly elliptic, unguiculate. Stamens with anthers 4–5 mm. long, equalling or half as long as the filaments, frequently incurved, dehiscing by apical pores. Carpels (8)10–12, with styles free at the apex and spreading; stigmas capitate. Drupelets 9–10 × 5·5–8 mm., flattened-cylindric, inserted near the base; embryo straight.

Mozambique. N: Quissanga, near Meluco, fl. & fr. 2.x.1948, *Barbosa* 2319 (LISC; LMJ). Z: Mocuba, fl. 1943, *Torre* (LISC).
Eastern districts of Kenya, Tanganyika and northern Mozambique In the shrub layer of evergreen forest and in evergreen scrub, 0–600 m.

O. kirkii has been confused (e.g. by Gilg in Engl., Bot. Jahrb. **33**: 245 (1903)) with the E. African *O. thomasiana*, a plant of coastal forests in which the leaves are usually smaller and auriculate at the base with long basal cilia and the flowers usually smaller and fewer in each inflorescence.

7. **Ochna mossambicensis** Klotzsch in Peters, Reise Mossamb. Bot. **1**: 88, t. 16 (1861).—Oliv., F.T.A., **1**: 317 (1868).—Gilg in Engl., Bot. Jahrb. **33**: 234, 244 (1903).—Sim, For. Fl. Port. E. Afr.: 28 (1909).—Brenan, T.T.C.L.: 384 (1949). Type: Mozambique, Sena, *Peters* (B, holotype†; K).
Ochna fischeri Engl., Bot. Jahrb. **17**: 78 (1893). Type from Tanganyika.
Ochna purpureocostata Engl., Pflanzenw. Ost-Afr. **C**: 273 (1895). Type from Tanganyika.
Discladium mossambicensis (Klotzsch) Van Tiegh. in Ann. Sci. Nat., Sér. 8, Bot. **16**: 351 (1902). Type as for *Ochna mossambicensis*.
Diporidium purpureocostatum (Engl.) Van Tiegh., tom. cit.: 356 (1902). Type as for *Ochna purpureocostata*.
Polythecium fischeri (Engl.) Van Tiegh., tom. cit.: 368 (1902). Type as for *Ochna fischeri*.

Bushy shrub or small tree up to 3(5) m. high or sometimes a rhizomatous shrublet c. 5–40 cm. high, evergreen, frequently galled, with bark rather rough, whitish-brown; branches rather stout, ± ascending, greenish-brown and flattened, without visible lenticels at first, becoming greyish-white and then greyish-purple and terete. Leaves petiolate; lamina (5·5)7–22·6 × (2)3·4–7·8(8·4) cm., obovate to oblanceolate or rarely oblong, obtuse to rounded (very rarely acute or acuminate) and sometimes shortly mucronate at the apex, with margin densely serrulate, cuneate (rarely truncate) at the base, coriaceous, with main and subsidiary lateral nerves almost equally prominent and densely reticulate tertiary venation more prominent above than below; petiole 1·5–8 mm. long, usually ± stout, sometimes swollen. Flowers ∞ in ± laxly branched panicles, terminating short lateral shoots; pedicels 1·5–3 cm. long in fruit, articulated about ¼ the distance from the base. Sepals 9–11(12·5) mm. long in flower, elliptic to elliptic-oblong, rounded, becoming red, 12–14 mm. long, flat, and spreading in fruit. Petals bright yellow, sometimes dark-veined, 10–17(22) × 7–13(19) mm., obovate or subcircular and shortly unguiculate. Stamens with anthers 4–7 mm. long, 3–5 times as long as the filaments, straight, dehiscing by apical pores. Carpels (6)8–10, with styles united almost to the apex with free ends spreading radially; stigmas slightly enlarged. Drupelets 8–9(10) × 6–8 mm., subglobose or flattened-ovoid-cylindric, inserted near the base; embryo straight.

Mozambique. N: Nampula, fl. & fr. 6.xi.1936, *Torre* 862 (COI; LISC). Z: Mocuba-Milange road, fl. & fr. 27.xi.1949, *Faulkner* K. 339 (2) (COI; K; SRGH). MS: Beira, Dondo, fl. & fr. 14.ix.1943, *Torre* 5893 (LISC).
Eastern districts of Kenya, Tanganyika and northern Mozambique. In *Brachystegia* or mixed deciduous woodland or among rocks, usually on sandy soils, 0–450 (800) m.
The record for Nyasaland (N.C.L.: 56) is probably due to a misidentification of *O. macrocalyx*.

Gomes e Sousa 4589, from the Palma Distr. of Niassa Prov., combines characters of *O. mossambicensis*, *O. kirkii* and *O. thomasiana* (a closely related species of coastal forests in East Africa). It may belong to a distinct species.

8. **Ochna natalitia** (Meisn.) Walp., Repert. **2**: 826 (1843).—Planch. in Hook., Lond. Journ. Bot. **5**: 655 (1846).—Gilg in Engl., Bot. Jahrb. **33**: 236 (1903).—Bak. f. in Journ. Linn. Soc., Bot. **40**: 37 (1911).—Phillips in Bothalia, **1**: 93 (1922).—Burtt Davy, F.P.F.T. **1**: 239 (1926). Type from Natal (Durban).
Diporidium natalitium Meisn. in Hook., Lond. Journ. Bot. **2**: 58 (1843).—Van Tiegh. in Ann. Sci. Nat., Sér. 8, Bot. **16**: 355 (1902). Type as above.
Ochna atropurpurea var. *natalitia* (Meisn.) Harv., F.C. **1**: 448 (1860).—Sim, For. Fl., Port. E. Afr.: 28 (1909). Type as above.
Ochna arborea sensu Sim, loc. cit. (1909) pro parte quoad descr. pars et tab.
Ochna chilversii Phillips, tom. cit.: 90 (1922). Syntypes from Natal and E. Cape Prov.

Bushy shrub or small tree 0·75–4·5 m. high (to 7 m. or more in Natal), sometimes branching below ground level, evergreen or deciduous, frequently galled, with bark rather rough, brown; branches ± ascending, whitish and flattened and frequently peeling at first, becoming purplish and terete with numerous lenticels (sometimes appearing lepidote). Leaves petiolate; lamina 3–12(14·8) × 1–3·5(5) cm., elliptic or oblong to oblanceolate (rarely obovate), rounded (more rarely obtuse to acute) and occasionally apiculate at the apex, with margin serrate with curved or straight teeth or almost entire, rounded or shallowly cordate (more rarely cuneate) at the base, coriaceous, with main and subsidiary lateral nerves almost equally prominent and tertiary venation prominent above but less so or almost plane below; petiole 1–2 mm. long, slender. Flowers (2)3–14(c. 20), in lax or ± condensed panicles or sometimes reduced to simple racemes or pseudumbels, terminal or on short axillary shoots; pedicels (0·8)1·2–2·5 cm. long in fruit, articulated in the lower ¼. Sepals 5–8(11) mm. long in flower, elliptic to elliptic-oblong, rounded, becoming red, 7–12(14) mm. long, ± convex and ± spreading in fruit. Petals bright yellow, sometimes dark-veined, (7)8–18(21) × 6–12(14) mm., obovate or obovate-oblong to suborbicular and unguiculate. Stamens with anthers (1·75)2–3 mm. long, ⅖–1(1⅓) times as long as the filaments, straight, dehiscing by apical pores. Carpels (6)8–13, with styles united almost to the apex with free ends spreading radially or recurved; stigmas flattened. Drupelets subglobose or flattened-ovoid-cylindric, inserted near the base, 6–11(12·5) × (4·5)5–7 mm.; embryo straight.

Mozambique. MS: Cheringoma, Inhaminga, fr. viii.1954, *Gomes e Sousa* 4254 (K; PRE; SRGH). SS: Massinga, near Quizugo, 110 m., fl. ix.1937, *Gomes e Sousa* 2029 (COI; K). LM: Vila Luís, fl. & fr. 2.x.1957, *Barbosa & Lemos* in *Barbosa* 7894 (COI; K; LISC; LMJ; SRGH).

North-eastern Cape Prov. to Beira and inland to the Transvaal and Swaziland. Deciduous woodland and in forest margins and scrub, usually on sandy soils, 0–300 m. (to 1375 m. in the Transvaal).

O. natalitia varies in habit, leaf shape and texture, and in size and numbers of floral parts. At the extreme south of its range there are large-flowered arboreal forms with elliptic curved-serrate leaves narrowing at both ends, occurring in open forest and forest margins (*O. chilversii* Phillips); but these are connected by numerous intermediates in Natal, Swaziland and the Transvaal with the form typical of our area, viz. a bushy shrub with flowers varying in size and oblong or elliptic-oblong leaves rounded at both ends. The latter form has frequently been confused with *O. arborea*, from which it can be easily distinguished by the number of carpels, the shape and insertion of the drupelets, the leaf-shape and less densely reticulate venation, and usually by the compound inflorescence.

9. **Ochna macrocalyx** Oliv., F.T.A. **1**: 319 (1868).—Gilg in Engl., Bot. Jahrb. **33**: 236 (1903).—Sim, For. Fl. Port. E. Afr.: 28 (1909).—Burtt Davy & Hoyle, N.C.L.: 56 (1936). TAB. **44** fig. A. Syntypes from Tanganyika and Nyasaland: Sotshi, *Kirk* (K); Manganja Mts. [? in Mozambique], *Meller* (K).

Ochna macrocarpa Engl., Bot. Jahrb. **17**: 77 (1893).—Gilg, loc. cit. Type from Tanganyika.

Ochna splendida Engl., op. cit. **28**: 434 (1900).—Gilg, loc. cit. Type from Tanganyika.

Diporidium macrocalyx (Oliv.) Van Tiegh. in Ann. Sci. Nat., Sér. 8, Bot. **16**: 355 (1902). Type as for *O. macrocalyx*.

Diporidium macrocarpa (Engl.) Van Tiegh., loc. cit. Type as for *Ochna macrocarpa*.

Polythecium splendidum (Engl.) Van Tiegh., tom. cit.: 368 (1902). Type as for *Ochna splendida*.

Ochna sp. nr. *O. macrocalyx* sensu Palgrave, Trees of Central Afr.: 29, tt. (1956).

Shrub up to c. 2 m. high but usually smaller, sometimes a rhizomatous shrublet with only leaves and flowers appearing above ground, evergreen or deciduous, with bark rough, grey-brown; branches spreading or ascending from stiff erect shoots, purplish or brown and often with prominent lenticels at first, becoming greyish-white, sometimes peeling. Leaves petiolate; lamina 6·8–18·8(24·6) × 2–4·7(7) cm., oblanceolate to narrowly obovate, acute to shortly acuminate (more rarely obtuse) at the apex, with margin densely serrulate, cuneate to rounded at the base, ± coriaceous, with main lateral nerves scarcely more prominent than subsidiary ones and densely reticulate tertiary venation prominent above but almost plane below; petiole 1–3 mm. long, stout or swollen. Flowers (2)3–9(14), in short axillary simple or compound racemes or fascicles, more rarely in condensed panicles; pedicels 1·3–4 cm. long in fruit, articulated in the lower ⅓. Sepals 12–20(25) mm. long in flower, elliptic to narrowly ovate, rounded or obtuse, becoming red, flat or slightly convex at the base only, 16–26 mm. long and eventually spreading but not navicular in fruit. Petals bright yellow to orange-yellow, sometimes dark-veined,

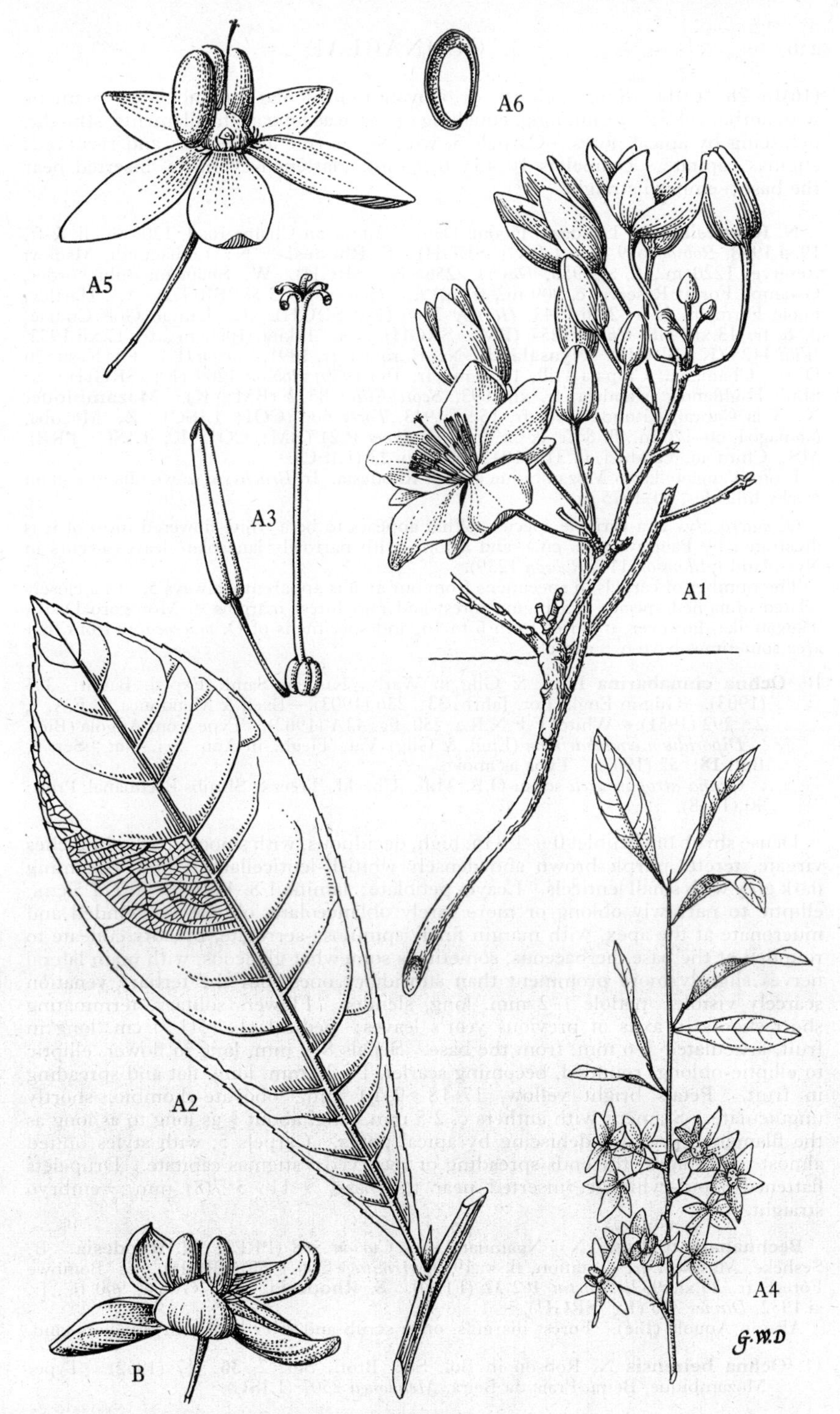

Tab. 44. A.—OCHNA MACROCALYX. A1, flowering shoot of dwarf form (×1), *Robson* 659; A2, leaf (×1), *Faulkner* P21; A3, stamen and gynoecium (×4), *Robson* 659; A4, fruiting branch of tall form (×½); A5, fruit (×1); A6, vertical section of drupelet (×1), all from *Carson* s.n. B.—OCHNA ROVUMENSIS, fruit (×1), *Kirk* 170.

(16)18–28 × (9)11–18 mm., elliptic or obovate to orbicular, unguiculate. Stamens with anthers (4·5)5–7 mm. long, equalling or twice as long as the filaments, straight, dehiscing by apical pores. Carpels 5, with styles free at the apex and recurved; stigmas capitate. Drupelets 10–13 × 6–9 mm., flattened-cylindric, inserted near the base; embryo straight.

N. Rhodesia. E: Lundazi, Tigone Dam, 3·2 km. on Chama road, 1200 m., fl. & fr. 19.xi.1958, *Robson* 659 (K; LISC; SRGH). **S. Rhodesia.** N: Lomagundi, Msukwi Reserve, 1220 m., fl. xi.1956, *Davies* 2256 (K; SRGH). W: Shangani-Bubi border, Gwampa Forest Reserve, c. 900 m., fr. ii.1957, *Goldsmith* 25/58 (SRGH). C: Hartley, Poole Farm, fl. & fr. 2.xii.1943, *Hornby* 2466 (K; SRGH). E: Umtali Golf Course, fl. & fr. 13.xi.1948, *Chase* 1257 (BM; SRGH). S: Bikita, 1000 m., fr. 17.xii.1953, *Wild* 4425 (K; SRGH). **Nyasaland.** N: Karonga, fr. 1891, *Carson* (K). C: Kasungu Distr., Chamama, Chipala Hill, 1050 m., fr. 16.i.1959, *Robson* 1209 (K; SRGH). S: Shire Highlands, Mandala, fr. xii.1893, *Scott Elliot* 8532 (BM; K). **Mozambique.** N: Vila Cabral, Litunde road, fr. 15.xii.1943, *Torre* 600 (COI; LISC). Z: Mocuba, Namagoa, 60–120 m., fl. & fr. ix–xi.1944, *Faulkner* P 21 (BM; COI; K; LISC; PRE). MS: Chimoio, Vanduzi, fl. 31.x.1942, *Salbany* 23 (LISC).

From Tanganyika to Mozambique and S. Rhodesia. In *Brachystegia* woodland and on rocky hillsides, 60–1675 m.

O. macrocalyx is a variable species. What appears to be a white-flowered form of it is illustrated by Palgrave (loc. cit.), and a form with narrowly lanceolate leaves occurs in Nyasaland (*Adamson* 116, *Robson* 1239).

The number of carpels in specimens from our area is apparently always 5. In a closely related unnamed species of fringing forest and rain forest margins in Morogoro Distr., Tanganyika, however, it varies from 6 to 10, and specimens of *O. macrocalyx* from that area sometimes have 6–8 carpels.

10. **Ochna cinnabarina** Engl. & Gilg in Warb., Kunene-Samb.-Exped. Baum: 305 (1903).—Gilg in Engl., Bot. Jahrb, **33**: 236 (1903).—Exell & Mendonça, C.F.A. **1**, 2: 292 (1951).—White, F.F.N.R.: 250, fig. 43A (1962). Type from Angola (Bié).

Diporidium cinnabarinum (Engl. & Gilg) Van Tiegh. in Ann. Sci. Nat., Sér. 8, Bot. **18**: 52 (1903). Type as above.

Ochna atropurpurea sensu O.B. Mill., Checkl. Trees & Shrubs Bechuanal. Prot.: 40 (1948).

Dense shrub or shrublet 0·6–1·8 m. high, deciduous, with smooth bark; branches virgate, terete, purple-brown and densely whitish-lenticellate at first, becoming dark grey with small lenticels. Leaves petiolate; lamina 1·8–4·3(6·5) × 0·7–1·5 cm., elliptic to narrowly oblong or more rarely oblanceolate, obtuse to rounded and mucronate at the apex, with margin finely spinulose-serrulate, broadly cuneate to rounded at the base, herbaceous, sometimes somewhat glaucous, with main lateral nerves slightly more prominent than subsidiary ones, but the tertiary venation scarcely visible; petiole 1–2 mm. long, slender. Flowers solitary, terminating short shoots in axils of previous year's leaves; pedicels 1–1·5(1·7) cm. long in fruit, articulated 3–6 mm. from the base. Sepals 8–9 mm. long in flower, elliptic to elliptic-oblong, rounded, becoming scarlet, 14–20 mm. long, flat and spreading in fruit. Petals bright yellow, 17–18 × 9–10 mm., obovate-rhombic, shortly unguiculate. Stamens with anthers c. 2·5 mm. long, about $\frac{3}{4}$ as long to as long as the filaments, straight, dehiscing by apical pores. Carpels 5, with styles united almost to the apex, the ends spreading or recurved; stigmas capitate. Drupelets flattened-ovoid-cylindric, inserted near the base, 9–11 × 5–7(8) mm.; embryo straight.

Bechuanaland Prot. N: Ngamiland, fr., *Curson* 568 (PRE). **N. Rhodesia.** B: Sesheke, Masese Forest Station, fl. xi.1959, *Armitage* 206/1959 (SRGH). S: Bombwe Forest, fr. 29.xi.1932, *Martin* 462/32 (FHO). **S. Rhodesia.** W: Gwaai, 900 m., fr. xi.1952, *Davies* 386 (K; SRGH).

Also in Angola (Bié). Forest margins, open scrub and river banks on Kalahari sand.

11. **Ochna beirensis** N. Robson in Bol. Soc. Brot., Sér. 2, **36**: 17 (1962). Type: Mozambique, Beira, Praia da Beira, *Mendonça* 2507 (LISC).

Shrub or small tree up to c. 5 m. high; branches terete or somewhat striate, reddish-brown at first, becoming pale brown or whitish, with numerous lenticels. Leaves petiolate; lamina 3·2–8·2 × 1·2–2·6 cm., rhombic-elliptic to oblong or oblanceolate, acute or shortly and bluntly acuminate at the apex, with margin sharply serrulate, narrowly cuneate to angustate at the base, subcoriaceous, with main and subsidiary lateral nerves equally prominent and the densely reticulate tertiary venation almost equally prominent on both sides; petiole 2–4 mm. long, slender. Flowers 3–5, in umbels terminating short axillary shoots; pedicels 2–2·5 cm. long in fruit, articulated 2–4 mm. from the base. Sepals becoming carmine-pink, 15–18 mm. long, convex and accrescent in fruit. Petals and stamens as yet unknown. Carpels 5. Drupelets flattened-ovoid-cylindric, inserted at the base, 11 × 7 mm.; embryo straight.

Mozambique. MS: Cheringoma, Chinizíua, fr. 22.x.1949, *Pedro & Pedrógão* 8865 (LMJ; PRE).

Known only from the Beira District of Mozambique. Deciduous woodland and evergreen littoral scrub, near sea-level.

O. beirensis is related to *O. macrantha* Bak. (*Diporidium greveanum* Van Tiegh.) from Madagascar, but differs in leaf-shape and venation.

12. **Ochna inermis** (Forsk.) Schweinf. apud Penzig in Atti Congr. Bot. Intern. Genova, 1892: 335 (1893); Arab. Pfl. Aegypt. Alger. & Jemen: 148 (1912). Type from the Yemen.

Euonymus inermis Forsk., Fl. Aegypt.-Arab.: 204 (1775). Type as above.

Ochna parvifolia Vahl, Symb. Bot. **1**: 33 (1790). Type as above.

Ochna rivae Engl. in Ann. R. Inst. Bot. Roma, **7**: 21 (1897). Type from Somalia.

Diporidium inerme (Forsk.) Van Tiegh. in Ann. Sci. Nat., Sér 8, Bot. **16**: 358 (1902). Type as for *O. inermis*.

Diporidium schimperi Van Tiegh., tom. cit.: 360 (1902). Type from Ethiopia.

Ochna pretoriensis sensu Phillips in Bothalia, **1**: 95 (1922) pro parte quoad specim. ex Messina et Moorddrift.—O.B. Mill., Checkl. Trees & Shrubs Bechuanal. Prot.: 41 (1948); in Journ. S. Afr. Bot. **18**: 58 (1952).

Ochna inermis var. *rivae* (Engl.) Chiov., Fl. Somal. **1**: 121 (1929). Type as for *O. rivae*.

Ochna rogersii Hutch., Botanist in S. Africa: 309, 317 (1946) *nom. nud.*

Ochna sp. sensu O.B. Mill. in Journ. S. Afr. Bot. **18**: 58 (1952).

Shrub or small tree 1–4·5(6) m. high, usually ± bushy, with smooth whitish bark; branches numerous, ascending, terete, brownish- to blackish-purple at first, becoming glaucous or whitish, with numerous lenticels. Leaves petiolate; lamina (1·4)1·7–4·8(7·5) × 0·9–2·8(3·1) cm., elliptic or obovate to subcircular or oblong, rounded to obtuse (rarely ± acute) and apiculate at the apex, with margin sharply curved-serrulate, cuneate to truncate (more rarely shallowly cordate) at the base, chartaceous to subcoriaceous, with main and subsidiary lateral nerves equally prominent and the ± densely reticulate tertiary venation almost equally prominent on both sides; petiole 1–2·5 mm. long, slender. Flowers 1(2), terminating short axillary shoots; pedicels 1·3–3 cm. long in fruit, articulated 1–3 mm. from the base. Sepals c. 5 mm. long in flower, elliptic, rounded, becoming crimson-red, (10)11–16 mm. long, convex and accrescent in fruit, eventually spreading and navicular. Petals golden-yellow, (7·5)9–11 × (4·5)5–6 mm., obovate and shortly unguiculate to subcircular with a longer claw. Stamens with anthers 1 mm. long, $\frac{1}{2}$–$\frac{1}{3}$ as long as the filaments, straight, dehiscing by apical pores. Carpels 5, with styles united almost to the apex, recurved at the ends; stigmas slightly capitate. Drupelets 9–10 × 5–7 mm., flattened-ovoid-cylindric, inserted near the base; embryo straight.

Bechuanaland Prot. SE: Kanye Distr., Pharing, 1220 m., fr. 12.xi.1948, *Hillary & Robertson* 476 (PRE). **S. Rhodesia.** W: Bulawayo, Matopos, fr. 18.xii.1950, *Knapman* 18 (SRGH). C: Chilimanzi Distr., Umvuma, Gwelo road, st. i.1951, *Kirkman* 7/51 (K; SRGH). E: Melsetter, Hot Springs, 610 m., fr. 28.xii.1948, *Chase* 1402 (BM; K; SRGH). S: Ndanga, Chipinda Pools, fr. 13.x.1951, *Wormald* 85/51 (K; SRGH). **Mozambique.** LM: Libombo Mts., fr. 9.i.1929, *Pole Evans* in Herb. Pret. 28766 (PRE).

From the Yemen, Eritrea and Ethiopia to northern Kenya and Uganda; apparently rare in Tanganyika and absent from N. Rhodesia and Nyasaland and the northern parts of S. Rhodesia and Mozambique; and recurring in southern S. Rhodesia, eastern Bechuanaland Prot., the northern Transvaal and SW. Mozambique. Dry *Acacia-Commiphora* or *Colophospermum mopane* scrub or among rocks, in stony or sandy ground, 425–1200 m. (100–1550 m. in E. Africa).

This widespread species is closely related to *O. hackarsii* Robyns & Lawalrée (a plant of damper habitats in Uganda and the eastern Congo) and *O. beirensis* N. Robson. In our area it has frequently been confused with *Ochna pretoriensis* Phillips, from which it can always be distinguished by the position of the pedicel articulation and by the leaf shape and venation.

13. **Ochna rovumensis** Gilg in Engl., Bot. Jahrb. **33**: 246 (1903).—Brenan, T.T.C.L.: 384 (1949). TAB. **44** fig. B. Type from Tanganyika (R. Rovuma).
Diporidium rovumense (Gilg) Van Tiegh. in Ann. Sci. Nat., Sér. 8, Bot. **18**: 52 (1903). Type as above.

Low shrub 1·5 m. high to small tree, deciduous, with bark smooth and grey with lighter patches; branches ± spreading or ascending, terete, whitish or pale brown and glabrous or puberulous at first, becoming greyish- or purplish-brown and glabrous, scarcely peeling, with small lenticels. Leaves petiolate; lamina 3–6·4(7) × 1·3–3·5 cm., broadly to rather narrowly elliptic, acute (sometimes mucronate) at the apex, with margin finely serrulate or subentire, cuneate to rounded at the base, subcoriaceous, with main and subsidiary lateral nerves equally prominent and densely reticulate tertiary venation slightly prominent on both sides; petiole 1·5–2 mm. long, slender. Flowers solitary, terminating short axillary shoots; pedicels 1·2–1·8 cm. long in fruit, glabrous or rarely pubescent, articulated within 1 mm. of the base. Sepals 9–13 mm. long in flower, elliptic, rounded, becoming red, 21–24(30) mm. long, convex and accrescent in fruit, eventually spreading and navicular. Petals bright yellow, c. 26 × 15 mm., obovate, unguiculate. Stamens with anthers 1·5–2 mm. long, $\frac{1}{4}$–$\frac{1}{2}$ as long as the filaments, straight, dehiscing by apical pores. Carpels 5, with styles free towards the apex with ends spreading; stigmas small, capitate. Drupelets c. 10 × 6 mm., ovoid-cylindric; embryo straight.

S. Rhodesia. N: Kariba, c. 600 m., fr. ii.1960, *Goldsmith* 20/60 (K; SRGH). **Nyasaland.** S: Sambani hills, fl. 1937, *Townsend* 246 (FHO). **Mozambique.** N: between Corrane and Meconta, fr. 18.x.1948, *Barbosa* 2501 (LISC; LMJ). MS: Chemba, Mt. Chiramba Dembe, fr. 24.i.1860, *Kirk* 170 (K).

Also in central and south Tanganyika. Rocky areas in river valleys.

Barbosa 2501 has a fawn pubescence on the young shoots, petioles and pedicels, and the leaf-apices are obtuse, not acute as in the other specimens. This specimen may belong to a different taxon (species or variety); but some pedicels of the S. Rhodesian specimen (*Goldsmith* 20/60) are very slightly puberulous. It seems advisable, therefore, to include *Barbosa* 2501 within *O. rovumensis* until more information is available on the frequency and distribution of pubescent specimens.

14. **Ochna barbosae** N. Robson in Bol. Soc. Brot., Sér, 2, **36**: 18 (1962). Type: Mozambique, Vila Luís, *Barbosa & Lemos* in *Barbosa* 7895 (COI; K; LISC, holotype; LMJ).
Ochna pretoriensis sensu Phillips in Bothalia, **1**: 95 (1922) pro parte quoad specim. Junod.

Shrub or small tree 0·5–3 m. high, occasionally galled, with bark smooth, brown to whitish; branches ascending, greenish-brown and ridged at first, becoming purplish-brown to whitish and ± ridged or terete, often peeling, with numerous small ± raised lenticels. Leaves petiolate; lamina 3–6·2 × 1–2·6 cm., elliptic-oblong to oblanceolate (rarely obovate), obtuse to rounded and sometimes mucronate at the apex, with margin usually bluntly serrulate to subentire, cuneate or rarely rounded at the base, subcoriaceous, with main and subsidiary lateral nerves almost equally prominent, forming almost a right angle with the midrib, and very densely reticulate tertiary venation slightly prominent on both sides; petiole (1·5)2–4 mm. long, slender. Flowers solitary, terminating short axillary shoots, scented; pedicels 0·8–2 cm. long in fruit, articulated at the base. Sepals 5–9 mm. long in flower, elliptic to elliptic-oblong, rounded, becoming red, 14–18 mm. long, convex and accrescent in fruit, eventually spreading and navicular. Petals bright yellow, sometimes dark-veined, 9–13 × 5–8 mm., obovate to subcircular, unguiculate. Stamens with anthers 1–1·5 mm. long, $\frac{3}{4}$ as long as or equalling the filaments, straight, dehiscing by apical pores. Carpels 5, with styles united almost to the apex but with ends spreading or recurved; stigmas slightly capitate. Drupelets 10–11 × 8–9 mm., flattened-ovoid-cylindric, inserted near the base; embryo straight.

S. Rhodesia. E; Chipinga Distr., Mutandawa Hills, fl. 13.ix.1958, *Phelps* 257 (BM; SRGH). S: Nuanetsi Distr., Chilojo escarpment, fl. 22.viii.1956, *Mowbray* 119 (K; SRGH). **Mozambique.** SS: Inhambane, Inharrime, Ponta Závora, fr. 16.x.1957,

Barbosa & Lemos in *Barbosa* 8071 (COI; K; LISC; LMJ). LM: Vila Luís, fr. 2.x.1957, *Barbosa & Lemos* in *Barbosa* 7895 (COI; K; LISC; LMJ).

Apparently confined to coastal districts of southern Mozambique and the low hills of south-eastern Rhodesia. Dunes and sandy soils, 0–565 m.

O. barbosae appears to be a southern derivative of *O. rovumensis*. Like that species it has solitary flowers with sepals enlarging and becoming convex in fruit, pedicels articulated at the base and leaves with widely-spreading lateral nerves and very densely reticulate venation. The sepals, however, are smaller in fruit and the leaf-shape is quite different. It had been confused with a species from the Transvaal, *O. pretoriensis* Phillips, which has smaller sepals in fruit and different leaf-shape and venation.

15. **Ochna angustata** N. Robson in Bol. Soc. Brot., Sér. 2, **36**: 20 (1962). Type: Mozambique, between Mocubela and Bajone, 7·7 km. from Mocubela, *Barbosa & Carvalho* in *Barbosa* 4265 (K, holotype; LMJ).

Shrub or small tree 1–5 m. high, sometimes much branched, with rough whitish bark; branches spreading, purplish-brown and striate at first, soon becoming whitish and terete with few lenticels. Leaves petiolate; lamina 4–9·5 × 1·5–3 cm., elliptic to oblong or rarely lanceolate, rounded or more rarely obtuse at the apex, with margin crenulate-serrulate to subentire, broadly to narrowly cuneate and narrowing towards the petiole at the base, subcoriaceous, with main lateral nerves slightly more prominent than subsidiary ones and the reticulate tertiary venation scarcely visible above and almost plane below; petiole 3–6 mm. long, slender. Flowers 3–4(5), in umbels terminating short axillary shoots; pedicels 1·3–3 cm. long in fruit, articulated within 3 mm. of the base. Sepals 4–5 mm. long in flower, elliptic, rounded, becoming orange-red, 9–11 mm. long, flat and spreading in fruit. Petals bright yellow, 11–12 × 6–8 mm., obovate or oblong-obovate, shortly unguiculate. Stamens with anthers 1·5–2 mm. long, as long as the filaments, straight, dehiscing by apical pores. Carpels 5(7), with styles almost completely united; stigmas slightly enlarged. Drupelets 5–6 × 5–6 mm., globose or subglobose, inserted near the base; embryo straight.

Mozambique. N: Between Corrane and Muatua, fr. 7.xi.1936, *Torre* 974 (COI; LISC). Z: Maganja da Costa, between Pebane and Mocubela, fr. 25.x.1942, *Torre* 4675 (LISC). MS: Cheringoma, between R. Urema and Durúndi, fl. 10.ix.1942, *Mendonça* 179 (LISC).

Known only from Mozambique from Beira northwards, in areas within c. 40 km. of the coast. Deciduous woodland and coastal scrub.

O. angustata is closely related to two species from W. Madagascar—*O. pervilleana* Baill. a species with white petals and acute or acuminate leaves, and *Diporidium* (=*Ochna*) *baronii* Van Tiegh. which has 1-flowered inflorescences. *O. pseudoprocera* Sleumer, from SE. Tanganyika, is also closely allied but differs in having sharply acuminate membranous leaves and larger flowers in 1–3-flowered inflorescences.

16. **Ochna glauca** Verdoorn in Bothalia, **6**: 232 (1951). Type from the Transvaal (Soutpansberg).

Shrub or small tree 1·5–2·5 m. high, usually straggling or sprawling, with smooth dark grey bark; branches spreading, brownish- to blackish-purple at first, becoming glaucous-purple-grey and vertically striate (eventually terete), with scattered lenticels. Leaves petiolate, glaucous; lamina 1·6–4·5 × 0·9–2·5 cm., elliptic or oblong to obovate or subcircular, obtuse to rounded or retuse at the apex, with margin finely serrulate, cuneate to rounded at the base, with main and subsidiary lateral nerves and the ± densely reticulate tertiary venation equally prominent on both sides; petiole 3–4 mm. long, slender. Flowers solitary, terminating short axillary shoots; pedicels 0·7–1·2 cm. long in fruit, articulated at the base. Sepals 4–5 mm. long in flower, elliptic, rounded, becoming brownish-red, 7–10 mm. long, flat and reflexed in fruit. Petals bright yellow, 8–11 × 4–6(8) mm., obovate, unguiculate. Stamens with anthers 1·5 mm. long (? or longer), ½ as long to as long as the filaments, straight, dehiscing by apical pores. Carpels 5, with styles united almost to the apex with ends recurved; stigmas slightly enlarged. Drupelets 8–9 × 5–6 mm., flattened-curved-cylindric, inserted near the base on a short stalk; embryo straight.

S. Rhodesia. W: Matopos, Silozwe Hill, 90 m., fr. x.1956, *Miller* 3707 (K; SRGH).

Apparently confined to the Matopo Hills and a hill in the Transvaal (Dongola Reserve, Soutpansberg). In cracks on granite kopjes or on granitic soils, 1370–1670 m.

O. glauca is most closely allied to an undescribed species from Tanganyika related to *O. ovata* F. Hoffm., but differs from it in several characters, notably the 1-flowered inflorescence and the smaller glaucous leaves. This last character also serves to distinguish it from the other 1-flowered species in the south of our area. In addition, it differs from *O. inermis* in several characters including the position of the pedicel articulation and from both the latter species and *O. pretoriensis* in the sepals which are flat and become reflexed in fruit.

17. **Ochna holstii** Engl. in Abh. Preuss. Akad. Wiss.: 69 (1894); Pflanzenw. Ost-Afr. **C**: 273 (1895).—Gilg in Engl., Bot. Jahrb. **33**: 234, 241 (1903).—Phillips in Bothalia, **1**: 93 (1922).—Burtt Davy, F.P.F.T. **1**: 238 (1926).—Hutch., Botanist in S. Afr.: 667 (1946).—Brenan, T.T.C.L.: 382 (1949). Type from Tanganyika (Usambara).

Ochna prunifolia Engl., Pflanzenw. Ost-Afr. **C**: 273 (1895).—Gilg, tom. cit.: 236 (1903).—Brenan, T.T.C.L.: 384 (1949). Syntypes from Kenya.

Ochna longipes Bak. in Kew Bull. **1897**: 247 (1897).—Burtt Davy & Hoyle, N.C.L.: 56 (1936).—Brenan in Mem. N.Y.Bot. Gard. **8**, 3: 234 (1953).—White, F.F.N.R.: 251 (1962). Type: Nyasaland, Mt. Malosa near Zomba, *Whyte* 429 (K).

Ochna shirensis Bak., loc. cit. Type: Nyasaland, Mt. Zomba and Mt. Malosa, *Whyte* 430 (K).

Ochna acutifolia Engl., Bot. Jahrb. **28**: 433 (1900). Type from Tanganyika (Uluguru).

Diporidium holstii (Engl.) Van Tiegh. in Ann. Sci. Nat., Sér. 8, Bot. **16**: 356 (1902). Type as for *Ochna holstii*.

Diporidium prunifolium (Engl.) Van Tiegh., loc. cit. Syntypes as for *Ochna prunifolia*.

Diporidium acutifolium (Engl.) Van Tiegh., loc. cit. Types as for *Ochna acutifolia*.

Ochna densicoma Engl. & Gilg in Engl., Bot. Jahrb. **33**: 241 (1903).—Brenan, T.T.C.L.: 382 (1949). Type from Tanganyika (Usambara).

Ochnella densicoma (Engl. & Gilg) Van Tiegh., op. cit., Sér. 8, **18**: 40 (1903). Type as for *Ochna densicoma*.

Biramella holstii (Engl.) Van Tiegh., tom. cit.: 41 (1903). Type as for *Ochna holstii*.

Biramella acutifolia (Engl.) Van Tiegh., tom. cit.: 42 (1903). Type as for *Ochna acutifolia*.

Ochna chirindica Bak. f. in Journ. Linn. Soc., Bot. **40**: 37 (1911).—Eyles in Trans. Roy. Soc. S. Afr. **5**: 419 (1916). Type: S. Rhodesia, Chirinda Forest, *Swynnerton* 106 (BM, holotype; K).

Ochna sp.—White, tom. cit.: 252 (1962) pro parte quoad specim. *White* 2806.

Tree 3–20 m. high (to 27 m. in East Africa), with bark smooth, grey or grey-brown; branches ± quadrangular, purplish-brown and glabrous or very rarely papillose-puberulous at first, becoming striate or slightly ridged and glabrous, with bark not exfoliating, with numerous small lenticels often somewhat elongated laterally. Leaves petiolate; lamina (3)5–12·2 × (1·2)1·5–3·9(4·3) cm., oblanceolate or obovate to elliptic or oblong, acute or acutely to obtusely acuminate at the apex, with margin densely curved-serrulate, cuneate at the base (more rarely narrowed to a rounded base), herbaceous to chartaceous, with numerous widely spreading lateral nerves (usually at almost 90° to the midrib) and densely reticulate tertiary venation prominent above (with main laterals rather more prominent than subsidiaries) and smooth below; petiole 1–3(3·5) mm. long, rather slender, grooved above. Flowers (5)7–14(20) in a ± elongated raceme, simple or very rarely branched at the base, with rhachis up to 20 mm. long, rarely pseudumbellate; pedicels 1·3–3·5(4·1) cm. long in fruit, articulated in the lower $\frac{1}{6}$ or at the base, glabrous or rarely papillose-puberulous. Sepals 6–9 mm. long in flower, oblong-elliptic, rounded, becoming pinkish-red to deep-red, 9–15(18) mm. long, flat and spreading in fruit. Petals pale to bright yellow, (8)10–12 × 3–6 mm., obovate, narrowed to a short claw. Stamens with anthers (1)1·5–2 mm. long, c. $\frac{1}{2}$ as long as the filaments, straight, dehiscing by longitudinal slits. Carpels 5(6), with styles completely united (or very rarely slightly free at the apex); stigma 5(6)-lobed or subglobose. Drupelets (8)9–12(14) × (5)6–7(9) mm., ± flattened-cylindric, rarely ovoid or subglobose, inserted at or near the base; embryo straight.

N. Rhodesia. N: Sunza Mt., 2000 m., fr. 9.i.1955, *Richards* 3963 (K). E: Nyika Plateau, Kangampande Mt., 2135 m., st. 8.v.1952, *White* 2793 (FHO; K). **S. Rhodesia.** W: Matopos, 1370 m., fr. 10.xi.1902, *Eyles* 1184 (BM; K; SRGH). C: Rusape, Mt. Dombo, fr. 31.xii.1950, *Munch* 361 (K; SRGH). E: Chimanimani Mts., 2100 m., fl. 26.ix.1906, *Swynnerton* 627 (BM; K). S: Victoria, 1200 m., *Mainwaring* in *Eyles* 2787

(PRE; SRGH). **Nyasaland.** N: Mzimba Distr., Mt. Hora, fr. 11.i.1950, *Krippner & Smith* N. 10 (BM); SE. slope of Mafinga Mts., 200 m., fr. 10.xi.1958, *Robson* 546 (BM; K; LISC; PRE; SRGH). C: Dedza, Chongoni Forest, fl. 8.xii.1960, *Chapman* 1081 (K; SRGH). S: Cholo Mt., fl. 20.ix.1946, *Brass* 17668 (BM; K; SRGH). **Mozambique.** Z: Serra do Gumi, fl. 18.x.1960, *Barbosa & Carvalho* 4485 (LMJ). MS: Serra da Gorongosa, Gogogo, fl. 29.ix.1943, *Torre* 5988 (LISC).

From the Sudan, Uganda, Congo (Kivu) and Kenya to Natal, Transvaal and NE. Cape Prov. (Tembuland). Evergreen rain-forest and in drier areas among rocks, 1160–2135 m. (915–2375 m. in East Africa).

O. holstii is a very variable species but one which defies attempts at subdivision. Forms with large leaves and long pedicels occur in scattered localities, e.g. Imatong Mts., Mt. Elgon, Usambaras, Mahali Mts. and the Chirinda-Vumba region; but they intergrade with the typical form and make it impossible to recognise *O. densicoma* as a distinct species. Forms with small leaves having an obtusely acuminate apex and a truncate or rounded base occur in drier habitats from S. Tanganyika to the Soutpansberg (Transvaal). These often have densely puberulous young shoots and inflorescences and whitish branches. Although they may appear to form a distinct taxon in some areas (e.g. S. Tanganyika and the Transvaal), completely intergrading forms occur in some intermediate regions (e.g. the Inyanga area).

Robson 546 is almost intermediate between *O. afzelii* and the *densicoma* type of *O. holstii*.

18. **Ochna oxyphylla** N. Robson in Bol. Soc. Brot., Sér, 2, **36**: 21 (1962). Type from Tanganyika (Uluguru).

Shrub or small tree (1)2–8 m. high, with bark fissured or rough, greyish- to reddish-brown; branches slender, ribbed, reddish-brown and puberulous or more rarely glabrous at first, becoming terete or somewhat fluted, purplish-brown, with numerous pale ± prominent lenticels. Leaves petiolate or subsessile, drying green or brown, occasionally with blue-green tinges; lamina 2–5·2(6·3) × 1–2 cm., elliptic to oblanceolate, acute or very acutely acuminate at the apex, with margin densely spinulose-curved-serrulate, rounded or more rarely broadly cuneate at the base, chartaceous, with numerous widely spreading lateral nerves and densely reticulate tertiary venation prominent above (with the laterals more prominent than the reticulation or not) and smooth below; petiole up to 1·5 mm. long or almost absent, grooved above. Flowers 1–3(5) in a condensed simple raceme with rhachis up to 2 mm. long or subumbellate; pedicels 1·1–2·4 cm. long in fruit, articulated at the base (or the terminal one within 2 mm. of the base), puberulous or glabrous, slender, not becoming reflexed. Sepals 5–7 mm. long in flower, oblong-elliptic, becoming deep pink or red, 9–11 mm. long, flat and spreading in fruit. Petals yellow, 7–8(10) × 3–4·5 mm., narrowly obovate. Stamens with anthers 1·3–2 mm. long, $\frac{3}{4}$–1$\frac{1}{4}$ times as long as the filaments, straight, dehiscing by longitudinal slits. Carpels 5, with styles completely united; stigma 5-lobed or subglobose. Drupelets 6–7 × 4 mm. (? immature), ovoid-cylindric, inserted at the base; embryo straight.

Nyasaland. N: Mafinga Mts., above Chisenga, fr. 22.xi.1952, *Angus* 839 (FHO; K).

Also in Tanganyika (Uluguru, Iringa and Mbeya Districts). In upland rain forest or fringing forest, c. 2000 m. (1650–2400 m. in Tanganyika).

An upland forest derivative of *O. holstii*, from which it differs essentially in size and numbers of parts.

19. **Ochna stolzii** Gilg ex Engl., Pflanzenw. Afr., **3**, 2: 480 (1921) in obs.—Gilg in Engl. & Prantl, Nat. Pflanzenfam., ed. 2, **21**: 68 (1925).—Sleumer in Notizbl. Bot. Gart. Berl. **12**: 69 (1934). Type from Tanganyika (Kyimbila).

Shrublet or shrub up to 1·2(2) m. high, bushy, with bark smooth, grey-brown; branches ascending, reddish-brown and puberulous at first, becoming terete and grey-brown with numerous pale lenticels. Leaves petiolate, drying ± metallic bluish-green; lamina 1·2–3·5 × 0·6–1·8 cm., elliptic to obovate, acute to obtuse or rounded at the apex, with margin serrulate, rounded or broadly cuneate at the base, chartaceous, with lateral nerves and reticulate tertiary venation prominent on both sides (usually less so below); petiole up to 1·5 mm. long or almost absent, grooved above. Flowers 1–3, fascicled, or rarely up to 7 in a simple raceme with rhachis up to 8 mm. long, scented; pedicels 0·8–1·5 cm. long in fruit, articulated within 1 mm. of the base (except the terminal one), puberulous or glabrous, becoming reflexed. Sepals 6 mm. long in flower, oblong-elliptic, becoming crimson-red, 8–10 mm. long, flat and spreading in fruit. Petals yellow, 9–10 × 4–5·5 mm., obovate, narrowed to a short claw. Stamens with anthers c. 1·5 mm. long, about $\frac{3}{4}$ as long as the filaments, straight, dehiscing by longitudinal slits. Carpels 5, with styles completely united; stigma subglobose. Drupelets 7–8 × 4–5 mm., ovoid-cylindric, inserted at the base; embryo straight.

Nyasaland. N: Nyika Mts., 2100–2400 m., fr. xi.1932, *Sanderson* 69 (BM).
Also in Tanganyika (Iringa, Kyimbila and Lupembe Districts). Hillside grassland and rain-forest remnants, 1500–2400 m.

O. stolzii appears to be related to *O. holstii*. In S. Tanganyika it may become very similar to *O. oxyphylla*, but its metallic-bluish-green-tinged leaves (usually obtuse to rounded at the apex) and recurved pedicels appear to be diagnostic.
Although the type specimen of *O. stolzii* (*Stolz* 2212) has been destroyed, Sleumer's extended description fits the material from Nyasaland and Tanganyika well, apart from the latter's smaller leaves. I have therefore identified our plant with *O. stolzii*; but a definite judgement on this matter will not be possible unless or until any isotypes or isopleths come to light.

20. **Ochna afzelioides** N. Robson in Bol. Soc. Brot., Sér. 2, **36**: 23 (1962). Type from Tan ganyika (Kigoma Distr.).

Shrub or tree up to 6 m. high, with bark grey, smooth; branches striate, lenticellate (but sometimes not in the first year), purplish- or yellow-brown, papillose-puberulous or glabrous at first, later always glabrous, eventually exfoliating in small scales or not. Leaves petiolate; lamina 4·5–9 × (1·8)2–2·9(3·3) cm., oblanceolate or rarely elliptic or obovate-oblanceolate, acute or acuminate at the apex, with margin densely spinulose-serrulate or curved-serrulate, cuneate to rounded at the base, herbaceous, with numerous widely spreading lateral nerves and densely reticulate tertiary venation prominent above (with main laterals more prominent than subsidiaries) and almost smooth below; petiole 1·5–2 mm. long, rather slender, grooved above. Flowers (5)7–10 in a ± elongated simple raceme with rhachis 9–20 mm. long; pedicels 1·4–2·6 cm. long in fruit, articulated in the lower $\frac{1}{4}$ (c. 1–5 mm. from the base), papillose-puberulous or glabrous. Sepals 5–7 mm. long in flower, elliptic, rounded, becoming scarlet, 7–9 mm. long, flat or somewhat convex and reflexed or rarely ascending in fruit. Petals pale yellow, c. 8–10 × 5–6 mm., obovate, narrowing to a short claw. Stamens with anthers 1·5 mm. long, about $\frac{1}{2}$ as long as the filaments, straight, dehiscing by longitudinal slits. Carpels 5–8, with styles completely united; stigma subglobose. Drupelets 6–8 × 4–5 mm., ovoid-cylindric, inserted below the centre; embryo somewhat curved.

N. Rhodesia. N: Abercorn Distr., top of Kambole Escarpment, 1650 m., fr. 1.ii.1959, *Richards* 10835 (K). W: Ndola Distr., Chichele Botanical Reserve, fr. 7.xii.1952, *White* 3827 (K). S: Pemba to Choma, km. 22·7, fr. 22.i.1960, *White* 6337 (FHO).
Also in W. Tanganyika (Kigoma and Bukoba Districts) and Katanga (Elisabethville). In deciduous woodland and evergreen forest edges, 1200–1650 m.

A species related to *O. holstii*, and *O. puberula* and probably *O. afzelii*. The Rhodesian and Congo specimens have ± densely papillose-puberulous young shoots and (sometimes) inflorescences, whereas those of the Tanganyikan specimens are glabrous.

21. **Ochna polyneura** Gilg in Engl., Bot. Jahrb. **33**: 240 (1903).—Brenan, T.T.C.L.: 383 (1949). Syntypes from SE. Tanganyika.
Ochna hylophila Gilg, tom. cit.: 242 (1903).—Brenan, tom. cit.: 382 (1949). Type from Tanganyika (Songea Distr.).

Shrublet or small tree, 0·3–6(8) m. high, with bark pale grey, rough and vertically fissured; branches smooth (rarely striate), reddish- to purplish-brown and sometimes papillose-puberulous at first, glabrous and whitish, with bark sometimes exfoliating in thin strips, usually with numerous rather faint lenticels. Leaves petiolate, bronze-tinged, drying bluish-green; lamina (4·3)5·5–10(12) × 1·6–3·4(3·8) cm., obovate to oblanceolate, obtuse to rounded (rarely very shortly and acutely acuminate) at the apex, with margin densely shallowly serrulate, rounded or truncate at the base, herbaceous, with lateral nerves numerous and widely spreading (usually at almost 90° to the midrib) and densely reticulate tertiary venation prominent above (with main laterals more prominent than subsidiaries) and almost smooth below; petiole 1–2 mm. long, rather stout. Flowers (5)6–10(14) in a ± elongated simple raceme with rhachis 3–25 mm. long, rarely subumbellate, primrose-scented; pedicels (1)1·8–2·8(3·3) cm. long in fruit, papillose-puberulous or glabrous, articulated usually in the lower ⅓ or ¼ (at least 1 mm. above the base). Sepals 5–6 mm. long in flower, oblong-elliptic, rounded, bronze-tinged, becoming pinkish-red to deep red, 10–12(15) mm. and spreading in fruit. Petals pale yellow, 9–13 × 5–8 mm., with obovate to subcircular limb c. twice as long as the claw. Stamens with anthers 1–2 mm. long, about ½ as long as the filaments, straight, dehiscing by longitudinal slits. Carpels 5(6–7) with styles completely united; stigma subglobose. Drupelets (7)8·5–10 × 5–6 mm., curved-cylindric to ovoid, inserted near the base; embryo straight.

S. Rhodesia. E: Umtali, Hondi Valley, fr. 21.xii.1948, *Chase* 1258 (BM; SRGH). **Nyasaland.** S: without precise locality, fr. 1895, *Buchanan* 47 (BM). **Mozambique.** N: Maniamba, fr. 20.xi.1934, *Gomes e Sousa* 599 (COI; LISC). T: Moatize, Zóbuè Mt., fr. 3.x.1942, *Mendonça* 618 (LISC). MS: Chimoio, Bandula, fl. xi.1923, *Honey* 778 (K; PRE).

Also in SE. Tanganyika. Open deciduous woodland, 500–1065 m.

A striking species characterised by smooth reddish-brown branches and bronze-tinged foliage and sepals. Towards the southern end of its range, glabrous forms appear which tend to approach *O. leptoclada* (apparently a derivative of *O. polyneura*). One of these (*Shinn* 191, from Rua, Mlanje) is almost exactly intermediate between these species, having the bluish-green foliage and pedicels articulated well above the base as in *O. polyneura*, but with the leaves glaucous and attenuate at the base and the stems whitish as in *O. leptoclada*.

22. **Ochna leptoclada** Oliv., F.T.A. **1**: 318 (1868).—Gilg in Engl., Bot. Jahrb. **33**: 233 (1903).—Bak. f. in Journ. Linn. Soc., Bot. **40**: 36 (1911).—Eyles in Trans. Roy. Soc. S. Afr. **5**: 419 (1916).—Burtt Davy & Hoyle, N.C.L.: 56 (1936).—Brenan, T.T.C.L.: 383 (1949); in Mem. N.Y.Bot. Gard. **8**, 3: 234 (1953).—White, F.F.N.R.: 251, fig. 43F (1962). Syntypes: Nyasaland, Manganja Hills, Margomero, *Meller* (K); Maravi country, 12° S; 34° E, *Kirk* (K).

Ochnella leptoclada (Oliv.) Van Tiegh. in Ann. Sci. Nat., Sér. 8, Bot. **16**: 344 (1902). Syntypes as above.

Ochna debeerstii De Wild. in Ann. Mus. Cong., Bot. Sér. 4, **1**: 88 (Jan. 1903); in Rev. Zool. Afr. **7,** Suppl. Bot.: 32 (1919).—Gilg, tom. cit.: 237 (Mar. 1903) pro parte quoad specim. cong. Syntypes from the Congo.

Ochnella debeerstii (De Wild.) Van Tiegh., op. cit., Sér. 8, Bot. **18**: 40 (1903). Syntypes as for *Ochna debeerstii.*

Shrub or rhizomatous shrublet up to 1(1·3) m. high (or sometimes a small tree?), with brown bark; shoots erect, ± branched, often caespitose, smooth, terete or angular, greyish-white, not lenticellate, with epidermis exfoliating in papyraceous strips. Leaves petiolate, glaucous, rarely bluish-green or metallic-tinged; lamina (4)6·5–12 × (1·2)2–4·5 cm., obovate to oblanceolate or rarely oblong-elliptic, rounded (more rarely obtuse or apiculate) at the apex, with margin entire or remotely spinulose-serrulate, attenuate and recurved at the base, chartaceous to subcoriaceous, with ± widely spreading lateral nerves and densely reticulate tertiary venation scarcely prominent above (with main and subsidiary laterals equally prominent) but not below; petiole 2·5–4 mm. long, rather stout, flattened above. Flowers 1–3(4), pseudumbellate, the pseudumbels usually forming narrowly cylindric or conic compound panicles; pedicels 0·9–2(2·8) cm. long in fruit, articulated at the base. Sepals 3–5 mm. long in flower, oblong-elliptic, rounded, soon turning reddish, becoming crimson, 9–13(15) mm. long and spreading in fruit. Petals bright yellow, 6–8(9) × 4–5(6) mm., obovate to elliptic, narrowed towards the base or shortly unguiculate. Stamens with anthers 1–2 mm. long, ⅔ as long to as long as the filaments, straight, dehiscing by longitudinal slits. Carpels 5, with styles completely united; stigma globose. Drupelets 6–8 × 5–6 mm., subglobose, inserted at the base; embryo straight.

N. Rhodesia. B: Kataba, fr. 30.x.1961, *Fanshawe* 6747 (K). N: Lake Mweru, near Kafulwe Mission, fr. 6.xi.1952, *White* 3630 (FHO; K). W: Ndola, fr. 16.xii.1955, *Fanshawe* 2659 (K). C: between Broken Hill and Bwana Mkubwa, fl. x.1906, *Allen* 299 (K; SRGH). E: Lundazi to Chama, 100 km., 1350 m., fl. 19.x.1958, *Robson* 172 (BM; K; LISC; PRE; SRGH). S: Mapanza, Choma, fr. 28.xi.1957, *Robinson* 2512 (K; PRE; SRGH). **S. Rhodesia.** N: Urungwe Reserve, Msukwe R., 1200 m., fr. xi.1956, *Davies* 2245 (K; SRGH). C: Salisbury, 1440 m., fr. 17.xii.1926, *Eyles* 4577 (K; SRGH). E: Inyanga Distr., 29 km. N. of Troutbeck, 1000 m., fl. 20.ix.1959, *Leach* 9422 (K; SRGH). **Nyasaland.** N: Stevenson Road, 1500–1800 m., fl. 1893–4, *Scott Elliot* 8257

(BM; K). C: Kota Kota Distr., Chia area, fr. 1.ix.1946, *Brass* 17477 (BM; K; SRGH). S: Road to L. Shirwa, fl. 14.x.1955, *Wiehe* N/655 (K; SRGH). **Mozambique.** N: Amaramba, Mandimba to Vila Cabral, 13 km., fl. 8.x.1942, *Mendonça* 653 (LISC). Z: between Mocuba and Namarroi, fr. 23.ix.1941, *Torre* 3465 (LISC). T: Maràvia, between Chicoa and Fíngoè, fr. 24.ix.1942, *Mendonça* 383 (LISC). LM: Manhiça, fr., *Gomes e Sousa* 53 (LISC).

Also in W. Tanganyika, E. Congo and the Sudan. Deciduous woodland on sandy soil, 480–1800 m.

O. leptoclada is quite distinct from *O. schweinfurthiana*, but young specimens of the latter may be easily confused with it. The leaves of *O. leptoclada* tend to be almost entire and glaucous, whereas those of *O. schweinfurthiana* are densely serrulate and not usually glaucous except when young, but these differences do not appear to be constant. Mr. Milne-Redhead has noticed another difference in the field, viz. that the fruits of young *O. schweinfurthiana* become pendulous whereas those of *O. leptoclada* remain erect. Only by further observation in the field can the constancy of these distinguishing features be established.

The specimen from Inyanga (*Leach* 9422) is said to be a tree 4 m. high and the Salisbury specimen (*Eyles* 4577) has narrowly elliptic leaves. The plant from Manhiça (*Gomes e Sousa* 53) was collected well to the south of the nearest S. Rhodesian and Mozambique localities, but is otherwise quite typical.

23. **Ochna puberula** N. Robson in Bol. Soc. Brot., Sér. 2, **36**: 25 (1962). FRONTISP. Type: Abercorn Distr., Kawimbe, *Richards* 10235 (K).
Ochna longipes sensu Nordlindh in Bot. Notis. **1948**: 32 (1948).
Ochna sp. aff. O. welwitschii sensu Suesseng. in Proc. & Trans. Rhod. Sci. Ass. **43**: 88 (1951).
Ochna sp.—White, F.F.N.R.: 252 (1962) pro parte excl. specim. *White* 2806.

Shrub or small tree 0·5–7·5 m. high (? or higher), with bark grey, smooth or ± reticulately fissured; branches ± quadrangular, reddish-brown and ± densely papillose-puberulous at first, becoming striate or shallowly fissured, with numerous raised brownish lenticels. Leaves petiolate, usually drying dark bluish-green; lamina (2·2)3·6–6·5(7·5) × 1–2·7(2·9) cm., obovate to oblanceolate or more rarely elliptic or elliptic-oblong, obtuse or obtusely acuminate to rounded (rarely acute) at the apex, with margin densely curved-serrulate, cuneate at the base or narrowed to a rounded or truncate base, herbaceous to chartaceous, rarely ± glaucous below, with numerous widely spreading lateral nerves and densely reticulate tertiary venation prominent above (with main and subsidiary laterals almost equally prominent) and less prominent or almost smooth below; petiole (0·5)1–1·5(2) mm. long, rather slender, grooved above. Flowers 2–8, in a ± condensed pseudumbellate raceme; pedicels 1–2·7(3·5) cm. long in fruit, articulated in the lower $\frac{1}{6}$ (except sometimes the terminal one) or at the base, papillose-puberulous. Sepals 3–6(7) mm. long in flower, oblong-elliptic, rounded, drying dark bluish-green, enclosing the developing drupelets and becoming orange-red to crimson, 6–10(14) mm. long, convex and eventually spreading in fruit. Petals bright yellow, (5)7–13 × 3–7·5 mm., obovate, narrowing to a short claw. Stamens drying dark bluish-green, with anthers 1–1·5 mm. long, $\frac{1}{3}$–$\frac{1}{2}$ as long as the filaments, straight, dehiscing by longitudinal slits. Carpels 5–7, with styles completely united (or very rarely free at the apex); stigma 5–7-lobed or subglobose. Drupelets 8–10 × 5–6 mm., ovoid-cylindric, inserted at or near the base; embryo straight.

N. Rhodesia. N: Abercorn Distr., Kawimbe, 1740 m., fl. & fr. 27.xi.1958, *Richards* 10235 (K). W: Kitwe, fl. 16.xi.1955, *Fanshawe* 2610 (K; SRGH). E: Fort Jameson Distr., Mfumu, Asamfa, 900 m., 6.i.1959, *Robson* 1052 (BM; K; LISC; PRE; SRGH). S: Mazabuka Distr., Siamambo Forest Reserve , near Choma, fr. 13.xii.1952, *Angus* 939 (FHO; K). **S. Rhodesia.** N: Sebungwe Distr., Kariangwe, fr. 20.xi.1951, *Lovemore* 196 (K; SRGH). W: Matobo, Farm Chesterfield, ± 1465 m., fl. x.1958, *Miller* 5469 (K; SRGH). C: Rusape, fl. & fr. xi.1952, *Dehn* 468/53 (K; SRGH). E: Melsetter, Rocklands, The View, ± 1370 m., fl. 7.x.1950, *Sturgeon* in GHS 30492 (K; SRGH).

Also in central and south Tanganyika. Deciduous woodland and wooded grassland, often among rocks, 800–1740 m. (to 2135 m. in East Africa).

The combination of puberulous pseudumbellate inflorescence and sepals enclosing the developing drupelets and then spreading will usually differentiate *O. puberula* from its near relatives. In eastern S. Rhodesia, however, there are forms in which the inflorescence tends to be more elongated and glabrous. These may be hybrids with *O. holstii*.

24. **Ochna cyanophylla** N. Robson in Bol. Soc. Brot., Sér. 2, **36**: 27 (1962). Type: S. Rhodesia, Mazoe Dam, *Stables* 14/56 (K, holotype; SRGH).

Ochna welwitschii sensu Nordlindh in Bot. Notis. **1948**: 33 (1948).
Ochna mechowiana sensu Palgrave, Trees of Central Afr.: 295, tt. (1956).

Small tree 3·5–6 m. high, with bark light brown, rather rough; branches striate to shallowly ribbed, yellow-brown with raised lenticels at first, later with epidermis exfoliating in thin papery strips. Leaves petiolate, dark bluish-green when dried; lamina 5·5–10·6 × 2·3–3·8(4·3) cm., obovate or more rarely oblanceolate or oblong-elliptic, obtuse to rounded at the apex, with margin densely crenulate-serrulate, cuneate or rarely truncate at the base, subcoriaceous with ± widely spreading lateral nerves and densely reticulate tertiary venation more prominent above than below, the main and subsidiary laterals almost equally prominent; petiole 1–2 mm. long, stout, grooved above. Flowers (1)3–7, in a condensed raceme with rhachis up to 5 mm. long or subumbellate, cowslip-scented; pedicels 1·3–2·1(2·4) cm. long in fruit, articulated within 1 mm. of the base. Sepals 4–6 mm. long in flower, oblong-elliptic, rounded, becoming deep cherry-red, 8–11(13) mm. long, convex and spreading in fruit. Petals deep yellow, 7–11 × 5–7 mm., obovate, narrowed to a short claw. Stamens with anthers 1–1·5 mm. long, $\frac{1}{3}$–$\frac{1}{2}$ as long as the filaments, straight, dehiscing by longitudinal slits. Carpels 6–8, with styles completely united; stigma subglobose. Drupelets flattened-cylindric, inserted at the base, 9–10(13) × 6·5–8 mm.; embryo slightly curved.

S. Rhodesia. N: Mazoe Dam, fr. i.1956, *Stables* 14/56 (FHO; K; SRGH). C: Salisbury, 1465 m., fr. xi.1919, *Eyles* 1843 b (K; PRE; SRGH). E: Inyanga, 1700 m., fr. 27.xi.1930, *F.N. & W.* 3187 (BM; PRE). S: Victoria Distr., fl. 1908, *Monro* 2159 (BM; K).

Also in south and central Tanganyika. Deciduous woodland, 1465–1700 m.

O. cyanophylla is superficially similar to *O. schweinfurthiana*, but can easily be distinguished from it by the dark green or bluish-green leaves, the number of carpels and the dark cherry-red colour of the sepals in fruit. It has also been confused with *O. afzelii* subsp. *mechowiana* which has longer petioles, leaves which do not turn bluish-green, smaller fruits, etc.

25. **Ochna katangensis** De Wild. in Ann. Mus. Cong., Bot. Sér. 4, **1**: 89, t. 33 figs. 5–6 (1903).—Gilg in Engl., Bot. Jahrb. **33**: 232 (1903). Type from the Congo (Katanga).
Ochna humilis Engl., Bot. Jahrb. **30**: 354 (1901) non Kuntze (1891).—Gilg in Engl., op. cit. **33**: 233 (1903).—Brenan, T.T.C.L.: 382 (1949). Type from SW. Tanganyika.
Ochnella humilis (Engl.) Van Tiegh. in Ann. Sci. Nat., Sér. 8, Bot. **16**: 344 (1902). Types as for *Ochna humilis*.
Campylochnella katangensis (De Wild.) Van Tiegh., op. cit. **18**: 60 (1903). Type as for *Ochna katangensis*.
Ochna hockii De Wild. in Rev. Zool. Afr. **7**, Suppl. Bot.: 33 (1919).—White, F.F.N.R.: 251, fig. 43J (1962). Type from the Congo (Katanga).
Ochna gracilipes sensu Exell & Mendonça, C.F.A. **1**, 2: 286 (1951) pro parte quoad. specim., *Gossweiler* 3035.

Shrublet up to c. 15 cm. high, but sometimes flowering at ground level, usually forming low cushions from a polycephalous woody rhizome with ± smooth or gnarled brown bark; branches brown, densely pustulose-lenticellate. Leaves petiolate, rarely somewhat glaucous; lamina (3·2)4·2–11(12) × 0·9–2·2(2·8) cm., narrowly elliptic or narrowly oblanceolate to linear-oblong, ± acute and shortly apiculate at the apex, with margin usually rather remotely curved-spinose-serrate, narrowly cuneate at the base, chartaceous, with main lateral nerves (especially the basal ones) ascending gradually to the margin and the ± densely reticulate reddish tertiary venation equally prominent on both sides or less prominent below; petiole c. 2 mm. long, rather stout; stipules c. 3–7 mm. long, linear-triangular, entire or ± divided, usually tardily deciduous. Flowers solitary or 2–5 in an umbellate or shortly racemose inflorescence, axillary on second-year shoots; pedicels 1·5–3·3 cm. long in fruit, articulated at (or the terminal one rarely up to 2 mm. above) the base. Sepals 5–8(10) mm. long in flower, elliptic, rounded, becoming carmine, 1·1–1·5 mm. long, ± convex and erect or ± spreading in fruit. Petals bright yellow to orange-yellow, 12–13(20) × 7–8(13) mm., obovate and unguiculate. Stamens with anthers (1·75)2–3 mm. long, $\frac{1}{3}$–$\frac{2}{3}$ as long as the filaments, straight, dehiscing by longitudinal slits. Carpels 5, with styles completely united and 5-lobed or globose stigma (rarely with free apices and separate stigmas). Drupelets

6–6·5(10) × 5–5·5(7) mm., subglobose (rarely flattened-obovoid), inserted near the base; embryo slightly curved.

N. Rhodesia. N: Abercorn Distr., Old Katwe Road, 1525 m., fl. & fr. 25.xi.1954, *Richards* 2345 (K). W: Mwinilunga Distr., 26 km. W. of Kabompo, fr. 11.ix.1930, *Milne-Redhead* 1103A (K). **Nyasaland.** N: Nyika Plateau, path from Rest House to waterfall on N. Rukuru R., 1850 m., fr. 29.x.1958, *Robson* 473 (BM; K; LISC; PRE; SRGH).

Also in Angola (Bié), the Congo (Katanga) and southern Tanganyika. In grassland and open woodland subjected to burning, 1500–2150 m.

The large deep-coloured flowers and relatively large anthers of this species are characteristic, as are the cushion-like habit and the long acute leaves with elongated lateral nerves. The Nyika Plateau specimen differs from the other ones examined in having larger fruits and bluish-green-tinged leaves with more densely serrate leaf margins.

26. **Ochna richardsiae** N. Robson in Bol. Soc. Brot., Sér. 2, **36**: 29 (1962). TAB. **45.** Type: N. Rhodesia, Abercorn Distr., Ballymain, *Richards* 7017 (K).

Shrublet up to 0·6 m. high, with brown bark; shoots erect, little-branched, shallowly striate, yellowish-white, densely lenticellate but not markedly pustulose, papillose-puberulous (at least in part). Leaves petiolate; lamina (6)8–12 × (1)1·2–2 cm., narrowly elliptic or narrowly oblong-elliptic, obtuse to subacute at the apex, with margin ± densely spinulose-serrulate, attenuate to cuneate or truncate at the base, herbaceous to chartaceous, with spreading or somewhat ascending lateral nerves (main and subsidiary almost equally prominent) and densely reticulate tertiary venation more prominent above than below; petiole c. 1 mm. long, stout, grooved. Flowers (1)2–5(6) in a condensed racemose or pseudumbellate axillary inflorescence; pedicels 1–2 cm. long in fruit, articulated at the base or within 1 mm. of it, glabrous or very sparsely puberulous. Sepals 5–8 mm. long in flower, elliptic or oblong, rounded, becoming crimson, 9–13 mm. long, slightly convex and accrescent or ± spreading in fruit. Petals orange to orange-yellow, 8–15 × 7–10(14) mm., rhombic-subcircular, unguiculate. Stamens with anthers 2·5 mm. long, $\frac{2}{3}$ as long to as long as the filaments, straight, dehiscing by longitudinal slits. Carpels 5, with styles completely united; stigma 5-lobed. Drupelets 5–7 × 4–5·5 mm., subglobose, inserted near the base; embryo slightly curved.

N. Rhodesia. N: Abercorn Distr., Lunzua valley, Ballymain, 1200 m., fl. 19.xi.1956, *Richards* 7017 (K). W: Mufulira, fr. 27.xii.1954, *Fanshawe* 1762 (K). C: Kapiri Mposhi, fr. 5.xii.1957, *Fanshawe* 4117 (FHO; K). S: Siamambo Forest Reserve, near Choma, st. vii.1952, *White* 3837 (FHO). **S. Rhodesia.** C: Domboshawa, fl. 5.x.1945, *Wild* 173 (SRGH).

Also in the Congo (Elisabethville). Deciduous woodland on sandy soil, c. 1200–1500 m.

Although *O. richardsiae* has large flowers and long narrow leaves like those of *O. humilis*, the habit and leaf venation of these two species are quite distinct.

A specimen from Macanga, near Furancungo, Tete (*Mendonça* 449), has similar larger anthers and may also belong here. The stem appears to be glabrous, however, and the material is insufficient to allow a more definite determination.

27. **Ochna gambleoides** N. Robson in Bol. Soc. Brot., Sér. 2, **36**: 30 (1962). Type: N. Rhodesia, Chadiza turn-off to Fort Jameson, 1·7 km., *Robson* 32 (BM; K, holotype; LISC; PRE; SRGH).

Small tree 3–7 m. high sometimes branching from the base, with slender trunk and silver-grey rectangularly fissured bark; branches stout, lenticellate, brown and ridged at first, becoming angular, whitish-brown and non-lenticellate. Leaves petiolate, ± glaucous; lamina 8·8–14·5 × 5–8·2 cm., oblong-elliptic to obovate, rounded or obtusely apiculate at the apex, with margin serrulate, broadly cuneate (more rarely subtruncate) to attenuate at the base, coriaceous, with widely spreading lateral nerves prominent on both sides (main more prominent than subsidiary) and densely reticulate tertiary venation more prominent above than below; petiole 5–20 mm. long, stout, flattened above. Flowers c. 7–16, in a ± condensed racemose inflorescence with rhachis up to c. 13 mm. long; pedicels 2·6–4 mm. long in fruit, articulated at the base. Flowering specimens not yet collected. Sepals orange-red to scarlet, 16–25 mm. long, convex and ascending in fruit. Stamens with anthers 2 mm. long when dry, dehiscing by longitudinal slits. Carpels 5(6), with styles completely united; stigma small, 5-lobed. Drupelets 10–13 × 8–10 mm., flattened-ovoid-subcylindric, inserted at the base; embryo straight.

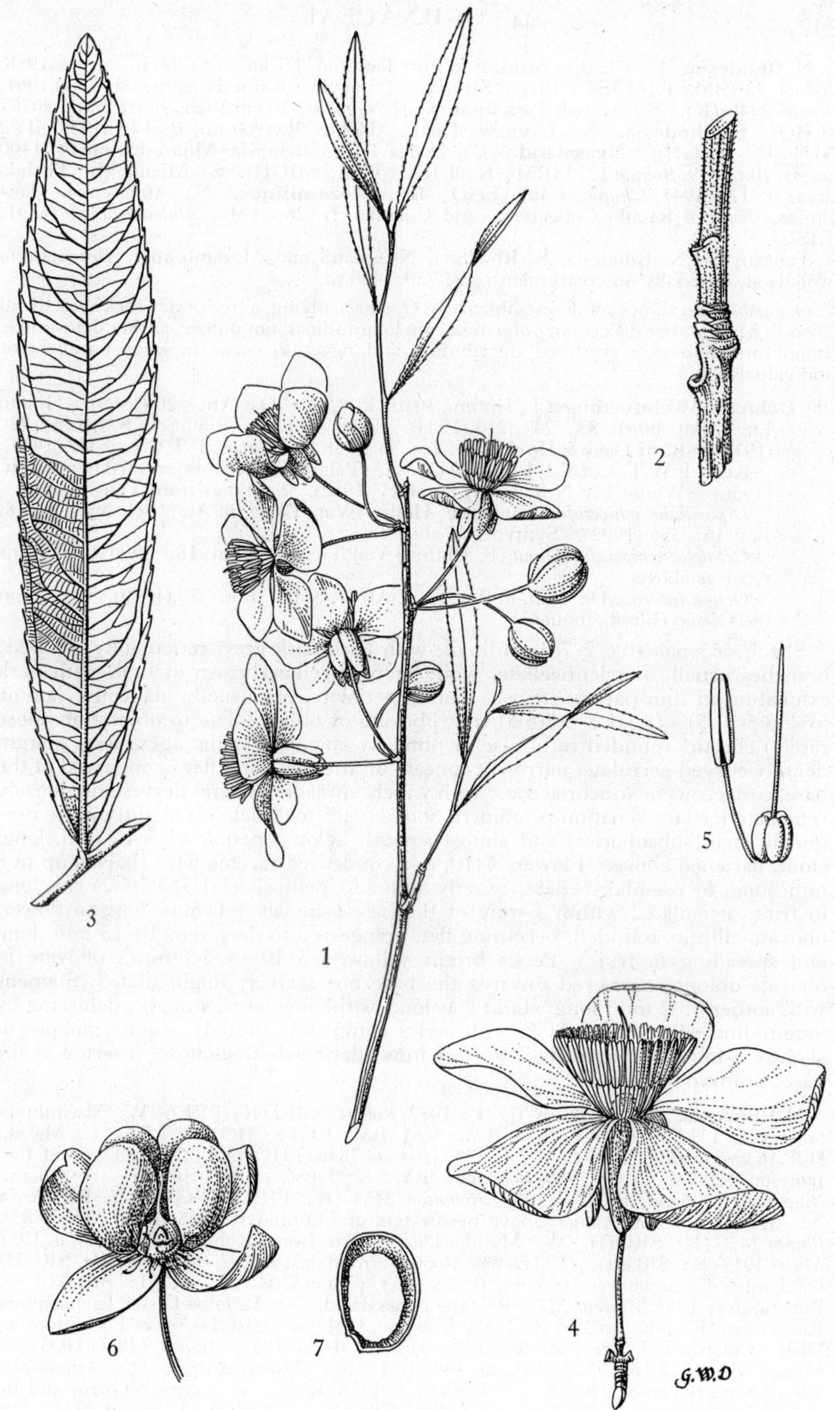

Tab. 45. OCHNA RICHARDSIAE. 1, flowering shoot (×1); 2, part of stem (×4), both from *Richards* 7017; 3, leaf (×1), *Fanshawe* 1762; 4, flower (×2); 5, stamen and gynoecium (×4), both from *Richards* 7017; 6, fruit (×4); 7, vertical section of drupelet (×4), both from *Fanshawe* 1762.

N. Rhodesia. E: Chadiza turn-off to Fort Jameson, 1·7 km., c. 1200 m., fr. 8.x.1958, *Robson* 32 (BM; K; LISC; PRE; SRGH). C? or S?: Kafue R. gorge, st. 25.vi.1961, *Angus* 2946 (K). S: Munali Pass, near Nega Nega, st. 20.viii.1928, *Burtt Davy* 20767 (FHO). **S. Rhodesia.** N: Urungwe Distr., Msukwe R., 900 m., fr. 17.xi.1953, *Wild* 4171 (K; SRGH). **Nyasaland.** C: Dedza Distr., Bembeke-Mua escarpment, 1400 m., fr. 20.i.1959, *Robson* 1273 (BM; K; LISC; PRE; SRGH). S: Mlanje Mt., Tuchila area, fr. 12.x.1947, *Chapman* 467 (FHO; K). **Mozambique.** N: Amaramba, Mandimba, between Régulo Congerenge and Cuamba, fr. 26.x.1948, *Andrada* 1450 (COI; LISC).

Tanganyika, N. Rhodesia, S. Rhodesia, Nyasaland and Mozambique. *Brachystegia* woodland, especially on escarpment soils, 750–1400 m.

O. gambleoides shows some resemblance to *O. gamblei* King, a tree of dry rocky woodland areas in Madras, but differs *inter alia* in having longitudinal, not porose, anther dehiscence. It appears to have a scattered distribution and never to occur in large numbers of individuals.

28. **Ochna schweinfurthiana** F. Hoffm., Beitr. Fl. Centr.-Ost-Afr.: 20 (1889).—Gilg in Engl., Bot. Jahrb. **33**: 234, 240 (1903).—Eyles in Trans. Roy. Soc. S. Afr. **5**: 420 (1915).—Burtt Davy & Hoyle, N.C.L.: 56 (1936).—Brenan, T.T.C.L.: 383 (1949). —Keay, F.W.T.A., ed. 2, **1**, 1: 223 (1954).—Palgrave, Trees of Central Afr.: 302, tt. (1956).—White, F.F.N.R.: 251, fig. 43H (1962). Syntypes from Tanganyika.

Diporidium schweinfurthianum (F. Hoffm.) Van Tiegh. in Ann. Sci. Nat., Sér. 8, Bot. **16**: 356 (1902). Syntypes as above.

Ochnella schweinfurthiana (F. Hoffm.) Van Tiegh., op. cit. **18**: 39 (1903). Syntypes as above.

Ochna suberosa De Wild. in Rev. Zool. Afr. **7**, Suppl. Bot.: 39 (1919). Type from the Congo (Elisabethville).

Shrub or small tree 2–7(9) m. high, with bark dark grey, reticulately fissured; branches usually not lenticellate, white or pale yellow-brown at first, with bark exfoliating in thin papery strips. Leaves petiolate, not usually glaucous; lamina 5·5–13·5(17·5) × (1·3)1·7–5·5(6·5) cm., obovate or oblanceolate to oblong or (more rarely) elliptic, rounded to obtuse or obtusely apiculate at the apex, with margin densely curved-serrulate, narrowly cuneate or attenuate and flat or incurved at the base, coriaceous or subcoriaceous, with widely spreading lateral nerves and densely reticulate tertiary venation prominent above (with main laterals slightly more prominent than subsidiaries) and almost smooth below; petiole (3)5–12 mm. long, stout, flattened above. Flowers 4–10, in a condensed raceme with rhachis up to 8 mm. long, or pseudumbellate, sweetly scented; pedicels (1)1·5–3·5(4·2) cm. long in fruit, articulated within 1 mm. of the base. Sepals 4–6 mm. long in flower, obovate-elliptic, rounded, becoming flat, orange-red to deep red, 10–15 mm. long and spreading in fruit. Petals bright yellow, 5·5–10 × 4–5·5 mm., obovate to obovate-oblong, narrowed towards the base but scarcely unguiculate. Stamens with anthers 1–2 mm. long, about ⅔ as long as the filaments, straight, dehiscing by longitudinal slits. Carpels 5, with styles completely united; stigma globose or slightly 5-lobed. Drupelets 7–9 × 6–7 mm., flattened-subglobose, inserted at the base; embryo straight.

N. Rhodesia. N: Abercorn, fr. 24.x.1952, *Robertson* 162 (K; PRE). W: Mwinilunga to Solwezi, 131 km., fr. 14.ix.1952, *White* 3261 (BM; COI; FHO; K; PRE). C: Makaka Hill, 16 km. north of Lusaka, fl. 8.ix.1947, *Brenan* 7836 (FHO; K). E: hill west of Fort Jameson, 1200 m., fr. 9.x.1958, *Robson* 35 (BM; K; LISC; PRE; SRGH). S: Choma, Mapanza, 1100 m., fr. 29.ix.1957, *Robinson* 2454 (K; PRE; SRGH). **S. Rhodesia.** N: Binga, Chizorira Range, above headwaters of Masumo R., 855 m., fr. 11.xi.1958, *Phipps* 1450 (K; SRGH). W: Matobo Distr., Farm Besna Kobila, 1370 m., fl. ix.1953, *Miller* 1919 (K; SRGH). C: Hartley, Poole Farm, fl. ix.1950, *Hornby* 3206 (K; SRGH). E: Umtali Commonage, 1095 m., fl. 2.x.1954, *Chase* 5304 (BM; COI; SRGH). S: Buhera, st. iv.1953, *Vincent* 222 (SRGH). **Nyasaland.** N: Mzimba Distr., Livingstonia, Kaziweziwe R., 1400 m., fr. 8.i.1959, *Robinson* 3111 (K; SRGH). C: Fort Manning Distr., Colansano Estate near Kambobo Hill, fr. 16.xii.1955, *Adlard* 138 (FHO). S: Mlanje Mt., fr. 27.ix.1957, *Chapman* 440 (BM; K). **Mozambique.** N: Amaramba, Freixo Nova (Cuamba), fr. 22.x.1948, *Andrada* 1438 (LMJ). Z: between Gúruè and Ile, 16 km. from Gúruè, fl. 19.x.1949, *Barbosa & Carvalho* 4522 (COI; LISC; LMJ). MS: Báruè, Vila Gouveia, fr. 1.xi.1941, *Torre* 3743 (LISC).

From Mali and Ghana to the Sudan and Uganda and southwards to S. Rhodesia and Mozambique. Also in Angola (Benguela). Deciduous woodland on sandy or stony soil, 750–1675 m. (to 2100 m. in Tanganyika).

Although very variable in shape and size, the leaves of *O. schweinfurthiana* are always narrowed into the petiole at the base and have a characteristic blunt serrulation. The size of the fruiting calyx and the smaller leaves usually without a glaucous bloom distinguish it from *O. gambleoides*. The almost constantly 5-carpellary ovary is a useful diagnostic character.

29. **Ochna afzelii** R.Br. ex Oliv., F.T.A. **1**: 319 (1868).—Gilg in Engl., Bot. Jahrb. **33**: 239 (1903).—Keay, F.W.T.A., ed. 2, **1**: 223 (1954).—White, F.F.N.R.: 251, fig. 43I (1962). Type from Sierra Leone.

Ochnella afzelii (R.Br. ex Oliv.) Van Tiegh. in Ann. Sci. Nat., Sér. 8, Bot. **16**: 345 (1902). Type as for *Ochna afzelii*.

Ochna rhodesica R.E.Fr., Wiss. Ergebn. Schwed. Rhod.-Kongo-Exped. **1**: 149, t. 12 figs. 1–2 (1914). Type: N. Rhodesia, Abercorn, S. end of Lake Tanganyika, *Fries* 1248 (UPS).

Ochna ituriensis De Wild. in Rev. Zool. Afr. **7**, Suppl. Bot.: B35 (1919). Type from the Congo (Ituri).

Shrub or tree (1·5)3–12 m. high, with bark pale grey-brown, flaking (? always); branches densely lenticellate, purplish- or dark brown (more rarely whitish), usually not exfoliating. Leaves petiolate; lamina 4–13 × 1·5–4·5 cm., oblanceolate to oblong-oblanceolate or rarely elliptic, rounded to shortly and ± obtusely acuminate at the apex, with margin densely curved-serrulate to crenulate, cuneate to attenuate at the base, herbaceous to chartaceous, with numerous widely spreading lateral nerves and densely reticulate tertiary venation prominent above (with main and subsidiary laterals almost equally prominent) but almost smooth below; petiole 2·5–5(8) mm. long, slender, grooved above. Flowers 2–6(8), in a condensed raceme with rhachis up to 4 mm. long, or pseudumbellate; pedicels 0·8–3·1(3·5) cm. long in fruit, articulated at or within 1 mm. of the base. Sepals 4–6(8) mm. long in flower, elliptic-oblong, rounded, becoming deep pink to scarlet, 6–14(18) mm. long, often ± convex below, and ± spreading in fruit. Petals white to lemon-yellow, 7–13 × 3–7 mm., obovate, scarcely unguiculate. Stamens with anthers bright orange, 1–1·5(2) mm. long, about ½ as long as the filaments, straight, dehiscing by longitudinal slits. Carpels (5)6–8, with styles completely united; stigma globose or slightly 5–8-lobed. Drupelets 6–8 × 4–6 mm., subglobose to subreniform, inserted below their centre; embryo somewhat curved.

From Guinea to the Sudan and Uganda and southwards to W. Tanganyika, the Congo, N. Rhodesia and Angola. In dry evergreen forest, fringing forest or plateau woodland, 900–1650 m.

The above range of *O. afzelii* includes those of *O. mechowiana* O. Hoffm. and *O. congoensis* (Van Tiegh.) Gilg, taxa which cannot be specifically separated from *O. afzelii* sens. str. Three subspecies can be recognised, however, one northern and eastern (subsp. *afzelii*), one south-central (subsp. *congoensis*), and one south-western with outliers in north-eastern N. Rhodesia and south-western Tanganyika (subsp. *mechowiana*).

Fruiting sepals 6–10 mm. long; petals white to pale yellow; leaves oblanceolate to oblong-elliptic:
- Leaves oblanceolate to oblong-oblanceolate; inflorescence-rhachis usually very short (but sometimes up to 4 mm. long) or absent; pedicels usually articulated at the base - - - subsp. *afzelii*
- Leaves elliptic to oblong (rarely oblong-oblanceolate); inflorescence-rhachis usually ± elongated, up to 5 mm. long; pedicels usually articulated above but within 1 mm. of the base - - - - - - - - - - - subsp. *congoensis*

Fruiting sepals (10)12–18 mm. long; petals lemon to bright yellow; leaves oblanceolate or rarely oblong; inflorescence-rhachis very short or absent; pedicels articulated at the base - - - - - - - - - - - - subsp. *mechowiana*

Subsp. **afzelii**

Inflorescence-rhachis very short or absent (more rarely up to 4 mm. long). Sepals in flower 4–6 mm. long, in fruit 6–10 mm. long. Petals white to pale lemon-yellow, 7–10 mm. long. Anthers 1–1·5 mm. long. Carpels (5)6–7(8). Drupelets 6–7 × 4–6 mm.

N. Rhodesia. N: Abercorn, L. Chila, 1550 m., fr. 6.xi.1958, *Robson* 502 (BM; K; LISC; PRE; SRGH). W: Ndola, fl. 29.x.1954, fr. 26.xii.1954, *Fanshawe* 1639 (FHO; K).

From Guinea to the Sudan and southwards to north-western Tanganyika, eastern Congo, N. Rhodesia and north-eastern Angola.

Subsp. **congoensis** (Van Tiegh.) N. Robson in Bol. Soc. Brot., Sér, 2, **36**: 32 (1962). Type from the Congo (Bas Congo).

Polyochnella congoensis Van Tiegh. in Ann. Sci. Nat., Sér. 8, Bot. **16**: 349 (1902). Type as above.

Ochna congoensis (Van Tiegh.) Gilg [ex Van Tiegh., loc. cit., *in synon.*] in Engl., Bot. Jahrb. **33**: 239 (1903). Type as above.

Ochna mechowiana sensu Exell & Mendonça, C.F.A. **1**, 2: 288 (1951) pro parte.

Inflorescence-rhachis usually ± elongated, up to 5 mm. long. Sepals in flower c. 4 mm. long, in fruit 6–10 mm. long. Petals white to pale yellow, 9–13 mm. long. Anthers 1–1·5 mm. long. Carpels (6)7–8. Drupelets c. 6 × 5 mm.

N. Rhodesia. W: Mwinilunga, fl. 1951, *Dunning* 28 (FHO).
In the Congo (Bas Congo), N. Rhodesia and Angola.

Subsp. **mechowiana** (O. Hoffm.) N. Robson in Bol. Soc. Brot., Sér. 2, **36**: 33 (1962). Type from Angola (Malange).

Ochna mechowiana O. Hoffm. in Linnaea, **43**: 123 (1881).—Gilg in Engl., Bot. Jahrb. **33**: 234 (1903).—R.E. Fr., Schwed. Rhod.-Kongo-Exped. **1**: 150, t. 3 fig. 3 (1914).—Exell & Mendonça, C.F.A. **1**, 2: 288 (1951) pro parte. Type as above.

Ochna welwitschii Rolfe in Bol. Soc. Brot. **11**: 84 (1893).—Hiern, Cat. Afr. Pl. Welw. **1**: 121 (1896).—Gilg, loc. cit. Type from Angola (Cuanza Norte).

Ochnella mechowiana (O. Hoffm.) Van Tiegh. in Ann. Sci. Nat., Sér. 8, Bot. **16**: 344 (1902). Type as for *O. afzelii* subsp. *mechowiana*.

Polyochnella welwitschii (Rolfe) Van Tiegh., loc. cit. Type as for *Ochna welwitschii*.

Inflorescence-rhachis very short or absent. Sepals in flower 5–8 mm. long, in fruit (10)12–18 mm. long. Petals lemon to bright yellow, 8–13 mm. long. Anthers 1–2 mm. long. Carpels (6)7–8. Drupelets 7–8 × 6 mm.

N. Rhodesia. N: Abercorn, fl. 30.ix.1949, *Bullock* 1114 (K; SRGH). W: Ndola, fl. & fr. 26.ix.1955, *Fanshawe* 2462 (K; SRGH).
In north-western Angola and the lower Congo with outliers in N. Rhodesia, Katanga and western Tanganyika.

30. **Ochna confusa** Burtt Davy & Greenway in Burtt Davy, F.P.F.T. **1**: 238 in clav., 239 (1926); in Kew Bull. **1926**: 239 (1926). Syntypes from the Transvaal.

Ochna leptoclada sensu Phillips in Bothalia, **1**: 94 (1921) pro parte quoad specim. Transvaal.

Ochna gracilipes sensu Brenan in Mem. N.Y.Bot. Gard. **8**, 3: 234 (1953).

Shrub up to c. 1(2) m. high or shrublet with woody rootstock, with bark brown becoming vertically striate; branches virgate, terete or slightly ridged, yellowish-white, densely pustulose-lenticellate. Leaves petiolate; lamina 4·4–9(11) × (1)1·5–2·1(2·9), elliptic to oblong-elliptic or oblanceolate, acute or shortly and bluntly acuminate at the apex, with margin densely serrulate with straight or incurving teeth, cuneate or attenuate at the base, herbaceous or chartaceous, with usually spreading main and subsidiary lateral nerves almost equally prominent and very densely reticulate tertiary venation more prominent above than below; petiole 2–3(4) mm. long, ± slender. Flowers (2)3–6 in a very shortly racemose or pseudumbellate axillary inflorescence; pedicels 0·9–1·2 cm. long in fruit, articulated at the base. Sepals 4–5 mm. long in flower, oblong-elliptic, rounded, becoming carmine- or brownish-red, 8–11 mm. long, flat and spreading in fruit. Petals bright yellow, 7–10 × 3–4 mm., broadly elliptic to suborbicular and unguiculate. Stamens with anthers 1–1·5 mm. long, $\frac{1}{3}$–$\frac{1}{2}$ as long as the filaments, straight, dehiscing by longitudinal slits. Carpels 5, with styles completely united; stigma 5-lobed. Drupelets 6·5–8 × 5–6·5 mm., subglobose, inserted near their base; embryo slightly curved.

N. Rhodesia. N: road to Isoka village from Abercorn road, 915 m., fr. 31.i.1955, *Richards* 4294 (K). **S. Rhodesia.** C: near Goromonzi, fr. 17.xi.1941, *Pardy* in GHS 8231 (SRGH). E: Chimanimani Mts., The Corner, 1525 m., fl. 9.x.1950, *Wild* 3547 (K; SRGH). **Nyasaland.** N: Mafinga Mts. above Chisenga, 2100 m., fl. 9.xi.1958, *Robson* 534 (BM; K; LISC; PRE; SRGH). C: Kota Kota Distr., Chintembwe,

1400 m., fl. & fr. 9.ix.1946, *Brass* 17588 (K; SRGH). **Mozambique.** MS: Manica, Ratanda (Mavita), fl. 26.x.1944, *Mendonça* 2621 (LISC).

Also in the eastern Transvaal, and southern Tanganyika. Stony hillsides, grassland or deciduous woodland, 915–2100 m.

O. confusa is probably a derivative of *O. afzelii* R.Br. ex Oliv.

31. **Ochna pygmaea** Hiern, Cat. Afr. Pl. Welw. **1**: 122 (1896).—Gilg in Engl., Bot. Jahrb. **33**: 233 (1903). Type from Angola (Huila).

Ochna leptoclada? sensu O. Hoffm. in Linnaea, **43**: 123 (1881).

Ochna dekindtiana Engl. & Gilg in Engl., Bot. Jahrb. **32**: 135 (1902).—Gilg, tom. cit.: 233 (1903). Type from Angola (Huila).

Ochnella pygmaea (Hiern) Van Tiegh. in Ann. Sci. Nat., Sér. 8, Bot. **16**: 345 (1902). Type as for *Ochna pygmaea.*

Ochnella dekindtiana (Engl. & Gilg) Van Tiegh., loc. cit. Type as for *Ochna dekindtiana.*

Diporidium hoepfneri Van Tiegh., tom. cit.: 358 (1902). Type from Angola (Cunene).

Ochna debeerstii sensu Gilg, tom. cit.: 237 (1903) pro parte quoad specim. angol.

Proboscella hoepfneri Van Tiegh., op. cit. **18**: 50 (1903). Type from Angola (Bié).

Proboscella emarginata Van Tiegh., op. cit. **18**: 50 (1903). Type from Angola (Huila).

Ochna hoepfneri (Van Tiegh.) Engl. & Gilg [ex van Tiegh., op. cit. **16**: 357 (1902) *in synon.*] in Warb., Kunene-Samb.-Exped. Baum: 303 (1903).—Gilg, tom. cit.: 233 (1903). Type as for *Diporidium hoepfneri.*

Ochna homblei De Wild. in Rev. Zool. Afr. **7**, Suppl. Bot.: 334 (1919). Type from the Congo.

Ochna gracilipes sensu Exell & Mendonça, C.F.A. **1**, 2: 286 (1951).—White, F.F.N.R.: 251, fig. 43G (1962).

Shrub or rhizomatous shrublet up to 0·6 m. high (? or higher), with bark brown and longitudinally ridged; branches virgate or ± widely spreading, usually markedly flattened and/or ridged, white or purplish, ± densely pustulose-lenticellate. Leaves petiolate or subsessile; lamina 3·1–9·5 × 0·6–2·6 cm., oblanceolate to linear-oblanceolate or more rarely narrowly oblong, obtuse (rarely subacute) to rounded at the apex, herbaceous or chartaceous, with margin densely serrulate with the curved teeth often overlapping the margin, attenuate at the base, with spreading or somewhat ascending main and subsidiary lateral nerves and densely reticulate tertiary venation prominent above but scarcely so below; petiole very short, 0·5–1 mm. long, stout. Flowers 2–4, in a subumbellate axillary inflorescence; pedicels 0·7–1·6 cm. long in fruit, articulated at or within 1 mm. of the base. Sepals 5–6 mm. long in flower, oblong, rounded, becoming carmine, (11)13–16 mm. long, flat, and spreading or reflexed in fruit. Petals bright yellow, 7–10 × 3–6 mm., obovate, shortly unguiculate. Stamens with anthers 1–1·5 mm. long, $\frac{1}{4}$–$\frac{1}{3}$ as long as the filaments, straight, dehiscing by longitudinal slits. Carpels 5, with styles completely united; stigma 5-lobed. Drupelets subglobose, 5–8 × 4–6·5 mm., inserted near the base; embryo slightly curved.

N. Rhodesia. B: Balovale South, fr. 1.xi.1952, *Gilges* 251 (K; PRE; SRGH). W: Mwinilunga Distr., near source of Matonchi R., fr. 7.x.1937, *Milne-Redhead* 2623 (BM; K). **S. Rhodesia.** W: 26 km. from Victoria Falls near Kazene, 900 m., fl. 2.x.1928, *Pardy* in GHS 4531 (FHO; SRGH). C: Chilimanzi, Mtao, fr. 25.ii.1951, *Greenhow* 28/51 (FHO; K; SRGH).

Also in Angola, SW. Africa (*fide* Engl. & Gilg, Kunene-Samb.-Exped. Baum: 303 (1903)) and the Congo. Open woodland and grassland on Kalahari sand plains, c. 1050–1250 m.

O. pygmaea is very closely related to *O. confusa* and, indeed, may not be specifically distinct from it. They can usually be distinguished, however, on leaf shape, petiole length and length of the sepals in fruit. Although none of these characters is sufficient by itself to separate the two species, when they are taken together it is usually quite easy to allocate a given specimen to one or other of them.

O. pygmaea was described in the same paper as *O. gracilipes* Hiern and has been regarded as a synonym of it. However, the type specimen of *O. gracilipes* has flowers with longer pedicels than in *O. pygmaea*, ovaries with 6–7 carpels, and broader leaves, characters all of which suggest that it may be a dwarf specimen of *O. afzelii* subsp. *mechowiana*. Although *O. pygmaea* is probably derived from this subspecies it seems preferable not to use a name (*O. gracilipes*) based on an apparently intermediate specimen.

2. BRACKENRIDGEA A. Gray

Brackenridgea A. Gray, Bot. U.S. Expl. Exped. **1**: 361, t. 42 (1854).
Pleuroridgea Van Tiegh. in Ann. Sci. Nat., Sér. 8, Bot. **16**: 198, 399 (1902).
Campylochnella Van Tiegh., tom. cit.: 198, 400 (1902).

Trees, shrubs or shrublets, glabrous, with yellow pigment under the bark. Leaves petiolate; lamina with margin glandular-serrulate or entire, and characteristic nervation of ascending primary laterals linked by secondary laterals (see specific descriptions); stipules markedly longitudinally striate, laciniate or deeply divided into linear segments, free, persistent on first-year shoots (at least in the African spp.). Inflorescence paniculate or fasciculate or reduced to a single flower, or fascicles secondarily aggregated into spikes or capitula, terminal or at the base of current year's growth; bracts striate, laciniate, persistent or deciduous; pedicels articulated at the base. Sepals (4)5, imbricate (usually quincuncial) in bud, white or pink in flower, persistent, enlarging and becoming red and coriaceous in fruit. Petals (4)5, white to pink, not or scarcely unguiculate, deciduous. Stamens (8)10–20(22), free; anthers yellow, dehiscing by longitudinal slits, deciduous; filaments ± slender, approximately equal in length to the anthers, persistent. Carpels (3–4)5–10, apparently free at the base, 1-ovulate; styles slender, gynobasic, completely united; stigmas terminal, scarcely enlarged. Fruit of 1 to several free black 1-seeded drupelets with fleshy mesocarp, inserted on the enlarged red receptacle. Seeds curved, without endosperm but with an internal projection of the endocarp round which the embryo develops; embryo curved, incumbent or accumbent, isocotylous.

A genus of 12 species, 4 in tropical Africa and Madagascar and the others in Malaysia from the Andaman Is., Perak and the Philippines to New Guinea, and also in Queensland and Fiji.

Stamens 10(11); petals 4–5 mm. long; sepals white or greenish in flower; tree or shrub usually at least 2 m. high - - - - - - - - 1. *zanguebarica*
Stamens (13–19)20(21–22); petals 6–8 mm. long; sepals pink in flower; shrub or rhizomatous shrublet usually 1 m. high or less - - - 2. *arenaria*

1. **Brackenridgea zanguebarica** Oliv. in Hook., Ic. Pl. **11**: 77, t. 1096 (1871).—Eyles in Trans. Roy. Soc. S. Afr. **5**: 419 (1916).—Engl., Pflanzenw. Afr. **3,** 2: 486 (1921).—Brenan, T.T.C.L.: 381 (1949). TAB. **46** fig. A. Type from Tanganyika.
Ochna alboserrata Engl., Bot. Jahrb. **17**: 75 (1893). Type from Kenya.
Pleuroridgea zanguebarica (Oliv.) Van Tiegh. in Journ. de Bot. **16**: 203 (1902); in Ann. Sci. Nat., Sér. 8, Bot. **16**: 399 (1902).—Bak. f. in Journ. Linn. Soc., Bot. **40**: 38 (1911). Type as for *Brackenridgea zanguebarica.*
Pleuroridgea alboserrata (Engl.) Van Tiegh. in Journ. de Bot. **16**: 204 (1902); in Ann. Sci. Nat., Sér. 8, Bot. **16**: 400 (1902). Type as for *Ochna alboserrata.*
Pleuroridgea lastii Van Tiegh. in Journ. de Bot. **16**: 204 (1902); in Ann. Sci. Nat., Sér. 8, Bot. **16**: 400 (1902) *nom. nud.*
Brackenridgea bussei Gilg in Engl., Bot. Jahrb. **33**: 273 (1903).—Engl., Pflanzenw. Afr. **3,** 2: 486 (1921). Lectotype from Tanganyika.
Pleuroridgea bussei (Gilg) Van Tiegh. in Ann. Sci. Nat., Sér. 8, Bot. **18**: 59 (1903). Lectotype as for *Brackenridgea bussei.*
Ochna praecox Sleumer in Fedde, Repert. **39**: 276 (1936). Type from Tanganyika (Lindi).

Shrub or small tree (0·3)2–10(15) m. high, with bark black or dark grey, very rough; branches striate, purplish with pale lenticels at first, becoming white or yellowish-white, not usually exfoliating. Leaves petiolate; lamina 2·7–7·7(9·2) × (0·9)1·1–2·6(3) cm., oblong or elliptic to oblanceolate or obovate, acute or shortly acuminate or mucronate to rounded at the apex, with margin densely glandular-serrulate, cuneate at the base, herbaceous, with primary lateral nerves ascending, interconnected by more widely spreading secondary laterals in turn linked by tertiary cross-veins, prominent above but less so or almost plane below; petiole 1–2·5 mm. long, rather slender, grooved above. Flowers solitary or 2–4 in fascicles, axillary or terminating short shoots, appearing before the leaves, scented; pedicels 1–2 cm. long in fruit. Sepals 3–4(5) mm. long in flower, oblong, rounded, reflexed after flowering, becoming crimson, 7–9(10) mm. long, narrowly oblong, flat and spreading in fruit. Petals white to creamy white, 4–5 mm. long, oblong-oblanceolate, narrowed at the base but not clawed. Stamens 10(11), with anthers yellow,

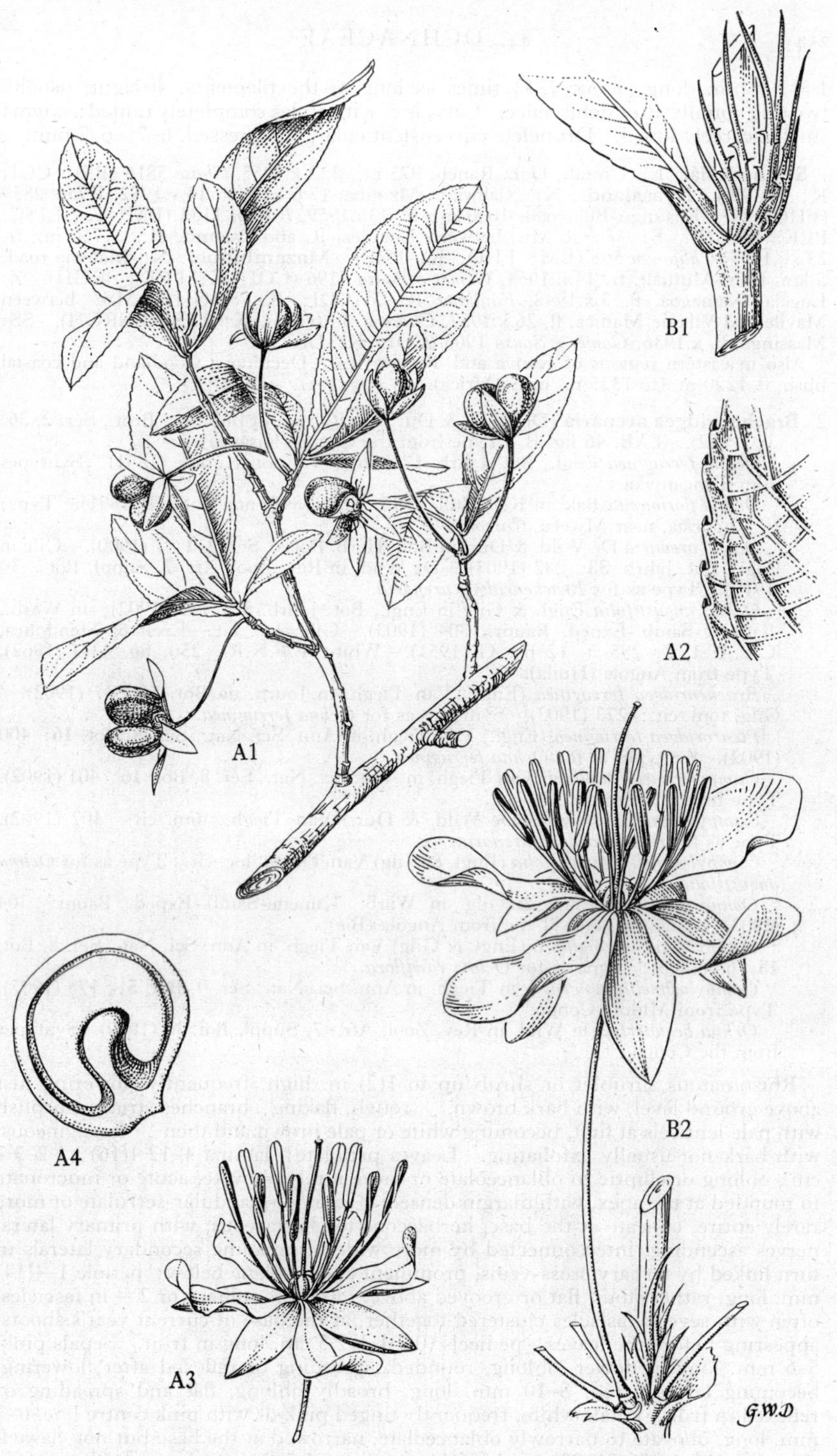

Tab. 46. A.—BRACKENRIDGEA ZANGUEBARICA. A1, fruiting branch (×1); A2, part of leaf showing marginal glands (×8), both from *Faulkner* 93 (b); A3, flower (×4) *Faulkner* 93; A4, vertical section of drupelet (×4) *Gomes e Sousa* 2204. B.—BRACKENRIDGEA ARENARIA. B1, stipule (×4) *Gilges* 245; B2, flower (×4) *Holmes* 1181.

1·5–2·5 mm. long, about $\frac{3}{4}$–$1\frac{1}{2}$ times as long as the filaments, straight, usually twisting spirally after dehiscence. Carpels 5, with styles completely united; stigma small, scarcely lobed. Drupelets curved-lenticular, compressed, 6–7 × 6–7 mm.

S. Rhodesia. E: Umtali, Dora Ranch, 975 m., fl. 2.x.1955, *Chase* 5816 (BM; COI; K; SRGH). **Nyasaland.** N: Njakwa to Mzimba, 17·6 km., st. 13.v.1952, *White* 2859 (FHO). C: Kasungu-Bua road, 1000 m., fr. 13.i.1959, *Robson* 1133 (BM; K; LISC; PRE; SRGH). S: Mlanje Mt., between Lukulezi R. and Palombe R., c. 760 m., fr. 23.xii.1957, *Chapman* 506 (BM; FHO; K; PRE). **Mozambique.** N: Malema road, 5 km. from Mutuali, fr. 14.ii.1954, *Gomes e Sousa* 4196 (COI; K; PRE; SRGH). Z: Lugela, Namagoa, fl. 3.x.1948, *Faulkner* K 93 (COI; K; SRGH). MS: between Mavila and Vila de Manica, fl. 26.x.1957, *Chase* 6730 (COI; K; LISC; SRGH). SS: Massinga, fl. x.1936, *Gomes e Sousa* 1903 (COI; K; LISC).

Also in eastern regions of Kenya and Tanganyika. Deciduous woodland and coastal bush, 0–1220 m. (to 1525 m. in E. Africa).

2. **Brackenridgea arenaria** (De Wild. & Dur.) N. Robson in Bol. Soc. Brot., Sér. 2, **36**: 37 (1962). TAB. **46** fig. B. Type from the Congo (Kisantu).

Ochna ferruginea Engl., Bot. Jahrb. **17**: 76 (1893) non Kuntze (1891). Syntypes from Tanganyika.

Ochna floribunda Bak. in Kew Bull. **1895**: 289 (1895) non Kuntze (1891). Type: N. Rhodesia, near Mweru, *Carson* 8 (K).

Ochna arenaria De Wild. & Dur. in Bull. Herb. Boiss., Sér. 2, **1**: 7 (1900).—Gilg in Engl., Bot. Jahrb. **33**: 232 (1903).—De Wild. in Rev. Zool. Afr. **7**, Suppl. Bot.: 30 (1919). Type as for *Brackenridgea arenaria.*

Ochna angustifolia Engl. & Gilg in Engl., Bot. Jahrb. **32**: 135 (1902); in Warb., Kunene-Samb.-Exped. Baum: 304 (1903).—Gilg, loc. cit.—Exell & Mendonça, C.F.A. **1**, 2: 285, t. 12 fig. G (1951).—White, F.F.N.R.: 250, fig. 43D (1962). Type from Angola (Huila).

Brackenridgea ferruginea (Engl.) Van Tiegh. in Journ. de Bot. **16**: 47 (1902).—Gilg, tom. cit.: 273 (1903). Syntypes as for *Ochna ferruginea.*

Pleuroridgea ferruginea (Engl.) Van Tiegh. in Ann. Sci. Nat., Sér. 8, Bot. **16**: 400 (1902). Syntypes as for *Ochna ferruginea.*

Campylochnella thollonii Van Tiegh. in Ann. Sci. Nat., Sér. 8, Bot. **16**: 401 (1902). Type from the Congo.

Campylochnella arenaria (De Wild. & Dur.) Van Tiegh., tom. cit.: 402 (1902). Type as for *Brackenridgea arenaria.*

Campylochnella angustifolia (Engl. & Gilg) Van Tiegh., loc. cit. Type as for *Ochna angustifolia.*

Ochna roseiflora Engl. & Gilg. in Warb., Kunene-Samb.-Exped. Baum: 304 (1903).—Gilg, loc. cit. Type from Angola (Bié).

Campylochnella roseiflora (Engl. & Gilg) Van Tiegh. in Ann. Sci. Nat., Sér. 8, Bot. **18**: 60 (1903). Type as for *Ochna roseiflora.*

Campylochnella pungens Van Tiegh. in Ann. Sci. Nat., Sér. 9, Bot. **5**: 178 (1907). Type from Middle Congo.

Ochna bequaertii De Wild. in Rev. Zool. Afr., **7**, Suppl. Bot. 30 (1919). Syntypes from the Congo.

Rhizomatous shrublet or shrub up to 1(2) m. high, frequently flowering just above ground level, with bark brown, ± rough, flaking; branches striate, purplish with pale lenticels at first, becoming white or pale brown and then ± ferrugineous, with bark not usually exfoliating. Leaves petiolate; lamina 4–12·4(16) × 1·2–3·7 cm., oblong or elliptic to oblanceolate or more rarely obovate, acute or mucronate to rounded at the apex, with margin densely ± curved-glandular-serrulate or more rarely entire, cuneate at the base, herbaceous to chartaceous, with primary lateral nerves ascending, interconnected by more widely spreading secondary laterals in turn linked by tertiary cross-veins, prominent above, plane below; petiole 1–4(14) mm. long, rather stout, flat or grooved above. Flowers solitary or 2–4 in fascicles, often with several fascicles clustered together, at the base of current year's shoots, appearing before the leaves; pedicels (0·9)1·3–2·5 cm. long in fruit. Sepals pink, 5–6 mm. long in flower, oblong, rounded, spreading or reflexed after flowering, becoming crimson-red, 8–10 mm. long, broadly oblong, flat and spreading or reflexed in fruit. Petals white, frequently tinged pink or with pink centre line, 6–8 mm. long, obovate to narrowly oblanceolate, narrowed at the base but not clawed. Stamens (13–19)20(21–22), with anthers yellow, 1·5–2 mm. long, $\frac{3}{4}$–$1\frac{1}{4}$ times as long as the filaments, straight, often twisting spirally after dehiscence. Carpels 5–7, with styles completely united; stigma small, scarcely lobed. Drupelets 6–8 × 6–7 mm., curved-lenticular, compressed.

N. Rhodesia. B: Balovale Distr., near Chavuma, fr. 13.x.1952, *White* 3485 (BM; FHO; K). N: Abercorn, fr. 22.ix.1949, *Bullock* 1051 (K; SRGH). W: Mwinilunga Distr., Matonchi Farm, fl. 1.ix.1930, *Milne-Redhead* 1013 (K). C: Broken Hill, fr. 17.xii.1907, *Kassner* 2203 (BM). S: Near Siamambo Forest Reserve, fl. 27.vii.1952, *Angus* 63 (FHO; K). **S. Rhodesia.** N: Mrewa road, Shawanoye R., fl. 7.ix.1931, *Jack* in GHS 5236 (SRGH). W: Wankie, fr. 10.xi.1949, *Davison* in *West* 3054 (K; SRGH). C: Beatrice, Progress Farm, 1220 m., fl. 14.ix.1923, *Eyles* 4535 (K; SRGH).

Also in the Congo, Ruanda Urundi, W. Tanganyika, Middle Congo, Angola and northern SW. Africa. Deciduous woodland and margins of seasonal swamps, on sandy soils, 915–1675 m.

B. arenaria can be easily distinguished from dwarf species of *Ochna* by the laciniate or fin-like stipules persistent on first year stems, the white petals and pink sepals, the usually glandular-serrulate leaves with distinctive venation, and the discoid drupelets.

3. OURATEA Aubl.

Ouratea Aubl., Hist. Pl. Guian. Fr. **1**: 397, t. 152 (1775).
Gomphia Schreb. in L., Gen. Pl. ed. 8, **1**: 291 (1789).

Trees or shrubs, usually completely glabrous. Leaves petiolate to sessile and amplexicaul; lamina with margin serrate to ciliate or entire; stipules entire, not striate, coriaceous, free or ± united intrapetiolarly, caducous or ± persistent. Inflorescence paniculate to racemose or umbellate or rarely reduced to 1–2 flowers, terminal or axillary, sometimes at the base of the current year's growth; bracts scale-like, caducous or ± persistent; pedicels articulated at or above the base. Sepals 5, quincuncial in bud, green or yellow in flower, persistent and usually enlarging and becoming varying shades of red and coriaceous in fruit (in Old World species). Petals 5, yellow or rarely white, not or scarcely unguiculate, deciduous. Stamens 10, diplostemonous, free; anthers yellow or orange, elongate and frequently narrowed upwards, often rugose, dehiscing by terminal pores, deciduous; filaments stout, much shorter than the anthers, persistent, or absent. Carpels 5–10, apparently free at the base, 1-ovulate; styles slender, gynobasic, completely united; stigma terminal, not enlarged. Fruit of 1 to several free black or brown 1-seeded drupelets with fleshy or coriaceous mesocarp, inserted on the enlarged red or purplish receptacle. Seeds straight or curved, without endosperm or an internal projection of the endocarp; embryo straight or ± curved, incumbent or accumbent, isocotylous or heterocotylous.

A genus of c. 150–200 species of tropical rain forest, which is represented by different subgenera in the Eastern and Western Hemispheres. Old World species (mostly in Africa and Madagascar) have ± curved embryos, persistent calyces and stipules frequently ± united, whereas those from the New World have straight embryos, usually deciduous calyces and free stipules.

Leaves with lateral nerves ± widely separate, linked by more slender secondary veins and reticulate tertiary venation; inflorescences terminal or in axils of foliage leaves (Sect. *Reticulatae*):
- Stipules, scale-leaves at the base of inflorescence (if present) and bracteoles soon deciduous; leaf margins shallowly to ± sharply serrulate; inflorescence with main rhachis ± obviously zig-zag:
  - Leaves coriaceous, with venation not or scarcely visible on smooth matte upper surface; petiole (4)6–12 mm. long, stout - - - - - 1. *densiflora*
  - Leaves chartaceous to subcoriaceous, with venation visible and prominent and lateral nerves usually ± depressed on ± glossy upper surface; petiole 2–7 mm. long, rather slender or enlarged towards the base:
    - Inflorescence-rhachis and young stems olive-green, relatively stout and ± markedly compressed; older stems yellowish- to orange-brown; petals 2·5–4 mm. broad; secondary and tertiary venation not or scarcely prominent above 2. *hiernii*
    - Inflorescence-rhachis and young stems soon becoming dark purplish-red or blackish, relatively slender and scarcely compressed; older stems reddish- to purplish-brown; petals 4–7 mm. broad; secondary and tertiary venation usually ± prominent above - - - - - - - - 3. *andongensis*
- Stipules and scale-leaves at base of inflorescence persistent; bracteoles persistent until anthesis; leaf margins spinulose towards the apex; inflorescence-rhachis not zig-zag - - - - - - - - - - - - - 4. *lunzuensis*

Leaves with lateral nerves and veins crowded together, parallel, only occasionally branching, not or scarcely differentiated in size, not reticulate; inflorescences in axils of deciduous scale-leaves at base of current growth (Sect. *Calophyllae*) 5. *welwitschii*

1. **Ouratea densiflora** De Wild. & Dur. in Ann. Mus. Cong., Sér. 3, Bot. **1**, 1: 37 (1901).—Gilg in Engl., Bot. Jahrb. **33**: 265 (1903).—De Wild., Pl. Bequaert. **4**: 455 (1928).—Robyns, Fl. Parc Nat. Alb. **1**: 616, t. 61 (1948).—Brenan, T.T.C.L.: 385 (1949).—F. W. Andr., Fl. Pl. Anglo-Egypt. Sudan, **1**: 188 (1950).—White, F.F.N.R.: 252, fig. 44 (1962). Type from the Congo.

Exomicrum densiflorum (De Wild. & Dur.) Van Tiegh. in Ann. Sci. Nat., Sér. 8, Bot. **16**: 339 (1902). Type as for *Ouratea densiflora*.

Monelasmum densiflorum (De Wild. & Dur.) Van Tiegh., op. cit., Sér. 8, Bot. **18**: 35 (1903). Type as for *Ouratea densiflora*.

Monelasmum coriaceum Van Tiegh., op. cit., Sér. 9, Bot. **5**: 171 (1907). Type from Upper Ubangi.

Shrub or tree 1–12 m. high, with bark dark grey, smooth except for raised transversely elongated lenticels, and spreading or rounded crown; branches yellow-green and striate at first, becoming purple-brown and polished. Leaves petiolate; lamina (10)12–32(37) × (4)5–12 cm., elliptic or elliptic-oblong to oblanceolate, obtuse to acute or shortly acuminate at the apex, with margin shallowly serrulate, cuneate to rounded or truncate at the base, coriaceous, with ascending main lateral nerves linked by c. 1–5 ± widely spreading secondary nerves and tertiary reticulations slightly prominent below but smooth above; petiole (4)6–10(12) mm. long, stout; stipules 5–6·5 mm. long, free or almost so, soon deciduous. Flowers solitary or in clusters of 2–4(6) in axils of deciduous bracts on a terminal or axillary paniculate inflorescence (shorter or longer than surrounding leaves), with (1)3–7(11) ± elongated spreading or ascending lateral branches; rhachis angular, striate; pedicels 7–15 mm. long in fruit, longer than the sepals, articulated in the lower ¼. Sepals 5–7 mm. long in flower, ovate to lanceolate, acute to subacute, becoming 8–9 mm. long, dark red and spreading in fruit. Petals yellow, (7)9–10 × (4)6–7 mm., obovate. Stamens with anthers 4·5–5 mm. long, orange-yellow, rugose and narrowed above. Carpels 5. Drupelets 7–9 × 4–6 mm., cylindric, not carinate.

N. Rhodesia. N: Lumangwe, fr. 14.xi.1957, *Fanshawe* 4004 (FHO). W: Mwinilunga Distr., Musera R., c. 16 km. W. of Kakomo, fr. 28.ix.1952, *Angus* 561 (FHO; K).

Margin of the Congo forests from Ubangi-Chari and the Sudan to western Tanganyika and N. Rhodesia. Damp evergreen forest and fringing forest, 470–1500 m.

2. **Ouratea hiernii** (Van Tiegh.) Exell in Journ. of Bot. **65**, Suppl. Polypet.: 59 (1927). Type from Angola (Pungo Andongo).

Ouratea reticulata var. *poggei* Engl., Bot. Jahrb. **17**: 81 (1893) pro parte excl. specim. angol. Syntypes from the Congo.

Monelasmum hiernii Van Tiegh. in Ann. Sci. Nat., Sér. 8, Bot. **16**: 328 (1902). Type as for *O. hiernii*.

Ouratea sibangensis Gilg in Engl., Bot. Jahrb. **33**: 267 (1903) pro parte quoad *Soyaux* 43.

Ouratea bukobensis Gilg, tom. cit.: 271 (1903); apud Mildbr. in Deutsch-Z.-Afr. Exped. 1907–1908, **2**: 559 (1913).—De Wild., Pl. Bequaert. **4**: 419 (1928).—Brenan, T.T.C.L.: 385 (1949) pro parte excl. syn. *O. floribunda*. Syntypes from Tanganyika (Bukoba).

Ouratea poggei (Engl.) Gilg, tom. cit.: 272 (1903) pro parte excl. syn. *O. reticulata* var. *andongensis* et specim. Angol.—Exell & Mendonça, C.F.A. **1**, 2: 296 (1951) pro parte excl. specim. *Gossweiler* 7570 et 13570 b. Syntypes as for *O. reticulata* var. *poggei*.

Monelasmum bukobense (Gilg) Van Tiegh., op. cit., Sér. 8, Bot. **18**: 36 (1903). Syntypes as for *Ouratea bukobensis*.

Ouratea nutans sensu Exell, loc. cit. pro parte quoad specim. *Gossweiler* 6471.

Ouratea sp.—Exell, loc. cit. pro parte quoad specim. *Gossweiler* 7986.

Ouratea andongensis sensu Exell & Mendonça, loc. cit. pro parte quoad specim. *Gossweiler* 5434 et 9961.

Shrub or small tree 2–6(12) m. high, with brown bark and spreading or rounded crown; branches rather stout, olive-green, ± compressed and angular or striate at first, becoming yellowish- to orange-brown and polished. Leaves petiolate; lamina (8)10·7–23 × (3)3·6–6·7(7) cm., oblanceolate or obovate to oblong or elliptic, acute or shortly and abruptly acuminate to rounded at the apex, with margin serrulate to undulate and sharply curved-serrate, cuneate or attenuate at the base, chartaceous to subcoriaceous, with ascending main veins linked by numerous ± widely spreading secondary veins and tertiary reticulations prominent above but less so or plane below; petiole (2)4–6(7) mm. long, ± stout, sometimes enlarged towards the base; stipules 3–5 mm. long, free or partly or completely united interpetiolarly, caducous. Flowers solitary or paired or more usually in clusters of 3–8 in axils of deciduous bracts on a terminal or axillary paniculate inflorescence (usually shorter than surrounding leaves), with 1–4(6) short or ± elongated spreading lateral branches; rhachis stout, olive-green, compressed, angular, striate; peduncle present or not; pedicels 7–16 mm. long in fruit, longer to shorter than the sepals, articulated at the base or in the lower $\frac{1}{3}$. Sepals 5–7(8) mm. long in flower, lanceolate, subacute, becoming 7–9(10) mm. long, red and ascending in fruit. Petals bright yellow, (5)7–11(13) × (2·5)3–4 mm., oblanceolate to obovate, with apex rounded or retuse. Stamens with anthers (3)4(6) mm. long, orange-yellow, ± rugose and narrowed above. Carpels 5. Drupelets 6–8 × 4–6 mm., ellipsoid-cylindric, ± compressed, not carinate.

N. Rhodesia. N: Chishimba Falls, fl. 10.ix.1958, *Fanshawe* 4787 (FHO; K). W: Mwinilunga Distr., Zambezi R., 7 km. N. of Kalene Hill Mission, fl. 23.ix.1952, *White* 3339 (FHO; K).

From Uganda and W. Tanganyika to Angola and Gaboon. Understorey of swamp forest and also drier evergreen forest, 850–1200 m.

O. hiernii and *O. bukobensis* respectively appear to constitute opposite ends of a cline. Thus *O. bukobensis* tends to have acute leaf-apices, pedicels articulated above the base, and long petals, and to grow in swamp-forest; whereas *O. hiernii* tends to have rounded or abruptly acuminate leaf-apices, pedicels articulated at or near the base, and shorter petals, and to grow in fringing forest. None of these differences is complete, however, and although the two populations may possibly be subspecifically distinct, it seems most appropriate to treat them as one species.

3. **Ouratea andongensis** (Hiern) Exell in Journ. of Bot. **65,** Suppl. Polypet.: 58 (1927).—Exell & Mendonça, C.F.A. **1,** 2: 296 (1951) pro parte excl. syn. *Monelasmum hiernii* et specim. *Welwitsch* 4605, *Gossweiler* 4472, 9961, 14107. Type from Angola (Pungo Andongo).

Gomphia reticulata sensu O. Hoffm. in Linnaea, **43**: 122 (1881).

Ouratea reticulata var. *poggei* Engl., Bot. Jahrb. **17**: 81 (1893) pro parte quoad specim. *Welwitsch* 4604.

Ouratea reticulata var. *andongensis* Hiern, Cat. Afr. Pl. Welw. **1**: 122 (1896) emend. excl. specim. *Welwitsch* 4605. Type as for *O. andongensis*.

Monelasmum andongense (Hiern) Van Tiegh. in Ann. Sci. Nat., Sér. 8, Bot. **16**: 328 (1902). Type as for *Ouratea andongensis*.

Ouratea poggei (Engl.) Gilg in Engl., Bot. Jahrb. **33**: 272 (1903) pro parte quoad specim. *Welwitsch* 4604.

Ouratea flava sensu White, F.F.N.R.: 252 (1962).

Shrub 2–6 m. high, spreading, evergreen, with bark light brown; branches slender, dark red or blackish, terete (or rarely somewhat angular) and striate at first, becoming reddish- to purplish-brown and polished. Leaves petiolate; lamina (8·5)10·3–16·2(18) × (2·6)3·1–5·1(5·9) cm., oblong to elliptic (rarely obovate or oblanceolate), acute to shortly and suddenly acuminate at the apex, with margin sharply serrulate, narrowly cuneate to rounded at the base, chartaceous to subcoriaceous, with ascending main lateral nerves linked by numerous widely spreading secondary veins and tertiary reticulations scarcely prominent on either surface; petiole 4–7 mm. long, rather slender, not enlarged towards the base; stipules 3–5 mm. long, free, caducous. Flowers solitary or in clusters of 2–7 in axils of deciduous bracts on a terminal paniculate inflorescence (shorter to longer than surrounding leaves), with 1–4 ± elongate spreading lateral branches; rhachis slender, olive-green to purplish-red, slightly compressed, angular, striate; basal bracts caducous; peduncle short; pedicels 6–12 mm. long in fruit, equalling or exceeding the sepals, articulated at the base or within 4 mm. of it (in the lower $\frac{1}{3}$). Sepals c. 6 mm. long in flower, lanceolate, subacute, becoming 7–10 mm. long, red and spreading in fruit. Petals yellow, 7–10 × 4–7 mm., obovate, with apex retuse or rounded. Stamens with anthers 4 mm. long, orange-yellow, rugose and narrowed above. Carpels 5. Drupelets (6)8–9 × (4)5–6 mm., ellipsoid to cylindric, slightly compressed, not carinate.

N. Rhodesia. W: Zambezi R., 7 km. N. of Kalene Hill Mission, fl. 21.ix.1952, *Angus* 510 (FHO; K).

Also in Angola and the Congo (Kasai). Fringing forest and evergreen thickets, 700–750 m.

O. andongensis has been confused with *O. hiernii*, but can easily be distinguished by several characters, notably the more slender blackish to purplish inflorescence-rhachis and the more prominent leaf venation.

4. **Ouratea lunzuensis** N. Robson in Bol. Soc. Brot., Sér. 2, **36**: 38 (1962). TAB. **47** fig. A. Type: N. Rhodesia, Abercorn Distr., Lunzua R., 30·4 km. west of Abercorn, above Falls, *Bullock* 3877 (K).
Ouratea sp. 1.—White, F.F.N.R.: 252 (1962).

Shrub or small tree 1·5–6 m. high, evergreen; branches slender, wiry, green and striate at first, becoming pale brown. Leaves petiolate; lamina 8–14 × 2·2–4·1 cm., elliptic to narrowly oblong, very acute or slightly acuminate at the apex, with margin spinulose-serrate to subentire, becoming entire towards the base, cuneate or attenuate at the base, subcoriaceous, with c. 8–16 ascending main lateral nerves (basal ones becoming submarginal) linked by numerous widely spreading secondary veins and tertiary reticulations prominent on both sides; petiole 2–3 mm. long, expanded towards the base; stipules 3–3·5 mm. long, free, persistent on current growth. Flowers solitary or paired (rarely 3) in axils of bracts (persistent at least until anthesis) on a terminal or axillary paniculate inflorescence shorter than the leaves, with 1–3 short lateral branches; rhachis slender, green, slightly angular, striate; basal bracts persistent; peduncle present or not; pedicels 8–20 mm. long in fruit, longer than (or rarely equalling) the sepals, articulated in the lower ¼ (1·5–5 mm. above the base). Sepals 5–7 mm. long in flower, oblong, rounded, becoming (5)6–9 mm. long, cherry-pink to red and ascending in fruit. Petals bright to orange-yellow, 8–10 × 4·5–5 mm., obovate, with apex retuse. Stamens with anthers 4·5–6 mm. long, yellow, rugose and narrowed above. Carpels 5. Drupelets 7 × 5 mm., cylindric-ellipsoid, slightly flattened, slightly dorsally carinate when dry.

N. Rhodesia. N: Lunzua Falls, fr. 26.x.1952, *Robertson* 173 (K; PRE).
Also in Tanganyika (Mpanda Distr.). Fringing forest, 840–1500 m.

O. lunzuensis appears to have a restricted distribution round the south-eastern end of Lake Tanganyika. It is allied to *O. warneckei* Gilg ex Engl., from the Usambaras, and an as yet unnamed species from Lindi Distr., S. Tanganyika (*Eggeling* 6425), but differs from both these species by its spinulose-serrate leaves and pedicels articulated well above the base. In addition, the flowers of *O. warneckei* are smaller, while those of the Lindi plant are larger than in *O. lunzuensis*.

5. **Ouratea welwitschii** (Van Tiegh.) Exell in Journ. of Bot. **65**, Suppl. Polypet.: 58 (1957).—De Wild., Pl. Bequaert. **4**: 520, 533 (1929).—Exell & Mendonça, C.F.A. **1**, 2: 293 (1951).—White, F.F.N.R.: 252 (1962). TAB. **47** fig. B. Type from Angola.
Ouratea affinis sensu Hiern, Cat. Afr. Pl. Welw. **1**: 122 (1896).
Rhabdophyllum welwitschii Van Tiegh. in Ann. Sci. Nat., Sér. **8**, Bot. **16**: 322 (1902). Type as for *Ouratea welwitschii*.
Rhabdophyllum umbellatum Van Tiegh., tom. cit.: 323 (1902). Type from Angola.
Rhabdophyllum penicellatum Van Tiegh., loc. cit. Type from Angola.
Ouratea subumbellata Gilg in Engl., Bot. Jahrb. **33**: 254 (1903). Type as for *O. welwitschii*.

Shrub or small tree (1)2–5 m. high, much branched, with bark grey-brown, closely longitudinally fissured and with transverse cracks; branches slender, green and striate at first, becoming pale or purplish-brown. Leaves petiolate; lamina (5)5·9–12·2(14·5) × (1·5)2·2–4(5) cm., elliptic to oblanceolate or oblong, acute (rarely subobtuse) or acutely acuminate (with acumen up to 1 cm. long) at the apex, with margin undulate-serrulate to (rarely) subentire, cuneate at the base, chartaceous to subcoriaceous, with lateral nerves and veins very numerous, close together and parallel or almost so, scarcely prominent; petiole (1)2–4(5) mm. long, slender or rather stout; stipules 1–3 mm. long, free, intrapetiolar, partly or completely united, soon deciduous or persisting for 1–2 years. Flowers 4–18, in compound, paniculate or pseudumbellate racemes 1·5–7 cm. long borne in axils of caducous scale leaves at the base of current growth; bracteoles caducous; peduncle variable in length, subterete or ± compressed; pedicels c. 10–23 mm. long in fruit, longer than the sepals, articulated in the lower ¼ (rarely at the base). Sepals 5·5–7 mm. long in flower, narrowly elliptic to oblong, acute, occasionally with a small subapical tooth, becoming deep pink or red, 7–8·5 mm. long and spreading in fruit. Petals yellow, 6–8 × 2–3 mm., elliptic-oblong. Stamens with

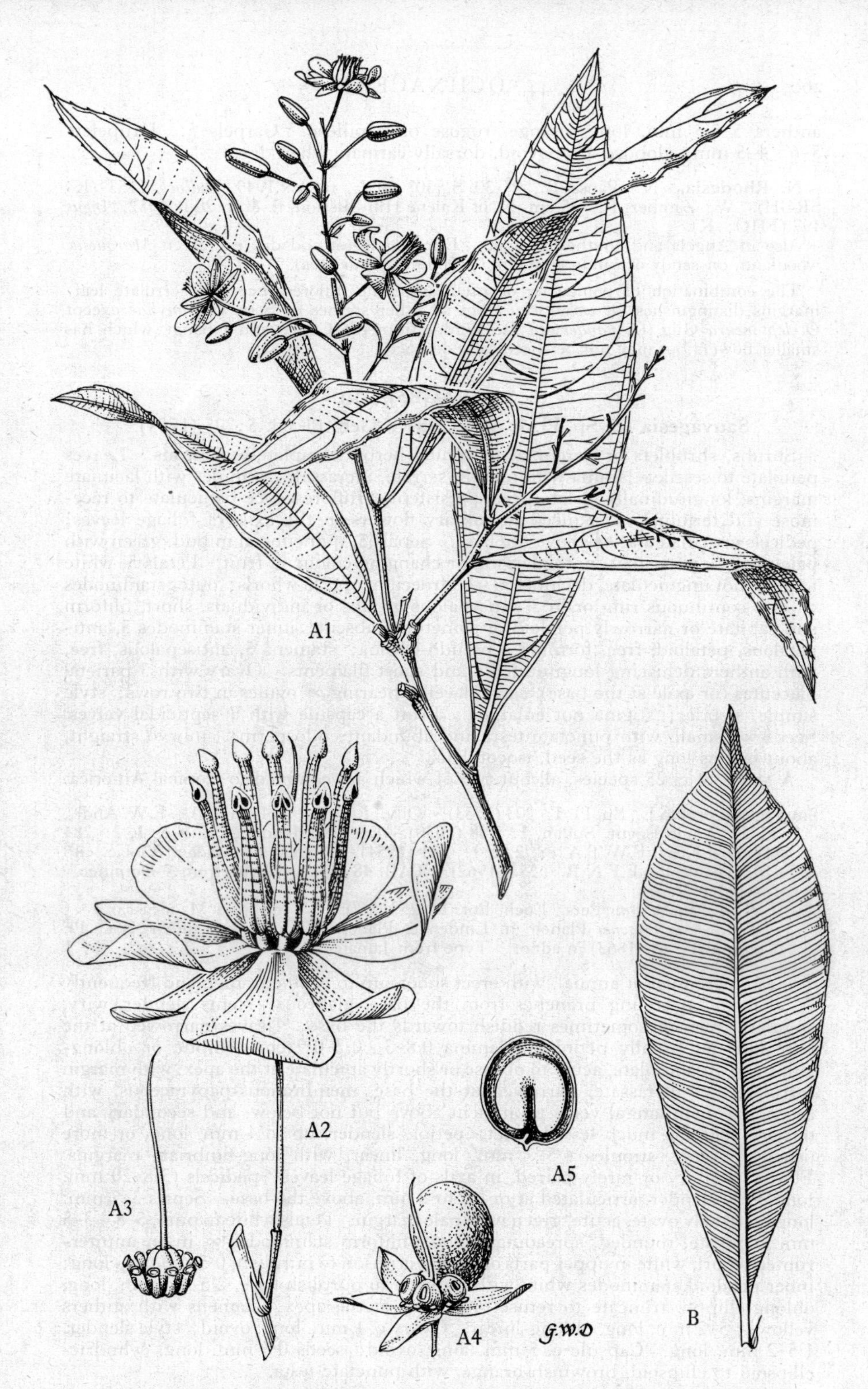

Tab. 47. A.—OURATEA LUNZUENSIS. A1, flowering branch (×1); A2, flower (×4); A3, stamen-filaments and gynoecium (×4), all from *Bullock* 3877; A4, fruit with one remaining drupelet (×2); A5, vertical section of drupelet (×2), both from *Robertson* 173. B.—OURATEA WELWITSCHII, leaf (×1) *Bullock* 1153.

anthers 5–5·5 mm. long, orange, rugose or papillose. Carpels 5. Drupelets 5–6 × 4–5 mm., globose or obovoid, dorsally carinate when dry.

N. Rhodesia. N: Pansa R., 9° 30′ S, 30° 30′ E, fr. 6.x.1949, *Bullock* 1153 (K; SRGH). W: Zambezi R., 6·5 km. N. of Kalene Hill Mission, fl. & fr. 20.ix.1952, *Angus* 497 (FHO; K).

Also in Angola and southern Congo. Fringing forest and dry evergreen *Marquesia* woodland, on sandy or sandy alluvium (c. 700 m. in our area).

The combination of paniculate or pseudumbellate inflorescence and serrulate leaf-margins distinguishes *O. welwitschii* from all other species of Sect. *Calophyllae* except *O. leptoneura* Gilg (*O. vanderystii* De Wild.), from the Congo and Ubangi, which has smaller flowers but may not be specifically distinct.

4. SAUVAGESIA L.

Sauvagesia L., Sp. Pl. **1**: 203 (1753); Gen. Pl. ed. 5: 95 (1754).

Shrubs, shrublets or perennial to annual herbs, completely glabrous. Leaves petiolate to sessile; lamina with margin serrate, incrassate; stipules with laciniate margins, longitudinally striate, free, persistent. Inflorescence paniculate to racemose and terminal or reduced to solitary flowers in the axils of foliage leaves; pedicels articulated at or above the base. Sepals 5, quincuncial in bud, green with pale margin, persistent, not enlarging or changing colour in fruit. Petals 5, white to pink, not unguiculate, deciduous. Androecium of (2)3 whorls; outer staminodes ∞ in a continuous ring or in 5 antisepalous groups or individuals, short, filiform and capitate or narrowly petaloid or sometimes absent; inner staminodes 5, antipetalous, petaloid, free, forming a pseudo-corona; stamens 5, antisepalous, free, with anthers dehiscing longitudinally and short filaments. Ovary with 3 parietal placentas (or axile at the base), each placenta bearing ∞ ovules in two rows; style simple, slender; stigma not enlarged. Fruit a capsule with 3 septicidal valves. Seeds ∞, small, with punctate testa and abundant endosperm; embryo straight, about half as long as the seed, isocotylous.

A genus of c. 25 species, all but two of which are confined to tropical America.

Sauvagesia erecta L., Sp. Pl. **1**: 203 (1753).—Oliv., F.T.A. **1**: 111 (1868).—F.W. Andr., Fl. Pl. Anglo-Egypt. Sudan, **1**: 188 (1950).—Exell & Mendonça, C.F.A. **1**, 2: 284 (1951).—Keay, F.W.T.A. ed. 2, **1**, 1: 231 (1954).—Perrier, Fl. Madag., Violac.: 46 (1955).—White, F.F.N.R.: 252 (1962). TAB. **48** fig. A. Type from " Domingo " (probably Haiti).

Sauvagesia nutans Pers., Ench. Bot. **1**: 253 (1805). Type from Madagascar.

Sauvagesia brownei Planch. in Linden & Planch., Trois Voy. Linden, Bot., Pl. Columb. **1**: 64 (1863) in adnot. Type from Jamaica.

Herb, perennial or annual, with erect shoots up to c. 60 cm. high and frequently ± elongate ascending branches from the base, glabrous; stems slender, wiry, angular, green or sometimes reddish towards the base. Leaves narrowed at the base but not usually petiolate; lamina 0·8–3 × 0·2–0·9 cm., elliptic or oblong-elliptic to oblanceolate, acute to obtuse or shortly apiculate at the apex, with margin serrulate and incrassate, narrowed at the base, membranous-papyraceous, with ascending main lateral veins prominent above but not below, and secondary and tertiary venation much less distinct; petiole slender, up to 4 mm. long, or more usually absent; stipules 4·5–7 mm. long, linear, with long-fimbriate margins. Flowers solitary or rarely paired, in axils of foliage leaves; pedicels (5)8–20 mm. long, very slender, articulated at or up to 2 mm. above the base. Sepals 3–7 mm. long, narrowly ovate, acute, green with pale margin. Petals white to pink, 5–8 × 3–5 mm., obovate, rounded, spreading. Outer filiform staminodes ∞ in an uninterrupted whorl, white in upper parts or wholly crimson to purplish, 0·5–1·5 mm. long; inner petaloid staminodes white with crimson to purplish base, 2·5–4·5 mm. long, oblong-elliptic, truncate to retuse or eroded at the apex; stamens with anthers yellow, 1·5–2 mm. long, oblong-linear. Ovary c. 1 mm. long, ovoid; style slender, 1·5–2 mm. long. Capsule c. 5 mm. long, ovoid; seeds 0·5 mm. long, cylindric-ellipsoid to ellipsoid, brownish-orange, with punctate testa.

N. Rhodesia. N: Mbereshi, 900 m., fl. & fr. 24.vi.1957, *Robinson* 2388 (K; SRGH). **Mozambique.** SS: Gaza, Bilene, S. Martinho, fl. & fr. 30.xi.1955, *Gonçalves-Sanches* 14 (LISC).

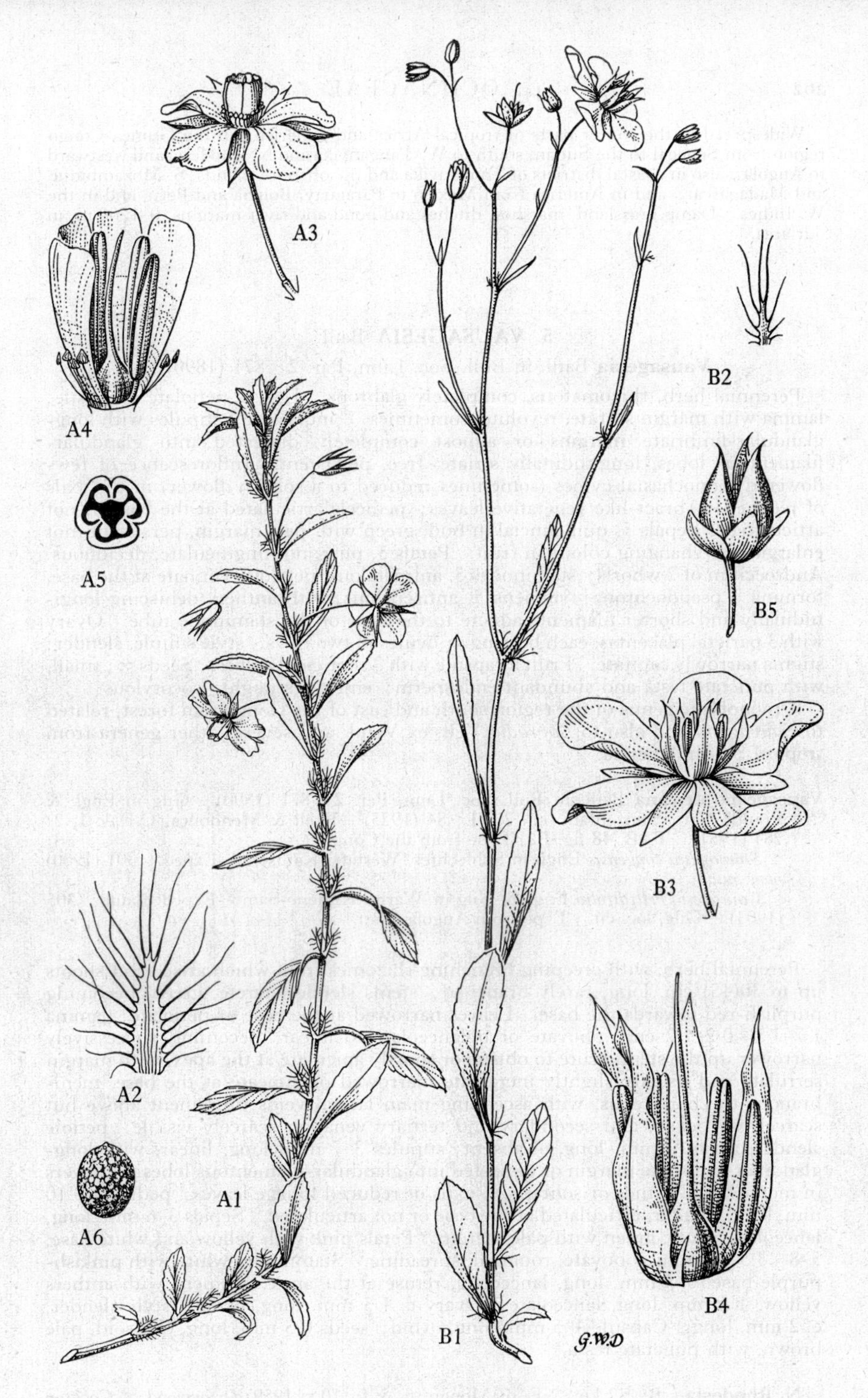

Tab. 48. A.—SAUVAGESIA ERECTA. A1, flowering shoot (×1); A2, stipule (×4); A3, flower (×2); A4, androecium and gynoecium with some outer staminodes, one inner staminode and one stamen removed (×8); A5, transverse section of ovary (×16), all from *Fanshawe* 3630; A6, seed (×20) *Walter* 28. B.—VAUSAGESIA AFRICANA. B1, flowering shoot (×1); B2, stipule (×4), both from *Milne-Redhead* 3030; B3, flower (×2); B4, androecium and gynoecium with one staminode and one stamen removed (×8), both from *Marks* C4 i.e. 56; B5, dehisced capsule (×2), *Gossweiler* 4110.

Widespread in the wetter parts of tropical Africa and America; in the Guinea-Congo region from Senegal to the Sudan, south to W. Tanganyika and N. Rhodesia and westward to Angola; also in coastal districts of Tanganyika and the offshore islands, S. Mozambique and Madagascar; and in America from Mexico to Paraguay, Bolivia and Peru, and in the W. Indies. Damp grassland, marshes, ditches and pond and river margins, 0–1200 m. in our area.

5. **VAUSAGESIA** Baill.

Vausagesia Baill. in Bull. Soc. Linn. Par. **2**: 871 (1890).

Perennial herb, rhizomatous, completely glabrous. Leaves petiolate to sessile; lamina with margin serrate, revolute, sometimes ± incrassate; stipules with long-glandular-fimbriate margins or almost completely dissected into glandular-filamentous lobes, longitudinally striate, free, persistent. Inflorescence of few-flowered monochasial cymes (sometimes reduced to a solitary flower) in the axils of reduced ± bract-like vegetative leaves; pedicels articulated at the base or not articulated. Sepals 5, quincuncial in bud, green with pale margin, persistent, not enlarging or changing colour in fruit. Petals 5, pink, not unguiculate, deciduous. Androecium of 2 whorls; staminodes 5, antipetalous, petaloid, connate at the base, forming a pseudocorona; stamens 5, antisepalous, with anthers dehiscing longitudinally and shorter filaments adnate to the base of the staminode tube. Ovary with 3 parietal placentas, each bearing ∞ ovules in two rows; style simple, slender; stigma narrowly capitate. Fruit a capsule with 3 septicidal valves. Seeds ∞, small, with punctate testa and abundant endosperm; embryo straight, isocotylous.

A monotypic genus of the region south and east of the Congo rain forest, related to *Sauvagesia* and also to *Lavradia* Vell. ex Vand. and several other genera from tropical South America.

Vausagesia africana Baill. in Bull. Soc. Linn. Par. **2**: 871 (1890).—Gilg in Engl. & Prantl, Nat. Pflanzenfam. ed. 2, **21**: 84 (1925).—Exell & Mendonça, C.F.A. **1**, 2: 284 (1951). TAB. **48** fig. B. Type from the Congo.

Sauvagesia congensis Engl. in Schlechter, Westafr. Kautschuk-Exped.: 301 (1900) *nom. nud.*

Vausagesia bellidifolia Engl. & Gilg in Warb., Kunene-Samb.-Exped. Baum: 305 (1903).—Gilg, loc. cit. Type from Angola (Bié).

Perennial herb, with creeping branching rhizomes from which arise erect shoots up to 30(45) cm. long, rarely branched; stems slender, terete, green, becoming purplish-red towards the base. Leaves narrowed at the base or petiolate; lamina 1·7–3·5 × 0·2–1·2 cm., obovate or oblanceolate to linear, becoming successively narrower up the stem, acute to obtuse or shortly apiculate at the apex, with margin serrulate and usually slightly incrassate, narrowed or cuneate at the base, membranous to chartaceous, with ascending main lateral veins prominent above but scarcely so below, and secondary and tertiary venation scarcely visible; petiole slender, up to 2 mm. long, or absent; stipules 3–7 mm. long, linear, with long-glandular-fimbriate margin or dissected into glandular-filamentous lobes. Flowers in monochasial cymes or solitary, in axils of reduced foliage leaves; pedicels 3–10 mm. long, slender, articulated at the base or not articulated. Sepals 3–6 mm. long, lanceolate, acute, green with pale margin. Petals pink with yellow and white base, 5–8 × 3·5–4·5 mm., obovate, rounded, spreading. Staminodes white with pinkish-purple base, 3–5 mm. long, lanceolate, retuse at the apex; stamens with anthers yellow, 1·5 mm. long, lanceolate. Ovary c. 1·5 mm. long, ovoid; style slender, c. 2 mm. long. Capsule 4–5 mm. long, ovoid; seeds 0·5 mm. long, ellipsoid, pale brown, with punctate testa.

N. Rhodesia. B: 32 km. NE. of Mongu, fl. & fr. 10.xi.1959, *Drummond & Cookson* 6308 (K; SRGH). N: Chinsali Distr., Lake Young, Shiwa-Ngandu, 1350 m., fl. & fr. 15.i.1959, *Richards* 10680 (K; SRGH). W: Mwinilunga Distr., bank of R. Kasompa, fl. 31.x.1937, *Milne-Redhead* 3030 (K).

Also in Angola (Lunda, Bié), the Congo and Uganda. Bogs, river and lake margins, and other damp habitats, 350–1500 m.

45. BURSERACEAE

By H. Wild

Trees or shrubs, usually secreting resin or oil. Leaves alternate, rarely opposite, bipinnate (but not in our area) or imparipinnate or 3-foliolate or 1-foliolate or very rarely simple; leaflet margins entire or variously serrate; stipules absent. Flowers unisexual or more rarely bisexual. Calyx 3–5-lobed, lobes imbricate or valvate. Petals 3–5 (rarely absent), free (or variously connate outside our area). Disk present (or sometimes apparently absent), often lobed. Stamens twice as many as the petals or of equal number (outside our area); filaments free or slightly connate at the base; anthers 2-thecous, opening longitudinally. Ovary superior, 2–5-locular; ovules (1) 2 in each loculus, axile. Fruit a drupe (the exocarp sometimes tardily splitting into 2–4-valves). Seeds without endosperm; cotyledons usually contorted, palmatilobed or conduplicate.

Calyx-lobes and petals 4; stamens 8; endocarp partially surrounded by a ± fleshy pseudaril - - - - - - - - - - - 1. *Commiphora*
Calyx-lobes and petals 3; stamens 6; pseudaril absent:
  Indumentum of inflorescences stellate or dendroid with some simple hairs; ovary 2(3)-locular - - - - - - - - - - - 2. *Dacryodes*
  Indumentum of inflorescences of simple hairs only; ovary 3-locular 3. *Canarium*

1. COMMIPHORA Jacq.

Commiphora Jacq., Hort Schoenbr. **2**: 66, t. 249 (1797) *nom. conserv.*

Dioecious (rarely monoecious) shrubs or trees, often spiny; bark papery or smooth, often secreting an odoriferous resin. Leaves usually grouped at the ends of the branches, alternate, petiolate or more rarely sessile, 1-foliolate or 3-foliolate or imparipinnate or rarely simple; leaf or leaflet margins entire or variously dentate, surface usually reticulately veined below. Inflorescences often appearing before the leaves, of axillary ± elongated paniculate sometimes dichasial cymes or ± reduced to abbreviated usually 1–2-flowered axillary cymes often borne on abbreviated side-shoots. Flowers small, unisexual or more rarely apparently bisexual. Calyx narrowly infundibuliform, cylindric, campanulate or broadly campanulate, with 4 valvate persistent lobes. Petals 4, valvate, straight, spreading or recurved at the apex and the latter usually somewhat hooded and mucronulate. Stamens 8 (vestigial in ♀ flowers), inserted just outside the disk, antisepalous ones longer; filaments subterete or broadened and flattened towards the base; anthers oblong, those of the shorter stamens usually mucronulate. Disk apparently absent (lining the calyx-tube) or of 8 small lobes alternating with the stamens or of 4 lobes opposite the longer stamens (not so well developed in ♀ flowers). Ovary (absent or rarely vestigial in ♂ flowers), 2(3)-locular with 2 ovules per loculus; style very short; stigma capitate, obscurely 4-lobed. Fruit an ovoid or ellipsoid or globose drupe; exocarp ± fleshy or occasionally leathery, resinous, splitting when ripe into 2 longitudinal valves (4 valves in a few species outside our area); mesocarp forming a pseudaril intimately clasping the endocarp or part of the endocarp, very variable in shape (rarely apparently absent); endocarp crustaceous or bony, smooth or rugose, enclosing one fertile loculus and a smaller abortive loculus. Seed with a straight embryo; cotyledons much folded.

Most species of *Commiphora* produce aromatic resins and, outside our area, *C. myrrha* (Nees) Engl. of Arabia and *C. molimol* Engl. of Somaliland, as well as perhaps some other species, produce the aromatic resin Myrrh, whilst *C. gileadensis* (L.) Christ. of Arabia and north-east Africa is the source of the medicinal resin Balm of Gilead. Species of *Commiphora* are also widely used to produce " live fences ". Truncheons are planted as a fence and these then proceed to grow.

All our species belong to Subgen. *Commiphora* with the exocarp eventually dividing into 2 rather than 4 valves. (See Wild in Bol. Soc. Brot., Sér. 2, **33**: 81 (1959)).

Commiphora acutidens Engl., Bot. Jahrb. **44**: 153 (1910), whose type was collected near Bulawayo, is not a *Commiphora* sp. but *Sclerocarya caffra* Sond.

Key to Flowering Specimens

Male or bisexual flowers should be examined, for in the female flowers the disk and stamens (staminodes) are less well developed, the calyx is usually smaller and the degree of lobing variable.

Calyx pubescent or glandular:
  Petals glabrous:
    Calyx glandular; leaves 1-foliolate (very rarely 3-foliolate)
      2. *pyracanthoides* subsp. *glandulosa*
    Calyx not glandular; leaves 3-foliolate or pinnate:
      Disk-lobes not developed or of 8 very small lobes; flowers appearing with the young or mature leaves:
        Leaves scabrous above - - - - - - - 16. *edulis*
        Leaves pilose or pubescent - - - - - - - 18. *serrata*
      Disk-lobes 4, well developed; flowers appearing with the leaves or before the leaves:
        Flowers appearing before the leaves; calyx lobed to about ⅓-way
          14. *africana* var. *rubriflora*
        Flowers appearing with the leaves; calyx lobed to at least half-way:
          Leaflets 2–4-jugate, oblong-lanceolate, glabrous, margin crenate-serrate
            4. *puguensis*
          Leaflets 3–6-jugate, lanceolate to narrowly ovate, pilose on the nerves and veins, margin entire - - - - - - - - 6. *karibensis*
  Petals hairy outside except at the margins:
    Disk-lobes absent or of 8 very small lobes; stamen-filaments flattened and broadened towards the base; calyx infundibuliform or infundibuliform-campanulate; flowers appearing with the young leaves:
      Bark pale bluish-yellow; calyx divided to almost half-way - 21. *caerulea*
      Bark brown and smooth or chestnut-brown and polygonal-reticulate, under layer green; calyx lobed to at most ⅓-way - - - - - 20. *angolensis*
    Disk-lobes 4; stamen-filaments slender, terete:
      Disk-lobes hairy:
        Leaflets rather sparsely pubescent when mature, nerves and veins prominently raised below, margins distinctly serrate-undulate - - 12. *pedunculata*
        Leaflets densely pilose or tomentose on both surfaces, nerves and veins not prominently raised, margins coarsely crenate or crenate-serrate
          13. *marlothii*
      Disk-lobes not hairy:
        Leaves imparipinnate; leaflets 2–10-jugate; calyx densely pilose:
          Leaflets entire - - - - - - - 9. *mollis*
          Leaflets serrate - - - - - - - 11. *ugogensis*
        Leaves 3-foliolate or rarely imparipinnate with 2 pairs of leaflets; calyx sparsely pilose with glands intermingled - - - - - 10. *mossambicensis*
Calyx glabrous:
  Disk-lobes not developed or of 8 very small lobes:
    Leaflets 3–8-jugate:
      Leaves quite glabrous; leaflets 3–4-jugate, margins entire or almost so; calyx campanulate - - - - - - - - 17. *zanzibarica*
      Leaves with hairs along the midrib and nerves; leaflets 4–8-jugate, margins crenate-serrate - - - - - - - - 18. *serrata*
    Leaves 3-foliolate or leaflets 2-jugate; calyx infundibuliform 19. *tenuipetiolata*
  Disk-lobes 4:
    Flowers appearing with the leaves (very rarely before the leaves); calyx lobed half-way or almost half-way; leaves 3-foliolate or pinnate with well-developed lateral leaflets:
      Young branches cinereous-pubescent - - - - - - 8. *neglecta*
      Young branches quite glabrous:
        Leaves 3-foliolate or pinnate and leaflets 2–3-jugate; lamina oblong-lanceolate or oblanceolate; calyx campanulate - - - - 7. *schlechteri*
        Leaves 3-foliolate; leaflets broadly elliptic to ovate or lanceolate; calyx campanulate-cylindric - - - - - - - 5. *pteleifolia*
    Flowers appearing before the leaves (very rarely with the leaves); calyx usually lobed to ⅓-way or less (sometimes up to halfway in *C. madagascariensis*); leaves simple or 3-foliolate with very small lateral leaflets:
      Young branches smooth and purplish - - - - - - - 3. *merkeri*
      Young branches smooth, green, brown, grey or yellowish:
        Young twigs and branches densely pubescent or tomentose - 14. *africana*

Young twigs and branches glabrous:
  Bark smooth, dark green or yellowish, peeling in yellowish strips — 15. *schimperi*
  Bark grey or brown or greenish-brown:
    Low spreading much branched shrub 1–2 m. tall; bark grey and papery — 2. *pyracanthoides* subsp. *pyracanthoides*
    Shrub or small tree up to 5 m. tall; bark dark brown or green-brown — 1. *madagascariensis*

Key to Fruiting Specimens

Before the pseudaril is examined dry fruits need boiling for about 15 mins. so that the exocarp can be removed without damage to the pseudaril.

Leaves usually 1-foliolate, occasionally with 2 additional much smaller lateral leaflets, usually fascicled on short side-shoots:
  Endocarp with a raised elongated hump on its flatter face:
    Young branches greyish or at least not purplish:
      Persistent calyx glabrous - - 2. *pyracanthoides* subsp. *pyracanthoides*
      Persistent calyx (and often the leaves also) glandular — 2. *pyracanthoides* subsp. *glandulosa*
    Young branches smooth, purplish - - - - - - - 3. *merkeri*
  Endocarp with no elongated hump on the flatter face - - 1. *madagascariensis*
Leaves all 3-foliolate with the lateral leaflets half the size of the terminal leaflet or more, or leaves pinnate:
  Leaves 3-foliolate; leaflet-margins crenate-serrate; endocarp lumpy, rugose; pseudaril apparently absent (rarely with the pseudaril covering $\frac{3}{4}$ or more of the endocarp and with its margin lobulate):
    Branchlets and leaves quite glabrous - - - - - - - 15. *schimperi*
    Branchlets and leaves pubescent or tomentose:
      Branchlets pubescent with very short white hairs or longer yellowish hairs; persistent calyx glabrous - - - - - - 14. *africana* var. *africana*
      Branchlets tomentose; persistent calyx pubescent 14. *africana* var. *rubriflora*
  Leaves all pinnate or some of them pinnate or if 3-foliolate then leaflets entire or if all leaflets 3-foliolate and crenate-serrate then endocarp never lumpy-rugose:
    Leaflets entire or with a very few crenations near the apex:
      Leaflets quite glabrous:
        Leaflets 3–4-jugate, oblong-lanceolate; pseudaril fleshy and cupular, clasping the lower third of the endocarp - - - - - 17. *zanzibarica*
        Leaves 3-foliolate or leaflets 2(3)-jugate and elliptic to obovate in shape:
          Leaflets broadly elliptic to ovate or lanceolate, apex acuminate; pseudaril with 4 arms - - - - - - - - 5. *pteleifolia*
          Leaflets elliptic to obovate, apex rounded; pseudaril covering about $\frac{2}{3}$ of the endocarp, very thin towards its upper margin - 19. *tenuipetiolata*
      Leaflets hairy:
        Leaflets scabrous above - - - - - - - - 16. *edulis*
        Leaflets not scabrous:
          Young branchlets and leaves glandular; leaves 3-foliolate; leaflets broadly ovate to subcircular or oblate - - - 10. *mossambicensis*
          Young branchlets and leaves not glandular; leaves pinnate, or if 3-foliolate then leaflets elliptic to obovate:
            Leaflets elliptic to oblong-elliptic, often whitish below; fruit pubescent — 9. *mollis*
            Leaflets lanceolate to narrowly ovate; fruit glabrous - 6. *karibensis*
    Leaflets crenate-serrate:
      Leaves glabrous or with a few scattered hairs:
        Leaves all 3-foliolate; leaflet-apices rounded or shortly apiculate 8. *neglecta*
        Leaves pinnate or if 3-foliolate then leaflet-apices acute:
          Young stems with a waxy bloom; leaflets lanceolate to oblong-lanceolate, apices acuminate; fruit not beaked - - - - - 4. *puguensis*
          Young stems without a waxy bloom; leaflets oblong-lanceolate to oblanceolate; apices acute; fruits beaked - - - - - 7. *schlechteri*
      Leaves hairy, at least on the nerves or midrib:
        Lateral leaflets oblong or obovate-oblong or oblong-elliptic; pseudaril fleshy and clasping the lower $\frac{1}{3}$ of the endocarp or with 3 (4) arms:
          Leaflets with hairs usually confined to the midrib; pseudaril fleshy and clasping the lower $\frac{1}{3}$ of the smooth endocarp - - - 18. *serrata*
          Leaflets hairy beneath at least on both midrib and nerves; pseudaril with (3) 4 arms; endocarp rugose:
            Fruit glabrous; branches spiny; leaflets 6–10-jugate - 11. *ugogensis*
            Fruit pubescent; branches not spiny; leaflets 2–5-jugate:

Leaflets sparsely pubescent, venation very prominently reticulate below 12. *pedunculata*
Leaflets densely pilose or tomentose on both sides, veins not prominently reticulate - - - - - - - - - 13. *marlothii*
Lateral leaflets elliptic, broadly elliptic, ovate, rotund, or obovate, or if oblong then pseudaril covering up to $\frac{2}{3}$ of the endocarp and with a thin upper margin:
Pseudaril with 4 fleshy arms; leaves with a few sparse hairs (or glabrous), always 3-foliolate - - - - - - - - 8. *neglecta*
Pseudaril covering about $\frac{2}{3}$ or more of the endocarp and its upper margin very thin; leaves ± pubescent (rarely glabrescent), 3-foliolate and/or pinnate:
Bark pale bluish-yellow: leaflets up to 8 × 4·7 cm., elliptic to oblong-elliptic - - - - - - - - - 21. *caerulea*
Bark brown and smooth or chestnut-brown and polygonal-reticulate; leaflets up to 4 × 2·3 cm., ovate, rotund or oblong-elliptic to oblong 20. *angolensis*

Sect. COMMIPHORA

Subsect. *Madagascarienses* (Engl.) Wild in Bol. Soc. Brot., Sér. 2, **33**: 81, t. 1 fig. B (1959).

Leaves simple (or 1-foliolate) or 3-foliolate with two very reduced lateral leaflets. Inflorescences of much reduced 1–2-flowered axillary cymes; flowers with short pedicels, clustered on abbreviated side-shoots. Calyx shortly divided or divided about half-way. Endocarp with one very convex face and a second only slightly convex face which lacks a pronounced hump or protuberance and remainder of surface smooth.

1. **Commiphora madagascariensis** Jacq., Hort. Schoenbr. **2**: 66, t. 249 (1797); in Engl. & Prantl, Nat. Pflanzenfam. ed. 2, **19a**: 436 (1931).—White, F.F.N.R.: 176, fig. 34C (1962). TAB. **49** fig. A. Type a cultivated plant, country of origin not known.

Balsamodendrum habessinicum O. Berg in Bot. Zeit. **20**: 161 (1862). Type from Ethiopia.

Balsamodendrum africanum var. *habessinicum* (O. Berg) Oliv., F.T.A. **1**: 325 (1868). Type as above.

Commiphora habessinica (O. Berg) Engl. in A. & C. DC., Mon. Phan. **4**: 10 (1883); Bot. Jahrb. **48**: 478, fig. 2Q (1912); in Engl. & Prantl, Nat. Pflanzenfam. ed. 2, **19a**: 436, fig. 205 (1931) " abyssinica ".—Troupin, F.C.B. **7**: 136, fig. 2A (1958). Type as above.

Commiphora subsessilifolia Engl., Bot. Jahrb. **34**: 303 (1904). Type from Tanganyika.

Commiphora salubris Engl., Bot. Jahrb. **54**: 294 (1917). Type from Tanganyika (Kyimbila).

Shrub or small tree up to c. 5 m. tall; bark smooth (dark brown, grey-brown or green-brown); branches often spiny, glabrous. Leaves 1-foliolate or 3-foliolate with 2 much smaller lateral leaflets; petiole very short (c. 1 mm. long) or considerably longer (up to 1·5 cm.), glabrous except for a minute tuft of hairs at the articulation just below the base of the sessile leaflet; leaflet-lamina up to 4 (6) × 1·5 (3) cm., elliptic or narrowly obovate-spathulate, apex acute or obtuse, margins finely crenate-serrate, base cuneate, glabrous on both sides. Flowers appearing before the leaves or with the young leaves in subsessile clusters on short side-shoots or spines. Calyx c. 2 mm. long, tubular, lobed to $\frac{1}{3}$–$\frac{1}{2}$-way, glabrous. Petals c. 4 mm. long. Disk-lobes 4. Stamen-filaments slender, subterete. Fruit c. 1·2 × 0·6 cm., ellipsoid, somewhat flattened, glabrous, apex somewhat apiculate; pseudaril with (3) 4 arms; endocarp c. 7 × 5 mm., smooth, with one face rather deeply convex and one shallowly convex.

N. Rhodesia. N: Lake Mweru, fr. 11.xi.1957, *Fanshawe* 3909 (K). **Mozambique.** N: Palma, fr. 17.ix.1948, *Andrada* 1362 (COI; LISC). Z: Mopeia, fr. 12.ix.1944, *Mendonça* 2045 (LISC).

Widely distributed in tropical Africa north of our area, also in India. A species of lake-shore thicket or in the undergrowth of forest patches.

Subsect. *Pyracanthoides* Wild in Bol. Soc. Brot., Sér. 2, **33**: 82, t. 1 fig. A (1959).

Leaves simple or 3-foliolate. Inflorescences usually of much reduced 1–2-

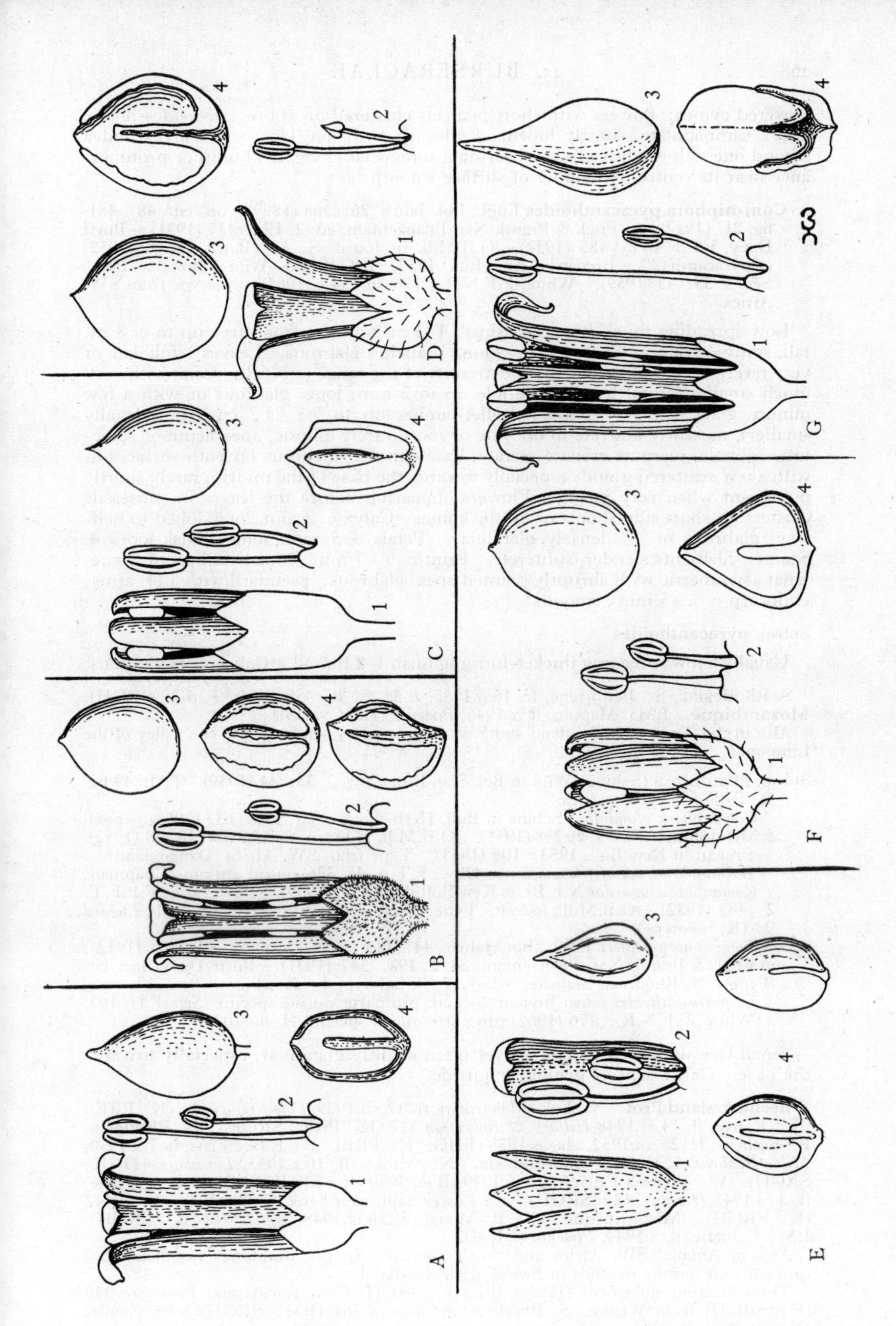

Tab. 49. COMMIPHORA. A.—C. MADAGASCARIENSIS. B.—C. PYRACANTHOIDES SUBSP. GLANDULOSA. C.—C. MERKERI. D.—C. PUGUENSIS. E.—C. PTELEIFOLIA. F.—C. KARIBENSIS. G.—C. SCHLECHTERI. 1, flower (×6); 2, stamens and disk-lobes (×6); 3, fruit (×2); 4, endocarp and pseudaril (×2). All magnifications approximate.

flowered cymes; flowers with short pedicels clustered on abbreviated side-shoots. Calyx campanulate, shortly lobed. Endocarp with one face very convex and a second one only slightly convex bearing a somewhat elongated hump or protuberance near its centre, remainder of surface smooth.

2. **Commiphora pyracanthoides** Engl., Bot. Jahrb. **26**: 368 (1899); op. cit. **48**: 481, fig. 2U (1912); in Engl. & Prantl, Nat. Pflanzenfam. ed. 2, **19a**: 437 (1931).—Burtt Davy, F.P.F.T. **2**: 485 (1932).—O.B.Mill. in Journ. S. Afr. Bot.: **18**: 38 (1952) (" pyracantha ").—Brenan in Kew Bull. **1953**: 104 (1953).—Wild in Bol. Soc. Brot., Sér. 2, **33**: 43 (1959).—White, F.F.N.R.: 176, fig. 34A (1962). Neotype from SW. Africa.

Low spreading much-branched shrub 1–2 m. tall or a small tree up to c. 8 m. tall, spiny, bark grey and papery; young branches glabrous. Leaves 1-foliolate or very rarely (? only at the northern extremity of the range in N. Rhodesia) with very much smaller lateral leaflets; petiole up to 2 mm. long, glabrous or with a few minute glands and ± pilose; leaflet-lamina up to 7·5 × 3·2 cm. (but usually smaller), narrowly obovate to obovate or more rarely elliptic, apex acute or sometimes obtuse, margins crenate-serrate, base cuneate, glabrous on both surfaces or with a few scattered glands especially towards the base of the midrib, rarely shortly pubescent when very young. Flowers appearing before the leaves in subsessile clusters on short side-shoots or on the spines. Calyx c. 2 mm. long, lobed to halfway, glabrous or ± densely glandular. Petals 4–5 mm. long. Disk-lobes 4. Stamen-filaments slender, subterete. Fruit c. 9 × 7 mm., broadly ellipsoid, somewhat asymmetric with abruptly pointed apex, glabrous; pseudaril with 3 (4) arms; endocarp c. 7 × 5 mm., smooth.

Subsp. **pyracanthoides**

Usually a low spreading thicket-forming bush 1–2 m. tall. Calyx quite glabrous.

S. Rhodesia. S: Beitbridge, fr. 16.ii.1955, *E.M. & W.* 458 (BM; LISC; SRGH). **Mozambique.** LM: Maputo, fl. x.1946, *Pimenta* (LISC; LM).
Also in the Transvaal, Zululand and SW. Africa. Dry sandy flats in the valley of the Limpopo.

Subsp. **glandulosa** (Schinz) Wild in Bol. Soc. Brot., Sér. 2, **33**: 44 (1959). TAB. **49** fig. B. Type from SW. Africa.
Commiphora glandulosa Schinz in Bull. Herb. Boiss., Sér. 2, **8**: 633 (1908).—Exell & Mendonça, C.F.A. **1**, 2: 298 (1951).—O.B.Mill. in Journ. S. Afr. Bot. **18**: 38 (1952). —Brenan in Kew Bull. **1953**: 106 (1953). Type from SW. Africa (Damaraland).
Balsamodendrum africanum sensu Oliv., F.T.A. **1**: 326 quoad specim. Chapman.
Commiphora lugardae N.E.Br. in Kew Bull. **1909**: 99 (1909).—Burtt Davy, F.P F.T. **2**: 485 (1932).—O.B.Mill., loc. cit. Type: Bechuanaland Prot., Kwebe Hills, *Lugard* 23 (K, holotype).
Commiphora seineri Engl., Bot. Jahrb. **44**: 145 (1910); op. cit. **48**: 480 (1912); in Engl. & Prantl. Nat. Pflanzenfam. ed. 2, **19a**: 347 (1931).—Burtt Davy, loc. cit. Type: N. Rhodesia, Sesheke, *Seiner* 57 (B, holotype †; K, photo).
C. pyracanthoides sensu Brenan, loc. cit. pro parte quoad specim. *Seiner* II, 103. —White, F.F.N.R.: 176 (1962) pro parte quoad specim. *Angus* 1038.

Small tree up to 8 m. tall. Leaves often slightly glandular, especially toward the base. Calyx densely glandular outside.

Bechuanaland Prot. N: Lake Makarikari, fr. 12.xii.1929, *Pole Evans* 2595 (K; PRE). SE: Kanye, fl. 14.xi.1948, *Hillary & Robertson* 517 (K; PRE; SRGH). **N. Rhodesia.** B: Sesheke, fr. 26.xii.1952, *Angus* 1038 (FHO; K; PRE). S: Kafue Flats, fl. 7.x.1930, *Milne-Redhead* 1223 (K). **S. Rhodesia.** N: Mtoko, fl. 10.x.1955, *Lovemore* 443 (K; SRGH). W: Victoria Falls, fr. 18.xi.1949, *Wild* 3080 (K; SRGH). C: Hartley, Poole, fr. 17.i.1945, *Hornby* 3129 (SRGH). S: Lower Sabi, west bank, fr. 29.i.1948, *Wild* 2442 (K; SRGH). **Mozambique.** T: R. Mouzi, fl. 26.ix.1948, *Wild* 2642 (K; SRGH). LM: Catuane, fl. x.1944, *Pimenta* (LISC).
Also in Angola, SW. Africa and the Transvaal. In dry deciduous woodland, occasionally on termite mounds in *Brachystegia* woodland.
Three fruiting specimens (*Phelps* 102 (K; SRGH) from Kariba and *Lovemore* 239 (K; SRGH) from Wankie, S. Rhodesia and *Banda* 361 (BM; SRGH) from Zomba District, Nyasaland) have slender petioles up to 2 cm. long and a tendency to produce very small lateral leaflets. They have the fruits of *C. pyracanthoides* and have glandular petioles and leaves. By analogy with the variation of leaf form in *C. madagascariensis* they are probably not a distinct species but forms of *C. pyracanthoides* subsp. *glandulosa.*

3. **Commiphora merkeri** Engl., Bot Jahrb. **44**: 144 (1909); op. cit. **48**: 480 (1912); in Engl. & Prantl, Nat. Pflanzenfam. ed. 2, **19a**: (1931) TAB. **49** fig. C. Type from Tanganyika.
Commiphora viminea Burtt Davy, F.P.F.T. **2**: XXI, 485 (1932).—Brenan in Kew Bull. **1953**: 104 (1953). Type from the Transvaal.

Small tree up to 6 m. tall; bark dark green, becoming rough and dark with age but peeling in thin yellowish strips; young branches smooth, rich purple, shallowly longitudinally furrowed, spiny. Leaves 1-foliolate; petiole up to 0·5 mm. long, often with a tuft of brownish hairs; leaflets up to 4 × 2·3 cm., narrowly obovate to obovate, apex rounded, margin crenate in the upper third, base cuneate, glabrous on both sides or with a few minute glands, glaucous. Flowers appearing before the leaves or with the young leaves in clusters on short side-shoots; pedicels up to 5 mm. long, very slender, glabrous. Calyx c. 2·5 mm. long, tubular, lobed about ⅓-way, glabrous. Petals (? immature) 4 mm. long. Disk-lobes 4. Stamen-filaments slender. Fruit c. 1 × 0·6 cm., ellipsoid, somewhat flattened, apex apiculate, glabrous; pseudaril with (3) 4 arms; endocarp c. 8 × 5 mm., smooth, with one face deeply convex and the other shallowly convex with an elongated hump at its centre.

S. Rhodesia. N: Urungwe, Nyanyanga R., fl. 23.xi.1956, *Mullin* 91/56 (SRGH). S: Birchenough Bridge, fr. 29.i.1948, *Wild* 2426 (K; SRGH). **Mozambique.** N: Montepuez, fr. 17.x.1942, fr. *Mendonça* 911 (LISC). SS: Vilanculos, Inhambane, fr. 7.ii.1941, *Torre* 3836 (LISC).
Also in the Transvaal and Tanganyika. Often on granite sand soils in mixed or *Colophospermum mopane* woodland.

Very little flowering material has been collected so far but the Tanganyika specimens appear to have somewhat smaller anthers and less deeply divided calyces than those from the Transvaal, southern Mozambique and S. Rhodesia. It may be that two subspecies are involved but more flowering material is needed to decide this.

Subsect. *Quadricinctae* (Engl.) Wild in Bol. Soc. Brot., Sér. 2, **33**: 83, t. 1 fig. C (1959).

Leaves 3-foliolate or pinnate. Inflorescences of more or less elongated dichasial cymes or much reduced usually 1–2-flowered axillary cymes. Calyx campanulate (often broadly so), divided to about half-way (rarely to more than half-way). Endocarp often with both faces rather shallowly convex or more strongly convex on one or both sides, remainder of surface smooth.

4. **Commiphora puguensis** Engl., Bot. Jahrb. **46**: 289 (1911); op. cit. **48**: 476 (1912); in Engl. & Prantl, Nat. Pflanzenfam. ed. 2, **19a**: 435 (1931).—White, F.F.N.R.: 177, fig. 34I (1962). TAB. **49** fig. D. Type from Tanganyika (Pugu Hills).
Commiphora kyimbilensis Engl., Bot. Jahrb. **54**: 293 (1917). Type from Tanganyika (Kyimbila).

Tree up to 10 (13) m. tall; bark smooth, grey, fluted; young stems with a waxy bloom, strigose-pilose. Leaves pinnate, sparsely strigose-pilose or glabrous; petiole up to 7 cm. long, slender; leaflets 2–4-jugate, membranous, up to 8·5 × 3 cm., lanceolate to oblong-lanceolate, apex acuminate, margin crenate-serrate, base broadly cuneate, glabrous; petiolules up to 2 mm. long. Flowers appearing with the leaves in axillary paniculate cymes up to 10 cm. long; panicle-branches cinereous-pilose or glabrescent; pedicels up to 2 mm. long, pilose. Calyx c. 1·5 mm. long, broadly campanulate, lobed about half-way, pilose. Petals 4 mm. long. Stamen-filaments slender, subterete. Disk-lobes 4. Fruit c. 1·4 × 1 cm., ellipsoid, mucronate, glabrous; pseudaril with 3 arms; endocarp c. 1·2 × 0·9 cm., smooth, with both faces very convex.

N. Rhodesia. N: Kalambo Falls, fr. 15.v.1936, *Burtt* 6025 (BM; K). **Nyasaland.** C: Dedza, Bambele, fl. 21.ii.1961, *Chapman* 1154 (SRGH).
Also in Tanganyika. Lake-shore thicket and on rocky hills with *Brachystegia* spp.

The leaves are highly aromatic and resinous. The branches have a somewhat weeping habit.

5. **Commiphora pteleifolia** Engl., Pflanzenw. Ost-Afr. **C**: 229 (1895); Bot. Jahrb. **48**: 469, fig. 1D (1912); in Engl. & Prantl, Nat. Pflanzenfam. ed. 2, **19a**: 433 (1931). —White, F.F.N.R.: 176 (1962). TAB. **49** fig. E. Type from Tanganyika.

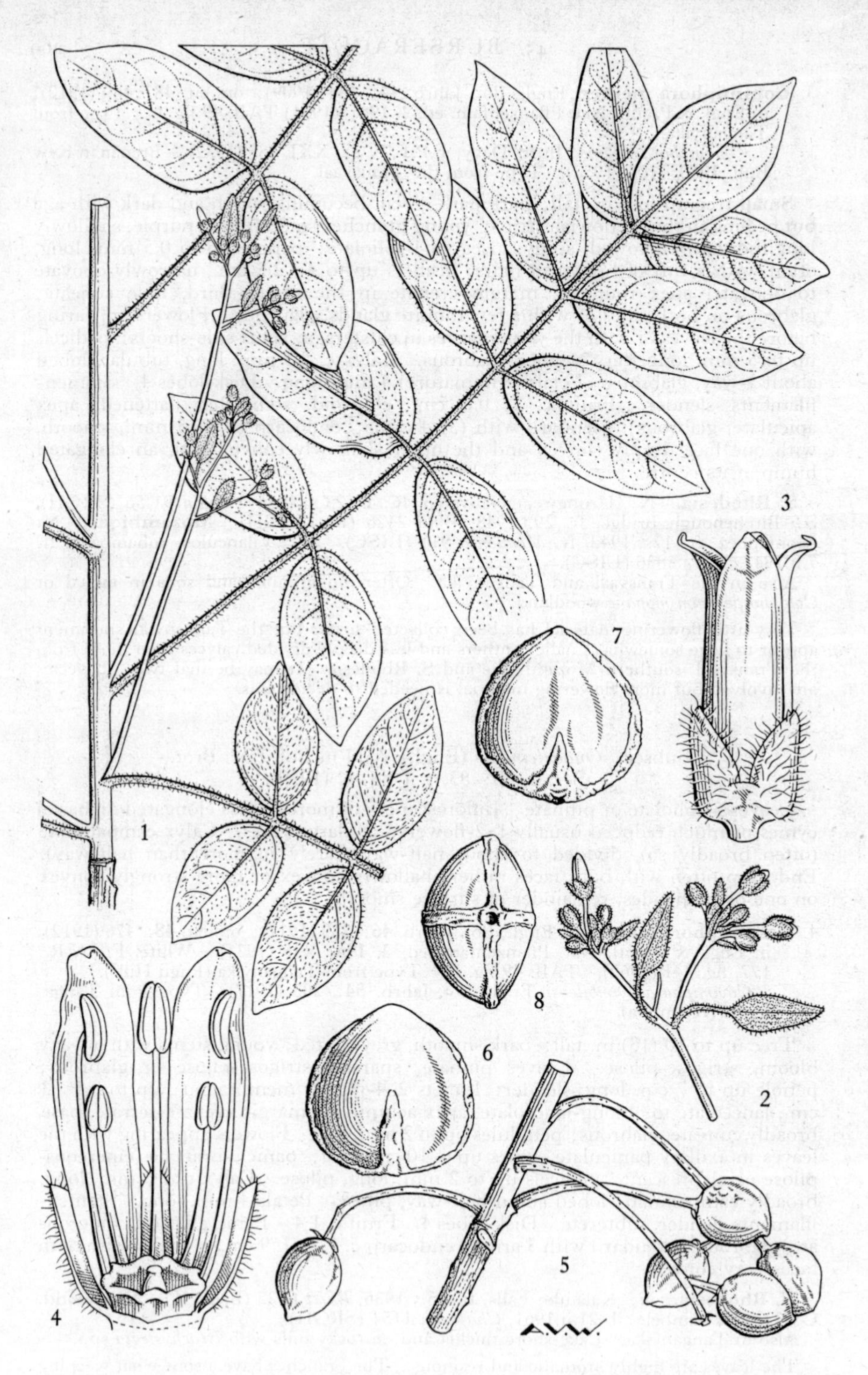

Tab. 50. COMMIPHORA KARIBENSIS. 1, branch with flowers (×1) *Wild* 4159; 2, part of inflorescence (×2) *Robinson* 367; 3, flower (×8) *Robinson* 367; 4, longitudinal section of male flower (×10) *Robinson* 367; 5, part of fruiting branch (×1) *White* 2391; 6, 7 & 8, three views of endocarp with pseudaril (×3) *White* 2312.

Straggling glabrous shrub or small tree up to 7 m. tall; bark smooth, dark green or grey, papery; branches spiny. Leaves 3-foliolate; petiole up to 6 cm. long, slender; leaflets subsessile, lamina up to 7 (9) × 4 (5) cm., broadly elliptic to ovate or lanceolate, with the terminal leaflet somewhat larger and broader than the laterals, apex acuminate, margin very finely crenate-serrate and entire towards the base, base cuneate. Flowers appearing before the leaves or with the young leaves, in small axillary clusters on short side-shoots or on the spines; pedicels c. 1 mm. long. Calyx c. 3 mm. long, campanulate-cylindric, lobed to almost half-way. Petals c. 4 mm. long (immature). Stamen-filaments slender, subterete. Disk-lobes 4. Fruit c. 1·4 × 0·8 cm., ellipsoid-ovoid, apiculate; pseudaril with 4 arms; endocarp c. 0·9 × 0·6 mm., smooth, with one face very convex and the other not so convex.

N. Rhodesia. N: Mkupa, fr. 7.x.1949, *Bullock* 1166 (K).
Also in Tanganyika. Lake shore thicket and woodland.

Appears to contain little or no resin and to be scarcely or not aromatic.

6. **Commiphora karibensis** Wild in Bol. Soc. Brot., Sér. 2, **33**: 83, t. 1 fig. C (1959). —White, F.F.N.R.: 177 (1962). TABS. **49** fig. F, **50**. Type: S. Rhodesia, Urungwe Distr., Msukwe R., *Wild* 4159 (K, holotype; PRE; SRGH).

Tree up to 14 m. tall; trunk longitudinally fluted; bark smooth, grey; branches striate with reddish-grey bark, pilose near the apex; lenticels orange. Leaves pinnate; petiole up to 4·5 cm. long, densely pilose; leaflets 3–6-jugate, lamina up to 7 × 3·3 cm., lanceolate to narrowly ovate, apex acute to acuminate, margin entire, base rounded or broadly cuneate, pilose on the nerves and veins; petiolule up to 1 mm. long, densely pilose; rhachis densely pilose. Flowers appearing with the leaves in axillary paniculate cymes up to 8 cm. long; panicle-branches glabrous; bracts up to 6 × 2·5 mm., foliaceous, lanceolate, apex acute, base cuneate, pilose; bracteoles 2–3 mm. long, caducous, subulate, pilose; pedicels up to 5 mm. long, glabrous. Calyx c. 1·75 mm. long, broadly campanulate, lobed $\frac{3}{4}$-way, pilose outside. Petals c. 4 mm. long. Stamen-filaments subterete. Disk-lobes 4. Fruit c. 1·2 × 1 cm., ovoid, glabrous; pseudaril with 4 arms; endocarp c. 9 × 9 mm., cordiform, smooth with both faces moderately convex.

N. Rhodesia. C: Kafue Gorge, fr. 20.iii.1952, *White* 2312 (FHO; K). S: Gwembe Valley, fr. 1.iv.1952, *White* 2391 (FHO; K). **S. Rhodesia.** N: Urungwe Distr., Mensa pan, fl. 25.xi.1952, *Lovemore* 307 (K; SRGH). W: Sebungwe Distr., Kariangwe, fl. 6.xi.1951, *Lovemore* 155 (K; SRGH).
Not known elsewhere. Dense woodland or thickets in the Zambezi and Kafue valleys.

7. **Commiphora schlechteri** Engl., Bot. Jahrb. **26**: 372 (1899); op. cit. **48**: 477 (1912); in Engl. & Prantl, Nat. Pflanzenfam. ed. 2, **19a**: 435 (1931).—Burtt Davy, F.P.F.T. **2**: 484 (1932).—Mogg in Macnae & Kalk, Nat. Hist. Inhaca I.: 147 (1958). TAB. **49** fig. G. Type: Mozambique, Lourenço Marques, *Schlechter* 11673 (B, holotype†; BM; COI; K; PRE).

Small glabrous tree up to 6 m. tall; bark smooth, grey-green, papery. Leaves 3-foliolate or pinnate; petioles up to 5 cm. long, leaflets 2–3-jugate, lamina up to 7 × 3·2 cm., the terminal leaflet rather larger than the rest, oblong-lanceolate or oblanceolate, apex acute, margin serrate-crenate, base cuneate and often somewhat asymmetric. Flowers appearing with the leaves in axillary paniculate cymes up to 11 cm. long; pedicels up to 2 mm. long. Calyx 1·5 mm. long, campanulate, lobed to at least half-way. Petals c. 4 mm. long. Stamen-filaments subterete, slender. Disk-lobes 4. Fruit c. 2·3 × 1·2 cm., narrowly ovoid, tapering to an acuminate apex, glabrous; pseudaril with 4 winged arms; endocarp c. 2 × 0·7 cm., smooth, both faces deeply convex.

Mozambique. SS: Massinga, fr. 26.ii.1955, *E. M. & W.* 646 (BM; LISC; SRGH). LM: Marracuene, fr. 3.iii.1949, *Barbosa* 2640 (K; LM; SRGH).
Also in the Transvaal and Natal. Coastal dunes.

8. **Commiphora neglecta** Verdoorn in Bothalia, **6**, 1: 214 (1951).—Mogg in Macnae & Kalk, Nat. Hist. Inhaca I.: 147 (1958). TAB. **51** fig. A. Type from the Transvaal.

Tree up to 9 m. tall; bark smooth, green, peeling; branches cinereous-pubescent when young, often spiny. Leaves 3-foliolate; petiole up to 4·5 cm. long, sparsely pilose; leaflets up to 6 × 3·5 cm., terminal usually very broadly obovate, the laterals smaller and broadly elliptic to rotund, apex shortly apiculate or rounded, margins

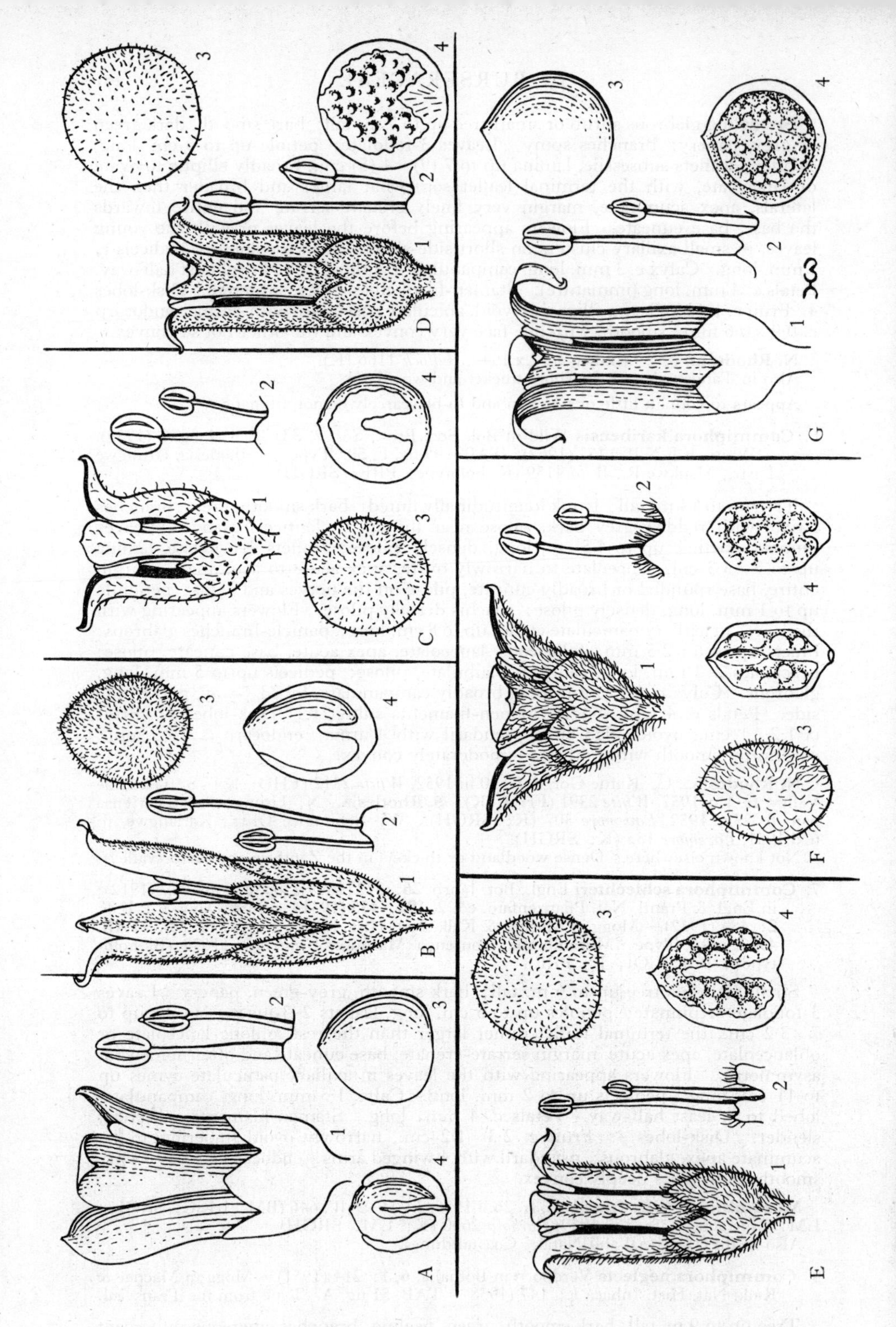

Tab. 51. COMMIPHORA. A.—C. NEGLECTA. B.—C. MOLLIS. C.—C. MOSSAMBICENSIS. D.—C. UGOGENSIS. E.—C. PEDUNCULATA. F.—C. MARLOTHII. G.—C. AFRICANA VAR. AFRICANA. 1, flower (×6); 2, stamens and disk-lobes (×6); 3, fruit (×2); 4, endocarp and pseudaril (×2). All magnifications approximate.

finely crenate-serrate but entire towards the base, base cuneate, glabrous or with a few scattered hairs. Flowers appearing with the leaves or before the leaves, in clusters in short side-shoots or (male) in short dichasial cymes; pedicels c. 2 mm. long, glabrous. Calyx 2 mm. long, campanulate, lobed to half-way. Petals 3·5 mm. long. Stamen-filaments subterete. Disk-lobes 4. Fruit c. 1·3 × 1·2 cm., subglobose, minutely apiculate, glabrous; pseudaril with 4 rather thick and slightly winged arms; endocarp c. 1 × 0·8 cm., smooth, both faces deeply convex.

Mozambique. SS: Guijá, Caniçado, fr. 22.v.1948, *Torre* 7908 (LISC). LM: Maputo, fr. 28.i.1947, *Hornby* 2531 (K; LMJ; PRE; SRGH).
Also in the Transvaal and Natal. Mixed bush or woodland.

9. **Commiphora mollis** (Oliv.) Engl. in A. & C. DC., Mon. Phan. **4**: 23 (1883); Bot. Jahrb. **48**: 472, fig. 10 (1912); in Engl. & Prantl, Nat. Pflanzenfam. ed. 2, **19a**: 435 (1931).—Burtt Davy, F.P.F.T. **2**: 484 (1932).—Brenan in Kew Bull. **1950**: 367 (1950).—Exell & Mendonça, C.F.A. **1**, 2: 298 (1951).—O.B.Mill. in Journ. S. Afr. Bot. **18**: 38 (1952).—Troupin, F.C.B. **7**: 134, fig. 2D (1958).—White, F.F.N.R.: 177, fig. 34H (1962). TAB. **51** fig. B. Type: Mozambique, Chiramba, *Kirk* (K, holotype).
Balsamodendrum molle Oliv., F.T.A. **1**: 326 (1868). Type as above.
Commiphora welwitschii Engl. in DC., Mon. Phan. **4**: 22 (1883); Bot. Jahrb. **48**: 473 (1912); in Engl. & Prantl, Nat. Pflanzenfam. ed. 2, **19a**: 435 (1931). Type from Angola.

Small tree up to c. 8 m. tall; trunk often fluted; bark smooth, dark greenish or dark grey; young branches densely pubescent, sometimes spiny. Leaves pinnate; petiole up to 5 cm. long, densely pubescent; leaflets 2–6-jugate, lamina up to 6 × 3 cm., elliptic or oblong-elliptic, apex acute or obtuse, margin entire, base rounded or broadly cuneate (terminal leaflet always cuneate), pubescent on both surfaces, sometimes distinctly paler and tomentose below; petiolules c. 0·5 mm., densely pubescent. Flowers appearing before the leaves or with the young leaves in clustered axillary dichasial cymes up to c. 2 cm. long; branches of inflorescence pubescent; pedicels slender, c. 2 mm. long; bracteoles caducous, c. 2 mm. long, subulate, brown, pubescent. Calyx c. 2·5 mm. long, campanulate, lobed at least half-way, densely pubescent. Petals up to 5 mm. long, pubescent at the back except towards the edges. Stamen-filaments slender, subterete. Disk 4-lobed. Fruit c. 1·1 × 0·9 cm., ellipsoid, densely pubescent; pseudaril with 4 arms; endocarp c. 9 × 8 mm., smooth, both faces deeply convex.

Caprivi Strip. Mpilila I., fr. 15.i.1959, *Killick & Leistner* 3406 (K; PRE; SRGH). **Bechuanaland Prot.** N: Bosoli, fr. 4.xii.1929, *Pole Evans* 2577 (K; PRE). SE: Macloutsie, fr. 14.iv.1931, *Pole Evans* 3221 (K; PRE). **N. Rhodesia.** B: Mankoya, fr. 23.ii.1952, *White* 2131 (FHO; K). W: Mwinilunga, fl. 2.xi.1955, *Holmes* 1304 (K). C: Broken Hill, fr. 20.v.1955, *Fanshawe* 2328 (K). S: Mazabuka, fr. ii.1933, *Trapnell* 1162 (K). **S. Rhodesia.** N: Urungwe Distr., Chavaru R., fr. i.1956, *Goodier* 27 (K; PRE; SRGH). W: Victoria Falls, fr. 18.xi.1949, *Wild* 3085A (K; SRGH). C: Beatrice, st. 4.iv.1929, *Eyles* 6328 (K; SRGH). E: Hot Springs, fl. 1.i.1949, *Chase* 1510 (BM; K; SRGH). S: Lower Sabi, W. bank, fr. 29.i.1948, *Wild* 2439 (K; SRGH). **Nyasaland.** N: Rumpi, fr. 30.iv.1952, *White* 2541 (FHO; K). C: Domira Bay, st. 2.vii.1936, *Burtt* 6019 (BM; K). S: Boadzulu I., fr. 14.iii.1955. *E.M. & W.* 888 (BM; LISC; SRGH). **Mozambique.** N: Mutuáli, Lioma Rd., fr. 20.iii.1954, *Gomes e Sousa* 4244 (K; LISC). T: Boroma, Sisitso, 15.vii.1950, *Chase* 2760 (BM; SRGH). MS: Chiramba, fr. 13.iv.1860, *Kirk* (K). LM: Magude, fr. 30.xi.1944, *Mendonça* 3169 (LISC).
Also in the Transvaal, SW. Africa, Angola, Congo and Tanganyika. In the hotter and drier types of woodland, common in the valleys of our lower altitude rivers and sometimes thicket forming.

Unusually variable in leaf-form and indumentum for a *Commiphora* sp.

Subsect. *Latifoliolatae* (Engl.) Wild in Bol. Soc. Brot., Sér. 2, **33**: 86, t. 1 fig. D (1959).

Leaves 3-foliolate or pinnate. Inflorescence of more or less elongated dichasial cymes. Calyx usually rather broadly campanulate, lobed about half-way. Endocarp discoid with two slightly convex faces. Pseudaril arms broad, with ± wavy margins, side-arms reaching almost to the apex of the endocarp, remainder of pericarp smooth.

10. **Commiphora mossambicensis** (Oliv.) Engl. in A. & C. DC., Mon. Phan. **4**: 26 (1883).—Sim, For. Fl. Port. E. Afr.: 29 (1909).—White, F.F.N.R.: 176, fig. 34B (1962). TAB. **51** fig. C. Type: Nyasaland, Shire R., 16° S., *Kirk* (K, holotype).
Protium ? mossambicense Oliv., F.T.A. **1**: 329 (1868). Type as above.
Commiphora fischeri Engl., Bot. Jahrb. **15**: 97 (1893); op. cit.: **48**: 471, fig. 1L (1912); in Engl. & Prantl, Nat. Pflanzenfam. ed. 2, **19a**: 435 (1931).—Bak. f. in Journ. Linn. Soc., Bot. **40**: 38 (1911).—R.E.Fr., Rhod. Schwed.-Kongo-Exped. **1**: 110 (1914).—Eyles in Trans. Roy. Soc. S. Afr. **5**: 388 (1916).—Steedman, Trees etc. S. Rhod.: 30, t. 27 (1933).—O.B.Mill. in Journ. S. Afr. Bot. **18**: 38 (1952). Type from Tanganyika.
Commiphora stolzii Engl., Bot. Jahrb. **54**: 292 (1917); in Engl. & Prantl, Nat. Pflanzenfam. ed. 2, **19a**: 435, fig. 203 (1931).—O.B.Mill., tom. cit.: 39 (1952). Type from Tanganyika.

Small tree occasionally reaching 10 m. tall; bark smooth, grey; young branches densely pubescent. Leaves 3-foliolate or more rarely pinnate with the leaflets 2-jugate; petioles up to 10 cm. long, pubescent; leaflet-lamina up to 7 × 8·5 cm., from broadly ovate to subcircular or oblate, apex subacute or abruptly acuminate, margins entire, pubescent or rarely almost glabrous and with minute golden glands, base truncate or shallowly cordate, rarely broadly cuneate; petiolules up to 1·8 cm. long, pubescent. Flowers in paniculate cymes up to c. 7 cm. long; branches of inflorescence sparsely pilose and glandular; pedicels clustered, up to 3 mm. long, glandular-pilose. Calyx c. 1·5 mm. long, broadly campanulate, lobed to rather more than half-way, sparsely pilose and glandular. Petals sparsely pilose outside. Stamen-filaments subterete. Disk 4-lobed. Fruit c. 1 cm. in diam., globose, sparsely pubescent and glandular; pseudaril with 4 rather flattened and wavy-margined arms; endocarp c. 9 mm. in diam., smooth, with one face very convex and the other shallowly convex.

Caprivi Strip. Mpilila I., st. 13.i.1959, *Killick & Leistner* 3365 (K). **Bechuanaland Prot.** N: Kazungula, fr. iv.1935, *Miller* B6 (BM; FHO). **N. Rhodesia.** N: Abercorn Distr., Lucheche R., fr. 23.iv.1936, *Burtt* 6024 (BM; K; PRE). S: Victoria Falls, fl. 19.xi.1949, *Wild* 3101 (K; SRGH). **S. Rhodesia.** N: Urungwe Distr., Zambezi escarpment, st. 31.iii.1955, *Lovemore* 424 (K; SRGH). W: Nyamandhlovu, fr. 9.i.1954, *Plowes* 1658 (K; SRGH). C: Salisbury, fl. ix.1919, *Eyles* 1926 (K; PRE; SRGH). E: Hot Springs, fl. 19.xi.1950, *Chase* 3093 (BM; PRE; SRGH). S: Sabi-Lundi Junction, Chitsa's Kraal, fr. 9.vi.1950, *Chase* 2370 (BM; SRGH). **Nyasaland.** N: Karonga, st. iii.1954, *Jackson* 1252 (K). **Mozambique.** T: Boroma, fl. 1891 *Menyhart* 755 (K).

Also in Tanganyika and Kenya. Common in *Brachystegia* woodland and often on stony hills.

The wood readily exudes resin and the leaves are aromatic with a somewhat peppery smell. The leaves are bright green and conspicuous when young.

Subsect. *Ugogenses* (Engl.) Wild in Bol. Soc. Brot., Sér. 2, **33**: 86, t. 1 fig. E (1959).

Leaves pinnate. Inflorescence of ± elongated dichasial cymes or reduced cymes on abbreviated fertile side-shoots. Calyx campanulate, lobed to less than half-way. Endocarp discoid with rather shallowly convex faces, surface lumpy-rugose. Pseudaril-arms broad and with ± wavy margins, side-arms reaching almost to the apex of the endocarp.

11. **Commiphora ugogensis** Engl., Bot. Jahrb. **34**: 314 (1904); op. cit. **48**: 483, fig. 3H (1912); in Engl. & Prantl, Nat. Pflanzenfam. ed. 2, **19a**: 438 (1931). —White, F.F.N.R.: 177, fig. 34J (1962). TAB. **51** fig. D. Type from Tanganyika.

Small tree up to 8 (15) m. tall; bark papery, green when young, rusty-red and flaking off when old; young branches pubescent, spiny. Leaves pinnate; petiole c. 1 cm. long, densely pubescent; leaflets 6–10-jugate, lamina up to 7·5 × 2·3 cm., lowest leaflets smaller and ovate, others narrowly oblong, terminal one obovate-oblong, apex obtuse or subacute, margins serrate, base rounded and often slightly asymmetric, pubescent on both surfaces or at least on the nerves beneath, venation prominently reticulate. Flowers in axillary sessile clusters on short side-shoots. Calyx c. 2 mm. long, campanulate, lobed somewhat less than half-way, densely pubescent outside. Petals c. 4·5 mm. long, densely pubescent outside. Stamen-filaments subterete. Disk 4-lobed. Fruit c. 1·8 cm. in diam., subglobose, somewhat flattened, glabrous; pseudaril with 4 wavy-margined arms; endocarp c. 1 cm. in diam., lumpy-rugose, with 2 moderately convex faces.

N. Rhodesia. S: Gwembe, 3 km. N. of Sinazongwe, st. 7.iv.1952, *White* 2625 (FHO; K). **S. Rhodesia.** W: Wankie, fr. 25.i.1952, *Lovemore* 238 (K; SRGH). N: Sebungwe Distr., Kariangwe, st. 6.xi.1951, *Lovemore* 154 (K; SRGH). E: Inyanga North, Lawley's Concession, st. 19.ii.1954, *West* 3375 (K; SRGH). **Mozambique.** N: Mossuril, between Maguema and Monapo, st. 5.v.1948, *Pedro & Pedrógão* 3151 (K; LMJ).

Also in Tanganyika. Riverine alluvium, sometimes thicket-forming.

Subsect. *Pedunculatae* (Engl.) Wild in Bol. Soc. Brot., Sér. 2, **33**: 87, t. 1 fig. F (1959).

Leaves pinnate. Disk hairy. Endocarp lumpy-rugose. Pseudaril arms almost reaching apex of endocarp.

12. **Commiphora pedunculata** (Kotschy & Peyr.) Engl. in A. & C. DC., Mon. Phan. **4**: 23 (1883); Bot. Jahrb. **48**: 487, fig. 3S (1912); in Engl. & Prantl, Nat. Pflanzenfam. ed. 2, **19a**: 438 (1931).—Keay, F.W.T.A. ed. 2, **1**, 2: 695 (1958).—White, F.F.N.R.: 177 (1962). TAB. **51** fig. E. Type from the Sudan.
Balsamodendrum pedunculatum Kotschy & Peyr., Pl. Tinn.: 11, t. 5B (1867). Type as above.

Shrub or small tree up to 5 m. tall; bark papery; young branches densely pubescent. Leaves pinnate with leaflets 2–5-jugate, or very occasionally 3-foliolate; petiole up to 3 cm. long, densely pilose; leaflet-lamina up to 8·5 × 3·7 cm., narrowly obovate-oblong to oblong, apex obtuse or subacute, margin coarsely serrate and undulate, base broadly cuneate (at least in the terminal one) or rounded and often asymmetric, sparsely pubescent, nerves and veins very prominently reticulate below; petiolule c. 0·5 mm. long, densely pilose. Flowers in axillary paniculate cymes up to 5 cm. long; branches of inflorescence densely pilose; pedicels c. 1 mm. long, clustered, densely pilose; bracteoles up to 5 mm. long, densely pilose, caducous. Calyx c. 2 mm. long, campanulate, lobed a little less than half-way, densely pilose. Petals 4 mm. long, densely pilose outside. Stamen-filaments subterete. Disk 4-lobed, sparsely pilose. Fruit c. 1·4 cm. in diam., subglobose, somewhat flattened, pubescent; pseudaril with 4 rather wavy-margined arms; endocarp c. 1 cm. in diam., lumpy-rugose, both faces moderately convex.

N. Rhodesia. N: Lake Mweru, st. xi.1930, *Allen* 12 (FHO). E: 32 km. E. of Kachalolo, fr. 24.iii.1955, *E.M. & W.* 1167 (BM; SRGH). **Nyasaland.** C: Domira Bay, st. 2.vii.1936, *Burtt* 6026 (BM; K).

Also in Ubangi-Chari, the Sudan and Nigeria. *Brachystegia* or mixed woodland.

13. **Commiphora marlothii** Engl., Bot. Jahrb. **44**; 155 (1909); op. cit. **48**: 485 (1912); in Engl. & Prantl, Nat. Pflanzenfam. ed. 2, **19a**: 438 (1931).—O.B.Mill. in Journ. S. Afr. Bot. **18**: 38 (1952).—Palgrave, Trees Centr. Afr.: 55 cum t. et 57 cum photo (1958).—White, F.F.N.R.: 177, fig. 34F–G (1962). TAB. **51** fig. F. Syntypes: S. Rhodesia, Matopos, *Marloth* 3397 (B†; K; PRE), 3402 (B†).

Small tree c. 5 m. tall; bark peeling, yellowish (green in the underlayer); young branches tomentose or densely pubescent. Leaves pinnate; petiole up to 10 cm. long, tomentose; leaflets 3–4-jugate, lamina c. 6 × 3·3 cm., oblong to obovate-oblong, apex acute or obtuse, margin coarsely crenate or crenate-serrate, base rounded (cuneate in terminal leaflet) and asymmetric, densely pilose or tomentose on both surfaces; petiolules c. 0·5 mm. long, tomentose. Flowers appearing with the leaves, in axillary paniculate cymes up to c. 15 cm. long; branches of inflorescence tomentose; pedicels clustered, very short, or the flowers subsessile; bracteoles c. 6 mm. long, lanceolate-subulate, densely pilose. Calyx c. 2 mm. long, campanulate, lobed less than half-way, densely pilose outside. Petals 2–3 mm. long, spreading, pilose outside. Stamen-filaments subterete. Disk-lobes 4, pilose. Fruit c. 1·2 × 0·8 cm., ellipsoid, minutely apiculate, pilose; pseudaril with (3) 4 arms; endocarp c. 8·5 × 5 mm., surface rather lumpy, both faces moderately convex.

Bechuanaland Prot. N: between Tutumi and Sebena, fr. 14.xii.1929, *Pole Evans* 2605 (K; PRE). **N. Rhodesia.** N: Mporokoso, 8 km. S. of Chiengi, fl. 5.xi.1952, *Angus* 717 (FHO; K; PRE). S: Victoria Falls, 4th Gorge, fr. 24.xi.1949, *Wild* 3194 (K; SRGH).

S. Rhodesia. N: Lomagundi, st. v.1926, *Eyles* in GHS 1244 (K; SRGH). W: Plumtree, Nata Reserve, fr. iii.1949, Davies 311 (SRGH). C: Salisbury, st. 30.iii.1929, *Eyles* 6316 (K; SRGH). E: Umtali, fr. 21.xi.1948, *Chase* 1274 (BM; K; SRGH). S. Victoria, fl., *Monro* 1573 (BM).

Also in the Transvaal. Open woodland, particularly on rocky hills.

Sect. AFRICANAE Engl., Bot. Jahrb. **48**: 462 (1912).—Wild in Bol. Soc. Brot., Sér. 2, **33**: 87, t. 1 fig. G (1959).

Leaves 3-foliolate. Inflorescence of short reduced cymes. Calyx broadly campanulate, lobed less than half-way. Disk-lobes 4 (5), often slightly bifid at the apex and the longer stamens arising from behind them. Endocarp always more or less rugose on the more convex side. Pseudaril apparently absent but probably united too intimately with the endocarp (and enclosing it completely) to be visible (but sometimes an obvious pseudaril with a lobulate margin and incompletely enclosing the endocarp is visible).

14. **Commiphora africana** (A. Rich.) Engl. in A. & C. DC., Mon. Phan. **4**: 14 (1883); Bot. Jahrb. **48**: 484, fig. 3N (1912); in Engl. & Prantl, Nat. Pflanzenfam. ed. 2, **19a**: 438 (1931).—Sim, For. Fl. Port. E. Afr.: 28 (1909).—Exell & Mendonça, C.F.A. **1,** 2: 300 (1951).—Keay, F.W.T.A. ed. 2, **1,** 2: 695 (1958).—Troupin, F.C.B. **7**: 134, fig. 2B, photo 1 (1958).—Wild in Bol. Soc. Brot., Sér. 2, **33**: 42 (1959).—White, F.F.N.R.: 176, fig. 34B (1962). Type from Senegambia.

Heudelotia africana A. Rich. in Guill., Perr. & Rich., Fl. Senegamb. Tent. **1**: 150, t. 39 (1831). Type as above.

Balsamodendrum africanum (A. Rich.) Arn. in Ann. Nat. Hist. **3**: 87 (1839).—Oliv., F.T.A. **1**: 325 (1868) pro parte excl. syn. *B. schimperi* et vars. Type as above.

Balsamea pilosa Engl., Bot. Jahrb. **1**: 41 (1880). Type from Zanzibar.

Commiphora pilosa (Engl.) Engl. in A. & C. DC., Mon. Phan. **4**: 12 (1883); Bot. Jahrb. **48**: 488, fig. 3U (1912); in Engl. & Prantl, Nat. Pflanzenfam. ed. 2, **19a**: 440 (1931).—R.E.Fr., Schwed. Rhod.-Kongo-Exped. **1**: 111 (1914).—Steedman, Trees etc. S. Rhod.: 31 (1933).—Gomes e Sousa in Soc. Estud. Col. Moçamb. **32**: 75 (1936).—Palgrave, Trees Centr. Afr.: 58 cum t., 60 cum photo (1958). Type as above.

Commiphora nkolola Engl., Bot. Jahrb. **34**: 308 (1904); op. cit. **48**: 490 (1912); in Engl. & Prantl, Nat. Pflanzenfam. ed. 2, **19a**: 440 (1931). Type: Mozambique, ? Madanda (" Mandandu "), *Busse* 528 (B, holotype †; K, fragment of holotype).

Commiphora sambesiaca Engl., Bot. Jahrb. **44**: 146 (1909); op. cit. **48**: 490, fig. 3W (1912); in Engl. & Prantl, Nat. Pflanzenfam. ed. 2, **19a**: 330 (1931).—Burtt Davy, F.P.F.T. **2**: 485 (1932). Type: N. Rhodesia, Katombora, *Seiner* 90 (B, holotype †; K, fragment of holotype).

Commiphora calcicola Engl., Bot. Jahrb. **44**: 147 (1909); op. cit. **48**: 490, fig. 3V (1912); in Engl. & Prantl, Nat. Pflanzenfam. ed. 2, **19a**: 440 (1931).—Burtt Davy, loc. cit. Type from SW. Africa (Damaraland).

Shrub or small tree c. 2–5 m. tall; bark smooth, dark green or yellowish, peeling in yellowish strips; young branches densely pubescent or tomentose with short whitish hairs or sometimes yellowish longer hairs, spiny. Leaves 3-foliolate; petiole up to 4·5 cm. long, but usually less, pubescent or pilose; terminal leaflet up to 8 × 5 cm., often smaller, obovate, apex acute or obtuse, base gradually cuneate, lateral leaflets about $\frac{1}{2}$–$\frac{3}{4}$ the size of the terminal one, elliptic to rotund, apex acute or rounded, margins (of all leaflets) coarsely crenate or crenate-serrate, base rounded or broadly cuneate, both surfaces pubescent or occasionally glabrous above. Flowers appearing before the leaves in axillary abbreviated clusters, often borne on the spines; pedicels up to 2 mm. long, pubescent or glabrous. Calyx c. 2 mm. long, campanulate, lobed to about $\frac{1}{3}$-way, glabrous or pubescent. Petals 3–5 mm. long, stamen-filaments subterete, slender above but broadening somewhat near the base. Disk-lobes 4, ± bifid at the apex. Fruit c. 1·2 cm. in diam., subglobose; pseudaril apparently absent or occasionally irregularly lobulate and covering about $\frac{3}{4}$ of the endocarp; endocarp c. 1 cm. in diam., subglobose but with one face more nearly hemispherical than the other, rugose, the two faces separated by a narrow wing-like rim.

Widely distributed through tropical Africa and also in the Transvaal and SW. Africa. In low-altitude and drier types of woodland and bush throughout our area; rather rare in *Brachystegia* woodland at higher altitudes in N. Rhodesia.

Var. **africana** TAB. **51** fig. G.

Calyx and pedicels quite glabrous.

Bechuanaland Prot. N: Nata R., fr. 20.iv.1931, *Pole Evans* 3345 (K; PRE). **N. Rhodesia.** B: Sesheke, fr. 28.xii.1952, *Angus* 1060 (FHO; K). N: Lake Mweru, fr. 5.xi.1952, *White* 3621 (FHO; K). W: Lake Kashiba, fr. 22.x.1957, *Fanshawe* 3804 (K). S: Victoria Falls, fr. 22.xi.1949, *Wild* 3151 (K; SRGH). **S. Rhodesia.** N: Urungwe Distr., Kariba, fr. 14.i.1956, *Phelps* 105 (BM; K; SRGH). W: Bulawayo, fl. & fr. x.1930, *Eyles* 6629 (K; SRGH). C: Hartley, Poole Farm, fl. & fr. 27.x.1953, *Sturgeon* in GHS 44208 (K; SRGH). S: Lower Sabi R., W. bank, fr. 29.i.1948, *Wild* 2437 (K; SRGH). **Nyasaland.** S: Fort Johnston, Nankumba, fr. 17.xi.1954, *Jackson* 1382 (FHO; K). **Mozambique.** N: R. Rovuma, fr. 8.iii.1961, *Kirk* (K). Z: Mocuba, Namagoa, fr. x–xi.1944, *Faulkner* 20 (COI; K; PRE). T: Boroma, fr. 1891–2, *Menyhart* 760 (K). MS: Gorongoza, fl. 24.ix.1953, *Chase* 5076 (BM; LISC; PRE; SRGH). SS: between Mabote and Zimane, fl. & fr. 2.ix.1944, *Mendonça* 1958 (LISC). LM: Maputo, Goba, fr. 23.xii.1944, *Mendonça* 3464 (LISC).

Distribution and ecology as for species.

Var. **rubriflora** (Engl.) Wild in Bol. Soc. Brot., Sér. 2, **33**: 42 (1959). Type from Tanganyika.

Commiphora rubriflora Engl., Bot. Jahrb. **30**: 336 (1901); op. cit. **48**: 490 (1912); in Engl. & Prantl, Nat. Pflanzenfam. ed. 2, **19a**: 440 (1931). Type: Tanganyika, Unyika, *Goetze* 1406 (B, holotype†; K).

Leaves and stems more densely pilose or tomentose than in var. *africana*. Calyx and pedicels pubescent.

N. Rhodesia. N: Kalambo Falls, fl. 2.xi.1951, *Robertson* 205 (K; PRE; SRGH). **S. Rhodesia.** W: Bulawayo, fl. x.1930, *Steedman* in *Herb. Eyles* 6626 (K; SRGH). E: Umtali, fl. 29.ix.1954, *Chase* 5303 (BM; SRGH). S: Fort Victoria, Umshandige Dam, fl. 6.x.1949, *Wild* 2884 (SRGH). **Mozambique.** N: between Muatua and Angoche, fl. & fr. 7.xi.1936, *Torre* 980 (COI).

Also in Tanganyika & Uganda. Ecology as for species.

Chase 5303 is a specimen somewhat intermediate between the two varieties as the hairs on the calyx are very sparse.

Although the pseudaril is apparently absent in *C. africana* and other members of Sect. *Africanae* (Engl.) Wild, there is evidence that it is not in reality so but that it has been absorbed into the tissue of the endocarp and is now indistinguishable from it. This is shown by some specimens (e.g., *Angus* 1060 (FHO; K) from Sesheke) which have a lobulate pseudaril enclosing about three-quarters of the endocarp. In mature fruits of this kind the naked endocarp is black and the pseudaril surface creamy white, as is the whole surface in typical specimens, but if a transverse section be cut through below the edge of the pseudarillar tissue no black or otherwise distinguishable lower layer can be seen.

15. **Commiphora schimperi** (O. Berg) Engl. in A. & C. DC., Mon. Phan. **4**: 13 (1883); Bot. Jahrb. **48**: 477, fig. 2N (1912); in Engl. & Prantl, Nat. Pflanzenfam. ed. 2, **19a**: 435, fig. 204 C–D (1931). TAB. **52** fig. A. Type from Ethiopia.

Balsamodendrum schimperi O. Berg in Bot. Zeit. **20**: 162 (1862). Type as above.

Balsamodendrum africanum sensu Oliv., F.T.A. **1**: 325 (1868) pro parte quoad specim. Schimper.

Commiphora betschuanica Engl., Bot. Jahrb. **44**: 149 (1909); op. cit. **48**: 478 (1912); in Engl. & Prantl, Nat. Pflanzenfam. ed. 2, **19a**: 435 (1931).—Burtt Davy, F.P.F.T. **2**: 484 (1932).—O.B.Mill. in Journ. S. Afr. Bot. **18**: 38 (1952). Type from Cape Province.

Shrub or small bushy tree, spiny, quite glabrous except for a few brown hairs at the apex and base of the petioles; bark smooth, peeling in yellowish strips. Leaves 3-foliolate; petiole up to 2·5 cm. long; terminal leaflet up to 2·6 × 1·4 cm., narrowly obovate or obovate, apex acute (often truncate to the north of our area), base gradually cuneate; lateral leaflets about half the size of the terminal ones, broadly obovate or ovate, apex acute, base broadly cuneate; margin of all leaflets coarsely crenate in the upper half. Flowers appearing before the leaves in axillary abbreviated clusters, often borne on the spines; pedicels c. 1 mm. long. Calyx 1·5 mm. long, lobed to $\frac{1}{3}$-way or less. Petals c. 3·5 mm. long. Stamen-filaments subterete. Disk-lobes 4, ± bifid at the apex. Fruits c. 1·3 × 0·8 cm., ellipsoid to ± globose, slightly apiculate at the apex; pseudaril apparently absent; endocarp c. 0·75 × 0·5 cm., ellipsoid, lumpy-rugose, with one face very deeply convex and the other shallowly so.

Bechuanaland Prot. N: Zambezi (*fide* O.B.Mill., loc. cit.). **S. Rhodesia.** N: Mazoe, fr. 5.iii.1928, *Eyles* 5846 (K; SRGH). W: Bulawayo, fl. 29.viii.1908, *Chubb* 319

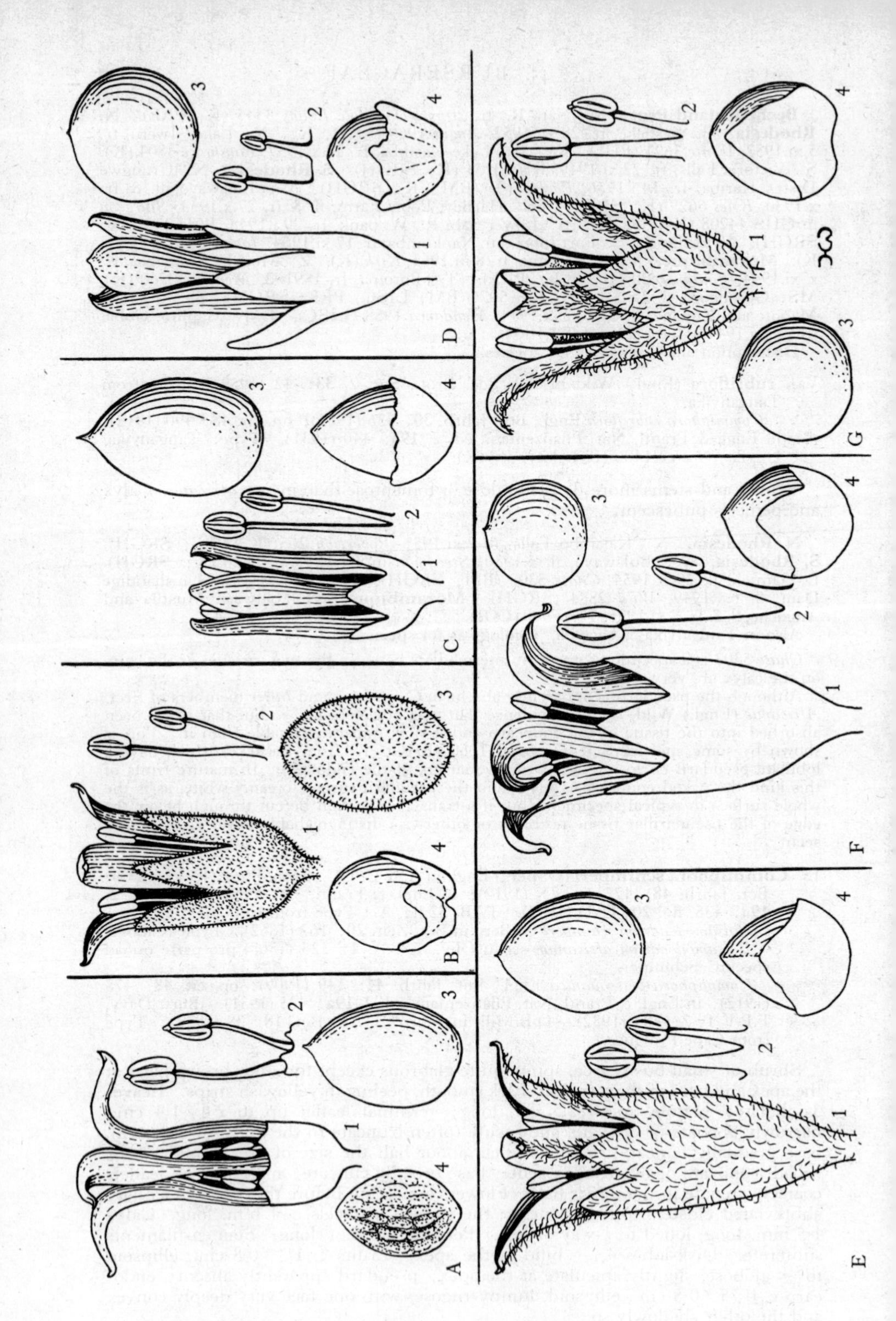

Tab. 52. COMMIPHORA. A.—C. SCHIMPERI. B.—C. EDULIS. C.—C. ZANZIBARICA. D.—C. SERRATA. E.—C. ANGOLENSIS. F.—C. TENUIPETIOLATA. G.—C. CAERULEA. 1, flower (×6); 2, stamens and disk-lobes (×6); 3, fruit (×2); 4, endocarp and pseudaril (×2). All magnifications approximate.

(BM). C: Marandellas Distr., Skipton, st. 5.xii.1946, *Wild* 1613 (K; SRGH). S: Zimbabwe, fr. 4.ii.1957, *Mullin* 23/51 (SRGH). **Mozambique.** SS: Vilanculos, fl. 31.viii.1942, *Mendonça* 44 (LISC).

Widely distributed from Natal, Transvaal and Northern Cape through East Africa to Ethiopia.

Rare in our area. Usually on rocky hillsides or outcrops in *Brachystegia* woodland or in mixed woodland at higher altitudes apparently than *C. africana*.

Very close to *C. africana* but with quite glabrous leaves. Our material, like that from the Union of S. Africa, has acute leaflets and was previously separated as *C. betschuanica*. Material to the north has more truncate or flabellate leaflets but exceptions can be found and this difference is scarcely of specific or even varietal significance.

Sect. SPONDIOIDES Engl., Bot. Jahrb. **48**: 453 (1912) emend. Wild in Bol. Soc. Brot., Sér. 2, **33**: 90 (1959).

Subsect. *Cupulares* Wild, tom. cit.: 90, t. 2 fig. D (1959).

Leaves pinnate. Inflorescence of ± elongated paniculate cymes. Calyx campanulate, lobed to about half-way. Pseudaril cup-like, enclosing c. $\frac{1}{3}$ of the endocarp, margin often rather sinuate.

16. **Commiphora edulis** (Klotzsch) Engl. in A. & C. DC., Mon. Phan. **4**: 22 (1883); Bot. Jahrb. **48**: 474, fig. 1S (1912); in Engl. & Prantl, Nat. Pflanzenfam. ed. 2, **19a**: 435 (1931).—Burtt Davy, F.P.F.T. **2**: 484 (1932).—O.B.Mill. in Journ. S. Afr. Bot. **18**: 38 (1952).—White, F.F.N.R.: 176, fig. 34E (1962). TAB. **52** fig. B. Type: Mozambique, Sena, *Peters* (B†, holotype; K).

Hitzeria edulis Klotzsch in Peters, Reise Mossamb. Bot. **1**: 89 (1861). Type as above.

Balsamea edulis (Klotzsch) Baill. in Bull. Soc. Linn. Par. **1877**: 122 (1877). Type as above.

Commiphora chlorocarpa Engl., Bot. Jahrb. **28**: 414 (1900); op. cit. **48**: 472 fig. 1N (1912); in Engl. & Prantl, Nat. Pflanzenfam. ed. 2, **19a**: 435 (1931). Type from Tanganyika (Uhehe).

Shrub or small tree up to 10 m. tall; bark pale grey, smooth, under layer green peeling; young branches densely pubescent. Leaves pinnate; petiole up to 8 (10) cm. long, pubescent; leaflets 2–4-jugate, up to 9·5 × 4 cm., oblong, narrowly ovate-elliptic or elliptic, apex acute or subacute, margin entire, base rounded (cuneate in terminal leaflet), scabrous above, harshly and densely pubescent below; petiolules up to 1 mm. long, densely pubescent. Flowers appearing with the young or mature leaves, in axillary paniculate cymes up to 15 cm. long; side-branches short; branches of inflorescence densely pubescent; bracteoles c. 4 mm. long, caducous, brown, pubescent; pedicels clustered, very short or flowers sessile. Calyx c. 3·5 mm. long, cylindric-campanulate, lobed to rather more than half-way, densely pubescent outside. Petals c. 5 mm. long. Stamen-filaments slender, broadening and flattening somewhat towards the base. Disk-lobes not developed. Fruit c. 2 × 1·2 cm., ellipsoid, pubescent; pseudaril cupular, very fleshy, clasping the lowest third of the endocarp, with 4 short lobes or undulations; endocarp c. 1·5 × 1 cm., ellipsoid, smooth, both faces moderately convex.

Bechuanaland Prot. N: Chobe R., Kasane, fl. x.1951, *Miller* 1193 (K; PRE). **N. Rhodesia.** C: Lusaka, fr., *Puffett* (PRE). E: Petauke Distr., Ndefu, fr. 19.iv.1952, *White* 2413 (FHO; K). S: Kazungula, fr. iii.1932, *Trapnell* 1058 (K). **S. Rhodesia.** N: Urungwe Distr., Ruwe R., fr. 21.ii.1956, *Phelps* 121 (K; PRE: SRGH). W: Doddieburn Ranch, fr. 24.iv.1924, *Davison* (PRE). E: Hot Springs, fr. 30.xii.1948, *Chase* 1472 (BM; K; SRGH). S: Lower Sabi, fr. 27.i.1948, *Wild* 2294 (K; SRGH). **Nyasaland.** S: Fort Johnston, Naukumba, fl. 20.xi.1954, *Jackson* 1392 (K). **Mozambique.** T: between Lupata and Tete, fr. ii.1859, *Kirk* (K). MS: Sena, fl. ix., *Peters* (B†; K).

Also in the Transvaal and Tanganyika. In thicket or mixed woodland or bush in the hotter and drier areas, common in the valleys of the Zambezi, Shire, Sabi and Limpopo.

The fruit is reported to be eaten by baboons.

17. **Commiphora zanzibarica** (Baill.) Engl. in A. & C. DC., Mon. Phan. **4**: 28 (1883); Bot. Jahrb. **48**: 468, fig. 1A (1912); in Engl. & Prantl, Nat. Pflanzenfam. ed. 2, **19a**: 433 (1931). TAB. **52** fig. C. Type from Zanzibar.

Balsamea zanzibarica Baill. in Adansonia, **11**: 180 (1874). Type as above.

Commiphora spondioides Engl., Bot. Jahrb. **26**: 371 (1899); op. cit. **48**: 468 (1912); in Engl. & Prantl, Nat. Pflanzenfam. ed. 2, **19a**: 433 (1931). Type:

Mozambique, Lourenço Marques, *Schlechter* 11559 (B, holotype †; BM; COI; K; PRE).

Tree up to 12 m. tall; bark smooth, grey, peeling with straw-coloured strips; young branches glabrous. Leaves pinnate; petiole up to 13 cm. long, glabrous; leaflets 3–4-jugate, up to 9·5 × 3 cm., oblong-lanceolate, apex acute or acuminate, margin entire or undulate or very sparsely denticulate, base rounded or broadly cuneate, glabrous; petiolules up to 8 mm. long, but usually less, often with a few hairs near the point of attachment to the rhachis. Flowers appearing with the leaves, in axillary paniculate cymes up to 25 cm. long; branches of inflorescence glabrous; bracteoles up to 1 mm. long, ovate; pedicels 1–3 mm. long, slender, glabrous. Calyx c. 2·5 mm. long, campanulate, tapering to the pedicel, lobed to half-way, glabrous. Petals c. 4 mm. long. Stamen-filaments somewhat broadened towards the base, disk-lobes not developed. Fruit c. 1·6 × 0·9 cm., ellipsoid, apiculate, glabrous; pseudaril cupular, very fleshy, clasping the lowest third of the endocarp, margin undulate; endocarp c. 1 × 0·8 cm., ellipsoid, smooth, both faces moderately convex.

S. Rhodesia. E: Chipinga, Msoloti, fr. 19.vi.1955, *Mowbray* 37 (SRGH). S: Ndanga, Chiribira Falls, fr. 9.vi.1950, *Chase* 2370 (SRGH). **Mozambique.** N: N. of Mandimba, fl. 11.xii.1941, *Hornby* 2404 (PRE). Z: between Mopeia and Campo, fr. 12.ix.1944, *Mendonça* 2046 (LISC). T: near Tete, fr. 8.vi.1947, *Hornby* 2729 (PRE; SRGH). MS: Marromeu, fr. 8.v.1942, *Torre* 4106 (LISC). LM: Lourenço Marques, 30.xi.1897, *Schlechter* 11559 (B†; BM; K).

Also in coastal Tanganyika and Zanzibar. Coastal thickets and low-altitude river valleys. Leaves slightly aromatic when crushed.

18. **Commiphora serrata** Engl. in A. & C. DC., Mon. Phan. **4**: 24 (1883); Bot. Jahrb. **48**: 476, fig. 2K (1912); in Engl. & Prantl, Nat. Pflanzenfam. ed. 2, **19a**: 435 (1931). TAB. **52** fig. D. Type from Tanganyika (Dar es Salaam).

Bush or small tree up to 8 m. tall; bark smooth, green; young branches glabrous or very sparsely hairy, sometimes spiny. Leaves pinnate; petiole up to 5·5 cm. long, glabrous or puberulous, leaflets 4–8-jugate, up to 5·2 × 1·6 cm., oblong-elliptic, apex acute, margin crenate-serrate, base rounded or broadly cuneate, pilose or glabrous on both surfaces except for the nerves and midrib which are sparingly hispidulous; petiolules up to 0·5 mm. long, glabrous or hispidulous. Flowers appearing with the young leaves, in axillary dichasial cymes c. 8 cm. long; branches of inflorescence pilose or glabrous; bracteoles c. 3 mm. long, subulate-lanceolate, margins ciliate; pedicels up to 5 mm. long, slender, pilose or glabrous. Calyx c. 3·5 mm. long, cylindric-campanulate, lobed to at least half-way, pilose or glabrous. Petals 4–5 mm. long. Stamen-filaments broadening towards the base. Disk-lobes probably not developed (male flowers not yet seen). Fruit c. 2 cm. in diam., globose, glabrous; pseudaril cupular, very fleshy, grasping the lowest third of the endocarp, upper margin 4-lobed; endocarp c. 1·3 × 1·1 cm., ellipsoid, smooth, both faces moderately convex.

Mozambique. N: Corrane, fl. 15.xi.1936, *Torre* 1012 (LISC). Z: Mocuba, Namagoa, fl. 20.x.1949, *Faulkner* 418(2)A (K; PRE). MS: Cheringoma, fl. 18.x.1944, *Mendonça* 2495 (LISC).

Also in Tanganyika. Coastal thickets and coastal woodland.

Forms with pilose and glabrous calyces grow together and intermediates are common.

Subsect. *Glaucidulae* (Engl.) Wild in Bol. Soc. Brot., Sér. 2, **33**: 93, t. 2F (1961).

Leaves 3-foliolate or pinnate. Inflorescences of axillary cymes usually shorter than the leaves but not usually reduced to subfasciculate cymose clusters. Calyx with a rather deep funnel-shaped tube and rather short lobes. Stamen-filaments flattened and broadened towards the base. Pseudaril enclosing about ⅔ of the endocarp, becoming very thin above.

19. **Commiphora tenuipetiolata** Engl., Bot. Jahrb. **48**: 483, 3L (1912); in Engl. & Prantl, Nat. Pflanzenfam. ed. 2, **19a**: 438 (1931).—Burtt Davy, F.P.F.T. **2**: 485 (1932). TAB. **52** fig. F. Syntypes from SW. Africa.

Bush or small tree up to 5 m. tall, usually glabrous except for the bracteoles; bark smooth, peeling in whitish strips, under layer dark green. Leaves 3-foliolate or pinnate with the leaflets 2-jugate (rarely 3-jugate); petiole up to 4·5 cm. long, sometimes sparsely pilose, slender and often twisted; terminal leaflet up to 5·8 × 3 cm., obovate to broadly obovate, apex rounded, base gradually cuneate, lateral leaflets up to 4 × 2·5 cm., elliptic, broadly elliptic or obovate, apex obtuse, base cuneate; all leaflets with margins entire or with a few crenations near the apex, glabrous, glaucous; petiolules up to 1 mm. long. Flowers appearing with the very young leaves, in axillary paniculate cymes up to 4 cm. long or the female inflorescences reduced to 1–2-flowered abbreviated cymes; branches of inflorescence very slender; bracteoles c. 3 mm. long, filamentous, pubescent; pedicels up to 3 mm. long, slender. Calyx c. 3–5 mm., long, infundibuliform, lobed to about ⅓-way. Petals 2–3 mm. long. Stamen-filaments flattened and broadened towards the base. Disk-lobes not developed. Fruit c. 1·2 cm. in diam., globose, somewhat flattened; pseudaril covering about ⅔ of the endocarp and becoming thin above; endocarp c. 1 × 0·75 cm., subcordiform, smooth, both faces moderately convex.

S. Rhodesia. S: about 80 km. N. of Beitbridge towards Gwanda, st. 8.v.1948, *Rodin* 4516 (K; PRE; SRGH).

Also in the Transvaal and SW. Africa. Low-altitude and drier types of woodland, often with *Colophospermum mopane*.

20. **Commiphora angolensis** Engl. in A. & C. DC., Mon. Phan. **4**: 24 (1883); Bot. Jahrb. **48**: 486 (1912); in Engl. & Prantl, Nat. Pflanzenfam. ed. 2, **19a**: 438 (1931).—Exell & Mendonça, C.F.A. **1**, 2: 300 (1951).—White, F.F.N.R.: 176, fig. 34D (1962). TAB. **52** fig. E. Syntypes from Angola.

C. rehmannii Engl. in A. & C. DC., Mon. Phan. **4**: 15 (1883); Bot. Jahrb. **48**: 483, fig. 3J (1912); in Engl. & Prantl, Nat. Pflanzenfam. ed. 2, **19a**: 438 (1931).—Burtt Davy, F.P.F.T. **2**: 485 (1932). Type from the Transvaal.

C. oliveri Engl. in A. & C. DC., Mon. Phan. **4**: 24 (1883); Bot. Jahrb. **48**: 483, fig. 3K (1912); in Engl. & Prantl, Nat. Pflanzenfam. ed. 2, **19a**: 438 (1931). Type: Bechuanaland Prot., South-Western Division, *Baines* (K, holotype).

C. kwebensis N.E.Br. in Kew Bull. **1909**: 98 (1909).—O.B.Mill. in Journ. S. Afr. Bot. **18**: 38 (1952). Syntypes: Bechuanaland Prot., Kwebe Hills, *Lugard* 86 (K), *Mrs. Lugard* 34 (K).

Bush or small tree up to 5 m. tall; bark on older specimens chestnut-brown, polygonal-reticulate on bole, flaking in scales 2 cm. across, greenish on branches and peeling in papery buff strips; young branches densely pubescent. Leaves pinnate with the leaflets 2–4-jugate or occasionally 3-foliolate; petiole up to 5 cm. long, pubescent; leaflets up to 6 × 2·3 cm., oblong, elliptic-oblong, ovate or rotund, apex acute or obtuse, margin crenate-serrate, base rounded or very broadly cuneate, slightly asymmetric, sparsely pubescent above, more densely so below, occasionally almost glabrous; petiolules up to 0·5 mm. long, pubescent. Flowers appearing with the young leaves, in axillary dichasial cymes up to 8 cm. long or the female inflorescences reduced to 1–2-flowered abbreviated cymes; branches of inflorescence pubescent; bracteoles up to 3 mm. long, filamentous, pilose; pedicels up to 5 mm. long, slender, pilose. Calyx c. 4 mm. long, infundibuliform, lobed to ⅓-way, pilose. Petals 2·5–4 mm. long, pilose outside except towards the margins. Stamen-filaments flattened and broadened towards the base. Disk-lobes not developed. Fruit 0·85–1·3 × 0·7 × 1 cm., ovoid-globose, minutely apiculate, glabrous; pseudaril covering up to ⅔ of the endocarp and becoming thin above; endocarp c. 1·1 × 0·8 cm., cordiform, smooth, both faces moderately convex.

Bechuanaland Prot. N: Kwebe Hills, fl. 1.xii.1897, *Lugard* 34 (K). SW: without precise locality, *Baines* (K). **N. Rhodesia.** B: Sesheke Distr., 1·5 km. W. of Katima, fr. 1.iii.1952, *White* 1997 (FHO; K). **S. Rhodesia.** W: Wankie Game Reserve, Dett Rd., fr. 22.ii.1956, *Wild* 4784 (K; PRE; SRGH). E: Hot Springs, fl. 16.xi.1952, *Chase* 4713 (BM; K; PRE; SRGH).

Also in Angola, Transvaal, northern Cape Prov. and SW. Africa. Kalahari Sands with *Baikiaea plurijuga* and in *Colophospermum mopane* woodland. Often thicket-forming.

21. **Commiphora caerulea** B. D. Burtt in Kew Bull. **1935**: 111 (1935). TAB. **52** fig. G. Type from Tanganyika.

Tree up to 12 m. tall; bark pale bluish-yellow, peeling in papery buff or straw-coloured strips; young branches greyish-pubescent. Leaves pinnate or more rarely 3-foliolate; petiole up to 5 cm. long, pubescent; leaflets 2–3-jugate, up to

8 × 4·7 cm., elliptic or oblong-elliptic, apex acute, margin crenate-serrate, base rounded or very broadly cuneate, asymmetric, upper surface with a few scattered hairs on the nerves and midrib or glabrescent, more densely pubescent below; petiolules up to 3 mm. long, pubescent. Flowers appearing with the very young leaves, in axillary dichasial cymes up to 2·5 cm. long or the female inflorescences reduced to very short 1–2-flowered cymes; branches of inflorescence greyish-pubescent. Calyx c. 2·5 mm. long, campanulate, lobed to almost half-way, densely pilose outside. Petals c. 3 mm. long, pilose outside except at the margins. Stamen-filaments broadened and flattened towards the base. Disk-lobes not developed. Fruit c. 1·5 cm. in diam., subglobose, somewhat flattened, glabrous; pseudaril covering about ¾ of the endocarp, and becoming thin above; endocarp c. 1 × 0·8 cm., subcordiform, smooth, both faces moderately convex.

S. Rhodesia. N: Kariba, fl. xi.1959, *Goldsmith* 52/59 (SRGH). **Nyasaland.** N: Rumpi, Njakwa Gorge, fr. 30.iv.1952, *White* 2533 (FHO; K; SRGH). S: Lake Nyasa, Boadzulu I., fr. 14.iii.1955, *E.M. & W.* 887 (BM; LISC; SRGH).

Also in Tanganyika. Thicketed ravines and rocky slopes.

2. DACRYODES Vahl

Dacryodes Vahl in Skrivt. Nat. Selsk. **4**: 116 (1810).

Dioecious trees. Leaves imparipinnate; leaflets entire. Inflorescences of axillary or terminal, elongated panicles; branches of inflorescence, calyx and corolla usually with a dense indumentum of stellate or dendroid hairs mixed with simple hairs. Flowers unisexual. Calyx rotate or broadly campanulate, divided almost to the base; lobes 3, valvate. Petals 3, valvate or sometimes slightly imbricate, incurved, somewhat hooded and usually mucronulate at the apex. Stamens 6, inserted just outside the disk, equal in size, with filaments broader towards the base (smaller and infertile in female flowers). Disk annular but slightly lobed between the stamens. Ovary (vestigial in male flowers) 2(3)-locular with 2 ovules per loculus; style very short, stigma 3–4-lobed. Fruit an ovoid or ellipsoid drupe; endocarp thin and cartilaginous. Seed large; cotyledons very much thickened and deeply folded or conduplicate, thus appearing palmately lobed.

Dacryodes edulis (G. Don) H. J. Lam in Bull. Jard. Bot. Buitenz., Sér. 3, **12**: 336 (1932).—Keay, F.W.T.A. ed. 2, **1**, 2: 696 (1958).—Troupin, F.C.B. **7**: 139 (1958). —White, F.F.N.R.: 177 (1962). TAB. **53** fig. B. Type from S. Tomé.
Pachylobus edulis G. Don, Gen. Syst. **2**: 89 (1832).—Hemsl. in Hook., Ic. Pl. **26**: tt. 2566–7 (1898).—Exell & Mendonça, C.F.A. **1**, 2: 304 (1951). Type as above.
Canarium edule (G. Don) Hook. f. ex Benth. in Hook., Niger Fl.: 285 (1849).—Oliv., F.T.A. **1**: 327 (1868). Type as above.

Evergreen tree up to 24 m. tall with a much-branched dense crown; trunk shallowly fluted; bark reddish-brown; young branches with a dense indumentum of ferruginous stellate or dendroid hairs. Leaves pinnate; petiole up to 7·5 cm. long, sparsely ferruginous-hairy, longitudinally striate; leaflets 5–8-jugate, up to 17 (25) × 6 (8) cm., pairs slightly overlapping as a rule, lamina oblong-lanceolate or, especially towards the petiole, ovate-lanceolate, apex long-acuminate, base often asymmetric, broadly cuneate, rounded or slightly cordate, glabrous above, very sparsely stellate-pubescent or glabrous below, nerves in 10–15 pairs, prominent below, anastomosing well within the margin; petiolules up to 1 cm. long. Inflorescences ferruginous-tomentose like the young branches, up to 40 cm. long; bracts up to 2 × 0·6 cm., elliptic, carinate; pedicels 1–2 mm. long. Calyx c. 3 mm. long, lobed nearly to the base, ovate-lanceolate, ferruginous-stellate-tomentellous. Petals c. 5 mm. long, with ovate-lanceolate lobes. Stamen-filaments c. 3 mm. long in male flowers; anthers ovoid, muticous. Ovary ovoid, glabrous or sometimes pubescent. Fruit (described from material outside our area) bright blue, up to 6 × 3 mm., oblong-ellipsoid.

N. Rhodesia. N: Shiwa Ngandu, fl. 7.viii.1938, *Greenway* 5578 (EA: K).

Also in Angola, Congo, Nigeria, Cameroons, Principe and S. Tomé. Swamp forest. Fruit edible.

N. Rhodesian material is characterised by the very dense indumentum on the inflorescences and young branches with a proportion of well-developed dendroid hairs as well as stellate hairs. These can also be seen, however, though less well developed, in some West African specimens.

3. CANARIUM L.

Canarium L., Herb. Amboin.: 10 (1754); Amoen. Acad. **4**: 121 (1759).

Dioecious trees. Leaves imparipinnate, leaflets in up to c. 24 pairs, usually entire. Inflorescences of axillary paniculate cymes, peduncle usually elongated; cymules surrounded by persistent or caducous bracts. Flowers unisexual. Calyx infundibuliform, campanulate or saucer-shaped, lobes 3. Petals 3, usually imbricate below but valvate towards the apex. Stamens 6, free or connate below, inserted on or outside the disk or sometimes, in male flowers, inserted on the disk which then resembles an androphore, smaller and infertile in female flowers. Disk variable in form, sometimes pubescent. Ovary (absent or vestigial in male flowers) 3-locular, loculi 2-ovulate; style ± elongated; stigma subcapitate, ± 3-lobed. Fruit a drupe; pericarp fleshy; endocarp crustaceous, ± trigonous, 3-locular with 3 seeds or 2 loculi sometimes abortive. Seed with 3-foliolate or palmatifid, plicate or conduplicate cotyledons.

Leaflets in 8–12 (23) pairs; calyx c. 1 cm. long - - - - 1. *schweinfurthii*
Leaflets in 4–5 (9) pairs; calyx 2–3 mm. long - - - - 2. *madagascariense*

1. **Canarium schweinfurthii** Engl. in A. & C. DC., Mon. Phan. **4**: 145 (1883).—Exell & Mendonça, C.F.A. **1**, 2: 305 (1951).—Keay, F.W.T.A. ed. 2, **1**, 2: 697 (1958).—Troupin, F.C.B. **7**: 144 (1958).—Leenhouts, Mon. Gen. Canarium in Blumea, **9**: 382, fig. 20 (1959).—White, F.F.N.R.: 173 (1962). TAB. **53** fig. A. Type from the Sudan.

Tree up to 30 (45) m. tall (but often less in our area); bark grey, rough; crown flattish and much branched; young branches densely ferruginous-pubescent. Leaves up to 50 cm. long; petiole up to c. 5 cm. long, ferruginous-pubescent; rhachis and petiole flattened above, semicircular in section below, often slightly winged at the edges towards the base; leaflets 8–12 (23)-jugate, up to 15 (18) × 4 (5·5) cm., oblong, apex acuminate, margins entire, base cordate, sparsely pubescent or glabrescent above, ± densely hairy beneath, especially on the nerves, nerves and veins moderately prominent above, more prominent below; petiolules 1–5 mm. long, densely pubescent. Flowers in dense panicles up to 30 cm. long; branches of inflorescence ferruginous-tomentellous; bracts caducous, 1–2 cm. long, ovate-triangular to lanceolate-triangular, ferruginously tomentellous; pedicels 1–5 mm. long, ferruginously tomentellous. Calyx c. 1 cm. long, infundibuliform, lobed about ½-way, ferruginous-tomentellous outside and within. Petals 0·6–1·2 cm. long, lanceolate, keeled, greyish-tomentellous outside, glabrous within. Stamens with filaments c. 0·4 mm. long, inserted on the cylindric disk, glabrous; anthers oblong (in female flowers c. 2 mm. long, with the thecae sparsely pubescent). Disk* cylindric, pubescent in male flowers. Ovary ovoid to obconic, glabrous; style c. 2 mm. long; stigma 3-lobed. Fruit purplish, up to 4 × 2 cm., plum-like, ellipsoid, usually mucronulate; endocarp almost as long as the fruit, hard and trigonously spindle-shaped.

N. Rhodesia. N: Abercorn Distr., near Chileshi village, fl. & fr. 30.xii.1955, *Kerfoot* 41 (FHO; K).

Also in Angola, Congo, Senegal, Guinea, Portuguese Guinea, Sierra Leone, Ivory Coast, Ghana, Nigeria, Togoland, Cameroons, Ubangi-Chari, Sudan and Tanganyika. Riverine forest and forest patches or remaining as isolated trees probably because of the destruction of other forest species.

The wood is said to secrete oil and is used for canoe making. The endocarp is used by children as spinning tops and the fruit is eaten by birds. The species is said to be very rare in N. Rhodesia (Kerfoot).

2. **Canarium madagascariense** Engl. in A. & C. DC., Mon. Phan. **4**: 111 (1883).—Perrier, Fl. Madag., Burserac.: 43, t. 10 (1946).—Leenhouts, Mon. Gen. Canarium in Blumea, **9**: 377 (1959). Type from Madagascar.

* The staminal tube is said to be formed from the disk, hence the apparent absence of the latter in TAB. **53** fig. A3.

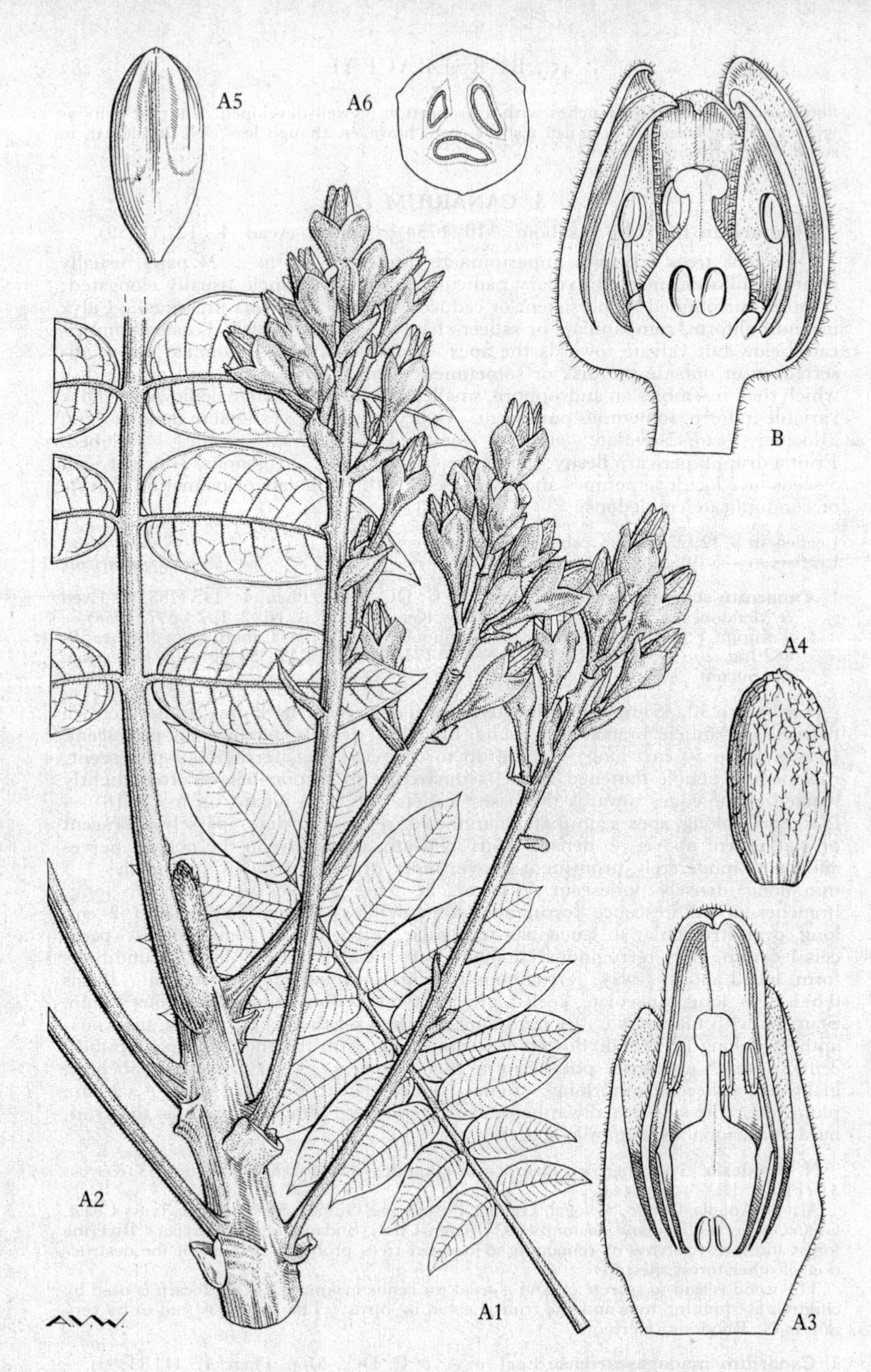

Tab. 53. A.—CANARIUM SCHWEINFURTHII. A1, leaf ($\times \frac{1}{3}$); A2, flowering branch ($\times 1$); A3, longitudinal section of female flower ($\times 5$); A4, fruit ($\times 1$), all from *Kerfoot* 41; A5, endocarp ($\times 1$) *Kennedy* 536; A6, transverse section of endocarp ($\times 1{\cdot}5$) *Welwitsch* 1877. B.—DACRYODES EDULIS. Vertical section of female flower ($\times 8$) *Greenway* 5571.

Canarium liebertianum Engl. in Notizbl. Bot. Gart. Berl. **2**: 270 (1899). Type from Tanganyika.

Laxly branched deciduous tree up to 12 (30) m. tall; bark brown, rough, fissured; branches spreading; crown round; young branches, petiole, leaf-rhachis and branches of inflorescence covered with a ferruginous or greyish-ferruginous indumentum of simple hairs. Leaves up to 25 (55) cm. long; petiole up to c. 7 cm. long; leaflets 4–5 (9)-jugate, up to 12 (16) × 4 (5·5) cm., the lower pairs smaller, ovate-oblong to oblong, apex bluntly acuminate, margins entire or undulate, base rounded, glabrous or almost so on both sides except for the midrib which is densely pilose below; nerves slightly raised below; petiolules c. 1·2 cm. long, hairy. Flowers in laxly paniculate inflorescences up to 20 cm. long (♀ inflorescence smaller), with clusters of 6–15 flowers in cymose glomerules; outer bracts broad and very caducous, inner bracts c. 3 mm. long, linear, ferruginous-pubescent; pedicels 1 (6) mm. long, calyx 2–3 mm. long, campanulate, shortly and obtusely lobed to about $\frac{1}{3}$ of the way or less, densely ferruginous-pubescent outside, glabrous within. Petals c. 5 × 3 mm., oblong, keeled, apex mucronulate, pubescent outside, glabrous within. Stamens 2–3 mm. long, with thickened filaments tapering to the apex, glabrous. Disk annular, very small. Female flowers not seen. Fruit up to 4·5 × 2·3 cm., ovoid-ellipsoid, plum-like; endocarp almost as large as the fruit, ovoid-ellipsoid and somewhat trigonous.

Mozambique. Z: Ile, between Mulevala and Nampero, fr. 31.v.1949, *Andrada* 1557 (LISC).

Also in Madagascar and Tanganyika. Forest patches and river banks.

Yields an aromatic gum.

According to Leenhouts (loc. cit.) our material should be referred to subsp. *madagascariense*, the only subsp. recorded so far from the mainland of Africa.

46. MELIACEAE

By F. White and B. T. Styles

Trees, shrubs or shrublets. Wood often scented. Indumentum of simple, glandular or stellate hairs. Leaves usually alternate, 2- or 3-pinnate, simply impari- or pari-pinnate, 3-foliolate, 1-foliolate or simple; leaflets entire, crenate or serrate. Stipules absent. Inflorescence usually axillary or in axils of fallen leaves, of cymose panicles or compound or simple cymes or flowers fasciculate. Flowers bisexual or unisexual, monoecious or dioecious, occasionally polygamous, actinomorphic, mostly 5-merous; sepals and petals dissimilar. Sepals small, 4–6, variously connate or almost free, the lobes imbricate or with open aestivation, never completely covering corolla in bud. Petals usually 4–5, free, valvate, imbricate or contorted. Disk intrastaminal, very variable, often developed from the gynophore, completely fused to base of staminal tube, or annular, cup-shaped or cushion-shaped and free from staminal tube and ovary or cushion-shaped and enveloping the base of the ovary. Stamens (5)8–10(20), rarely completely free, usually partly or completely fused to form a staminal tube, usually bearing appendages; anthers 2-thecous, dehiscing longitudinally, connective usually apiculate beyond anther-lobes. Ovary superior, (2)4–5(20)-locular, with axile placentation; ovules 1–∞ per loculus; style 1; style-head expanded, capitate, globose, ovoid, cylindric, discoid or coroniform, entire or shallowly lobed, only partly stigmatic, sometimes (*Turraea*) functioning as a *receptaculum pollinis*. Fruit a loculicidal or septifragal capsule or a drupe. Seeds usually arillate, winged or with a corky outer integument; endosperm present or absent.

A family of 50 genera and 800 species, almost confined to the tropics and subtropics of both hemispheres.

Most authors have placed *Ptaeroxylon* Eckl. & Zeyh. in or near Sapindaceae. Harms placed it in Meliaceae in both editions of Die Natürlichen Pflanzenfamilien. Leroy (in Journ. Agr. Trop. Bot. Appl., **7**: 455–6 (1960)) has recently shown convincingly that it is best treated as part of a small family, Ptaeroxylaceae, closely related to Sapindaceae. It will be dealt with in the second part of this volume.

That unisexual flowers occur in the Meliaceae has long been known to some workers but the fact is not often mentioned in text-books and floras. They are found in many more genera than has previously been supposed. All the specimens we have examined of all our species of *Trichilia*, *Pseudobersama* and *Ekebergia* are either male or female and the plants themselves presumably dioecious, and not polygamous as various authorities suggest. *Entandrophragma caudatum*, *E. stolzii*, *Khaya nyasica* and *Lovoa swynnertonii* have male and female flowers in the same inflorescence. The sterile vestiges of the non-functional sex in all these genera are always well-developed and, on a superficial examination, can easily be mistaken for fertile organs. In all cases the antherodes are indehiscent and lack pollen. The pistillodes may have vestigial loculi and minute vestigial ovules, but they are always much more slender than the fertile gynoecia. In dioecious species the overwhelming majority of herbarium specimens are male, and of some species the female plant has been collected only once. Field observations and experimental studies of pollination are needed for all species.

The term "aril" in this account refers to a red or yellow fleshy structure intimately associated with the seed. Within the family its morphological nature varies greatly (see Harms in Engl. & Prantl, Nat. Pflanzenfam., ed. 2, **19 b1**: 25–6, 1940) and in many cases is not known.

Azadirachta indica A. Juss., the " Nim " or " Neem Tree ", a native of India and Burma, is planted for forestry purposes at Ndola and probably elsewhere; *Cedrela ordorata* L., the " Central American Cedar ", widespread in Central and S. America, is planted as an avenue tree in Blantyre; *Swietenia mahagoni* Jacq., the " Mahogany ", native of Florida and the W. Indies, is grown in the Municipal Garden at Lourenço Marques; *Swietenia humilis*, " Mahogany ", native of the Pacific Coast of Central America, differing from *S. mahagoni* in its almost sessile leaflets with long filiform apices, ciliate calyx and petals, and bright brown (not dark chestnut-brown) seeds, is grown at Zomba Investigational Farm; *Toona ciliata* M. J. Roem., the " Toon Tree " or " Red Cedar ", native of tropical Asia, is widely planted as an avenue tree and is sometimes self-sown in our area; *Toona serrata* (Royle) M. J. Roem., native of tropical Asia, with serrate leaflets, is planted on the lower slopes of Mt. Mlanje.

Leaves paripinnate; fruit a large septifragal capsule; seeds winged or with a thick corky outer integument:
- Capsule elongate, more than 3 times as long as broad; seeds winged at one end only; staminal tube without appendages or with paired deltoid-acuminate appendages:
  - Flowers 5-merous; staminal tube without appendages; capsule more than 12 cm. long, opening by 5 valves, columella deeply indented distally with the imprints of 3 or more seeds per loculus; seeds attached by the seed end to the distal end of of the columella and winged towards the base of the capsule — 2. **Entandrophragma**
  - Flowers 4-merous; staminal tube with paired deltate-acuminate appendages; capsule less than 6 cm. long, columella slightly indented proximally with the imprints of 1–2 seeds per loculus; seeds attached by the wing end to the distal end of the columella - - - - - - - - - - - 3. **Lovoa**
- Capsule more or less spherical, not or scarcely longer than broad; seeds winged more or less equally all round the margin or unwinged; staminal appendages subcircular:
  - Calyx-lobes and staminal appendages overlapping; ovules 12–16 per loculus; capsule woody with a well-developed woody columella with 4–5 sharp ridges; seeds 6 or more per loculus, broadly transversely ellipsoid, narrowly winged ± all round the margin; trees of evergreen forest - - - - - - 1. **Khaya**
  - Calyx-lobes and staminal appendages not overlapping; ovules 2–4 per loculus; capsule leathery with a vestigial columella; seeds 2–4 per loculus, large, pyramidal or tetrahedral with angular margins, unwinged but with a thick corky outer integument; trees of mangrove swamps and coastal scrub - 4. **Xylocarpus**

Leaves imparipinnate, pinnately 3-foliolate, 1-foliolate, simple or 2- or 3-pinnate; fruit a drupe or a small loculicidal capsule; seeds neither winged nor with a corky outer integument, often arillate:
- Fruit a capsule; seeds arillate or with a fleshy testa:
  - Leaves pinnate, very rarely 3-foliolate or 1-foliolate; flowers unisexual; filaments usually united in lower half only, or if completely united then staminal tube without appendages; capsule with 2–5 valves:

Anthers and antherodes glabrous; ovary (2)3(4)-locular; capsule with 2(3)4 leathery, rather thin valves, lacking antler-shaped appendages - **5. Trichilia**
Anthers and antherodes puberulous; ovary (4)5-locular; capsule with (4)5 thick woody valves bearing branched antler-like appendages - **6. Pseudobersama**
Leaves always simple; flowers bisexual; filaments fused to apex of staminal tube and bearing paired appendages alternating with the anthers or completely or partly fused to form a frill continuing the staminal tube; fruit with (4)5–10(20) valves **8. Turraea**
Fruit a drupe; seeds usually exarillate:
Leaves 2- or 3-pinnate; leaflets crenate or serrate - - - - - **9. Melia**
Leaves 1-pinnate; leaflets entire:
Indumentum stellate; flowers bisexual; petals valvate-induplicate; filaments fused in lower half; anthers inserted between a pair of deltate-acuminate appendages - - - - - - - - - - - **7. Lepidotrichilia**
Indumentum of simple hairs; flowers unisexual; petals imbricate; filaments fused almost or quite to the apex; appendages absent - - **10. Ekebergia**

1. **KHAYA** A. Juss.

Khaya A. Juss. in Mém. Mus. Hist. Nat. Par. **19**: 249 (1830).

Trees, usually of large size. Leaves paripinnate; leaflets entire. Flowers monoecious, but with well-developed vestiges of opposite sex, and with little external difference between the sexes, borne in large much-branched panicles. Calyx lobed almost to base, lobes 4–5, subcircular, imbricate. Petals 4–5, free, much longer than the calyx in bud, contorted. Staminal tube urceolate, glabrous inside and out, bearing 8 or 10 included anthers or antherodes inside towards the apex and terminated by 8 or 10 subcircular, emarginate or irregularly lobed appendages alternating with the anthers or antherodes. Disk cushion-shaped, fused to the base of the ovary but free from the staminal tube. Ovary 4–5-locular, each loculus with 12–16 or more pendulous ovules in 2 rows; style short; style-head discoid, almost completely blocking entrance to staminal tube; margin crenellate, upper surface with minute central papilla with 4–5 radiating, (?) stigmatic ridges. Pistillode similar to ovary but much narrower; loculi well-developed, but vestigial ovules minute. Fruit an upright, almost spherical woody septifragal capsule, opening by 4–5 valves from the apex; columella not extending to apex of capsule, with 4–5 sharp ridges, hard and woody; seed scars white, conspicuous. Seeds 6 or more per loculus, broadly transversely ellipsoid, narrowly winged more or less all round margin.

A small genus with 5 species in tropical Africa and a sixth in Madagascar and the Comoros. The differences separating the species are slight and inconstant and further work may result in their being reduced to subspecific rank or complete synonymy. *Khaya* is one of the more important timber-producing genera in Africa and is one of the sources of " African Mahogany ". According to Caudwell, " Variation and Taxonomic Categories in the Genus *Khaya* " (unpublished thesis, Oxford University), all the flowers of *K. nyasica* are male or female. In the other African species functionally bisexual flowers can occur as well.

Khaya nyasica Stapf ex Bak. f. in Journ. Linn. Soc., Bot. **40**: 42, t. 1 fig. 4 (" 3 in text. sphalm.") (1911).—Eyles in Trans. Roy. Soc. S. Afr. **5**: 388 (1916).—Steedman, Trees etc. S. Rhod.: 32–3 (1933).—Staner in Bull. Jard. Bot. Brux. **16**: 217 (1941).—Brenan, T.T.C.L.: 317 (1949).—Gomes e Sousa, Dendrol. Mozamb., **1**: 174 cum tab. (1951); op. cit. **5**: 161 cum tab. (1960).—Pardy in Rhod. Agr. Journ. **50**: 153 cum photogr. (1953).—Garcia in Contr. Conhec. Fl. Moçamb. **2**: 143 (1954).—Williamson, Useful Pl. Nyasal.: 72 (1956).—Palgrave, Trees of Central Afr.: 222 cum photogr. et tab. (1957).—Staner & Gilbert, F.C.B. **7**: 178, t. 21 (1958).—White, F.F.N.R.: 181, fig. 35F–H (1962). TAB. **54.** Type: Mozambique, Lower Umswirizwi R. and Chirinda Forest, fl. x.1904, *Swynnerton* 15 (BM, holotype; K; SRGH).

Khaya senegalensis sensu Oliver, F.T.A. **1**: 388 (1868) pro parte quoad plant. mossamb.—Sim, For. Fl. Port. E. Afr.: 27, t. 21 (1909).

Khaya zambesiaca Stapf ex Engl., Pflanzenw. Afr. **3,** 1: 803 (1915) *nom. nud.*

Large evergreen tree up to 40 (60 or more) m. tall; bole of large trees markedly buttressed at the base to a height of 3–4 m., very straight and reaching a considerable height before branching, up to 4 m. or more in diam. above the buttresses; bark with a smooth surface but exfoliating in scales the size of a half-crown, mottled

Tab. 54. KHAYA NYASICA. 1, flowering branchlet (×⅔) *Andrada* 1469; 2, vertical section of flower (×7) *Faulkner* 369; 3, fruit (with 1 valve removed) and seed (×⅔) *Burtt Davy* 21037.

grey and brown. Leaves paripinnate; petiole and rhachis 8–36 cm. long, glabrous; leaflets up to 17 × 7 cm., usually much smaller, opposite or subopposite, 2–7(8)-jugate, usually oblong-elliptic, sometimes lanceolate-oblong, apex cuspidate, base cuneate to obtuse, slightly asymmetric, glabrous, coriaceous, upper surface glossy; petiolules 5–15 mm. long. Flowers white, sweet-scented, in large many-flowered axillary panicles; peduncle 3–14 cm. long; pedicels 2–3 mm. long; bracts minute, 1 mm. long, squamiform, persistent. Calyx 1–1·5 mm. long, lobed almost to the base, lobes subcircular, ciliate. Petals up to 5·5 × 3 mm., elliptic, somewhat hooded, glabrous. Staminal tube 4·5–5·5 mm. long. Anthers 1 mm. long in male flowers; antherodes 0·5 mm. long and not producing pollen in female flowers. Ovary 1·5–2 mm. in diam.; style less than 1 mm. long; pistillode similar to ovary but much more slender. Capsule 4–6 cm. in diam., dehiscing by 4–5 valves. Seeds 1–2 × 1·5–3 cm., 6–12 per loculus.

N. Rhodesia. N: Kawambwa, fl. viii.1957, *Fanshawe* 3632 (K; SRGH). W: Kasaria Forest Reserve, Ndola, fl. ix.1951, *Holmes* 202 (FHO). C: between Luangwa Bridge and Rufunsa, fr. ix.1947, *Brenan & Greenway* 7816 (FHO; K). E: Fort Jameson, fr. iv.1952, *White* 2461 (BR; FHO; K; ND). S: Mululu, Chibilwabilwa Valley, Gwembe, fr. ix.1961, *Bainbridge* 592 (FHO). **S. Rhodesia.** N: Lomagundi, fl., *Eyles* 6193 (SRGH). C: Greenwood Park, Salisbury, fl. x.1937, *Finlay* M148/37 (BM; FHO) cult. E: Chirinda, fr. x.1947, *Wild* 2211 (K; SRGH). **Nyasaland.** N: near Rumpi Boma, fr. vi.1953, *Chapman* 123 (FHO). C: Namikokwe R., st. ix.1929, *Burtt Davy* 21686 (FHO). S: Malabri Hill near Limbe, fl., *Townsend* 100–104 (FHO). **Mozambique.** N: Nampula, fl. & fr. xi.1948, *Andrada* 1469 (COI; LISC). Z: Mocuba, Namagoa, fr. x.1944, fl. vi.1946, *Faulkner* 17 (BM; COI; FI; K; PRE; SRGH). T: Moatize, fr. iv.1948, *Mendonça* 4126 (LISC). MS: Inhandoa R., Gorongosa, fl. x.1956, *Gomes e Sousa* 4326 (COI; FHO; K; LMJ; PRE; SRGH). SS: Vilanculos, Macovane, fr. v.1947, *Hornby* 2713 (SRGH).

Also in Tanganyika and the Congo. In riverine forest at low and medium altitudes; nearly always near water.

The well-known " Big-tree " in the Chirinda forest near Chipinga in Southern Rhodesia, which is this species, is more than 60 m. high and is probably the tallest tree in the Flora Zambesiaca area. The bark is astringent in taste, reminding one of quinine, hence the African name "Umbaba" (Southern Rhodesia) meaning "to be bitter". The timber is reddish in colour with a handsome grain, hard but fairly easily worked, takes a fine finish and polishes well; it weathers well above ground and is untouched by termites and borers (*Swynnerton*). It is considered an excellent wood for furniture and general cabinet making. Over large parts of Mozambique, Nyasaland and N. Rhodesia it is the most important indigenous timber species, but supplies have been sadly depleted by exploitation. Artificial regeneration in its natural habitat in Nyasaland has given promising results so far. It is under trial in many parts of the tropics and in Florida it grows twice as rapidly as the indigenous pines.

Khaya senegalensis (Desr.) A. Juss. has recently been introduced at Lake Bangweulu for trial for forestry purposes. It and *K. grandifoliola* C.DC. are under trial in Nyasaland.

2. ENTANDROPHRAGMA C.DC.

Entandrophragma C.DC. in Bull. Herb. Boiss. **2**: 582, t. 12 (1894).

Trees, usually of large size. Leaves paripinnate, leaflets entire. Flowers (? always) monoecious, borne in panicles. Calyx lobed in upper half or almost to base, lobes 5, aestivation open. Petals 5, free, much longer than the calyx in bud, contorted. Staminal tube cup-shaped (in our area), margin entire with 10 anthers borne on very short filaments. Disk (in our area) cushion-shaped, enveloping the base of the ovary and connected to the staminal tube by 10 short ridges. Ovary 5-locular, each loculus with 4–12 pendulous anatropous ovules in 2 rows; style short; style-head discoid with 5 radiating (?) stigmatic grooves on the upper surface, not completely blocking the entrance to the staminal tube. Fruit a pendulous elongate woody septifragal capsule, cigar-shaped, fusiform, cylindric or claviform, opening by 5 valves from the apex or base or from the apex and base simultaneously; columella extending to the apex of the capsule, 5-angled or 5-ridged, softly woody, deeply indented with the imprints of seeds; seed-scars conspicuous or inconspicuous; seeds 3–8 per loculus, attached by the seed-end to the distal part of the columella, and winged towards the base of the capsule.

10 or 11 species confined to tropical Africa, where they are emergents from tropical rain forest, montane rain forest and drier forest types. The timber of the

rain forest species is exported in quantity and is highly prized for cabinet making and veneers.

Dissection of herbarium material shows that the flowers of *E. caudatum* and *E. stolzii* are unisexual with well-developed vestiges of the other sex. For *E. delevoyi* the evidence is inconclusive.

Capsule clavate, apex rounded; valves pale brown, closely lenticellate, dehiscing from the apex and ultimately breaking away from the base of the capsule; points of attachment of seeds to the columella small and inconspicuous; leaflets with filiform tips; inflorescence contracted, up to 2 cm. in diam.; petals densely puberulous outside 1. *caudatum*

Capsule cylindric, apex pointed; valves black, sparsely lenticellate, dehiscing from the base of the capsule; points of attachment of seeds to the columella large and conspicuous, covered with cottony hairs; leaflets without filiform tips; inflorescence lax, more than 3 cm. in diam.; petals glabrous:

Petiolules 1–2(3·5) cm. long, slender; leaflets shortly acuminate; venation not prominent - - - - - - - - - - - - - 2. *delevoyi*

Petiolules 0·3–0·8 cm. long, stout; leaflets rounded at the apex; venation prominent on both surfaces - - - - - - - - - 3. *stolzii*

1. **Entandrophragma caudatum** (Sprague) Sprague in Kew Bull. **1910**: 180 (1910); in Hook., Ic. Pl. **31**: t. 3023 (1915).—Bremek. & Oberm. in Ann. Transv. Mus. **16**: 420 (1935).—Harms in Notizbl. Bot. Gart. Berl. **14**: 443–4 (1939).—O. B. Mill, in Journ. S. Afr. Bot. **18**: 39 (1952).—Pardy in Rhod. Agr. Journ. **52**: 515 cum photogr. (1955).—Williamson, Useful Pl. Nyasal.: 55 (1956).—Palgrave, Trees of Central Afr.: 218 cum photogr. et tab. (1957).—White, F.F.N.R.: 180, fig. 35C–E (1962). TAB. **55** fig. B. Type from the Transvaal.

Pseudocedrela caudata Sprague in Kew Bull. **1908**: 163 (1908). Type as above.

Entandrophragma utile sensu Gomes e Sousa, Dendrol. Mozamb. **1**: 161–9, cum tab. (1951).

Large deciduous tree up to 30 m. tall; bole up to 2 m. in diam., not buttressed; bark grey or grey-brown, exfoliating in large irregular scales up to 12·5 cm. in diam., leaving buff patches and producing a mottled appearance. Leaves paripinnate; petiole and rhachis up to 25 cm. long, puberulous; leaflets up to 11 × 3·5 cm., usually opposite, sometimes alternate, (5)6–7(8)-jugate, ovate to lanceolate, tapering gradually from near the base to a narrowly acuminate apex with a filiform tip, base of distal leaflets very asymmetric, of the proximal less so, upper surface dull, lower surface glabrous except for the pubescent midrib and a few hairs on the secondary nerves; petiolules 1–2·5 cm. long. Flowers pale green, in contracted panicles up to 20 × 2 cm., borne in leaf-axils and towards the base of the current year's growth; peduncles up to 14 cm. long; pedicels 2–4 mm. long, densely puberulous. Calyx 1·1–2 mm. long, tomentellous, lobed almost to the base. Petals 5–6 mm. long, densely puberulous outside. Staminal tube 3–4 mm. long, densely puberulous inside, glabrous outside. Capsule 15–21 × 5–5·5 cm., clavate, apex rounded; valves pale brown, closely lenticellate, apex thickened and recurved, dehiscing from the apex and ultimately breaking away from the base of the capsule; points of attachment of seeds to columella small and inconspicuous. Seeds (including wing) 9–10 × 2·5 cm.

Bechuanaland Prot. N: Tsessebe Camp, st. iv.1931, *Pole Evans* 3250 (FHO; K; PRE). **N. Rhodesia.** B: Katongo Forest Reserve, Sesheke, fr. i.1952, *White* 1977 (FHO; K). E: between Yakobe and Mpelembe, Luangwa Valley, fr. viii.1938, *Greenway & Trapnell* 5639 (FHO). S: near Sinazongwe, Gwembe Valley, fl. ix.1955, *Bainbridge* 126 (FHO; K). **S. Rhodesia.** N: Chicomba R., Sebungwe, st. xi.1951, *Lovemore* 170 (SRGH). W: Matopos National Park, fl. x.1949, *West* 3116 (SRGH). E: Hot Springs, near Odzi R., Melsetter, fl. x.1948, *Chase* 1189 (BM; K; SRGH). S: Chipinda Pools, Ndanga, fl. x.1951, *Mullin* 118/51 (FHO; SRGH). **Nyasaland.** S: near Bennis's Village, Ncheu, st. xi.1957, *Jackson* 2121 (FHO; SRGH). **Mozambique.** LM: Namaacha, between Porto Henrique and Chengalene, fl. viii.1948, *Gomes e Sousa* 3784 (COI; K; PRE).

Also in the Transvaal and Natal. Often associated with *Baikiaea plurijuga* as an emergent from dense " mutemwa " thicket on Kalahari Sands. Elsewhere most often reported as an emergent from other thicket types on deep well-drained soil in low-lying river valleys, or as a constituent of open woodland on rocky slopes.

E. caudatum is the Royal tree of Barotseland, from which canoes are made for the Paramount Chief. The wood is dark brown with an attractive figure. It is valued for

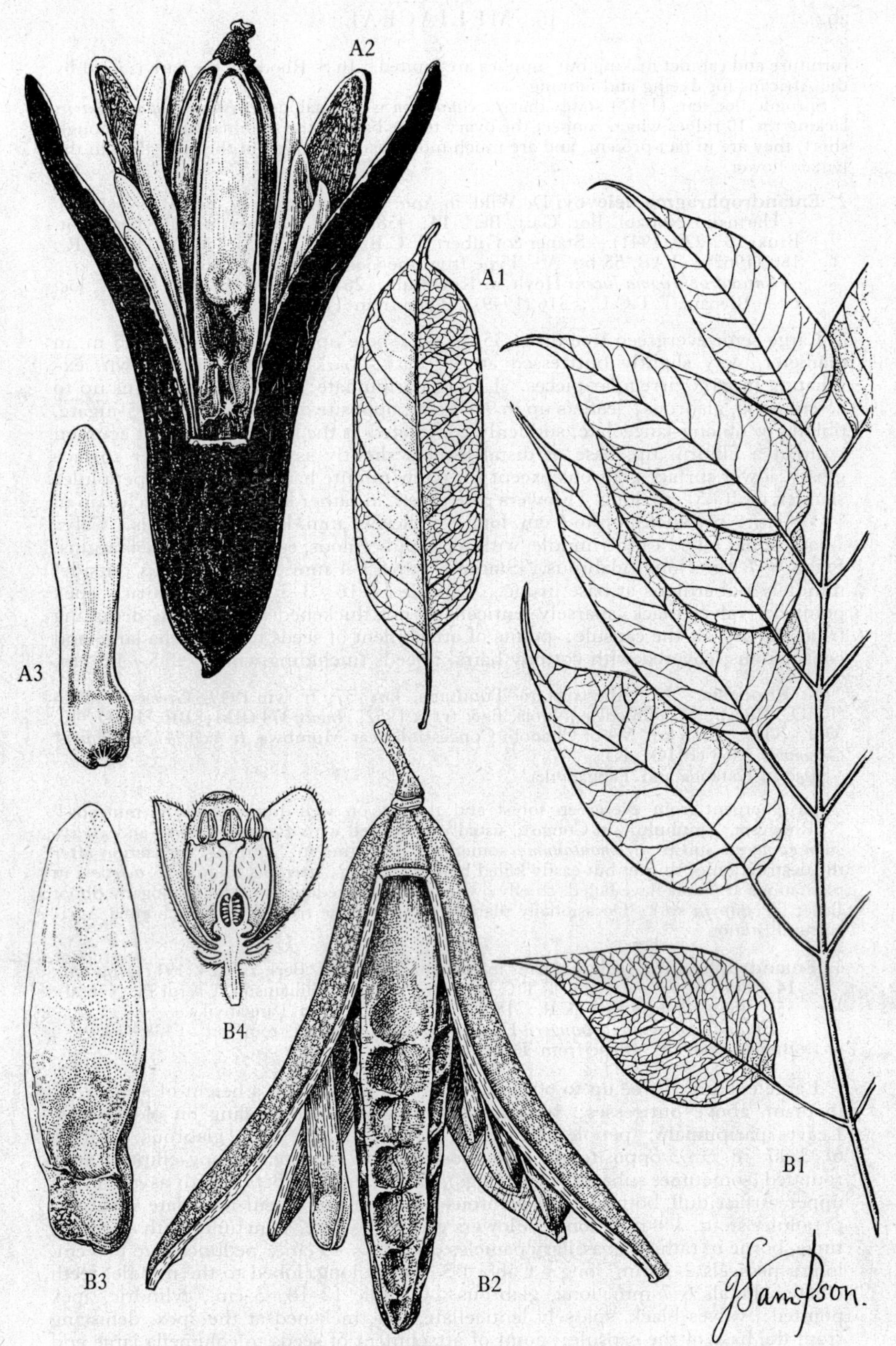

Tab. 55. A.—ENTANDROPHRAGMA DELEVOYI. A1, leaflet ($\times\frac{2}{3}$); A2, fruit (with 1 valve partly removed) ($\times\frac{2}{3}$); A3, seed ($\times\frac{2}{3}$). All from *Angus* 374. B.—ENTANDROPHRAGMA CAUDATUM. B1, leaf ($\times\frac{2}{3}$) *Angus* 985; B2, fruit (with 1 valve removed) ($\times\frac{1}{3}$) *Burtt Davy* 20573; B3, seed ($\times\frac{2}{3}$) *Brenan* 7772; B4, vertical section of flower ($\times 5$) *Martin* 307.

furniture and cabinet making but supplies are limited. In S. Rhodesia the bark is used by the Africans for dyeing and tanning.

Sprogue, loc. cit. (1915) states that *E. caudatum* is anomalous in *Entandrophragma* in lacking the 10 ridges which connect the ovary to the base of the staminal tube. Although short, they are in fact present, and are much more easily observed in the male than in the female flower.

2. **Entandrophragma delevoyi** De Wild. in Ann. Soc. Sci. Brux. **47,** Sér. B: 78 (1927).—Harms in Notizbl. Bot. Gart. Berl. **14**: 438 (1939).—Staner in Bull. Jard. Bot. Brux. **16**: 229 (1941).—Staner & Gilbert, F.C.B. **7**: 181 (1958).—White, F.F.N.R.: 180 (1962). TAB. **55** fig. A. Type from the Congo.

Entandrophragma lucens Hoyle in Kew Bull.: 267 cum tab. (1932).—Harms, loc. cit.—Brenan, T.T.C.L.: 316 (1949). Type from Tanganyika.

Large semi-evergreen tree up to 35 m. tall; bole up to 20 m. tall and 1·5 m. in diameter, very slightly buttressed at the base; bark smooth, grey-brown, exfoliating in large, irregular pieces. Leaves paripinnate; petiole and rhachis up to 25 cm. long, glabrous; leaflets up to 9 × 4 cm., opposite or subopposite, 3–5-jugate, oblong to oblong-lanceolate, suddenly contracted at the apex into a short acumen without a filiform tip, base of distal leaflets slightly asymmetric, upper surface glossy, lower surface glabrous except for a few minute hairs on midrib; petiolules slender, 1–2(3·5) cm. long. Flowers pale green, in rather lax axillary panicles up to 15 × 8 cm.; peduncles up to 6 cm. long; pedicels 2 mm. long, puberulous. Calyx 1 mm. long, lobed to the middle, with teeth puberulous, especially on the margins. Petals 5–7 mm. long, glabrous. Staminal tube 3–4 mm. long, glabrous outside, minutely puberulous at base inside. Capsule 14–16 × 3–3·5 cm., cylindric; apex pointed; valves black, sparsely lenticellate, not thickened at the apex, dehiscing from the base of the capsule; points of attachment of seeds to columella large and conspicuous, covered with cottony hairs. Seeds (including wing) 7–8·5 × 1·5 cm.

N. Rhodesia. N: Abercorn to Tunduma, km. 37, fr. viii.1949, *Greenway* 8376 (FHO; PRE). W: Chichele, Ndola, fl. & fr. ix.1952, *Angus* 374 (BM; BR; FHO; K; MO; ND). S: 8 km. W. of Chanobi Concession near Mumbwa, fr. ix.1947, *Brenan & Greenway* 7863 (FHO; K).
Also in Katanga and Tanganyika.

An emergent from evergreen forest and thicket on well-drained soils (" muteshi " N. Rhodesia, " muhulu " in Congo); usually associated with *Parinari excelsa* and *Syzygium guineense* subsp. *afromontanum*; sometimes persisting in " chipya " vegetation after the destruction of thicket but easily killed by fierce fires. Attempts to grow *E. delevoyi* in plantations at Ndola have failed, chiefly owing to the depredations of the Mahogany Shoot Borer (*Hypsipyla* sp.). Occasionally planted as an avenue tree, as at Kanchomba Agricultural Station.

3. **Entandrophragma stolzii** Harms in Notizbl. Bot. Gart. Berl. **7**: 224 (1917); op. cit., **14**: 439 (1939).—Brenan, T.T.C.L.: 316 (1949).—Williamson, Useful Pl. Nyasal.: 55 (1956).—White, F.F.N.R.: 180 (1962). Type from Tanganyika.

Entandrophragma deiningeri Harms, loc. cit. (1917); tom. cit.: 438 (1939).—Brenan, loc. cit. Type from Tanganyika.

Large evergreen tree up to 60 m. tall; bole buttressed to a height of 4 m., 2 m. in diam. above buttresses; bark greyish-brown, rough, scaling on older trees. Leaves paripinnate; petiole and rhachis up to 35 cm. long, glabrous; leaflets 6 × 3–17 × 8 cm., opposite or subopposite, 5–6(7)-jugate, oblong-elliptic, apex rounded (sometimes subacute in saplings), base of distal leaflets slightly asymmetric; upper surface dull, both surfaces glabrous and with prominent reticulate venation; petiolules stout, 3–8 mm. long. Flowers greenish-white, sometimes with a reddish tinge, borne in rather lax axillary panicles up to 25 × 15 cm.; peduncles up to 9 cm. long; pedicels 2–3 mm. long. Calyx 1·5–2 mm. long, lobed to the middle, teeth ciliate. Petals 6–7 mm. long, glabrous. Capsule 12–18 × 3 cm., cylindric, apex pointed; valves black, sparsely lenticellate, not thickened at the apex, dehiscing from the base of the capsule; point of attachment of seeds to columella large and conspicuous, covered with cottony hairs. Seeds (including wing) 6–7·5 × 1·5 cm.

Nyasaland. N: Misuku Forest, fr. x.1953, *Chapman* 214 (FHO).
Also in Tanganyika. In montane forest.

E. stolzii may be a synonym of *E. excelsum* (Dawe & Sprague) Sprague from Uganda and the Eastern Congo. Better material is needed before this can be established. The wood is of high quality, but too scarce in our area to be of much commercial importance.

3. LOVOA Harms

Lovoa Harms in Engl., Bot. Jahrb. **23**: 164 (1896).

Large trees. Leaves paripinnate; leaflets entire. Flowers monoecious, borne in large panicles. Calyx lobed almost to the base; lobes 2+2, imbricate, the outer larger than the inner. Petals 4, free, much longer than the calyx in bud, imbricate. Staminal tube shortly cylindric or cup-shaped, margin entire or with very short broad teeth or with paired deltate-acuminate appendages alternating with the anthers; anthers 8. Disk short, broad, cushion-shaped, enveloping the base of the ovary, but free from the staminal tube. Ovary 4-locular, each loculus with 2 rows of 2 ovules; style short; style-head capitate, obscurely 4-lobed. Fruit a pendulous elongate tetragonal woody septifragal capsule, dehiscing from the apex or from the apex and base simultaneously; columella extending to the apex of the capsule, 4-ridged, softly woody, each face slightly indented with the imprints of 1–2 seeds. Seeds 2(4) per loculus, but only 1–2 fertile, attached to the apex of the columella by the winged end of the seed and leaving inconspicuous scars on falling.

A small genus with 2 or 3 species, confined to evergreen forest in tropical Africa. *L. trichilioides* Harms yields the well-known timber " African walnut ".

Lovoa swynnertonii Bak. f. in Journ. Linn. Soc., Bot. **40**: 41, t. 3 (1911).—Eyles in Trans. Roy. Soc. S. Afr. **5**: 390 (1916).—Steedman, Trees etc. S. Rhod.: 33 (1933).—Staner in Bull. Jard. Bot. Brux. **16**: 249 (1941).—Brenan, T.T.C.L.: 317 (1949).—Eggeling & Dale, Indig. Trees Uganda Prot.: 193 (1952).—Staner & Gilbert, F.C.B. **7**: 195 (1958).—Dale & Greenway, Kenya Trees and Shrubs: 269 (1961). TAB. **56**. Type: S. Rhodesia, Chirinda, fl. ii.1906, *Swynnerton* 16 (BM, holotype; K).

Large evergreen tree occasionally reaching a height of 50 m.; bole fluted or slightly buttressed at the base to a height of 2 m., tall and straight, sometimes 30 m. to first branch, slender, up to 2 m. in diam.; bark similar to that of *Khaya nyasica* but lacking its bitter taste. Leaves paripinnate, petiole and rhachis up to 30 cm. long, minutely puberulous; leaflets up to 10 × 3·5 cm., opposite or alternate, 3–5(6)-jugate, up to 10 × 4 cm., more or less oblong-elliptic or lanceolate-elliptic, slightly falcate, apex shortly and bluntly acuminate, base markedly asymmetric, upper surface glossy, lateral nerves closely spaced in c. 16 pairs, lower surface glabrous except for a few hairs on nerves when young; petiolules 5–10 mm. long. Flowers white, in large axillary panicles; peduncles 6–9 cm. long; inflorescence-axes densely puberulous; bracts very small, less than 1 mm. long, puberulous; pedicels ± 1 mm. long. Calyx 1 mm. long, puberulous, especially on the margins. Petals 2·5–3·5 mm. long, elliptic, hooded, glabrous inside and outside. Staminal tube 1–1·5 mm. long, appendages 1–1·5 mm. long; anthers inserted between the appendages on short filaments, exserted. Ovary 1 mm. in diam., subglobose, minutely papillose; style 1 mm. long. Capsule up to 5·5 × 2 cm.; valves brownish-black outside with scattered white lenticels, separating first from the apex and remaining attached to the base for some time before falling. Seeds (including wing) up to 4·5 × 1 cm., falling from the columella without leaving conspicuous scars.

S. Rhodesia. E: Chirinda Forest, fl. i.1948, *McGregor* 16/48 (FHO; SRGH). **Mozambique.** MS: Chimoio, Garuso, st. iv.1935, *Gilliland* 1816 (BM; FHO).

Also in the Congo, Uganda, Kenya and Tanganyika. In our area known only from the Garuso forests of Mozambique and the Chirinda forest in S. Rhodesia, where it prefers well-drained slopes of banks of streams.

In S. Rhodesia the timber is known as " Brown Mahogany ". The heart-wood is grey-brown and handsome, but frequently cross-grained and difficult to work. It has been used a great deal for outdoor work, being very durable and untouched by insects (*Swynnerton*).

4. XYLOCARPUS Koen.

Xylocarpus Koen. in Naturforsch. **20**: 2 (1784).

Trees of mangrove swamps and coastal scrub, often with pneumatophores or ribbon-like buttresses. Leaves paripinnate, leaflets entire. Flowers probably monoecious, borne in compound cymes. Calyx lobed to or beyond the middle,

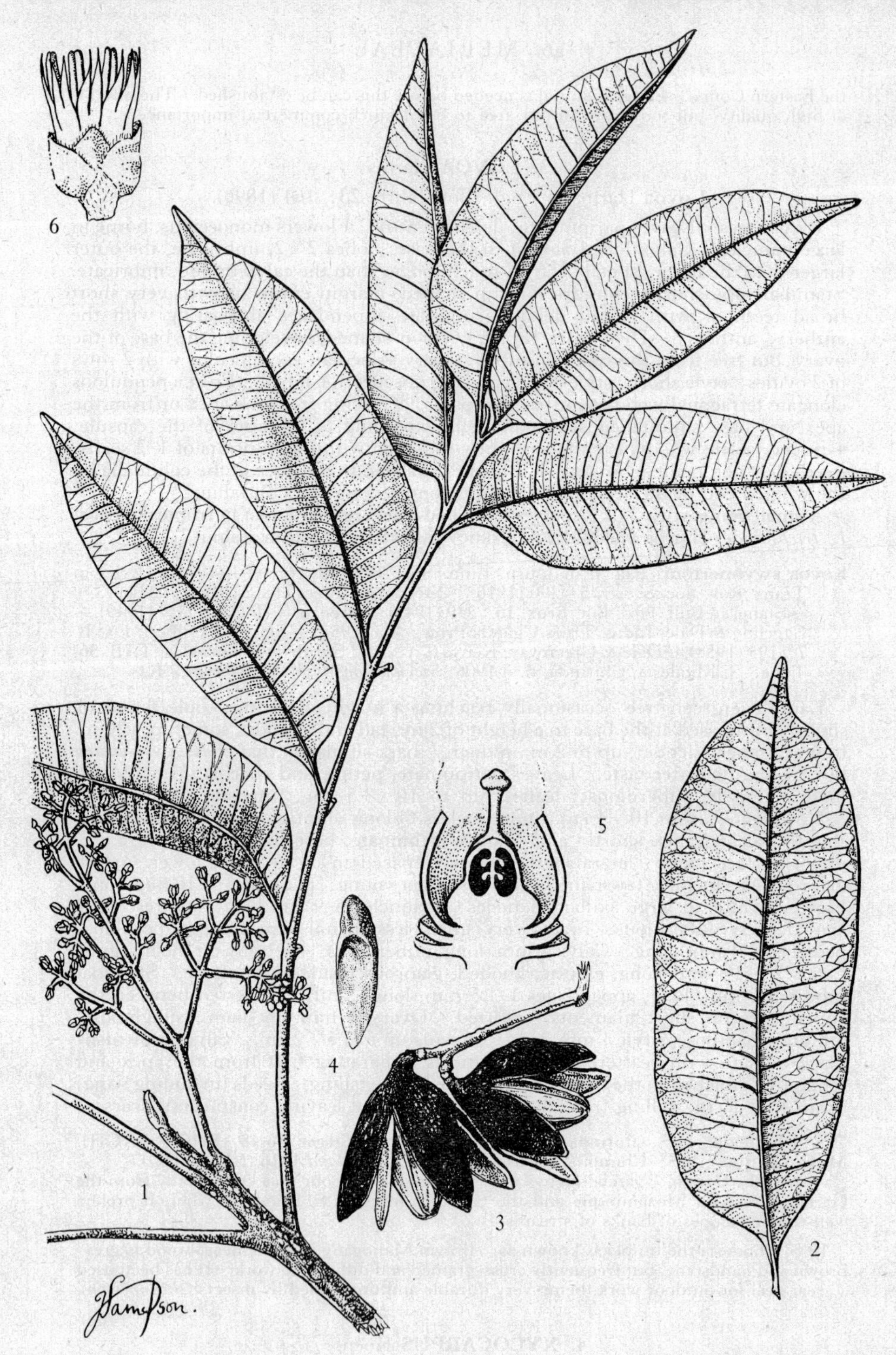

Tab. 56. LOVOA SWYNNERTONII. 1, flowering branchlet ($\times \frac{2}{3}$) *McGregor* 16/48; 2, leaflet from sapling ($\times \frac{2}{3}$) *Jack* in GHS 5958; 3, fruits ($\times \frac{2}{3}$) *Wild* 2210; 4, seed ($\times \frac{2}{3}$) *Wild* 2210; 5, vertical section of gynoecium and base of staminal tube ($\times 10$) *English* 1/48; 6, flower with corolla removed ($\times 7$) *English* 1/48.

lobes 4, aestivation open. Petals 4, free, much longer than the calyx in bud, contorted. Staminal tube urceolate, bearing 8 included anthers inside towards the apex and terminated by 8 appendages alternating with the anthers, appendages subcircular, retuse or shallowly and irregularly 2(3)-lobed. Disk large, red, cushion-shaped, situated beneath or engulfing and fused to the ovary, free from the staminal tube. Ovary 4-locular, each loculus with 2–4 ovules; style short; style-head discoid, almost completely blocking the entrance to the staminal tube, margin crenellate, upper surface with a minute central papilla with 4 radiating, (?) stigmatic grooves. Fruit a large almost spherical leathery septifragal capsule, dehiscing by 4 valves; septa thin, ultimately breaking down. Seeds 8–16, large, pyramidal or tetrahedral, with angular margins due to mutual compression, outer side somewhat rounded, attached by the apex to the placenta and so forming a spherical mass; outer integument thick, corky.

A small genus with two species widespread in the mangrove swamps and coastal scrub of the Old World. Noamesi (a revision of the *Xylocarpeae*, unpublished thesis; see " Dissertation Abstracts " **19**, no. 7 (1959)) maintains a third species, *X. mekongensis* Pierre, and suggests that it may have arisen through hybridization between the other two. Intermediates between *X. granatum* and *X. moluccensis* appear to be widespread and locally numerous in Asia and the western Pacific. The two specimens of *X. mekongensis* that Noamesi cites from the African mainland (including *Gomes e Sousa* 3800 from Mozambique) in our opinion belong to *X. granatum*. More information on the distribution and biology, including the occurrence of hybrids, is needed for both species in our area. In the 17th century Rumphius observed that certain individuals, although flowering profusely, never produced fruit. More recently Mrs. H. G. Faulkner has made similar observations in East Africa. It would appear that the flowers are dioecious, but evidence from herbarium material is so far inconclusive.

The mahogany-like timber is valued for furniture and carpentry, but is of too small a size and too scarce in Africa to be of much commercial importance.

Leaflets drying orange-brown, elliptic, oblong-elliptic or obovate-elliptic, apex rounded, obtuse or emarginate; disk fused to lower half of ovary; fruit up to 20 cm. in diam.; bark smooth, flaking; surface roots forming ribbon-like pneumatophores 1. *granatum*

Leaflets drying yellow-green; ovate or ovate-lanceolate, tapering from near the base to a subacuminate apex; disk free from the ovary; fruit up to 8 cm. in diam.; bark rough and longitudinally fissured; surface roots not ribbon-like - - 2. *moluccensis*

1. **Xylocarpus granatum** Koen. in Naturforsch. **20**: 2 (1784).—A. Juss. in Mém. Mus. Hist. Nat. Par. **19**: 244 (1830).—Parkinson in Indian Forester, **60**: 138, t. 15 (1934). —Merr., Interpr. Rumph. Herb. Amboin.: 306 (1917).—Ridl. in Kew Bull. **1938**: 288 (1938).—Burtt Davy in Journ. S. Afr. Bot. **6**: 31 (1940). TAB **57** fig. A. Type from eastern Asia.

Carapa obovata Bl., Bijdr.: 179 (1825).—Baill. in Grandid. Hist. Nat. Pl. Madag. **3**: t. 260 (1886).—Battiscombe, Trees, etc. Kenya Col., ed. 2: 103 (1936). Type from Java.

Carapa moluccensis sensu Oliv., F.T.A. **1**: 337 (1868).—Sim, For. Fl. Port. E. Afr.: 27, t. 16 (1909).

Xylocarpus benadirensis Mattei in Boll. Ort. Bot. Palermo, **7**: 99 (1908).—Ridl., loc. cit., descr. ampl.—Brenan, T.T.C.L.: 322 (1949).—Gomes e Sousa, Dendrol. Moçamb., **2**: 103 cum tab. (1949).—Dale & Greenway, Kenya Trees and Shrubs: 276 (1961). Type from Somalia.

Xylocarpus obovatus (Bl.) A. Juss., loc. cit.—Chiovenda, Fl. Somala, **2**: 131 (1932). Glover, Prov. Check-List Brit. & It. Somal.: 187 (1947). Type as above.

Medium-sized crooked much-branched evergreen tree up to 10 m. tall (taller elsewhere); bark smooth and yellowish, or brown and green and flaking; surface roots laterally compressed and forming a spreading network of ribbon-like pneumatophores with the upper edges protruding above the mud and suggesting a mass of snakes. Leaves paripinnate, drying orange-brown; petiole and rhachis up to 8·5 cm. long, glabrous; leaflets up to 12 × 5 cm., usually much smaller, opposite, 1–2(3)-jugate, elliptic, oblong-elliptic or obovate-elliptic, apex usually rounded, rarely obtuse or emarginate, base narrowly or broadly cuneate, glabrous, coriaceous, venation prominent on both sides; petiolules 2–5 mm. long. Flowers whitish or pale pink, in lax racemes of (2)3-flowered cymes; peduncle plus rhachis 4–7 cm. long; bracts minute, usually caducous. Calyx about 3 mm. long, glabrous, lobed

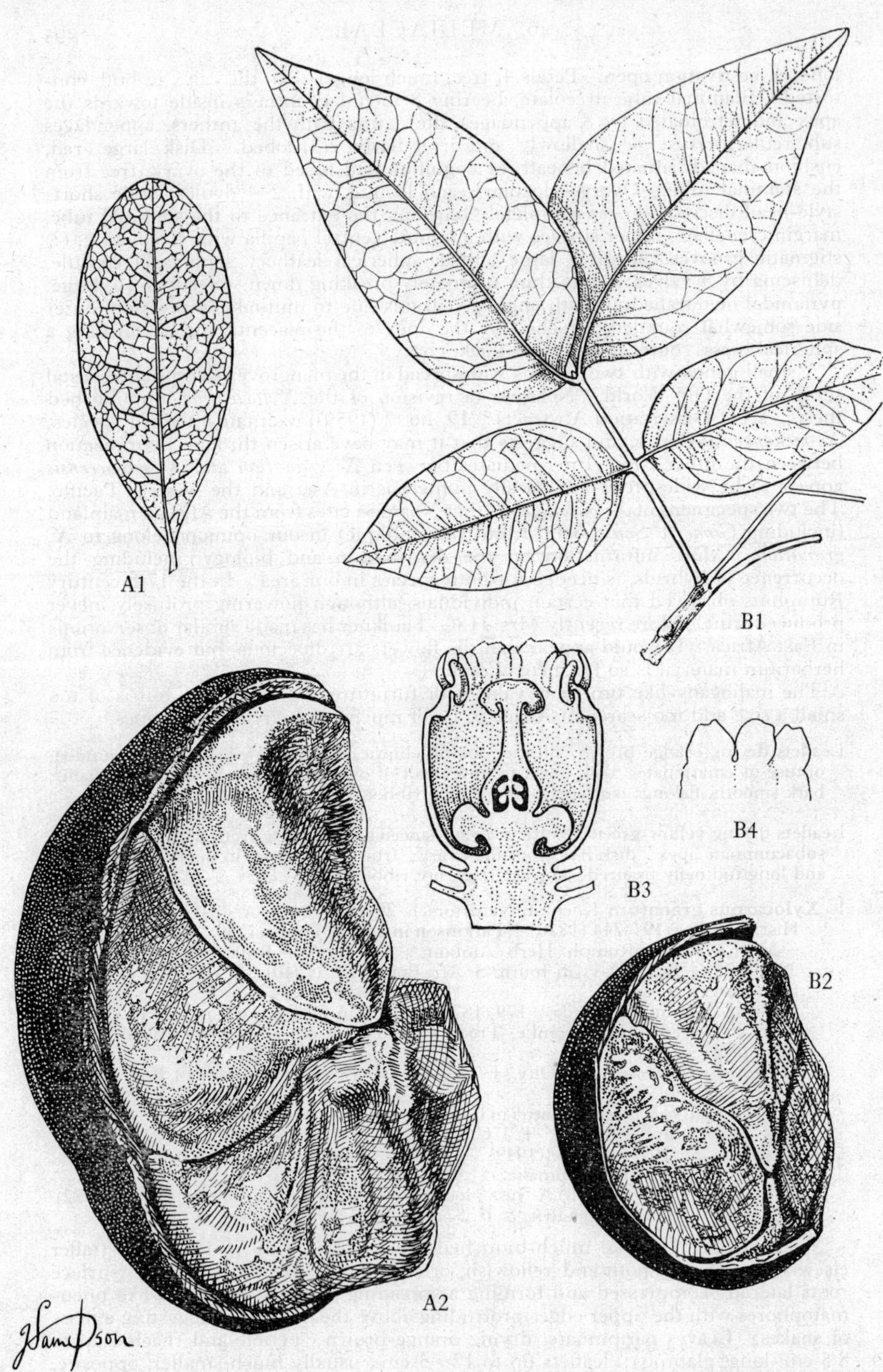

Tab. 57. A.—XYLOCARPUS GRANATUM. A1, leaflet (× ⅔) *Faulkner* 1881; A2, part of fruit (× ⅔) *Tanner* 2460. B.—XYLOCARPUS MOLUCCENSIS. B1, leaf (× ⅔) *Faulkner* 1512; B2, part of fruit (× ⅔) *Greenway* 1446; B3, section of gynoecium and staminal tube (× 10) *Faulkner* 1512; B4, appendages of staminal tube (× 10) *Faulkner* 1512.

to the middle, lobes rounded. Petals 5–6·5 × 2·5 mm., glabrous. Staminal tube 4–5 mm. long, glabrous. Ovary less than 1 mm. in diam.; style 1·5 mm. long; disk fused to the lower half of the ovary. Fruit large, up to 20 cm. in diam., obscurely 4-sulcate. Seeds 4–8 cm. long.

Mozambique. N: R. Tari, Cabo Delgado, fr. xi.1959, *Gomes e Sousa* 4523 (COI). Z: Quelimane, *Sim* 21074 (PRE, not seen). MS: Beira, fr. iv.1894, *Kuntze* (K). LM: R. Maputo, fl. xi.1948, *Gomes e Sousa* 3895 (COI; K; MO; PRE; SRGH).
Also in Somalia, Kenya, Tanganyika, Mafia, Pemba, Zanzibar, Madagascar and throughout most of the Old World tropics to Australia, Fiji and Tonga. In tidal mud of mangrove swamps, especially towards their upper limits.

The ribbon-like roots are illustrated by Watson (in Malayan Forest Records, **6**: t. 37 (1928)) and by Graham (in Journ. E. Afr. & Uganda Nat. Hist. Soc. **36**: 157 (1929)).
Ridley (loc. cit.) maintained the African plant, *X. benadirensis*, distinct from the Asiatic because of its dentate not retuse staminal appendages. This trifling character is too variable (even within the same flower) to be of taxonomic value. On Mafia I. a decoction of the crushed fruits is drunk as an aphrodisiac (*Greenway* 5373).

2. **Xylocarpus moluccensis** (Lam.) M. J. Roem., Syn.—Monogr. Hesper.: 124 (1846).—Parkinson in Indian Forester, **60**: 142, t. 16 (1936).—Merr., Interpr. Rumph. Herb. Amboin.: 307–8 (1917).—Ridl. in Kew Bull.: 291 (1938).—Brenan, T.T.C.L.: 322 (1949).—Dale & Greenway, Kenya Trees and Shrubs: 276 (1961). TAB. **57** fig. B. Type from Eastern Asia.
Carapa moluccensis Lam., Encycl. Méth. Bot. **1**: 621 (1785).—Baill. in Grandid., Hist. Nat. Pl., Madag., **3**: t. 259 (1886). Type as above.

Medium-sized much-branched spreading semi-evergreen tree up to 13 m. tall (much taller elsewhere), sometimes flowering as a shrub; bark rough like an *Ulmus*; ribbon-like buttresses absent. Leaves paripinnate, drying yellow-green, petiole and rhachis up to 16 cm. long, glabrous; leaflets up to 12 × 6 cm., opposite, 2–3(4)-jugate, ovate or ovate-lanceolate, tapering from near the base to a subacuminate apex, base variable, asymmetric, glabrous, subcoriaceous, venation closer and less prominent than in *X. granatum*. Flowers white or greenish-white in lax racemes or panicles of 3–7-flowered cymes; peduncle plus rhachis 5–14 cm. long; bracts minute, usually caducous. Calyx c. 2 mm. long, glabrous, lobed to beyond the middle, the lobes of irregular outline, often apiculate. Petals up to 7 × 4 mm., glabrous. Staminal tube 4–5·5 mm. long, glabrous. Ovary less than 1 mm. in diam., style 1 mm. long; disk situated beneath ovary. Fruit up to 8 cm. in diam., obscurely 4-sulcate. Seeds 3·6–7 cm. long.

Mozambique. N: Goa I., *Gomes e Sousa* 4265 (PRE).
Also in Kenya, Tanganyika, Mafia, Pemba, the Mascarenes, Madagascar and throughout most of the Old World tropics to Australia, Fiji and Tonga, but apparently absent from India except for the Andamans. In coastal scrub just above high-water mark on sandy soil, seashore cliffs and rocks near surf.

5. TRICHILIA Browne

Trichilia Browne, Hist. Jam.: 278 (1756).—L., Syst. Nat. ed. 10, **2**: 1020 (1759) *nom. conserv.*

Trees, often of large size, or rarely shrublets. Leaves imparipinnate, rarely 3-foliolate or 1-foliolate, alternate, leaflets entire. Flowers unisexual and apparently dioecious but with little external difference between those that are functionally male and those that are functionally female, borne in cymes or cymose panicles. Calyx cupuliform with 5 minute teeth, or deeply lobed with 5 deltate or circular lobes. Petals 5, free, imbricate, much longer than the calyx in bud, linear or linear-oblong. Filaments 10, completely united to form a staminal tube bearing the anthers on its more or less entire margin or united only in the lower half and then anthers inserted between a pair of deltate appendages; anthers glabrous; antherodes smaller than the anthers and not producing pollen. Disk always present (in our area), either completely fused to the base of the staminal tube or cup-shaped and free from it. Ovary small, with 2–3(4) loculi, each with 2 collateral or superposed ovules; style elongate, much longer than the ovary, distally expanded to form a capitate coroniform or ovoid style-head, only part of which is stigmatic; stigmatic surface entire or deeply 2–3-lobed. Pistillode similar to the gynoecium but narrower, usually with vestigial ovules. Fruit a loculicidal capsule with (2)3(4) ± leathery rather thin valves. Seeds large, each partly or almost completely covered by a bright red aril, or the aril apparently absent and then the testa fleshy.

About 15 species occur in tropical Africa. More than 240 species have been described from America but many of these are synonyms. Recent work by Bentvelzen (in Act. Bot. Neerl. **11**: 11 (1962)) has shown that the small Asiatic genus, *Heynia* Roxb. is congeneric.

Vermoesen (in Rev. Zool. Afr. **10**, Suppl. Bot.: 16 (1922)), observed that in the Congo some trees produced fruits; in others all the flowers fell after anthesis. Medley Wood recorded the same phenomenon from Natal in his field notes for Nos. 5612 & 5615 in 1894. Vermoesen assumed that the fertile flowers were bisexual and suggested that some species were polygamous, but that in others fertile flowers were produced at one season and male flowers on the same individual at another season. All the flowers we have examined are either male or female and the species appear to be dioecious. Careful field observations are needed to confirm this.

Filaments united only at the base or for less than half of their length, villous in the upper half inside; anthers inserted between a pair of deltate appendages:
  Calyx lobed to beyond the middle, lobes imbricate; filaments united for approximately half of their length; disk completely fused to the base of the staminal tube (Sect. *Trichilia*):
    Large trees; appendages nearly $\frac{3}{4}$ as long as the anthers; disk glabrous, thin, with 10 deltate teeth alternating with the filaments of the staminal tube; seed black, almost completely concealed by the scarlet aril:
      Capsule without a stipe; leaflets distinctly broadest near apex, drying dark brown, apex nearly always acute or acuminate, rarely rounded or emarginate, lateral nerves in 8–9(12) widely spaced pairs, lower surface glabrous or with a few strigose hairs, very rarely sparsely to densely pilose; large tree of evergreen forest patches in higher rainfall areas - - - - - 1. *dregeana*
      Capsule with a stipe (4)6–10 mm. long; leaflets not or scarcely broadest near the apex, drying olive-green or pale brown, apex nearly always rounded or emarginate, very rarely acute or apiculate or subacuminate, lateral nerves in (10–12)13–16(19) closely spaced pairs, lower surface densely puberulous with short curly hairs especially on nerves; medium-sized tree, especially of riparian woodland in regions of lower rainfall - - - - - - - - 2. *emetica*
    Suffrutex; appendages scarcely $\frac{1}{8}$ as long as the anthers; disk densely setulose, annular, thick and fleshy; seed without an aril but with a fleshy orange testa 3. *quadrivalvis*
  Calyx cupuliform with minute teeth; filaments united only at the base; disk cupuliform, almost free from the staminal tube (Sect. *Apotrichilia*) - - 4. *capitata*
Filaments completely united; anthers inserted on the rim of the staminal tube which is minutely puberulous in the upper half inside; appendages absent (Sect. *Moschoxylum*) 5. *prieuriana*

1. **Trichilia dregeana** Sond. in Harv. & Sond., F.C. **1**: 246 (1860).—Monro in Trans. Rhod. Sci. Ass. **7**: 67 (1908). Type from S. Africa.

Trichilia dregeana var. *oblonga* Sond., loc. cit.—Harv., Thes. Cap. **1**: 49, t. 76 (1859).* Type from S. Africa.

Trichilia dregei E. Mey [ex Drège, Zwei Pflanz.-Docum.: 227 (1843) *nom. nud.*] ex C.DC. in A. & C.DC., Mon. Phan. **1**: 657 (1878). Type from S. Africa.

Trichilia dregei var. *oblonga* C.DC., op. cit.: 658 (1878). Type from S. Africa.

Trichilia strigulosa Welw. ex C.DC., loc. cit.—Exell & Mendonça, C.F.A. **1, 2**: 314–15 (1951). Type from Angola.

Trichilia vestita C.DC. in Bull. Herb. Boiss. **4**: 428 (1896). Type from Angola.

Trichilia splendida A. Chev. in Bull. Soc. Bot. Fr., **58**, mém. 8: 147 (1911).—Staner in Bull. Jard. Bot. Brux. **16**: 159, t. 4. et fig. 9 (1941).—Keay, F.W.T.A. ed. 2, **1**: 705 (1958).—Staner & Gilbert, F.C.B. **7**: 165 (1958). Type from Guinée.

Trichilia chirindensis Swynnerton & Bak. f. in Journ. Linn. Soc., Bot. **40**: 39–41 (1911).—Eyles in Trans. Roy. Soc. S. Afr. **5**: 390 (1916).—Steedman, Trees etc. S. Rhod.: 33 (1933).—Brenan, T.T.C.L.: 319 (1949). Type: S. Rhodesia, Chirinda Forest, fl. x.1906, *Swynnerton* 1 (BM, holotype; K; SRGH).

Trichilia grotei Harms in Notizbl. Bot. Gart. Berl. **7**: 230 (1917).—Brenan, loc. cit. Type from Tanganyika.

Trichilia umbrosa Vermoes. in Rev. Zool. Afr. **10**, Suppl. Bot.: 53 (1922). Type from the Congo.

* Dated 1859 on the title-page but probably not published until 1860.

Trichilia schliebenii Harms in Notizbl. Bot. Gart. Berl. **11**: 1070 (1934).—Brenan, loc. cit. Type from Tanganyika.

Large evergreen tree 25–40 m. tall; bole slightly buttressed, up to 2 m. in diam.; bark grey and smooth like a *Ficus*. Leaves imparipinnate; petiole and rhachis up to 26 cm. long, glabrous or densely pilose; leaflets up to 21 × 8·5 cm., opposite or alternate, (2)3–4(5)-jugate, obovate to oblanceolate or oblanceolate-elliptic, nearly always distinctly broadest near the apex, apex nearly always acute or acuminate, rarely rounded or emarginate, base rounded or cuneate, upper surface drying dark brown, lateral nerves in 8–9(12) pairs, lower surface glabrous or with a few strigose hairs, very rarely sparsely to densely pilose; petiolules up to 1 cm. long. Flowers dirty white, in axillary cymes or cymose panicles. Pedicels up to 1 cm. long. Calyx usually 5–7 mm. long, strigulose-tomentellous, lobed to half-way or more, lobes subcircular, imbricate. Petals (1·3)1·4–2·4(2·6) cm. long, linear, tomentellous on both surfaces. Filaments usually 1–1·5 cm. long, united for about half their length, sparsely puberulous outside, densely villous in the upper half inside. Appendages deltate, nearly $\frac{3}{4}$ as long as the anthers; anthers 2 mm. long, slightly apiculate, antherodes a little smaller, not producing pollen. Disk glabrous, thin, with 10 deltate teeth alternating with the filaments. Ovary 3(4)-locular; style usually 8–10 mm. long, columnar, densely setulose-puberulous almost to the apex; stylehead capitate, with a crateriform apical (?)stigmatic region. Pistillode with vestigial ovules. Capsule c. 3 × 3 cm., obovoid-globose, without a stipe, slightly sulcate, surface transversely wrinkled, fulvous-tomentellous, opening by 3(4) valves. Seed black, almost completely concealed by the scarlet aril.

N. Rhodesia. N: Kawimbe, Abercorn, fl. x.1956, *Richards* 6415 (K; SRGH). W: Ndola, fr., *Trapnell* 1400 (K). **S. Rhodesia.** E: Chirinda, fl. x.1947, *Wild* 2057 (K; SRGH). **Nyasaland.** N: Mugesse Forest Reserve, fl. xi.1952, *Chapman* 50 (FHO; K). C: Nchisi Mountain, fl. ix.1929, *Burtt Davy* 21250 (FHO). S: Zomba, fr. i.1932, *Clements* 179 (FHO). **Mozambique.** MS: Chimoio, st. iv.1948, *Simão* 3901 (LISC).
Also in Guinée, Uganda, Kenya, Tanganyika, Congo, Angola, Transvaal, Natal and Cape Prov. In patches of evergreen forest in regions of higher rainfall. It is one of the commonest of the larger trees in Chirinda Forest, S. Rhodesia.

The timber is pink, easily worked, polishes well and is used by Africans for dishes, pillows and other carved work, a previous boiling being said to prevent subsequent attack by *Bostrichidae* (Swynnerton). Fat from the seed is used in the same way as in *T. emetica*. The aril is eaten by Africans and baboons. On Kilimanjaro the seeds are used as a bait for the barbel fish and the fat from the seeds for soap-making and anointing the body (*Haarer* 1729).
In the majority of specimens the leaves are glabrous or, at most, have a few strigose hairs on the lower leaflet surface. A few specimens have densely pilose leaf-rhachides and sometimes lower leaf-surfaces as well. This variant (*T. strigulosa, T. vestita*) occurs sporadically through most of the range of the species.

2. **Trichilia emetica** Vahl, Symb. Bot. **1**: 31 (1790)—Oliv., F.T.A. **1**: 335 (1868).—C.DC. in A. & C.DC., Mon. Phan. **1**: 660 (1878).—Gürke in Engl., Pflanzenw. Ost-Afr. **C**: 231 (1895).—Gibbs in Journ. Linn. Soc., Bot. **37**: 435 (1906).—Monro in Trans. Rhod. Sci. Ass. **7**: 67 (1908).—Sim, For. Fl. Port. E. Afr., 26, t. 15 (1909).—R.E.Fr., Wiss. Ergebn. Schwed. Rhod.-Kongo-Exped. **1**: 111 (1914).—Eyles in Trans. Roy. Soc. S. Afr. **5**: 390 (1916).—Vermoes. in Rev. Zool. Afr. **10,** Supple. Bot.: 34 (1922).—Staner in Bull. Jard. Bot. Brux. **16**: 175 (1941).—Brenan, T.T.C.L.: 318 (1949).—O. B. Mill. in Journ. S. Afr. Bot. **18**: 40 (1952).—Pardy in Rhod. Agr. Journ. **51**: 492 cum photogr. (1954).—Palgrave, Trees of Central Afr.: 226 cum tab. et photogr. (1957). TAB. **58** fig. B.—Type from Arabia.
Mafureira oleifera Bertol. in Mem. Acc. Sci. Bol. **2**: 269, t. 12 (1850). Type: Mozambique, *Fornasini* (P).
Trichilia umbrifera Swynnerton & Bak. f. in Journ. Linn. Soc., Bot. **40**: 39 (1911). Type: Mozambique, Lower Umswirizwi, 300 m., fl. xi.1905, *Swynnerton* 148 (BM; K).
Trichilia roka Chiov., Fl. Somala, **2**: 131 (1932) *nom. illegit.*—Brenan in Mem. N.Y. Bot. Gard. **8, 3**: 235 (1953).—Garcia in Contr. Conhec. Fl. Moçamb. **2**: 142 (1954).—Williamson, Useful Pl. Nyasal.: 119 (1956).—Keay, F.W.T.A. ed. 2, **1**: 705 (1958).—Staner & Gilbert, F.C.B. **7**: 163 (1958). Same type as *T. emetica*.
Trichilia chirindensis sensu Garcia, loc. cit.

Medium-sized handsome evergreen tree 8–20(25) m. tall, with a wide umbrageous crown when growing in the open; bark dark grey or dark brown, rough or smooth, foliage very dark green, glossy. Leaves imparipinnate; petiole and rhachis up to 28 cm. long, tomentellous or densely puberulous; leaflets up to 15·5 × 5 cm., usually smaller, opposite or alternate, (3)4–5-jugate, elliptic or oblong-elliptic, rarely narrowly elliptic or lanceolate-elliptic, not or scarcely broadest near the apex, apex nearly always rounded or emarginate, very rarely acute, apiculate or subacuminate, base rounded or cuneate, upper surface drying olive-green or pale brown, lateral nerves in (10–12)13–16(19) pairs, lower surface densely puberulous with short curly hairs, especially on the nerves; petiolules up to 5 mm. long. Flowers pale green, pale yellow-green or pale yellow, fragrant, borne in congested cymes in leaf-axils or towards the base of the current-year's shoot. Pedicels very short. Calyx usually 3·5–5 mm. long, tomentellous, lobed almost to the base, lobes subcircular, imbricate. Petals 7(10)–15·5(16) mm. long, linear, tomentellous on both surfaces. Filaments usually 7–10 mm. long, united for about ½ of their length, sparsely puberulous outside, densely villous in the upper half inside. Appendages deltate, nearly ¾ as long as anthers; anthers 2 mm. long, slightly apiculate, antherodes a little smaller, not producing pollen. Disk glabrous, thin, with 10 deltate teeth alternating with the filaments. Ovary (2)3-locular; style usually 6–8 mm. long, columnar, densely setulose-puberulous almost to the apex; style-head capitate with a crateriform apical (?)stigmatic region. Pistillode with vestigial ovules. Capsule 1·8 × 1·8–2·5 × 2·5 cm., obovoid-globose, with a long stipe (0·4)0·6–1 cm. long, slightly sulcate, surface transversely wrinkled, fulvous-tomentellous, opening by (2)3-valves; seed black, almost completely concealed by the scarlet aril.

Caprivi Strip. Mpilila I., Chobe R., st. i.1959, *Killick & Leistner* 3388 (K; M; SRGH). **Bechuanaland Prot.** N: Kasane Rapids, Chobe R., fl. vii. 1950, *Robertson & Elffers* 52 (K; PRE). **N. Rhodesia.** B: Mwandi (old Sesheke), fr. xii.1952, *Angus* 1028 (BR; FHO; K; NDO). N: Samfya, Lake Bangweulu, fl. ix.1952, *Angus* 358 (BM; BR; COI; FHO; K; MO; PRE). C: 10 km. E. of Kafue Station, fr. x.1909, *Rogers* 8428 (SRGH). E: Fort Jameson, fl. xi.1951, *Gilges* 102 (PRE; SRGH). S: Munieke R., near Mapanza Mission, fl. viii.1952, *Angus* 131 (BM; BR; FHO; K; MO; NDO). **S. Rhodesia.** N: Nyanyanya R., Urungwe, fl. viii.1956, *Mullin* 49/56 (K; SRGH). W: Wankie, fl. iv.1954, *Levy* 1130 (K; PRE; SRGH). C: Lower Gwelo Reserve, fr. vi.1949, *Harvie* 5/49 (FHO; SRGH). E: Odzi R., Melsetter, fr. xii.1948, *Chase* 1459 (BM; K; SRGH). S: Ndanga, fl. x.1951, *Kirkham* 48/51 (SRGH). **Nyasaland.** N: Lukoma, fl. viii.1887, *Bellingham* (BM). C: Chia area, Kota Kota, fl. ix.1946, *Brass* 17503 (BM; K; SRGH). S: Tuchila R., near Mt. Mlanje, fl. viii.1957, *Chapman* 414 (BM; FHO; K). **Mozambique.** N: Montepuez, Cabo Delgado, fl. viii.1948, *Andrada* 1312 (COI; LISC). Z: Mocuba, Namagoa Estate, Quelimane, fr. i.1949, fl. ix.1949, *Faulkner* 388 (COI; K; SRGH). T: Mutarara, between Anacuaze and Dôa, fr. vi. 1949, *Barbosa* 3197 (LISC; LM; LMJ). MS: Cheringoma, fl. ix.1945; *Simão* 1 (LISC; LM). SS: Chibuto, fr. ii.1959, *Barbosa & Lemos* 8395 (COI; K; LMJ). LM: between Lourenço Marques and Matola, fr. ii.1953, *Barbosa* 5286 (LISC).

Widespread in Africa from Senegal to the Red Sea and through East and Central Africa to the Congo, Natal and the Transvaal; also in the Yemen. Usually in riparian forest and woodland or as an emergent from lake-shore thicket; more rarely in savanna woodland (on deep soils) or as an emergent of thickets away from water. It is replaced in evergreen forest, in regions of higher rainfall within our area, by *T. dregeana.*

T. emetica is one of the most widespread and better known African trees. Over much of Northern Rhodesia and parts of the adjoining territories it is known by the Lozi name "Musikili" or similar variants. Many uses have been recorded. Oil from the seeds is widely used by Africans for cooking, soap-making and for anointing their bodies and as a hair oil. In 1859 Kirk recorded that it was exported from Inhambane. 5000 metric tons of seed and 200 tons of oil were exported from Mozambique in some years before 1928, but since then the trade has greatly declined. The oil, which cannot be economically refined for edible purposes, is still used locally in the manufacture of soap. Great quantities of seeds are taken by the Africans of northern Nyasaland across the border to a soap factory at Kyela in Tanganyika. The bole is used for making dug-out canoes. The timber is light, pinkish-grey and makes excellent furniture, planking and boat material. In the Zambezi Valley a root infusion is drunk to facilitate labour in pregnancy (*Plowes* 1999). The seeds are edible. It is frequently planted as a shade tree in towns.

T. emetica and *T. dregeana* have been much confused in herbaria, although the notes of several collectors testify to their distinctness in the field. They differ in ecology and in several taxonomic characters, few of which, however, are absolutely diagnostic. Although most characters occasionally break down, genuine intermediates (at least in our area) are rare and, if all the key characters are used together, most specimens can be assigned to one or other species without hesitation. The key needs to be used intelligently. The leaf

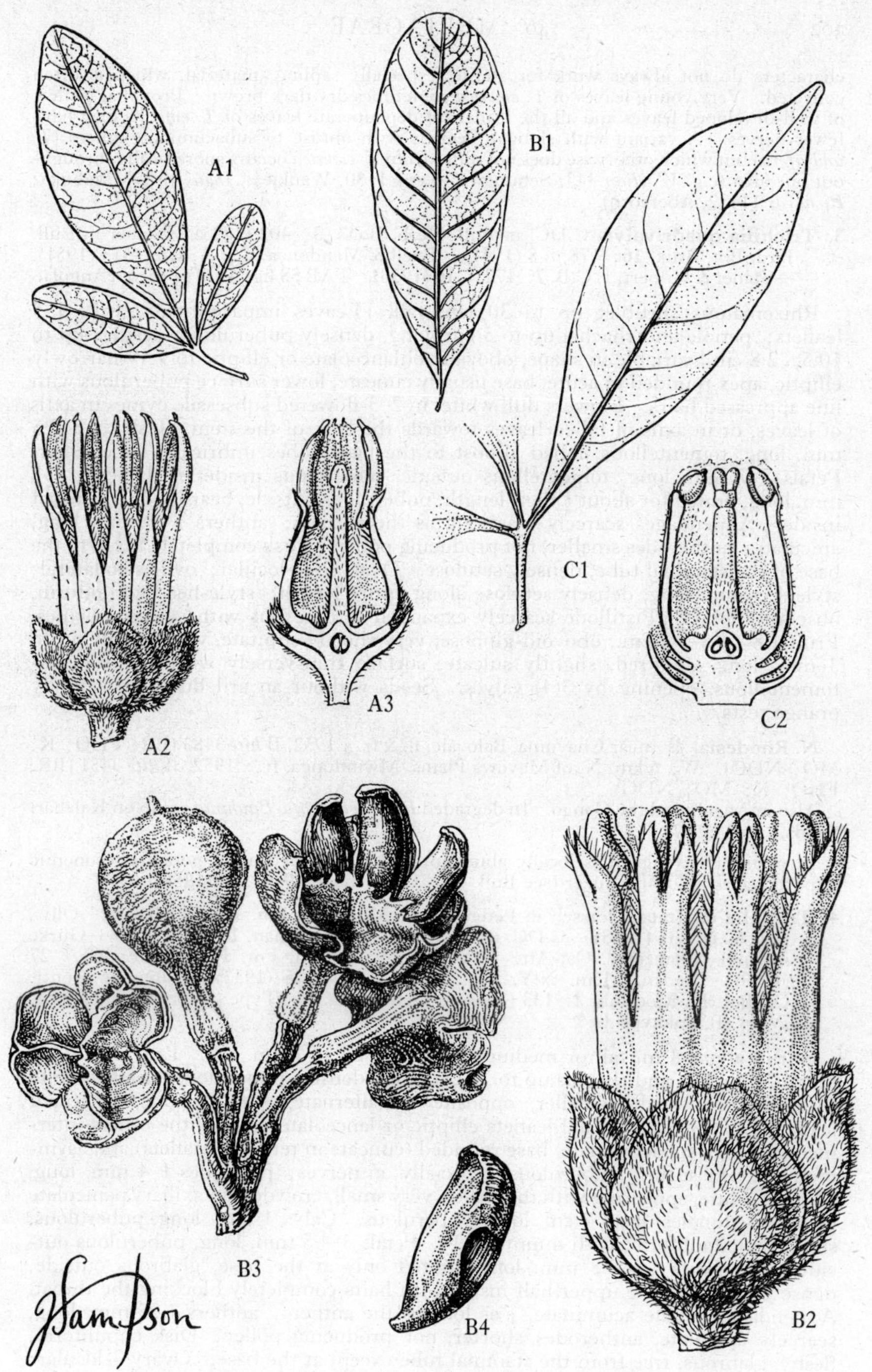

Tab. 58. A.—TRICHILIA QUADRIVALVIS. A1, leaf ($\times\frac{2}{3}$) *White* 3451; A2, male flower with corolla removed ($\times 7$) *White* 3451; A3, half of male flower with corolla removed ($\times 7$) *White* 3451. B.—TRICHILIA EMETICA. B1, leaflet ($\times\frac{2}{3}$) *Torre* 7663; B2, flower ($\times 7$) *Angus* 358; B3, fruits ($\times\frac{2}{3}$) *Barbosa & Lemos* 8395; B4, seed ($\times 1\frac{1}{3}$) *Barbosa & Lemos* 8395. C.—TRICHILIA PRIEURIANA. C1, leaflet ($\times\frac{2}{3}$) *Hoyle* 1101; C2, vertical section of gynoecium and androecium ($\times 7$) *Hoyle* 1101.

characters do not always work for sterile, especially sapling, material, which is often collected. Very young leaves of *T. emetica* sometimes dry dark brown. Proximal leaflets of well-developed leaves and all the leaflets of depauperate leaves of *T. emetica* may have fewer nerves. A variant with elongate leaflets with obtuse to subacuminate apices (*T. umbrifera*) but which otherwise does not differ from *T. emetica* occurs sporadically throughout our area (e.g. *Whellan* 443, Sebungwe; *Levy* 1130, Wankie; *Angus* 131, Mazabuka; *Richards* 11432, Abercorn).

3. **Trichilia quadrivalvis** C.DC. in Bull. Herb. Boiss. **3**: 402 (1895).—Staner in Bull. Jard. Bot. Brux. **16**: 178, t. 8 (1941).—Exell & Mendonça, C.F.A. **1**, 2: 313 (1951). —Staner & Gilbert, F.C.B. **7**: 172, t. 19 (1958). TAB **58** fig. A. Type from Angola.

Rhizomatous shrublet up to 30 cm. tall. Leaves imparipinnate, with 1–5 leaflets; petiole and rhachis up to 5 cm. long, densely puberulous; leaflets up to 10·5 × 2·8 cm., variable in shape, obovate, oblanceolate or elliptic to very narrowly elliptic, apex rounded to acute, base usually cuneate, lower surface puberulous with fine appressed hairs. Flowers dull white, in 2–5-flowered subsessile cymes in axils of leaves, or in axils of fallen leaves towards the base of the stem. Calyx 1·5–2·5 mm. long, tomentellous, lobed almost to the base, lobes imbricate, subcircular. Petals 5–7 mm. long, tomentellous outside, puberulous inside. Filaments 4–5 mm. long, united for about ½ their length, puberulous outside, bearded at the throat inside. Appendages scarcely ⅛ as long as the anthers; anthers 1·75 mm. long, apiculate; antherodes smaller, not producing pollen. Disk completely fused to the base of the staminal tube, densely setulose. Ovary 3(4)-locular; ovules collateral; style 4–5 mm. long, densely setulose along entire length; style-head coroniform, obscurely lobed. Pistillode scarcely expanded at base but with vestigial ovules. Fruit 1·8 × 1·8–2·2 cm., obovoid-globose, very shortly stipitate, with a stipe up to 3 mm. long, dull red, slightly sulcate, surface transversely wrinkled, fulvous-tomentellous, opening by 3(4) valves. Seeds without an aril but with a fleshy orange testa.

N. Rhodesia. B: near Chavuma, Balovale, fl. & fr. x.1952, *White* 3483 (BR; FHO; K; MO; NDO). W: 6 km. N. of Mayowa Plains, Mwinilunga, fr. x.1952, *White* 3451 (BR; FHO; K; MO; NDO).

Also in Angola and the Congo. In degraded *Baphia obovata-Bauhinia* scrub on Kalahari Sands.

In parts of Kwango it is locally abundant and is regarded as of potential economic importance as an oil-producer (see Bull. Agr. Cong. Belg. **49**: 1304 (1958)).

4. **Trichilia capitata** Klotzsch in Peters, Reise Mossamb. Bot. **1**: 120 (1861).—Oliv., F.T.A. **1**: 335 (1868).—C.DC. in A. & C.DC., Mon. Phan. **1**: 707 (1878).—Gürke in Engl., Pflanzenw. Ost-Afr. **C**: 231 (1895).—Sim, For. Fl. Port. E. Afr.: 27 (1909).—Brenan, Mem. N.Y. Bot. Gard., **8**, 3: 235 (1953).—Garcia in Contr. Conhec. Fl. Moçamb. **2**: 143 (1954). TAB. **59** fig. A. Type: Mozambique, Sena, *Peters* (B, holotype †).*

Shrub about 3 m. tall or medium-sized tree up to 15 m. tall. Leaves imparipinnate; petiole and rhachis up to 21 cm. long, densely puberulous; leaflets up to 11·5 × 5 cm., usually smaller, opposite or alternate, 4–6(7)-jugate, proximal leaflets broadly ovate, distal leaflets elliptic or lanceolate-elliptic, the others intermediate, apex subacuminate, base rounded (cuneate in terminal leaflets) and asymmetric, lower surface puberulous, especially on nerves; petiolules 1–4 mm. long. Flowers white, appearing with the leaves, very small, crowded, in axillary paniculate cymes; peduncles up to 7 cm. long, puberulous. Calyx 1 mm. long, puberulous, shallowly cupuliform, with minute teeth. Petals 3–4·5 mm. long, puberulous outside. Filaments 1·5–2·5 mm. long, united only at the base, glabrous outside, densely villous in the upper half inside with hairs completely blocking the throat. Appendages deltate-acuminate, ¾ as long as the anthers; anthers 1·25 mm. long, scarcely apiculate, antherodes shorter, not producing pollen. Disk cupuliform, fleshy, glabrous, free from the staminal tube except at the base. Ovary 3-locular; style 2 mm. long, glabrous; style-head ovoid, with 3 erect stigmatic lobes; ovules collateral. Pistillode similar to the gynoecium but narrower and apparently without vestigial ovules. Capsule c. 1·5 × 1·5 cm., globose, surface smooth, not or scarcely wrinkled, tomentellous, glabrescent, opening by 3 thinly woody valves. Seeds dull dark red, less than ⅓ covered by an orange aril forming a cushion at the apex.

* Tete, *Kirk* (K) was compared with the type by Oliver in 1865.

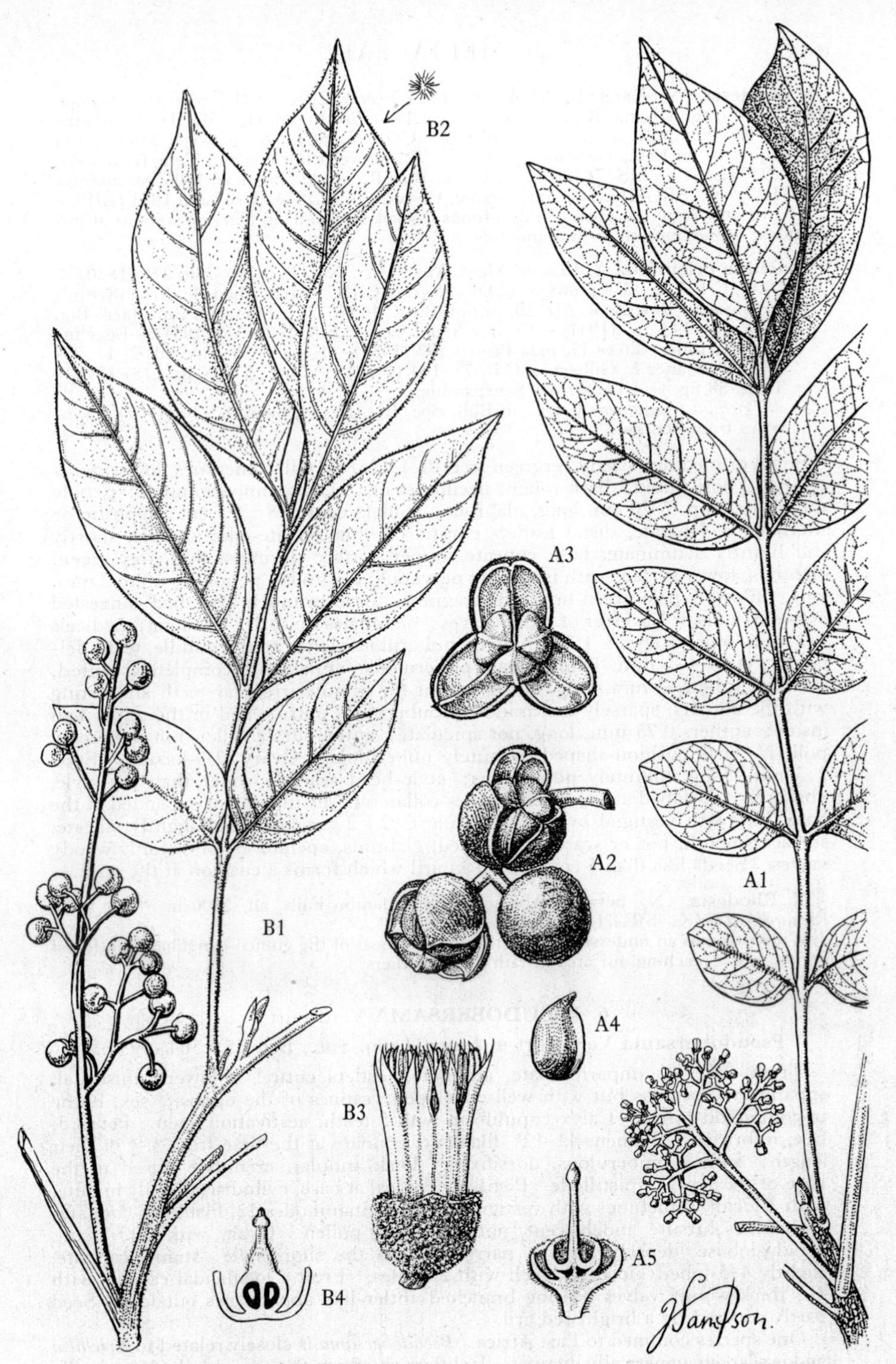

Tab. 59. A.—TRICHILIA CAPITATA. A1, flowering branchlet ($\times \frac{2}{3}$) *Phelps* 22; A2, infructescence ($\times \frac{2}{3}$) *Chase* 2608; A3, open fruit ($\times \frac{2}{3}$) *Phelps* 56; A4, seed ($\times 1\frac{1}{2}$) *Phelps* 56; A5, vertical section of pistillode and disk ($\times 7$) *Kirk* s.n. B.—LEPIDOTRICHILIA VOLKENSII. B1, fruiting branchlet ($\times \frac{2}{3}$) *Chapman* 269; B2, stellate hair ($\times 20$); B3, flower with corolla removed ($\times 7$) *Brass* 17800; B4, vertical section of gynoecium ($\times 7$) *Brass* 17800.

S. Rhodesia. N: Kafefe R., Mtoko, fl. i.1953, *Phelps* 22 (K; PRE; SRGH). **Nyasaland.** S: Lower Mwanza R., Chikwawa, fr. viii.1946, *Brass* 18006 (K; SRGH). **Mozambique.** N: Nampula, fl. xii.1936, *Torre* 776 (COI; LISC). Z: between Mopeia and Massingire, fr. vii.1942, *Torre* 4464 (LISC). T: between Boroma and Tete, fr. vi.1941, *Torre* 2930 (LISC). MS: Chibabava, Lower Buzi R., alt. 130 m., fl. xii.1906, *Swynnerton* 1033 (BM; K; SRGH). SS: Inhambane, Govuro, fr. ix.1944, *Mendonça* 1983 (LISC).

Known only from our area. In deciduous thicket or woodland fringing streams at low altitudes; sometimes on termite mounds.

5. **Trichilia prieuriana** A. Juss. in Mém. Mus. Hist. Nat. Par. **19**: 236, 276 (1830).—Oliv., F.T.A. **1**: 334 (1868).—C.DC. in A. & C.DC., Mon. Phan. **1**: 678 (1878).—Vermoes. in Rév. Zool. Afr. **10,** Suppl. Bot.: 46 (1922).—Staner in Bull. Jard. Bot. Brux. **16**: 143, t. 7 (1941).—Exell & Mendonça, C.F.A. **1,** 2: 312 (1951).—Eggeling & Dale, Indig. Trees Uganda Prot.: 197 (1952).—Keay, F.W.T.A. ed. 2, **1**: 704 (1958).—Staner & Gilbert, F.C.B. **7**: 164 (1958).—White, F.F.N.R.: 181 (1962). TAB. **58** fig. C. Type from Senegambia.

Trichilia senegalensis C.DC. in Bull. Soc. Bot. Fr. **54,** mém. 8: 10 (1907). Syntypes from Senegal.

Small or medium-sized evergreen tree 10–15(25) m. tall, sometimes flowering as a shrub; bole fluted; bark rough, peeling easily. Leaves imparipinnate; petiole and rhachis up to 20 cm. long, glabrous; leaflets up to 18 × 6·5 cm., opposite or alternate, 3–4-jugate, distal leaflets elliptic or oblanceolate-elliptic, apex shortly and bluntly acuminate, base cuneate, upper surface drying a dull grey-green, glabrous, lower surface with indistinct nervation, glabrous; petiolules up to 2 mm. long. Flowers pale green or white, fragrant, borne in much branched congested cymes in axils of leaves or of fallen leaves; peduncles up to 1·5 cm. long; pedicels 2–3 mm. long. Calyx 1–1·5 mm. long, puberulous, lobed to the middle, teeth deltate. Petals 4–7 mm. long, linear, puberulous. Filaments completely united, staminal tube 5–7 mm. long, entire except for minute irregular teeth alternating with the anthers, sparsely and minutely puberulous outside and in the upper half inside; anthers 0·75 mm. long, not apiculate; antherodes smaller, not producing pollen. Disk cushion-shaped, minutely puberulous. Ovary (2)3-locular. Style 2–3 mm. long, minutely puberulous; style-head scarcely wider than the style, obscurely 2–3-lobed at the apex; ovules collateral. Pistillode not expanded at the base, but with vestigial ovules. Capsule c. 2 × 2 cm., globose, slightly sulcate, surface smooth, not or scarcely wrinkled, glabrous, opening by 3(4) thinly woody valves. Seeds less than ⅓ covered by an aril which forms a cushion at the apex.

N. Rhodesia. N: between Abercorn and Kalambo Falls, alt. 1500 m., fl. ix.1956, *Richards* 6206 (K; SRGH).

Widespread as an understorey tree throughout most of the guineo-congolian rain forest region. Just reaching our area in rain forest outliers.

6. PSEUDOBERSAMA Verdcourt

Pseudobersama Verdcourt in Journ. Linn. Soc., Bot. **55**: 504 (1956).

Trees. Leaves imparipinnate, alternate, leaflets entire. Flowers unisexual, apparently dioecious but with well-developed vestiges of the opposite sex, borne in compound cymes. Calyx cupuliform with 5 teeth, aestivation open. Petals 5, free, imbricate. Stamens 11–12; filaments connate at the base from ⅓–⅔ of their length; anthers puberulous, dorsifixed. Disk annular, crenulate, fused to the base of the ovary or pistillode. Pistillode conical at base, cylindric, distally hirsute, with 3 loculi, sometimes with vestigial ovules. Staminodes 12, filaments connate; antherodes hirsute, indehiscent, not producing pollen. Ovary with (4)5 loculi, ovoid-globose, densely hirsute, narrowed into the short style; stigma capitate, slightly 4–5-lobed; loculi 5, each with 2 ovules. Fruit a loculicidal capsule with 4–5 thick woody valves bearing branched antler-like appendages outside. Seed partly covered by a bright red aril.

One species confined to East Africa. *Pseudobersama* is closely related to *Trichilia* and is scarcely generically distinct. It differs chiefly in the capsule which is usually 5-valved (not 2–4-valved). The valves in *Trichilia* are never as thick and woody. The curious appendages of *Pseudobersama* are distinctive, but smaller unbranched appendages occur in some American species of *Trichilia*, e.g. *T. trinitensis* A. Juss.

As in *Trichilia* male plants appear to be much more numerous than female. We have seen only one female flowering specimen collected in our area. According to Verdcourt (loc. cit.) the ovary is 5-locular with axile placentation. In the only ovary seen by us 5 intrusive placentae almost meet in the middle but are not fused. The discrepancy may reflect different stages of development.

1. **Pseudobersama mossambicensis** (Sim) Verdcourt in Journ. Linn. Soc., Bot. **55**: 504, t. 1 (1956).—Gomes e Sousa, Dendrol. Mozamb. **5**: 184 cum tab. (1960).—Dale & Greenway, Kenya Trees and Shrubs: 270 (1961). TAB. **60**. Type: Mozambique, without locality, *Sim* 5204.*

   Bersama mossambicensis Sim, For. Fl. Port. E. Afr.: 34, t. 23 (1909).—Brenan, T.T.C.L.: 324 (1949). Type as above.

Small or medium-sized evergreen tree up to 20 m. tall, sometimes flowering as a shrub 2–4 m. tall. Leaves imparipinnate; petiole and rhachis up to 30 cm. long, minutely puberulous towards base; leaflets up to 15 × 6 cm., usually smaller, 4–8(9)-jugate, alternate or subopposite, elliptic or oblong-elliptic, apex shortly and bluntly acuminate, base cuneate, asymmetric, the proximal leaflets smaller and proportionally broader, young leaflets densely pubescent, glabrous above when mature, lower surface with conspicuous tufts of hairs in nerve-axils and prominent open reticulate venation. Flowers white, 3–12, in lax to subcapitate compound cymose inflorescences, each part of the inflorescence bearing a simple 3-flowered cyme; peduncle 1–6 cm. long; bracts 1–3 mm. long, subulate, puberulous; pedicels 0·5–1 mm. long. The flowers are functionally either male or female and apparently dioecious, but the differences between them are slight. Calyx c. 3 mm. long, lobed to about the middle, lobes deltate. Petals 4·5–6 mm. long, glabrous except for minute papillae. Filaments about 3·5 mm. long, united for $\frac{1}{3}$–$\frac{2}{3}$ of their length, densely hairy towards the apex; anthers 1·5 mm. long, hairy. Pistillode 3 mm. long. Antherodes 1 mm. long. Ovary 3 × 3 mm.; style 1·5 mm. long. Capsule 3–4·5 cm. in diam., very woody, red, densely covered with lobed antler-shaped appendages 7 mm. long, dehiscing by (4)5(6) valves, which remain connate at the base; valves 4–6 mm. thick; stipe about 5 mm. long; seeds 2 per loculus, 7 × 5 mm., purple-red, aril bright red, confined to adaxial part of seed, and forming a cushion at the apex.

Mozambique. MS: Dondo, Beira, fl. xii.1943, *Torre* 6328 (LISC). SS: Chibuto, fr. vii.1944, *Torre* 6777 (LISC). LM: Maputo, Magala, fr. vii.1948, *Mendonça* 4509 (LISC).

Also in Kenya, Tanganyika and Natal. Ecology imperfectly known; reported from " dense mixed forest with *Brachystegia spiciformis, Dialium schlechteri, Albizia, Millettia, Afzelia* ".

7. LEPIDOTRICHILIA Leroy

Lepidotrichilia Leroy in Compt. Rend. Acad. Sci. **247**: 1025 (1958); in Journ. Agr. Trop. Bot. Appl. **5**: 673 (1958).

Trees. Leaves imparipinnate, leaflets entire, stellate-puberulous. Flowers bisexual, borne in large cymose panicles. Calyx cupuliform with 5 minute teeth. Petals 5, free, much longer than the calyx in bud, induplicate-valvate. Filaments 10, fused in the lower half to form a staminal tube; anthers inserted between a pair of deltate-acuminate appendages which slightly exceed them in length. Disk absent but base of staminal tube apparently slightly modified for nectar secretion. Ovary (in our area) with 2(3) loculi, each loculus with 1 ovule; style-head capitate, surmounted by 2–3 erect stigmatic lobes. Fruit drupaceous, indehiscent.

A small genus with 1 species in Africa and 3 in Madagascar.

1. **Lepidotrichilia volkensii** (Gürke) Leroy, comb. nov.† TAB. **59** fig. B. Type from Tanganyika.

   Trichilia volkensii Gürke in Engl., Bot. Jahrb. **19,** Beibl. 47: 33 (1894); in Engl.,

* This has not been definitely traced. The only specimen of this species in the Pretoria Herbarium collected by Sim in Mozambique (No. 21178) is possibly the type specimen subsequently re-numbered.

† Leroy first proposed this name in Compt. Rend. Acad. Sci. **247**: 1026 (1958) but did not give a full reference to the basionym. He has kindly consented to validate the name here.

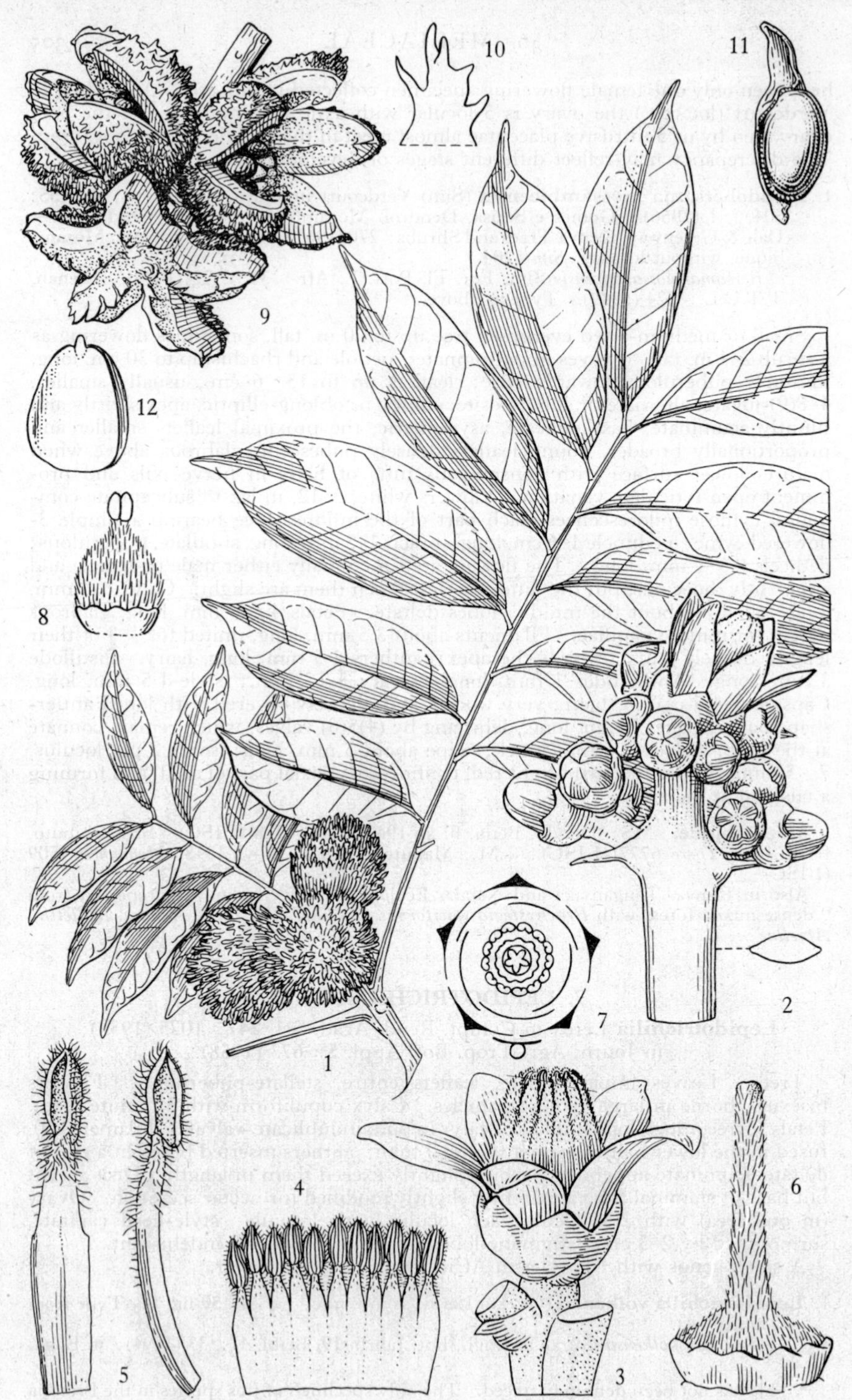

Tab. 60. PSEUDOBERSAMA MOSSAMBICENSIS. 1, branchlet showing immature fruit ($\times \frac{1}{2}$); 2, male inflorescence ($\times 3$); 3, male flower ($\times 6$); 4, staminal tube, opened out, from within ($\times 6$); 5, stamens ($\times 12$); 6, vestigial ovary from male flower ($\times 12$); 7, floral diagram of female flower; 8, ovary ($\times 4$); 9, infructescence ($\times \frac{2}{3}$); 10, appendage from surface of capsule, ($\times 4$); 11, seed, with aril ($\times 2$); 12, cotyledons ($\times 4$). From Journ. Linn. Soc., Bot. 55: t. 1 (1956)

Pflanzenw. Ost-Afr. **C**: 232 (1895).—Brenan T.T.C.L.: 319 (1949); in Mem. N.Y. Bot. Gard. **8**, 3: 235 (1953).—Eggeling & Dale, Indig. Trees Uganda Prot.: 198 (1952).—Staner & Gilbert, F.C.B. **7**: 160 (1958).—Dale & Greenway, Kenya Trees and Shrubs: 271 (1961). Type as above.

Trichilia buchananii C.DC. in Bull. Herb. Boiss. **2**: 580 (1894).—Lebrun, Ess. For. Reg. Mont. Cong. Or. **2**: 111 (1935). Type: Nyasaland, *Buchanan* 155 (BM; G, holotype; K).

Trichilia volkensii var. *buchananii* (C.DC.) P.-Sermolli in Webbia, **7**: 333 (1950). —White, F.F.N.R.: 182 (1962). Type as above.

Trichilia volkensii var. *genuina* P.-Sermolli, tom. cit.: 334 (1950). Type as for *T. volkensii*.

Small or medium-sized evergreen tree 5–20 m. tall, sometimes flowering as a shrub; bole fluted; bark smooth, grey. Leaves imparipinnate; petiole and rhachis up to 23 cm., long, densely stellate-pubescent; leaflets very variable in shape and size, up to 15 × 6 cm., usually much smaller, opposite or alternate, 3–4-jugate, distal leaflets more or less elliptic, apex acute to shortly acuminate, base markedly asymmetric, upper surface drying pale yellow-green, lower surface puberulous with small stellate hairs and minute red and black glands; petiolules up to 1 cm. long. Flowers creamy-white, becoming darker yellow with age, fragrant, in contracted cymose panicles in leaf-axils; peduncles up to 10 cm. long; pedicels 1–2·5 mm. long. Calyx 2 mm. long, cupuliform, with minute teeth, scurfy-stellate-pubescent. Petals 4–5 mm. long, densely puberulous outside. Filaments 3–4 mm. long, puberulous at the base outside, sparsely villous in the upper half inside. Style 1·5–2 mm. long. Fruit small, up to 1·5 × 1 cm., encrusted with stellate scales, 2–3-locular, with 1 dark brown or black exarillate seed in each loculus.

Nyasaland. N: Mugesse Forest, Misuku Hills, fl. x.1953, *Chapman* 172 (FHO; K). C: Nchisi Mt., fl. ix.1929, *Burtt Davy* 21206 (FHO; K). S: Cholo Mt., fl. ix.1946, *Brass* 17765 (BM; K; SRGH).

Also in Ethiopia, the Sudan, Uganda, Kenya, Tanganyika and the Congo. In montane forest.

This species has been collected in the Mafinga Mts. very near the N. Rhodesian border and almost certainly occurs on the Rhodesian side.

8. TURRAEA L.

Turraea L., Mant. Pl. Alt.: 150, 237 (1771).

Shrubs or small trees, sometimes scrambling. Leaves simple, alternate, sometimes fasciculate, usually entire, rarely shallowly lobed. Flowers bisexual, solitary or fasciculate or in axillary or terminal cymes or false racemes. Calyx cupuliform, with (4)5(6) teeth or lobes, sometimes almost entire-margined, persistent in fruit. Petals (4)5, much longer than the calyx, linear-spathulate to linear. Staminal tube cylindric, sometimes expanded distally, terminated by simple or 2-lobed free or partly or completely fused appendages opposite to or alternating with the anthers. Anthers (8)10(20), apiculate, inserted on the rim of the staminal tube or inside the tube towards the apex, sessile or with short filaments. Disk always present (in our area), sometimes vestigial, partly or completely fused to the base of the staminal tube. Ovary small, with (4)5–10(20) loculi, each with 2 anatropous or almost campylotropous superposed ovules. Style elongate, distally expanded to form an ovoid, globose, cylindric or conical style-head of which the proximal part serves as a *receptaculum pollinis* and the distal part forms a disk- or cushion-shaped stigmatic surface. Fruit a small leathery or woody loculicidal capsule with (4)5–10(20) valves. Seeds large, with a large or small, orange or red aril; testa black or red, shining.

A palaeotropical genus with about 70 species, concentrated in Madagascar and East Africa, and rapidly diminishing in number to the east and to the west.

The flowers of *T. fischeri* are said to be pollinated by honey birds (*Burtt* 3336); those of species with very long staminal tubes are probably pollinated by moths. Stigmatic papillae are confined to the distal part of the expanded style-head. In *T. vogelioides* (a living specimen from Tanganyika in the Oxford Botanic Garden was examined) the anthers dehisce before the flower opens and pollen is deposited on the proximal part of the style-head which serves as a *receptaculum pollinis* as in certain *Rubiaceae* and *Apocynaceae*. At this stage the stigmatic surface is pale green; a day or two later it turns yellow and is presumably then receptive. This interpretation of the pollination mechanism requires confirmation in the field for *T. vogelioides* and other species.

Staminal tube densely bearded at the throat, or at least densely hairy with retrorse hairs in the upper half; appendages shorter than the anthers, connate at the base to form a frill which continues the staminal tube beyond the insertion of the filaments:
Staminal tube distally expanded, bearded at the throat; filaments as long as or longer than the anthers, geniculate; appendages usually 2-lobed; disk with a free, toothed margin; ovary and capsule ± 10-locular; pedicels 1–1·5 cm. long:
Flowers borne in terminal and axillary, congested cymes or false racemes; peduncles 3–4·5 cm. long; bracts up to 1·5 cm. long; calyx exceeding 5 mm. in length, tomentellous, distinctly lobed; capsule puberulous - - 1. *robusta*
Flowers borne in fascicles; peduncles less than 5 mm. in length; bracts up to 5 mm. long; calyx less than 3 mm. in length, puberulous, with indistinct teeth; capsule glabrous:
Branchlets of flowering specimens stout, usually exceeding 4 mm. in diam.; leaves rounded to acute; flowers usually produced when the plant is leafless, in subsessile fascicles in axils of the fallen leaves; staminal appendages sparsely ciliate or glabrous on the margin, usually deeply and regularly 2-lobed 2. *nilotica*
Branchlets of flowering specimens slender, usually less than 3 mm. in diam.; leaves acuminate; flowers usually produced with the leaves, in fascicles in the leaf-axils or terminating lateral shoots; staminal appendages densely ciliate, usually shallowly and irregularly 2-lobed - - - - 3. *zambesica*
Staminal tube not distally expanded, not bearded but densely hairy inside in the upper half; filaments much shorter than the anthers, not geniculate; appendages entire or slightly emarginate, scarcely lobed; disk completely fused to the staminal tube, margin not toothed; ovary and capsule 5–6-locular; inflorescence axillary, 1–4-flowered, pedicels 2·5–2·75 cm. long - - - - - - - 4. *holstii*
Staminal tube not bearded at the throat, glabrous in the upper half inside; appendages much longer than the anthers (except in *T. fischeri*) and usually occurring in pairs between them, completely free or sometimes united at the base, not fused to form a frill (except in *T. obtusifolia*); disk completely fused to staminal tube, margin not toothed:
Appendages usually shorter than the anthers, occasionally just a little longer; staminal tube distinctly curved - - - - - - - - - - 5. *fischeri*
Appendages much longer than the anthers; staminal tube straight:
Style-head just protruding from the staminal tube; ovary 5-locular, glabrous or setose-hairy; capsule 5-locular:
Ovary densely setose-hairy; style-head conical-cylindric from a broad base; filaments not fused; staminal appendages 3–3·5 mm. long - 6. *mombassana*
Ovary glabrous; style-head obovoid-cylindric, broadest at apex; filaments fused beyond the insertion of the appendages to form a short frill continuing the staminal tube; staminal appendages 2 mm. long - - - - 7. *obtusifolia*
Style-head exserted for 1 cm. or more; ovary 10–12-locular, puberulous; capsule 10–12-locular:
Leaves with conspicuous tufts of hairs in axils of secondary nerves beneath; style-head conical-cylindric tapering from a broad base; capsule fulvous-tomentellous 8. *wakefieldii*
Leaves without conspicuous tufts of hairs in axils of secondary nerves beneath; style-head globose; capsule dark brown or black, glabrous - 9. *floribunda*

1. **Turraea robusta** Gürke in Engl., Bot. Jahrb. **19,** Beibl. 47: 34 (1894); in Engl., Pflanzenw. Ost-Afr. **C**: 231 (1895).—Bak. f. in Journ. of Bot. **41**: 12 (1903).—R.E.Fr., Wiss. Ergebn. Schwed. Rhod.-Kongo-Exped. **1**: 111 (1914).—Brenan, T.T.C.L.: 321 (1949); in Mem. N.Y. Bot. Gard. **8,** 3: 234 (1953).—Eggeling & Dale, Indig. Trees Uganda Prot.: 200 (1952).—Dale & Greenway, Kenya Trees and Shrubs: 275 (1961).—White, F.F.N.R.: 182 (1962). TAB. **61** fig. C. Type from Tanganyika.

Turraea volkensii Gürke, loc. cit.—Gürke in Engl., loc. cit.—Bak. f., loc. cit. Type from Tanganyika.

Turraea goetzei Harms in Engl., Bot. Jahrb. **28**: 415 (1900).—Bak. f., loc. cit.—Brenan, T.T.C.L.: 320 (1949). Type from Tanganyika.

Turraea sacleuxii C.DC. in Ann. Conserv. Jard. Bot. Genève, **10**: 130 (1907). Type from Kenya.

Turraea squamulifera C.DC., tom. cit.: 133. Type from Kenya.

Turraea nilotica sensu Staner in Bull. Jard. Bot. Brux. **16**: 127, t. 3 (1941)—Staner & Gilbert, F.C.B. **7**: 156 (1958).

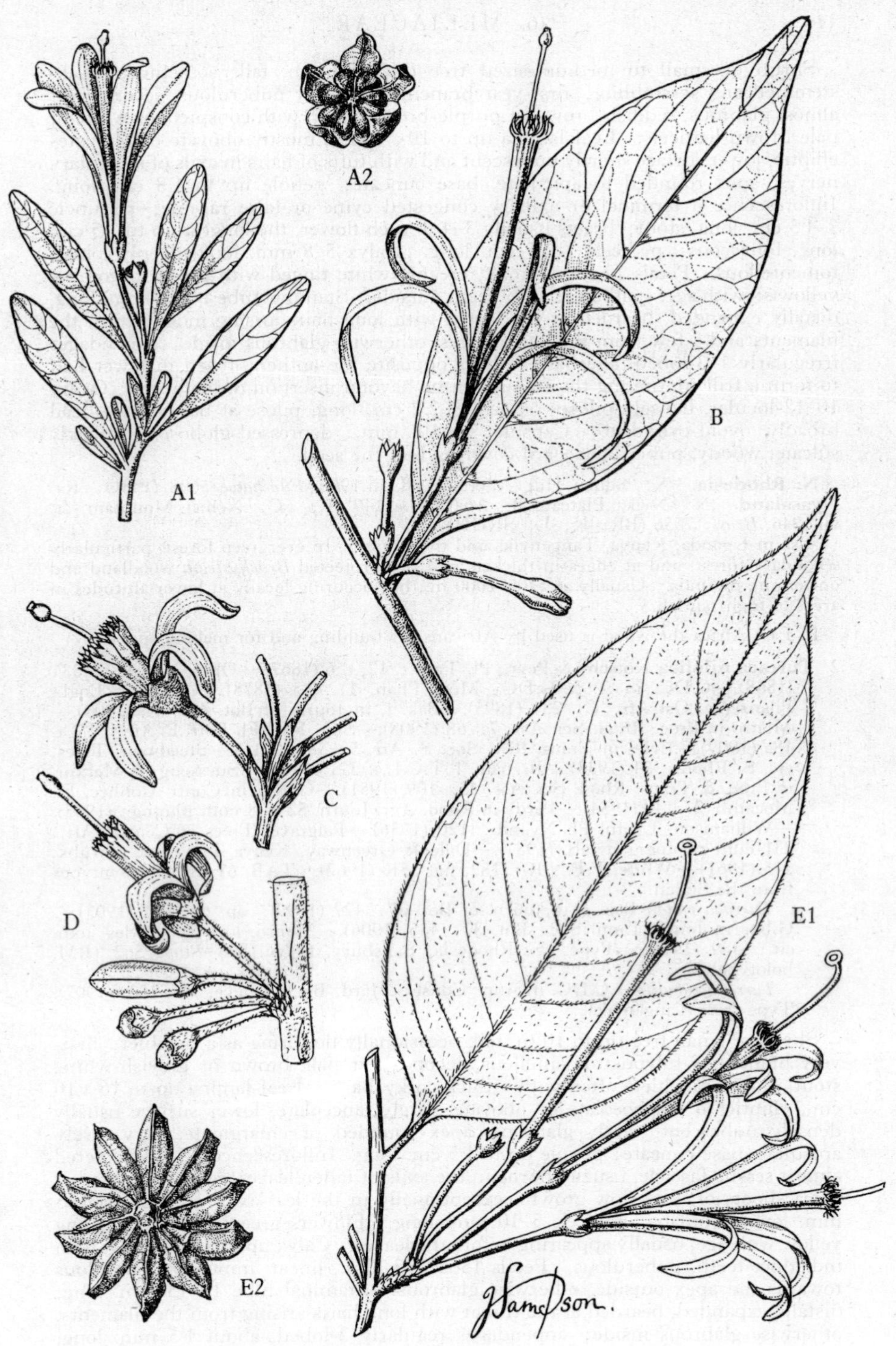

Tab. 61. A.—TURRAEA OBTUSIFOLIA. A1, flowering shoot (×1) *Sousa* 141; A2, fruit (×1) *Torre* 7105. B.—TURRAEA WAKEFIELDII, flowering shoot (×1) *Barbosa & Lemos* 8548. C.—TURRAEA ROBUSTA, flower (×1) *Richards* 4342. D.—TURRAEA NILOTICA, inflorescence (×1) *Angus* 168. E.—TURRAEA FLORIBUNDA. E1, flowering shoot (×1) *Wild* 4333; E2, fruit (×1) *Watt & Brandwyk* 1095.

Shrub or small to medium-sized tree (1)2–8(16) m. tall, sometimes weak-stemmed and scrambling; first-year branchlets densely puberulous, second-year almost glabrous, reddish-brown or purple-brown, often with conspicuous white or pale brown lenticels. Leaf-lamina up to 10 × 7 cm., mostly obovate or obovate-elliptic, lower surface shortly pubescent and with tufts of hairs in axils of secondary nerves, apex rounded to apiculate, base cuneate; petiole up to 1·8 cm. long. Inflorescence a terminal or axillary congested cyme or false raceme; peduncle 3–4·5 cm. long, stout; bracts usually 3–4 to each flower, the longest up to 1·5 cm. long, foliaceous; pedicels 1–1·5 cm. long. Calyx 5–8 mm. long, deeply lobed, tomentellous. Petals 12–14 × 4 mm., creamy white tinged with green, becoming yellowish with age, spathulate, pubescent outside. Staminal tube 1·2–1·4 cm. long, distally expanded, bearded at the throat with long hairs arising mostly from the filaments and a few from the appendages, otherwise glabrous inside; appendages irregularly 1–3-lobed, alternate with or opposite the anthers, fused in lower half to form a frill continuing the staminal tube beyond insertion of filaments. Ovary 10–12-locular, densely pilose; style 1·5–2·2 cm. long, pilose at base; style-head broadly ovoid-cylindric. Capsule 6 × 10 mm., depressed-globose, shallowly sulcate, woody, puberulous; aril covering half the seed.

N. Rhodesia. N: Sunzu Hill, Abercorn, fl. iii.1960, *Fanshawe* 5608 (FHO; K). **Nyasaland.** N: Nyika Plateau, fl., *McClounie* 149 (K). C: Nchisi Mountain, fr. vii.1946, *Brass* 17056 (BR; K; L; SRGH).

Also in Uganda, Kenya, Tanganyika and the Congo. In evergreen forest, particularly secondary forest, and at edges in thickets and fire-protected *Brachystegia* woodland and on termite mounds. Usually at 1200–2000 m., but occuring locally at lower altitudes in areas of high rainfall.

In East Africa the wood is used by Africans for building and for making spoons.

2. **Turraea nilotica** Kotschy & Peyr., Pl. Tinn.: 12, t. 6 (1867).—Oliv., F.T.A. **1**: 331 (1868).—C.DC. in A. & C.DC., Mon. Phan. **1**: 445 (1878).—Gürke in Engl., Pflanzenw. Ost-Afr. **C**: 231 (1895).—Bak. f. in Journ. of Bot. **41**: 11 (1903).—Monro in Proc. Rhod. Sci. Ass. **7**: 68 (1908).—Sim, For. Fl. Port. E. Afr.: 25, t. 18a (1909).—Eyles in Trans. Roy. Soc. S. Afr. **5**: 388 (1916).—Steedman, Trees, etc. S. Rhod.: 34 (1933).—Brenan, T.T.C.L.: 321 (1949).—Suesseng. & Merxm. in Proc. & Trans. Rhod. Sci. Ass. **43**: 109 (1951).—Garcia in Contr. Conhec. Fl. Moçamb. **2**: 140 (1954).—Pardy in Rhod. Agr. Journ. **52**: 38 cum photogr. (1955).—Williamson, Useful Pl. Nyasal.: 120 (1956).—Palgrave, Trees of Central Afr.: 231 cum photogr. et tab. (1957).—Dale & Greenway, Kenya Trees and Shrubs: 275 (1961).—White, F.F.N.R.: 182, fig. 35K (1962). TAB. **61** fig. D. Syntypes from the Sudan.

Turraea randii Bak. f. in Journ. of Bot. **37**: 427 (1899); op. cit.: 11 (1903).—Gibbs in Journ. Linn. Soc., Bot. **37**: 435 (1906).—Monro, loc. cit.—Eyles tom. cit.: 389 (1916). Type: S. Rhodesia, Salisbury, fl. vii.1898 *Rand* 562 (BM, holotype).

Turraea tubulifera C.DC. in Ann. Conserv. Jard. Bot. Genève, **10**: 133, (1907). Type from Tanganyika.

Shrub or small tree up to 10 m. tall, occasionally flowering as a shrublet; first-year branchlets fulvous-tomentellous, second-year pale brown or greyish-white, stout, older branchlets often with a thick corky bark. Leaf-lamina up to 16 × 10 cm., elliptic to oblanceolate or obovate, rarely lanceolate, lower surface usually densely pubescent, rarely glabrous, apex rounded or emarginate, very rarely apiculate, base cuneate; petiole up to 1·5 cm. long. Inflorescence a 5–12-flowered almost sessile fascicle, usually borne in the axils of fallen leaves, rarely terminating short shoots of very slow growth, exceptionally in the leaf-axils; bracts up to 5 mm. long, subulate; pedicels 5–10 mm. long. Flowers greenish-white, turning yellow with age, usually appearing before the leaves. Calyx up to 3 mm. long, with indistinct teeth, puberulous. Petals 15–22 × 3 mm., linear, minutely puberulous towards the apex outside, otherwise glabrous. Staminal tube 10–15 mm. long, distally expanded, bearded at the throat with long hairs arising from the filaments, otherwise glabrous inside; appendages regularly 2-lobed, about 1·5 mm. long, alternating with the anthers, fused in lower half to form a frill continuing staminal tube beyond the insertion of filaments, glabrous outside or rarely with a few marginal cilia. Ovary 10-locular, glabrous; style 2–2·5 cm. long, glabrous or pilose at base, style-head ovoid-cylindric. Capsule 7 × 15 mm., depressed-globose, shallowly sulcate, leathery, glabrous; aril covering about half of the seed.

N. Rhodesia. N: Mpika, fl., *Fanshawe* 2475 (BR; K). W: Ndola, fl. viii.1953, *Fanshawe* 226 (BR; FHO; K). C: 10 km. E. of Lusaka, fl. ix.1955, *King* 170 (K). E: Lundazi, Tigone Dam, fr. xi.1958, *Robson* 665 (K). S: between Mazabuka and Nega NegaHills, fl. viii. 1952, *Angus* 168 (BM; BR; FHO; K; MO; ND; PRE). **S. Rhodesia.** N: Chipoli, Mazoe, fr. ix.1958, *Moubray* 26 (SRGH). W: 40 km. S. of Bulawayo, fl. vi.1947, *Keay* in FHI 21332 (FHO). C: Enkeldoorn, fl. ix.1947, *Johnston* (K; SRGH). E: Umtali, fr. xii.1951, *Chase* 4220 (SRGH). S: Lundi R., Ndanga, *Mowbray* 51 (SRGH). **Nyasaland.** S: Zomba, fr., *Clements* 618 (FHO). **Mozambique.** N: between Monapo and Quixaxe, fr. x.1948, *Barbosa* 2470 (BR; LM). Z: Mocuba, Namagoa, fl. viii.1944, *Faulkner* 29 (BR; COI; FI; G; K; PRE). T: Angonia, fl. vii.1949, *Andrada* 1785 (LISC). MS: Ilha de Chiloane, fl. 1884–5, *Carvalho* (COI). SS: Chibuto, fr. x.1957, *Barbosa & Lemos* 7988 (COI; LISC; LMJ).

Also in the Sudan, Uganda, Kenya, Tanganyika, Zanzibar and the northern Transvaal. Usually in open savanna woodland, especially on the more fertile soils dominated by *Acacia* and *Combretum*, also on termite mounds, 60–1525 m. Recorded in error from the Victoria Falls by Oliver (loc. cit.).

The leaves when green are said to be edible by cattle, but poisonous when dried (*Greenway* 801).

3. **Turraea zambesica** [Sprague & Hutch. ex Hutch., Botanist in S. Afr.: 481 (1946) *anglice tantum descr*.—O. B. Mill. in Journ. S. Afr. Bot. **18**: 40 (1952).—White, F.F.N.R.: 182 (1962)] Styles & White in Bol. Soc. Brot., Sér. 2, **36**: 71, t. 1 (1962). Type: N. Rhodesia, Victoria Falls, *Hutchinson & Gillett* 3493 (BM; K, holotype; LISC; SRGH).

   Turraea nilotica sensu Bremek. & Oberm. in Ann. Transv. Mus. **16**: 420 (1935).—O. B. Mill., loc. cit.—Wild, Guide Fl. Vict. Falls: 151 (1952).

Shrub or small slender tree up to 4 m. tall; first-year branchlets fulvous-puberulous, second-year glabrous, grey-brown, slender, older branchlets not corky. Leaf-lamina up to 10 × 5 cm., more or less elliptic but tapering rapidly to each end, lower-surface quite glabrous except for a few hairs on the nerves, apex usually shortly acuminate, base cuneate; petiole up to 1·3 cm. long. Inflorescence a 3–7-flowered almost sessile axillary or terminal fascicle; bracts 3 mm. long, subulate; pedicels 4–10 mm. long. Flowers greenish-white, becoming yellow with age, fragrant, usually present with the leaves. Calyx up to 3 mm. long, with indistinct teeth, puberulous. Petals 15–22 × 3 mm., linear, minutely puberulous towards the apex outside, otherwise glabrous. Staminal tube 10–15 mm. long, distally expanded, bearded at the throat with long hairs arising from filaments, otherwise glabrous inside or minutely puberulous at the base; appendages fused almost to the apex and forming a frill about 1·5 mm. long beyond the insertion of the filaments, densely ciliate, irregularly lobed. Ovary 10-locular, glabrous; style 2–2·5 cm. long, glabrous or pilose at base; style-head ovoid-cylindric. Capsule 7 × 12 mm., depressed-globose, shallowly sulcate, leathery, glabrous; aril covering about half the seed.

Caprivi Strip. Lisikili, 24 km. E. of Katima Mulilo, fl. vii.1952, *Codd* 7094 (K; SRGH). **Bechuanaland Prot.** N: Serondela, Chobe R., fl. v.1950, *Miller* B/1042 (K; PRE; SRGH). **N. Rhodesia.** B: without locality, fl. viii.1921, *Borle* 303 (PRE). C: Chingombe, fr. ix.1957, *Fanshawe* 3744 (K). S: Katombora, Zambezi R., fl. & fr. viii.1947, *Brenan* 7741 (BR; FHO; K). **S. Rhodesia.** N: Rekomitje R., Urungwe, fr. viii.1959, *Goodier* 592 (SRGH). W: Sebungwe R., Wankie, fl. v.1955, *Plowes* 1840 (K; LISC; SRGH). **Mozambique.** T: Tete, Sisito, R. Zambeze, fl. vii.1950, *Chase* 2630 (BM; BR; K; SRGH).

Known only from our area, where it appears to be confined to the Zambezi Valley and its tributaries. In riverine woodland and thicket, 250–975 m.

4. **Turraea holstii** Gürke in Engl., Bot. Jahrb. **19**, Beibl. 47: 35 (1894); in Engl., Pflanzenw. Ost-Afr. **C**: 231 (1895).—Bak. f. in Journ. of Bot. **41**: 10 (1903).—Staner in Bull. Jard. Bot. Brux. **16**: 126, t. 2 (1941).—Brenan, T.T.C.L.: 321 (1949).—Eggeling & Dale, Indig. Trees Uganda Prot.: 200 (1952).—Staner & Gilbert, F.C.B. **7**: 155 (1958).—Dale & Greenway, Kenya Trees and Shrubs: 274 (1961). Type from Tanganyika.

   Turraea abyssinica var. *longipedicellata* Oliv., F.T.A. **1**: 331 (1868). Type from Ethiopia.

   Turraea usambarensis Gürke in Engl., loc. cit.—Bak. f., tom. cit.: 12 (1903). Type from Tanganyika.

   Turraea laxiflora C.DC. in Ann. Conserv. Jard. Bot. Genève, **10**: 127 (1907). Type from Tanganyika.

Small tree up to 15 m. tall, more rarely a shrub, sometimes scrambling; first-year branchlets shortly pubescent, second-year glabrous, mostly grey-brown, sometimes purplish, slender. Leaf-lamina up to 8 × 3 cm., lanceolate-elliptic to elliptic, lower surface almost glabrous except for a few hairs on the nerves and conspicuous tufts of short hairs in axils of the secondary nerves, apex shortly and somewhat bluntly acuminate, base cuneate; petiole up to 8 mm. long. Inflorescence an axillary 1–4-flowered cyme; peduncle 8–20 mm. long; bracts very small, 1 mm. long, squamiform; pedicels 2·5–2·75 cm. long, very slender. Flowers white, pendulous, jasmine-scented. Calyx 2–3·5 mm. long, puberulous, especially on the teeth and margin. Petals 18–25 × 3 mm., linear-spathulate, puberulous outside. Staminal tube 15–20 mm. long, cylindric, densely hairy in the upper half inside; appendages opposite the anthers, entire and truncate or emarginate, scarcely lobed, fused for the greater part of their length to form a frill 1–1·5 mm. long beyond the insertion of the filaments. Ovary 5–6(7)-locular, glabrous; style 2–3 cm. long; style-head ovoid or narrowly ovoid. Capsule depressed-globose, 7 × 10–15 mm., shallowly sulcate, woody, glabrous; aril almost completely covering seed.

Nyasaland. N: Lwanjati Hill, Champila, fl. i.1956, *Chapman* 276 (BR; FHO; K). C: Mulunduni, Dedza, fl. iii.1961, *Chapman* 1194 (FHO).

Also in Ethiopia, Sudan, Uganda, Kenya, Somaliland, Congo and Tanganyika. In montane forest.

5. **Turraea fischeri** Gürke in Engl., Bot. Jahrb. **14**: 308 (1891); in Engl., Pflanzenw. Ost-Afr. **C**: 231 (1895).—Bak. f. in Journ. of Bot. **41**: 11 (1903).—Milne-Redh. in Kew Bull. **1936**: 475 (1936).—Brenan, T.T.C.L.: 320 (1949).—Eggeling & Dale, Indig. Trees Uganda Prot.: 200 (1952). Syntypes from Tanganyika.

Subsp. **eylesii** (Bak. f.) Styles & White in Bol. Soc. Brot., Sár. 2, **36**: 72 (1962). Type: S. Rhodesia, Matopos, *Eyles* 29 (BM, holotype; K; SRGH).

Turraea eylesii Bak. f., op. cit. **43**: 45 (1905).—Monro in Proc. Rhod. Sci. Ass. 7: 68 (1908).—Eyles in Trans. Roy. Soc. S. Afr. **5**: 388 (1916). Type as above.

Shrub or small tree up to 8 m. tall; first-year branchlets glabrous, reddish-brown or grey. Leaf-lamina up to 10 × 6 cm., ovate to elliptic, lower surface glabrous, apex bluntly acuminate, base cuneate to rounded; petiole up to 1 cm. long. Inflorescence an axillary 1–2(3)-flowered fascicle; peduncle up to 5 mm. long; bracts small, up to 2 mm. long, squamiform; pedicels 10–15 mm. long. Calyx 2–4 mm. long, shortly puberulous. Petals 25–32 × 4 mm., linear to linear-spathulate, puberulous outside. Staminal tube 15–25 mm. long, white, distinctly curved and gradually narrowed from apex to base, glabrous inside except for a few scattered hairs near the base; appendages variable, 1–1·5 mm. long, 2-fid or in pairs, alternating with the anthers, glabrous outside. Ovary 9–10-locular, densely puberulous with short, weak hairs; style 3–3·5 cm. long, softly puberulous in lower half; style-head narrowly ovoid-cylindric. Capsule 8 × 13 mm., depressed-globose, shallowly sulcate, leathery, puberulous; aril covering half of seed.

S. Rhodesia. W: Matopos, fl. xii.1947, *Hodgson* 15/48 (FHO; K; SRGH).

Confined to granite hills in the SW. corner of S. Rhodesia.

Subsp. *fischeri* is known only from rocky hills in northern and central Tanganyika, with small outlying populations in Uganda. It differs from subsp. *eylesii* in having pubescent leaves, and flowers in terminal (3)4–7-flowered inflorescences produced when the plant is leafless. The root is used as a vermifuge (*Koritschoner* 1643).

6. **Turraea mombassana** Hiern ex C.DC. in A. & C.DC., Mon. Phan. **1**: 439 (1878).—Gürke in Engl., Pflanzenw. Ost-Afr. **C**: 231 (1895).—Bak. f. in Journ. of Bot. **41**: 10 (1903).—C.DC. in Ann. Conserv. Jard. Bot. Genève, **10**: 125 (1907).—Brenan, T.T.C.L.: 320 (1949).—Dale & Greenway, Kenya Trees and Shrubs: 274 (1961). Type from Kenya.

Turraea cuneata Gürke in Engl., loc. cit.—Bak. f., loc. cit. Type from Tanganyika.
Pittosporum jaegeri Engl., Bot. Jahrb. **43**: 372 (1909). Type from Tanganyika.
Pittosporum spathulifolium Engl., loc. cit. Type from Tanganyika.
Turraea mombassana var. *cuneata* (Gürke) Engl., Pflanzenw. Afr. **3, 1**: 813 (1915).—Brenan, loc. cit. Type as above.

Shrub up to 4 m. tall, sometimes scrambling; first-year branchlets densely puberulous with spreading hairs, second-year more sparsely so, purplish-brown or reddish. Leaves mostly in fascicles. Leaf-lamina very variable in shape and size, up to 7 × 3·5 cm., but usually much smaller, rhombic to rhombic-spathulate, but outline often irregular, lower-surface glabrous except for a few scattered hairs on the nerves and conspicuous tufts of short hairs in the nerve-axils, apex subacuminate to emarginate, base narrowly cuneate, decurrent almost to the base of the petiole; petiole very short or absent. Inflorescence with peduncle less than 1 mm. long or flowers (1)4–6-fasciculate; bracts very small, 1 mm. long, subulate; pedicels 4–10 mm. long. Flowers pure white. Calyx 2·5–5 mm. long, teeth pubescent. Petals 4–5 × 0·5 cm., linear-spathulate, glabrous outside. Staminal tube (2·5)3·8–5 cm. long, cylindric, glabrous inside; appendages in pairs alternating with the anthers, 3–3·5 mm. long, glabrous outside, filaments not fused at the base. Ovary 5(6)-locular, densely setose-hairy; style 4–5·5 cm. long; style-head conical-cylindric from a broad base, exserted up to 5 mm. beyond the base of the staminal appendages. Capsule 5 × 10 mm., depressed-globose, very shallowly sulcate, leathery, glabrous; aril covering c. $\frac{1}{6}$ of seed.

Nyasaland. C: Chintembwe Mission, near Nchisi, fl. ii.1959, *Robson* 1713 (BM; K; LISC; SRGH).

An East African species known in our area only from a single gathering. In undergrowth of montane forest, 1370 m. alt. In East Africa from 0–2500 m.

T. cuneata is an extreme variant not sufficiently distinct to be kept up as a species or as a variety.

7. **Turraea obtusifolia** Hochst. in Flora, **27**: 296 [962] (1844).—Sond. in Harv. & Sond., F.C. **1**: 245 (1860).—Oliv., F.T.A. **1**: 331 (1868).—C.DC. in A. & C.DC., Mon. Phan. **1**: 440 (1878).—Gürke in Engl., Pflanzenw. Ost-Afr. **C**: 231 (1895).—Bak. f. in Journ. of Bot. **41**: 10 (1903).—Monro in Proc. Rhod. Sci. Ass. **7**: 68 (1908).—Eyles in Trans. Roy. Soc. S. Afr. **5**: 388 (1916).—Garcia in Contr. Conhec. Fl. Moçamb. **2**: 140 (1954). TAB. **61** fig. A. Type from Natal.
Turraea obtusifolia var. *microphylla* C.DC., loc. cit.—Bak. f., loc. cit.—O. B. Mill. in Journ. S. Afr. Bot. **18**: 40 (1952). Type from Cape Prov.
Turraea obtusifolia var. *matopensis* Bak. f. in Journ. of Bot. **43**: 45 (1905).—Eyles, loc. cit. Type: S. Rhodesia, Matopos, *Eyles* 154 (BM, holotype; SRGH).
Turraea oblancifolia Bremek. in Ann. Transv. Mus. **15**: 245 (1933). Type from the Transvaal.
Turraea mombassana sensu Schinz in Mém. Herb. Boiss. **10**: 45 (1900).

Shrub up to 3 m. tall, sometimes scrambling; first-year branchlets puberulous with spreading hairs, second-year more sparsely so, reddish-brown or grey. Leaves mostly in fascicles; lamina very variable in shape and size, up to 5 × 2·5 cm., usually smaller, oblanceolate to narrowly obovate, lower surface glabrous, apex subacuminate to emarginate, sometimes shallowly 3-lobed, base decurrent almost to the base of the petiole; petiole very short or absent. Inflorescence axillary, 1–3-flowered; peduncles 5–8 mm. long; bracts very small, 1 mm. long, subulate. Flowers pure white. Calyx 3–5 mm. long, lobes pubescent. Petals 3–3·6 × 0·5 cm., linear to linear-spathulate, glabrous outside. Staminal tube 2·5–3·2 cm. long, cylindric, glabrous inside; appendages in pairs, alternating with the anthers, 1·5–2 mm. long, glabrous outside; filaments fused beyond the insertion of the appendages to form a short frill continuing the staminal tube. Ovary 5-locular, glabrous; style 2·7–3·3 cm. long; style-head broadly obovoid-cylindric, widest at the apex, exserted up to 3 mm. beyond the base of the staminal appendages. Capsule 5 × 10–13 mm., depressed-globose, shallowly sulcate, leathery, glabrous; aril vestigial, confined to the adaxial surface; testa red.

Bechuanaland Prot. N: Francistown, fl. ii.1926, *Rand* 79 (BM). SE: Kanye, fl. & fr. i.1941, *Miller* B/266 (K; PRE). **S. Rhodesia.** W: Matopos, fl., xii.1947, *Hodgson* 16/48 (FHO; SRGH). E: Umtali, fr. ii.1953, *Chase* 4791 (SRGH). S: 112 km. S. of Gwanda, fl. ii.1955, *Plowes* 1769 (BR; K; SRGH). **Mozambique.** SS: Guijá, near Mabalane, fr. vi.1959, *Barbosa & Lemos* 8621 (K; LISC; LMJ; M; PRE; SRGH). LM: Ponta do Ouro, fl. xii.1948, *Gomes e Sousa* 3942 (COI; K; PRE).

Also in the Transvaal, Natal and Cape Prov. Often on kopjes or granite hills in S. Rhodesia. Recorded from sand dunes near Lourenço Marques, 0–1525 m.

Closely related to but quite distinct from *T. mombassana*, which has a more northerly distribution.

The variety *matopensis* and *T. oblancifolia* are based on specimens with narrower, almost entire leaves. Such forms occur most frequently towards the northern limit of the species range. Plants from the Cape Prov. usually have broader lobed leaves. Although this variation shows some correlation with geography it does not appear to be sufficiently definite to warrant taxonomic recognition.

8. **Turraea wakefieldii** Oliv. in Hook., Ic. Pl. **15**: t. 1489 (1885).—Gürke in Engl., Pflanzenw. Ost-Afr. **C**: 231 (1895).—Bak. f. in Journ. of Bot. **41**: 12 (1903).—Brenan, T.T.C.L.: 320 (1949).—Dale & Greenway, Kenya Trees and Shrubs: 275 (1961). TAB. **61** fig. B. Type from Kenya.
Turraea junodii Schinz in Mém. Herb. Boiss. **10**: 45 (1900).—Bak. f., loc. cit. Type: Mozambique, Delagoa Bay, fl. 1890, *Junod* 118 (G).
Turraea breviracemosa C.DC. in Ann. Conserv. Jard. Bot. Genève, **10**: 129 (1907). Type from Tanganyika.
? *Turraea cylindrica* Sim, For. Fl. Port. E. Afr.: 25, t. 18b (1909). Type: Mozambique, locality uncertain, *Sim* 373.*
Turraea schlechteri Harms in Notizbl. Bot. Gart. Berl. **7**: 228 (1917). Type: Mozambique, Manica, Inhamadzi, *Schlechter* 12069 (BR; G; K; L; P; PRE; Z).

Shrub or small tree up to 7 m. tall, sometimes scrambling; first-year branchlets sericeous-tomentellous, second-year almost glabrous, reddish-brown or purple. Leaves sometimes crowded at the ends of slow-growing short shoots; lamina up to 8 × 4 cm., broadly ovate to lanceolate-elliptic, coriaceous, upper surface shining, lower surface glabrous except for tufts of hairs in the nerve-axils, apex shortly and bluntly subacuminate, base cuneate; petiole up to 1·5 cm. long. Inflorescence subfasciculate, 1–4-flowered, sessile or subsessile; bracts 2 mm. long, squamiform; pedicels 8–10 mm. long. Flowers white. Calyx 5 mm. long, densely puberulous. Petals 5–6 × 0·15 cm., linear, sparsely puberulous outside. Staminal tube 4·5–5 cm. long, cylindric, glabrous inside; appendages in pairs alternating with the anthers, 3·5 mm. long, glabrous outside. Ovary 10–12-locular, shortly pubescent; style 5·5–6·5 cm. long; style-head conical-cylindric, tapering from a broad base. Capsule 7·5 × 12 mm., depressed-globose to cylindric, deeply ribbed and sulcate, woody, fulvous-tomentellous; aril covering about half of seed.

Mozambique. SS: between Mavume and Mapinhame, fr. ii.1939, *Gomes e Sousa* 2222 (COI; K; LISC). LM: Marracuene, fl. & fr. vi.1959, *Barbosa & Lemos* 8548 (COI; K; LISC; LMJ; PRE; SRGH).

Confined to the low-lying coastal parts of East Africa. In secondary forest, 20–150 m.

9. **Turraea floribunda** Hochst. in Flora, **27**: 297 (1844).—C.DC. in A. & C.DC. Mon. Phan. **1**: 445 (1878).—Bak. f. in Journ of Bot. **41**: 12 (1903).—Brenan, T.T.C.L.: 320 (1949).—Eggeling & Dale, Indig. Trees Uganda Prot.: 200 (1952).—Garcia in Contr. Conhec. Fl. Moçamb. **2**: 140 (1954).—Staner & Gilbert, F.C.B. **7**: 154 (1958).—Dale & Greenway, Kenya Trees and Shrubs: 274 (1961). TAB. **61** fig. E. Type from Natal.
Turraea heterophylla sensu Sond. in Harv. & Sond., F.C. **1**: 245 (1860).

Deciduous shrub or small tree up to 10(13) m. tall, sometimes scrambling; first-year branchlets shortly pubescent, second-year purple-brown and glabrous. Leaf-lamina up to 14 × 7 cm., ovate to lanceolate, densely setose when young, more sparsely so later, except on nerves beneath, apex acuminate, base subtruncate, rounded or broadly cuneate; petiole up to 1·5 cm. long. Inflorescence a 2–7-flowered false raceme or flowers subfasciculate; peduncle 3–8 mm. long; bracts 2 mm. long, squamiform; pedicels 5–17 mm. long. Calyx 2·5–5 mm. long, densely puberulous to tomentellous. Petals (3·8)4·5–5·2 × 0·3 cm., greenish-white, linear-spathulate, sparsely puberulous outside. Staminal tube (2·5)3·6–4·5 cm. long, pure white, cylindric, glabrous inside; appendages in pairs alternating with the anthers, 3 mm. long, glabrous outside. Ovary 10-locular, puberulous; style (4)4·8–5·5 cm. long; style-head globose. Capsule 1·3–2·4 × 2 cm., obovoid-cylindric, globose or depressed-globose, deeply ribbed and sulcate, woody, glabrous; aril covering about half of seed.

S. Rhodesia. W: Matopos, fl. xi.1951, *Plowes* 1339 (SRGH). C: Chilimanzi, Shasha R., fr. vii.1951, *Seward* 48/51 (SRGH). E: Chipinda, Umsiliswe R., fr. iii.1957, *Goodier* 229 (SRGH). S: Belingwe, Mt. Buhwe, fl. xii.1953, *Wild* 4333 (K; LISC; SRGH). **Nyasaland.** S: Malosa, fl. xi., *Whyte* (K). **Mozambique.** Z: Morrumbala, fr. v.1943, *Torre* 5323 (LISC). MS: Dondo, Chiluvo Hills, fl. xi.1923, *Honey* 758 (K; PRE). LM: Costa do Sol, fl. xi.1940, *Torre* 1950 (LISC).

* *Sim* 21106 (PRE) labelled " Turraea cylindrica " in Sim's hand may well be *Sim* 373. See footnote p. 305.

Also in Uganda, Kenya, Tanganyika, Congo, Swaziland, Natal and Cape Prov. In evergreen forest and on kopjes, 15–1830 m.

9. MELIA L.

Melia L., Sp. Pl. **1**: 384 (1753); Gen. Pl. ed. 5: 182 (1754).

Trees or shrubs. Indumentum of simple, glandular, and stellate hairs. Leaves 2- or 3-pinnate; leaflets usually rather deeply crenate or serrate. Flowers bisexual. Calyx lobed almost to the base, lobes 5(6), lanceolate, imbricate. Petals 5(6), free, much longer than the calyx in bud, imbricate (sometimes only distinctly so at the apex). Staminal tube cylindric; anthers 10–12, shortly apiculate, alternating with a pair of narrowly deltate appendages. Disk annular, crenulate, free from the ovary and staminal tube. Ovary 4–8-locular, each loculus with 2 superposed ovules (the lower pendulous, the upper directed upwards); style elongate, shorter than the staminal tube; style-head scarcely wider than the style, coroniform, with 4–8 erect stigmatic lobes. Fruit drupaceous with 4–8 loosely united pyrenes, each with 1–2 seeds.

A small genus confined to the Old World tropics; with c. 6 poorly defined species.

1. **Melia azedarach** L., Sp. Pl. **1**: 384 (1753).—Oliv., F.T.A. **1**: 332 (1868).—C.DC. in A. & C.DC., Mon. Phan. **1**: 451 (1878).—Sim, For. Fl. Port. E. Afr.: 26 (1909).—Eyles in Trans. Roy. Soc. S. Afr. **5**: 389 (1916).—Brenan, T.T.C.L.: 317 (1949).—Exell & Mendonça, C.F.A. **1**, 2: 317 (1951).—O. B. Mill. in Journ. S. Afr. Bot. **18**: 39 (1952).—Garcia in Contr. Conhec. Fl. Moçamb. **2**: 141 (1954).—Staner & Gilbert, F.C.B. **7**: 173 (1958).—White, F.F.N.R.: 181 (1962). Type from Asia.

Medium-sized rapidly growing short-lived deciduous tree up to 15 m. tall, sometimes flowering as a shrub; bark grey-brown, smooth. Leaves usually 2-, rarely 3-pinnate; petiole and rhachis up to 40 cm. long; leaflets up to 5·5 × 2·5 cm., opposite or subopposite, more or less lanceolate, apex acuminate or subacuminate, base asymmetric, margin rather deeply crenate or serrate, sparsely puberulous, glabrescent; petiolules up to 7 mm. long. Flowers sweet-scented, in large many-flowered axillary cymose panicles. Calyx 2·5 mm. long, densely stellate-puberulous. Petals up to 8 × 3 mm., pale lilac, spathulate, sparsely puberulous outside with simple hairs, glabrous inside except for a few hairs towards apex. Staminal tube up to 7 mm. long, dark purple, glabrous outside, hairy inside, especially in the upper half; appendages c. 1 mm. long. Ovary less than 1 mm. in diam., 5(7)-locular; style 4·5 mm. long. Drupe up to 2 × 1·5 cm.

N. Rhodesia. N: Kafulwe, Lake Mweru, fl. ii.1957, *Richards* 9433 (K). W: Dola Hill, fl. ix.1958, *Savory* 4 (FHO). E: Ndefu, Luangwa Valley, fl. iv.1952, *White* 2414 (FHO; K). **S. Rhodesia.** N: Umvukwe Mts., fr. iv.1948, *Rodin* 4422 (K; SRGH). C: Ruwa R., Salisbury, fl. ix.1946, *Wild* 1254 (SRGH). **Nyasaland.** S: Kundwelo Village, Palombe Plain, fl. vii.1956, *Newman & Whitmore* 287 (SRGH). **Mozambique.** N: Nampula, fl. & fr. i.1937, *Torre* 1170 (COI; LISC). MS: Muchino, fl. & fr. vii.1949, *Barbosa & Carvalho* 3464 (LM). SS: between Vila Luísa and Manhiça, fr. xi.1950, *Carvalho* 1038 (LM). LM: between Moamba and Lourenço Marques, fr. ii.1948, *Torre* 7425 (LISC).

Native of India. Widely planted in tropics and subtropics for ornament and for building poles, and known as " Persian Lilac ".

Most of the above records are from cultivated specimens but the species has been recorded as naturalized in S. Rhodesia (*Rodin* 4422), and Mozambique (Sim, loc. cit.). In India the wood is used for furniture, agricultural implements and cigar-boxes.

10. EKEBERGIA Sparrm.

Ekebergia Sparrm. in Svensk. Vet. Akad. Handl. **40**: 282, t. 9 (1779).

Small or medium-sized trees. Leaves imparipinnate; leaflets entire. Flowers dioecious, with little external difference between the sexes, but with well-developed vestiges of the opposite sex, borne in axillary cymose panicles or extra-axillary towards the base of the current year's growth. Calyx lobed in upper half, lobes (4)5, deltate or rounded. Petals (4)5, free, much longer than the calyx in bud, imbricate. Staminal tube shortly cylindric, margin more or less entire or shallowly divided to form 8–10 teeth each bearing an anther or antherode; appendages absent; anthers less than 1 mm. long, ± oblong, scarcely apiculate; antherodes more than 1 mm. long, narrowly pyramidal, apiculate, without pollen-sacs Disk annular, free from the ovary but fused at the base to the base of the staminal tube, glabrous outside, densely setulose-hairy on margin and inner face. Ovary 2–5-locular, each loculus with 2 superposed ovules; style very short, style-head capitate, surmounted by 2–5 indistinct erect stigmatic lobes. Vestigial ovary similar to the functional ovary but more slender, with loculi and sometimes with vestigial ovules. Drupe with 2–4 pyrenes. Seeds mostly one per loculus; testa fleshy and resembling an aril.

A small genus with 3 or 4 species, confined to the African mainland. According to Leroy (in Compt. Rend. Acad. Sci. **247**: 861 (1958)) the Madagascan species belong to a distinct genus, *Astrotrichilia*, bι.t the differences between the two genera appear to be slight.

Previous authors, e.g. Harms (in Engl. & Prantl, Nat. Pflanzenfam. **19 bl**: 120 (1940)), Andrews (in Fl. Pl. Anglo-Egypt. Sudan, **2**: 327 (1952)) and Corner (in Phytomorphology, **3**: 471 (1953)) mention the presence of an aril. So far as one can judge from herbarium material the greater part of the testa is modified to form a yellow somewhat fleshy sarcotesta, but a distinct aril does not seem to occur.

Second-year branchlets usually less than 6 mm. in diam., smooth, with scattered leaf-scars, closely lenticellate with large whitish lenticels; leaflets tapering to an acuminate or subacuminate apex, lower surface never drying whitish, without globose wax particles; medium-sized tree of evergreen (usually riparian) forest - - - - 1. *capensis*
Second-year branchlets usually more than 7 mm. in diam., rough, with thick corky bark and crowded leaf-scars, lenticels inconspicuous; leaflets not tapering, apex broadly rounded to subtruncate, emarginate or with a minute apiculus, lower surface nearly always drying whitish because of a covering of minute globose wax particles; small tree of open woodland - - - - - - - - 2. *benguelensis*

1. **Ekebergia capensis** Sparrm. in Svensk Vet. Akad. Handl. **40**: 282, t. 9 (1779).—Sond. in Harv. & Sond. F.C. **1**: 247 (1860).—C.DC. in A. & C.DC., Mon. Phan **1**: 641 (1878).—Monro in Trans. Rhod. Sci. Ass. **7**: 67 (1908).—Burtt Davy, F.P.F.T. **1**: 487 (1926).—Steedman, Trees etc. S. Rhod.: 32 (1933).—Chalk et al. in Chalk & Burtt Davy, For. Trees Brit. Emp. **3**: 51, t. 7, fig. 10 (1935).—White, F.F.N.R.: 433 (1962). TAB. **62**. Type from Cape Prov.

Trichilia rueppelliana Fresen. in Mus. Senckenb. **2**: 278 (1837). Type from Ethiopia.

Trichilia ekebergia E. Mey. [ex Drège, Zwei Pflanz.-Docum.: 227 (1843) (" ekebergia ") *nom. nud.*] ex Sond. in Harv. & Sond., tom. cit.: 246 (1860). Syntypes from S. Africa.

Ekebergia rueppelliana (Fresen.) A. Rich., Tent. Fl. Abyss. **1**: 105 (1847).—Oliv., F.T.A. **1**: 333 (1868).—C.DC., tom. cit. 643 (1878).—Gürke in Engl., Pflanzenw. Ost-Afr. **C**: 231 (1895).—Steedman, loc. cit.—Lebrun, Ess. For. Reg. Mont. Cong. Or. **2**: 112 (1935).—Brenan, T.T.C.L.: 315 (1949).—Eggeling & Dale, Indig. Trees Uganda Prot.: 174 (1952).—Staner & Gilbert, F.C.B. **7**: 208 (1958).—Dale & Greenway, Kenya Trees and Shrubs: 267, t. 54 (1961). Type as for *Trichilia rueppeliana*.

Ekebergia meyeri Presl [Bot. Bemerk.: 25 (1844) *nom. nud.*] ex C.DC., tom. cit.: 642 (1878).—Sim, For. Fl. Port. E. Afr.: 26, t. 20 (1909).—Bak. f. in Journ. Linn. Soc., Bot. **40**; 38 (1911).—Eyles in Trans. Roy. Soc. S. Afr. **5**: 389 (1916).—Burtt Davy, loc. cit.—Steedman, loc. cit.—Brenan, loc. cit.—Gomes e Sousa, Dendrol. Mozamb., **1**: 157 cum tab. (1951).—Garcia in Contr. Conhec. Fl. Moçamb. **2**: 141 (1954).—White, tom. cit.: 180 (1962). Type from S. Africa.

Ekebergia buchananii Harms in Engl., Bot. Jahrb. **23**: 164 (1896).—Brenan, tom. cit.: 314 (1949). Type: Nyasaland, *Buchanan* 39 (B, holotype†; K).

Medium-sized evergreen or semi-evergreen tree up to 30 m. tall but usually less; bole slightly buttressed or fluted at base, up to 1 m. d.b.h.; second-year branchlets slender, usually less than 0·6 cm. diameter, smooth, with scattered leaf-scars, closely lenticellate with large whitish lenticels. Leaves imparipinnate, usually entirely glabrous, sometimes pubescent, rarely tomentose; petiole and rhachis up to 35 cm. long; leaflets up to 14·5 × 6 cm., usually much smaller, opposite or sub-opposite, 3–5(6)-jugate, subsessile or shortly petiolulate, lanceolate to oblong-lanceolate, tapering to an acuminate or subacuminate apex, base asymmetric, lower surface never drying whitish. Flowers dioecious, white or pinkish-white, sweet-scented, borne in many-flowered cymose panicles. Calyx 2 mm. long, sparsely to densely puberulous. Petals 4–5 mm. long, elliptic-oblong, densely puberulous on both surfaces. Staminal tube 2 mm. long, puberulous outside, densely bearded

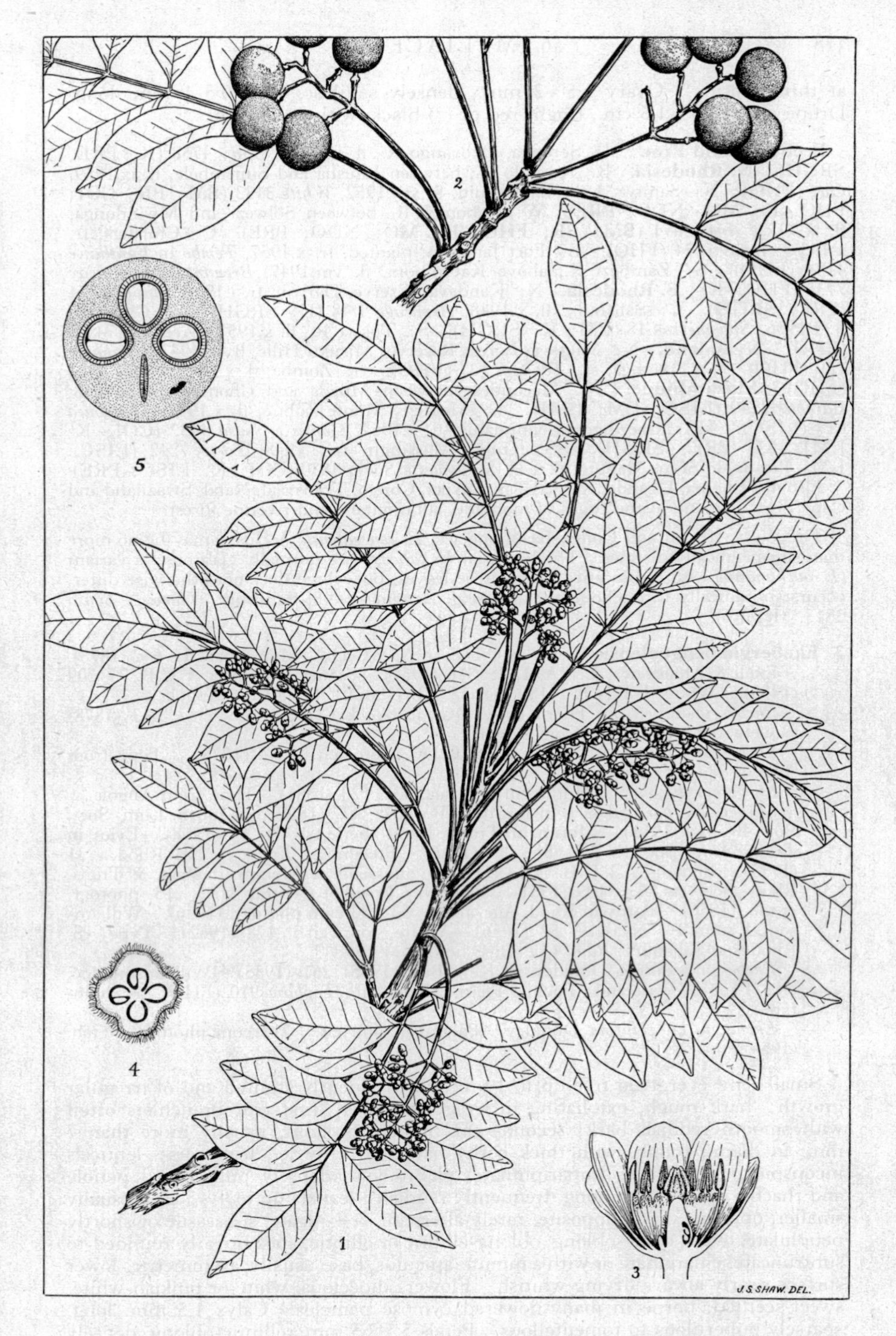

Tab. 62. EKEBERGIA CAPENSIS. 1, flowering branchlet ($\times\frac{2}{3}$); 2, fruiting branchlet ($\times\frac{2}{3}$); 3, half flower ($\times 8$); 4, ovary in transverse section ($\times 20$); 5, fruit in transverse section, showing aborted loculi ($\times 1\frac{1}{2}$). From For. Trees & Timbers Brit. Emp.

at throat inside. Ovary 1·5 × 2 mm., densely setulose, style 0·5–1 mm. long. Drupe about 1·5 × 1·5 cm., bright red or (?) black, with 2–4 pyrenes.

Bechuanaland Prot. N: Seronga, Okovango R., fl. ix.1954, *Story* 4758 (K; PRE; SRGH). **N. Rhodesia.** B: Kwando R., between Imusha and Sinjembele, fl. ix.1959, *Guy* (FHO). N: Samfya, Lake Bangweulu, fl. viii.1952, *White* 3133 (BM; BR; COI; FHO; K; MO; NDO; PRE). W: Kabompo R., between Solwezi and Mwinilunga, fl. ix.1952, *Angus* 471 (BM; BR; FHO; K; MO; NDO; PRE). C: Chilanga, fr. x.1957, *Angus* 1774 (FHO). E: Fort Jameson, planted, fr. x.1957, *Tembo* in *Fanshawe* 4184 (FHO). S: Zambezi R., above Katombora, fl. viii.1947, *Brenan & Greenway* 7735 (FHO; K). **S. Rhodesia.** N: Kandeya Reserve, Darwin, fr. i.1960, *Phipps* 2274 (PRE; SRGH). C: Salisbury, fl. x.1948, *Armitage* 9/48 (K; SRGH). E: Chirinda, fl. x.1906, *Swynnerton* 18 (BM; K; SRGH). S: Tokwe R., fl. x.1951, *Greenhow* 74/51 (FHO). **Nyasaland.** N: Mugesse Forest Reserve, Misuku Hills, fl. ix.1953, *Chapman* 159 (FHO). C: Kasunga, st., *Topham* 137 (FHO). S: Zomba, fl. x.1948, *Vigne* 4881 (FHO). **Mozambique.** N: Macondes, between Mueda and Chomba, fl. ix.1948, *Barbosa* 2249 (LISC; LM; LMJ). T: Moatize, Monte Zóbuè, fl. x.1942, *Mendonça* 617 (LISC). MS: Cheringoma, Chiniziúa, fl. x.1957, *Gomes e Sousa* 4432 (COI; K; LMJ; M; PRE; SRGH). SS: Guijá, Massingir, fr. i.1948, *Torre* 7242 (LISC). LM: Picada do Sêco, Maputo, fl. x.1948, *Gomes e Sousa* 3830 (COI; K; LISC; PRE).

Also in Ethiopia, Uganda, Kenya, Tanganyika, Congo, Transvaal, Natal, Swaziland and Cape Prov. In and at the edges of montane, submontane and riverine forest.

The West African and Congo Basin species, *E. senegalensis* A. Juss., may be no more than subspecifically distinct. The leaves of *E. capensis* are normally glabrous; a variant (*E. buchananii*) with pubescent to tomentose leaves, but which does not otherwise differ, occurs sporadically throughout our area (e.g. *Armitage* 9/48, Salisbury; *Gomes e Sousa* 3813, Maputo).

2. **Ekebergia benguelensis** Welw. ex C.DC. in A. & C.DC., Mon. Phan. **1**: 642 (1878).—Exell & Mendonça, C.F.A. **1, 2**: 316 (1951).—Staner & Gilbert, F.C.B. **7**: 209 (1958).—White, F.F.N.R.: 180, fig. 35A–B (1962). Type from Angola.

Ekebergia welwitschii Hiern ex C.DC. in A. & C.DC., tom. cit.: 643 (1878) (" welwitichii "). Type from Angola.

Ekebergia fruticosa C.DC. in A. & C.DC., tom. cit.: 644 (1878). Type from Angola.

Ekebergia discolor O. Hoffm. in Linnaea, **43**: 123 (1881). Type from Angola.

Ekebergia arborea Bak. f. in Journ. of Bot. **37**: 427 (1899); in Journ. Linn. Soc., Bot. **40**: 38 (1911).—Monro in Trans. Rhod. Sci. Ass. **7**: 67 (1908).—Eyles in Trans. Roy. Soc. S. Afr. **5**: 389 (1916).—Steedman, Trees etc. S. Rhod.: 31 (1933).—Brenan, T.T.C.L.: 314 (1949).—Suesseng. & Merxm. in Proc. & Trans. Rhod. Sci. Ass. **43**: 109 (1951).—Palgrave, Trees of Central Afr.: 215, photogr. et tab. (1956).—Pardy in Rhod. Agr. Journ. **53**: 53 cum photogr. (1956).—Williamson, Useful Pl. Nyasal.: 52 (1956).—White, tom. cit.: 178 (1962). Type: S. Rhodesia, Salisbury, *Rand* 612 (BM, holotype).

Ekebergia velutina Dunkley in Kew Bull. **1935**: 261 (1935).—Williamson, loc. cit. Type: Nyasaland, Tuchila Plain, Mlanje, fl., *Topham* 910 (FHO; K, holotype).

Ekebergia cf. *velutina* Dunkley.—Palgrave, tom. cit.: 212 cum photogr. et tab. (1956).

Small semi-evergreen tree up to 10 m. tall, frequently stunted and of irregular growth; bark rough, exfoliating in irregular scales; first-year branchlets often with smooth reddish bark; second-year branchlets stout, usually more than 7 mm. in diam., rough, with thick corky bark and crowded leaf-scars; lenticels inconspicuous. Leaves imparipinnate, glabrous to densely pubescent; petiole and rhachis up to 20 cm. long, frequently reddish; leaflets up to 9 × 5 cm., usually smaller, opposite or subopposite, rarely alternate, 3–4-jugate, subsessile or shortly petiolulate, ovate, ovate-oblong, oblong-elliptic or elliptic, apex broadly rounded to subtruncate, emarginate or with a minute apiculus, base usually asymmetric, lower surface nearly always drying whitish. Flowers dioecious, white or pinkish-white, sweet-scented, borne in many-flowered cymose panicles. Calyx 1·5 mm. long, sparsely puberulous to tomentellous. Petals 5 × 2·5 mm., elliptic-oblong, densely puberulous on both surfaces. Staminal tube 2–3 mm. long, puberulous outside, densely bearded at throat inside. Ovary 1·5 × 2–2·5 mm., densely setulose; style 0·5 mm. long. Drupe c. 15 × 15 mm., bright red, with 2–4 pyrenes.

N. Rhodesia. B: Mankoya, st. ii.1952, *White* 2114c (FHO). N: Lake Chila, Abercorn, fr. xi.1952, *White* 3663 (BR; FHO; K; MO; NDO). W: 5 km. E. of Mumbezi

R., fl. ix.1930, *Milne-Redhead* 1132 (BR; K; PRE). C: Broken Hill Forest Reserve, fl. ix.1947, *Brenan & Trapnell* 7900 (BR; FHO; K). E: Lundazi, fl. ix.1929, *Conservator of Forests* 87 (FHO). S: Muckle Neuk, 20 km. N. of Choma, fl. x.1954, *Robinson* 925 (K; SRGH). **S. Rhodesia.** C: Domboshawa, Salisbury, fl. ix.1954, *Meyer* 5 (K; PRE; SRGH). E: Chikore, Chipinga, fl. ix.1947, *Whellan* 270 (BR; K; SRGH). S: Makaholi, Fort Victoria, fr. x.1952, *Davies* 383 (SRGH). **Nyasaland.** N: Rumpi to Mzimba km. 23, st. v.1952, *White* 2862 (FHO). C: Dzalanyama Forest, Dedza, fl. x.1953, *Adlard* 43 (FHO). S: Mt. Mlanje, fl. x.1948, *Vigne* 4882 (FHO). **Mozambique.** N: Maniamba, Metangula, fl. i.1942, *Hornby* 2563 (PRE). Z: between Alto Ligonha and Alto Molócuè, fl. x.1949, *Barbosa & Carvalho* 4408 (K; LM). T: Angónia, fl. vii.1949, *Andrada* 1783 (COI; LISC).

Also in Angola, Congo and Tanganyika. Usually in and at the edge of *Brachystegia* woodland on infertile soils, but ascending higher than the upper limit of these woodlands on mountains.

Although the differences separating *E. capensis* and *E. benguelensis* are slight, most specimens can be assigned to one or other species without difficulty. A few specimens that are atypical for one character have been collected but genuine intermediates have not been detected. The synonyms of *E. benguelensis* were originally separated on the basis of number of loculi and amount and distribution of the indumentum. A detailed study made by one of us (B.T.S.) has shown that these characters vary widely within restricted populations and show no correlation with geographic or ecological features.

47. DICHAPETALACEAE

By A. R. Torre

Shrublets or erect or scandent shrubs or rarely trees. Leaves alternate, subsessile or petiolate, simple, entire, often provided with glands at the base and below the apex; stipules caducous or persistent. Inflorescence of cymes, fascicles or glomerules, often axillary. Flowers actinomorphic or zygomorphic, bisexual or rarely unisexual by abortion. Sepals 5, ± connate below. Petals 5, equal (in actinomorphic flowers) or unequal (in zygomorphic flowers), often 2-dentate, 2-lobed or 2-fid at the apex, sometimes united with the stamens at the base. Stamens (4)5, all fertile or 2–3 fertile and 2 staminodes; anthers introrse. Disk annular or often divided into 5 squamulous hypogynous glands, distinct or connate. Ovary superior or semi-inferior, composed of 2–3 carpels, 2–3-locular, with 2 pendulous ovules in each loculus; style simple, 2–3-lobulate or 2–3-fid at the apex. Fruits drupaceous, with 3 (often reduced to 2 or 1) ± separate mericarps, 1–2-seeded. Seeds without endosperm; cotyledons containing starch.

Flowers actinomorphic; petals equal; fertile stamens 5; peduncle rarely adnate to the petiole - - - - - - - - - **1. Dichapetalum**
Flowers zygomorphic; petals unequal; fertile stamens 2–3; peduncle always adnate to the petiole - - - - - - - - - - - - **2. Tapura**

DICHAPETALUM Thou.

Dichapetalum Thou., Gen. Nov. Madag.: 23 (1806).

Flowers actinomorphic. Stamens all fertile. Disk sometimes divided into 5 squamulous glands. Fruits drupaceous, dry or rarely fleshy with (1)2–3 mericarps; mericarps obtuse or acuminate. Otherwise characters of the family.

A pantropical genus with about 190 species in tropical Africa.

Stipules ± linear or linear-lanceolate, sometimes 2-fid to 2-partite at the apex, deciduous or ± persistent:
Inflorescence sessile; leaves elliptic or oblong or oblong-obovate, acute or caudate at the apex:
Flowers up to 5 mm. long, in dense glomerules; stipules up to 4 mm. long, caducous; leaves glabrous on both surfaces:
Leaves ± elliptic, 5–17 × 2–4 cm., cuneate at the base; petiole c. 10 mm. long; flowers distinctly petiolate - - - - - - - 3. *thouarsianum*

Leaves oblong, 7–18 × 3·5–6 cm., cordate at the base, shining on the upper surface; petiole 2–4 mm. long; flowers sessile - - - - - - - 2. *whitei*
Flowers c. 10 mm. long, in few-flowered cymes; stipules linear-lanceolate, c. 15 × 2 mm., ± persistent; leaves oblong-obovate, obtuse or rounded at the base, acute at the apex, pubescent on the lower surface - - - - 1. *macrocarpum*
Inflorescence clearly pedunculate; peduncle 3–40 mm. long:
Leaves glabrous or glabrescent or pubescent only on the midrib:
Erect shrublets; cymes few- or many-flowered, 2–9 cm. long; peduncle 1·5–4 cm. long; leaves sessile or shortly petiolate; petiole 1–3(6) mm. long:
Stems, peduncles and pedicels glabrous; young leaves glabrous, 7–16 × 0·5–2·5 cm., narrowly oblanceolate - - - - - - - 13. *bullockii*
Stems, peduncles and pedicels pubescent; young leaves ± pubescent, 5–13 × 2–4·5 cm., oblong or subelliptic; fruit obovoid, c. 4 × 2·5 cm. 12. *cymosum*
Erect or climbing shrubs; cymes up to 2(2·5) cm. long, shortly pedunculate; peduncle less than 1 cm. long; leaves distinctly petiolate; petiole 3–8 mm. long:
Shrubs erect or sometimes climbing; calyx and corolla c. 4·5 mm. long; leaves ovate-elliptic or elliptic, 7–15 × 3·5–6 cm., rounded at the base, obtuse or rounded at the apex; petiole 4–10 mm. long - - 11. *zambesianum*
Shrubs climbing with ± patent branches; calyx 2–4 mm. long; leaves oblong or elliptic, 5–12 × 2·5–5 cm., rounded or cuneate at the base, obtuse or acute at the apex; petiole 2–5 mm. long:
Petals entire or emarginate, c. 4 mm. long; sepals ovate, shorter than the petals - - - - - - - - - - - 5. *crassifolium*
Petals deeply 2-lobed, 4–4·5 mm. long; sepals lanceolate, ± equal to the petals - - - - - - - - - - - 10. *barbosae*
Leaves pubescent, rarely glabrescent when mature (*D. stuhlmannii*), 4–14 × 1·5–7 cm.; petiole up to 10 mm. long; stipules linear or 2-fid; cymes subsessile or with long peduncles; corolla 3–7 mm. long; shrublets, shrubs or climbing shrubs:
Corolla up to 5 mm. long; leaves with indumentum not arachnoid; stipules linear or linear lanceolate:
Cymes subsessile, few-flowered; pedicels c. 2 mm. long; leaves subelliptic, rounded or subcordate at the base, 4–10 × 2–5 cm., sometimes glabrescent and shining on the upper surface when mature; climbing shrubs 4. *edule*
Cymes distinctly pedunculate, many-flowered, 1·5–2·5 cm. long; pedicels 1–3 mm. long; leaves oblong-obovate, subelliptic or oblong, rounded at the base:
Peduncle adnate to the petiole in its lower part; leaves not discolorous:
Leaves 2·5–5 × 1–2 cm., oblong or oblong-elliptic; stipules 1–2 mm. long, linear, caducous; pubescence cinereous; style 3-lobed or -partite to the base - - - - - - - - - - - 9. *mendoncae*
Leaves 4–8 × 2–4·5 cm., ± oblong-obovate or subelliptic; stipules 2–3 mm. long, linear, caducous; pubescence fulvous; style 3-lobed 8. *deflexum*
Peduncle free for the whole of its length, axillary; leaves ± discolorous, densely tomentose on the lower surface:
Stipules c. 2 mm. long, linear, very caducous; leaves 4–14 × 2–7 cm., oblong-obovate, rounded at the apex, tomentose above, or eventually glabrescent; lateral nerves 3–4 pairs - - - - 15. *stuhlmannii*
Stipules 6–7 mm. long, linear-lanceolate; leaves 5–12 × 2·5–4·5 cm., oblong or subelliptic, rounded or obtuse at the base, acute or rounded and mucronulate at the apex, tomentose above; lateral nerves 5–7 pairs 14. *rhodesicum*
Corolla 6–7 mm. long; leaves discolorous with an arachnoid indumentum on the lower surface, upper surface glabrous when mature; stipules 2-fid, linear; climbing shrubs - - - - - - - - 16. *thonneri* var. *ellipticum*
Stipules 4-partite, with subulate laciniae, persistent; leaves subsessile, cordate at the base:
Upper surface of leaf with sparse long hairs, lower surface with long hairs on the nerves and a sparse pubescence of short hairs - - - - 6. *mossambicense*
Upper surface of leaf with sparse long fulvous hairs, lower surface with long hairs on the nerves and a dense fulvo-cinereous pubescence - - - 7. *aureonitens*

1. **Dichapetalum macrocarpum** Engl., Bot. Jahrb. **46**: 565 (1912).—M. B. Moss in Kew Bull. **1928**: 121 (1928). Type from Tanganyika (Lindi Distr.).

Shrub; branchlets tomentose with spreading hairs. Leaf-lamina 4–10(14) × 2–5 cm., elliptic or oblanceolate, glabrescent above (except for the midrib and lateral nerves which are pilose with long hairs), densely hispid-tomentose below, acuminate and mucronate at the apex, rounded at the base and often slightly asymmetric; lateral nerves c. 6–9 pairs, prominent below; stipules lanceolate or linear-lanceolate, c. 15 × 2 mm., ± persistent and with appressed hairs especially on the outside.

Cymes subsessile, few-flowered; bracts up to 5 mm. long, hairy; pedicels 2 mm. long, pilose. Sepals c. 6 mm. long, free almost to the base, densely tomentose outside, glabrous within. Petals c. 8 × 4 mm., obovate, 2-fid for c. ¼ their length, rounded at the apex, tomentose outside. Ovary covered with very long white woolly hairs; style glabrescent, shortly 3-lobed. Fruit 2–3-lobed, c. 2·5 cm. long, densely covered with long stiff hairs.

Mozambique. N: Cabo Delgado, fl. & fr. 12.i.1912, *Allen* 148 (K).
In dry *Brachystegia* woodland. Known only from Tanganyika and Mozambique.

The vernacular name is " Chibwayajika ". According to Engler (loc. cit.) the fruit is a deadly poison.

2. **Dichapetalum whitei** Torre in Bol. Soc. Brot., Sér. 2, **36**: 67, t. 1 (1962). Type: N. Rhodesia, Mwinilunga, *White* 3364 (BM, holotype; COI; FHO; K).
Dichapetalum sp. 1.—White, F.F.N.R.: 185 (1962).

Climbing shrub; young branchlets densely brownish-red-tomentose. Leaf-lamina 7–18 × 3·5–6 cm., oblong-elliptic or oblong-obovate, papyraceous, glabrous on both surfaces and shining above, ± acuminate and subcaudate at the apex, rounded or subcordate at the base; lateral nerves c. 6 pairs; petiole 2–4 mm. long and c. 1·5 mm. in diam.; stipules caducous. Inflorescence a very dense many-flowered sessile glomerule; flowers subsessile, rufous when dried. Sepals c. 2·5 mm. long, pubescent. Petals c. 5 mm. long, free, glabrous, 2-fid for ½ their length, with a long claw; squamulous glands c. 0·5 mm. long. Stamens 4–5 mm. long. Ovary sparsely puberulous; style c. 5 mm. long, glabrous, shortly 3-lobed.

N. Rhodesia. W: Mwinilunga, N. of Kalene Hill Mission, fl. 25.ix.1952, *White* 3364 (BM; COI; FHO; K).
Not known from elsewhere. In deciduous woodland of *Parinari*, *Sarcocephalus*, *Albizia*, etc.

3. **Dichapetalum thouarsianum** Roem. & Schult. in L., Syst. Veg. ed. nov. **5**: 324 (1819).—Descoings, Fl. Madag., Dichapetalac.: 6 (1961). Type from Madagascar.

Shrub; young branchlets puberulous or glabrescent; lenticels sparse or numerous on the young and old branches. Leaf-lamina 5–11 × 2–4 cm. oblong or subelliptic, acute or shortly caudate at the apex, cuneate at the base; lateral nerves 5–6 pairs, conspicuous above, prominent below; petioles 2–10 mm. long; stipules c. 2 mm. long, linear, caducous. Cymes c. 7-flowered, axillary, subglobose, sessile; pedicels 1–1·5 mm. long, pubescent. Sepals c. 2 mm. long, pubescent outside. Petals c. 4 × 1·4 mm. glabrous, narrowed below, 2-fid. Ovary woolly-tomentose; style c. 4 mm. long, glabrous, shortly 3-lobed. Fruit (unripe) villous.

Mozambique. SS: Gaza, Chipenhe, fr. 23.x.1957, *Barbosa & Lemos* 8047 (BR; COI; K; LISC; LMJ; SRGH). LM: Marracuene, Vila Luisa, Rikatla, fl. x.1917, *Junod* 108 (G; LISC; PRE).
Also in Madagascar. In deciduous woodland.

4. **Dichapetalum edule** Engl., Bot. Jahrb. **46**: 571 (1912).—M. B. Moss in Kew Bull. **1928**: 121 (1928).—Engl. & Krause in Engl. & Prantl, Nat. Pflanzenfam. ed. 2. **19c**: 7 (1931). Type from Tanganyika (Namguru-Tal).
Dichapetalum deflexum sensu P. Lima in Broteria, Sér. Bot. **19**, 3: 133 (1922) (" reflexum ").

Shrub c. 1·5 m. tall; branchlets tomentose when young with dense rusty-brown hairs, older branches glabrescent. Leaf-lamina 4–10 × 2–5 cm., elliptic or ovate-oblong, pubescent on both surfaces when young, later glabrescent above, acute or obtuse and mucronate at the apex, subcordate or rounded at the base; lateral nerves 6–7 pairs, slightly prominent above, very prominent below; petiole 2–4 mm. long, with fulvous hairs; stipules linear, c. 5 mm. long, deciduous. Cymes shortly pedunculate or subsessile; pedicels c. 2 mm. long, hirsute; bracteoles small, hirsute. Sepals c. 2 mm. long, retroflexed, tomentose outside, glabrous inside. Petals c. 3·5 mm. long, spathulate or obovate, 2-lobed, glabrous; squamulous glands short. Ovary ovoid-globose, tomentose; style c. 4 mm. long, glabrous, shortly 3-lobed. Fruit ovoid-oblong, densely fulvous-pilose, apex obliquely acuminate.

Mozambique. N: Cabo Delgado, Palma, fl. 17.ix.1948, *Andrada* 1365 (BR; LMJ; LISC).
Also in Tanganyika. In deciduous bush.

The vernacular name is " Mtosh ". According to Engler (loc. cit.) the pericarp of the ripe fruit is eaten.

5. **Dichapetalum crassifolium** Chod. in Bull. Herb. Boiss., Sér. B, **3**: 672 (1895). Type from Angola (Cuanza Norte).

Scandent shrub up to 1·5 m. tall, glabrous; bark of young and old branches with sparse lenticels. Leaf-lamina 7–12 × 3–4 cm., elliptic or oblong, glabrous on both surfaces, obtuse or acute or shortly caudate at the apex, rounded, subrounded or cuneate at the base; lateral nerves 5–7 pairs, slightly impressed above, prominent below; petiole 4–10 mm. long, pubescent or glabrous; stipules c. 2 mm. long, linear, early deciduous. Cymes c. 1·5 cm. long, axillary; peduncle c. 5 mm. long (up to 2 cm. long in the type), velutinous; pedicels c. 1·5 mm. long. Sepals c. 3 mm. long, ovate, pubescent. Petals c. 4 mm. long, obovate, apex entire, obtuse, glabrous. Stamens adnate to the petals at the base. Ovary subglobose, velutinous; style c. 2·5 mm. long, glabrous, shortly 3-lobed. Fruit asymmetric, laterally rostrate, 2 cm. long (type), sericeous-velutinous.

N. Rhodesia. W: Mwinilunga, fl. 23.ix.1955, *Holmes* 1208 (K).
Also in Angola.

Note: the specimen cited has the leaves cuneate and smaller than those of the type and shorter cymes (c. 1·5 cm. long).

6. **Dichapetalum mossambicense** (Klotzsch) Engl., Pflanzenw. Ost-Afr. **C**: 235 (1885) pro parte; in Engl. & Prantl, Nat. Pflanzenfam. **3,** 4: 349 (1906); Bot. Jahrb. **46**: 572 (1912) pro parte.—M. B. Moss in Kew Bull. **1928**: 120 (1928).—Engl. & Krause in Engl. & Prantl, Nat. Pflanzenfam. ed. 2, **19c**: 6 (1931).—Brenan, T.T.C.L.: 130 (1949). Type: Mozambique, Sena, *Peters* (B, holotype†).
Chailletia mossambicensis Klotzsch in Peters, Reise Mossamb. Bot. **1**: 108, t. 19 (1861).—Oliv., F.T.A. **1**: 342 (1868) pro parte.—Sim, For. Fl. Port. E. Afr.: 30 (1909). Type as above.

Scandent shrub up to 2 m. tall; branches, branchlets and peduncles terete, covered with long reddish-yellow bristly hairs and also short hairs. Leaves subsessile or very shortly petiolate; lamina 6–15 × 4–6·5 cm., oblong-elliptic, broadly acute or obtuse, thinly pilose above and with scattered appressed long hairs, with paler sparse short pubescence below but with long hairs on the nerves, mucronate or shortly apiculate at the apex, distinctly cordate at the base; lateral nerves 5–7 pairs, prominent below, conspicuous above; petiole up to 4 mm. long, reddish-yellow-villous; stipules c. 1 cm. long, persistent, pinnatipartite, pilose, with filiform segments. Cymes 2–5 cm. long, axillary; bracts setaceous; pedicels slender, pilose, articulated above. Sepals c. 3 mm. long, ovate, whitish-tomentose outside, at length recurved. Petals c. 4 mm. long, obovate, free, 2-lobed for about half their length, narrowed at the base. Ovary villous; style c. 2 mm. long, glabrous, shortly 3-lobed. Fruit c. 1·5 cm. long, villous, oblong.

Mozambique. MS: Sena, rivers of Sena, *Peters* (B†; K; P).
Also in Tanganyika.

The vernacular name is " Chibwaya-dume ". According to Engler (loc. cit.) the fruit is a deadly poison.

7. **Dichapetalum aureonitens** Engl., Bot. Jahrb. **46**: 573 (1912).—M. B. Moss in Kew Bull. **1928**: 119 (1928). Type from Tanganyika.
Chailletia mossambicensis sensu Oliv., F.T.A. **1**: 342 (1868) pro parte.
Dichapetalum mossambicense sensu Engl., Pflanzenw. Ost-Afr. **C**: 235 (1895) pro parte.—P. Lima in Bol. Soc. Brot. Sér. 2, **2**: 138 (1922).

Scandent shrub; branchlets laxly villous with rusty brown hairs intermixed with a short dense indumentum. Leaf-lamina 7–14 × 2·5–8 cm., oblong to slightly obovate, with sparse long appressed hairs on both surfaces, denser on the midrib, intermixed with a short dense woolly indumentum below, tapering to a fine point at the apex, cordate at the base; lateral nerves c. 7 pairs, impressed above, prominent below; petiole c. 2 mm. long; stipules c. 10 mm. long, pinnatipartite, brown-villous. Cymes 3–7 cm. long, axillary, pedunculate, laxly branching, villous; bracts c. 5 mm. long, linear, villous; pedicels shortly hairy. Sepals c. 2 mm. long, oblong, at length recurved, white-woolly outside, glabrous and dark coloured inside. Petals about the same length as the sepals, obovate, 2-fid.

Stamens slightly longer than the petals. Ovary dense and with a long white woolly indumentum; style shorter than the stamens, shortly 3-lobed. Fruits oblong-cylindric, covered with dense rusty-brown hairs.

Mozambique. N: between Mucojo and Quiterajo, fl. & fr. 12.ix.1948, *Andrada* 1339 (BR; COI; K; LISC; LMJ); R. Msalu, i.1912, *Allen* 146 (K).
Also in Tanganyika. Evergreen coastal bush.

Vernacular name: " Quicuaia " (Cabo Delgado, Palma).

8. **Dichapetalum deflexum** (Klotzsch) Engl., Pflanzenw. Ost-Afr. **C**: 235 (1895); Bot. Jahrb. **46**: 575 (1912).—M. B. Moss in Kew Bull. **1928**: 128 (1928).—Engl. & Krause in Engl. & Prantl, Nat. Pflanzenfam. ed. 2, **19c**: 6 (1931).—Brenan., T.T.C.L.: 130 (1949). Type: Mozambique, Mossuril, Cabaceira, *Peters* (B, holotype†).
Chailletia deflexa Klotzsch in Peters, Reise Mossamb. Bot. **1**: 109, t. 20 (1861).—Oliv., F.T.A. **1**: 343 (1868) " reflexa ".—Sim, For. Fl. Port. E. Afr.: 30 (1909). Type as above.

Erect scandent shrub; branches and branchlets covered with a short rusty or yellowish tomentum. Leaf-lamina 4–8 × 2–4·5 cm., oblong-elliptic or elliptic, acute or shortly acuminate at the apex, rounded or obtuse at the base, softly pilose-pubescent at first, later glabrescent above, sparsely pubescent below; petiole 2–3·5 mm. long, deflexed, rusty-pubescent; stipules c. 3 mm. long, linear-subulate, caducous. Flowers in axillary shortly pedunculate cymes 1–2·5 cm. long; peduncle adnate to the petiole at the base, yellowish-tomentose; bracts subulate; pedicels c. 2 mm. long, pubescent; calyx lobes c. 3 mm. long, linear-acute, whitish- or yellowish-tomentose outside. Petals c. 4 mm. long, spathulate, 2-fid for about $\frac{2}{3}$ of their length, much narrowed to the base. Ovary tomentose; style c. 2·5 mm. long, glabrescent, shortly 2–3-lobed. Fruit oblique, tomentose.

Mozambique. N: Moçambique, Mossuril, Cabaceira, fl. 1884, *Carvalho* (COI). SS: between Vilanculos and Macovane, fl. 24.xi.1942, *D'Orey* 14 (LISC).
Also in Tanganyika. In bush.

Vernacular name " Nadula-Nicane " (Moçambique). Poisonous shrub.

9. **Dichapetalum mendoncae** Torre in Bol. Soc. Brot., Sér. 2, **36**: 67 (1962). Type: Moçambique, Massinga, *Mendonça* 1895 (LISC, holotype).

Climbing shrub with patent cinereous-pubescent branchlets. Leaf-lamina 2–5 × 1–2 cm., oblong or oblong-lanceolate, sparsely pubescent on both surfaces, acute and apiculate at the apex, rounded at the base; lateral nerves 5–7 pairs; petiole 2–3 mm. long, pubescent; stipules 2 mm. long, linear, deciduous. Cymes c. 1·5 cm. long, axillary, many-flowered, shortly pedunculate, with the peduncle adnate to the petiole; bracts c. 1 mm. long, subulate, caducous. Flowers pedicellate; pedicels c. 2 mm. long, cinereous-pubescent. Calyx c. 2·5 mm. long, cinereous-pubescent outside; lobes reflexed. Petals c. 3 mm. long, glabrous, deeply 2-fid for about $\frac{1}{2}$ their length, with claw narrowed. Stamens c. 2·5 mm. long. Ovary covered with woolly hairs; style glabrous deeply 3-partite or 3-lobed.

Mozambique. SS: Massinga, Inhachengo, fl. 30.viii.1944, *Mendonça* 1895 (BR; COI; K; LISC; LM; SRGH).
Known only from Mozambique. In mixed woodland.

10. **Dichapetalum barbosae** Torre in Bol. Soc. Brot., Sér. 2, **36**: 68, t. 2 (1962). Type: Mozambique, Cabo Delgado, Macomia, *Barbosa* 2035 (LISC, holotype).

Scandent shrub; branches and branchlets terete, pubescent; bark longitudinally striate, covered with small and numerous lenticels. Leaves 5–12 × 2–5 cm., oblong-elliptic or oblong, papyraceous, glabrous on both surfaces, subacute or obtuse at the apex, subrounded or obtuse at the base; lateral nerves 5–6 pairs, slightly conspicuous above, prominent below; petiole c. 5 mm. long, pubescent or glabrescent; stipules c. 2 mm. long, subulate, caducous. Inflorescence cymose, short, axillary; peduncle 1–3 mm. long, free; pedicels c. 1 mm. long, articulated at the base of calyx; bracts small, caducous. Sepals c. 3 mm. long, oblong, tomentose. Petals c. 4 mm. long, lanceolate, pubescent, glabrous at the apex, 2-lobed. Stamens c. 4 mm. long. Ovary tomentose; style c. 3·5 mm. long, pubescent, shortly 3-lobed. Fruit c. 1·5 × 1 cm., subellipsoid, densely and shortly whitish-tomentose.

Mozambique. N: Cabo Delgado, Macomia, between Ingoane and Quiterajo, fl. 12.ix.1948, *Barbosa* 2085 (COI; K; LISC). Z: Maganja da Costa, fr. (immat.) viii.1908, *Sim* 5705 (PRE). MS: Cheringoma, Serração de Cardoso Lopes, about 70 km. from Beira, fr. 17.ix.1943, *Torre* 5907 (LISC).

Known only from Mozambique. In dry bush and on margins of rivers.

11. **Dichapetalum zambesianum** Torre in Bol. Soc. Brot., Sér. 2, **36**: 69, t. 3 (1962). Type: Mozambique, Bajone, *Barbosa & Carvalho* 4272 (LMJ, holotype).

An erect or scandent shrub; young branches sparsely pubescent, later glabrous; bark longitudinally striate, lenticellate. Leaf-lamina 7–15 × 3·5–6 cm., glabrous on both surfaces, chartaceous, oblong-elliptic, obtuse to acute at the apex, rounded at the base; lateral nerves 7–9 pairs, slightly conspicuous above, prominent below; petiole 5–10 mm. long and c. 2 mm. diam., sparsely pubescent, later glabrescent; stipules c. 2 mm. long, subulate, caducous. Inflorescence a many-flowered axillary cyme, shortly pedunculate; peduncle 3–6 mm. long; pedicels c. 1 mm. long, articulated at the base of the calyx. Sepals c. 4 mm. long, oblong, tomentose. Petals c. 4 mm. long, with pubescent claw, glabrous at the apex, 2-lobed. Stamens c. 4 mm. long. Ovary tomentose; style c. 4·5 mm. long, pubescent, apex glabrous, shortly 3-lobed. Fruit c. 2 cm. long, subglobose, densely and shortly whitish-tomentellous.

Mozambique. N: Cabo Delgado, between Macomia and Chai, fr. 30.ix.1948, *Barbosa* 2280 (K; LISC; LMJ). Z: Bajone, between Murroa and Namuera, fl. 2.x.1949, *Barbosa & Carvalho* 4272 (LISC; LM; LMJ).

Known only from Mozambique. In deciduous and secondary woodland.

12. **Dichapetalum cymosum** (Hook.) Engl. in Eng. & Prantl, Nat. Pflanzenfam. **3,** 4: 349 (1896).—Eyles in Trans. Roy. Soc. S. Afr. **5,** 4: 392 (1916).—Engl. & Krause in Engl. & Prantl, Nat. Pflanzenfam ed. 2, **19c**: 6 (1931).—Wild, Comm. Rhod. Weeds: t. 50 (1955). TAB. **63.** Type from the Transvaal.
Chailletia cymosa Hook., Ic. Pl. **6**: t. 591 (1943). Type as above.

Stems simple from a creeping woody rhizome, pubescent, striate. Leaves subsessile; lamina 5–13 × 2–4·5 cm., narrowly oblong to oblanceolate, rounded and minutely mucronate at the apex, narrowed or cuneate at the base, papyraceous when young, glabrous on both surfaces, pale green below; lateral nerves c. 6 pairs, tertiary nerves markedly reticulate and prominent below; stipules c. 4 mm. long, linear, pilose. Cymes up to 5 cm. long, c. 8-flowered; peduncle sparsely hairy or glabrous, partially adnate to the stem for c. 5 mm. above the axil; pedicels ± 2 mm. long; bracts c. 2 mm. long. Sepals c. 7 mm. long, narrow, free almost to the base, tomentose at the apex, becoming glabrous towards the base. Petals c. 7 mm. long, narrowly oblong, 2-fid for ⅔ their length, glabrous. Stamens as long as petals. Ovary villous; style c. 6 mm. long, deeply 3-divided. Fruit 4 × 2·5 cm., obovoid.

Bechuanaland Prot. N: Ngamiland, between Natal and Iamasetse, *Pole Evans* 4631 (PRE). **S. Rhodesia.** W: Nyamandhlovu Distr., fl. 25.ix.1953, *Plowes* 1635 (K; SRGH), fr. 18.xi.1953, *Plowes* 1643 (K; SRGH).

Also in the Transvaal. On Kalahari Sand.

Very near to *D. venenatum* Engl. & Gilg. from which it is sometimes separable only with difficulty. Extremely poisonous to cattle.

13. **Dichapetalum bullockii** Hauman in Bull. Jard. Bot. Brux. **28,** 1: 74 (1958).—White, F.F.N.R.: 185 (1962). Type: N. Rhodesia, Mkupa, *Bullock* 1196 (BR, holotype).

Shrublet c. 40 cm. tall; stems many, erect, virgate, glabrous, from a woody root-stock. Leaves sessile or shortly petiolate; lamina 7–16 × 0·5–2·5 cm., narrowly lanceolate or linear, papyraceous, glabrous on both surfaces, obtuse or rounded at the apex, cuneate at the base; lateral nerves prominent below; stipules linear, c. 3 mm. long, deciduous; petiole up to 7 mm. long. Inflorescence a few-flowered axillary cyme up to 9 cm. long; peduncle 1–3 cm. long, 2-forked. Flowers 5–8 mm. long; pedicels c. 4 mm. long, glabrous; bracts and bracteoles c. 2 mm. long, linear, deciduous. Calyx campanulate; sepals 4–5 mm. long, linear, acute tomentellous. Petals 5–6 mm. long, linear, 2-lobed to ½ of their length; squamulous glands minute. Stamens free. Ovary lanate; style 4–5 mm. long, pilose, 3-fid. Fruit (unripe) oblong-obovoid.

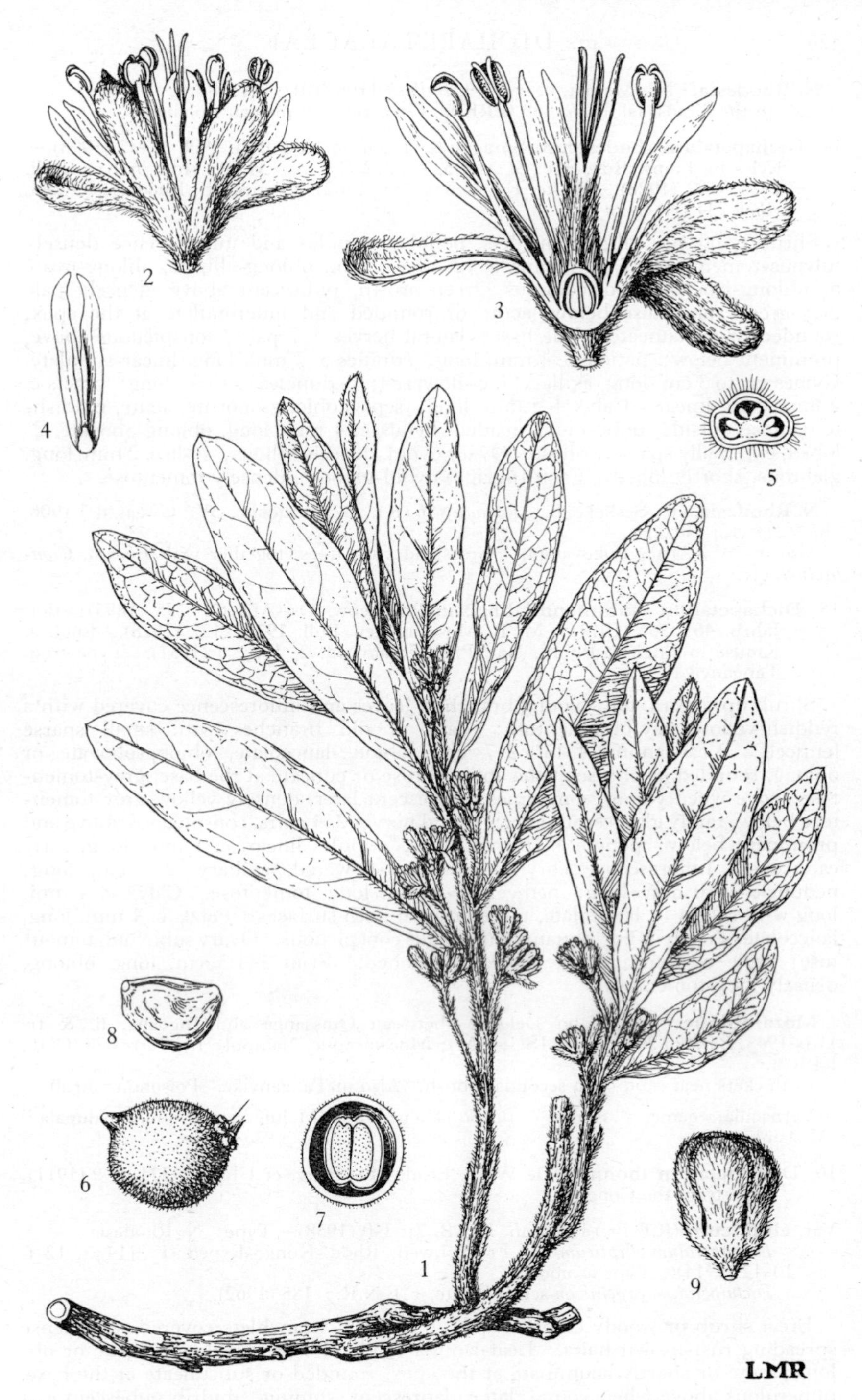

Tab. 63. DICHAPETALUM CYMOSUM. 1, flowering shoot ($\times \frac{2}{3}$) *Plowes* 1635; 2, flower ($\times 4$) *Plowes* 1635; 3, vertical section of flower ($\times 6$) *Moss & Ottery* 11935; 4, petal ($\times 4$) *Plowes* 1635; 5, transverse section of ovary ($\times 10$) *Plowes* 1635; 6, fruit ($\times \frac{2}{3}$) *Leemann* s.n.; 7, transverse section of fruit ($\times \frac{2}{3}$) *Leemann* s.n.; 8, seed ($\times \frac{2}{3}$) *Leemann* s.n.; 9, flower-bud ($\times \frac{2}{3}$) *Plowes* 1635.

N. Rhodesia. N: Mkupa, fl. ix.1949, *Bullock* 1196 (BR; EA; K).
Also in the Transvaal (*Codd* 869 (PRE)). Gregarious in plateau woodlands.

14. **Dichapetalum rhodesicum** Sprague & Hutch. in Kew Bull. **1908** : 433 (1908).—Eyles in Trans. Roy. Soc. S. Afr. **5,** 4: 322 (1916).—M. B. Moss in Kew Bull. **1928**: 122 (1928).—White, F.F.N.R.: 185 (1962). Types: S. Rhodesia, Gwaai, *Allen* 234 (K, holotype).

Shrub c. 60 cm. tall; branchlets, petioles, stipules and inflorescence densely fulvous-tomentose. Leaf-lamina 5–12 × 2–4·5 cm., oblong-elliptic, oblong-ovate or oblong-lanceolate, discolorous, green-brown, pubescent above, densely pale grey-green tomentose below, acute or rounded and mucronulate at the apex, rounded or subcuneate at the base; lateral nerves 5–7 pairs, conspicuous above, prominent below; petiole 3–4 mm. long; stipules 5–7 mm. long, linear-subulate. Cymes up to 3 cm. long, axillary, few-flowered; peduncles c. 1 cm. long; bracts c. 2 mm. long, linear. Calyx 4–5 mm. long; sepals oblong-obovate, acute, reddish-tomentose outside, puberulous inside. Petals c. 4 mm. long, oblong-obovate, 2-lobed, externally sparsely pilose. Ovary ovoid, densely villous; style c. 2 mm. long, glabrous, shortly lobed. Fruit (unripe) ovoid-globose, densely tomentose.

N. Rhodesia. B: Sesheke, fl., *Macaulay* 86 (K). **S. Rhodesia.** W: Gwaai, fl. i.1906, *Allen* 234 (K; SRGH).
Also in SW. Africa (Okovango). Grassland with trees, together with *Burkea*, *Combretum*, etc.

15. **Dichapetalum stuhlmannii** Engl. in Pflanzenw. Ost-Afr. **C**: 235 (1895); Bot. Jahrb. **46**: 566 (1912).—M. B. Moss in Kew Bull. **1928**: 122 (1928).—Engl. & Krause in Engl. & Prantl, Nat. Pflanzenfam. ed. 2, **19c**: 6 (1931). Type from Tanganyika.

Shrub up to 4 m. tall; young branches, leaves and inflorescence covered with a reddish-yellow-grey indumentum; bark of old branches with small sparse lenticels. Leaf-lamina 4–14 × 2–7 cm., oblong-lanceolate, oblong-obovate or oblong, rounded or obtuse at the apex, obtuse or cuneate at the base, grey-tomentose above when young, sometimes glabrescent later, densely yellow-grey tomentose below, rarely glabrescent later; lateral nerves 3–4 pairs, conspicuous above and prominent below; petiole 4–10 mm. long; stipules linear, c. 2 mm. long, early caducous. Inflorescence corymbose, many-flowered, axillary, 2–3 cm. long; peduncle 3–10 mm. long; pedicels 2–3 mm. long, tomentose. Calyx c. 4 mm. long with sepals ± lanceolate, tomentose on both surfaces. Petals c. 4 mm. long, lanceolate, shortly 2-fid; squamulous glands conspicuous. Ovary subovoid, tomentose; style c. 4 mm. long, glabrous, 3-lobed. Fruit 1–1·5 cm. long, oblong, densely grey-tomentose.

Mozambique. N: Cabo Delgado, between Quissanga and Ingoane, fl. & fr. 11.ix.1948, *Barbosa* 2052 (K; LISC; LM); Moçambique, Nampula, fr., *Torre* 678 (COI; LISC).
In thickets near sand or in secondary bush. Also in Tanganyika. Poisonous shrub.

Vernacular names "Sunho" (Cabo Delgado), "Melule", "Nodula-malupala" (Mozambique).

16. **Dichapetalum thonneri** De Wild., Études Fl. Bang. et Ubang: 224, t. 9 (1911). Type from the Congo.

Var. **ellipticum** (R.E.Fr.) Hauman, F.C.B. **7**: 340 (1958). Type: N. Rhodesia.
Dichapetalum ellipticum R.E.Fr., Schwed. Rhod.-Kongo-Exped. **1**: 114, t. 12 f. 10–12 (1914). Type as above.
Dichapetalum argenteum sensu White, F.F.N.R.: 185 (1962).

Erect shrub or woody climber up to 4 m. tall; branchlets covered with dense spreading rust-brown hairs. Leaf-lamina 5–14 × 2–7 cm., oblong-elliptic or oblong, acute or shortly acuminate at the apex, rounded or subcuneate at the base, puberulous above when young, later glabrescent, shining; midrib pubescent and covered with a whitish-tomentose indumentum below; lateral nerves 5–8 pairs, slightly visible above, prominent below; petiole 3–6 mm. long with spreading rust-brown hairs; stipules c. 5 mm. long, linear or divided to the base, deciduous. Cymes axillary, many-flowered; peduncle c. 5 mm. long, densely rust-brown-pubescent; pedicels c. 2 mm. long; bracts linear pubescent. Sepals c. 7 mm. long,

Tab. 64. TAPURA FISCHERI VAR. PUBESCENS. 1, branchlet with inflorescences (×⅔); 2, flower (×8); 3, vertical section of calyx and ovary (×8); 4, corolla opened to show 4 petals, 2 stamens and 2 staminodes (×10); 5, lateral petal and stamen from inside (×10); 6, lateral petal from outside (×10); 7, branchlet with fruits (×⅔); 8, fruit (×3); 9, transverse section of fruit (×3); 10, 2-locular fruit (×3); 11, vertical section of apex of fruit (×3); 12, inflorescence adnate to petiole (×4). 1–6 and 12 from *Simão* 1173, 7–11 from *Kirk* s.n.

oblong or obtuse, densely pubescent, glabrous outside and dark-coloured within. Petals 6·5–7 × c. 3 mm., oblong-lanceolate, 2-fid, narrowed at the base; squamulous glands very small. Ovary densely whitish-woolly; style c. 7 mm. long, shortly 2-lobed. Fruit c. 12 mm. long, ovoid, covered with dense rust-brown hair.

N. Rhodesia. N: Mpika, fl. & fr. 31.i.1955, *Fanshawe* 1911 (K); Kasama Distr., Malole Rocks, fl. 1.iii.1960, *Richards* 4250 (SRGH).
Also in Upper Katanga and Angola. In gallery forest and evergreen thickets.

2. TAPURA Aubl.

Tapura Aubl., Pl. Guian. **1**: 126, t. 48 (1775).

Small trees. Leaves entire; stipules minute, caducous. Flowers zygomorphic, in small axillary shortly pedunculate glomerules; peduncle adnate to the petiole. Sepals 5, unequal, united. Petals 5, adnate at the base to the stamens to form an unequal 5-lobed tube, 2 of the lobes 2-fid, 3 entire. Stamens 2–3 fertile, 2 sterile. Ovary sessile, 2–3-locular; style filiform, shortly 2–3-fid at the apex. Fruit drupaceous.
A tropical American and tropical African genus.

Tapura fischeri Engl., Pflanzenw. Ost-Afr. **C**: 423 (1895). Type from Tanganyika.

Var. **pubescens** Verdcourt & Torre in Bol. Soc. Brot., Sér, 2, **36**: 69 (1962). TAB. **64**. Type: Mozambique, Mossurize, *Simão* 1173 (EA, holotype).

Small tree; old branches glabrous; young and floriferous branches pubescent; bark with very small lenticels. Leaf-lamina 4–7 × 1·5–3 cm., subelliptic, glabrescent on the upper surface when old, except for the midrib, sparsely pubescent when old on the lower surface, acute or obtuse at the apex, broadly cuneate at the base; lateral nerves 4–6 pairs, slightly prominent below, conspicuous above; petiole 4–7 mm. long; stipules linear, very small, caducous. Infloresencec a c. 15-flowered pseudumbel; peduncle adnate to the petiole; pedicels c. 2 mm. long, pubescent; bracts very small. Sepals c. 1 mm. long, subcircular or ovate, pubescent, 2 outside ones smaller. Petals with 2 longer ones connate below up to half-way, c. 2·5 mm. long, 2-fid, with hooded pubescent lobes, 3 shorter ones pubescent. Fertile stamens 2, anthers enclosed in the hood before anthesis. Ovary glabrous or puberulous; style c. 2 mm. long, 2–3-lobed. Fruit c. 4 mm. long, ovoid, glabrescent.

Nyasaland. S: Shire Highlands, fl. ix.1881, *Buchanan* 30 (K). **Mozambique.** T: between Tete and Lupata, fr. vii.1859, *Kirk* (K). MS: Cheringoma, Inhamitanga, fl. 30.x.1944, *Simão* 237 (LISC); Mossurize, Machaze, Mecufi, *Simão* 1173 (EA, holotype; LISC; LM).
Known only from Nyasaland and Mozambique. In woodland with some evergreen trees.
A specimen *Burtt* 6006 (K) from N. Rhodesia (Kalambo Falls) cited by White (F.F.N.R.: 185 (1962)) may belong to this species but it is sterile and not identifiable with certainty. If it is *T. fischeri* it is the type variety, not var. *pubescens*.

48. OLACACEAE

By J. G. Garcia

Trees or shrubs, sometimes armed with spines. Leaves alternate, simple, entire, penninerved, exstipulate. Flowers actinomorphic, bisexual, in axillary fascicles, cymes or racemes. Calyx cupuliform, often accrescent, with the base free or more or less adnate to the ovary, and the margin entire or lobed. Petals 3–5(6), valvate in bud, sometimes connate in pairs. Stamens as many as or more numerous than the petals, free or with the filaments more or less adnate to the petals, all fertile or some sterile. Disk present or inconspicuous. Ovary free or more or less immersed, usually 3–5-locular at the base or nearly to the apex; style short, columnar; stigma capitate, sometimes lobulate; ovules (2)3–5, pendulous from the apex of a central placenta. Fruit drupaceous, ellipsoid to globose. Seed with abundant fleshy endosperm; embryo minute, apical.

Stamens, or stamens and staminodes, more numerous than the petals:
Staminodes absent; petals bearded within - - - - - - **1. Ximenia**
Staminodes present; petals glabrous within - - - - - - **2. Olax**
Stamens as many as and opposite to the petals - - - - - **3. Strombosia**

1. XIMENIA L.

Ximenia L., Sp. Pl. **2**: 1193 (1753); Gen. Pl., ed. 5: 500 (1754).

Shrubs or small trees up to 6 m. tall, usually spiny. Leaves obtuse to retuse, mucronulate. Flowers axillary, solitary or fasciculate or in pedunculate cymes. Calyx 4(5)-lobed, not accrescent. Petals 4(5), linear-oblong, later recurved, glabrous to pubescent outside, densely bearded within. Stamens usually 8, 4 opposite the petals and 4 alternating with them, free. Ovary free, lageniform, usually 4-locular nearly to the apex; placenta central, columnar; ovules 2–4, apical, pendulous, anatropous; style short; stigma minute, capitate. Fruit drupaceous with a large stone.

Flowers in pedunculate cymes - - - - - - - 1. *americana*
Flowers fasciculate or solitary - - - - - - - - 2. *caffra*

1. **Ximenia americana** L., Sp. Pl. **2**: 1193 (1753).—Sim, For. Fl. Port. E. Afr.: 30 (1909).—Engl., Pflanzenw. Afr. **3**, 1: 82, fig. 46 (1915).—Eyles in Trans. Roy. Soc. S. Afr. **5**: 345 (1916).—Burtt Davy & Hoyle, N.C.L.: 57 (1936).—Louis & Léonard, F.C.B. **1**: 256 (1948).—Brenan, T.T.C.L.: 387 (1949).—Exell & Mendonça, C.F.A. **1**, 2: 322 (1951).—Wild, Guide Fl. Vict. Falls: 151 (1953).—Keay, F.W.T.A. ed. 2, **1**, 2: 646 (1958).—White, F.F.N.R.: 42 (1962). TAB. **65** fig. B. Type from tropical America.
Ximenia americana var. *microphylla* Welw. ex Oliv., F.T.A. **1**: 346 (1868).—O. B. Mill., B.C.L.: 12 (1949). Type from Angola.
Ximenia rogersii Burtt Davy, F.P.F.T. **2**: xxxv, 454 (1932). Holotype: S. Rhodesia, Bulawayo, *Rogers* 22569 (K).

Shrub or small tree up to 5 m. tall, glabrous, usually spiny; branches divaricate or more rarely ascendent. Leaf-lamina 2–8 × 1–4 cm., oblong-elliptic, obtuse to retuse at the apex, coriaceous; midrib impressed above, prominent below; lateral nerves 3–6 pairs, subinconspicuous on both surfaces; petiole 3–6 mm. long, canaliculate, puberulous or pubescent above. Flowers in pedunculate racemose or umbelliform cymes; pedicels 3–7 mm. long; buds ellipsoid or oblong-obovoid. Calyx 4(5)-lobed, with ± slightly ciliate margins. Petals 4(5), 5–10 × 1–2·5 mm., glabrous outside, densely bearded within, apiculate. Stamens 8; filaments 2–4 mm. long; anthers 2–4 × 0·4–0·8 mm. Ovary c. 3 × 1·5 mm.; style 1–2 mm. long, columnar, caducous. Fruit c. 2·5 cm. in diam., drupaceous, yellow when ripe, ellipsoid or subglobose, edible. Seed c. 1·5 cm. in diam.

Bechuanaland Prot. N: Ngamiland, Kwebe Hills, fl. vii.1897, *Lugard* 3 (K). SW: Ghanzi, fr. 22.vii.1955, *Story* 5061 (K; PRE). SE: Mochudi, st. i–iv.1914, *Harbor* in *Rogers* 6629 (K). **N. Rhodesia.** B: Sesheke Distr., Seembe Pool, fl. 15.ix.1947, *Rea* 108 (K). N: Luangwa Valley, Mwuniamazi R., fl. 1.x.1933, *Michelmore* 618 (K). W: Kitwe, fl. 3.ii.1957, *Fanshawe* 2990 (K). C: Chilanga, Quien Sabe, fl. 19.ix.1929, *Sandwith* 13 (K). S: Mazabuka, Choma, fl. 28.vii.1952, *Angus* 65 (K); Livingstone, fl. 26.viii.1947, *Brenan & Greenway* 7752 (EA). **S. Rhodesia.** N: Urungwe, Kariba, fl. & fr. 14.i.1956, *Phelps* 103 (K). W: Victoria Falls, fl. & fr. 12.xi.1919, *Shantz* 406 (K); Bulawayo, fl. 21.ii.1922, *Eyles* 3312 (BOL; SRGH). C: Hunyani, fl. 15.ix.1946, *Wild* 1243 (K; SRGH). E: Umtali, fl. 20.ix.1948, *Chase* 978 (K). S: Sabi-Odzi Junction, fl. 5.ix.1945, *McGregor* 49/47 (SRGH). **Mozambique.** N: Moçambique, Imala, fl. & fr. 23.x.1948, *Barbosa* 2564 (LISC; LMJ). Z: between Mopeia and Campo, fl. 12.ix.1944, *Mendonça* 2042 (COI; K; LISC). T: Tete, 17 km. from Changara on road to Mtoko, fl. 28.ix.1948, *Wild* 2662 (K; SRGH). MS: Mossurize, between the Catholic Mission and R. Mossurize, fl. 9.vi.1942, *Torre* 4287 (BR; COI; K; LISC). SS: Guijá, between Mabalane and Mapai, fr. 4.xi.1944, *Mendonça* 2755 (LISC; LM; SRGH). LM: between Lourenço Marques and Namaacha, fl. 16.x.1940, *Torre* 1794 (COI; K; LISC).
Widespread in the tropics. Dry woodland, grassland with trees and coastal bush.

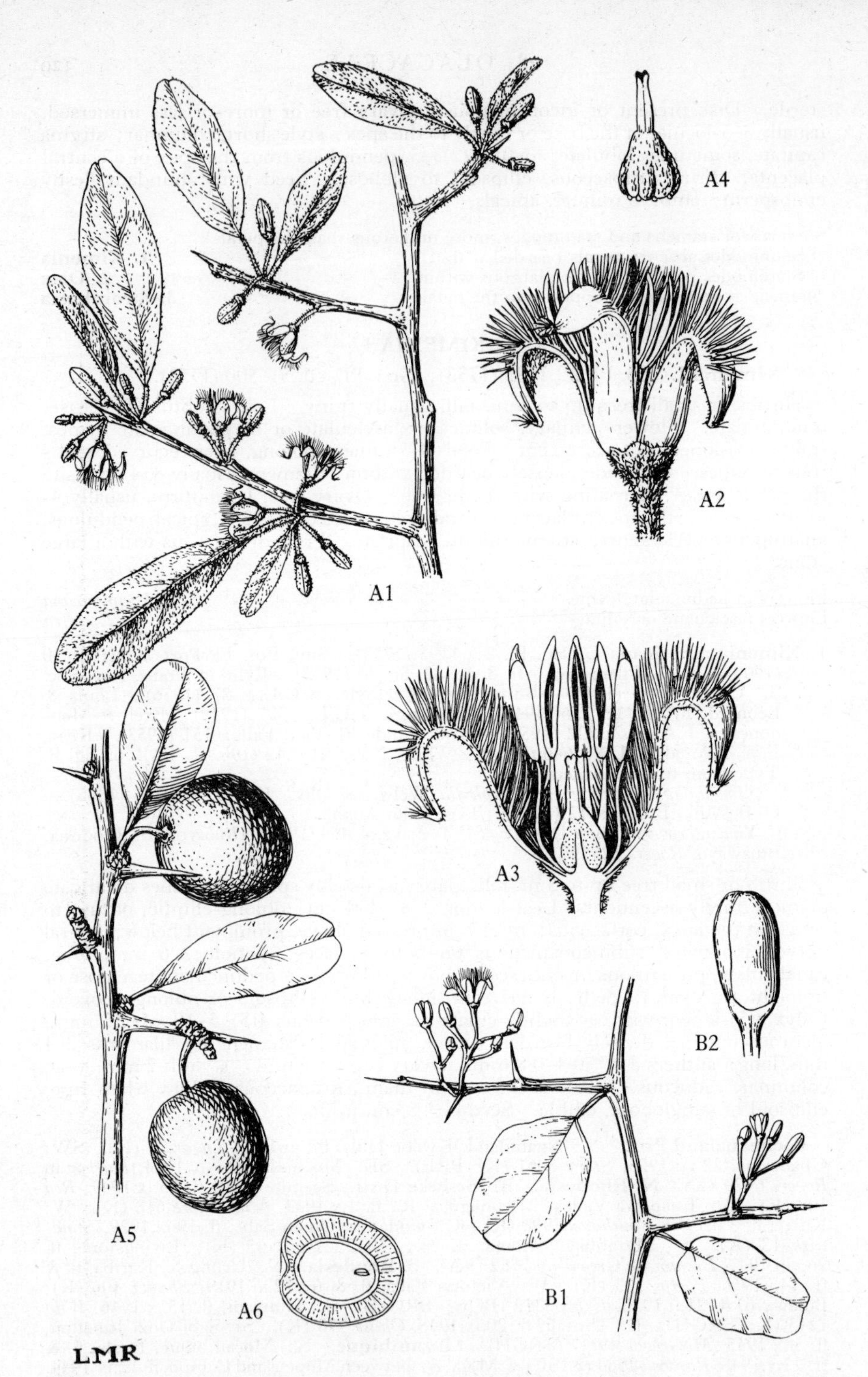

Tab. 65. A.—XIMENIA CAFFRA VAR. CAFFRA. A1, flowering branch (× ⅔) *Torre* 1032; A2, flower (× 3) *Torre* 1032; A3, vertical section of flower (× 4) *Torre* 1032; A4, ovary (× 4) *Mendonça* 3151; A5, fruiting branch (× ⅔) *Mendonça* 3151; A6, longitudinal section of fruit (× ⅔) *Mendonça* 3151. B.—XIMENIA AMERICANA. B1, flowering branch (× ⅔) *Mendonça* 2042; B2, bud (× 4) *Mendonça* 2042.

Ximenia caffra Sond. in Linnaea, **23**: 21 (1850).—Sim, For. Fl. Port. E. Afr.: 30 (1909).—Eyles in Trans. Roy. Soc. S. Afr. **5**: 345 (1916).—Burtt Davy, F.P.F.T. **2**: 453, fig. 72 (1932).—Burtt Davy & Hoyle, N.C.L.: 57 (1936).—Louis & Léonard, F.C.B. **1**: 257 (1948).—Brenan, T.T.C.L.: 388 (1949).—O. B. Mill., B.C.L.: 13 (1949).—Exell & Mendonça, C.F.A. **1,** 2: 333 (1951).—Wild, Guide Fl. Vict. Falls: 151 (1953).—White, F.F.N.R.: 42 (1962). Type from S. Africa.
Ximenia americana sensu Bak. f. in Journ. Linn. Soc., Bot. **40**: 43 (1911).

Shrub or small tree up to 6 m. tall, armed with spines; young branches and leaves densely rusty-tomentose to glabrous. Leaf-lamina 2·5–8 × 1·2–4 cm., elliptic or oblong-elliptic, obtuse to retuse at the apex, coriaceous; midrib impressed above, prominent below; lateral nerves 4–5 pairs, very slightly or not at all prominent on both surfaces; petiole 5–6 mm. long, canaliculate. Flowers solitary or fasciculate in the axils of the leaves; pedicels 3–7 mm. long; buds oblong-obovoid. Calyx 4(5)-lobed, ± pubescent. Petals 4(5), 6–12 × 1·5–2·5 mm., externally pubescent or almost glabrous, densely bearded within, shortly cucullate at the apex, recurved. Stamens 8; filaments 2·5–4 mm. long; anthers 2·5–4·5 × 0·6–0·8 mm. Ovary c. 3 × 1·5 mm.; style up to 1 mm. long, columnar, caducous. Fruit c. 2·5 × 1·8 cm., drupaceous, bright scarlet when ripe, ellipsoid, edible. Seed c. 1·5 × 1 cm.

Young branches and leaves ± tomentose - - - - - - - - var. *caffra*
Young branches and leaves glabrous - - - - - - - var. *natalensis*

Var. **caffra.** TAB. **65** fig. A.
Ximenia americana var. *caffra* (Sond.) Engl., Pflanzenw. Afr. **3,** 1: 82 (1915).

Bechuanaland Prot. N: Ngamiland, Kwebe Hills, fl. 17.xii.1897, *Lugard* 58 (K). SW: Chukuda Pan, 375 km. NW. of Molepolole, st. 20.vi.1955, *Story* 4949 (K; PRE). SE: Mahalapye, Western Lephephe, fr. 13.i.1958, *de Beer* 554 (SRGH). **N. Rhodesia.** B: near Senanga, ± 1040 m., fl. 30.vii.1952, *Codd* 7259 (K; PRE). N: Luangwa Valley, fl. 6.x.1933, *Michelmore* 643 (K). C: Lusaka, fl. & fr. 31.x.1956, *Angus* 1433 (BM). E: Petauke, st. 26.v.1952, *White* 2884 (K). S: Mazabuka, Kafue Flats, fl. 7.x.1930, *Milne-Redhead* 1222 (K). **S. Rhodesia.** N: Sinoia, fr. 2.xi.1931, *Rattray* 385A (BM; SRGH). W: Victoria Falls, fl. ix.1905, *Allen* 159 (K). C: Salisbury, alt. 1450 m., fl. 15.x.1931, *Brain* 6248 (COI; SRGH). E: Chirinda, fl. 18.x.1947, *Wild* 2022 (K; SRGH). S: Nuanetsi, fl. ix.1955, *Davies* 1623 (K; SRGH). **Mozambique.** N: between Porto Amélia and Ancuabe, fl. 24.viii.1948, *Andrada* 1297 (COI; LISC; LMJ). Z: Mocuba, Namagoa, fl. & fr. x–xi, *Faulkner* P 41 (COI; K). T: Zumbo, R. Mudzi, 15 km. from the Rhodesian Border, fl. 25.ix.1948, *Wild* 2614 (K). MS: Inhaminga, Metenguere, fl. 15.x.1947, *Pimenta* 210 (SRGH). SS: Macia, Muianga, st. 11.vii.1947, *Pedro & Pedrógão* 1440 (K). LM: Marracuene, Bobole, fl. 2.x.1957, *Barbosa & Lemos* in *Barbosa* 7928 (LISC).

Also in Angola, Congo, Tanganyika and S. Africa. Dry woodland and grassland with trees.

Var. **natalensis** Sond. in Linnaea, **23**: 21 (1850).—Brenan, T.T.C.L.: 388 (1949). Type from S. Africa.

N. Rhodesia. N: Abercorn, fr. 16.xi.1952, *Angus* 768 (K); Kawambwa, fl. 15.x.1947, *Brenan & Greenway* 8018 (EA); Kasama, fl. 11.x.1960, *Robinson* 3927 (K), W: Solwezi, fl. 16.ix.1952, *White* 3262 (K). **Nyasaland.** S: Mposa, Chikala Hills, Zomba, fr. 3.ii.1955, *Jackson* 1456 (K). **Mozambique.** N: Cabo Delgado, between Quissanga and Ingoane, fl. 11.ix.1948, *Barbosa* 2053 (K; LISC; LMJ); Moçambique, between Corrane and Liupo, fl. 15.ix.1936, *Torre* 1032 (COI; LISC). Z: between Mocuba and Inhamacurra, fl. 6.x.1941, *Torre* 3603 (COI; K; LISC; LM; SRGH).

Also in Tanganyika and S. Africa. Dry woodland and grassland with trees.

Intermediates between these two varieties sometimes occur in our area.

2. OLAX L.

Olax L., Sp. Pl. **1**: 34 (1753); Gen. Pl., ed. 5: 20 (1754).

Shrubs or small trees up to 6(12) m. tall; young branches yellowish-green, rugulose or scabridulous, narrowly 4-winged. Leaves distichous. Flowers solitary and axillary, or in axillary solitary or fasciculate racemes; pedicels articulate at the base; buds oblong. Calyx short, entire, cupuliform, undulate and/or inconspicuously denticulate at the margin. Petals 5(6) or 3, valvate, connate in the bud, later free, recurved-apiculate inside at the top. Stamens 3–6; staminodes 3–6;

filaments ± attached to the corolla; sterile anthers filiform, usually bifid and longer than the fertile ones. Disk circular. Ovary superior, lageniform or conical, usually 3-locular at the base; ovules usually 3, pendulous from the apex of a central placenta; style included; stigma capitate, sometimes 3–4-lobed. Fruit drupaceous, glabrous, sometimes enclosed nearly to the apex by the accrescent calyx. Seed with a minute apical embryo surrounded by fleshy endosperm.

Petiole 3–8 mm. long - - - - - - - - - - 1. *dissitiflora*
Petiole never more than 2 mm. long:
  Leaf-lamina 4–8 cm. long - - - - - - - - - 2. *obtusifolia*
  Leaf-lamina 9–14 cm. long - - - - - - - - - 3. *gambecola*

1. **Olax dissitiflora** Oliv., F.T.A. **1**: 350 (1868).—Eyles in Trans. Roy. Soc. S. Afr. **5**: 345 (1916).—Burtt Davy, F.P.F.T. **2**: 454 (1932).—Burtt Davy & Hoyle, N.C.L.: 56 (1936).—Brenan, T.T.C.L.: 387 (1949).—O. B. Mill., B.C.L.: 12 (1949).—Wild, Guide Fl. Vict. Falls: 151 (1953).—White, F.F.N.R.: 41 (1962). TAB. **66.** Type: Mozambique, Sena, *Kirk* (K, lectotype).
Olax stuhlmanni Engl. in Notizbl. Bot. Gart. Berl. **2**: 283 (1899).—Bake. f. in Journ. Linn. Soc., Bot. **40**: 43 (1911). Syntypes from Tanganyika (*Stuhlmann* 8562) and Mozambique: Lourenço Marques, *Schlechter* 11620, of which the former is the lectotype.

Shrub or small tree up to 6–12 m. tall; crown rounded, low, spreading; twigs pendulous, flexuous; bark smooth, light grey. Leaf-lamina dull deep green, 1·5–6 × 0·5–2·5 cm., elliptic-lanceolate to ovate, obtuse or acuminate and usually mucronulate at the apex, cuneate or rounded at the base, decurrent into the petiole; midrib impressed above, somewhat prominent below; venation inconspicuous, widely reticulate; petiole 3–8 mm. long. Flowers axillary, solitary, or in short axillary racemes, sweet-scented; buds broadly clavate, 5–7 mm. long. Calyx membranous, truncate, with margin somewhat undulate. Petals 5, white, 6–8 × 0·8–1·2 mm., valvate, shortly cucullate at the apex. Fertile stamens 3–4; filaments adnate to the petals; anthers 1·2–1·6 × 0·6–0·8 mm.; staminodes 5; filaments similar to those of the fertile stamens; anthers 1·5–2·5 mm. long, divaricate; ovary c. 1·2 × 1·4 mm.; style up to 4 mm. long, cylindric; stigma capitate, 3–4-lobed. Fruit 7–9 mm. long, ellipsoid or obovoid, drupaceous, red, crowned with the persistent base of the style and enveloped nearly to the apex by the accrescent calyx.

Bechuanaland Prot. N: Ngamiland, st. 19.vi.1937, *Erens* 256 (SRGH). **N. Rhodesia.** S: Zambezi bank, Katambora, fr. 5–11.xi.1949, *West* (SRGH); Lusito, fl. & fr. 25.ix.1959, *Fanshawe* 5215 (K); Livingstone, fl. 18.xii, *Rogers* 7462 (BM; K; SRGH); Victoria Falls, fr. 20.xi.1949, *Wild* 3114 (K; SRGH). **S. Rhodesia.** N: Mtoko, Musicheri R., Fungwi Reserve, fl. 11.x.1955, *Lovemore* 444 (K; LISC; SRGH). W: Matopo Hills, fl. x.1905, *Gibbs* 263 (BM; K). C: Selukwe, fl. xi.1948, *Hodgson* 427/48 (SRGH). E: Umtali, fr. 6.xi.1949, *Chase* 1815 (COI; K; LISC; SRGH). S: Gwanda, fl. xi.1956, *Davies* 2159 (K; LISC; SRGH). **Nyasaland.** C: Salima Bay, fl. 22.ix.1935, *Galpin* 15020 (K); Dowa, Ngora Hill, fl. 21.x.1954, *Adlard* 188 (COI; K). S: Mlanje, between Nampende and Nandiwo, st. 17.vii.1958, *Chapman* W/615 (SRGH). **Mozambique.** N: Cabo Delgado, Quissanga, Maate, fl. 11.ix.1948, *Barbosa* 2036 (EA; LISC; LMJ). Z: Alto Molócuè, between Mugema and R. Ligonha, fl. 14.x.1949, *Barbosa & Carvalho* in *Barbosa* 4426 (K; LMJ; SRGH). T: between Chicoa and Chioco, fl. 25.ix.1942, *Mendonça* 427 (BR; COI; K; LISC; LM; SRGH). MS: Sena, fl. x.1858, *Kirk* (K); Manica, Chimanimani, fl. 15.x.1950, *Chase* 2919 (BM; COI; LISC). SS: Guijá, Massingir, Algés, fl. & fr. 19.xi.1957, *Barbosa* 8192 (COI; K; LISC; LMJ). LM: Maputo, Santaca, fl. 27.ix.1948, *Gomes e Sousa* 3851 (COI; K; LISC; PRE).
Also in Tanganyika and the Transvaal. Deciduous woodland.

2. **Olax obtusifolia** De Wild., Ann. Mus. Cong. Belg., Bot. Sér. 4, **1**: 177, t. 40, fig. 6–11 (1903); Contr. Fl. Katanga: 54 (1921).—Th. & H. Dur., Syll. Fl. Cong.: 97 (1909).—R.E.Fr., Wiss. Ergebn. Schwed. Rhod.-Kongo-Exped. **1**: 24 (1914).—Louis & Léonard, F.C.B. **1**: 261 (1948).—White, F.F.N.R.: 42 (1962). Type from the Congo.

Shrub or small tree up to 6 m. tall. Young shoots rust colour, rugulose; bark later exfoliating. Leaves very shortly stalked; lamina 4–8 × 2–4 cm., ovate-lanceolate, blunt or subacute at the apex, rounded or obtuse at the base, dull, light green, chartaceous; midrib somewhat conspicuous above, prominent below; lateral nerves 6–10 pairs, very slightly raised on both surfaces; petiole up to 1 mm.

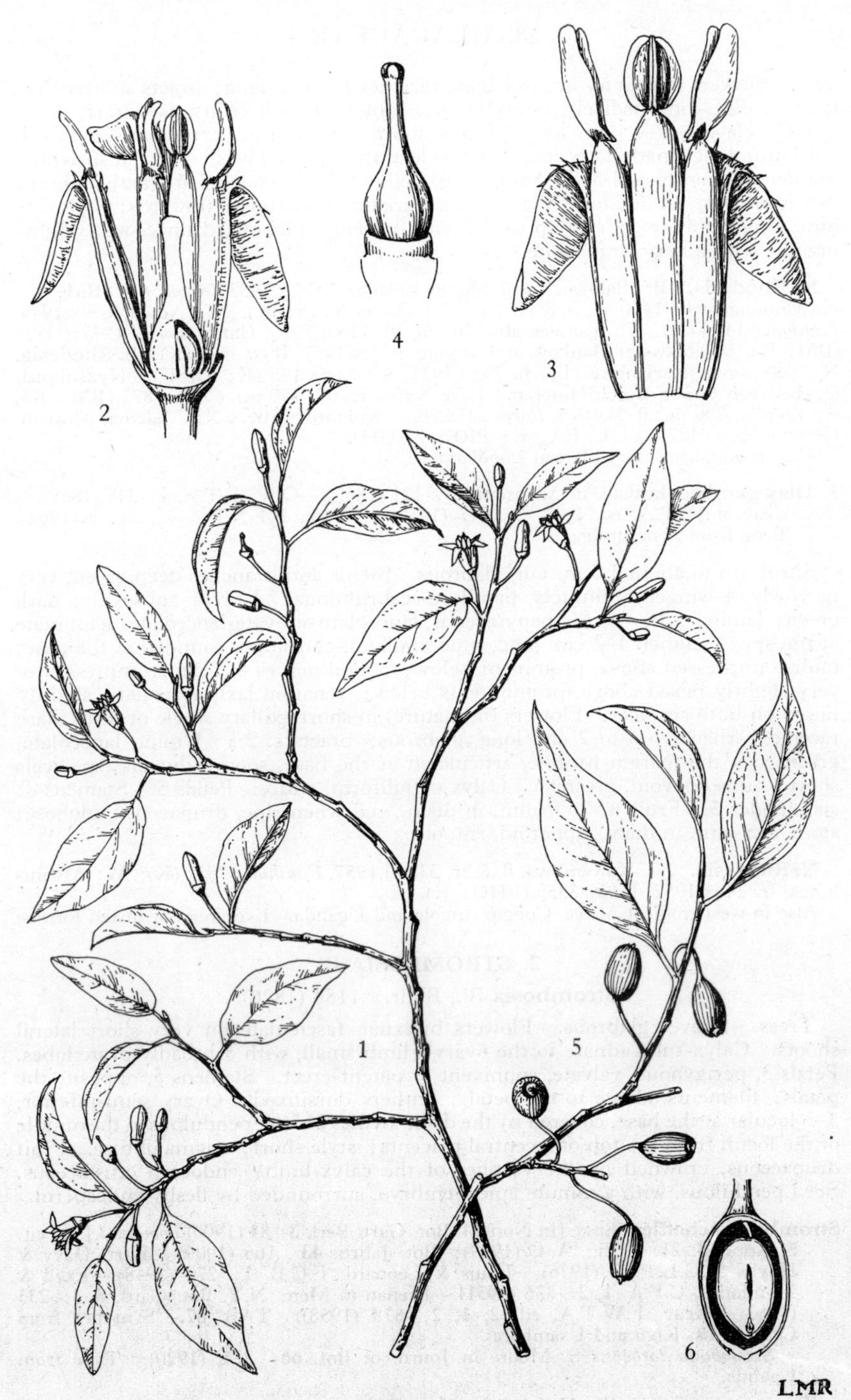

Tab. 66. OLAX DISSITIFLORA. 1, flowering branch ($\times \frac{2}{3}$) *Mendonça* 427; 2, vertical section of flower ($\times 6$) *Mendonça* 427; 3, pair of petals with stamen and staminodes ($\times 8$) *Mendonça* 427; 4, ovary and disk ($\times 6$) *Mendonça* 427; 5, fruiting branch ($\times \frac{2}{3}$) *Barbosa & Lemos* 8168; 6, longitudinal section of fruit ($\times 2$) *Barbosa & Lemos* 8168.

long. Flowers in solitary or fasciculate racemes 1–3 cm. long; bracts shorter than the 1–2 mm. long pedicels, rarely longer, sometimes foliaceous, caducous; buds broadly clavate, 6–9 mm. long. Calyx short, cupuliform, accrescent. Petals 3, 7–10 mm. long, spathulate, apiculate inside at the top, yellowish or greenish-white. Stamens 3; anthers 1·5–2·5 mm. long, yellow; staminodes 5–6; anther-thecae 2–3 mm. long, undulate, white. Ovary gradually attenuate upwards; style 2–4 mm. long, capitate. Fruit up to 2·5 cm. in diam., globose, drupaceous, yellow-orange, with a very large stone.

N. Rhodesia. B: Balovale, near Shingi Village, fl. 14.x.1952, *Angus* 630 (BM; K). N: Chilongowelo, 1450 m., fl. & fr. 4.xi.1954, *Richards* 2152 (K). W: Ndola, fl. 9.x.1949, *Fanshawe* 1416 (K). C: Lusaka, alt. 1200 m., fl. 12.vii.1930, *Hutchinson & Gillett* 3594 (BM; K). S: between Mankoya and Mongu, fl. 1.x.1957, *West* 3472 (K). **S. Rhodesia.** N: Sebungwe, Kariangwe Hill, fl. 20.xi.1951, *Lovemore* 198 (K; LISC). **Nyasaland.** C: between Grand Beach Hotel and Lake Nyasa Hotel, 450 m., *Chase* 3873 (BM; K); S: Zomba, 900 m., fl. xi.1915, *Purves* 258 (K). **Mozambique.** N: Malema, Mutuáli, *Gomes e Sousa* 4275 (COI; EA; K; PRE; SRGH).

Also in Katanga. Dry plateau woodland.

3. **Olax gambecola** Baill. in Adansonia, **3**: 121 (1863).—Oliv., F.T.A. **1**: 348 (1868).—Keay, F.W.T.A. ed. 2, **1**, 2: 647 (1958).—White, F.F.N.R.: 42, fig. 9 (1962). Type from Senegambia.

Shrub up to about 1·5 m. tall, glabrous. Stems and branches deep green, very narrowly 4-winged, minutely papillose-scabridulous. Leaves subsessile, dark green; lamina c. 10 × 4 cm., papyraceous, lanceolate or ovate-lanceolate, acuminate at the apex (acumen 1–2 cm. long, mucronulate), cuneate to rounded at the base; midrib impressed above, prominent below; lateral nerves 4–7 pairs, impressed or very slightly raised above, prominulous below; venation laxly reticulate, slightly raised on both surfaces. Flowers (immature) in short axillary single or fasciculate racemes; rhachis up to 2 cm. long, glabrous; bracts c. 2·5 × 1 mm., lanceolate, green with the margin hyaline, articulated at the base, soon caducous; pedicels short; buds obovoid, greenish. Calyx cupuliform, entire. Petals 5. Stamens 3; staminodes 5. Fruit up to 9 mm. in diam., red when ripe, drupaceous, globose; stone c. 5 mm. in diam., spheroid, smooth.

N. Rhodesia. N: Kawambwa, fl. & fr. 23.viii.1957, *Fanshawe* 3537 (K); W: Mwinilunga, fr. 24.ix.1952, *White* 3356 (FHO; K).

Also in west tropical Africa, Congo, Angola and Uganda. Evergreen or mixed forests.

3. STROMBOSIA Bl.

Strombosia Bl., Bijdr.: 1154 (1826).

Trees. Leaves glabrous. Flowers bisexual, fasciculate on very short lateral shoots. Calyx-tube adnate to the ovary; limb small, with 5 broadly ovate lobes. Petals 5, perigynous, valvate, connivent or patent-erect. Stamens 5, opposite the petals; filaments adnate to the petals; anthers dorsifixed. Ovary semi-inferior, 3–5-locular at the base, covered by the disk; ovules 3–5(6), pendulous in the middle of the loculi from the top of a central placenta; style short; stigma obtuse. Fruit drupaceous, crowned by the vestiges of the calyx-limb; endocarp crustaceous. Seed pendulous, with a minute apical embryo, surrounded by fleshy endosperm.

Strombosia scheffleri Engl. [in Notizbl. Bot. Gart. Berl. **3**: 84 (1900) *nom. nud.*] op. cit. **5,** Append. 21: 4, fig. A–C (1909); Bot. Jahrb. **43**: 166 (1909).—Burtt Davy & Hoyle, N.C.L.: 57 (1936).—Louis & Léonard, F.C.B. **1**: 270 (1948).—Exell & Mendonça, C.F.A. **1,** 2: 336 (1951).—Brenan in Mem. N.Y. Bot. Gard. **8,** 3: 235 (1953).—Keay, F.W.T.A. ed. 2, **1**, 2: 678 (1958). TAB. **67.** Syntypes from Cameroons, Kivu and Usambara.

Strombosia toroensis S. Moore in Journ. of Bot. **66**: 224, (1920). Type from Uganda.

Tree up to 30 m. tall. Young branches flexuous, green, compressed, later pale brown, cylindric. Leaf-lamina 7–20 × 4–12 cm., subcoriaceous, oblong or ovate-oblong, subacute at the apex, obtuse at the base, glabrous, light green, somewhat shining above; lateral nerves in 4–8 pairs, prominulous below, arcuate-ascendent; veins loosely reticulate, slightly raised below; petiole 1–3 cm. long, canaliculate. Inflorescence a short cyme or fascicle, axillary or extra-axillary, multiflorous.

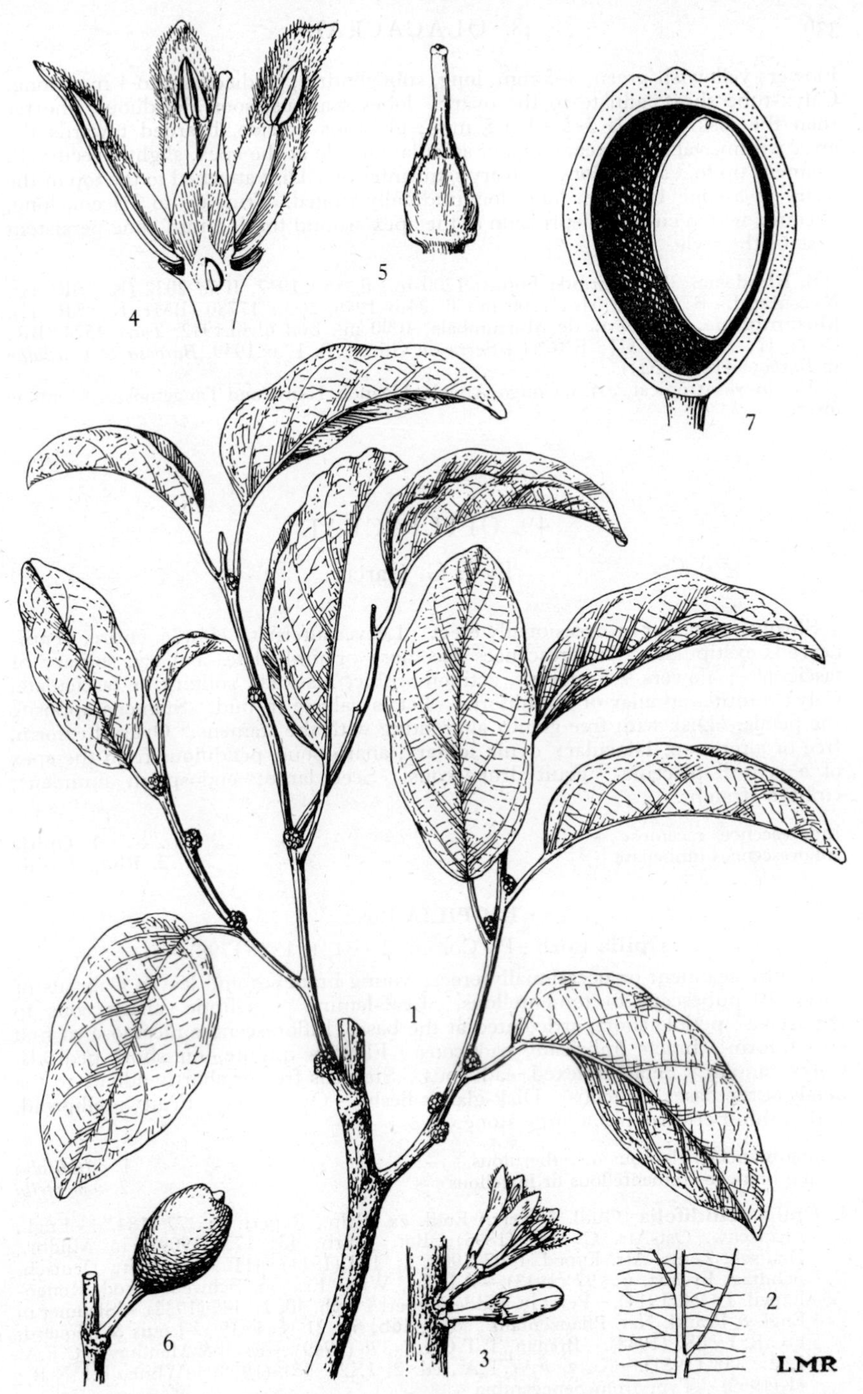

Tab. 67. STROMBOSIA SCHEFFLERI. 1, branch with young inflorescences (× ⅔) *Torre* 4524; 2, lower surface of leaf showing venation (× 2) *Fries* 1718; 3, flower and bud (× 2) *Maitland* 431; 4, vertical section of flower (× 6) *Greenway & Eggeling* 7057; 5, ovary with surrounding disk (× 6) *Greenway & Eggeling* 7057; 6, fruit (× ⅔) *Gardner* 1330; 7, longitudinal section of fruit (× 2) *Gardner* 1330.

Flowers yellowish-green, 3–5 mm. long, subcylindric; pedicels up to 4 mm. long. Calyx-tube short, adnate to the ovary; lobes scarious, soon deciduous, shorter than the disk. Petals 3–5 × 1–1·5 mm., glabrous outside, bearded towards the apex within, valvate, linear-oblong, apiculate inside at the apex, slightly recurved. Stamens up to 3·5 mm. long. Ovary semi-inferior. Disk attached to the top of the ovary in a cone 1–2 mm. long, longitudinally ridged. Fruit up to 2·5 cm. long, obconic, with a circular depression at the apex around the 1–2 mm. long persistent base of the style.

S. Rhodesia. E: Chirinda Forest, 1200 m., fl. 18.x.1947, *Wild* 2032 (K; SRGH). **Nyasaland.** S: Cholo Mt., 1200 m., fl. 24.iv.1946, *Brass* 17780 (BM; K; SRGH). **Mozambique.** Z: Serra de Morrumbala, 1000 m., bud 6.viii.1942, *Torre* 4524 (BR; COI; K; LISC; LMJ; SRGH); Serra de Gúruè, fl. 17.ix.1949, *Barbosa & Carvalho* in *Barbosa* 4120 (LMJ).

Also in west tropical Africa, Congo, Angola, Uganda, Kenya and Tanganyika. Montane forest.

49. OPILIACEAE

By J. G. Garcia

Shrubs, scandent or occasionally erect. Leaves alternate, simple, entire, penninerved, exstipulate. Inflorescences racemose or umbellate, axillary, solitary or fasciculate; flowers 4–5-merous, actinomorphic, bisexual, solitary or fasciculate. Calyx minute, annular or cupuliform. Petals valvate in bud. Stamens opposite the petals. Disk with free glands alternating with the stamens. Ovary fusiform, free or almost so, 1-locular; ovule solitary, anatropous, pendulous from the apex of a central placenta. Fruit drupaceous. Seed large; endosperm abundant; embryo straight.

Inflorescences racemose - - - - - - - - - - **1. Opilia**
Inflorescences umbellate - - - - - - - - - **2. Rhopalopilia**

1. OPILIA Roxb.

Opilia Roxb., Pl. Corom. **2**: 31, t. 158 (1798).

Shrubs, scandent or occasionally erect; young branches and leaves glabrous or variously pubescent to tomentellous. Leaf-lamina ± elliptic, chartaceous to coriaceous; petiole short, articulated at the base. Inflorescences racemose, at first strobiliform; bracts imbricate, caducous. Flowers minute, on short pedicels. Calyx annular. Petals reflexed, caducous. Stamens free or shortly adnate to the petals at the base, caducous. Disk-glands fleshy. Ovary free. Fruit ± ellipsoid, with a thin pericarp and a large stone.

Young branches glabrous or puberulous - - - - - - 1. *celtidifolia*
Young branches tomentellous or hirtellous - - - - - - 2. *tomentella*

1. **Opilia celtidifolia** (Guill. & Perr.) Endl. ex Walp., Repert. **1**: 377 (1842).—Engl., Pflanzenw. Ost-Afr. **C**: 168 (1895); Bot. Jahrb. **43**: 173 (1909); in Mildbr., Deutsch. Zentr.-Afr.-Exped. 1907–1908, **2**: 193 (1911).—H. Mey., Mitt. Deutsch. Schutzg. Ergänz. **6**: 97 (1913).—R.E.Fr., Wiss. Ergebn. Schwed. Rhod.-Kongo-Exped. **1**: 24 (1914).—Peter in Fedde, Repert., Beih. **40,** 2: 147 (1932).—Sleumer in Engl. & Prantl, Nat. Pflanzenfam., ed. 2, **16b,** fig. 21 K (1935).—Louis & Léonard, F.C.B. **1**: 282 (1948).—Brenan, T.T.C.L.: 396 (1949).—Exell & Mendonça, C.F.A. **1,** 2: 338 (1951).—Keay, F.W.T.A., ed. 2, **1,** 2: 651 (1958).—White, F.F.N.R.: 41 (1962). Type from Senegambia.

Groutia celtidifolia Guill. & Perr. in Guill., Perr. & A. Rich., Fl. Senegamb. Tent. **1**: 101, t. 22 (1831). Type as above.

Opilia amentacea sensu Oliv., F.T.A. **1**: 352 (1868).—Bak. f. in Journ. Linn. Soc., Bot. **40**: 43 (1911).

Shrub, scandent or occasionally erect; young branches glabrous or puberulous, longitudinally furrowed. Leaf-lamina 5–10 × 2–4 cm., lanceolate to elliptic, acute

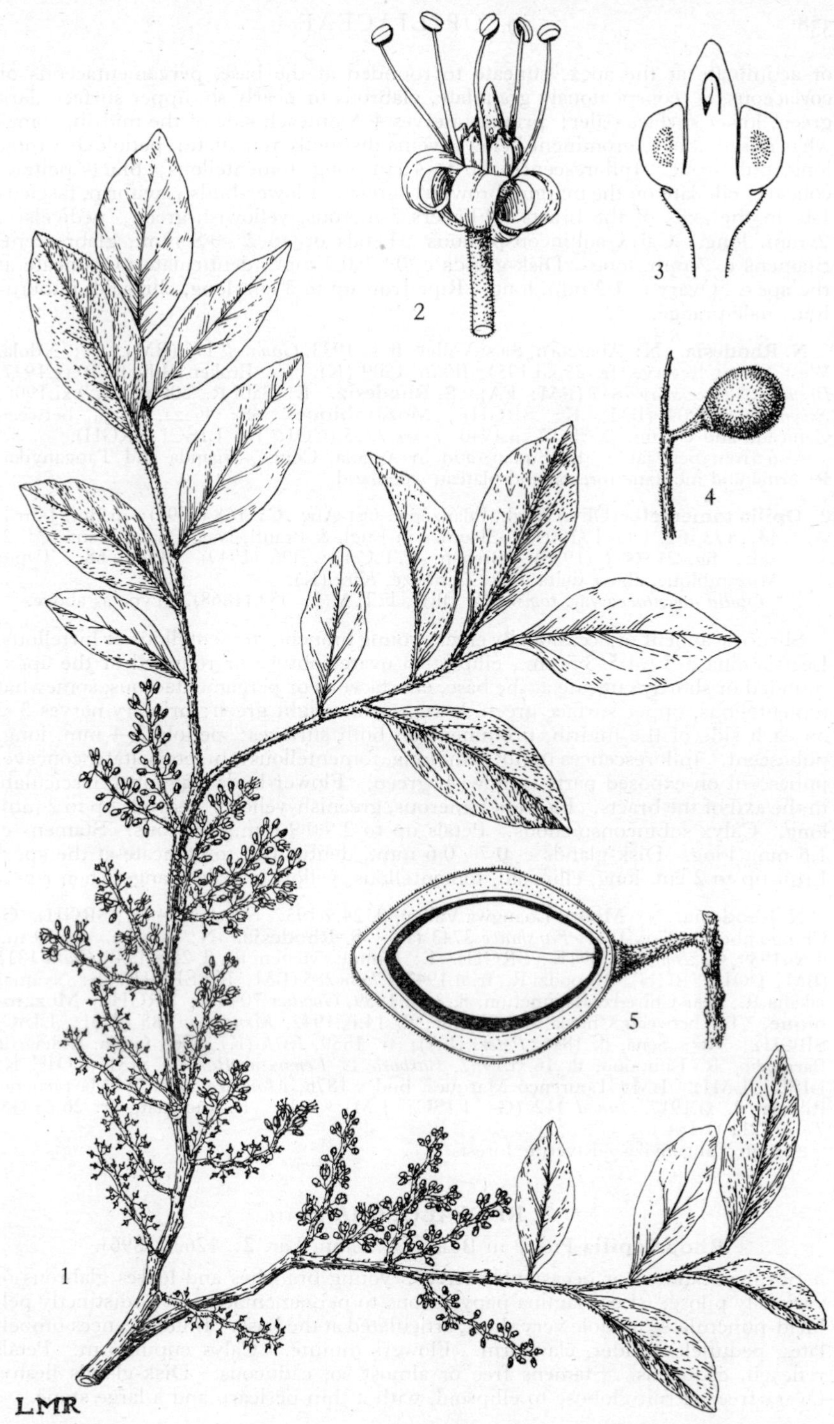

Tab. 68. OPILIA TOMENTELLA. 1, flowering branch (×⅔) *Goldsmith* 53/59; 2, flower (×10) *Wild* 5283; 3, vertical section of disk and ovary (×16) *Wild* 5283; 4, fruit (×⅔) *Barbosa & Lemos* 8397; 5, longitudinal section of fruit (×2) *Barbosa & Lemos* 8397.

or acuminate at the apex, cuneate to rounded at the base, pergamentaceous or coriaceous, ± conspicuously granulate, glabrous or nearly so, upper surface dark green, lower surface paler; primary nerves 4–8 on each side of the midrib, somewhat raised above, prominent below; veins distinctly reticulate; petiole 3–8 mm. long, pubescent. Inflorescences up to 4 cm. long, tomentellous; bracts peltate, concave, ciliolate on the margin, brownish-green. Flower-buds claviform, fasciculate in the axils of the bracts. Flowers 5-merous, yellowish-green; pedicels c. 2 mm. long. Calyx subinconspicuous. Petals up to 2 × 0·8 mm., glabrescent. Stamens c. 2 mm. long. Disk-glands c. 0·8 × 0·7 mm., denticulate to truncate at the apex. Ovary c. 1·2 mm. long. Ripe fruit up to 3 cm. long, ellipsoid, puberulous, pale orange.

N. Rhodesia. N: Abercorn, Saisi Valley, fl. ix.1933, *Gamwell* 173 (BM). W: Ndola, West Forest Reserve, fr. 25.xii.1951, *White* 1809 (K). C: Broken Hill, fl. 10.ix.1947, *Brenan & Greenway* 7847 (BM; EA). **S. Rhodesia.** E: Sabi R., 300 m., fl. 9.xi.1906, *Swynnerton* 1204 (BM; K; SRGH). **Mozambique.** SS: Gaza, Guijá, between Caniçado and Chamusca, fl. 12.xii.1940, *Torre* 2365 (COI; K; LISC; SRGH).

Also from Senegal to the Sudan, and in Angola, Congo, Uganda and Tanganyika. Riverine and montane forests, and plateau woodland.

2. **Opilia tomentella** (Oliv.) Engl., Pflanzenw. Ost-Afr. **C**: 168 (1895); in Bot. Jahrb. **43**: 173, fig. 1 G–J (1909).—Sleumer in Engl. & Prantl, Nat. Pflanzenfam., ed. 2, **16b**: fig. 21 G–J (1935).—Brenan, T.T.C.L.: 396 (1949). TAB. **68**. Type: Mozambique, above delta of R. Zambeze, *Kirk* (K).

Opilia amentacea var. *tomentella* Oliv., F.T.A. **1**: 352 (1868). Type as above.

Shrub, scandent or occasionally erect; young branches tomentellous or hirtellous. Leaf-lamina 3–7 × 1·5–3·5 cm., elliptic to ovate, obtuse or rounded at the apex, rounded or shortly cuneate at the base, chartaceous or pergamentaceous, somewhat tomentellous, upper surface green, lower surface light green; primary nerves 3–5 on each side of the midrib, prominent on both surfaces; petiole 2–4 mm. long, pubescent. Inflorescences up to 3 cm. long, tomentellous; bracts peltate, concave, pubescent on exposed parts, yellowish-green. Flower-buds claviform, fasciculate in the axil of the bracts. Flowers 5-merous, greenish-yellow; pedicels up to 2 mm. long. Calyx subinconspicuous. Petals up to 2 × 0·9 mm., glabrous. Stamens c. 1.6 mm. long. Disk-glands c. 0·7 × 0·6 mm., denticulate to truncate at the apex. Fruit up to 2 cm. long, ellipsoid, tomentellous, yellow or pale orange when ripe.

N. Rhodesia. N: Mpika, Luangwa Valley, fl. 24.x.1957, *Savory* 234 (K; SRGH). C: Chingombe, fl. 27.ix.1957, *Fanshawe* 3743 (K). **S. Rhodesia.** N: Kariba, ± 750 m., fl. xi.1959, *Goldsmith* 53/59 (K; SRGH). E: Umtali, Meneni R., fl. 25.x.1948, *Chase* 1315 (BM; COI; SRGH); Impodzi R., fr. ii.1947, *Chase* 285 (BM; K; SRGH). S: Nyamasakana R., near Chiredzi R. junction, fl. 3.xii.1959, *Goodier* 709 (K; SRGH). **Mozambique.** T: between Chicoa and Fíngoè, fl. 14.ix.1942, *Mendonça* 385 (COI; LISC; SRGH). MS: Sena, fl. 1846, *Peters* (EA); fr. 1859, *Kirk* (K). SS: Guijá, Aldeia da Barragem, R. Limpopo, fl. 16.xi.1957, *Barbosa & Lemos* in *Barbosa* 8151 (COI; K; LISC; LMJ). LM: Lourenço Marques, bud x.1876, *Monteiro* 57 (K); Marracuene, Rikatla, fl. xi.1917, *Junod* 112 (G; LISC; LM); Sabié, R. Incomáti, fr. 26.i.1948, *Torre* 7211 (LISC).

Also in Tanganyika. Riverine forests.

2. RHOPALOPILIA Pierre

Rhopalopilia Pierre in Bull. Soc. Linn. Par. **2**: 1263 (1896).

Shrubs scandent or occasionally erect; young branches and leaves glabrous or variously pilose. Leaf-lamina papyraceous to pergamentaceous, ± distinctly pellucid-punctulate; petiole very short, articulated at the base. Inflorescences umbellate; peduncle slender, claviform. Flowers minute. Calyx cupuliform. Petals reflexed, caducous. Stamens free or almost so, caducous. Disk-glands fleshy. Ovary free. Fruit globose to ellipsoid, with a thin pericarp and a large stone.

Young branches glabrous - - - - - - - - - - - 1. *umbellulata*
Young branches fulvous-pilose - - - - - - - - - - 2. *marquesii*

1. **Rhopalopilia umbellulata** (Baill.) Engl., Bot. Jahrb. **43**: 175 (1909).—Brenan, T.T.C.L.: 397 (1949). TAB. **69**. Type from Zanzibar.

Opilia umbellulata Baill. in Adansonia, **8**: 199 (1867). Type as above.

Tab. 69. RHOPALOPILIA UMBELLULATA. 1, flowering branch (× ⅔) *Andrada* 1353; 2, flower (× 10) *Andrada* 1353; 3, flower with petals and stamens removed (× 20) *Andrada* 1353; 4, vertical section of ovary (× 40) *Andrada* 1353; 5, longitudinal section of fruit (× 2) *Faulkner* 681; 6, fruit (× ⅔) *Faulkner* 681.

Shrub, scandent or occasionally erect; young branches glabrous. Leaves subsessile; lamina 4–8 × 1·5–3·5 cm., lanceolate or oblong-lanceolate, cuneate or obtuse at the base, papyraceous or chartaceous, glabrous, upper surface green, lower surface paler; primary nerves 5–7 on each side of the midrib, slightly prominent on both surfaces; petiole up to 3 mm. long, glabrous. Inflorescences usually geminate, but sometimes solitary or fasciculate, shortly pubescent; peduncle up to 12·5 mm. long, with the thickened apex bracteate. Flowers 5-merous, greenish-yellow; pedicels up to 5 mm. long, capillaceous. Calyx minute. Petals c. 1·6 × 0·8 mm., puberulous outside, glabrous within. Stamens c. 1·4 mm. long. Disk-glands c. 0·6 × 0·5 mm., denticulate ot subtruncate at the apex. Ovary c. 1 mm. long. Fruit up to 10 mm. in diam. when ripe, subglobose, glabrous, yellowish.

Mozambique. N: Cabo Delgado, R. Msálu, fl. ix.1911, *Allen* 9 (K); Palma, fl. 16.ix.1948, *Andrada* 1353 (COI; LMJ). Z: Pebane, Mualama, fl. 7.x.1949, *Barbosa & Carvalho* in *Barbosa* 4332 (EA; LMJ).

Also in Kenya and Tanganyika. Low-altitude open woodland.

2. **Rhopalopilia marquesii** (Engl.) Engl., Bot. Jahrb. **43**: 176, fig. 1 B–C (1909); Pflanzenw. Afr. **3,** 1: 75, fig. 39 B–C (1915).—Sleumer in Engl. & Prantl, Nat. Pflanzenfam., ed. 2, **16b**; fig. 21 B–C (1935).—Louis & Léonard, F.C.B. **1**: 837 (1948).—Exell & Mendonça, C.F.A. **1,** 2: 339 (1951).—White, F.F.N.R.: 41 (1962). Type from Angola.

Opilia umbellulata var. *marquesii* Engl. in Notizbl. Bot. Gart. Berl. **2**: 282 (1899). Type as above.

Shrub, scandent or occasionally erect, with the young branches slender, ± densely fulvous-pilose. Leaves subsessile; lamina 2–8 × 1–3·5 cm., lanceolate or ovate-lanceolate, acute or acuminate at the apex, rounded or subcordate at the base, chartaceous or pergamentaceous, glabrescent except for the midrib ± densely pubescent above, pilose especially on the midrib and main nerves below, upper surface pale green, lower one paler; primary nerves 3–7 on each side of the midrib, slightly raised on both surfaces; petiole up to 2 mm. long, densely tawny-pilose. Inflorescences solitary, puberulous; peduncle up to 14 mm. long, with the thickened apex bracteate. Flowers 5-merous, greenish-yellow; pedicels up to 3·5 mm. long. Calyx minute. Petals c. 1·5 × 0·7 mm., puberulous outside, glabrous within. Stamens c. 1·5 mm. long. Disk-glands up to 0·7 × 0·6 mm., denticulate or subtruncate at the apex. Ovary up to 1·5 mm. long. Fruit up to 11 mm. long when ripe, ellipsoid, longitudinally sulcate, puberulous, pale orange.

N. Rhodesia. W: Mufulira, fl. 13.viii.1954, *Fanshawe* F. 1461 (K; SRGH); Mwinilunga, Zambezi R., fr. 21.ix.1952, *White* 3314 (BM; K). Type from Angola.

Also in Angola and Congo. Riverine forests.

50. ICACINACEAE

By E. J. Mendes

Trees or shrubs, sometimes climbing, or lianes. Leaves simple, alternate or opposite, penninerved or rarely palmatinerved; stipules absent (in our area). Inflorescences axillary, supraxillary or terminal. Flowers bisexual or unisexual by abortion, (4)5-merous, actinomorphic. Calyx small or absent. Petals usually valvate, free or ± united, or united into a long tube. Stamens free or adnate to the petals; anthers 2-thecous, opening lengthwise. Ovary superior, 1-locular (in our area); ovules pendulous from near the apex, usually 2. Style simple or stigma sessile. Fruit drupaceous, 1-seeded. Seed usually with copious endosperm; embryo usually small, ± straight.

Tropics and subtropics.

Leaves opposite; inflorescence repeatedly forked - - - - - - **1. Cassinopsis**
Leaves alternate:
Calyx and corolla present; flowers bisexual; style filiform:
Petals united nearly to the apex; filaments adnate to the petals; leaves drying green **2. Leptaulus**

Petals free or united only at the base; filaments free; leaves drying blackish:
Flowers in terminal panicles; filaments subulate; fruit with a lateral fleshy appendage - - - - - - - - - - - **3. Apodytes**
Flowers in supraxillary fascicles; filaments broadened below; fruit with no lateral appendage - - - - - - - - - **4. Rhaphiostylis**
Calyx absent; flowers unisexual; stigma sessile; endocarp spiny on the inner face; stems twining - - - - - - - - - **5. Pyrenacantha**

1. CASSINOPSIS Sond.

Cassinopsis Sond. in Harv. & Sond., F.C. **1**: 473 (1860).

Tall shrubs or small trees. Leaves opposite, petiolate. Inflorescences cymose, alternating from node to node. Flowers bisexual, inconspicuous. Calyx 5-fid. Petals slightly joined at the base, imbricate. Stamens 5, inserted on the base of the corolla, alternating with the petals. Ovary sessile, 1–2 ovuled; style short. Fruit a nearly dry drupe.

A south and tropical African and Madagascan genus of 4 species.

Leaves entire, margin recurved, apex obtuse; inflorescences axillary, more than half as long as the subtending leaf; drupes terete, elongated; unarmed shrub or small tree 1. *tinifolia*
Leaves ± serrulate, rarely entire, apex acute; inflorescences subinterpetiolar, scarcely exceeding the petiole; drupes ovoid, subcompressed; spiny shrub - 2. *ilicifolia*

1. **Cassinopsis tinifolia** Harv., Thes. Cap. **2**: 44, t. 168 (1863).—Engl., Pflanzenw. Afr. **3,** 2: 253 (1921).—Sleumer in Engl. & Prantl, Nat. Pflanzenfam. ed. 2, **20b**: 348 (1942).—Wild in Proc. & Trans. Rhod. Sci. Ass. **43**: 57 (1951).—Goodier & Phipps in Kirkia, **1**: 58 (1961). Syntypes from S. Africa (Natal and Zululand).

Straggling unarmed shrub or small tree, 4 m. in height; bark smooth, grey. Leaf-lamina 3–9 × 2–5 cm., apex bluntly acuminate, margins entire and recurved, broadly cuneate at base, dark green, glossy above. Flowers in much-forked cymes, often 3 cm. across, on axillary peduncles c. 3 cm. long. Calyx puberulous. Petals white. Fruit an elongated longitudinally corrugated drupe, c. 8 × 4·5 mm., black at maturity.

S. Rhodesia. E: Umtali Distr., Stapleford, on hillock N. of Ruperi Peak, Nyamkwarara Valley, 1675 m., fl. 28.ix.1952, *Chase* 4636 (BM; COI; K; LISC; SRGH). **Mozambique.** MS: Báruè, Vila Gouveia, Serra de Choa, fl. 7.ix.1943, *Torre* 5853 (K; LISC; LM).

Also known from Zululand, Natal and Pondoland. Margin of mountain forest, wooded gullies and riverine vegetation.

2. **Cassinopsis ilicifolia** (Hochst.) Kuntze, Rev. Gen. Pl. **3,** 2: 36 (1898).—Sleumer in Notizbl. Bot. Gart. Berl. **15**: 228 (1940); in Engl. & Prantl, Nat. Pflanzenfam. ed. 2, **20b**: 348 (1942). TAB. **70.** Type from S. Africa (Cape).
Hartogia ilicifolia Hochst. in Flora, **27**: 305 (1844). Type as above.
Cassine ilicifolia Hochst., loc. cit., *in synon.* Type as above.
Cassinopsis capensis Sond. in Harv. & Sond., F.C. **1**: 474 (1860).—Engl., Pflanzenw. Afr. **3,** 2: 252 (1921).—Burtt Davy, F.P.F.T. **2**: 451 (1932). Syntypes from S. Africa (Orange Free State and Cape).

A glabrous spiny scrambling shrub, 5 m. high; branches alternate, terete, more or less flexuous, shining, green; spines straight, very acute, c. 2·5 cm. long, subinterpetiolar, in age apparently at the forks of the older branches. Leaf-lamina 4–6 × 2–2·5 cm., shining above, narrowly ovate, subacuminate and acute at the apex, margin spinulose-serrulate or subentire, broadly cuneate at the base (smaller on the flowering branches); petiole 0·5–0·8 cm. long. Flowers white, in subinterpetiolar dichotomous cymes. Petals 3 times as long as the pubescent calyx-lobes. Fruit yellow-orange at maturity, c. 1 cm. long, slightly compressed, tipped by the persistent style.

S. Rhodesia. E: Inyanga Distr., Nyamziwa Falls, 1645 m., fr. 14.i.1951, *Chase* 3658 (BM; BR; COI; LISC; SRGH); Umtali, N.E. of Inodzi homestead, 1220 m., fl. 27.x.1957, *Chase* 6750 (COI; K; SRGH).

Also known from the Transvaal and Natal southwards to the Cape. In wooded kloofs and stream banks.

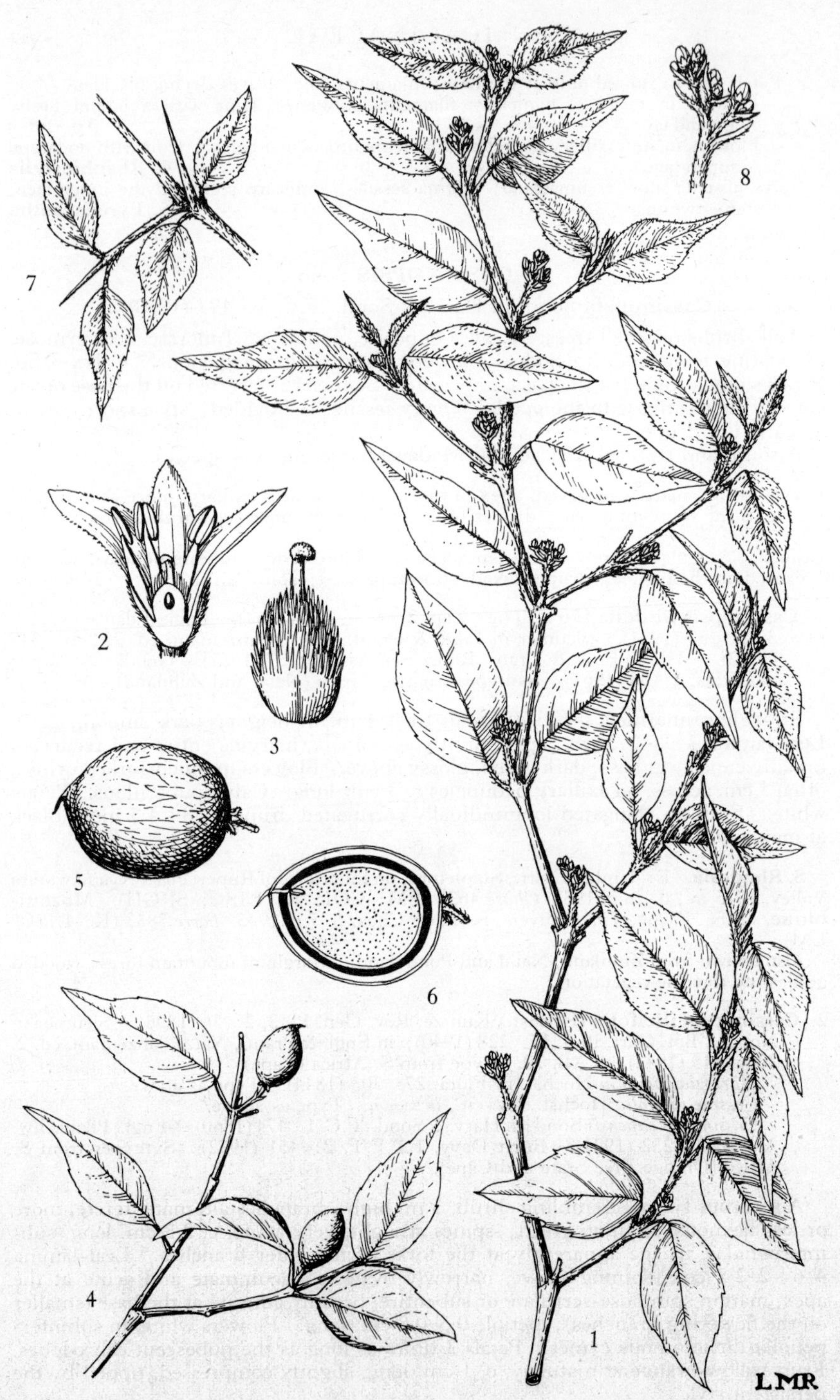

Tab. 70. CASSINOPSIS ILICIFOLIA. 1, branch with flowers ($\times \frac{2}{3}$) *Chase* 6750; 2, vertical section of flower ($\times 6$) *Chase* 6750; 3, ovary ($\times 16$) *Chase* 6750; 4, branchlet with fruits ($\times \frac{2}{3}$) *Chase* 3658; 5, fruit ($\times 2$) *Chase* 3658; 6, vertical section of fruit ($\times 2$) *Chase* 3658; 7, branchlet with spines ($\times \frac{2}{3}$) *Cheadle* 598; 8, cyme ($\times 3$) *Chase* 6750.

2. LEPTAULUS Benth.

Leptaulus Benth. in Benth. & Hook., Gen. Pl. **1**: 351 (1862).

Shrubs or small trees. Leaves alternate, entire, rather coriaceous. Flowers bisexual, in subumbellate cymes or solitary. Calyx deeply 5-lobed; lobes slightly imbricate. Petals 5, united nearly to the apex, valvate with inflexed tips. Stamens 5, alternating with the petals and adnate to them for most of their length. Ovary with a somewhat excentric style; stigma funnel-shaped. Fruit an oblong-ellipsoid drupe.

A tropical African genus of c. 6 species.

Leptaulus zenkeri Engl., Bot. Jahrb. **43**: 179 (1909).—Sleumer in Engl. & Prantl, Nat. Pflanzenfam. ed. 2, **20b**: 358 (1942).—Exell & Mendonça, C.F.A. **1,** 2: 341 (1951). —Keay, F.W.T.A. ed. 2, **1,** 2: 637 (1958).—Boutique, F.C.B. **9**: 262, fig. 4B (1960).—White, F.F.N.R.: 221 (1962) (" Leptaulis "). TAB. **71.** Syntypes from the Cameroons.

Evergreen shrub or small tree, up to 6 m. high; bark pale grey-brown; twigs sparsely puberulous or glabrous. Leaves petiolate; lamina 3–12 × 1·2–5 cm., elliptic to narrowly elliptic, apex abruptly caudate-acuminate with a linear obtuse or slightly spathulate acumen $\frac{1}{3}$–$\frac{1}{6}$ the length of the lamina, broadly cuneate at base. Flowers white, drying chrome-yellow, (1)3–5 in axillary shortly pedunculate cymes. Calyx-lobes narrowly triangular, 1·5–2 mm. long. Petals united into a tube c. 5 mm. long; lobes triangular, c. 1·5 mm. long. Ovary 1–1·5 mm. long; style as long as the corolla tube, pubescent-hirsute throughout, persisting after anthesis. Drupe c. 1·5 × 0·6 cm., oblong-ellipsoid. Seed c. 1 × 0·4 cm., oblong-ellipsoid.

N. Rhodesia. N: Abercorn, Chinakila, young fr. 29.iii.1960, *Fanshawe* 5584 (K); W: Mwinilunga Distr., near Kalene Hill Mission, island in Zambezi R., fl. 21.ix.1952, *White* 3312 (BM; BR; COI; EA; K).

Also from Nigeria to the Congo and Angola (Cabinda); in forest-relics (muteshi), in understorey of dry evergreen forest and in evergreen riverine forest.

3. APODYTES E. Mey. ex Arn.

Apodytes E. Mey. ex Arn. in Hook., Lond. Journ. Bot. **3**: 155 (1840).

Trees or shrubs. Leaves alternate, entire, penninerved. Flowers in terminal panicles, bisexual, 5-merous. Calyx small, 5-fid. Petals free, valvate, linear, glabrous. Stamens alternating with the petals and shortly adnate to their bases; filaments subulate. Ovary 1-locular, with a lateral appendage; style excentric, more or less oblique; ovules 2, pendulous. Fruit a crustaceous drupe with a lateral fleshy appendage.

A genus of c. 15 species from tropical and subtropical Africa, Madagascar, Réunion, Mauritius, Indomalaysia (incl. Borneo), China (Hainan) and Australia (Queensland).

Apodytes dimidiata E. Mey. [ex Drège, Cat. Pl. Exs. Afr. Austr.: 19 (1838) *nom. nud.*— ex Bernh. in Linnaea, **12**: 136 (1838) *nom. nud.*] ex Arn. in Hook., Lond. Journ. Bot. **3**: 155 (1840).—Sond. in Harv. & Sond., F.C. **1**: 235 (1860).—Oliv., F.T.A. **1**: 355 (1868) pro parte excl. specim. *Schimper*.—Engl., Pflanzenw. Ost-Afr. **C**: 248 (1895).—Burkill, List Pl. Brit. Central Afr.: 239 (1897).—Schinz & Junod in Mém. Herb. Boiss. **10**: 32 (1900).—Sim, For. Fl. Port. E. Afr.: 31 (1909).—Bak. f., Journ. Linn. Soc., Bot. **40**: 43 (1911).—R.E.Fr., Schwed. Rhod.-Kongo-Exped. **1**: 130 (1914).—Eyles in Trans. Roy. Soc. S. Afr. **5**: 405 (1916).—Engl., Pflanzenw. Afr. **3,** 2: 256 (1921).—Burtt Davy, F.P.F.T. **2**: 451, fig. 71 (1932).—Burtt Davy & Hoyle, N.C.L.: 47 (1936).—Sleumer in Engl. & Prantl, Nat. Pflanzenfam. ed. 2, **20b**: 367, fig. 103 D–E (1942).—Hutch., Botanist in S. Afr.: 532 (1946).—Brenan, T.T.C.L.: 251 (1949).—Exell & Mendonça, C.F.A. **1,** 2: 343 (1951).—Williamson, Useful Pl. Nyasal.: 19 (1955).—Cufod. in Bull. Jard. Bot. Brux. **28,** Suppl.: 487 (1958).—Mogg. in Macnae & Kalk, Nat. Hist. Inhaca I., Moçamb.: 9 (1958).— Boutique, F.C.B. **9**: 274 (1960).—Goodier & Phipps in Kirkia, **1**: 58 (1961).— White, F.F.N.R.: 221 (1962). Type from S. Africa.

An evergreen tree up to 5–15 m. or a shrub; young branches purplish-green with appressed pale brown hairs; older twigs dark grey with pale lenticels; sap-wood very narrow; heart-wood extremely hard and pink or pale violet in colour; bark smooth, grey. Leaf-lamina 2–15 × 1·2–8 cm., ovate-elliptic or broadly elliptic,

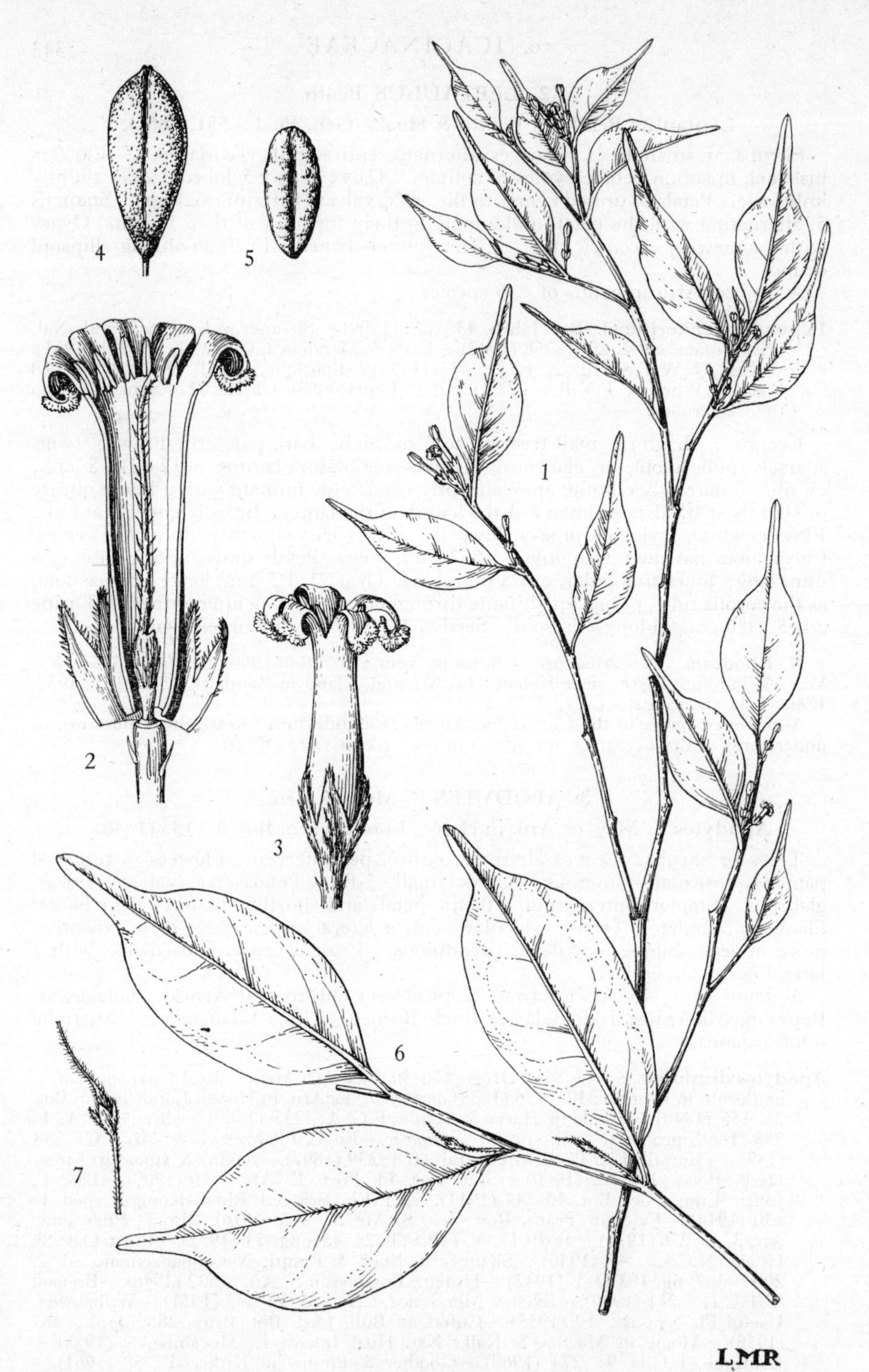

Tab. 71. LEPTAULUS ZENKERI. 1, branchlet with flowers (× $\frac{2}{3}$) *White* 3312; 2, opened flower (× 8) *White* 3312; 3, flower (× 6) *White* 3312; 4, fruit (× 2) *White* 3312; 5, seed (× 2) *White* 3312; 6, branchlet with young fruits (× $\frac{2}{3}$) *Fanshawe* 5584; 7, immature fruit (× 2) *Fanshawe* 5584.

slightly glossy or ± glaucous or pruinose, subcoriaceous, apex emarginate or obtuse or acute or shortly acuminate, margin entire or sometimes undulate and slightly recurved, midrib slightly impressed above and prominent beneath, venation obscure; petiole 1–3 cm. long, flattened and channelled above. Flowers 2 mm. long in bud, sessile or shortly pedicellate, with very minute or obsolete bracts and bracteoles, in terminal, loose or subthyrsiform many-flowered sweet-scented panicles. Calyx 0·3–0·5 mm. long, 5-fid; lobes deltate. Petals white, drying black, 5 × 1 mm, linear. Anthers linear, bilobed-sagittate at the base. Ovary ovoid. Fruit 0·6–0·7 × 0·5–0·7 × c. 0·3 cm., ovoid-reniform, subcompressed, black, with a persistent style and a fleshy lateral red (drying black) appendage.

From Angola to Ruanda-Urundi and Ethiopia and southwards to the Cape. Widespread from sea-level in coastal evergreen bush to fringing forest in *Brachystegia-Julbernardia* woodland and mountain grassland.

Subsp. **dimidiata.** TAB. **72.**

Ovary pubescent. Fruit and its appendage sparsely pubescent.

N. Rhodesia. N: Abercorn Distr., Lake Chila, fr. 15.xi.1952, *White* 3657 (BM; BR; COI; EA; K). W: Luanshya, seedlings 21.ix.1955, *Fanshawe* 2453 (K); Solwezi Distr., Solwezi, fl. 10.ix.1952, *Angus* 391 (BM; BR; K). C: Chisamba, fl. 10.ix.1958, *Brownjohn* in *Fanshawe* 4868 (K). S: Mazabuka Distr., Choma to Namwala, mile 30, fl. 22.vi.1952, *White* 2955 (BR; K). **S. Rhodesia.** N: Umvukwes Distr., Mazoe Ranch, 1525 m., fr. 16.xii.1952, *Wild* 3982 (K; LISC; SRGH). W: Bulawayo Distr., Matopos, Diana's Pool, fl. & fr. 8.iv.1950, *Orpen* 45/50 (SRGH). C: Makoni Distr., 37 km. E. of Rusape, fl. xi.1957, *Miller* 4750 (COI; SRGH). E: Umtali Distr., Commonage, 1000 m., fl. 22.xi.1951, *Chase* 4209 (BM; COI; LISC; SRGH). S: Chibi Distr., Native Reserve, fr. 7.iii.1951, *Kirkham* 24/51 (SRGH). **Nyasaland.** N: Vipya, Chikangawa, fl. 19.x.1957, *Adlard* 255 (K; SRGH). S: Mboma Valley, fl. 20.xi.1943, *Hornby* 2924 (K). **Mozambique.** N: Moçambique, Malema, fl., *Serrano* 12 (LMJ). T: Macanga, between Furancungo and Casula, fl. 14.x.1943, *Torre* 6025 (BR; K; LISC; LM; SRGH). MS: Gorongosa, fl. 29.ix.1943, *Torre* 5980 (BR; K; LISC; LM; SRGH). SS: Gaza, Bilene, fl. 8.xii.1950, *Andrada* 1994 (BM; LM; LMJ). LM: between Manhiça and Marracuene, fl. 9.xii.1940, *Torre* 2257 (BR; K; LISC; LM; SRGH).

From Angola to Kenya and southwards to the Cape.

Subsp. **acutifolia** (Hochst. ex A. Rich.) Cufod. in Bull. Jard. Bot. Brux. **28**, Suppl.: 487 (1958). Type from Ethiopia.

Apodytes acutifolia Hochst. ex A. Rich., Tent. Fl. Abyss. **1**: 92 (1847).—Oliv., F.T.A. **1**: 355 (1868) pro parte quoad specim. Schimper.—Engl., Pflanzenw. Afr. **3**: 256 (1921). Type as above.

Apodytes dimidiata var. *acutifolia* (Hochst. ex A. Rich.) Boutique, F.C.B. **9**: 274 (1960). Type as above.

Ovary glabrous. Fruit and its appendage glabrous.

Nyasaland. S: Zomba Hill, fl. ii.1901, *Sharpe* 113 (K).

Also from Ruanda-Urundi, Ethiopia, Congo (Haut-Katanga and Kivu), Kenya, Tanganyika and Zanzibar.

Some specimens from N. Rhodesia (Kaloswe, fl. 24.xii.1930, *Hutchinson & Gillett* 4070 (BM; COI; K; LISC; SRGH)) and from S. Rhodesia (Matopos Distr., Mtchabesi Valley, fl. 1.v.1952, *Plowes* 1446 (SRGH); Matopos Distr., Besna Kobila, 1435 m., fl. i.1953, *Miller* 1449 (SRGH) and Melsetter Distr., Chimanimani Mts., 1675 m., fr. 1.ii.1958, *Hall* 283 (BM; SRGH)) have slightly pubescent ovaries or fruits and appendages, suggesting the possibility of the occurrence of natural hybrids between the two subspecies in the boundary zone of their respective areas.

4. RHAPHIOSTYLIS Planch. ex Benth.

Rhaphiostylis Planch. ex Benth. in Hook., Niger Fl.: 259 (1849).

Shrubs or lianes. Leaves alternate, petiolate, entire. Flowers bisexual, pedicellate. Inflorescences supra-axillary. Calyx small, 5-fid. Petals 5, valvate, free. Stamens 5; filaments compressed below, filiform above, alternipetalous. Ovary with an excentric style. Fruit with a persistent style.

A tropical African genus with c. 7 species.

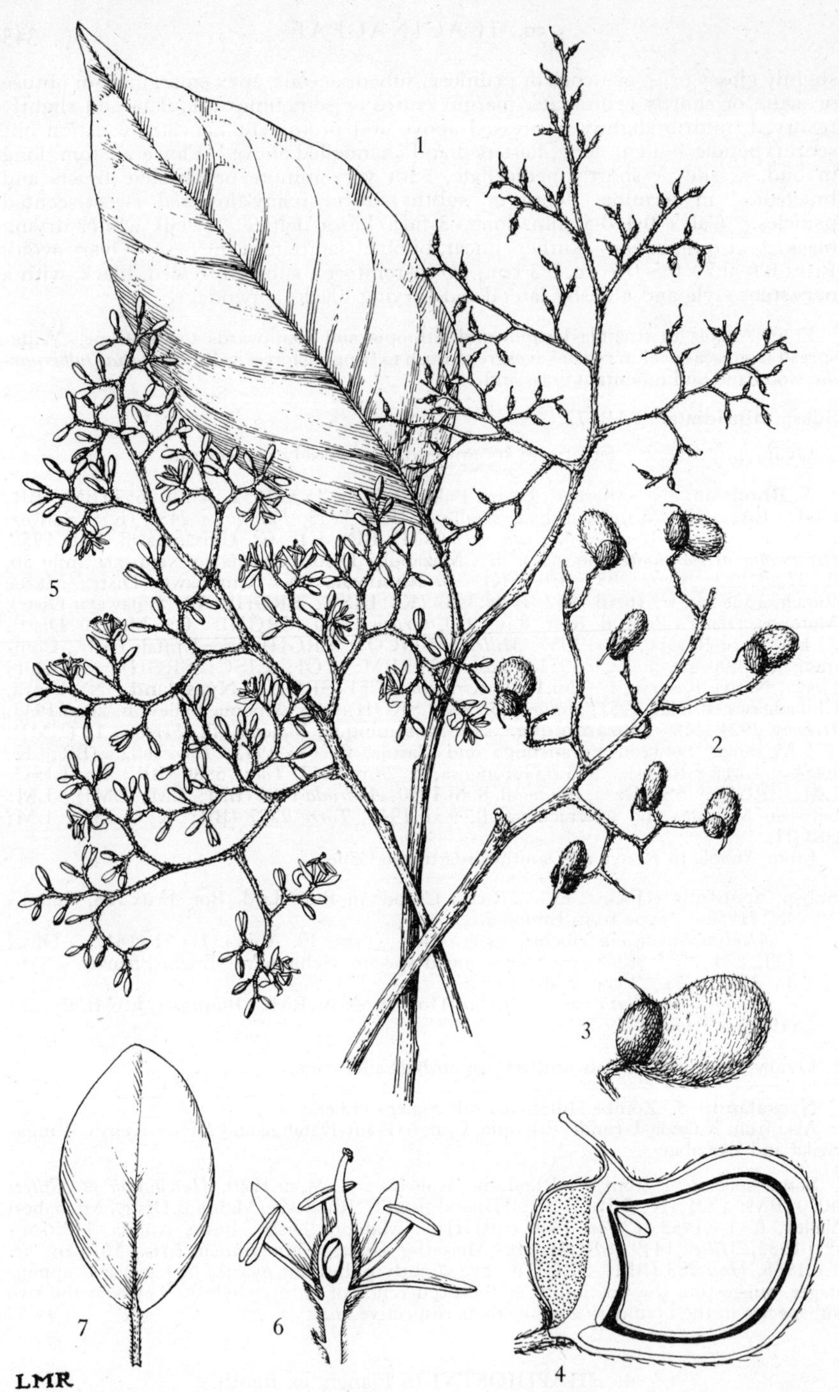

Tab. 72. APODYTES DIMIDIATA SUBSP. DIMIDIATA. 1, branchlet with leaf and young fruits (× ⅔) *Holmes* 1262; 2, branchlet with fruits (× ⅔) *White* 3657; 3, fruit (× 2) *White* 3657; 4, longitudinal section of fruit (× 4) *White* 3657; 5, inflorescence (× ⅔) *Torre* 5980; 6, vertical section of flower (× 4) *Torre* 5980; 7, leaf (× ⅔) *Barbosa & Balsinhas* 4886.

Rhaphiostylis beninensis (Hook. f. ex Hook.) Planch. ex Benth. in Hook., Niger Fl.: 259, t. 28 "Apodytes beninensis" (1849).—R.E.Fr., Wiss. Ergebn. Schwed. Rhod.-Kongo-Exped. **1**: 130 (1914).—Engl., Pflanzenw. Afr. **3,** 2: 256 (1921), "Raphiostyles".—Sleumer in Engl. & Prantl, Nat. Pflanzenfam. ed. 2, **20b**: 368 (1942).—Exell & Mendonça, C.F.A. **1,** 2: 343 (1951).—Keay, F.W.T.A. ed. 2, **1,** 2: 638 (1958).—Boutique, F.C.B., **9**: 275 (1960).—White, F.F.N.R.: 221 (1962). TAB. **73.** Type from Liberia.

Apodytes beninensis Hook. f. ex Hook., Ic. Pl. **8**: t. 778 (1848).—Oliv., F.T.A. **1**: 355 (1868). Type as above.

Ptychopetalum cuspidatum R.E.Fr., Wiss. Ergebn. Schwed. Rhod.-Kongo-Exped. **1**: 23, fig. 3 (1914). Type: N. Rhodesia, Mano R., near Lake Bangweulu, *Fries* 762 (UPS, holotype).

A glabrous straggling evergreen shrub or woody liane; bark smooth, dark grey; young branches dark purple. Leaf-lamina 7–15 × 2·5–5·5 cm., glossy dark green, pale below, drying blackish, narrowly lanceolate to elliptic, apex abruptly or gradually acuminate with obtuse acumen up to 2 cm. long, cuneate or slightly rounded at the base; midrib impressed above, prominent beneath; petiole c. 0·4 cm. long, wrinkled. Flowers c. 6 mm. long in bud, sweet-scented, in 4–many-flowered, supra-axillary fascicles; pedicels glabrous, ± equalling the flowers. Petals white. Fruit markedly reticulate-wrinkled, oblique on the stipe, broad-ovoid, red at maturity, drying glossy black.

N. Rhodesia. N: Luwingu Distr., Chiluvi I., L. Bangweulu, fl. & fr. 13.x.1947, *Brenan* 8095 (BM; BR; EA; K). W: Kitwe, fl. 2.vii.1955, *Fanshawe* 2358 (EA; K; LISC); Kitwe, forest nursery, 12 mth. seedlings 29.viii.1959, *Fanshawe* 5160 (K). **S. Rhodesia.** E: Chirinda, st. x.1947, *Chase* 498 (K; SRGH). **Mozambique.** N: Cabo Delgado, Macondes, between Mueda and Chomba, fl. 25.ix.1948, *Pedro & Pedrógão* 5355 (EA; LMJ).

Also from Senegambia to the Congo and Angola. In semi-deciduous and evergreen thickets, especially on sandy loam.

5. PYRENACANTHA Wight

Pyrenacantha Wight in Hook., Bot. Misc. **2**: 107 (1831) *nom. conserv.*

Lianes or scandent shrubs, dioecious or monoecious. Leaves alternate, petiolate. Inflorescences spicate or racemose. Flowers unisexual by abortion. Calyx absent. Petals valvate, forming a 4(5)-lobed corolla, persisting in fruit. Male flowers isomerous, with stamens alternating with the petals and a vestigial ovary. Female, flowers with isomerous vestigial staminodes alternating with the petals and an ovoid ovary crowned by a sessile discoid or multiradiate stigma. Drupes broadly ovoid, usually compressed and margined; endocarp fragile, having on the inner face prickles or warts which penetrate, sometimes deeply, into the seed; albumen copious, fleshy, bearing in the centre the embryo with its large, thin, foliaceous cotyledons.

A genus of tropical and subtropical Africa, Madagascar, India, Ceylon, Indochina and the Philippines, of c. 20 species.

Inflorescences borne on stems with no leaves or with undeveloped ones; leaves usually deeply 3–5(7)-lobed, deeply cordate or cordate-auriculate at the base, with hydathodes at the ends of the secondary nerves - - - - - 1. *kaurabassana*
Inflorescences (or rarely the solitary flower) ± supra-axillary to well-developed leaves without hydathodes:
 Leaf-lamina clearly palmately 5–7-nerved and deeply cordate at the base; margin doubly dentate, teeth gland-like; petioles longer than 2·5 cm. - 2. *grandiflora*
 Leaf-laminae all penninerved or some of them subpalmately 3-nerved; petioles up to 2 cm. long:
  Leaves sparsely hispid with ± appressed hairs (the smaller ones hooked) above, densely hispid with patent straight hairs below; margin with gland-like denticles; frequently some leaf-laminae clearly 3(5?)-lobate - - - - 3. *scandens*
  Leaves glabrous above, very sparsely and appressedly hispid (hairs all straight) below; margin sometimes irregularly sublobulate towards the apex, never beset with gland-like teeth - - - - - - - - - - - - - - 4. *kirkii*

1. **Pyrenacantha kaurabassana** Baill. in Adansonia, **10**: 272 (1872).—C. H. Wright in Kew Bull. **1906**: 17 (1906).—Bak. f. in Journ. Linn. Soc., Bot. **40**: 43 (1911).—

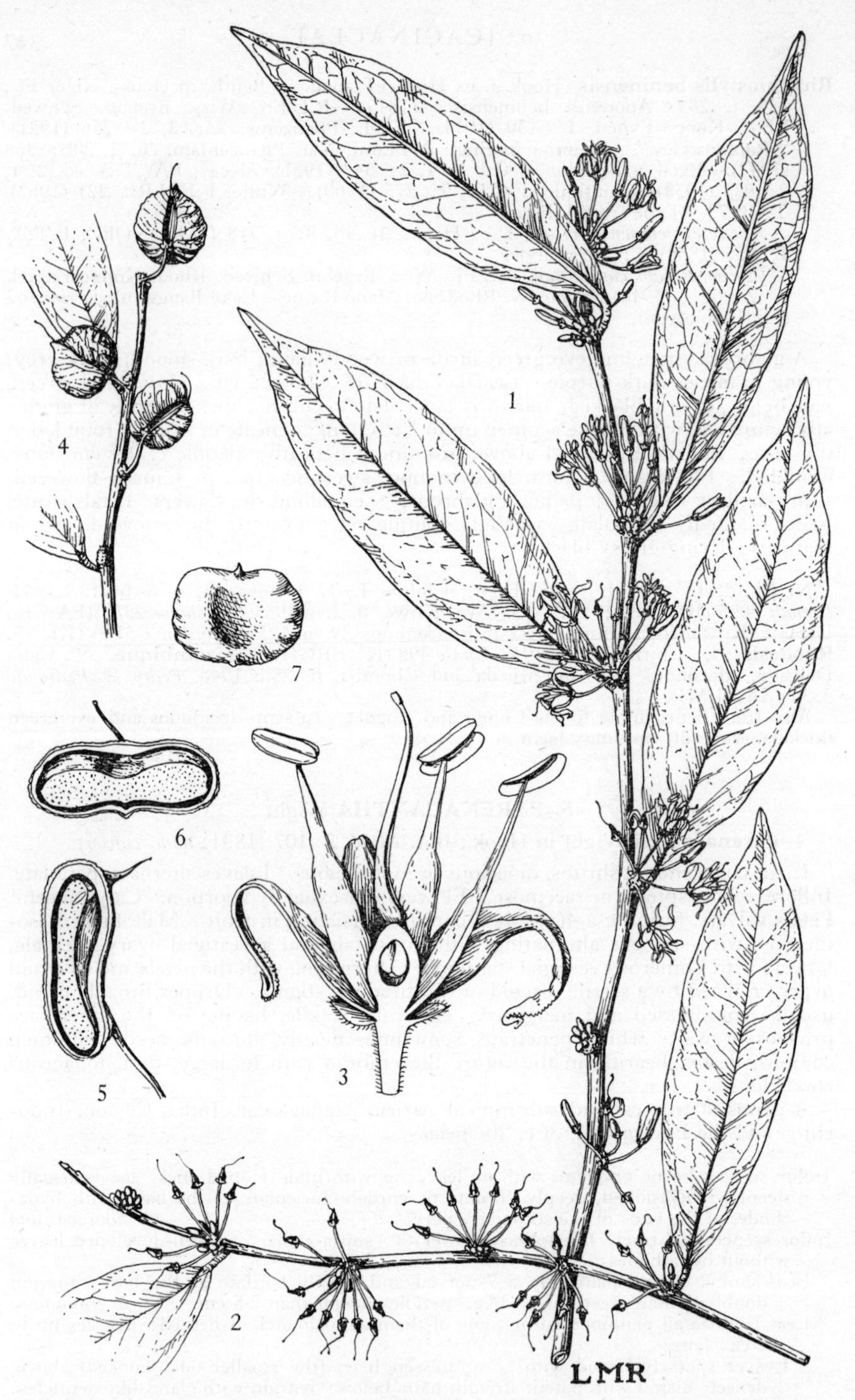

Tab. 73 RHAPHIOSTYLIS BENINENSIS. 1, branchlet with flowers (× ⅔) *Fanshawe* 2358; 2, branchlet with young fruits (× ⅔) *Fanshawe* 2358; 3, vertical section of flower (× 6) *Fanshawe* 2358; 4, branchlet with fruits (× ⅔) *Brenan & Greenway* 8095; 5, longitudinal section of fruit (× 2) *Brenan & Greenway* 8095; 6, transverse section of fruit (× 2) *Brenan & Greenway* 8095; 7, seed (× ⅔) *Brenan & Greenway* 8095.

Sleumer in Engl. & Prantl, Nat. Pflanzenfam. ed. 2, **20b**: 385 (1942) (" kamassana "). TAB. **74.** Type: Mozambique, Tete, Kaurabassa (" Kamassa " = lapsus calami!), *Kirk* (K, holotype).

Chlamydocarya sp.—Burkill, List Pl. Brit. Central Afr.: 239 (1897).

Pyrenacantha menyharthii Schinz in Denkschr. Math.-Nat. Kl. Akad. Wiss. Wien, **78**: 427 (1905).—Gomes e Sousa in Bol. Soc. Est. Col. Moçamb. **32**: 78 (1936). Type: Mozambique, Tete, Boroma, *Menyharth* 819 (K; WU, holotype).

Several trailing or twining stems from a perennial tuberous root-stock; stems olive green, hispid. Leaf-lamina 7–9 × 5–7 cm., very variable in shape, ovate-triangular to subpentagonal, subentire to deeply 3–5(7)-lobed, lobes rounded or subacute and mucronulate, deeply cordate to sagittate-cordate at the base, hirsute above, densely hirsute below, basal nerves 3–5(7)-palmate, secondary nerves mostly ending in well-marked ovoid hydathodes (1 mm. long with annular edge); petiole 4–6 cm. long. Inflorescences pedunculate, borne at the nodes and appearing before the often abortive subtending leaves; peduncles hispid. Male spikes 3–4 cm. long, very dense, on slender peduncles 4–12 cm. long; flowers numerous, sessile, subtended by bracteoles. Female spikes 1–1·5 cm. long, subcapitate; flowers fewer, sessile; rhachis bracteolate. Fruits 1·8–2 × 1·1–1·3 × 0·6–0·8 cm., yellow-orange at maturity, hirsute, crowded in spikes up to 6 cm. long and 3·5 cm. broad, on robust peduncles up to 2·5 cm. long.

S. Rhodesia. N: Lomagundi, 1100 m., fl. x.1920, *Eyles* 2680 (K; SRGH). S: Nuanetsi Distr., Tswiza, 415 m., fl. 31.x.1955, *Wild* 4714 (K; SRGH). **Nyasaland.** S: Ncheu Distr., 610 m., fr. 2.xi.1943, *Barker* 524 (EA). **Mozambique.** N: Amaramba, Cuamba, fl. 14.x.1942, *Mendonça* 817 (BR; LISC; LMJ). Z: Mocuba, Lugela, Namagoa, fl. & fr. ix–x.1943, *Faulkner* 101 (BR; K; PRE; SRGH). T: Tete, Caorabassa, fl. ix.1958, *Kirk* (K); Tete, Massanga, R. Luenha, 520 m., fr. 25.ix.1948, *Wild* 2618 (K; SRGH). MS: Madanda, 120 m., fr. 5.xii.1906, *Swynnerton* 2067 (BM). LM: Maputo, Goba, st. 23.xi.1944, *Mendonça* 3071 (LISC).

Also in Tanganyika. Among rocks and on termite-mounds in bushland and dry woodland.

The natives use the root-stock, heated, as a poultice (*Faulkner* in sched.).

2. **Pyrenacantha grandiflora** Baill. in Adansonia, **10**: 271 (1872).—Sleumer in Engl. & Prantl, Nat. Pflanzenfam. ed. 2, **20b**: 385 (1942).—Hutch., Botanist in S. Afr.: 311, fig. excl. B (1946). Type from S. Africa (Natal).

A twiner or sometimes a rambling shrub, usually monoecious; stems rugose with circular whitish lenticels. Leaf-lamina 5–14 × 4–11 cm., ovate, broadly-ovate or subcircular, rounded or subacute and mucronulate at the apex, margin doubly dentate with gland-like and hairy teeth, cordate or subcordate at base, hispid on both sides, basal nerves 5–7, palmate; petiole 2·5–5 cm. long. Male flowers* numerous, subsessile in supra-axillary rather loose spikes up to 16 cm. long with peduncles up to 3 cm. long; rhachis bracteolate. Female flowers (rarely bisexual) solitary near the leaf-axil*, with pedicels 1–2 mm. long or subsessile. Mature fruits not seen.

Mozambique. LM: Maputo, Goba, fl. 22.ix.1944, *Mendonça* 3016 (K; LISC; LM). Also from Transvaal and Natal. Bushland and dry woodland.

3. **Pyrenacantha scandens** Planch. ex Harv., Thes. Cap. **1**: 14, t. 23 (1859).—Sleumer in Engl. & Prantl, Nat. Pflanzenfam. ed. 2, **20b**: 385, fig. 111 A–B (1942). Syntypes from S. Africa.

Climber; young branches twining, densely setose with retrorse bristles; bark grey rough. Leaves with very hairy petioles; lamina 5–10 × 2·5–6·5 cm., variable in shape, that of the leaves in the lower part of the branches clearly 3(5?)-lobed, that of the others ovate to narrowly ovate; apex obtuse to acute and mucronate, margin with prominent gland-like hairy teeth, base broadly cuneate, ± rounded or rarely cordate; the upper surface sparsely hispid with appressed hairs (the smaller ones hooked) pointing mainly to the margin, the lower densely hispid with simple ± patent straight hairs. Inflorescences supra-axillary to proportionately smaller leaves. Male spikes c. 5 cm. long, on hairy peduncles c. 2·5 cm. long; flowers numerous, rather loose, sessile, subtended by bracteoles. Female flowers in short,

* In some specimens the 1–2 lower flowers (bisexual or female?) of the " male " inflorescence may develop fruits (only seen immature).

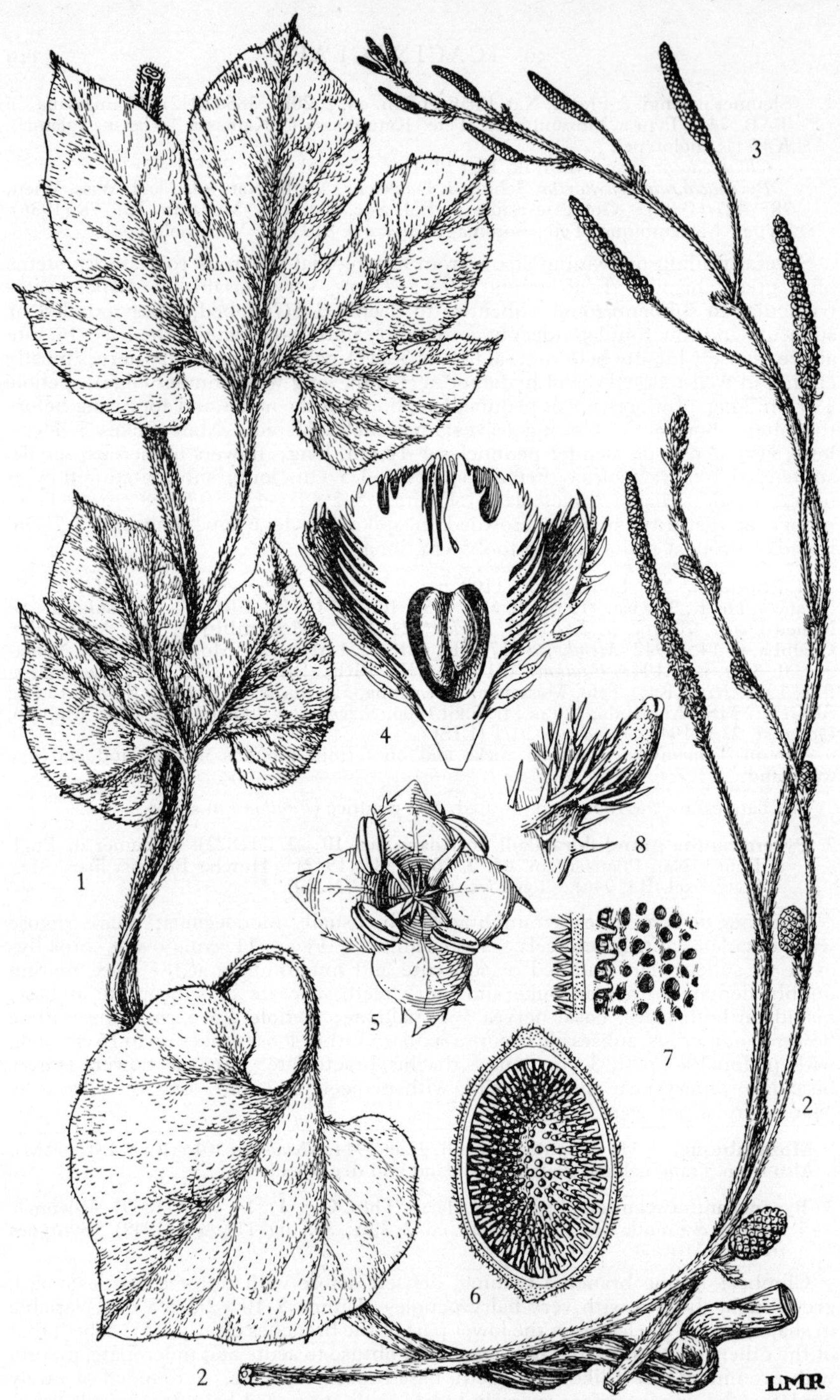

Tab. 74. PYRENACANTHA KAURABASSANA. 1, leaves ($\times\frac{2}{3}$) *Faulkner* P 101; 2, female inflorescences on stem ($\times\frac{2}{3}$) *Wild* 4714; 3, male inflorescences on stem ($\times\frac{2}{3}$) *Wild* 4714; 4, vertical section of female flower ($\times 16$) *Wild* 4714; 5, male flower from above ($\times 10$) *Wild* 4714; 6, vertical section of fruit with seed removed ($\times 2$) *Menyhart* 819; 7, seed surface with pits into which the prickles of the inner face of the exocarp penetrate ($\times 4$) *Menyhart* 819; 8, hydathode at the end of a secondary nerve ($\times 16$) *Faulkner* 101.

oblong or subcapitate pedunculate few-flowered spikes. Drupe c. 1·8 cm. long, slightly compressed, hispid.

Mozambique. SS: Gaza, Chipenhe, Chirindzi forest, fl. 13.x.1957, *Barbosa & Lemos* 8034 (K; LISC).
From Mozambique to the Cape.

4. **Pyrenacantha kirkii** Baill. in Adansonia, **10**: 272 (1872).—Sleumer in Engl. & Prantl, Nat. Pflanzenfam. ed. 2, **20b**: 385 (1942). Type: Mozambique, Zambesia, near Chamo R. junction with Zambezi R., *Kirk* (K, holotype).

Twining climber, stems hairy. Leaf-lamina 5–12 × 2–6 cm., papery, narrowly to broadly elliptic or rhombic, ± bluntly acuminate, margin subentire, subundulate or irregularly sublobulate towards the apex, ± narrowly rounded at the base, upper surface glabrous and glossy, lower surface sparsely appressed-hispid; venation penninerved, midrib prominent on both surfaces; petiole 0·5–1 cm. long, ± flexuous, hispid. Inflorescences supra-axillary. Male racemes c. 9 cm. long on peduncles c. 1 cm. long; flowers numerous, rather loose, on pedicels 0·5–1 mm. long, with bracteoles on the rhachis. Female spikes c. 1 cm. long, on peduncles c. 2·5 cm. long, with fewer sessile flowers, with bracteoles on the rhachis. Fruits 2–5 together per inflorescence, bright orange at maturity.

N. Rhodesia. N: Nchelenge, L. Mweru, *Fanshawe* 2970 (BR; K). **Mozambique.** Z: Lower Chire Valley, Chamo junction with Zambeze, fl. & immat. fr. 1.1862, *Kirk* (K). MS: Dondo, fl. 31.xii.1943, *Torre* 6320 (BM; BR; LISC; LM; SRGH).
Known only from our area. Climber of semi-deciduous forest and woodland.

P. kirkii is very near the Indian *P. volubilis* Wight (in Hook., Bot. Misc. **2**: 107 (1831)) and the different structure of the inflorescences may eventually prove insufficient justification for separating the two taxa specifically; more material, both African and Indian, is necessary to decide this.

Note: **Desmostachys planchonianus** Miers in Ann. Mag. Nat. Hist., ser. 2, **9**: 399 (1852).—Although there is a specimen of this species in the Kew Herbarium (" Herbarium Hookerianum "), collected by Forbes, with a manuscript indication on the sheet that it came from Mozambique, and was so recorded by Oliver (F.T.A. **1**: 353 (1868)), Engler (Pflanzenw. Afr. **3,** 2: 257 (1921)) and Perrier de la Bâthie (Fl. Madag. Com. **119**: 22 (1952)), we consider that it does not occur on the continent of Africa since no other collector has ever found it there up to the present date.
In view of the frequency of the species in the Madagascan region we think that the specimen in question was collected either in Santa Maria Island or, more probably, in the Comoro Islands where Forbes collected between 21.xii.1822 and 30.i.1823 (Exell & Hayes in Kirkia, **1**: 135–136 (1961)).
The species should therefore probably be excluded from our flora.

